REPRODUCTION IN DOMESTIC ANIMALS
FOURTH EDITION

REPRODUCTION
— IN —
DOMESTIC ANIMALS
FOURTH EDITION

Edited by

PERRY T. CUPPS

Department of Animal Science

University of California, Davis

Davis, California

ACADEMIC PRESS, INC.

Harcourt Brace Jovanovich, Publishers

San Diego   New York   Boston

Toronto   London   Sydney   Tokyo

ACADEMIC PRESS, INC.
San Diego, California 92101

United Kingdom Edition published by
ACADEMIC PRESS LIMITED
24-28 Oval Road, London NW1 7DX

LIBRARY OF CONGRESS CATALOG CARD NUMBER: 90-39853

ISBN   0-12-196575-9   (alk. paper)

PRINTED IN THE UNITED STATES OF AMERICA
90 91 92 93   9 8 7 6 5 4 3 2 1

This edition is dedicated to

Harold Harrison Cole

1897–1978

Contents

CHAPTER 18

Nutritional Influences on Reproduction

C. L. FERRELL

CHAPTER 19

Genetic Variation and Improvement in Reproduction

G. E. BRADFORD, J. L. SPEAROW, and J. P. HANRAHAN

CHAPTER 20

Influence of Infectious Diseases on Reproduction

R. H. BONDURANT

Contributors

Numbers in parentheses indicate the pages on which the authors' contributions begin.

Gary B. Anderson (279), Department of Animal Science, University of California, Davis, Davis, California 95616

Janice M. Bahr (555), Department of Animal Sciences, University of Illinois, Urbana, Illinois 61801

R. L. Baldwin (385), Department of Animal Science, University of California, Davis, Davis, California 95616

R. H. BonDurant (637), Department of Reproduction, School of Veterinary Medicine, University of California, Davis, Davis, California 95616

George R. Bousfield (25), Department of Biochemistry and Molecular Biology, The University of Texas M. D. Anderson Cancer Center, Houston, Texas 77030

G. E. Bradford (605), Department of Animal Science, University of California, Davis, Davis, California 95616

Hubert R. Catchpole (361), Department of Histology, College of Dentistry, University of Illinois at Chicago, Chicago, Illinois 60680

Patrick W. Concannon (517), Department of Physiology, College of Veterinary Medicine, Cornell University, Ithaca, New York 14853

Peter F. Daels (413), Department of Clinical Sciences, College of Veterinary Medicine, Cornell University, Ithaca, New York 14853

M. A. Driancourt (119), Station de Physiologie de la Reproduction, I.N.R.A., 37380 Nouzilly, France

Philip J. Dziuk (471), Department of Animal Sciences, Animal Genetics, University of Illinois, Urbana, Illinois 61801

C. L. Ferrell (577), United States Meat Animal Research Center, Clay Center, Nebraska 68933

P. F. Flood (315), Department of Veterinary Anatomy, Western College of Veterinary Medicine, University of Saskatchewan, Saskatoon, Saskatchewan S7N 0W0, Canada

William F. Ganong (1), Department of Physiology, School of Medicine, University of California, San Francisco, San Francisco, California 94143

D. L. Garner (251), Department of Animal Science, University of Nevada, Reno, Reno, Nevada 89557

J. P. Hanrahan (605), Agriculture and Food Development Authority, Belclare, Tuam County, Galway, Ireland

Donald M. Henricks (81), Department of Animal Science, Clemson University, Clemson, South Carolina 29634

John P. Hughes (413), Department of

Reproduction, School of Veterinary
Medicine, University of California, Davis,
Davis, California 95616

Larry Johnson (173), Department of
Veterinary Anatomy, College of
Veterinary Medicine, Texas A&M
University, College Station, Texas 77843

Patricia A. Johnson (555), Department of
Poultry and Avian Sciences, Cornell
University, Ithaca, New York 14853

D. R. Lindsay (491), School of Agriculture,
The University of Western Australia,
Perth 6009, Western Australia

J. C. Mariana (119), Station de Physiologie
de la Reproduction, I.N.R.A., 37380
Nouzilly, France

P. Mauleon (119), Station de Physiologie de
la Reproduction, I.N.R.A., 37380
Nouzilly, France

P. S. Miller (385), Department of Animal
Science, University of Nebraska, Lincoln,
Nebraska 68583

D. Monniaux (119), Station de Physiologie
de la Reproduction, I.N.R.A., 37380
Nouzilly, France

Katherine H. Moore (25), Department of

Biochemistry and Molecular Biology,
The University of Texas M. D. Anderson
Cancer Center, Houston, Texas 77030

T. J. Robinson (443), Department of Animal
Science, University of Sydney, Sydney
2006, Australia.

B. P. Setchell (221), Waite Agricultural
Research Institute, Department of
Animal Sciences, The University of
Adelaide, Glen Osmond, South Australia
5064, Australia

J. N. Shelton (443), John Curtin School of
Medical Research, Australian National
University, Canberra, A. C. T., 2601,
Australia

J. L. Spearow (605), Department of Animal
Physiology, University of California,
Davis, Davis, California 95616

George H. Stabenfeldt (413), Department
of Reproduction, School of Veterinary
Medicine, University of California, Davis,
Davis, California 95616

Darrell N. Ward (25), Department of
Biochemistry and Molecular Biology,
The University of Texas M. D. Anderson
Cancer Center, Houston, Texas 77030

Preface

The objectives of the fourth edition of *Reproduction in Domestic Animals* are (1) to summarize our current basic knowledge of the physiological and biochemical mechanisms regulating reproduction with emphasis on domestic species; and (2) to review and evaluate the current literature reflecting our understanding of these phenomena. The book is designed to be useful as a text for upper division undergraduate students and graduate students, and a reference source for research scientists and professional workers in fields related to the production of animal products. Information on "companion" animals is included as they form an important segment of the animal population.

Efforts have been made to keep the length of the book to a minimum, including the elimination of Chapter 1, "Historical In-troduction," and the incorporation of methods of hormone assay into the chapters dealing with the respective hormones. Since the publication of the third edition, assays have become more standardized and the decision was made to decrease space used to describe them.

Recently, important discoveries have made it possible to extend our understanding of the mechanisms controlling reproductive phenomena. They have provided better control of reproduction than was possible only a short time ago. They have also resulted in changes in the nomenclature of some of the phenomena related to reproduction. No attempt has been made here to standardize terms as this should be accomplished by a biological commission selected from several fields in the biological sciences.

REPRODUCTION IN
DOMESTIC ANIMALS
FOURTH EDITION

CHAPTER 1

Role of the Nervous System in Reproduction

WILLIAM F. GANONG

I. Introduction

The nervous system is involved to varying degrees in almost every aspect of the physiology of reproduction. Reflexes integrated at various levels of the nervous system are involved in sperm transport, parturition, and lactation. Copulation itself is made up of a series of reflexes and reaction patterns integrated into a coordinated whole, and sexual behavior is manifestly a subject for psychological and neurophysiological investigation. Another major aspect of neural involvement is the regulation of gonadal function by the brain through hypothalamic regulation of anterior pituitary gonadotropin secretion. The brain exercises a control-ling influence on the amount and type of pituitary gonadotropic hormones secreted into the circulation. The hormones act on the gonads to bring about, in both sexes, the state of readiness in the reproductive organs and the maturation of the germ cells necessary for successful procreation.

Such preparation would, of course, be in vain if it were not associated in both sexes with appropriate sexual behavior. This behavior is known to be dependent on an adequate level of circulating gonadal steroids. Thus, the gonads are involved in a kind of "feedback" mechanism. The brain controls the secretion of the gonadotropins via the hypophyseotropic hormones; the gonadal hormones are secreted in response to these tropic hormones; and the gonadal secretions act back on the brain to initiate the behavior necessary for successful reproductive performance (Fig. 1).

The brain not only regulates gonadotropin secretion in adulthood but is also responsible for the timing and coordination of the increase in the secretion of the gonadotropins that brings about sexual maturation. Puberty occurs when the episodic secretion of gonadotropin-releasing hormone (GnRH) is no longer held in check by neural mechanisms.

The actions of gonadal hormones on the brain also play a key role in the development and differentiation of hypothalamic function. In rats, both the adult pattern of gonadotropin secretion and the sexual behavior depend on the pattern of sex steroid secretion during infancy. In other species hormones exert similar inductive effects during fetal life, and in primates, there are also effects of early steroids on brain function even though the cyclic release of gonadotropins is unaffected by neonatal exposure

to androgens. Thus, brain–endocrine interrelations determine the development and sexual differentiation of the individual as well as reproductive capacity once sexual maturity has been attained.

The role of the nervous system in several reproductive processes and various aspects of brain–endocrine interactions are discussed elsewhere in this treatise. Neural mechanisms involved in parturition are discussed in Chapter 10, and the neuroendocrine reflex responsible for oxytocin-induced milk letdown is described in Chapter 11. The effects of light and other environmental stimuli on gonadotropin secretion are discussed in various chapters. In this chapter, the basic neural substrates of copulation are briefly considered, and attention is focused on the brain–gonad relationship in adulthood, the mechanisms regulating the onset of puberty, and the inductive effects of sex steroids on the brain early in life.

To keep this chapter as short as possible, emphasis has been placed on reviews rather than original papers as references. Additional references to original work published before 1977 can be found in the previous editions of this chapter (Ganong, 1959; Ganong and Kragt, 1969; Ganong, 1977).

II. Neural Substrates of Mating Behavior

Mating behavior may legitimately be divided into two components. It includes, first, activity consequent to the urge to copulate—the interest in or drive to sexual congress. Second, it includes the act of copulation itself. Sexual interest and the instinctual mating drive basic to the preservation of the species depend on neural circuits in the lim-

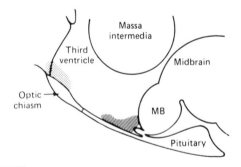

Figure 1 Sites at which estrogens act on the hypothalamus. Estradiol implants in infundibular region (striped area) near the piutitary inhibit gonadotropin secretion, producing ovarian atrophy, whereas estradiol implants above the optic chiasma (dotted area) induce estrous behavior. From Ganong (1989)

bic lobe of the brain and hypothalamus (Goy and Goldfoot, 1974). Copulation itself is made up of a collection of reflexes and reaction patterns, including erection, the necessary postural adjustments, the pelvic thrusts in the male, the lordotic adjustment of the pelvis in the female, ejaculation, and orgasm.

The reflex arcs and centers in the nervous system controlling the motor patterns of the sexual act have been studied in considerable detail. It is known, for instance, that most of the postural adjustments for coitus in both the male and female are integrated at the spinal level (Bard, 1940). In dogs with spinal cord transections, stimulation of the genitalia leads to erection and pelvic thrusts in males, and perineal stimulation produces elevation of the pelvis in the females.

Erection may be initiated in humans by purely psychic stimuli, but the reaction is primarily a reflex one, initiated by genital stimulation and integrated in the sacral segments of the spinal cord. The efferent pathway is parasympathetic. The motor fibers pass to the genitalia in a relatively well-defined bundle, and since these fibers are also involved in ejaculation, the bundle has come to be called the "nervus erigens." The vascular engorgement responsible for erection is produced in part by closure of the so-called small sluice channels within the corpora cavernosa, but the main factor involved is arterial dilation with consequent compression of the venous drainage (see Ganong, 1989).

Ejaculation in the male is initiated by stimulation of the glans penis, the adequate stimulus being gentle friction and the efferent pathway, the internal pudendal nerve. It is appropriatedly divided into two parts, emission and ejaculation proper (see Ganong, 1989). The first event, emission, is the delivery of semen into the urethra. This is primarily a sympathetic response integrated in the upper lumbar segments of the spinal cord and produced by the impulses that reach the smooth muscle of the vas deferens and associated organs via fibers in the hypogastric plexus. Ejaculation proper follows emission and is the expulsion of the seminal fluid from the urethra. This response is primarily parasympathetic, but it also involves a contraction of somatic musculature, particulary the bulbocavernosus muscles, which aids the expulsion. It is integrated in the upper sacral and lower lumbar portion of the spinal cord, and the motor fibers pass through the internal pudendal nerves and the nervus erigens. Ejaculation can still occur after sympathectomy or treatment with the modern sympathetic blocking drugs used to treat hypertension, but the ejaculate is "dry" because there is no contraction of the musculature around the bladder neck, and the ejaculate spills into the bladder.

Genital and other changes occurring during intercourse in human females have been studied in considerable detail (Masters and Johnson, 1966). Orgasm regularly occurs in less than one-half of the female population. When it does, there are rhythmic contractions of the vaginal wall. Impulses also travel via the pudendal nerves and produce rhythmic contractions of the bulbocavernosus and other pelvic muscles. The contractions may aid sperm transport but are clearly not essential for it.

Uterine contractions may occur in response to a spinal reflex during coitus in the female. A neuroendocrine reflex involving the posterior pituitary may also be involved. There is considerable evidence that genital stimulation during coitus initiates reflex release of oxytocin from the posterior pituitary (Fitzpatrick, 1966; see also Chapter 8).

Some investigators have argued that the oxytocin acts on the uterus to initiate a series of contractions that facilitate the transport of sperm from the vagina to the fallopian tubes. However, there are great variations in the rate of sperm transport in different mammalian species, and Fitzpatrick (1966) concluded, after a thorough review of this subject, that there was no proof that oxytocin secretion is an essential physiological component in mating.

III. Regulation of the Secretion of Pituitary Gonadotropins by the Nervous System

A. *Mechanism by Which the Nervous System Regulates Pituitary Secretion*

1. *Neurovascular Control of the Anterior Piutitary*

There is abundant evidence that the brain regulates gonadotropin secretion. Sexual cycles in animals are correlated with changes in seasons, an observation that is difficult to explain except in terms of the intermediation of the nervous system between the environment and the endocrine system. Temperature and rainfall changes may be responsible in part for seasonal variations, but in birds and mammals, fluctuation in the incident light is the major environmental factor involved. In certain mammalian species (e.g., the cat, rat, ferret, and mink), ovulation occurs only after copulation, and this reflex ovulation occurs in response to afferent stimuli that converge in the hypothalamus from the genitalia, the eyes, the nose, and other organs (Clegg and Doyle, 1967). In humans, hypothalamic disease is associated with amenorrhea and hypogonadism,

or, alternatively, precocious puberty. Similarly, lesions of the hypothalamus produce gonadal atrophy and inhibition of gonadotropin secretion in adult rats, cats, dogs, monkeys, sheep and other species, and in young animals, they produce precocious puberty (Reichlin, 1985).

The possibility that hypothalamic control of the anterior pituitary is exerted by nerve fibers to the gland deserves mention. There are sympathetic fibers from the superior cervical ganglion that reach the anterior lobe along blood vessels. Parasympathetic fibers also reach the gland by way of the greater superficial petrosal nerve. However, complete sympathectomy does not prevent ovulation in the rabbit or pregnancy in other species (Harris, 1955). Simple section of the pituitary stalk in female laboratory animals, provided it does not infarct the pituitary or interfere with its revascularization by the portal vessels (see below), permits a return of normal estrus cycles in a relatively short period of time. This interval is too short a period for regeneration of nerve fibers. Thus, it seems clear that nerve fibers to the adenohyphphysis do not play any important role in the control of reproductive function.

There is, however, a unique vascular connection between the brain and the anterior pituitary. The blood supply to the hypothalamus and pituitary in mammals is derived from the carotid arteries and the circle of Willis, the anastomotic arterial ring at the base of the brain. Branches of these vessels form a primary capillary plexus on the external surface of the median eminence (ventral portion of the hypothalamus overlying the pituitary) and the neurohypophysis. In addition, some of the capillaries form loops that penetrate the median eminence. From the primary plexus and the loops, blood is channeled into the sinusoidal portal hypophyseal vessels, which pass down the pituitary stalk and end in capillaries in the anterior pituitary (Fig. 2). The portal vessels that originate from the median eminence are referred to as "long portal vessels," and those that originate from the neurohypophysis are referred to as "short portal vessels." The portal hypophyseal system is a true portal system that begins and ends in capillaries without going through the heart, and it provides a vascular pathway by which substances released in the median eminence can be transported in high concentration directly to the anterior pituitary gland.

The portal hypophyseal system is a constant anatomical feature in higher vertebrates (Harris, 1955). There has been some discussion about the possibility that some of the blood in it flows from the pituitary to the brain, but certainly most if not all the flow is from the brain to the pituitary (Reichlin, 1985). In many species of mammals and in all birds studied, the portal vessels provide essentially all the blood supply reaching the anterior pituitary (Reichlin, 1985). The importance of the vascular connection to the hypothalamus is demonstrated by the observation that section of the pituitary stalk inhibits gonadotropin secretion. The portal vessels have a marked tendency to regrow (Harris, 1955), but if regrowth is prevented by the insertion of a plate of inert material between the hypothalamus and the pituitary, stalk section leads to marked gonadal atrophy and an increase in prolactin secretion. Pituitary transplants "take" particularly well in the anterior chamber of the eye and under the capsule of the kidney. Such transplants are associated with gonadal atrophy, but if the pituitary is retransplanted back to the median eminence region, the portal vessels regrow and there is a recrudesence of gonadal function.

These observations plus the presence in extracts of hypothalamic tissue of substances that stimulate the secretion of gonadotropins and other pituitary hormones led to the hypothesis that neurons in the hypothalamus with endings in the external layer of the median eminence secrete hypophyseotropic hormones that enter the portal vessels and regulate anterior pituitary function (Fig. 3). This neurovascular hypothesis, which was pioneered and popularized by Harris (1955), has been established by the isolation, characterization, and measurement in portal blood of six hypothalamic hypophyseotropic hormones: corticotropin-releasing hormone (CRH), thyrotropin-releasing hormone (TRH), gonadotropin-releasing hormone (GnRH); prolactin-inhibiting hormone (PIH); so-

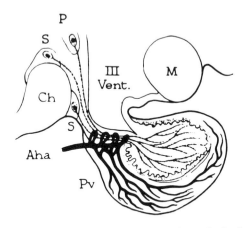

Figure 2 Hypophyseal portal vessels. Left: Sagittal section of the hypothalamus showing a branch of the anterior hypophyseal artery (Aha) breaking up into capillary loops that penetrate the median eminence. The loops drain into the portal vessels that end in capillaries in the anterior piutitary. Ch, optic chiasm; S, supraoptic nucleus; p, paraventricular nucleus; M, mammillary bodies; III VENT, third ventricle. Right: Detail of capillary loops penetrating the median emince.

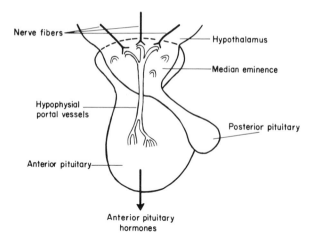

Figure 3 Diagrammatic summary of neurovascular control of anterior pituitary secretion. After Harris (1955).

matostatin, and growth-hormone–releasing hormone (GRH). There may be a prolactin-releasing hormone (PRH) as well but its existence has not been established (see below). The actions of these hormones are summarized in Fig. 4. Note that two of the hormones (CRH and GnRH) each regulate the secretion of two anterior pituitary hormones, whereas growth hormone is dually regulated and prolactin may be as well (see Ganong, 1989). The structures of the six hypophyseotropic hormones are shown in Fig. 5. All are peptides except PIH, which is the

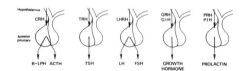

Figure 4 Effects of hypophyseotropic hormones on the secretion of anterior piutitary hormones. From Ganong (1989).

catecholamine, dopamine. TRH is a tripeptide, and GnRH is a decapeptide. Somatostatin is a tetradecapeptide (SS14), but preprosomatostatin also gives rise to an N-terminal extended polypeptide containing 28 amino acid residues (SS28) and a polypeptide containing 12 amino acid residues ([SS28]1–12). Both are found with SS14 in many tissues. CRH contains 41 amino acid residues, and GRH contains 44.

The structure and localization of the neurons secreting each of the hypophyseotropic hormones have been studied in detail (Everitt and Hökfelt, 1986). The parvocellular portion of the paraventricular nucleus on each side of the hypothalamus contains the cell bodies of the CRH- and TRH-secreting neurons, with the latter located medial to the former. The cell bodies of GnRH-secreting neurons are located in a band of tissue running through the hypothalamus, but most of them are found in the medial preoptic area. The cell bodies of the dopamin-

TRH　　　　　　(pyro)Glu-His-Pro-NH$_2$

LHRH　　　　　(pyro)Glu-His-Trp-Ser-Tyr-Gly-Leu-Arg-Pro-Gly-NH$_2$

┌S————————————————S┐

Somatostatin　　Ala-Gly-Cys-Lys-Asn-Phe-Phe-Trp-Lys-Thr-Phe-Thr-Ser-Cys

CRH　　　　　　Ser-Glu-Glu-Pro-Pro-Ile-Ser-Leu-Asp-Leu-Thr-Phe-His-Leu-Leu-Arg-Glu-Val-Leu-Glu-Met-Ala-Arg-Ala-Glu-Gln-Leu-
　　　　　　　　Ala-Gln-Gln-Ala-His-Ser-Asn-Arg-Lys-Leu-Met-Glu-Ile-Ile-NH$_2$

GRH　　　　　　Tyr-Ala-Asp-Ala-Ile-Phe-Thr-Asn-Ser-Tyr-Arg-Lys-Val-Leu-Gly-Gln-Leu-Ser-Ala-Arg-Lys-Leu-Leu-Gln-Asp-Ile-Met-
　　　　　　　　Ser-Arg-Gln-Gln-Gly-Glu-Ser-Asn-Gln-Glu-Arg-Gly-Ala-Arg-Ala-Arg-Leu-NH$_2$

PIH　　　　　　Dopamine

Figure 5 Structure of hypophyseotroic hormones in humans. Somatostatin, shown here as the tetradecapeptide form (SS14), exists in two additional forms. From Ganong (1989).

ergic neurons that secrete PIH are located in the arcuate nuclei. The cell bodies of the GRH neurons are also located in the arcuate nuclei, and the cell bodies of the somatostatin-secreting neurons that regulate anterior pituitary function are located in the periventricular nuclei above the optic chiasm (Fig. 6). All these neurons project to the external layer of the median eminence. The endings of the dopaminergic and GnRH-secreting neurons terminate close to each other in the lateral portions of the median eminence, but the termination of the rest of the neurons is diffuse.

The genes for CRH, TRH, GnRH, GRH, and somatostatin have been cloned and the structures of the preprohormones for each are known (Mayo *et al.*, 1986). Like the vasopressin- and oxytocin-secreting neurons in the supraoptic and paraventricular nuclei and other peptide-secreting neurons in general (Alberts *et al.*, 1983), hypophyseotropic hormones are formed in the cell bodies of the neurons that secrete them. From the ribosomes, they enter the endoplasmic reticulum, are processed in the Golgi apparatus, and become concentrated in secretory granules that migrate by axoplasmic flow from the cell body to the axon terminals (Fig. 7). Processing occurs in the secretory granules as they migrate, so that the final secretory products are stored in the granules in the endings. Release of these granules probably occurs by calcium-dependent exocytosis in response to action potentials reaching the endings.

It is important to note that the hypophyseotropic hormones are not unique to the hypothalamus and that they also exist in most instances in the endings of neurons in many other parts of the brain. This is particularly true in the case of CRH, TRH, and somatostatin. Dopamine is an established neurotransmitter in the striatum, the nucleus accumbens, and many other parts of the brain; and CRH, TRH, and somatostatin are also widely distributed. In addition, many hypophyseotropic hormones are found in nonneural tissues, and they are found in invertebrate animals that have no piuitary gland (see Ganong, 1985). Thus, it seems clear that when the anterior pituitary and its neurovascular control system evolved, existing brain peptides and transmitters were pressed into service to control the pituitary.

What regulates the secretion of the hormone-secreting neurons? The hormones of the anterior pituitary, adrenal cortex, thyroid, and gonads feed back to inhibit and, in some instances, to stimulate the secretion of the hypophyseotropic hormones. In addition, neural pathways converge on the hormone-secreting neurons in the hypothalamus and impulses in these pathways regulate secretion. The principal synaptic transmitters in the parts of the hypothalamus concerned with neuroendocrine control include norepinephrine, dopamine, epinephrine, serotonin, acetylcholine, γ-aminobutyric acid (GABA), glutamate, and aspartate. Most if not all of these substances affect the secretion of hypophyseotropic hormones, and consequently anterior pituitary secretion (Weiner and Ganong, 1978).

2. GnRH

GnRH was originally isolated as the luteinizing hormone- (LH) releasing hypophyseotropic hormone of the hypothalamus, and is still often called luteinizing-hormone–releasing hormone (LHRH). Many investigators initially felt that there was a separate follicle-stimulating hormone- (FSH) releasing hormone (FSHRH, or FRH). However, LHRH was found to have FSH releasing activity, and LH- and FSH-releasing activity increased or decreased in

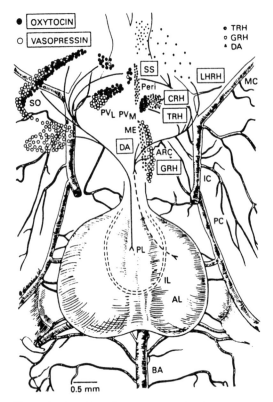

Figure 6 Location of neurons secreting hypothalamic hormones. The hypophyseotropic hormone neurons are shown on the right and the vasopressin- and oxytocin-secreting neurons on the left. AL, anterior lobe; ARC, arcuate nucleus; BA, basilar artery; IC, internal carotid; IL, intermediate lobe; MC, middle cerebral; ME, median eminence; PC, posterior cerebral; peri, periventricular nucleus; PL, posterior lobe; PV$_L$, PV$_M$, lateral and medial portions of the paraventricular nucleus; SO, supraoptic nucleus; SS, somatastatin. Courtesy of L. W. Swanson and E. T. Cunningham, Jr., from Ganong (1989).

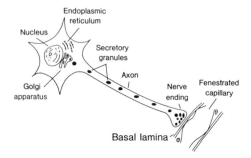

Figure 7 Typical peptidergic neuron secreting a neural hormone into the bloodstream. From Ganong (1986).

parallel when the molecule was modified. In addition, no FSH-releasing hormone has as yet been isolated from hypothalamic tissue. This has led many to accept the hypothesis that there is only one gonadotropin-releasing hormone, and increasingly, LHRH has come to be known as GnRH. The proportions of FSH and LH in pituitary effluent vary from time to time, but this variation may be due to variations in the concentrations of gonadal steroids or inhibin perfusing the gland at the time of stimulation.

PreproGnRH contains GnRH near its signal sequence plus a long C-terminal peptide that has been called the GnRH-associated peptide, or GAP peptide (Nikolics *et al.*, 1988). In some preparations, the GAP peptide has prolactin-inhibiting activity, and there has been speculation about its physiological role. However, the secretion of prolactin is often not inversely proportional to the secretion of LH, and there is considerable evidence against the idea that the GAP protein is a physiologically significant PIH (McNeilly, 1987).

When measured at intervals of 2 h or more, LH secretion is relatively flat and tonic in males. In females, there is a prominent midcycle surge in LH secretion with a sharp rise in the blood level that causes ovulation (see Chapter 4). There is also a midcycle surge in FSH secretion that is not as large as the LH increase, and in rats, but not in some other species, there is also a midcycle prolactin surge.

Several years ago, it was discovered that when LH was measured at frequent intervals in ovariectomized animals, plasma levels rose and fell in an episodic, rhythmic pattern. Peaks occurred every 60–120 min. For this reason, Knobil and associates called the LH secretory rhythm circhoral, that is,

having cycles about 1 h in length (Knobil, 1980). Subsequently, episodic secretion was also demonstrated in castrated male animals, and at a lower amplitude, in intact males and females. It was also demonstrated in experiments on monkeys and sheep that circhoral secretion of LH is due to circhoral secretion of GnRH. Thus, the secretion of GnRH is driven by a GnRH pulse generator in the mediobasal hypothalamus.

The significance of episodic secretion of GnRH becomes apparent when one compares steady infusion of GnRH with pulse infusions. Like many other receptors, GnRH receptors downregulate when the concentration of their ligand is high for a period of time and upregulate when the concentration of ligand is low. Furthermore, the GnRH receptors change their sensitivity at a rapid rate. Therefore, constant infusion of GnRH first stimulates but then rapidly inhibits LH secretion as its receptors become unresponsive. This phenomenon is so regular and reproducible that GnRH and its synthetic agonist analogs can be used to prevent rather than facilitate fertility by simply administering them at a steady rate or giving long-acting analogs (Marshall and Kelch, 1986; Marshall *et al.*, 1988). Given the rapid development of receptor unresponsiveness, the necessity of having GnRH secreted episodically rather than constantly becomes apparent.

The significance of differences in the amplitude and the frequency of GnRH pulses has now been studied in considerable detail (Marshall and Kelch, 1986). The frequency is increased by estrogen and decreased by testosterone and progesterone. In humans, frequency increases in the late follicular phase of the menstrual cycle, peaking with the production of the LH surge (Fig. 8).

During the luteal phase of the menstrual cycle, frequency declines due to the action of progesterone, and at the end of the luteal phase, when progesterone drops, frequency increases again.

The exact location and nature of the pulse generator in the hypothalamus is still unknown. However, the generator is affected by neurotransmitters in hypothalamic pathways. Norepinephrine and possibly epinephrine increase GnRH pulse frequency, and opioids such as β-endorphin and enkephalins reduce GnRH pulse frequency. It has been argued that some cases of hypothalamic amenorrhea are due to ex-

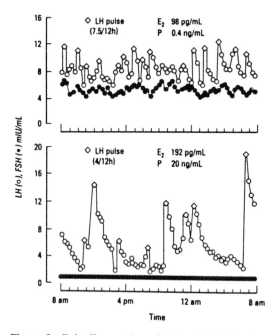

Figure 8 Episodic secretion of LH and FSH during the follicular stage (top) and the luteal stage (bottom) of the menstrual cycle. The numbers above each graph indicate the numbers of LH pulses per 12 h and the plasma estradiol (E_2) and progesterone (P) concentrations at these two times of the cycle. From Marshall and Kelch (1986).

cess activity in opioid circuits; and in a few trials, the opioid antagonist naltrexone has been reported to increase pulse frequency toward normal and bring about clinical improvement.

3. Prolactin-Regulating Hormones

The evidence is now overwhelming that the principal physiologic PIH is dopamine secreted by neurons with their cell bodies in the arcuate nuclei. This does not rule out the possibility of other prolactin-inhibiting hormones. The GAP peptide is the most recent of these to receive attention. However, as noted above, there is very little good evidence that it is a physiologic prolactin regulator, and at present no other PHI has been proved to exist and play a role under normal conditions.

Hypothalamic extracts sometime stimulate rather than inhibit prolactin secretion, and there is physiologic reason to expect that a PRH is secreted during stress-induced increases in prolactin secretion and suckling. However, no single PRH has been isolated and characterized. Instead, there are a number of hypothalamic peptides that can increase prolactin secretion (Shin *et al.*, 1987). TRH has potent prolactin-stimulating activity, but its physiologic role is uncertain because in many circumstances in which prolactin secretion is increased, there is no evidence of increased thyroid function or a rise in circulating TSH. Vasoactive intestinal polypeptide (VIP) also has prolactin-stimulating activity and is found in relatively high concentrations in portal hypophyseal blood. However, it has not been proved to be the active hormone in conditions in which prolactin secretion is increased. PreproVIP also contains a related peptide, PHI, and this compound also increases prolactin secretion. Whether any or all of these or other peptides are physiologic prolactin-releasing hormones remains to be determined by future research.

4. Feedback Control by Ovarian Steroids and Inhibin

Testosterone inhibits LH secretion by a direct action on the anterior pituitary, and it also acts on the hypothalamus to inhibit GnRH secretion. Estrogens inhibit LH secretion except for the period just before the ovulatory surge in females, when positive feedback predominates and the estrogens stimulate rather than inhibit GnRH secretion. Like testosterone, the negative feedback effect of estrogen is probably exerted at both the anterior pituitary and hypothalamic level. Estrogens also inhibit FSH secretion. The site of the positive feedback effect of estrogen apparently varies with the species. In rats, it appears to be in the anterior hypothalamus, rostral and superior to the arcuate nuclei, the site of the negative feedback. However, in monkeys, it appears to be in the same area as the site of negative feedback (Knobil, 1980).

Androgens do not exert as strong an inhibitory effect on FSH secretion as they do on LH secretion, and in humans with marked atrophy of the seminiferous tubules but normal Leydig cells, circulating FSH is elevated (see Ying, 1988). These observations led to a search for a gonadal secretion that specifically inhibited FSH secretion and eventually to the isolation and chemical characterization of a family of polypeptides called inhibins. These substances are secreted by the Sertoli cells in the male and the granulosa cells in the female, and they act directly on the anterior pituitary to inhibit FSH secretion. They are heterodimers that are produced from three precursor polypeptides: the α subunit precursor, the

β_A subunit precursor, and the β_B subunit precursor (Fig. 9). The α subunit and the β_A subunit are connected by disulfide bonds to form inhibin A, and the α subunit and the β_B subunit are joined to form inhibin B. The two dimers are equally active.

An interesting unexpected discovery was the observation that homodimers also form from the subunits, and that these homodimers stimulate rather than inhibit FSH secretion. These compounds have been named activins. The existence of $\beta_A\beta_A$ has been established, and $\beta_B\beta_B$ may also exist. In addition, $\beta_A\beta_B$ is formed. At present, the physiological role of the activins is unknown. It appears that inhibins and activins are also produced outside the reproductive system and that they may be involved in the functions of the brain and the hematopoietic system (Ying, 1988).

An additional structurally unrelated polypeptide that inhibits FSH secretion has been isolated from follicular fluid and named follistatin (Ying, 1988). Its physiological role is unknown at this time.

B. Control of GnRH Secretion in the Adult Male

In adult male mammals, gonadotropin secretion is generally steady. In some species, there is some seasonal variation in gonadal activity, but there are no regular cycles like those that occur in the female.

In males, lesions of the arcuate region of the hypothalamus cause diffuse testicular atrophy with loss of both Leydig cells and seminiferous tubules, indicating that the secretion of LH and that of FSH are both compromised. There is no convincing evidence that lesions in other parts of the hypothalamus selectively inhibit FSH secretion. If the mediobasal hypothalamus is

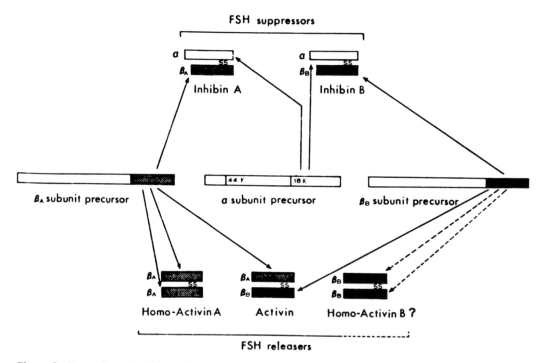

Figure 9 Formation of inhibins and activins from three precursor polypeptides. From Ying (1988).

isolated by knife cuts that separate it from the rest of the brain, testicular function is maintained (Halász, 1969). Thus, the mechanisms responsible for regulating testicular function are relatively autonomous and reside in the mediobasal hypothalamus and adjacent anterior pituitary.

A diagram summarizing current views on the hormonal interactions responsible for the control of testicular function in the adult is shown in Fig. 10. Testosterone secreted by Leydig cells of the testes enters the general circulation. In addition to maintaining secondary sex characteristics, it exerts a negative feedback effect on the anterior pituitary and hypothalamus to decrease LH secretion. In addition, testosterone passes in a paracrine fashion to the seminiferous tu-

bules, where it plays an important role in maintaining spermatogenesis (see Chapter 5). The Sertoli cells in the seminiferous tubules produce inhibins, and these act directly on the anterior pituitary to inhibit FSH secretion. In this fashion, steady levels of gonadotropin secretion are maintained with production of relatively constant amounts of testosterone and maintenance of spermatogenesis.

C. Control of GnRH Secretion in the Adult Female

The changes in circulating hormones that occur during the sexual cycle in spontaneously cycling adult females are summarized in Fig. 11. The human menstrual cycle

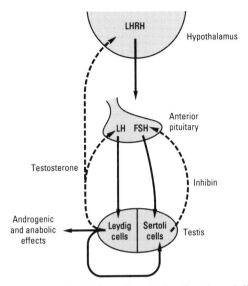

Figure 10 Regulation of testicular function. Solid arrows indicate excitatory effects, dashed arrows indicate inhibitory effects. From Ganong (1989).

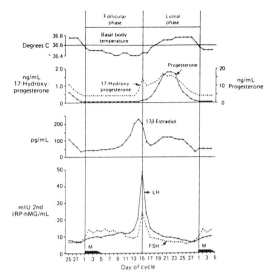

Figure 11 Typical basal body temperature and plasma hormone concentrations during a normal 28-day menstrual cycle. M, menstruation. From Midgley (1973).

is illustrated, but the changes in other mammals are similar, with major variation only in cycle length. Early in the follicular phase of the cycle, FSH is slightly elevated, producing follicular growth. Later in this phase, there is a fall in FSH secretion due to the negative feedback effect of estrogen. As the dominant follicle or follicles near ovulation, the negative feedback effect of estrogen is overridden by the positive feedback effect. The result is increased frequency and amplitude of GnRH pulses, with an additive, crescendo effect leading to the LH surge that produces ovulation (Knobil and Neill, 1987).

Estrogen secretion increases throughout the early stages of the follicular phase of the cycle, reaching a peak in humans one day before ovulation, then declining. The estrogen is coming from the growing dominant follicle, but the reason for the decline close to midcycle is unknown.

The exact mechanism by which estrogen produces its positive feedback is still unknown, but the characteristics that produce the effect have been analyzed in detail in monkeys. Both time and amplitude appear to be involved. In ovariectomized monkeys, only negative feedback was observed when circulating estrogen was increased 300% for 24 h (Knobil, 1980). However, when it was increased 300% for 3 days, a brief decline in LH secretion was followed by a burst of secretion that resembled the midcycle LH surge. Thus, it is the rising estrogen titer that triggers the LH surge, and the subsequent fall that is responsible for the return of LH secretion to low levels. There is a midcycle surge in FSH secretion as well, and this surge primes immature follicles for growth in the next succeeding cycle. In rats, but not in humans, there is also a midcycle prolactin surge.

After ovulation, the ruptured follicle is rapidly converted to a corpus luteum and the corpus luteum begins to secrete estrogen and progesterone. Progesterone inhibits LH secretion, and progestins are used for this purpose in oral contraceptive pills. The combination of estrogen and progesterone produced by the corpus luteum causes circulating LH and FSH to fall to low levels. Finally, the corpus luteum regresses (luteolysis), and as circulating estrogen and progesterone levels decline, LH and FSH levels rise and a new cycle begins.

It is obvious from the preceding comments that luteal regression is the key event in the reproductive cycle. The rest of the sequential events follow logically after luteolysis. Prostaglandins appear to be involved in luteolysis, possibly by inhibiting the cyclic AMP response to LH; in some species, oxytocin produced in the corpus luteum contributes to its decline (see Ganong, 1989). However, the details of the mechanism responsible for regression of the corpus luteum are unknown.

The feedback mechanisms involved in the neuroendocrine control of the sexual cycle in females are summarized in Fig. 12. Inhibins are produced by the granulosa cells of the ovarian follicles as well as the Sertoli cells of the testes, and they are present not only in follicular fluid but in the circulation. They presumably play a part in the overall regulation of FSH secretion, but their exact role in neuroendocrine control in females remains to be determined.

The neural pathways and transmitters involved in producing the cyclic release of gonadotropins have been studied in considerable detail in female experimental animals,

particularly in rats. In rats, there is abundant evidence that release of norepinephrine in the medial preoptic area from the endings of ascending noradrenergic pathways is necessary for the GnRH surge to occur (see reviews by Weiner and Ganong, 1978, and Ramirez *et al.*, 1984). Some investigators have argued that ascending epinephrine-secreting neurons are also involved (see Steele and Ganong, 1986), but the evidence for this is controversial.

An intriguing question is how release of norepinephrine is made specific, given the fact that noradrenergic neurons in the brainstem have multiply-branched axons that are distributed to many different parts of the brain. One possibility for which there is now considerable evidence is that the pattern of norepinephrine release depends on the particular combination of excitatory and

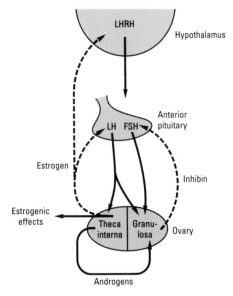

Figure 12 Regulation of ovarian function. Solid arrows indicate excitatory effects, dashed arrows indicate inhibitory effects. From Ganong (1989).

inhibitory receptors on the noradrenergic terminals. Neurons release transmitters that act on these presynaptic receptors to increase or decrease the release of norepinephrine. It may be that in the medial preoptic area, angiotensin II functions in this fashion to cause the increase in norepinephrine release that triggers secretion of GnRH and consequently the LH surge. Evidence for this hypothesis includes the observation that intraventricular angiotensin II produces increased LH secretion in proestrus female rats, that intraventricular administration of the angiotensin II receptor blocking drug, saralasin, or the angiotensin-converting enzyme inhibitor, enalaprilat, blocks the spontaneous LH surge and ovulation, and that the stimulatory effects of angiotensin II are blocked by adrenergic blocking drugs and catecholamine depletion (Steele and Ganong, 1986).

On the other hand, opioid transmitters are involved in pathways that inhibit overall LH secretion by reducing the amplitude and frequency of the GnRH pulses. In humans with hypothalamic amenorrhea, the opiate antagonist naloxone seems to be of benefit in restoring menstrual cycles (see above).

The site of positive feedback of estrogen to produce the ovulatory surge in rats appears to be separate from the negative feedback site. This was clearly demonstrated by Halász and associates (Halász, 1969), who showed that knife cuts separating the preoptic area from the mediobasal hypothalamus abolished ovarian cycles and produced a state of constant vaginal cornification. Lesions in the preoptic, suprachiasmatic, and related anterior hypothalamic areas produce a similar condition in rats and guinea pigs (Flerko, 1966). The ovaries of these animals are full of follicles, but no ovulation

occurs and no corpora lutea are seen. The vaginal epithelium becomes cornified and remains in that state. The condition produced by the lesions has been called the "constant estrus syndrome." The term is unfortunate since the "estrus" is confined to the vagina; the rats do not mate. The term "constant vaginal cornification syndrome" seems more appropriate. Enough estrogen is secreted to maintain the uterus and cornify the vaginal mucosa in these rats. FSH and LH are secreted, since FSH is necessary for follicular development and LH is necessary for estrogen secretion from the follicles. The follicles are capable of ovulation since injections of LH promptly produce luteinization. However, endogenous LH is not secreted in sufficient amounts to produce ovulation. The lesions in the suprachiasmatic and preoptic area that produce this syndrome exert their effects through the basal arcuate area (Flerko, 1966).

It is worth noting that a variety of other procedures produce the constant vaginal cornification syndrome, apparently by acting on the anterior hypothalamic LH regulating region. It can be produced, for example, by constant exposure to light (Critchlow, 1963). It also appears when female rats reach adulthood if they are injected with androgen in infancy (see Section V).

The situation in the monkey is different from the rat. In this species, knife cuts that create hypothalamic islands containing only the median eminence and the supraoptic, ventromedial, arcuate, and premammillary nuclei are associated with regular menstrual cycles (Krey et al., 1975). On the other hand, stalk section in the monkey produces gonadal atrophy, and injection of antibodies to GnRH causes gonadotropin secretion to decrease promptly to low levels (McCormack and Knobil, 1975). Therefore, the medioba-sal hypothalamus is necessary for both the negative feedback and positive feedback effects of estrogen.

Impulses in afferent pathways converging on the hypothalamus trigger the ovulation-producing increases in LH secretion in animals that ovulate after copulation (Brooks, 1940). In spontaneously ovulating species, the effects of light and other environmental stimuli on reproductive function indicate that in these animals, impluses in afferents from sense organs also affect gonadotropin secretion. In rabbits, ovulation can be produced by stimulation of the uterine cervix with a glass rod. However, genital stimulation is not essential for ovulation since this occurs following coitus after local anesthesia of the vagina and neighboring regions. Ovulation can also occur after a great variety of sensory receptors have been destroyed. Neither removal of the olfactory bulbs nor destruction of the vestibular apparatus and cochlea by itself blocks copulation-induced ovulation. Blinding is also ineffective. In cats, which also ovulate after coitus, complete sympathectomy does not alter the response. It appears probable, therefore, that many stimuli coverage on the hypothalamus and that no single afferent pathway is essential for LH release.

In rats, lesions just above the optic chiasm block production of constant vaginal cornification by exposure to constant light (Critchlow, 1963). In ferrets, a reflexly ovulating species that can be brought into estrus in the winter by exposure to extra light, the effect of light is mediated via the optic nerve as far as the hypothalamus. However, interruption of the optic pathways beyond the hypothalamus has no effect on the response to light (Clark et al., 1949). It is now known that the fibers that are primarily responsible for the effect of light on reproductive function in mammals leave the optic chiasm in the midline and enter the suprachiasmatic nuclei and related areas (Reiter, 1980; Youngstrom and Nunez, 1986). In birds, light not only acts by way of the eye, but also penetrates the skull and appears to exert a direct effect on the diencephalon or the pituitary to increase gonadotropin secretion (Benoit, 1962). The eyes are certainly the major receptors in mammals, but light does penetrate to the region of the diencephalon in rats, rabbits, dogs, and sheep (Ganong et al., 1963).

D. Control of Prolactin Secretion

The net neural influence on adrenocorticotropic hormone (ACTH), TSH, growth hormone, and the gonadotropins is stimulatory, and section of the pituitary stalk causes their secretion to decrease. However, secretion of prolactin is tonically inhibited by the central nervous system, and the section of the pituitary stalk or transplantation of the pituitary to a distant site is associated with increased secretion of the hormone (McLeod, 1976).

This tonic inhibition is due to the continued secretion of the prolactin-inhibiting hormone, dopamine, into the portal hypophyseal circulation. The concentration of dopamine in portal hypophyseal blood is much greater than it is in simultaneously collected peripheral blood, whereas there is no portal vein–peripheral difference in norepinephrine or numerous other substances (Weiner and Ganong, 1978). Furthermore, lesions that destroy the cell bodies of the dopaminergic neurons in the arcuate nuclei produce a marked increase in prolactin secretion. The effects of dopamine on prolactin secretion are mediated by D_2 receptors that act via G_i, the inhibitory GTP binding protein in the membrane, to inhibit intracellular generation of cyclic AMP (see Ganong, 1989). Drugs that block D_2 receptors produce marked increases in prolactin secretion. The major tranquilizers used in psychiatry for the treatment of psychoses are also D_2 receptor blockers, and the plasma concentration of prolactin has been used as an index of the degree of D_2 inhibition and consequent tranquilizer effectiveness. Conversely, long-acting dopamine agonists such as bromocryptine have been used to treat patients with prolactin-secreting tumors of the anterior pituitary gland.

Hypothalamic extracts contain prolactin-releasing as well as prolactin-inhibiting activity (McLeod, 1976). As noted above, no single prolactin-releasing hormone has been isolated and identified, but a number of neuropeptides found in the hypothalamus, including VIP, PHI, and TRH, may function as prolactin-releasing hormones (Shin et al., 1987). Drugs that increase serotonin release in the brain increase prolactin secretion. There is a considerable amount of serotonin in the median eminence, but the bulk of the evidence indicates that serotonin acts by releasing one or more prolactin releasing hormones into the portal vessels (Shin et al., 1987).

Many different stimuli that act via the nervous system affect prolactin secretion. Prolactin secretion increases throughout pregnancy, then falls after delivery unless the young are suckled. Suckling is a potent stimulus to prolactin secretion, although the magnitude of the prolactin response gradually decreases with repeated suckling over long periods of time. Prolactin secretion is also increased by stimulation of the nipples in nonlactating women (Frantz, 1978). These prompt, relatively marked increases

produced by breast stimulation cannot be explained by a decrease in dopamine secretion, and are probably due to secretion of one or more of the hypothalamic peptides with prolactin-releasing activity into the portal vessels. Prolactin secretion is increased by various stressful stimuli and by strenous exercise. In women but not in men, it is increased by sexual intercourse, and the increase seems to correlate with the occurrence of orgasm (Frantz, 1978). Prolactin secretion is increased during sleep. The increase starts at the onset of sleep and persists throughout the sleep period, peaking 5–7 h after onset. Prolactin secretion is also increased by estrogens (see Yen, 1978), but this effect is due, at least in part, to a DNA-mediated direct effect on the lactotropes (Maurer, 1982).

In some species, increased prolactin secretion and pseudopregnancy can be induced by copulation with a sterile male and by mechanical or electrical stimulation of the cervix (Everett, 1966). Prolactin plays an important role in the development of pseudopregnancy (McLeod, 1976).

Visual and tactile stimulation such as seeing or touching eggs or young in the nest may lead to prolactin release in birds. The prolactin facilitates subsequent incubation, crop sac development, and broody behavior in ringed doves and domestic pigeons. Prolactin release in response to seasonal and environmental changes has also been suggested to the responsible for premigratory deposition of fat in passerine species of birds. The mechanism regulating prolactin secretion in birds is different from that in mammals (Kragt and Meites, 1965). It is interesting that hypothalamic extracts from birds stimulate rather than inhibit the release of prolactin in vitro. Hypothalamic extracts from parent pigeons actively secreting crop milk have been shown to be more effective in stimulating prolactin release by pigeon pituitaries than extracts from young pigeons after no crop-gland stimulation.

There is some information about the afferent paths to the hypothalamus that are involved in the regulation of prolactin secretion. Lesions of the reticular formation in the brainstem inhibit lactation (Beyer and Mena, 1969). This effect may be due to interruption of the milk ejection reflex (see Chapter 11), but the possibility that such lesions also reduce the prolactin response to suckling also should be considered. There is, in addition, evidence that olfactory nerve stimulation can inhibit prolactin secretion. This may be the basis of the "Bruce" effect, the interruption of pregnancy that occurs in mice exposed to a cage recently occupied by a male mouse of another strain (Bruce and Parkes, 1963).

IV. Neural Components in the Regulation of the Onset of Puberty

In addition to the part it plays in the endocrine and behavioral aspects of reproduction, the brain is involved in the initiation of puberty. In all mammals of both sexes, there is a period that starts just after birth in which the gonads are quiescent. This period ends with rapid growth of the gonads, development of secondary sex characteristics, and production of sperm in the male and sexual cycles in the female. Technically, puberty is the time at which these changes have progressed to the point that reproduction is possible, but the term is often used in a broader sense to refer to the whole process of sexual maturation. In males, and to a less extent, at least in some species, in females, the gonads are active in utero and at the time of birth before they become quiescent (see Grumbach and Kaplan, 1990). The length of the quiescent period as a percent of total life span is greater in primates and humans than in rats. The quiescent period has survival value in that it delays the onset of reproduction until the parents are better able to care for the young, and presumably it evolved on this basis. However, what is the mechanism that produces the quiescent period? This question has been addressed in studies that have been carried out primarily in rats (Ojeda *et al.*, 1986), sheep (Foster *et al.*, 1986), rhesus monkeys (Plant, 1988), and humans (Grumbach and Kaplan, 1990).

In female rats, the vaginal canal does not become patent until the time of puberty. In some rats, the vaginal smear indicates estrus when opening occurs, whereas in others the first estrus occurs 1–2 days later. The average normal age of vaginal opening is about 35 days. In male rats, the testes become mature and sperm heads appear in the seminiferous tubules at about 40 day of age. Motile sperms appear later, and puberty occurs about 50 days of age (Bloch *et al.*, 1974).

The failure of puberty to occur earlier in female rats is not due to unresponsiveness of the tissues to gonadal steroids, or to unresponsiveness of the gonads to gonadotropin. Ovaries of immature animals function in the adult manner when transplanted into adults, and injections of gonadotropins after the age of 20 days can cause ovulation and corpus luteum formation. Precocious vaginal opening and cornification of the vaginal epithelium can also be produced by injections of gonadal steroids. The pituitaries of immature animals contain gonadotropins, and these can be released by appropriate hypothalamic hypophyseotropic hormones. In addition, the pituitaries of immature animals are capable of supporting normal estrous cycles when transplanted under the hypothalamus in hypophysectomized adult female rats (Jacobsohn, 1966). This is good evidence that the mechanism holding puberty in check resides in the central nervous system.

Additional evidence for central control is the observation that lesions of the hypothalamus in female rats produce precocious puberty (Gellert and Ganong, 1960). Lesions in the anterior hypothalamus produce premature vaginal opening followed by the constant vaginal estrus. This is consistent with the finding that anterior hypothalamic lesions in adults produce constant estrus in which abundant estrogen is produced but there is no ovulation, and consequently progesterone is low (see above). However, lesions in the posterior tuberal area, just in front of the mammillary bodies, produce premature vaginal opening followed by regular estrous cycles—that is, they cause true precocious puberty.

In males rats it has not been possible to produce clear-cut precocious puberty with hypothalamic lesions. Since it takes 40 days for sperm to develop from spermatogonia in the rat, it would be difficult to produce any significant acceleration in the time of appearance of mature sperm even if spermatogenisis were stimulated early in life. A search for early activation of the Leydig cells might be more productive. In humans, the length of the spermatogenic cycle is 74 days, but sexual maturation does not normally occur for approximately 14 years. Consequently, there is plenty of time for precocious puberty to manifest itself, and precocious puberty in boys is a well-established syndrome (Jolly, 1955).

In rhesus monkeys, sexual maturation occurs at $2\frac{1}{2}$–3 years of age. In male monkeys, as well as rats and humans, testosterone secretion is increased *in utero,* and this increase is responsible for the development of male external genitalia. This increase is associated with increased LH secretion and with episodic secretion of LH, presumably indicating that episodic secretion of GnRH occurs at the adult rate. Thus, all components of the hypothalamic–pituitary–gonadal system are capable of normal function in utero (Plant, 1988). In females, episodic LH secretion also occurs in the neonatal period, but its frequency is somewhat slower than it is in the adult. After birth, the levels of LH and FSH in plasma fall to low, fixed values in males (Fig. 13) and females. Removal of the gonads during the first 6 months of life causes increased plasma levels of the gonadotropins, but from 6 to 30 months of age, this response disappears (Fig. 13). It reappears at the time of puberty.

The cause of the quiescent period in this species has been investigated by injecting the excitatory amino acid *N*-methyl-D-aspartate (NMDA) in a pulsatile fashion (Plant, 1988). Pulsatile injection of GnRH produces ovulatory menstrual cycles in females, and increased secretion of testosterone in males. Similar increases in testosterone secretion are produced in perpubertal males by injection of NMDA. This substance activates glutamate receptors, and glutamate is a general excitatory transmitter in the brain. The excitation produced by injected NMDA is general, and includes increased secretion of ACTH, prolactin, and growth hormone. The NMDA experiment suggests that the neurons responsible for the pulsatile release of GnRH are mature and capable of stimulation before puberty. Therefore, the normal failure of puberty to occur is probably

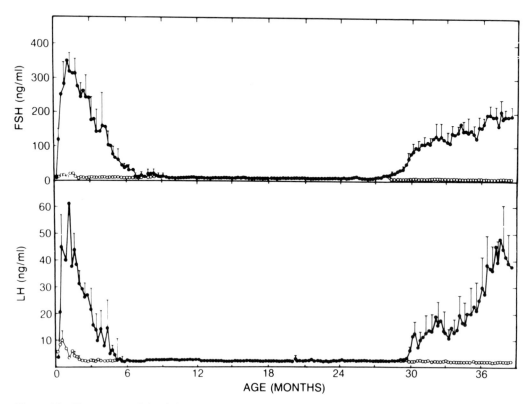

Figure 13 Time course of circulating FSH (top) and LH (bottom) in intact male monkeys (open circles) and monkeys orchidectomized at 1 week of age (closed circles). Note that gonadotropin secretion declines in the absence of testicular steroids during the quiescent period, then rises again about 6 months before the expected time of puberty in normal monkeys. From Plant (1985).

due to some sort of event proximal to the GnRH generator. In female monkeys, knife cuts behind the suprachiasmatic nuclei do not accelerate the onset of puberty. However, anterior hypothalamic lesions appear to accelerate the onset of menstruation, and as in rats, lesions of the posterior hypothalamus have a similar effect (Terasawa *et al.,* 1984). Thus, in some way, neural mechanisms are involved in keeping the pulse generator under restraint.

Although adult rhesus monkeys in cap-

tivity have regular menstrual cycles throughout the year, wild monkeys have a definite breeding period: ovulatory cycles occur in the fall, and births occur in spring. During the nonbreeding period, ovulatory cycles can be induced by intermittent injection of GnRH (Plant, 1988). Thus, the GnRH pulse generator appears to become quiescent during the nonbreeding season in a fashion that may be analogous to the quiescent period that occurs before puberty. In monkeys, food intake also has an effect on the onset of puberty, and malnutrition inhibits the GnRH pulse generator, delaying the onset of puberty.

Precocious puberty occurs in children with tumors or infections of the diencephalon (Jolly, 1955); indeed, precocious puberty is the most common endocrine manifestation of hypothalamic disease. Other cases of precocious puberty are "constitutional," meaning no brain abnormality can be identified. Sexual development in "true" precocious puberty includes spermatogenesis in the male and ovulation in the female, and is normal in all respects except its timing. It differs from "precocious pseudopuberty," the condition in which secondary sexual characteristics develop without maturation of germ cells due to excessive secretion of gonadal steroids in children with conditions such as ovarian and adrenal tumors. True sexual precocity may occur in very young children, and cases of regular menstruation in 2-year-old girls have been reported.

In boys, the syndrome of gonadotropin-independent precocious puberty (familial male precocious puberty) is also seen. In this condition, the gonads develop early independent of gonadotropin secretion. The condition responds to combined treatment with an antiandrogen and a drug that inhibits the conversion of androgen to estrogen (Laue et al., 1989). The pattern of inheritance is autosomal dominant, but the nature of the genetic defect is unknown.

Pineal tumors have also been reported to cause precocious puberty in humans (Jolly, 1955). For unknown reasons, precocity in association with such tumors is found almost exclusively in males. The syndrome is generally associated with lesions that destroy the pineal body, but pineal tumors, because of their anatomical position, also compress the hypothalamus early in their growth. Thus, the precocity may well be secondary to hypothalamic damage. In addition, some pineal tumors secrete human chorionic gonadotropin (hCG) (Sklar et al., 1981), and excess secretion of this gonadotropin could cause precocity.

What, then, is the mechanism responsible for suppressing the occurence of puberty during the prepubertal period? One hypothesis that has received considerable attention is that during this period, the negative feedback effect of gonadal steroids is magnified and puberty occurs when feedback sensitivity declines. However, the data on the effect of gonadectomy in monkeys indicate that this hypothesis is incorrect: gonadectomy between 6 and 30 months of age has no effect on gonadotropin secretion (Fig. 13). Similarly, there is no marked increase in circulating gonadotropins during the quiescent period in agonadal children (Grumbach and Kaplan, 1990). Others have argued that the pineal holds puberty in check. The pineal hormone melatonin may have some antigonadotropic effects, but pinealectomy has no stimulating effect on gonadotropin secretion in immature monkeys after gonadectomy (Plant, 1988). A third possibility is that the size of the animal is in some way monitored and when a critical size is reached, puberty occurs. This hypothesis was given impetus by observations that women who lost weight through jogging, competitive athletics, strenuous dancing, or anorexia nervosa became amenorrheic, and that they regained regular cycles when they regained the lost weight (Frisch and Revelle, 1970). However, this "somatometer" hypothesis has not been proved. Instead, it appears most likely that puberty is held in check by some sort of a neural mechanism that keeps the GnRH pulse generator quiescent. This would explain the available data, including the observation that hypothalamic lesions, which would disrupt the inhibitory mechanism, cause the development of early puberty. However, the exact mechanism of the inhibition remains to be determined.

V. Effects of Hormones on the Development and Differentiation of the Brain

The brain resembles the reproductive organs in that its development is determined by the hormonal environment early in life. The female pattern of gonadotropin secretion and sex behavior is innate, but the male pattern develops at puberty if the brain is briefly exposed to androgen during fetal or neonatal life. Most of the experiments demonstrating these actions have been performed in rats, a species in which the young are particularly immature at birth. If testes are transplanted to infant female rats, they do not ovulate when they mature (Gorski, 1971). Instead, they develop constant vaginal cornification. Ovulation can be produced in such animals by injection of LH or, after progesterone priming, by hypothalamic stimulation. Thus the rats are able to secrete LH at a steady level like the male, but they are unable to produce the peaks of LH secretion necessary for ovulation. In males castrated at birth, transplanted ovaries show the female pattern of cyclic ovulation with luteinization of the ruptured follicles; however, if the castrated rats with transplants are treated with androgens early in life, the male pattern of gonadotropin secretion develops. A single dose of androgen as

small as 10 μg testosterone on the fifth day of life in the female is capable of producing the male pattern of gonadotropin secretion in adulthood. Estrogens have no effect or they produce constant vaginal cornification. Consequently, it appears that the cyclic pattern of gonadotropin secretion seen in the female is the innate pattern, and that exposure to androgen early in life converts this pattern to the steady pattern of gonadotropin secretion seen in the male.

Pfeiffer (1936), who did much of the early work in this field, believed that androgens act on the pituitary to make the pattern of pituitary secretion that of the male. However, pituitaries transplanted from male fetal rats to hypophysectomized female adult rats maintain normal estrous cycles. Thus, the sex of the pituitary is not fixed, but depends on the sex of the brain under which it is located (Harris, 1964).

The early exposure of the brain to hormones also determines the pattern of sexual behavior that develops in adulthood. Female rats treated with testosterone when they are 5 days old do not behave sexually as females when they reach adulthood; instead, they attempt to mount other females with greater than normal frequency and show increased male sexual behavior. Conversely, males castrated at birth show increased female sexual behavior, although they also continue to act as males (Harris, 1964).

The similarity of the action of androgen on brain development to its action on the development of the external genitalia is striking. In many species, androgen from the fetal testis causes the undifferentiated genital anlagae to develop into male external genitalia. In the absence of androgen, female external genitalia develop regardless of genetic sex (Ganong, 1989). It is worth noting, however, that the androgen effects on the genitalia occur earlier than the androgen effects on the brain. Therefore, it is possible to have normal genital development and abnormal brain development (Harris, 1964).

The neural effects of exposure to steroids are most easily studied in rats because in this species the changes can be produced by treatment after birth. In other species, treatment must be given *in utero*, usually by treating the mother. For example, female pseudohermaphrodite offspring of rhesus monkeys treated during pregnancy with androgens have been shown to have abnormal sexual behavior in adulthood (Goy and Goldfoot, 1974). There is also evidence for masculinization of behavior in girls exposed to high levels of androgen *in utero* (Money, 1973). However, in primates, the pattern of gonadotropin secretion is not changed by early exposure to steroids, and sexually mature androgenized monkeys and humans can menstruate regularly if the secretion of androgens is suppressed by administration of glucocorticoids that inhibit the release of ACTH.

References

Alberts, B., Bray, D., Lewis, J., Raff, M., Roberts, K., and Watson, J. D. (1983). "The Molecular Biology of the Cell." Garland Publishing, New York.

Bard, P. (1940). *Res. Publ. Ass. Res. Nerv. Ment. Dis.* **20**, 551.

Benoit, J. (1962). *Gen. Comp. Endocrinol. Suppl.* **1**, 254.

Beyer, C., and Mena, F. (1969). *In* "Physiology and Pathology of Adaptation Mechanisms: Neural-Neuroendocrine Humoral" (E. Bajusz, ed.), p. 310. Pergamon, Oxford.

Bloch, G. J., Masken, C. L., Kragt, C. L., and Ganong, W. F. (1974). *Endocrinology* **94**, 947.

Brooks, C. M. (1940). *Res. Publ. Ass. Res. Nerv. Ment. Dis.* **20**, 525.

Bruce, H. M., and Parkes, A. S. (1963). *In* "Advances in Neuroendocrinology" (A. V. Nalbandov, ed.), p. 282. University of Illinois Press, Urbana.

Clark, W. E. L., McKeown, T., and Zuckerman, S. (1949). *Proc. R. Soc.* **B126**, 449.

Clegg, M. T., and Doyle, L. L. (1967). *In* "Neuroendocrinology" (L. Martini and W. F. Ganong, eds.), Vol II, p. 1. Academic Press, New York.

Critchlow, B. V. (1963). *In* "Advances in Neuroendocrinology" (A. V. Nalbandov, ed.), p. 377. University of Illinois Press, Urbana.

Everett, J. W. (1966). *In* "The Pituitary" (G. W. Harris and B. T. Donovan, eds.), Vol. II, p. 166. University of California Press, Berkeley.

Everitt, B. J., and Hökfelt, T. (1986). *In* "Neuroendocrinology" (S. L. Lightman and B. J. Everitt, eds.), pp. 5–31. Blackwell Scientific Publishers, Oxford.

Fitzpatrick, R. J. (1966). *In* "The Pituitary" (G. W. Harris and B. T. Bonovan, eds.), Vol. III, p.453. University of California Press, Berkeley.

Flerko, B. (1966). *In* "Neuroendocrinology" (L. Martini and W. F. Ganong, eds.), Vol. I, p. 613. Academic Press, New York.

Frantz, A. (1978). *New Engl. J. Med.* **298**, 201–207.

Frisch, R. E., and Revelle, R. (1970). *Science* **169**, 397–399.

Foster, D. L., Karsch, F. J., Olster, D. H., Ryan, K. D., and Yellon, S. M. (1986). *Recent Prog. Horm. Res.* **42**:331–378.

Ganong, W. F. (1959). *In* "Reproduction in Domestic Animals" (H. H. Cole and P. T. Cupps, eds.), p. 195. Academic Press, New York.

Ganong, W. F. (1977). *In* "Reproduction in Domestic Animals," 3rd ed. (H. H. Cole and P. T. Cupps, eds.), pp. 49–77. Academic Press, New York.

Ganong, W. F. (1985). *In* "Perspectives on Behavioral Medicine" (R. B. Williams, Jr., ed.), Vol. II, pp. 25–38. Academic Press, New York.

Ganong, W. F. (1986). *In* "Neuroregulation of Autonomic, Endocrine, and Immune Systems" (R. C. A. Fredrickson, H. C. Hendrie, J. M. Higten, and M. M. Aprison, eds.), pp. 223–241. Martinus-Nijhof, Boston.

Ganong, W. F. (1989). "Review of Medical Physiology," 14th edition, Appleton & Lange, Norwalk/San Mateo, California.

Ganong, W. F., and Kragt, C. L. (1969). *In* "Reproduction in Domestic Animals" (H. H. Cole and P. T. Cupps, eds.), Vol. II, P. 155. Academic Press, New York.

Ganong, W. F., Shepherd, J. R., Van Brunt, E. E., and Clegg, M. T. (1963). *Endocrinology* **72**, 962.

Gellert, R. T., and Ganong, W. F. (1960). *Acta Endocrinol.* **33**, 569–576.

Gorski, R. (1971). *In* "Frontiers in Neuroendocrinology, 1971" (L. Martini and W. F. Ganong, eds.), p. 237. Oxford University Press, New York.

Goy, R. W., and Goldfoot, D. A. (1974). *In* "The Neurosciences Third Study Program" (F. O. Schmidt and F. G. Worden, eds.), p. 571. MIT Press, Cambridge, Massachusetts.

Grumbach, M. M., and Kaplan, S. L. (1990). *In* "Control of the Onset of Puberty" (M. M. Grumbach, P. Sizonenko, and M. Oubert, eds.). p. 1–62. Williams & Wilkins, Baltimore.

Halász, B. (1969). *In* "Frontiers in Neuroendocrinology, 1969" (W. F. Ganong and L. Martini, eds.), p. 307. Oxford University Press, New York.

Harris, G. W. (1955). "Neural Control of the Pituitary Gland." Williams & Wilkins, Baltimore.

Harris, G. W. (1964). *Endocrinology* **75**, 627.

Jacobsohn, D. (1966). *In* "The Pituitary" (G. W. Harris and B. T. Donovan, eds.), Vol. II, p. 1. University of California Press, Berkeley.

Jolly, H. (1955). "Sexual Precocity." Charles C. Thomas, Springfield, Illinois.

Knobil, E. (1980). *Recent Prog. Horm. Res.* **36**, 53.

Knobil, E., and Neill, J. D., eds. (1987). "The Physiology of Reproduction," 2 volumes. Raven Press, New York.

Kragt, C. L., and Meites, J. (1965). *J. Endocrinol.* **76**, 1169.

Krey, L. C., Butler, W. R., and Knobil, E. (1975). *Endocrinology* **96**, 1073.

Laue, L., Kensinger, D., Rescovitz, O. H., Hench, K. D., Barnis, K. M., Loriaux, D. L., and Cutler, G. B., Jr. (1989). *N. Engl. J. Med.* **320**, 496–502.

Marshall, J. C., and Kelch, R. P. (1986). *N. Engl. J. Med.* **315**, 1459–1468.

Marshall, L. A., Monroe, S. E., and Jaffe, R. B. (1988). *In* "Frontiers in Neuroendocrinology" (L. Martini and W. F. Ganong, eds.), Vol. 10, pp.239–278. Raven Press, New York.

Masters, W. H., and Johnson, V. (1966). "Human Sexual Response." Little, Brown, Boston.

Maurer, R. A. (1982). *J. Biol. Chem.* **257**, 2133–2136.

Mayo, K. E., Evans, R. M., and Rosenfeld, G. M. (1986). *Annu. Rev. Physiol.* **48**, 431–436.

McCormack, J. T., and Knobil, E. (1975). *Endocrinology* **96**, A104.

McLeod, R. (1976). *In* "Frontiers in Neuroendocrinology" (L. Martini and W. F. Ganong, eds.), Vol. 4, pp. 169–194. Raven Press, New York.

McNeilly, A. S. (1987). *J. Endocrinol.* **115**, 1–5.

Midgley, A. R. (1973). *In* "Human Reproduction" (E. S. E. Hafez and T. N. Evans, eds.), pp. 201–236. Harper and Row, New York.

Money, J. (1973). *In* "Frontiers in Neuroendocrinology, 1973" (W. F. Ganong and L. Martini, eds.), p. 249. Oxford University Press, New York.

Nicoliks, K., Seeburg, P. H., and Ramachandran, J. (1988). *In* "Frontiers in Neuroendocrinology" (L. Martini and W. F. Ganong, eds.), Vol. 10, pp. 153–166. Raven Press, New York.

Ojeda, S. R., Urbanski, H. F., Ahmed, C. E. (1986). *Recent Prog. Horm. Res.* **42**, 385–440.

Pfeiffer, C. A. (1936). *Am. J. Anat.* **58**, 195.

Plant, T. M. (1985). *Endocrinology* **116**, 1341–1350.

Plant, T. M. (1988). *In* "Frontiers in Neuroendocrinology" (L. Martini, and W. F. Ganong, eds.), Vol 10, pp. 215–238. Raven Press, New York.

Ramirez, V. D., H. H. Feder, and C. H. Sawyer (1984). *In* "Frontiers in Neuroendocrinology" (L. Martini and W. F. Ganong, eds.), Vol. 8, pp. 27–84. Raven Press, New York.

Reichlin, S. (1985). *In* "Williams Textbook of Endocrinology" (J. D. Wilson and D. W. Foster, eds.), 7th ed., pp.492–567. Saunders, Philadelphia.

Reiter, R. J. (1980). *Endocrinol. Rev.* **1**, 109–131.

Shin, S. H., Papas, S., and Obonsawin, M. C. (1987). *Can J. Physiol. Pharmacol.* **65**, 2036–2043.

Sklar, C. A., Conte, F. A., Kaplan, S. L., and Grumbach, M. M. (1981). *J. Clin. Endocrinol. Metab.* **53**, 656–660.

Steele, M. K., and Ganong, W. F. (1986). *In* "Frontiers in Neuroendocrinology" (W. F. Ganong and L. Martini, eds.), Vol. 9, pp. 99–114. Raven Press, New York.

Terasawa, E., Noonan, J. J., Nass, T. E., and Loose, M. D. (1984). *Endocrinology* **115**, 2241–2250.

Weiner, R. W., and Ganong, W. F. (1978). *Physiol. Rev.* **58**, 905–976.

Yen, S. S. C. (1978). *In* "Reproductive Endocrinology" (S. S. C. Yen and R. B. Jaffe, eds.), pp. 152–170. Saunders, Philadelphia.

Ying, S.-Y. (1988). *Endocrinol. Rev.* **9**, 267–293.

Youngstrom, T. G., and Nunez, A. A. (1986). *Brain Res. Bull.* **17**, 485–492.

Gonadotropins

DARRELL N. WARD, GEORGE R. BOUSFIELD, and KATHERINE H. MOORE

I. Introduction

We have learned much about the processes of reproduction in domestic animals since the first edition of this monograph was published just 30 years ago. The progress in the interim is dramatically illustrated by the advances in our understanding of the chemistry of pituitary gonadotropins and related glycoprotein hormones that have been recorded during this period. Thirty years ago the isolation and characterization of this group of hormones were the major focus of effort. Since then these hormones have been characterized for their amino acid composition, carbohydrate composition, protein sequences, oligosaccharide sequences, and the sequences of the related genes. It will be the task of this chapter to summarize this progress.

In the third edition of this monograph Sherwood and McShan (1977) provided a useful review of the gonadotropins as of 12 years ago. Since then various authors have presented reviews that have been utilized in the preparation of the present chapter: see, for example, reviews by Pierce and Parsons (1981), Gordon and Ward (1985), Ryan *et al.* (1987), and Ward *et al.* (1989). For convenience, reviews will be cited whenever possible, but for subjects that have not been reviewed we will cite the pertinent literature.

II. Structural Features of the Gonadotropins

A. *Glycoprotein Hormones*

The two pituitary gonadotropins, follicle-stimulating hormone (FSH) and luteinizing hormone (LH), found in the pituitary glands of all vertebrates, and the placental gonadotropin, chorionic gonadotropin (CG), found only in primates and equids, along with another pituitary hormone, thyroid-stimulating hormone (TSH), make up the glycoprotein hormone family. These hormones are each composed of a common α subunit that is noncovalently associated with a hormone-specific β subunit. Because of the structural relatedness of TSH to the gonadotropins, its structure will be included in the discussion below. A comparative approach to gonadotropin structure–function

has proved fruitful in the past and is likely to remain so. When three-dimensional structures are available, comparisons between all the glycoprotein hormones will yet be necessary in order to understand how the receptors recognize the correct hormone.

The primary structures, some of which were available in the last edition, have been revised in light of new information provided by improvements in protein sequencing techniques, and by the development of DNA sequencing from which the protein sequence can be deduced. There have also been primary structure determinations on more examples of glycoprotein hormones both by protein sequencing, where supplies permit, and by cDNA sequencing for species such as the mouse and the rat where supplies of hormone preclude protein sequencing. As a result we better understand the structural relationships of these hormones. Progress has been made in elucidating the structures of the oligosaccharides found on the gonadotropins of several species. The nature of the microheterogeneity of glycoprotein hormones is now understood to arise not only from different stages of completion of oligosaccharides, but also from the fact that several different oligosaccharide structures are found on a single gonadotropin. Higher orders of structure remain elusive. There is little agreement on the disulfide bond placements in the α subunit and only partial agreement on the disulfide placements in the β subunit. However, the recent crystallization of hCG (Harris *et al.*, 1989) marks significant progress toward an eventual three-dimensional structure of the glycoprotein hormones. For the present we will have to make do with extrapolations from our one-dimensional structures.

Most techniques of biochemical analysis

require that the protein under investigation be torn apart by various chemical and enzymatic means and the resulting components analyzed separately. What follows will be a discussion of the three major components of the glycoprotein hormones: the oligosaccharide moieties, the α subunit, and the β subunit. It is important not to lose sight of the fact that the functional hormones consist of all three parts, each of which is essential for biologic activity and each of which is for practical purposes biologically inactive. At the end of this section, each of the gonadotropins will be discussed as complete entities in an effort to integrate some of the information derived from studies of the parts—but first the parts.

B. Common Features

1. Glycoprotein Hormone Gene Structure

The genes for several α subunits as well as for at least one example of each of the β subunits have been cloned and sequenced. Figure 1 shows the organization of the coding sequences for the mRNAs for the genes that have been examined. Figure 2 compares the mature messenger RNAs for each subunit. The genes for all of the pituitary hormones are found on different chromosomes. For example, in humans the α subunit gene is found on chromosome 6, the FSHβ gene is located on chromosome 11, the TSHβ gene is located on chromosome 1, and the LHβ gene is on chromosome 19. Also on chromosome 19 are the six CGβ genes (Graham et al., 1987), which are believed to be derived from a duplication of the LHβ gene (Fiddes and Talmadge, 1984).

The genes illustrated in Fig. 1 are drawn to the same scale to emphasize that the α

subunit gene is larger than the β subunit genes. The former consists of four exons separated by three introns of varying size, the largest of which, intron A, ranges from 6.5 to 13.4 kilobases in size. Exon 1 consists of untranslated sequences, exon 2 encodes the leader sequence and first 9 amino acids of the mature α subunit, exon 3 encodes residues 10–71 (actually the splice site splits the codon for residue 10 between exons 2 and 3), and exon 4 encodes the C-terminal residues 72–96 and some additional untranslated sequences.

The other interesting features of the α subunit gene, not shown in the figure are the regulatory elements that are only just beginning to be elucidated (Jameson et al., 1989; Nilson et al., 1989). In the pituitary, the α subunit gene is expressed in two cell types, gonadotropes and thyrotropes. In primates and equids, the α subunit gene is expressed in the same pituitary cells and is also expressed in the placenta. The upstream region of the human α gene contains two 18-base-pair elements that confer responsiveness to cAMP (called cAMP response element or CRE) in addition to a single upstream regulatory element (URE). The alpha subunit gene from species lacking a placental gonadotropin (such as cattle) has only a single CRE (Nilson et al., 1989). Transgenic mice containing the human α subunit gene express the α subunit in the placenta as well as in the pituitary, whereas transgenic mice possessing the bovine α subunit gene express it only in the pituitary (Nilson et al., 1989). These studies illustrate that comparative studies are useful at the gene level as well as at the protein level.

The β-subunit genes are similar in that they are much smaller than the α-subunit genes and all possess three exons (except the mouse TSHβ gene, which has two additional exons that code for additional 5′ untranslated sequences). The first exon of the LHβ and hCGβ genes encodes some untranslated sequences and part of the leader sequence. The second exon encodes the rest of the leader sequence and residues 1–41, and the remaining amino acids are encoded in the third exon. The primary differences

between the LH and CG β subunit genes are in the lengths of the untranslated sequences. The CGβ gene has a longer 5′ untranslated region and a shorter 3′ untranslated region. The latter is shorter because of the C-terminal extension, which arose by two changes in the LHβ gene. One change was a single-base deletion that eliminated

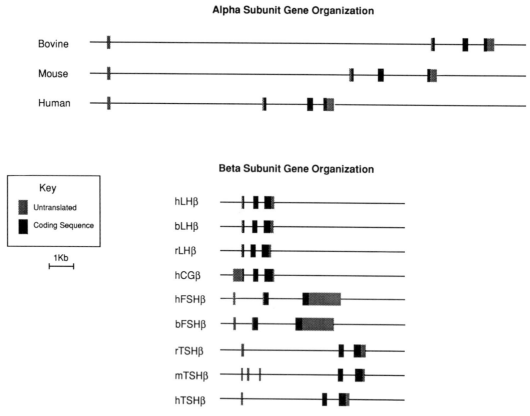

Figure 1 Comparison of the glycoprotein hormone α and β subunit genes. The genes are drawn to the same scale to illustrate the large differences in size between the α and β subunit genes, which are primarily accounted for by the larger introns in the α subunit genes. Based on genomic DNA sequences of bovine α (Goodwin *et al.*, 1983), mouse α (Gordon *et al.*, 1988), human α (Fiddes and Goodman, 1981), hLHβ (Talmadge *et al.*, 1984b), bLHβ (Virgin *et al.*, 1985), rLHβ (Jameson *et al.*, 1984), hCGβ (Talmadge *et al.*, 1984b), hFSHβ (Watkins *et al.*, 1987; Jameson *et al.*, 1988), bFSHβ (Kim *et al.*, 1988), rTSHβ (Croyle *et al.*, 1986), mTSHβ (Gordon *et al.*, 1988), and hTSHβ (Guidon *et al.*, 1988; Wondisford *et al.*, 1988).

the LHβ stop codon allowing readthrough into the 3' untranslated region and resulted in an additional 16 residues. The second change was a two base insertion that continued the open reading frame to code for the additional 8 amino acid residues. Together, these two mutations in the ancestral LHβ gene resulted in a shorter untranslated region in the third exon of the eCGβ gene (Talmadge *et al.*, 1984b). In the FSHβ and TSHβ genes the first exon encodes only untranslated sequences. The second exon encodes the leader sequence plus the first 34 or 35 residues of the mature protein. The remaining amino acids are encoded in the third exon. Bovine and human FSHβ have a relatively long 3' untranslated region, which may be a general feature of FSHβ genes.

2. Carbohydrate

The glycosylation of the glycoprotein hormones distinguishes them from the other pituitary hormones, although glycosylated forms of such hormones as prolactin and growth hormone have been reported in recent years (Markoff *et al.*, 1989; Sinha and Lewis, 1986). Variations in gonadotropin molecular weight have been correlated with the reproductive cycle (Bogdanov and Nansel, 1978). Subsequently these were attributed to changes in carbohydrate structure. A related functional role for the carbohydrate moieties has been suspected, and much work has been dedicated to discovering this role. Progress has been made in determining some of the oligosaccharide structures (Baenziger and Green, 1988;

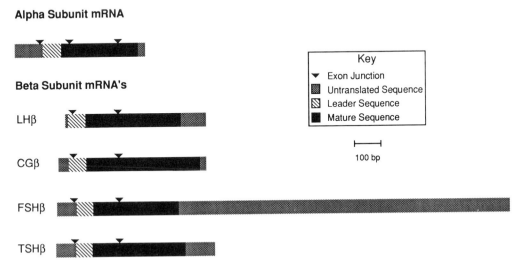

Figure 2 Comparison of the mature mRNAs for the glycoprotein hormone subunits. The mRNAs are drawn to illustrate the 3'- and 5'-untranslated regions, exon junctions, leader sequence and mature sequence. Adapted from Fig. 1.

Grotjan, 1989), and there is a growing body of evidence for carbohydrate involvement in signal transduction (Governman *et al.*, 1982; Liu *et al.*, 1984; Matzuk *et al.*, 1989; Sairam, 1980)

a. Carbohydrate Distribution Unlike other hormones now known to be glycoproteins, in which a single protein chain is glycosylated, both the α and β subunits of the glycoprotein hormones are glycosylated. Figure 3 shows the patterns of glycosylation for the α and β subunits of the glycoprotein hormones. For the α subunits, the pattern is very consistent: two N-linked oligosaccharides attached to asparagine 56 and 82. Perhaps the invariant glycosylation pattern results because the α-subunit oligosaccharides are essential for biological activity. Note that the human glycosylation pattern is slightly different. Because of a 12-base deletion in the second exon, there is a four amino acid deletion near the human α-subunit N-terminus, which results in the numbering of the glycosylation sites being 52 and 78. However, for the sake of clarity, all the numbering of the glycoprotein hormone sequence will be based on ovine LH, the first LH that was sequenced. All the other sequences will be aligned by their half-cystines, which are highly conserved. This permits comparisons of homologous parts of the hormones to be made more easily. The reader should remain aware that frameshifts in numbering systems for individual hormones will be encountered in the original literature for this reason. Thus, in Fig. 3, with the half-cystines in register the human glycosylation sites line up with those in all the other alpha subunits. A similar convention will be followed with the N-linked glycosylation sites in the β subunits below.

Free-α subunit is found in all glycoprotein hormone-producing tissues. It differs from α subunit that is part of a hormone in that it may be O-glycosylated at Thr[43]. In addition, the oligosaccharides on free-α tend to differ from those on other α subunits by being more complex and more completely processed (Blithe and Nisula, 1985). The functional significance of free-α is not known. There are reports that the α subunit stimulates differentiation of lactotropes *in vitro* (Bégeot *et al.*, 1984). Secretion of free-α results from the higher rate of biosynthesis in some systems, especially in the term placenta (Boime *et al.*, 1982), as well as its resistance to intracellular degradation (Peters *et al.*, 1984). Even in *in vitro* systems in which α subunit is synthesized at nearly the same rate as β subunit (Peters *et al.*, 1984) the uncombined β subunits are degraded and only a small amount is secreted, whereas most of the α subunit is secreted either as part of a hormone heterodimer or in the uncombined state (Corless *et al.*, 1987; Peters *et al.*, 1984).

Patterns of glycosylation on the β subunit are more variable than those on the α subunit, perhaps because the oliogsaccharide is

Figure 3 Patterns of glycosylation for the α and β subunits of the glycoprotein hormones and for free α subunit. The N-linked and O-linked glycosylation sites are illustrated for each of the known glycoprotein hormones. The protein sequences are represented by the solid bars, aligned with the half-cystines in register. The position in the sequence where the glycosylation site is found in each subunit is indicated by the numbers(s) below each glycosylation site. (Note: Numbering is based on oLH for convenience. Where the actual numbers differ significantly they are given in parentheses.)

Patterns of Alpha Subunit Glycosylation

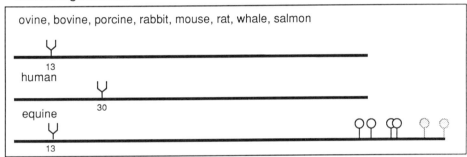

Patterns of Beta Subunit Glycosylation

Luteinizing Hormone

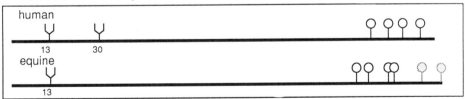

Follicle-Stimulating Hormone

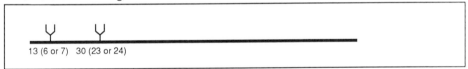

Chorionic Gonadotropin

Thyroid-Stimulating Hormone

less essential for biological activity. There are one or two N-linked glycosylation sites at positions 13 or 30, or both. Most LH β subunits have only a single glycosylation site at Asn^{13}. Human LH and equine LH (see below) are two exceptions. Human LH lacks the glycosylation site Asn^{13} but has the Asn^{30} site and is glycosylated there. The sequence at Asn^{13} in all other LH β subunits is -Asn-Ala-Thr-, which meets the requirement for N-glycosylation signal sequence: -Asn-X-Thr/Ser-, where the X means any amino acid can be substituted and the third position can be either Thr or Ser. In human LHβ the sequence at Asn^{13} is -Asn-Ala-Ile- (see Fig. 8). The substitution of Ile for Thr prevents glycosylation. Glycosylation is made possible at position 30 by the substitution of Asn for Thr resulting in the sequence -Asn-Thr-Thr-. Equine LHβ has recently been found to possess a C-terminal glycopeptide extension (Bousfield et al., 1985b) that heretofore had been considered to be characteristic of only the chorionic gonadotropins. This "characteristic structure," however, was based on only one example in humans where both hLH and hCG had been sequenced. N-Linked glycosylation of eLHβ occurs at Asn^{13} just as in all other LH β subunits. Preliminary data lead to the proposed sites for O-linked glycosylation indicated in Fig. 3. In the cases where two Ser residues or where Thr and Ser were adjacent, it was not possible to determine the exact position of the oligosaccharide, and these are indicated by the shaded symbols representing the O-linked oligosaccharides. Unfortunately, there is no recognized signal sequence for O-linked glycosylation. Glycosylation of Thr and Ser usually occurs in regions rich in Pro as well as Thr and Ser. The C-terminal extension of eLHβ is rich in all three of these amino acids. FSH β subunits

possess both N-linked glycosylation sites Asn^{13} and Asn^{30}, whereas TSHβ is glycosylated only at Asn^{30}. The chorionic gonadotropins show two different patterns of glycosylation for the two known examples: hCG and eCG. Human CGβ is N-glycosylated at both Asn^{13} and Asn^{30}. In addition it is O-glycosylated at Ser residues 121, 127, 132, and 138 (Birken and Canfield, 1977; Kessler et al. 1979; Keutmann and Williams, 1977). The predicted sequence from cDNA of baboon CG suggests an identical pattern of glycosylation. In contrast, there is a single N-linked glycosylation site in eCGβ at Asn^{13} and there are at least five O-glycosylation sites in the C-terminal extension. It is interesting to note that eCG is more heavily glycosylated than hCG. While hCG has 31% carbohydrate, eCG has 45% carbohydrate. Part of the difference in the carbohydrate contents results from more extensive branching in the equine oligosaccharides (Bahl and Wagh, 1986). This has hampered structure determination for the latter, and only partial structures are available at present.

b. Oligosaccharide Structures The story of the glycoprotein oligosaccharides is one of increasing complexity, reviewed by Grotjan (1989). The first structure proposed for hCG was for a single N-linked oligosaccharide structure (structure 5 in Fig. 4). The assumption was that all the other heterogeneity was due to incomplete processing (Kessler et al., 1979). At the same time, Endo et al. (1979) reported five charged and three neutral oligosaccharide structures. One structure was the same as structure 5, and others were similar in that they were missing only one residue such as the fucose or one or both of the terminal sialic acid residues. Some variations were

N-linked Oligosaccharide Structures found in Ovine, Bovine, and Human Gonadotropins

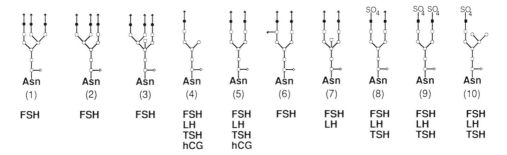

Asn (1)	Asn (2)	Asn (3)	Asn (4)	Asn (5)	Asn (6)	Asn (7)	Asn (8)	Asn (9)	Asn (10)
FSH	FSH	FSH	FSH LH TSH hCG	FSH LH TSH hCG	FSH	FSH LH	FSH LH TSH	FSH LH TSH	FSH LH TSH

O-linked Oligosaccharide Structures found in Equine and Human Gonadotropins

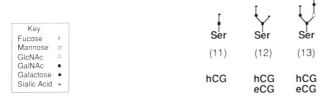

Key	
Fucose	▷
Mannose	○
GlcNAc	□
GalNAc	■
Galactose	●
Sialic Acid	▲

Ser (11)	Ser (12)	Ser (13)
hCG	hCG eCG	hCG eCG

Figure 4 Some of the N- and O-linked oligosaccharide structures found on the glycoprotein hormones (Baenziger and Green, 1988; Grotjan, 1989). Only the completed structures are represented. Many other oligosaccharides are found that are incomplete versions of these structures, mainly lacking terminal residues such as sulfate, sialic acid, and fucose.

consistent between subunits; for example, there was no fucose on any of the oligosaccharides isolated from the α subunit (Kessler *et al.*, 1979). More recent investigations suggest additional carbohydrate heterogeneity (Blithe and Nisula, 1985; Grotjan and Cole, 1989). Parsons and Pierce (1980) found that bLH was sulfated on its terminal sugars, which prevented exoglycosidases from hydrolyzing terminal sugars, thus hindering structural characterization. A recent series of studies from Baenziger's laboratory has proposed a variety of oligosaccharide structures from pituitary glycoprotein hormones (Baenziger and Green, 1988). They characterized the sulfated oligosaccharides on bLH (Green *et al.*, 1985). They next ex-

amined the oligosaccharides on the ovine, bovine, and human pituitary hormones and found that each hormone possessed several oligosaccharide structures (Green and Baenziger, 1988a, 1988b). The complete structures are shown in Fig. 4. Some hormones, such as oFSH, had 16 charged structures, accounting for 68% of the oligosaccharides released by peptide N-glycanase digestion. They did not analyze the numerous neutral oligossacharides that were released. The other characterized structures were variations on these lacking one or more sialic acid residues, or lacking sulfate or lacking fucose. Based on these results, the oligosaccharides fell into three categories: sialylated, sulfated and hybrid sulfated–sialylated. The figure shows the qualitative distribution of the oligosaccharide structures. For more detailed description of this distribution, see the recent review by Baenziger and Green (1988). The distributions of the structures varied from species to species and from hormone to hormone. For example, hCG and hFSH had essentially only sialylated oligosaccharides, and most species of LH possessed predominantly sulfated oligosaccharides or hybrids. Ovine FSH, on the other hand, had a little bit of everything. Figure 5, based on the report of Baenziger and Green (1988), summarizes the biosynthetic buildup of these structures. (See also Section III, page 55, on biosynthesis of the gonadotropins for further comment.)

c. Carbohydrate Function Carbohydrate influences glycoprotein hormone biosynthesis and action. Aspects of glycoprotein hormone biosynthesis that have been demonstrated (or speculated) to be influenced by carbohydrate include folding, intracellular survival, differential packaging of LH and FSH by the same cell, and secretion (Green et al., 1986). Carbohydrate affects circulating glycoprotein hormones by influencing their lifetime in circulation and by modulating their specific activities (Chappel et al., 1983). It is the effect of oligosaccharide variation on gonadotropin biologic activity that can best be studied on isolated hormone. With regard to this effect, in vitro studies have suggested that deglycosylated gonadotropins can act as antagonists. However, in vivo experiments do not support this idea.

Chemically or enzymatically deglycosylated CG, LH, FSH, and TSH bind their receptors with a higher apparent affinity than do native hormones (Berman et al., 1985; Calvo et al., 1986; Governman et al., 1982; Sairam and Schiller, 1979). However, the activities of deglycosylated glycoprotein hormones are much lower in assays that measure a target cell response such as steroidogenesis (Berman et al., 1985; Calvo et al., 1986; Governman et al., 1982; Sairam and Schiller, 1979). By deglycosylating α or β subunit and then recombing the deglycosylated subunit with the intact complementary subunit, it was demonstrated that only the α-subunit carbohydrate was required for these responses (Governman et al., 1982; Liu et al., 1984; Sairam, 1980). In the course of these studies, some investigators reported

---

Figure 5 Biosynthetic pathway leading to completed N-linked oligosaccharide structures shown in Fig. 4. Based on the pathway for N-linked glycosylation (Kornfeld and Kornfeld, 1985) and pathways suggested by structures on glycoprotein hormones (Baenziger and Green, 1988). (1) Oligosaccharyl transferase catalyzed transfer of dolicol-linked $Glc_3Man_9GlcNAc_2$ oligosaccharide to asparagine at glycosylation site -Asn-X-Ser/Thr-. (2) α-Glucosidase I removes terminal glucose. (3) α-Glucosidase II removes remaining two glucose residues. (4) Endoplasmic reticulum α-1,2-mannosidase removes one mannose residue. (5) Golgi α-glucosidase I removes another three mannose residues producing the precursor for N-acetylglucosamine transferase I. (6) N-Acetylglucosamine transferase I adds N-acetylglucosamine, creating precursor for S-1 type having only one completed branch or leading to continued processing; (a) addition of N-acetylgalactosamine diverts processing to complex bi- or triantennary oligosaccharide resulting in this single-branch sulfated structure. (7) Golgi α-mannosidase II removes two mannose residues; (a) addition of N-acetylgalactosamine could lead to this structure. (8) N-Acetylglucosamine transferase II adds second N-acetylglucosamine, leading to further processing which produces the following oligosaccharide structures: (a) triantennary sialylated, (b) biantennary sialylated, (c) hybrid sialylated–sulfated, and (d) disulfated.

Rough Endoplasmic Reticulum

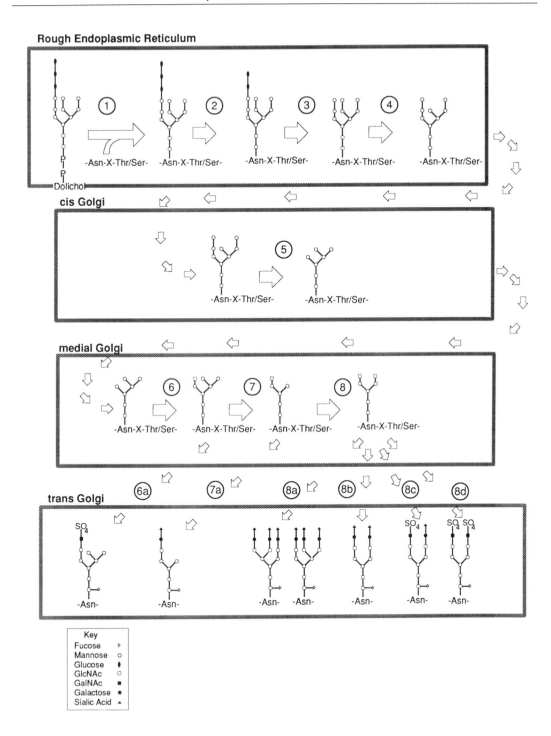

cis Golgi

medial Golgi

trans Golgi

Key
Fucose ▷
Mannose ○
Glucose ●
GlcNAc □
GalNAc ■
Galactose ●
Sialic Acid ▲

that the deglycosylated gonadotropins acted as antagonists to the native hormones (Sairam, 1983). It was proposed that the deglycosylated hormone occupied receptors and formed inactive hormone–receptor complexes, which prevented native hormone from binding to receptors and activating the target cells. On the other hand, Liu *et al.* (1984) showed deglycosylated hormone had a faster on-rate than native hormone, but the off-rates of both were comparable. This probably contributes to the poor *in vivo* antagonism of the deglycosylated hormone.

Deglycosylated hCG was administered to normal cycling women in an attempt to interfere with LH maintenance of postovulatory corpora lutea (Patton *et al.*, 1988). However, the deglycosylated hormone preparation had no effect on luteal phase except to elevate progesterone levels, prompting the investigators to suggest that agonist activity in the preparation interfered with antagonist activity. More recently, the activities of native, desialylated, and deglycosylated hCG were compared in male cynomolgus monkeys (Liu *et al.*, 1989). hCG was deglycosylated and the preparation was passed over a concanavalin A–Sepharose column to remove any partially deglycosylated hCG. In rat testis Leydig cells *in vitro*, a clear antagonism of hCG stimulation of cAMP production was observed with the deglycosylated hCG. The serum half-life of deglycosylated hCG was 23 min—23 times that of asialo-hCG, which is rapidly cleared by a galactose-binding lectin in liver (Morell *et al.*, 1971)—but less than one-tenth that of native hCG. However, the early (48 h) responses to stimulation by hCG, asialo-hCG, and deglycosylated hCG were identical. Again there was no detectable *in vivo* antagonistic activity in deglycosylated hCG. After 48 h, testosterone levels of hCG treated monkeys were still elevated while those of asialo-hCG and deglycosylated hCG were back to normal levels, suggesting carbohydrate affects long-acting hormone stimulation *in vivo.*

A similar set of results has accumulated for FSH. Although no *in vivo* studies have been completed, circulating forms of hFSH with *in vitro* antagonist effects have been reported (Dahl *et al.*, 1988). In a review of FSH biosynthesis and secretion, Chappel and colleagues summarized evidence by a number of investigators who had demonstrated physicochemical changes in FSH in response to changing physiologic conditions (Chappel *et al.*, 1983). These changes in apparent molecular size and distribution of isoelectric forms were attributed principally to variations in FSH sialic acid content. They noted that desialylation of FSH shortened its circulatory half-life but elevated its potency in receptor binding assays. A model was proposed in which FSH was modified by neuraminidase action, which converted relatively more acidic, long-acting, low-potency forms into more basic, more potent, short-lived forms. However, when the FSH assay was changed, the FSH activity profile changed. Hsueh and colleagues have developed an *in vitro* steroidogenesis assay sensitive enough to detect FSH in serum (Jia and Hsueh, 1986). In a subsequent study the more acidic forms of FSH (Chappel's less-potent, long-acting forms) were determined to be the active circulating forms. Moreover, the more basic forms, which were released in response to a GnRH antagonist, were found to be FSH antagonists in the steroidogenesis bioassay for FSH (Dahl *et al.*, 1988). These contradictory results emphasize our relative ignorance of the structure and function of glycoprotein hormone carbohydrate.

C. The Alpha Subunits

1. General

Two lines of evidence lead to the concept of the α subunit of the glycoprotein hormones as a common subunit: (1) chemical similarity between TSHα and LHα and (2) results of subunit recombination experiments which produced active hybrids of LH and TSH subunits (Liao and Pierce, 1970). The discovery that there is only a single gene for the α subunits in humans, cattle, horses, mice, and rats provides molecular confirmation of this concept (Chin et al., 1981; Fiddes and Goodman 1981; Godine et al., 1982; Goodwin et al., 1983; Stewart et al., 1987). This does not mean that the α subunits are identical. The striking qualitative differences for the carbohydrate compositions of the porcine α subunits (Maghuin-Rogister et al., 1975) suggest that the oligosaccharides distinguish α subunit from one hormone from α subunit of another hormone. However, as mentioned above, little information is yet available on oligosaccharide structures from subunits. When it became possible to chemically deglycosylate glycoprotein hormones it was found that the carbohydrate on the α subunit, but not that on the β subunit, was required for biologic activity (Governman et al., 1982; Liu et al., 1984; Sairam, 1980). Deglycosylated hormones bound receptors as well as or better than native gonadotropins, but responses requiring signal transduction, such as steroidogenesis, were only partially stimulated or even inhibited. Boime and colleagues have recently demonstrated that the Asn^{56}

position possessed the critical oligosaccharide for signal transduction (Matzuk et al., 1989). However, the role of the carbohydrate in signal transduction is unknown. It has been suggested that the carbohydrate is directly involved, binding a lectin essential for the hormone–receptor complex signaling (Ryan et al., 1987). It has also been suggested that absence of the carbohydrate results in subtle changes in conformation of the hormone that prevents signal transduction by the hormone–receptor complex (Rebois and Fishman, 1984). Little else is known of the carbohydrate structures except that differences in the susceptibility to exoglycosidase digestion have been reported with the Asn^{82} oligosaccharide more susceptible than the Asn^{56} oligosaccharide (Miura et al., 1988; Ronin et al., 1987).

2. Amino Acid Sequences

A generic mature α subunit mRNA and the corresponding amino acid sequences of the α subunits are shown in Fig. 6. The leader sequences are indicated for bovine, mouse, rat, and human alpha subunits, which have been deduced from the cDNA or genomic DNA sequences. A cDNA sequence for the equine α subunit also exists; however, the clone lacks the leader sequence and the first three residues at the amino terminus (Stewart et al., 1987).

The leader sequences that have been determined consist of 24 amino acid residues and are 76.5% identical. This is only slightly less than the degree of identity observed for the mature protein, 81.3%. Data for the predicted leader sequences of the human, mouse, and bovine alpha subunit agree with protein sequence data for these pre-α sub-

Alpha Subunit Sequences

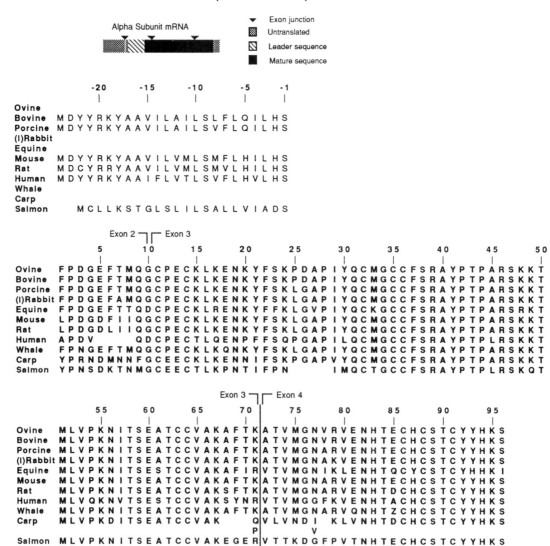

Figure 6 Compilation of the reported glycoprotein hormone α-subunit amino acid sequences. The original references to the studies indicated here are cited in several recent reviews (Pierce and Parsons, 1981; Ward and Bousfield, 1989; Ward *et al.,* 1989). Blank areas indicate deletions (gene level) compared to the other species. The two residues below the carp sequence represent genetic alleles at these positions. Exon junctions are indicated. (Single-letter amino acid code: A, alanine; B, either asparagine or aspartic acid; C, cysteine; D, aspartic acid; E, glutamic acid; F, phenylalanine; G, glycine; H, histidine; I, isoleucine; K, lysine; L, leucine; M, methionine; N, asparagine; P, proline; Q, glutamine; R, arginine; S, serine; T, threonine; V, valine; W, tryptophan; Y, tyrosine; Z, either glutamine or glutamic acid.) Glycosylation sites are indicated by an asterisk. Exon boundaries are indicated.

units (Birken *et al.*, 1981; Giudice *et al.*, 1979), but all the rest are predictions from DNA sequences.

Amino-terminal heterogeneity is extremely common in the α subunits that have been examined (Ward *et al.*, 1989). The first six residues at the amino terminus are frequently missing. The pFSHα subunit N-terminus was recently reexamined and found to have the same sequence as ovine, bovine, rabbit, and sheep α subunits (Sugino *et al.*, 1989). Prior to that time, the N-terminal amino acid was thought to be Thr[7]. When only the protein sequences were known, the sequence differences between alpha subunits were used to argue that the commom subunits were slightly different. Since then sequences have been refined and we now know them to be identical within a species. This is a consequence of the fact that there is only a single α-subunit gene (Chin, 1986). The only N-terminal sequence difference that has turned out to be real results from a 12-base-pair (four amino acid) deletion in the human α subunit gene. This was found by comparison of the human and bovine α-subunit cDNA sequences (Chin *et al.*, 1983). The gap in the protein sequences in Fig. 6 reflects the missing portion of the gene. Most of the mammalian α subunits average 93% identical. The two exceptors are human and equine α subunits, which average only 74% and 78% identical, respectively. A nonmammalian gonadotropin, carp gonadotropic hormone (GTH), has recently been sequenced that averages 73% identical with all mammalian α subunits. Two deletions, one of three residues and one single amino acid deletion, have occured in the carp α subunit, and there could be a missing glycosylation site that is otherwise remarkably conserved

with respect to the other α subunit sequences.[1]

Creation of "hybrid hormones" in which the α subunit of one species is combined with the β subunit of another species is readily done and has proven very informative (Reichert *et al.*, 1973). By this means it has been shown that alpha subunits are for the most part interchangeable. That is, the α subunit from any glycoprotein of any species can be recombined with the β subunit from any other glycoprotein hormone and the hybrid will take on the characteristics of the hormone from which the β subunit was obtained with little modulation by the α subunit (Strickland and Puett, 1981). Exceptions include oLHβ:cGTHα[1] (Burzawa-Gerard and Fontaine, 1976) and any recombinant of eLHβ or eCGβ with a nonequine α subunit. Often when recombination experiments don't yield active products, it is assumed that the subunits did not recombine. In the case of the equine β subunit recombinants, hybrids were determined to have formed but they were inactive. On the other hand, equine α subunit enhanced the potency of β subunits from normally less active hormones such as oLH and pLH (Bousfield *et al.*, 1985a).

1. The sequence in Fig. 6 shows an Asp residue at postion 56 instead of Asn. Since this is the glycosylation site that is essential for signal transduction, the loss of this glycosylation site in the carp hormone might be expected by this substitution. However, the assignment of Asp at this position might result from the chemistry involved in the sequence determination. If the dansyl– Edman procedures were used then the hydrolysis step would convert any Asn to Asp. (The details of this sequence study have not yet been published.) Certainly it is intriguing, since nonmammalian gonadotropins are inactive in mammalian bioassay systems, and this may be related to this structural feature.

3. Disulfide Bonds

There are 10 half-cystines in the α subunit that form 5 disulfide bonds. Figure 7 shows some of the proposals for the disulfide bonds that have been made. It can be seen that there is no agreement between the first three structures. The last two proposals are in complete agreement; however, the same procedures of partial reduction were followed in both cases. Disulfide bond placement is difficult because of the clustering of the half-cystines; there are two pairs of adjacent half-cystines, and one pair has a single residue in between. The clustering prevents proteolytic selective cleavage between disulfide bonds (one of the approaches to disulfide placement). Also, this is further complicated by dismutation (disulfide scrambling) under basic conditions commonly used for proteolysis. The first two proposals imply that the N- and C-terminal halves of the α subunit are folded over to each other. The assignment in the latter two proposals suggest that the N- and C-terminal halves fold independently to form disulfide-folded structures held together by the disulfide bond between half-cystines 32 and 64.

The primary structure of a protein determines the three-dimensional structure. There should be enough information in a subunit to refold following complete reduction of the disulfide bonds. This is true for ovine, bovine, and human alpha subunits, but only partly true for the equine α subunits. The latter form a significant amount of high-molecular-weight aggregates rather than forming monomeric α subunit.

4. Role of the Common Subunit

Individual subunits are essentially inactive from a physiological standpoint. When recombined they form an active dimer with a specific function and activity. Hormone specificity is determined by the β subunit. The potency of the hormone in most cases is also dictated by the potency of the hormone from which the beta subunit was obtained. Thus, hCG is one of the most active hormones in most measures of LH activity in mammalian bioassay systems. We refer here to "specific activity," for example, potency/unit weight. Hybrids made with hCGβ and the alpha subunits of ovine, bovine, and porcine LH are as active as hCG even though the hormones from which they were obtained are much less active than hCG (Bousfield et al, 1985a; Strickland and Puett, 1981). An exception of the potency determination rule is equine α subunit. Hydrids of equine LHα and the β subunits of ovine and porcine LH are much more active than oLH or especially pLH (Bousfield et al., 1985a). Examination of the amino acid sequences in Fig.6 shows these to be highly conserved for all α subunits. Equine α subunit has a His-Tyr transposition at positions 87 and 93. The C terminus of the α subunit is known to be required for biologic activity (Cheng et al., 1973). Removal of the C-terminal residues by carboxypeptidase digestion eliminates the ability to associate with TSHβ. This α derivative is able to associate with LH β, but the heterodimer is inactive. Since the C terminus is required for biologic activity, any substitutions in the C-terminal region might be expected to have effects on biologic activity. Both human and equine α subunits are the most divergent from all the other mammalian α subunits. The differences derive from scattered single amino acid substitutions in either subunit. In the human α subunit there are no detectable consequences of these substitutions. However, in the horse, the tendency to greater hydrophobicity as evidenced by the ten-

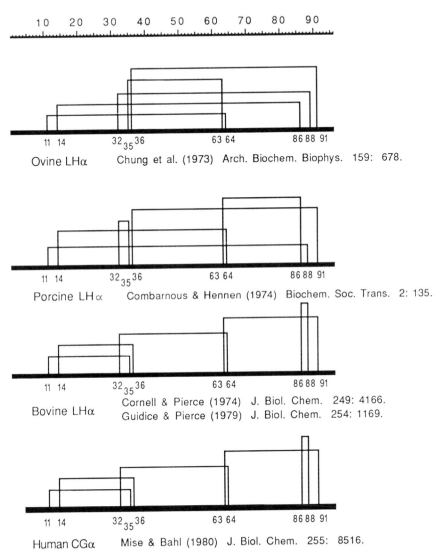

Figure 7 Disulfide proposals for the glycoprotein hormone α subunits.

dency to aggregate following disulfide bond reduction and the potency enhancement of hybrid LH molecules are two consequences of these substitutions.

D. The Beta Subunits

1. General

The β subunit is called the hormone-specific subunit because this subunit determines what hormonal activities the hormone will have. There are at least three β-subunit genes in all mammals, and primates have a fourth β-subunit gene. All of the glycoprotein hormone genes are found on seperate chromosomes except the hLH/hCGβ gene cluster, which is on chromosome 19 and consists of a single copy of the hLHβ gene and six copies of the hCGβ gene (Graham *et al*, 1987). Most of the latter appear to be pseudogenes, which lack transcriptional regulatory elements, but only two or three of them (gene 5, gene 3, and possibly gene 1) appear to be translated (Talmadge *et al*, 1984a). Comparison of the sequences of hCGβ and hLHβ (Fig. 8) shows them to be different proteins even if the C-terminal extension on hCGβ is ignored. In the horse, there may be a different organization of the LH and CG β-subunit encoding regions. The amino acid sequences of the two subunits are identical. Therefore, it is possible that a single gene might encode both LH and CG β subunits. Thus, in the case of the equine β-subunit gene, there may be gene regulatory elements that permit expression of the gene in both the pituitary and in the placenta, such as those that have been identified in the human α subunit gene (Nilson *et al*, 1989). Resolution of these possi-

bilities will require cloning of the equine genes.

2. Amino Acid Sequences

Comparison of the sequences in Fig. 8 shows differences and similarities between the β subunits. The sequences are aligned by placing their half cystines in register. This allows comparisons of highly conserved regions to be made even though TSH and FSH are shorter by six or seven amino acids at the N terminus. Two single amino acid gaps are inserted in the LH/CGβ and FSHβ sequences to accommodate two insertions in the TSHβ long loop (β38–57) sequence. Recent revisions of pFSH,hFSH, and hLH sequences have allowed these three sequences to be put in proper register with the rest of the LH/CGβ and FSHβ sequences. From the predicted sequence from the hLHβ gene (Talmadge *et al.*, 1984b), the sequence Met-Met was found at positions 41–42. This resolved a discrepancy between sequence proposals for hLHβ as to whether one (Shome and Parlow, 1973) or two methionines (Sairam and Li, 1975; Ward *et al.*, 1973) separated by an arginine (Closset *et al.*, 1973) were present in this region and thereby eliminated a gap that had to be made in some sequence compilations (Pierce and Parsons, 1981). There were some additional residues in pFSHβ and hFSHβ that were removed by reevaluation of the protein sequences (Shome *et al.*, 1988; Sugino *et al.*, 1989) and from the sequence predicted from the cDNA (Kato, 1988) that put these structures in register with all the other known FSHβ sequences (Ward and Bousfield, 1989).

The leader sequences for all the LH β subunits are similar to those of the other LHs yet different from those for FSH and

TSH β subunits. And the same is true for the leader sequences for both of those β subunits. The LH/CG β subunits have an additional six or seven amino acids at the N terminus that are lacking in TSHβ and FSHβ. N-Terminal heterogenity has been observed in FSHβ resulting in uncertainty of the N-terminal amino acid in some cases. Where unambiguous data exist, the cleavage sites of the leader sequences for the α subunits (Ser), LH (Ala), CG (Ala), and TSH β subunits (Ser) are strictly conserved. These observations are consistent with the preference of signal peptidases to cleave after amino acids with small side chains (Jackson and Blobel, 1980). Therefore, we have assumed the same cleavage point for all FSHβ sequences.

3. Disulfide Bonds

There are 12 half-cystines in the β subunit that form six disulfide bonds. Figure 9 shows some of the proposals for the disulfide bonds that have been made. The assignments for oLHβ and for hCGβ are the same (Mise and Bahl, 1981; Tsunasawa et al., 1973), which supports the belief that the disulfide bonds will be the same for all four β subunits. There is good agreement for two disulfides: 93–100 and 26–110. These exemplify two types of structures stabilized by disulfide bonds. One type results in long-distance interactions between the N- and C-terminal halves of the subunit. The other type forms loops. Two disulfide bonds, one at 93–100 and the other at 38–57, form loops. It is interesting to note that these two loops have been found to be part of the receptor region of gonadotropins (Ryan et al., 1987). The uncertainties in the disulfide bond assignments make disulfide determination an important problem. However, it is a difficult problem for the same reasons noted above for the α-subunit disulfides.

4. Beta Subunit Contributions to Hormone Structure

There is not much known about the interaction of the α and β subunits, other than the fact that the functional hormone is a combination of the two subunits. Early circular dichroism studies indicated that a conformational shift occurred when LH was dissociated into α and β subunits. Because most of the differences in the spectra were attributable to the α subunit, it was proposed that this subunit was more flexible than the β subunit (Jirgensons and Ward, 1970). This suggested that the common α subunit could conform to the structure of the more rigid β subunit in a complementary manner different for each β subunit to produce a unique hormone structure for each glycoprotein hormone. Subsequent studies followed the kinetics of change in the CD spectra during reassociation of subunits and indicated that there was an initial rapid change, which was interpreted as an initial loose binding of the subunits. This interpretation was supported by ultracentrifugation data that indicated that dimerization was complete by 2 h. There followed slower changes, followed by observing changes in one or more buried tyrosines as the hormone conformation finally achieved its native state (Bewley, 1979; Garnier, 1978). The advent of radioligand assays provided a relatively quick assay for biologic activity that permitted the kinetics of α–β association to be followed (Reichert et al., 1973). The generation of biologically active hormone was relatively slow, with maximum recovery of biologic activity requiring 8–10 h. The kinetics of the recovery of biologic ac-

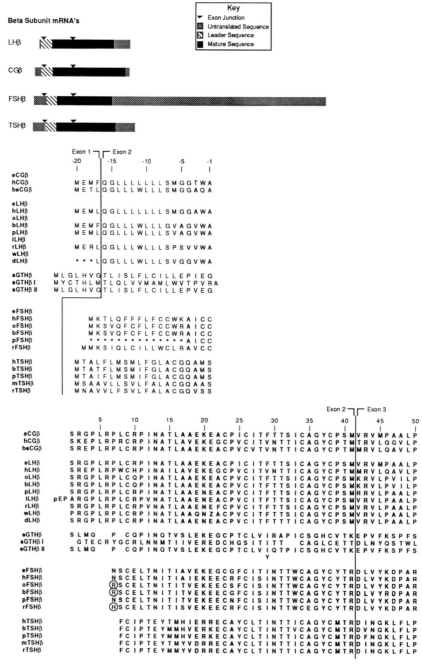

Beta Subunit Sequences

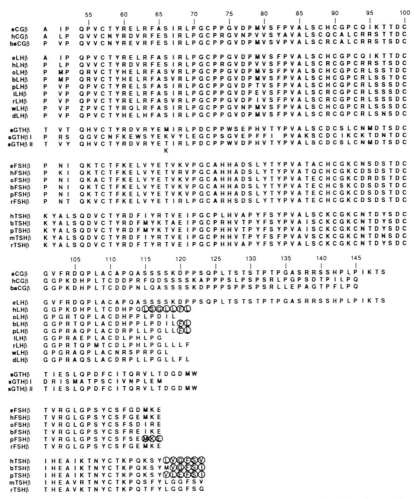

Figure 8 Compilation of the reported glycoprotein hormone β subunit amino acid sequences. The original references to the studies indicated here are cited in several recent reviews (Pierce and Parsons, 1981; Ward and Bousfield, 1989; Ward *et al.*, 1989). For the single letter amino acid code, see the legend to Fig. 6. The practice has evolved to designate species of origin by a lower case prefix for the species (e.g., oLH) as listed above. The ba designates baboon; d, dog; e, equine; h, human; o, ovine, b, bovine; p, porcine; r, rat; m, mouse; w, whale; s, salmon; (GTH for gonadotropic hormone, since in lower vertebrates the LH, FSH, and TSH functions appear not yet evolved). Since rabbit hormones would be confused with rat, their designation is with a "l" for lagomorph. Asterisk, cDNA did not include the entire coding sequence. Circled residues are those predicted from the DNA sequence that have not yet been observed in the protein. Exon junctions are indicated.

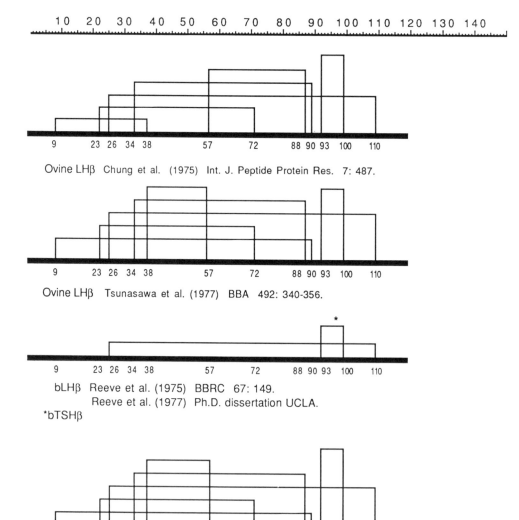

Figure 9 Disulfide proposals for the glycoprotein hormone β subunits.

tivity suggested a multistep process involving an initial association of α and β subunits followed by additional steps that convert this inactive α–β dimer into an active hormone. Sometimes the native conformation is not attained, as suggested by the existence of inactive interspecific hybrids that have been discussed above. For further reading, Garnier has provided a good view of these and many other studies of glycoprotein hormone subunit assembly (Garnier, 1978).

E. Follicle-Stimulating Hormone (Follitropin)

FSH, as its name implies, stimulates follicular development in the ovary. Its targets in the female are the granulosa cells of ovarian follicles. Along with LH, FSH prepares follicles for ovulation and luteinization. FSH induces the LH receptor in granulosa cells so that they become responsive to LH. The effects of FSH on granulosa cells have been studied extensively *in vitro* (Hsueh *et al.,* 1984). FSH stimulates the production of the steroid hormone, progesterone, and stimulates aromatase, which can be assayed by following the conversion of testosterone to estrogen (see assay section below) as well as stimulating the production of plasminogen activator (Beers and Strickland, 1978). Studies on induction of follicular development by exogenous gonadotropins of women and domesticated animals have been stimulated by the development of *in vitro* fertilization and embryo transfer. These studies should lead to a better understanding of how FSH and LH work *in vivo* together to promote follicular development. The practical demands of clinician and veterinarian are leading to a more thorough understanding of hormone action.

In the male, FSH acts on the Sertoli cells found in the seminiferous tubules of the testis. The end result of FSH stimulation is spermatogenesis. However, since the effects of FSH are mediated through the Sertoli cells, the mechanism of FSH stimulation of spermatogenesis remains unclear. Even FSH action on the Sertoli cell is not well characterized. In immature rat testes, cAMP production increases in response to FSH stimulation (Rao and Ramachandran, 1975). However, in older rats, the Sertoli cells are much less sensitive to FSH stimulation in terms of measurable changes in cAMP (Dorrington *et al.,* 1974).

One of the problems in studying FSH is the unavailability of purified preparations of the hormone. Table I shows the yields of FSH from various species. Except for human and equine FSH, the yields are very low, often less than 1% of the yield of the corresponding LH. Consequently, most investigators have to make do with crude FSH preparations that are usually contaminated with LH. In fact, most FSH preparations used for physiological or clinical studies are so crude that the FSH cannot be directly examined. This is unfortunate, since extent of sialylation can affect the length of time the hormone survives in the circulation, which has a direct effect on the *in vivo* potency of FSH (Chappel *et al.,* 1983). In addition, FSH can be nicked in either the α subunit in the vicinity of residues 46–50 or the β subunit in the 38–57 loop. These nicks are known to inactivate the hormone (Ward *et al.,* 1989).

FSH consists of a 92–96 amino acid residue α subunit and a 111–112 residue β subunit. The predicted molecular weights for the protein based on the amino acid sequences are around 23,000. Most molecualr weight estimates for FSH are around 33,000

(Table I). The difference in mass is accounted for by four oligosaccharide chains, two on the alpha subunit and two on the beta subunit. Human and bovine FSH possess mainly sialylated oligosaccharides (55–83%), corresponding to structures 1, 2, and 4 in Fig 4. They have very little of the sulfated oligosaccharides (7–13%). Ovine FSH, on the other hand has roughly equal amounts of all three types of oligosaccharide structures (Baenziger and Green, 1988).

The activities of FSH from various species are listed in Table I. NIH-FSH-S1, a common reference standard (see below), is estimated to be about 1% FSH. Thus, the

Table I
Yield, Potencies, and Molecular Size of Gonadotropins

Hormone and species	Yields (mg/kg pituitary)	Relative potency	Molecular weight by sedimentation analysis	Predicted mass of protein minus carbohydrate	Refs.[g]
LH					
Ovine	30–300	1.9–4.5[a]	28,000–32,500	23,486	1
Bovine	60–180	1.5–2.1	25,200–30,000	23,746	2
Porcine	270	0.7–3.9	27,000–34,000	23,149	3
Equine	16–110	5.3–6.2	33,500	26,674	4
Rabbit	93	1.5		23,117	5
Chicken	126	0.085			6
Turkey	197				7
Human	400	8.0–10.7	34,000	23,391	8
FSH					
Ovine	5–25	43–133[b]	32,700–33,800	22,977	9
Bovine	1.3	47–164		23,316	10
Porcine	71	81		23,152	11
Equine	7–67	90–110	33,200	23,211	12
Turkey	25				13
Human	2.9–200	390–428	32,600	22,674	14
CG	(mg/liter)				
Equine	2[e]	10,500–15,800[c]	28,000–68,500	26,674	15
Human	—[f]	12,000–18,500[d]	37,700	25,719	16

[a]Potency versus NIH-LH-S1.

[b]Potency versus NIH-FSH-S1

[c]Potency versus 1st Intl. Serum Gonadotropin Std.

[d]Potency versus 1st Intl. hCG Reference.

[e]Obtained from pregnant mare serum.

[f]Obtained from pregnancy urine, comparable volume figures are not meaningful.

[g]References: 1, Sherwood and McShan (1977); 2, Sherwood and McShan (1977); 3, Sherwood and McShan (1977); 4, Bousfield and Ward (1984), Guillou and Combarnous (1983), Landefeld and McShan (1974a), Licht et al. (1979), Sherwood and McShan (1977); 5, Ward, et al. (1979); 6, Bousfield (1981), Ishii and Furuya (1975); 7, Burke, et al. (1979); 8, Liu and Ward (1975); Sherwood and McShan (1977); 9, Liu and Ward (1975); Sherwood and McShan (1977); 10, Cheng (1976), Grimek et al. (1979), Liu and Ward (1975); 11, Closset and Hennen (1978); 12, Bousfield and Ward (1984), Guillou and Combarnous (1983), Landefeld and McShan (1974b), Licht et al. (1979), Rathnam et al. (1978); 13, Burke et al. (1979); 14, Sherwood and McShan (1977); 15, Moore (1978), Sherwood and McShan (1977); 16, Sherwood and McShan (1977).

potencies indicated in Table I are a rough estimate of the purity of the preparations. Absence of sufficient quantities of FSH prevent investigations into the nature of different relative potencies between hormones, such as the high activity of hFSH versus all other FSH preparations, which typically have potencies of ~100. We have compared human, equine, and porcine FSH in a chicken receptor assay and have found the eFSH to have a higher affinity for the FSH receptor than pFSH (Gordon *et al.*, 1989). Combarnous and colleagues have reported that eFSH is a "superagonist" in stimulating plasminogen activator production by rat granulosa cells and sertoli cells (Combarnous *et al.*, 1984). More such studies can be done as the provenance problem becomes better resolved.

F. Luteinizing Hormone (Lutropin or Interstitial Cell Stimulating Hormone)

Luteinizing hormone is so named because it causes ovulation in the female ovary resulting in the transformation of the pre-ovulatory follicles into corpora lutea. It stimulates the thecal cells of the ovary to produce precursors to estrogen early in the follicular phase when the follicular granulosa cells possess only FSH receptors and are unresponsive to LH. Under the combined influences of FSH and estrogen, the granulosa cells acquire LH receptors and become responsive to LH stimulation. Following ovulation, the luteal cells secrete progesterone in response to LH stimulation, or, in the case of primates, in response to CG stimulation also.

In the male, LH stimulates the interstitial Leydig cells to produce testosterone. Testosterone in turn is involved with spermatogenesis and maintenance of the male reproductive tract and secondary sex characteristics. Because Leydig cells are easy to procure from rat tests and because steroid radioimmunoassays are convenient to perform, testosterone production by rat Leydig cells *in vitro* is a popular LH bioassay (see below).

Highly purified hormone has been obtained from ovine, bovine, porcine, rabbit, chicken, turkey, whale, and equine pituitaries (Table I). Sequences for additional examples of LH β subunits have been provided by molecular biology techniques for rat and dog LH β subunits. The rat α subunit has also been cloned. Because of the relative abundance of LH, it has been characterized much more extensively than FSH. The subunit nature of the glycoprotein hormones was first determined on LH. At first this was thought to be a simple homodimer (Li and Starman, 1964), but our laboratory determined that the two subunits were different (Ward *et al.*, 1966). This was confirmed when the first gonadotropin sequence was reported for ovine LH (Liu *et al.*, 1972a, 1972b; Sairam *et al.*, 1972a, 1972b). Many of the chemical modification studies were done on LH (Gordon and Ward, 1985). The first proposal for disulfide bond placements were performed on LH (Ward *et al.*, 1973).

LH consists of a 96 residue α subunit, and most have a 117–121 residue β subunit. Recently eLHβ was found to have a C-terminal extension generally considered to be characteristic of CG β subunits (see below). The molecular weights for LH by equilibrium sedimentation range from 28,000 to 34,000 (Table I). The masses predicted from the amino acids sequences range from 23,486 to 26,647, with the rest of the mass accounted for by three N-linked oligosaccharides, two on the alpha and one on the beta subunit. Most LH β subunits are glycosylated at Asn[13]. Human LHβ is the exception, being glycosylated at Asn[30] instead. The oligosaccharide structures found on ovine and bovine LH are primarily biantennary. Over one third of the structures are neutral. Of the charged structures, most are the sulfated types (Structure 9, Fig. 4). Human LH has more hybrid sulfated/sialylated and sialylated oligosaccharides in addition to the sulfated oligosaccharides.

When the hLHβ gene was cloned and sequenced it was found that the gene encoded seven more amino acids at the C terminus than had been found in determining the amino acid sequence of the protein (Talmadge *et al.*, 1984b). The well-documented C-terminal heterogeneity of hLH preparations was blamed for missing these residues in the protein sequence (Keutmann *et al.*, 1979). Recently, experiments using site-directed mutagenesis have shown that these seven C-terminal residues along with a few others are responsible for hLHβ not being secreted from transfected cell lines (Boime, 1989; Corless *et al.*, 1987). Only when α subunit is present is the β subunit secreted, and then only as the LH dimer. Perhaps there is a reason why these residues were not seen in the secreted protein. Perhaps the pituitary has some processing enzymes not present in the CHO cells used in the transfection experiments. In any case, this serves as an illustration that analysis of the protein is as important as analysis of the gene.

G. Chorionic Gonadotropin (Choriogonadotropin)

Chorionic gonadotropins (CG) are the most accessible gonadotropins and have been used as a model for investigating LH and FSH activities *in vivo*. Human CG is used as a substitute for LH and equine CG (usually referred to as PMSG, i.e., pregnant mare serum gonadotropin) is used as a substitute for FSH even though it is known to have both LH and FSH activities in the usual rodent reproductive physiology models. This substitution is in turn related to the scarcity of purified FSH noted above.

1. Human Chorionic Gonadotropin

First discovered in the urine of pregnant women and almost simultaneously in the urine of some cancer patients (Hertz, 1980), hCG is one of the most extensively characterized gonadotropins. Its role is to rescue the corpus luteum from degenerating and to stimulate it to continue to produce progesterone to maintain pregnancy during the first trimester until the placenta is sufficiently developed to take over that function.

Human CG is the prototype of chorionic gonadotropin. It consists of a 92 amino acid residue α subunit and a 145-residue β subunit. The additional 24 amino acids at the C terminus of the β subunit were thought to be characteristic of hCG until eLH was found to have a similar C-terminal extension. The role of the C-terminal extension is unknown. It might play a role in prolonging the survival of the hormone in circulation (Birken and Canfield, 1978). Certainly it plays no role in α–β interaction and is not involved in receptor binding (Bousfield and Ward, 1986; El-Deiry *et al.*, 1989; Matzuk *et al.*, 1989). The C-terminal extension of hCGβ can be removed by mild acid hydrolysis, and the resulting β subunit derivative is capable of recombining with α subunit with full restoration of receptor binding activity (Bousfield and Ward, 1986). In fact, since the β subunit is desialylated by the mild acid hydrolysis, the hCG derivative is 2.5 times as active as native hCG in receptor binding.

It is assumed that LH and hCG share a common receptor (Roche and Ryan, 1985). However, hCG–receptor interactions are studied in species lacking a chorionic gonadotropin, such as the rat. The LH receptor has recently been cloned in the rat (McFarland *et al.*, 1989) and in the pig (Loosfelt *et al.*, 1989). The existence of the receptor clones means that the question of a common LH/CG receptor or a CG-specific receptor can now be addressed in humans and in horses. Although there is a common receptor, the response to hCG and LH stimulation in terms of steroid production by rat testis Leydig cells differs somewhat. Here we would caution the reader to question the significance of differences between hCG and LH found in a sex that does not need a CG, in a species that does not possess a CG. It is fairly well agreed that hCG binds to rat LH receptors longer than commonly available animal LH preparations such as oLH (Garfink *et al.*, 1976; Niswender *et al.*, 1985; Roche and Ryan, 1985). Niswender and colleagues have been investigating this phenomenon in ovine corpora luteal cells, another species that possesses no CG. The long-term hCG receptor occupancy appears to result from hCG–receptor complexes being anchored in the cell membrane so that they cannot migrate to coated pits where they would be internalized. Ovine LH–receptor complexes, on the other hand, are more mobile, and are rapidly internalized. Early speculation was that this might be a function of the C-terminal extension, the greatest difference between hCG and most LHs. However, when hLH was compared with hCG and oLH for mobility of the hormone–receptor complexes, it was found to be intermediate between the two hormones. Even though more closely related than oLH to hCG, hLHβ is still only 80% identical to

hCGβ. The differences between the two hormones that may be significant may be the distribution of the N-linked oligosaccharides. There are two N-linked oligosaccharides on hCGβ at positions 13 and 30. Ovine LH has one at position 13 while hLH has one at position 30 (Fig. 3). Currently site-directed mutagenesis is being used to determine with modified hCG if the glycosylation site at position 30 is the one responsible for anchoring the hormone–receptor complex in the cell membrane (Niswender, 1989). This example is shown to illustrate our ignorance about the role of carbohydrate in the expression of biologic activity in gonadotropins and some of the inappropriate conclusions that are readily made by the lack of homologous model systems for the study of CG. It would be very useful if a convenient-sized animal could be found that possessed a chorionic gonadotropin. In that regard, the isolation of a guinea pig chorionic gonadotropin has been reported (Bambra *et al.*, 1984). If this report is confirmed, the guinea pig might afford such a model system.

2. Equine Chorionic Gonadotropin (Pregnant Mare Serum Gonadotropin)

The discovery of hCG in pregnancy urine set off a search for similar gonadotropic activities in the blood and urine of most domesticated animals (Hertz, 1980). In pregnant mare serum a long-acting gonadotropin was discovered and named after its source, hence the designation pregnant mare serum gonadotropin (PMSG). It was originally believed that PMSG was produced by the uterus, rather than by the placenta as was the case for hCG. However, in the early 1970s, Allen and associates (Allen *et al.*, 1973; Allen and Moor, 1972) demonstrated that the endometrial cups that produce

PMSG were of placental origin and a proper designation for this hormone is equine chorionic gonadotropin. The other reason to change the name PMSG to eCG is a structural one. Equine CG is structurally related to hCG in that both possess a glycosylated C-terminal extension on their β subunits (Fig. 8). We will use the designation eCG throughout, but the reader should be aware that the early designation was PMSG.

The function of eCG is unclear. Perhaps misled by the traditional use of eCG as an FSH substitute in rats, physiologists have looked for FSH effects in the mare. Squires and Ginther (1975) reported that eCG had no FSH activity in the mare. Our laboratory has demonstrated that the equine FSH receptor can distinguish eFSH from LH better than the rat receptor (Moore and Ward, 1980). Given the placental origin and CG structure of eCG, it would perhaps be better to look for a CG effect in the pregnant mare. Allen and asssociates have compared the activities of eLH, eFSH, and eCG in binding assays (Stewart and Allen, 1979). The activities of both pituitary hormones are very high in receptor-binding assays compared with LH and FSH from other species and are very high compared with that of eCG. As a result, they questioned whether eCG had any functional role in the mare.

Equine CG is an interesting hormone because it is the only hormone that can be isolated from serum. All other gonadotropins are isolated from pituitaries or from urine. Carbohydrate heterogeneity resulting in incomplete oligosaccharide structures in most gonadotropins is blamed on extraction of partially processed forms of hormone. Alternatively, oligosaccharide degradation could be blamed for the carbohydrate heterogeneity observed in hCG. In fact, the growing impression is that this carbohydrate heterogeneity has important physiological significance and/or correlation. It is perhaps significant that oligosaccharide heterogeneity has been reported for eCG (Bahl and Wagh, 1986). Changes in oligosaccharides on circulating forms of FSH in response to physiologic changes during the reproductive cycle have been suggested by changes in molecular size estimates and have been attributed to changes in carbohydrate by isoelectric focusing and by neuraminidase digestions of pituitary extracts (Bogdanov and Nansel, 1978; Chappel et al., 1983). Equine CG is the only circulating form of gonadotropin that can be purified, and the observation of carbohydrate heterogeneity supports the idea that there is a physiologic significance to changes in carbohydrate. Unfortunately, the oligosaccharides for eCG are very complex and the structures have not been completely defined. Complete characterization of the oligosaccharides along with definition of the role of eCG in the pregnant mare would provide useful information on the role of carbohydrate in gonadotropin function.

H. Prolactin (PRL)

Prolactin is not classified as a gonadotropin by most authors. However, it has a definite role in the reproduction of domestic animals at least in the late stage—that of suckling the young. Prolactin has been shown to have a synergistic effect on sperm production, and is known to modulate both ovarian steroid production and testicular and prostatic processes, although often the mechanisms by which PRL exerts these effects on the accessory sex tissues are not yet known. Of course, the effect of PRL on the mammary gland and thus the role of PRL on lactation has long been known and indeed is the source of the name for this hormone.

Prolactin and growth hormone are the two most abundant hormones in the pituitary, considered either on a weight basis or a molar basis. Prolactin is present in the pituitary as a simple protein, that is, constituted only of amino acids for the most part. However, glycosylated forms have been described that constitute a small percentage of the prolactin present. Prolactin is secreted by cells in the pituitary called *lactotropes* (earlier nomenclature, mammotrophs or lactotrophs) The principal role of prolactin in all mammalian species is to stimulate the production of milk. However, McNeilly (1987) has provided a thoughtful analysis of the correlation of prolactin levels and control of gonadotropin secretion. In the rat, prolactin plays a role in the development of the corpora lutea and LH receptors, but this function for prolactin has not been shown in domestic animals, with the possible exception of cattle (Bartosik *et al.*, 1967), sheep (Denamur *et al.*, 1973) and rabbits (Spies *et al.*, 1968). Indeed, prolactin has been implicated in over 85 different biological functions or responses (Nicoll and Bern, 1972), but only in the case of lactation does this role seem to be a singular one. Thus, we may regard prolactin as a rather cosmopolitan hormone with broad specificity, often functioning in conjunction with other hormones. The placenta of domestic animals and humans contain a hormone that, like prolactin, stimulates mammary tissues *in vitro* or responds like prolactin in radioreceptor assays (Talamantes and Ogren, 1988), and is thus designated placental lactogen. In most of those species for which amino acid sequences are known (Talamantes and Ogren, 1988), *placental lactogen*

is highly homologous to growth hormone (e.g., about 96%). Prolactin shows some sequence homology to growth hormone and placental lactogens (of the order of 60–70% in the human), but it will be beyond the scope of this chapter to explore the structural relationships of prolactin and the growth hormone–placental lactogen family. It should be noted, however, that the recently described sequence of bovine placental lactogen (Schuler *et al.*, 1988) is about 70% homologous with bovine prolactin and only 46% homologous to the bovine growth hormone. Thus, the chemical similarities of placental lactogen in the domestic animals such as sheep and cow may be more closely similar to prolactin than to growth hormone, that is, they may differ from those patterns found for placental lactogens in the human, nonhuman primates, and rodents.

Growth hormones characteristically have four half-cystine residues that form a large disulfide loop and a small disulfide loop in their structure. The prolactin molecule has six half-cystines that form two small disulfide loops and one large disulfide loop (Fig. 10). For a detailed discussion of the similarities of prolactin and growth hormone structure and function, the review by Nicoll *et al.* (1986) is recommended. The review also includes a consideration of human placental lactogen.

The amino acid sequences of the pituitary hormone that are presently available (Lehrman *et al.*, 1988; Watahiki *et al.*, 1989) are summarized in Fig 10 using the single-letter amino code (see Fig. 5 legend). The prolactin forms from the lower vertebrates (salmon, carp, and tilapia) are foreshortened on the N-terminus by approximately 12 residues, but this removes the small N-terminal disulfide loop present in the other prolactin structures. In this respect these

```
                10                  20                  30                  40                  50
                |                   |                   |                   |                   |
cPRL       L P I C P I G S V N C Q V S L G E L F D R A V K L S H Y I H Y L S S E I F N E F D E R Y A Q G R G F
wPRL       I P I C P S G A V N C Q V S L R D L F D R A V I L S H Y I H N L S S E M F N E F D K R Y A Q G R G F
mPRL       L P I C S A G D - - C Q T S L R E L F D R V V I L S H Y I H T L Y T D M F I E F D K Q Y V Q D R E E
rPRL       L P V C S G G D - - C Q T P L P E L F D R V V M L S H Y I H T L Y T D M F I E F D K Q Y V Q D R E F
pPRL       L P E C P S G A V N C Q V S L R D L F D R A V I L S H Y I H N L S S E M F N E F D K R Y A Q G R G F
bPRL       T P V C P N G P G N C Q V S L R D L F D R A V M V S H Y I H D L S S E M F N E F D K R Y A Q G R G F
oPRL       T P V C P N G P G D C Q V S L R D L F D R A V M V S H Y I H N L S S E M E N E F D K R Y A Q G K G F
ePRL       L P I C P S G A V N C Q V S L R E L F D R A V I L S H Y I H N L S S E M F N E F D K R Y A Q G R G F
hPRL       L P I C P G G A A R C Q V T L R D L F D R A V V L S H Y I H N L S S E M F S E F D K R Y T H G R G F
sPRL                   I G L S D L M E R A S Q R S D K L H S L S T S L T K D L D S H F P P M G R V
caPRL                  V G L N D L L R R A S E L S D K L H S L S T S L T N D L D S H F P P V G R V
tiPRL-188              V P I N E L L E R A S Q H S D K L H S L S T T L T Q E L D S H F P P I G R V
tiPRL-177              V P I N D L I Y R A S Q Q S D K L H A L S T M L T Q E L G S E A F P I D R V


                60                  70                  80                  90                  100
                |                   |                   |                   |                   |
cPRL       I T K A V N G C H T S S L T T P E D K E Q A Q Q I H H E D L L N L V V G V L R S W N D P L I H L A S
wPRL       I T K A I N S C H T S S L Q T P E D K E Q A Q Q I H H E V L V S L I L G V L R S W N D P L Y H L V T
mPRL       M V K V I N D C P T S S L A T P E D K E Q A L K V P P E V L L N L I L S L V Q S S S D P L F Q L I T
rPRL       I A K A I N D C P T S S L A T P E D K E Q A Q K V P P E V L L N L I L S L V H S W N D P L F Q L I T
pPRL       I T K A I N S C H T S S L S T P E D K E Q A Q Q I H H E V L N L I L R V L R S W N D P L Y H L V T
bPRL       I T M A I N S C H T S S L P T P E D K E Q A Q Q T H H E V L H S L I L G L L R S W N D P L Y H L V T
oPRL       I T M A I N S C H T S S L P T P E D K E Q A Q Q T H H E V L M S L I L G L L R S W N D P L Y H L V T
ePRL       V T K A I N S C H T S S L S T P E D K E Q A Q Q I H H E D L L N L I L R V L K S W N D P L Y H L V S
hPRL       I T K A I N S C H T S S L A T P E D K E Q A Q Q M N Q K D F L S L I V S I L R S W N E P L Y H L V T
sPRL       M M P R P S M C H T S S L Q T P K D K E Q A L K V S E N E L I S L A R S L L A W N D P L L L L S S
caPRL      M M P R P S M C H T S S L Q V P N D K D Q A L K V P E D P L L S L A R S L L L A W S D P L A L L S S
tiPRL-8    I M P R P A M C H T S S L Q T P I D K D Q A L Q V S E S D L M S L A R S L L Q A W S D P L V V L S S
tiPRL-7    L A - - - - - C H T S S L Q T P T D K E Q A L Q V S E S D L L S L A R S L L Q A W S D P L E V L S S


                110                 120                 130                 140                 150
                |                   |                   |                   |                   |
cPRL       E V Q R I K E A P D T I L W K A V E I E E Q N K R L L E G M E K I V G R V H S G - H A G N E I Y S H
wPRL       E V R G M Q E A P D A I L S R A I Q E E E E N K R L L E G M E K I V G Q V H P G - V K E N E V Y S V
mPRL       G V G G I Q E A P E Y I L S R A K E I E E Q N K Q L L E G V B K I I S Q A Y P E - A K G N G I Y F V
rPRL       G L G G I H E A P D A I I S R A K E I E E Q N K R L L E G I E K I I G Q A Y P E - A K G N E I Y L V
pPRL       E V R G M K G A P D A I L S R A I Q E E E E N K R L L E G M E K I V G Q V H P G - I K E N E V Y S V
bPRL       E V R G M K G A P D A I L S R A I E I E E E N K R L I I E G M E M I F G Q V I P G - A K E T E P Y P V
oPRL       E V R G M K G V P D A I L S R A I E I E E E N K R L L E G M E M I F G Q V I P G - A K E T E P Y P V
ePRL       E V R G M Q E A P E A I L S K A I E I E E Q N R R L L E G M E K I V G Q V Q P R - I K E N E V Y S V
hPRL       E V R G M Q E A P E A L L S K A V E I E E Q T K R L L R G M E L I V S Q V H P R - T K E N E I Y P V
sPRL       E A P T C P H - P S N G D I S S K I R E L Q D Y S K S L G D G L D I M V N K M G P S S Q Y I S S I P
caPRL      E A S S L A H - P E R N T I D S K T K E L Q E N I N S L G A G L E H V F N K M D S T S D N L S S L P
tiPRL-8    S A S T L P H - P A Q S S I F N K I Q E M Q Q Y S K S L K D G L D V L S S K M G S P A Q A I T S L P
tiPRL-7    S T N V L P Y - S A Q S T L S K T I Q K M Q E H S K D L K D G L D I L S S K M G P A A Q T I T S L P


                160                 170                 180                 190                 200
                |                   |                   |                   |                   |
cPRL       S D G L P S L Q L A D E D S R L F A F Y N L L H C H R R D S H K I D N Y L K V L K C R L - - I H D S N C
wPRL       W S G L P S L Q M A D E D T R L F A F Y N L L H C L R R D S H K I D S Y L K L L K C R I - - I Y N S N C
mPRL       W S G L P S L Q G V D E E S K I L S L R N T I R C L R R D S H K V D N F L K V L R C Q I - - A H Q N N C
rPRL       W S Q L P S L Q G V D E E S K D L A F Y N N I R C L R R D S H K V D N Y L K F L R C Q I - - V H K N N C
pPRL       W S G L P S L Q M A D E D T R L F A F Y N L L H C L R R D S H K I D N Y L K L L K C R I - - I Y N S N C
bPRL       W S G L P S L Q T K D E D A R Y S A F Y N L L H C L R R D S S K I D T Y L K L L N C R I - - I Y N N N C
oPRL       W S G L P S L Q T K D E D A R H S A F Y N L L H C L R R D S S K I D T Y L K L L N C R I - - I Y N N N C
ePRL       W S G L P S L Q M A D E D S R L F A F Y N L L H C L R R D S H K I D N Y L K L L K C R I - - V Y D S N C
hPRL       W S G L P S L Q M A D E E S R L S A Y Y N L L H C L R R D S H K I D N Y L K L L K C R I - - I H N N N C
sPRL       F K G G D - L G N - D K T S R L I N F H F L M S C F R R D S H K I D S F L K V L R C R A T K M R P E T C
caPRL      F Y T N S - L G E - D K T S R L V N F H F L L S C F R R D S H K I D S F L K V L R C R A K K - R P E M C
tiPRL-8    Y R G G T N L G H - D K I T K L I N F N F L L S C F R R D S H K I D S F L K V L R C R A A K M Q P E M C
tiPRL-7    F I E T N E I G Q - D K I T K - - - - - L L S C F R R D S H K I D S F L K V L R C R A A N M Q P Q V C
```

Figure 10 Compilation of the amino acid sequences for prolactin taken mainly from Watahiki *et al.* (1989). The ePRL is from Lehrman *et al.* (1988). In addition to the abbreviations used in Fig. 8, ti designates tilapia (single-letter amino acid codes, see Fig. 6).

fish prolactins are similar to growth hormone disulfide loops (e.g., 58–174 and 191–199), but note that the lower vertebrate prolactins have one or two additional residues in the small C-terminal disulfide loop. The significance of this difference is not known.

Prolactin secretion is under the control of the hypothalamus, but, unlike the several other pituitary hormones under hypothalamic control, prolactin secretion is under tonic inhibition. Thus, when the pituitary is released from hypothalamic control, prolactin secretion and release rise. This was first shown by pituitary transplant to the kidney capsule (Everett, 1954). Dopamine and gamma-aminobutyric acid are two substances that show a prolactin inhibiting effect. Of these, dopamine is about 100-fold more effective on a molar basis. However, these amino derivatives do not account for all the prolactin-inhibiting activity in the hypothalamus, suggesting the existence of an as yet unidentified prolactin-inhibiting factor (PIF).

Thyrotropin-releasing hormone (TRH), as noted above, was the first hypothalamic peptide to be discovered and completely characterized based on its ability to stimulate thyrotropin (TSH) release, but shortly thereafter this simple tripeptide (pyroglutamyl-histidyl-prolinamide) was also shown to stimulate prolactin release *in vitro* and *in vivo* (Bowers *et al.*, 1971; Tashjian *et al.*, 1971).

Many reports have been made about possible physiological roles of PRL for an augmenting or ancillary role to other hormones. It is possible more important roles may yet be described, while some of the suggested effects may be no more than pharmacologic curiosities. At the present time prolactin may be regarded as having a sig-

nificant role in lactation for domestic animals with a supporting role for many other endocrine actions. It has been shown to modify (suppress) LH and FSH secretion during lactation (McNeilly, 1987).

III. Biosynthesis of the Gonadotropins

The biosynthesis of the glycoprotein hormones involves both the protein biosynthesis and the carbohydrate application to the protein structure as co- and posttranslational modifictions. The past decade has seen an impressive expansion of our knowledge of both these processes. The sequence of several of the genes for the α and β subunits, which are located on separate genes in separate chromosomes, have been elucidated in this period. The section on structure (see above) provides a diagrammatic summary of these gene structures (Figs. 1 and 2). For a recent review, see the monograph *Glycoprotein Hormones: Structure, Synthesis, and Biological Function* (Chin and Boime, 1990). The factors involved in the genetic control of transcription in both the pituitary and chorion are under intense investigation in several laboratories, but a detailed review of this molecular biologic aspect of biosynthesis will not be undertaken in the present chapter. The interested reader is referred to several detailed references on this subject (Chin, 1985; Chin and Boime, 1990; Habener, 1987).

The transcription process for the α and β subunits of the pituitary glycoprotein hormones is influenced by the steroid hormones levels (particularly progesterone, estradiol, and testosterone) acting through their respective receptor proteins and in conjunction with the action of initiation factors, promotors, etc. The "second messenger," cAMP, interacts specifically at sites on the genome designated cyclic-AMP responsive elements (CRE) to stimulate transcription. [See particularly the monograph by Chin and Boime (1990).]

At the anatomic and biochemical level the hypothalamo–anterior pituitary system plays a key role in the circulating glycoprotein hormone levels. The first hypophysiotropic peptide chemically characterized was the thyrotropin-releasing hormone (TRH). The characterization of this structurally simple tripeptide (described on page 55) in 1970–1971 ushered in a new era for our understanding of the control of the pituitary biosynthesis and release of its hormones. That discovery was followed shortly by the isolation and characterization of LHRH (luteinizing hormone releasing hormone). An interesting chronicle of these early events and the competitive spirit in science is provided by "The Nobel Duel" (Wade, 1981). Since then, a series of neuropeptides derived from the hypothalamo–pituitary axis have been shown, but these peptides are not exclusively products of the hypothalamus. They may be produced in other areas of the brain, the gut, etc. Somatostatin (inhibits growth hormone release), corticotropin-releasing factor, growth-hormone releasing factor, dopamine (inhibits prolactin release, although a separate prolactin inhibiting factor, PIF, may also exist), and inhibin [inhibits follitropin, (or FSH), release] are all factors that have been shown to be involved in the control of pituitary hormones. Inhibin is produced by the granulosa cells in the ovary (or Sertoli cells in the testis). There are other nonsteroidal hormones in ovarian follicular fluid whose physiologic role is yet to be shown. Examples of these are activin

(stimulates follitropin release and also functions as an erythroid differentiation factor—it is structurally related to the β subunit of inhibin), follistatin (structurally dissimilar to inhibin but with the same action, suppression of follitropin release, at least at pharmacologic levels), luteinization inhibitor, oocyte meiosis inhibitor, and other possible hormones from follicular fluid. [For comprehensive reviews see Steinberger and Ward (1988) and Ying (1988).] In a recent report follistatin has been shown to serve as a specific binding protein for activin; thus follistatin may be the solubilized receptor for activin or its structural equivalent, the erythroid differentiation factor (Nakamura et al., 1990). Since LHRH stimulates the release of both FSH and LH, the role of inhibin is to modulate FSH levels. Physiologically these levels are known to fluctuate independent of each other, as will be detailed in later chapters.

Once the genetic controls of the pituitary hormones initiate transcription of the required mRNA, protein translation occurs on the polysomes bound to the rough endoplasmic reticulum (RER). The completed linear subunit from this synthetic step carries a "signal peptide," and has no formed disulfide bonds. The signal peptide (also called the leader sequence) at the N terminus of the subunit precursor is characteristic of the particular subunit. For example, the LH/CG-beta subunit prehormone has a 20 amino acid residue leader sequence that is highly homologous throughout the known mammalian species (Fig. 8). (The leader sequence on the salmon gonadotropic hormone beta subunit has 23 residues but still maintains a considerable homology. This is the only available known leader sequence from the lower vertebrates at this time.) The alpha subunits currently known all have a 24-residue leader sequence (Fig. 6), again with a high homology among the species. The TSH β signal peptide for those known is comprised of 20 residues (Fig. 8). The known FSH beta leader sequences have 18 residues for the known cases (Fig. 8), but these have sometimes been reported as 19 or 20 residues due to uncertainties about the exact cleavage point of the signal protease. In the authors' opinion this cleavage site will be shown to be between cysteine and asparagine, arginine, or histidine (see Fig. 8 for the FSH β subunit leader sequences). This requires an 18-residue leader sequence.

The signal peptide, which may range from 15 to 30 amino acid residues, characteristically has a very hydrophobic cluster of amino acids 5–20 residues toward the N terminus from the site of cleavage (processing) by the signal peptidase. This step of the protein processing probably occurs *cotranslationally* at about the time the C-terminal portion of the peptide chain coded by the mRNA is being completed. Thus, the signal peptide removal marks the removal of the completed protein chain from the ribosomes and release of the linear protein (or disulfide cross-linked protein) into the cisternal space of rough endoplasmic reticulum (RER). [For review of the elements of the signal hypothesis, see Jackson and Blobel (1980).]

In the comments that follow, the reader is alerted to the fact that the processes related to glycosylation have only been studied to a limited degree for the glycoprotein hormones, and thus much of what we will propose is based by analogy to other systems. The reasons for this are twofold. First, the gonadotropes or thyrotropes of the pituitary are not readily obtained for study. Second, there are few suitable cell lines that provide reasonable systems to study this glycosylation. In this respect, those that have received the most extensive study are trophoblast-derived cell lines, particularly the JAR or BeWo human choriocarcinoma cell lines, originally islolated by Dr. Roland Patillo of the Medical College of Wisconsin. The JAR cell line, for example, has been through several hundred passages and maintained an essentially constant production of hCG throughout. This property is important for their utility to study biosynthesis. However, hCG is somewhat unusual among the glycoprotein hormones in that, besides an extra N-linked glycosylation site on the β subunit, it also has O-linked glycosylation sites on the C terminus (*vide supra*). This O-linked glycosylation may be characteristic among all the chorionic gonadotropins. However, we have shown it is not a unique feature, since equine pituitary LH also has this C-terminal extension on the β subunit, and, as in the chorionic gonadotropins, it is O-glycosylated (see structural summaries above).

The signal peptide removal in the RER is probably involved in the transfer mechanism from the RER cisternae along the membrane to the Golgi apparatus (GA), a cellular organelle that we will discuss later. This signal peptide removal is probably the final step in the cotranslational processing that occurs in the RER. However, for the glycoprotein hormones, there is another important cotranslational event (N-glycosylation) that also occurs in the RER. N-Glycosylation takes place at two sites on the α subunit and one or two sites on the β subunit. (See the diagrammatic summary, p. 31, Fig. 3.) O-Glycosylation, if it occurs, happens later as a posttranslational event in the GA.

N-Glycosylation (see Fig. 5 summary) is mediated through a "high-mannose" dolichol phosphate intermediate, enzymatically targeted to specific N-glycosylation sites. These sites have the structural requirement (single letter amino acid code) on the peptide chain -N-X-S (or T)-, where X may be any amino acid. Not all such N-glycosylation sites are glycosylated in those proteins that contain this structural feature, but in the glycoprotein hormones all potential N-glycosylation sites are, in fact, N-glycosylated. The carbohydrate moiety transferred from the dolichol phosphate intermediate has the general composition $glucose_3$ $mannose_9$ N-acetylglucosamine$_2$. The two N-acetylglucosamine residues form the link to the asparagine amide nitrogen in the N-glycosylation site (steps 1–3, Fig. 5). On the nonreducing end of this characteristic disaccharide attachment to the asparagine nitrogen amide is a mannose residue linked through position 1 to position 4 of the last N-acetylglucosamine residue. This mannose residue, in turn, is linked (through positions 3 and 6) to two additional mannose residues. This generates a two-armed branch point in the carbohydrate, which generates a "biantennary" structure (see diagram Fig. 5, pp. 34–35). In some of the more complex carbohydrate structures associated with the glycoprotein hormones a "triantennary" branching is required (see diagrammatic examples above). Whether a biantennary or triantennary carbohydrate will be produced in the "mature" glycoprotein will be determined in the GA, but a structure that provides for this possibility is contained in the high-mannose form or precursor form generated in the RER. Characteristically the

three glucose residues in a linear array are attached to the longest mannose chain on one arm of the triantennary high mannose precursor from dolichol phosphate. While the glycoprotein is still in the RER the three glucose residues are removed; α-glucosidase I removes the terminal glucose residue, then α-glucosidase II removes the remaining two glucose residues. (No mature glycoprotein from vertebrates contain glucose in their carbohydrate moiety with the exception of the recently reported case with inhibin (Ward et $al.$, 1988). Also in the RER, specific glycosidases remove 0, 1, or 3, of the 9 mannose residues added from the dolichol phosphate precursor (see review, Lodish, 1988).

The final step in the RER processing of the glycoprotein hormone subunit is probably one of loading (via the signal peptidase and its postulated chain of events) into a transport vesicle involved in the movement of the partially processed "high-mannose" carbohydrate-bearing subunit. Disulfide bond formation begins in the RER cisternae. For the α it is completed in the RER, but at least the final disulfide closures on the beta may possibly be accomplished in the GA along with the final posttranslational processing of the carbohydrate moieties to their several complex, mature forms. (The sequence of events at this point is somewhat conjectural.) The putative RER transport vesicle delivery of the α,β precursor to the GA has not been studied in the glycoprotein hormone system but is drawn from electron microscopy studies in other cell types.

The combination of the α and β subunits is initiated in the RER with the high-mannose forms of the subunits. It is not yet clear whether all of the disulfide bonds are formed here or if they are completed in the

GA. Ruddon and co-workers have detected discrete intermediates in this process (see discussion below).

However, the α,β dimeric precursor of the glycoprotein hormones may get from the RER to the GA, it is in the GA that various posttranslational events occur that have an important bearing on the biochemical and physiological properties of the mature protein hormone produced.

The Golgi apparatus is a cellular organelle that is best examined with the electron microscope. It is a structure in close juxtaposition with the nucleus. This organelle or complex appears to be composed of parallel, flattened saccules, vesicles, and vacuoles. The portion nearest the nucleus is designated the "cis Golgi," and it is here the protein from the RER enters the GA, as followed by pulse-chase labeling. Thus, the high-mannose form (nine, eight, or six mannose residues at this point) enters the Golgi for posttranslational processing. In the cis Golgi, α-1,2-mannosidase may remove an additional three mannose residues, resulting in the structure that is the substrate for N-acetylglucosamine transferase I, which adds N-acetylglucosamine. This structure can be further processed to yield a single-branched sulfated structure or go on to be processed into biantennary or triantennary oligosaccharide. The deciding factor is whether the pituitary enzyme N-acetylgalactosamine transferase adds N-acetylgalactosamine or not. If it does, then the monosulfated structure results. If not, then two more mannose residues are removed. The resulting oligosaccharide may be galactosylated and processed as monosialylated oligosaccharide or processed further by addition of a second N-acetylglucosamine. The latter structure is the precursor or triantennary si-alylated oligosaccharides, as well as biantennary sialylated, or hybrid sulfated–sialylated, or bisulfated oligosaccharides.

The glycoprotein hormone subunits are then moved to the "medial Golgi" for addition of specific sugars by specific glycosyl transferases. The substrates for these sugar transferases are synthesized in the cytoplasm (e.g., CMP-neuraminic acid, UDP-galactose, UDP-N-acetyl glucosamine, UDP-N-acetyl galactosamine, GDP-fucose, and the sulfate donor 3′-phosphoadenosine-5′-phosphosulfate, or PAPS.) There are selective transport proteins that facilitate each of these substrates' movement into the Golgi, where the glycosyl transferases bring about the buildup of hybrid intermediate forms of the glycosylated moieties on the subunits. [For review of the topography of glycosylation see Hirschberg and Snider (1987).] The "hybrid" intermediates in the formation of the carbohydrate moieties of the glycoprotein hormones are finally processed to the mature secreted form of the carbohydrate in the "trans Golgi." This is designated the "complex" carbohydrate moiety after addition of N-acetylglucosamine or N-acetylgalactosamine, galactose, N-acetyl neuraminic acid (also called sialic acid), fucose, or sulfate groups. The hCG-producing cells of the human chorion lack the enzyme system sulfotransferase in their Golgi apparatus for the addition of sulfate to the carbohydrate (Green et al., 1984). In the pituitary there is an N-acetylgalactosamine transferase in the Golgi apparatus that adds N-acetylgalactosamine to the peripheral ends of the carbohydrate moieties (Smith and Baenziger, 1988). The sulfation in the pituitary glycoprotein hormones is on the 4-position of these residues. Thus, the lack of sulfation in the human chorionic gonadotropin is a consequence of the absence of these two relatively unique transferases for PAPS and *N*-acetylgalactosamine. This leads to a heavier concentration of neuraminic acid termini on the hCG carbohydrate compared to the carbohydrate on pituitary glycoprotein hormones. This, in turn, is a factor in the longer *in vivo* half-life of hCG compared to the pituitary LH, TSH, or FSH. The longer half-life leads to a much longer presentation of hCG to the LH/CG receptor than is the case for LH. Moreover, the structural differences in LH and CG lead to a much longer binding to the receptor than the binding time of LH on the receptor. Thus, although the CG and LH interact with the same receptor, their net physiological (or pharmacological) effect may be considerably different.

In the conversion of the carbohydrate moieties on the glycoprotein hormone subunits from the high-mannose type to the mature complex type several degrees of structural buildup are possible. Thus it is possible to have some biantennary structures with two, one, or zero sulfate groups. Where there is no sulfate group it is possible that a neuraminic acid and galactose residue will complete the buildup of the antennary branch. In a brilliant series Baenziger and colleagues have studied the sulfate labeling patterns and possible combinations to produce the complex carbohydrate structural buildup that leads to a complex and heterogeneous set of carbohydrate moieties on LH, FSH, TSH, and CG. The first three were studied in bovine, ovine, and human synthetic systems and the CG in human cell systems. These studies culminated in two outstanding reports (Green and Baenziger, 1988a,1988b). Their extensive series of studies was reviewed comprehensively (Baenziger and Green, 1988), and this re-

view is recommended for a detailed statement of the structures and diversity (heterogeneity) encountered in the carbohydrate portion of these hormones.

As the carbohydrate moieties of the pituitary glycoprotein hormones reach their completed stages, the disulfide bond formation is also completed. It is likely that some of the early disulfide bond formation is completed in the RER with perhaps the last one-third of the disulfide bonds being completed in the GA. Strickland *et al.* (1985) showed that the glycosylation of the glycoprotein hormones was important for the proper folding of the subunits to allow the correct disulfide bonds to form. The available data do not allow us to rule out the converse, that some disulfide bonds may need to be formed to insure the proper folding. Perhaps the best data available on these points comes from a brilliant series by Ruddon and his colleagues. Their work has been with the JAR choriocarcinoma cell line to study the production of hCG, or other cell lines with selective production of the α or β subunits of hCG.

Ruddon *et al.* (1987, 1989) have identified a precursor form of the α (pα) that combines with only one of two β precursor forms (designated pβ1 and pβ2). These precursor forms were isolated by immunopurification and further characterized by gel electrophoresis and/or other methods such as sulfhydryl labeling. From their studies it is clear that the pα precursor has all the disulfide bonds formed before it combines with the β subunit to produce the α–β dimeric form, which is a precursor to the intact, mature hCG that is completed in the GA for transport out of the cell and into the circulation. There are no detectable storage forms of hCG in the cytoplasm of the trophoblastic cell. The situation for the β-subunit precursors is more complex. The pβ1 form has only half of its disulfides formed, and will not combine with the pα form. The pβ2 form may or may not require closure of all the remaining disulfide bonds before it produces the correct α–β dimeric intermediate. The presence of free β subunit with fully oxidized disulfide bonds, but probably incorrectly formed, complicates the interpretation at this point. Since the RER contains membrane-bound protein disulfide isomerase, it is even possible that all disulfide closures occur before vesicular transport from the RER to the GA, but the point remains for future clarification.

After processing the carbohydrate of the α–β dimer in the GA to the several forms of complex carbohydrate (mature hormone), the hormone is transported to storage vacuoles in the cytoplasm (pituitary cells) or directly to the cellular exterior (human chorion cells). The degree of sulfation among the pituitary hormones varies widely depending on the species of the pituitary cell donor, and the type of hormone (FSH, LH, or TSH). It has been postulated that sulfation may be involved with storage and secretion of these pituitary hormones. There is as yet no experimental support for this concept. (Some aspects of the storage and control of glycoprotein hormone release will be discussed in other chapters.)

In the case of hCG compared to the human pituitary glycoprotein hormones only the later are sulfated (to widely varying degrees). It should be noted that this does not hold for equine chorionic gonadotropin (eCG), for which our lab has been able to detect sulfation both in eCG and the eFSH or eLH. We have not examined eTSH. Thus between species marked differences in the degree and form of carbohydrate posttranslational processing are found. The

question of the heterogeneity of these carbohydrate forms as well as protein heterogeneity has been extensively examined in a monograph edited by Keel and Grotjan (1989).

There are multiple forms of the glycoprotein hormones that are for the most part attributable to the degree of processing of the carbohydrate during biosynthesis. These forms have been designated in a variety of ways, but usually as "isoforms" of the particular hormone. The designation "isohormones" has also been used. The relative concentration of these isoforms in the serum has been shown to correlate with physiological status of the individual in the case of all the pituitary hormones; see, for example, the review of the FSH isoforms (Chappel *et al.*, 1983). It is beyond the scope of the present chapter to consider the physiological and structural correlation of these isoforms, although some reference to the physiological parameters will be noted in other chapters of the text.

As summarized above, the biosynthesis of the glycoprotein hormones involves all the complexities of the protein biosynthetic mechanisms provided in nature from the genomic transcription, mRNA translation, cotranslational processing, to the final posttranslational processing. The numerous intermediates in the posttranslational processing in the Golgi apparatus, several of which may become final forms of the complex carbohydrate moieties, produce a significant heterogeneity in the secreted hormone. This heterogeneity is reflected in several isoforms of that particular glycoprotein hormone. These isoforms, to a degree, also correlate with physiological state of the individual domestic animal, person, or whatever. Thus the protein biosynthesis of glycoprotein hormones is a very complex process subject to many physiological controls, few of which are fully understood.

IV. Receptors

It is important to have a basic understanding of how protein hormones, including the gonadotropins and prolactin, mediate their effects in their target cells. The key to the specific action of any hormone on a target cell is the receptor for that hormone (Tepperman, 1980). The receptors for the gonadotropins and prolactin themselves are proteins. Because LH, hCG, FSH, and prolactin cannot diffuse through the lipid bilayer of the cell membrane of the target cells, their receptors must be located on the surface of the cells. In contrast, steroids can readily pass through cell walls, and thus their receptors are found within the target cells. In order for a cell surface protein to be classified as a receptor, it must exhibit a strict structural specificity for the hormone in question. The proposed receptor must be saturable by the hormone and have an affinity for the hormone that is within the range of physiological concentration of the hormone. Also, the receptor should be located on those tissues where a biological effect of the hormone has been demonstrated. Finally, there must be a mechanism for the reversal of the biological effect initiated by the binding of hormone to its receptor (Tepperman, 1980).

A. The LH Receptor

The receptor for LH also recognizes chorionic gonadotropin, as expected from their similar and/or identical amino acid sequences. Receptors for LH/hCG have been demonstrated on a variety of tissues in the reproductive system, including Leydig cells, granulosa cells and luteal cells (Ascoli and Segaloff, 1989). Until recently, there has not been agreement as to the nature of the LH/hCG receptor, beyond its identification as a protein. Much of the earlier data held that the LH/hCG receptor was composed of a number of disulfide-bound subunits of differing molecular weights; however, more current information indicated that the receptor was a single protein (Ascoli and Segaloff, 1989). The issue was recently clarified with the cloning and expression of the LH receptor using recombinant DNA methodology (Loosfelt et al., 1989; McFarland et al., 1989). One group isolated the cDNA for the LH receptor from a rat luteal DNA library (McFarland et al., 1989), the other from a porcine testis cDNA library (Loosfelt et al., 1989). The sequences are compared in Fig. 11. Both groups reported that the receptor was a single polypeptide chain. The final length of the rat LH receptor was slightly shorter (674 amino acids) than the functional pig LH receptor (699 amino acids). Both of these receptors have a long extracellular sequence, presumably the hormone binding site, that contains six potential N-linked glycosylation sites. Also both species'

Figure 11 Compilation of reported LH receptor sequences. The rat LH receptor was deduced from a rat luteal cell cDNA library (McFarland et al., 1989), the porcine LH receptor from porcine testis cDNA library (Loosfelt et al., 1989). The signal peptide is reported for the rat LH receptor and is indicated by negative numbering. The porcine LH receptor was aligned with the rat sequence for maximum homology. The proposed transmembrane segments are underlined (single-letter amino acid codes, see Fig. 6).

LH RECEPTOR SEQUENCE

```
        -25       -21        -16        -11         -6         -1 1          5           10          15          20
         |         |          |          |          |          |            |           |           |           |
Rat    M G R R V P A L R Q L L V L A V L L L K P S Q L Q S R E L S G S R C P G P C D C A P D G A L R C P G
Porcine M R R R S L A L R L L L A L L L L P P P L P Q T L L G A P             C P E P C S C R P D G A L R C P G

         25         30         35         40         45         50         55         60         65         70
          |          |          |          |          |          |          |          |          |          |
Rat    P R A G L A R L S L T Y L P V K V I P S Q A F R G L N E V V K I E I S Q S D S L E R I E A N A F D N
Porcine P R A G L S R L S L T Y L P I K V I P S Q A F R G L N E V V K I E I S Q S D S L E K I E A N A F D N

         75         80         85         90         95        100        105        110        115        120
          |          |          |          |          |          |          |          |          |          |
Rat    L L N L S E L L I Q N T K N L L Y I E P   A F T N L P R L K Y L S I C N T G I R T L P D V T K I S S
Porcine L L N L S E I L I Q N T K N L V Y I E P G A F T N L P R L K Y L S I C N T G I R K L P D V T K I F S

        125        130        135        140        145        150        155        160        165        170
          |          |          |          |          |          |          |          |          |          |
Rat    S E F N F I L E I C D N L H I T T I P G N A F Q G M N N E S V T L K L Y G N G F E E V Q S H A F N G
Porcine S E F N F I L E I C D N L H I T T V P A N A F Q G M N N E S I T L K L Y G N G F E E I Q S H A F N G

        175        180        185        190        195        200        205        210        215        220
          |          |          |          |          |          |          |          |          |          |
Rat    T T L   S L E L K E N I Y L E K M H S G A F Q G A T G P S I L D I S S T K L Q A L P S H G L E S I Q
Porcine T T L I S L E L K E N A H L K K M H N D A F R G A R G P S I L D I S S T K L Q A L P S Y G L E S I Q

        225        230        235        240        245        250        255        260        265        270
          |          |          |          |          |          |          |          |          |          |
Rat    T L I A L S S S Y S L K T L P S K E K F T S L L V A T L T Y P S H C C A F R N L P K K E Q N F S F S
Porcine T L I A T S   S Y                                       C C A F R N L P T K E Q N F S F S

        275        280        285        290        295        300        305        310        315        320
          |          |          |          |          |          |          |          |          |          |
Rat    I F E N F S K Q C E S T V R K A D N E T L Y S A I F E E N E L S G W D Y D Y G F C S P L T L Q C A P
Porcine I F K N F S K Q C E S T A R R P N N E T L Y S A I F A E S E L S D W D Y D Y G F C S P K T L Q C A P

        325        330        335        340        345        350        355        360        365        370
          |          |          |          |          |          |          |          |          |          |
Rat    E P D A F N P C E D I M G Y A F L R V L I W L I N I L A I F G N L T V L F V L L T S R Y K L T V P R
Porcine E P D A F N P C E D I M G Y D F L R V L I W L I N I L A I M G N V T V L F V L L T S H Y K L T V P R

        375        380        385        390        395        400        405        410        415        420
          |          |          |          |          |          |          |          |          |          |
Rat    F L M C N L S F A D F C M G L Y L L L I A S V D S Q T K G Q Y Y N H A I D W Q T G S G C G A A G F F
Porcine F L M C N L S F A D F C M G L Y L L L I A S V D A Q T K G Q Y Y N H A I D W Q T G N G C S V A G F F

        425        430        435        440        445        450        455        460        465        470
          |          |          |          |          |          |          |          |          |          |
Rat    T V F A S E L S V Y T L T V I T L E R W H T I T Y A V Q L D Q K L R L R H A I P I M L G G W L F S T
Porcine T V F A S E L S V Y T L T V I T L E R W H T I T Y A I Q L D Q K L R L R H A I P I M L G G W L F S T

        475        480        485        490        495        500        505        510        515        520
          |          |          |          |          |          |          |          |          |          |
Rat    L I A T M P L V G I S N Y M K V S I C L P M D V E S T L S Q V Y I L S I L I L N V V A F V V I C A C
Porcine L I A M L P L V G V S S Y M K V S I C L P M D V E T T L S Q V Y I L T I L I L N V V A F I I I C A C

        525        530        535        540        545        550        555        560        565        570
          |          |          |          |          |          |          |          |          |          |
Rat    Y I R I Y F A V Q N P E L T A P N K D T K I A K K M A I L I F T D F Y C M A P I S F F A I S A A F K
Porcine Y I K I Y F A V Q N P E L M A T N K D T K I A K K M A V L I F T D F T C M A P I S F F A I S A A L K

        575        580        585        590        595        600        605        610        615        620
          |          |          |          |          |          |          |          |          |          |
Rat    V P L I T V T N S K I L L V L F Y P V N S C A N P F L Y A I F T K A F Q R D F L L L L S R F G C C K
Porcine V P L I T V T N S K V L L V L F Y P V N S C A N P F L Y A I F T K A F R R D F F L L L S K S G C C K

        625        630        635        640        645        650        655        660        665        670
          |          |          |          |          |          |          |          |          |          |
Rat    R R A E L Y R R K E F S A Y T S N C K N G F P G S S K P S Q A T L K L S T V H C Q Q P I P P R A L T
Porcine H Q A E L Y R R K D F S A Y     C K N G F T G S N K P S R S T L K L T T L Q C Q Y S T V M D K T C

        675
          |
Rat    H
Porcine Y K D C
```

receptors possess seven predicted transmembrane segments, which are similar to
other receptors which have been shown to
be coupled to G proteins, the adenylyl cyclase second messenger system, and cAMP
(see below). The transmembrane segments
are predicted by hydropathy analysis, identifying hydrophobic regions of the protein
sequence. Figure 12 illustrates how the LH/
hCG receptor may be placed in the cell
membrane. The transmembrane segment
contains loops exposed to the cytoplasm,
which may be important in receptor function. One of the receptors that is also linked
to G protein is the β_2-adrenergic receptor.
Recently, the sites of interaction of this receptor with G protein were localized to one
of these cytoplasmic loops and a portion of
the cytoplasmic tail close to the cell membrane (O'Dowd et al., 1988). The amino acid
sequences of these regions are well conserved between the β_2-adrenergic receptor
and the LH receptor, implying that these re-

gions also may be involved in G-protein
binding to LH receptors.

Finally, both proteins contain a carboxy-
terminal domain that extends into the cell
cytoplasm, and may be a further site for the
regulation of hormone-receptor function,
as it contains potential phosphorylation
sites. In other G-protein-linked receptors,
phosphorylation of the receptor leads to uncoupling of the receptor from G protein
and subsequent systems (Sibley et al., 1988).
It is interesting to speculate that one mechanism for the regulation of LH at the target
cell could be phosphorylation of the receptor, induced by the phosphorylating enzymes stimulated by the hormone receptor
complex. The phosphorylated receptor
would be ineffective in stimulating further
cAMP production, shutting off the initiating
signal for biological activity.

In general, throughout their sequences
the rat LH receptor and porcine LH receptor exhibit 84% homology. The major dif-

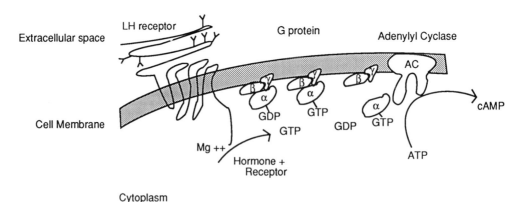

Figure 12 Summary of the proposed mechanism for signal transduction between the LH hormone–
receptor complex and adenyl cyclase. Once a hormone binds to a receptor, in the presence of Mg^{2+}, the
exchange of GTP for GDP occurs. The activated G_a dissociates from the G_{bg} dimer and activated adenylyl
cyclase, which produces cAMP from ATP. Indicated on the LH receptor are the six potential N-linked
glycosylation sites (Y). Not depicted is a predicted extracellular loop of adenylyl cyclase, which may also
be glycosylated.

ference in the length of the two receptors is accounted for by a 24-amino acid insert in the extracellular portion of the rat LH receptor compared to the porcine LH receptor. The most homologous region is the transmembrane portion, while the least similar region is the intracellular domain.

Also isolated from the porcine testis cDNA library were shorter forms of the LH receptor cDNA. All of these variants contained the proposed extracellular region, but were missing either the transmembrane domains or intracellular region. It is currently not known what role these variants may have; however, they account for 40% of the cDNA clones isolated.

B. The FSH Receptor

The receptor for FSH has been studied in both males and females. The major location for FSH receptor in males is the seminiferous tubule, specifically on Sertoli cells (Reichert and Dattatreyamurty, 1989). In females the granulosa cell FSH receptor has been extensively studied (Hsueh et al., 1984). In both sexes, the action of FSH through its receptor is mediated through the cAMP second messenger system, induced by adenylate cyclase and G protein.

Current evidence indicates that the FSH receptor may be an oligomeric glycoprotein, consisting of four disulfide-linked monomers (Reichert and Dattatreyamurty, 1989). However, the FSH receptor has yet to be cloned and the sequence deduced. This is an important distinction, as the structure of the LH/hCG receptor was clarified by using recombinant techniques. Studies of the LH receptor based on solubilization studies of tissues also had indicated that its structure was an oligomeric protein (Roche and Ryan, 1985). The current model proposed for the

LH receptor is consistent with other receptors known to couple to G protein and act through the adenylyl cyclase system (Loosfelt et al., 1989; McFarland et al., 1989). Since FSH and its receptor also mediate their effects through this second messenger system, it is likely that the receptor structure will fit into the class also. It is possible that the multimeric model currently proposed for the structure of the FSH receptor is a result of proteases clipping the receptor during its extraction from cells. However, a definitive answer awaits further research.

C. Prolactin Receptor

As is the case with the LH/hCG receptor, the prolactin receptor (Fig. 13) has been recently sequenced through isolation of the cDNA that encodes it (Boutin et al., 1988; Edery et al., 1989; Okamura et al., 1989). The prolactin receptor is found in mammary tissue, liver, and in certain species, the corpus luteum (Hsueh et al., 1984; Kelly et al., 1989). Unlike LH and FSH, where the second messenger systems involved in signal transduction is known, no second messenger has been identified for prolactin (Kelly et al., 1989).

The receptor for prolactin is a single protein with a long extracellular domain, a single transmembrane segment, and a segment that extends into the cell cytoplasm. Two forms of the prolactin receptor have been identified, differing by the length of this intracellular domain. A short form of the prolactin receptor was isolated from rat liver (Boutin et al., 1988), while the long form of receptor was isolated from rabbit mammary tissue (Edery et al., 1989), human hepatoma cells, and breast cancer cells (Boutin et al., 1989). The extracellular and transmembrane domains of these different forms of

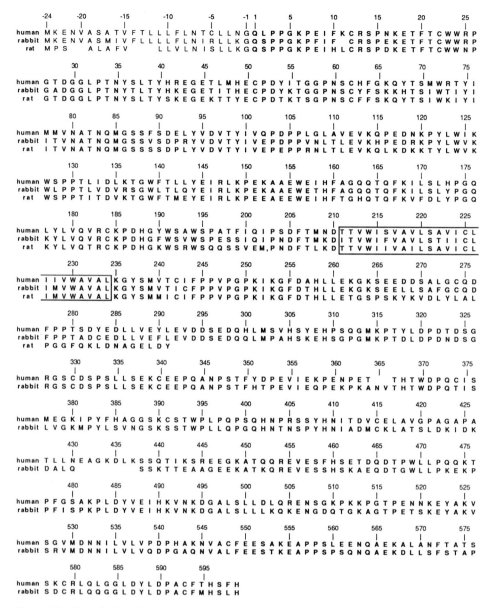

Figure 13 Compilation of reported prolactin receptor sequences. The human PRL receptor was deduced from hepatoma and breast cancer cDNA libraries, the rabbit receptor from mammary gland cDNA, and rat PRL receptor from a liver cDNA library. The proposed signal peptide is indicated by negative numbering, and the boxed region is the proposed transmembrane domain. The signal peptide and transmembrane domain are predicted from hydropathy analysis. The sequences are aligned for maximum homology (single-letter amino acid codes, see Fig. 6).

prolactin receptor are similar with 70–80% sequence homology between rat, human, and rabbit receptors. The major difference in the receptors seems to be in the intracellular region. Here, the short and long forms are similar for about half of the rat prolactin receptor sequence (the short version), but the remaining 25 amino acids in the rat prolactin receptor have no homology with any region of either the human or rabbit receptors. Both forms of receptors have been isolated in rats, the long form in the ovary and the short form in the liver, with alternative splicing of the receptor mRNAs proposed as being the likely mechanism involved in the generation of these different forms (Kelly *et al.*, 1989). The only protein that is similar in sequence to the prolactin receptor is the growth-hormone receptor. The prolactin and growth-hormone receptors have regions of similarity in the extracellular domain as well as the cytoplasmic domain (Edery *et al.*, 1989).

The different lengths of the cytoplasmic domain may have significance in the function of prolactin at those target tissues. The cytoplasmic region of the short prolactin receptor is similar in structure to other receptors which act as transport proteins, carrying ligands from blood to other biological fluids (Kelly *et al.*, 1989). The similarity of the long prolactin receptor with the growth hormone receptor may indicate that they have a similar, yet-to-be-determined mechanism of action (Edery *et al.*, 1989).

V. Intracellular Responses: G Protein and Second Messengers

As stated above, the effects of the gonadotropins and presumably prolactin are mediated by second messenger systems. The effects of the gonadotropins are mediated through at least two systems, the cyclic AMP (cAMP) system and the inositol phosphate system. The mechanism of action of prolactin has yet to be identified. There is a complex of proteins termed the G proteins (from their interaction with guanyl nucleotides), which function as an intermediate between the influence of hormone binding to receptor and activation of one of the messenger systems (here the adenylyl cyclase). The G proteins are associated with the cell membrane and are signal transducers in a number of systems including vision, neurotransmitters, and the gonadotropins (Neer and Clapham, 1988).

A. G Protein

The G protein (also called N protein) is actually a complex of three proteins, termed Gα, Gβ, and Gγ. This complex of proteins earned their collective name of G protein through their ability to bind guanine nucleotides. The constituent proteins are named in order of size: Gα is approximately 39–45 kDa, Gβ approximately 35–36 kDa, and Gγ approximately 7–10 kDa. The major types of Gα isolated thus far are Gαs (stimulating), Gαi (inhibitory), Gαt (transducing—rhodopsin system), and Gαo (other). These different Gα subunits have been identified by their function, except for Gαo, for which no function has been determined. One of the chemical characteristics used to differentiate the different types of Gα is the ability of either cholera toxin or pertussis toxin to add an ADP ribose group to the Gα protein. This functions to irreversibly activate the system. Generally cholera toxin is active on Gαs, and pertussis toxin is active on Gαi and Gαo (Gilman, 1987; Neer and Clapham, 1988).

The Gβγ complex does not seem to be specific for a particular type of Gα protein. Experiments have shown that the Gβγ component of a Gαs can associate with all other types of Gα. In addition, the Gβγ dimer may have additional functions besides being a reservoir for Gα subunits. Some of these actions include inhibition of brain cAMP phosphodiesterase, and activation of ion channels (Neer and Clapham, 1988).

After a gonadotropin binds to its receptor, the G protein complex is affected through a change in affinity of the G protein for Mg^{2+} (Hunzicker-Dunn and Birnbaumer, 1985). In the resting state the three subunits (α,β,γ) of the G protein are clustered together with GDP bound to the Gα protein. After the hormone binds to receptor, the affinity of the G complex for Mg^{2+} increases, catalyzing the exchange of GTP for GDP on the Gα protein. The Gα–GTP complex is now the "active" form, and the active Gα–GTP may actually dissociate from the Gβγ complex and enter the cytoplasm (Neer and Clapham, 1988). The active Gα–GTP interacts with other cellular systems such as adenylyl cyclase, initiating a second messenger signal. One of the controls of the system is the Gα action as a GTPase. Due to this endogenous activity of the Gα protein, Gα–GTP will revert to Gα–GDP, the inactive form, and reassociate with the Gβγ complex turning this part of the system off, until the appropriate hormone again binds to the receptor.

The gonadotropin's binding to a receptor, including the activation of G protein, leads to the production of an intracellular second messenger. The messenger system most heavily involved in the actions of gonadotropins on gonadal cells is the cAMP system. Evidence also exists that an additional second messenger system may be important in some cells, the inositol phosphate, diacylglycerol (DG) system.

B. Adenylate Cyclase and cAMP

The major second messenger system involved in the action of LH and FSH is the adenylate cyclase, cAMP system. The pathway is partially illustrated in Fig 12. Adenylate cyclase is the enzyme that produces cAMP and is closely assiciated with the G protein of the cell membrane. Adenylyl cyclase has recently been sequenced, and through hydropathic analysis a structure has been proposed that places the enzyme in the cell membrane with two large domains in the cytoplasm (Krupinski, et al., 1989).

When the Gα is activated by hormone binding to receptor and subsequent exchange of GDP for GTP, the activated Gα stimulates the catalytic activity of adenylate cyclase, producing cAMP from ATP (Hunzicker-Dunn and Birnbaumer, 1985). The cAMP then acts as the second messenger, activating protein kinase A by binding to the regulatory subunits, releasing the catalytic portion. It is through this cascade of events that a single hormone–receptor unit can amplify its signal. Up to several hundred molecules of cAMP are produced from the action of one activated Gα sununit (Gilman, 1987). The catalytic portion of protein kinase A then activates other enzymes by the addition of phosphate to either serine or threonine (Hunzicker-Dunn and Birnbaumer, 1985). The cAMP pathway is terminated through the action of the enzyme phosphodiesterase, which is regulated by calcium (Bolander, 1989). Under the influence of calcium, phosphodiesterase hydrolyzes cAMP into the inactive metabolite, AMP. The removal of cAMP favors the reassociation of the regulatory and catalytic subunits of protein kinase A. The Gαi has the opposite effect of Gαs. If Gαi is stimulated, it binds to adenylate cyclase, inhibiting its activity and preventing the production of cAMP (Casey and Gilman, 1988).

Actions of the gonadotropins modulated through cAMP include steriod production, induction of receptors for gonadotropins, and changes in cell shape (Hsueh *et al.*, 1984).

C. Inositol Phosphates

An additional second messenger system that may be induced by the gonadotropins is the diacylglycerol–inositol phosphate system. This second messenger system also is controlled by G protein; however, the Gα subunit (Gαp) is a different type than the one involved in activation of adenylyl cyclase (Berridge, 1987; Berridge and Irvine, 1989). Activation of this Gαp stimulates the action of phospholipase C, which cleaves inositol phosphate from phosphoinositol, producing two messenger molecules, inositol phosphate and diacylglycerol (DG). The inositol phosphates are primarily involved in controlling the calcium levels within the cell by redirecting intracellular calcium and opening calcium channels in the cell membrane, allowing calcium to enter the cell. Diacylglycerol is involved in the activation of protein kinase C, which also requires calcium. Protein kinase C is another protein-phosphorylating enzyme found in most cells. The importance of protein kinase C in the actions of the gonadotropins has also been studied through the use of phorbol esters. These are compounds isolated from the seed oil of the croton plant, and enhance the functions of protein kinase C. Recently the roles of both inositol phosphate and DG

have been investigated in LH induction of steroidogenesis, in which inositol phosphate increased progesterone release, and DG decreased progesterone production (Sadighia *et al.*, 1989).

VI. Assays

A critical step in indentifying hormones and understanding their functions in the animal and interactions with target cells is their detection and measurement. One of the basic experiments in endocrinology is to remove a gland, note changes in the animal, then replace the gland and verify that conditions return to normal. One example of this approach is the effect of growth hormone on growth. If the pituitary gland is removed from a young animal, it stops growing. When pituitary extracts, and later purified growth hormone were given, the animal resumed growth. This is an example of an *in vivo* (in the living) assay, one involving an entire animal (Fig. 14). Other general types of assays for hormones are *in vitro* (in glass) assays, where either tissue pieces, isolated cells, or cell membranes taken from animals are used . A third type of assay is the immunological assay, either radioimmunoassay (RIA) or enzyme-linked immunoassay (ELISA), in which an antibody specifically recognizes the desired hormone. Each of these assay types has advantages and disadvantages.

A. In Vivo Assays

These assays generally were the first ones developed for hormones, as a function that was possibly regulated by a hormone was identified and then used as an assay to monitor the purification of that hormone. The

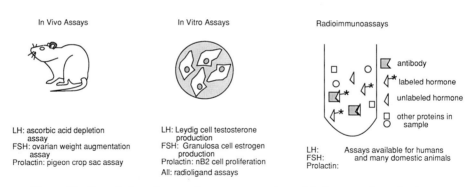

In Vivo Assays

In Vitro Assays

Radioimmunoassays

□ antibody

◁* labeled hormone

◁ unlabeled hormone

□ other proteins in
○ sample

LH: ascorbic acid depletion
   assay
FSH: ovarian weight augmentation
   assay
Prolactin: pigeon crop sac assay

LH: Leydig cell testosterone
   production
FSH: Granulosa cell estrogen
   production
Prolactin: nB2 cell proliferation
All: radioligand assays

LH: Assays available for humans
FSH: and many domestic animals
Prolactin:

Figure 14   Compilation of types of assays commonly used in hormone investigations.

major disadvantage of *in vivo* assays is expense, as generally only one or two experimental tests can be conducted per animal. Another disadvantage is lack of precision, as animal to animal variance is usually high. However, *in vivo* assays remain an important experimental tool as the normal environment for hormones is not a test tube, but a live animal. Testing hormone preparations in live animals is extremely important in understanding their functions in live animals, as there may be unknown factors that may modify the action of the hormone, which cannot be predicted by other methods.

The most common *in vivo* assay used to study LH is the depletion of ascorbic acid in the ovaries of pseudopregnant rats (Parlow, 1961). In this assay a pseudopregnant rat is treated with the hormone preparation, and depletion of ascorbic acid in the heavily luteinized ovary is indicative of LH content in the sample.

The standard *in vivo* assay for FSH is the ovarian augmentation test, described by Steelman and Pohley (1953) Similar to the test for LH, rat ovaries were collected after administration of the test substance, the end

point in this assay being the increase in ovarian weight. Steelman and Pohley minimized LH effects in this assay by including hCG in the test preparation, thus overwhelming any influence of LH in the test sample on ovarian weight gain. We now know that this increase in ovarian weight is due to follicular development induced by FSH.

The classical *in vivo* assay for prolactin is the pigeon crop sac assay, in which increase in crop sac weight is correlated with prolactin potency (Nicoll, 1967).

B. In Vitro *Assays*

In Vitro assays also are a valuable tool for understanding the relationship between the structure and function of hormones. These assays are artificial systems in that they are conducted in glass test tubes or plastic culture dishes, but are valuable in that they use animal tissue membranes or intact cells to gauge the potency of hormone preparations. One major advantage of these assays over *in vivo* assays is that many more experimental tests can be conducted per experimental animal, as the tissue is distributed into multiple culture dishes, and only a small amount of tissue is required per test. The precision of *in vitro* assays is usually much higher than *in vivo* assays.

There are essentially two types of *in vitro* assays, those that measure binding of a ligand to a membrane as an indicator of potency, and those that measure a cell's response to the ligand as the end point. Both types of assays have been developed for the gonadotropins and prolactin. These assays have been used to determine structural features of hormones that are required for binding to receptors and which features are important for activation of second messenger systems and cellular function. It has become clear through these types of experiments that these functional domains can be very distinct in gonadotropins (Matzuk *et al.*, 1989), allowing identification of specific structures of the hormones involved in receptor binding, and that are involved in determining biological activity. Recently, *in vitro* assays have been used to monitor gonadotropin concentrations in serum, facilitating experiments studying the regulation of biologically active hormone in circulation. (Dahl *et al.*, 1988; Dufau *et al.*, 1976).

The *in vitro* assays for LH utilize tissue from the *in vivo* target organ, the ovaries and testis. Testicular homogenates are a good source of LH receptor, and have an advantage in that tissue can be collected and frozen for future use (Glenn *et al.* 1981). MA-10 cells, derived from a Leydig cell tumor, also have been used to measure the binding affinity of modified hCG (Matzuk *et al.*, 1989). *In vitro* steroidogenesis assays for LH typically employ isolated Leydig cells in which testosterone production is under the control of LH/hCG (Dufau *et al.*, 1976). MA-10 cells also are used for steroidogenesis assays (Pereira *et al.*, 1987).

In vitro assays for FSH have also utilized testicular tissue for binding assays, from species ranging from horses (Moore and Ward, 1980) to chickens and rats (Gordon *et al.*, 1989). Three types of FSH bioassay have been reported, the production of cAMP in response to FSH in dispersed Sertoli cells (Rao and Ramachandran, 1975), activation of the enzyme plasminogen activator in granulosa cells (Beers and Strickland, 1978), and estrogen production by granulosa cells, controlled by FSH sensitive aromatase (Hsueh *et al.*, 1984).

Rabbit mammary glands are the most common source of tissue used in the prolactin radioreceptor assay (Shiu *et al.*, 1973). In addition, porcine mammary tissue and adrenal cortical tissue have been used in binding assays (Atkinson *et al.*, 1988). Also, the Nb2 cell line is used in prolactin bioassays, as mitogenesis increases under the influence of prolactin (Tanaka *et al.*, 1980).

C. Radioimmunoassay

The development of radioimmunoassay (RIA) for the gonadotropins and prolactin has greatly facilitated the study of the role of these hormones in reproduction. RIA was first applied to the measurement of insulin (Berson *et al.*, 1956; Yalow, 1978) but has since been applied to many aspects of endocrinological investigation, including reproduction. It is through RIA that patterns of hormones in blood were investigated, leading to greater understanding of reproductive cyclicity.

The components of an RIA are (1) an antibody that is specific for the hormone to be detected, (2) labeled hormone—commonly ^{125}I is used as the label—covalently added to the protein, and (3) the unknown sample.

Through competition of the labeled hormone and the hormone in the sample for binding to the antibody, the amount of hormone in the unknown can be quantified.

One of the fundamental aspects of RIA is that immunologically active hormone, that material recognized by the antibody, is the detected material. However, immunologically active hormone may not always be biologically active. The comparison or hormone patterns and isolated gonadotropins by RIA and bioassay has provided insights into physiological states that alter gonadotropin potency, and chemical determinants that are involved in biological action.

The technology of RIA as applied to the study of reproduction in domestic livestock is still evolving. Assay systems are being developed that use enzyme-catalyzed color markers, in place of ^{125}I- or ^{3}H-labeled ligands. While traditional RIAs are easily performed in laboratories, they are not easily performed in production locations due to the need for specialized equipment. Enzyme-linked assays show the promise of being accurate, precise, and reproducible, allowing the producer (or investigator) to conduct assays with a minimum amount of equipment (Bretzlaff et al., 1989).

VII. Reference Preparations for Bioassay and Radioimmunoassay

The availability of a suitable reference preparation upon which to evaluate potency of the particular hormone for comparison from experiment to experiment and from laboratory to laboratory has always been an important consideration. Currently, this problem generally has an appropriate solu-

tion, although in particular instances this may not be the case. For example, during the early stages in the isolation of a newly discovered hormone activity, and the presumptive pure hormone responsible for this activity, it may not be clear what a suitable reference preparation may be. As a recent example of this situation, the evolution of inhibin from a vague entity to a well-characterized hormone can be traced by reference to early and late reviews (Channing et al., 1978; Steinberger and Ward, 1988; Ying, 1988). Even so, the "ideal" reference standard for inhibin has not been provided. In the early stages of investigation of the gonadotropins, this same problem existed.

(From this point in this section we will be discussing only reference preparations for the glycoprotein hormones unless otherwise noted.)

In the early history of the isolation and characterization of the glycoprotein hormones as well as the physiological studies at that time, it was necessary to make definitions of animal units relating to a particular activity (e.g., "rat unit," "mouse unit," etc.) These were arbitrarily defined by the investigator in terms of some in vivo response. The utility of the "unit" was invariably compromised by many factors such as the astuteness of the investigator to define the unit, whether the definition was subsequently supported by a suitable reference preparation to serve as a yardstick for subsequent comparisons, the comparative conditions for animal handling in different laboratories, the actual animal variance, and finally, how effective the statistical analyses were for the results of the comparative assays. It should be readily apparent why such animal units were generally unsuitable, difficult to reproduce, and very insensitive. For the gonadotropin research, Armour

Laboratories before World War II and for a period shortly thereafter provided their own reference preparations to alleviate this situation somewhat, but these references were not always available to all researchers. The literature at that time frequently referred to Armour house standard 264-151-X (an FSH reference) and Armour 227-80 (an LH reference). The Endocrinology Study Section of the Division of Research Grants at the National Institutes of Health (NIH) undertook to provide a greater variety of reference preparations to allow researchers to make better comparisons from laboratory to laboratory. This effort was initiated in the 1950s and expanded to the issuance by NIH of a contract to produce these reference preparations. The first contract went to Emory University in 1959. Reichert and Wilhelmi summarized the preparations provided in two reports (Reichert and Wilhelmi, 1973, 1978). This is perhaps the most important set of reference preparations for endocrinologists, certainly in terms of their widespread usage. For the past several years the NIH contracts for these reference preparations and also rat pituitary preparations have been held by Dr. Albert F. Parlow and his colleagues in Torrence, California. Another important body for the preparation and dissemination of reference preparations has been the National Institute for Biological Standards and Control in the United Kingdom, and the affiliated World Health Organization (WHO) International Laboratory for Biological Standards. We will say more about the WHO program below.

In addition to the formal programs noted above, other specialized reference programs have supplemented the above with preparations developed for more specific applications. Examples of these are the materials made available from the Contraceptive Development Branch, National Institute of Child Health and Human Development, NIH, or from the U.S. Department of Agriculture, in conjunction with the above organizations. For the majority of the available hormone preparations from the above sources, these may be obtained by formal request to the Hormone Distribution Officer, NIDDK, NIH in Bethesda, Maryland, or by application to the National Hormone and Pituitary Program (NHPP) currently at the University of Maryland School of Medicine in Baltimore. The latter entity (NHPP) is under contract to the NIH for the storage, vialing and handling, and distribution of these reference (and other) hormone preparations. The actual production of these hormone preparations has been managed by contracts for human and animal materials. These contracts have been provided by both the National Institutes of Health and the U.S. Department of Agriculture. The available animal hormone preparations of interest to readers of the current volume are from rat, ovine, bovine, and porcine hormones, including suitable antisera for biological studies, for radioimmunoassay, and various research activities. (The materials are not available for commercial use.) Details for the available reference preparations and procedures for requesting the materials are carried as announcements twice a year in *Endocrinology* and other publications of The Endocrine Society (USA).

The second major source of reference preparations that may be of value to the readers of this volume is the National Institute of Biological Standards and Control, a World Health Organization International Laboratory for Biological Standards. Their latest catalog may be obtained by request to the foregoing at Blanche Lane, South

Mimms, Potters Bar, Hertfordshire, UK, EN6 3QG. The type of reference preparation this group provides ranges from viral and bacterial vaccines, serum factors (e.g., heparin and clotting factors), cytokines, and antibiotics. The preparations of direct concern to the readers of this volume will be the WHO reference standards for LH (and subunits), FSH, human chorionic gonadotropin (and subunits/RIA), equine serum gonadotropin (eCG), and TSH. A "standard" represents a defined amount of a given preparation that is used to define the international unit (IU) for that substance. The amount of a preparation that defines

an IU may be in terms of mass, but in most instances it is in terms of total contents of an ampoule provided. A batch of ampoules may comprise 500–5000 vials. When that supply is gone, a second batch is made and vialed and labeled 2nd International Standard (and so forth, as the case may be). Each preparation distributed contains a flier providing the assay data and handling instructions for that preparation. From the availability of the individual preparations (i.e., the limited number of vials) it is appropriate to prepare an in-house reference carefully calibrated against the International Standard. The in-house standard may then be

Table II
Reported Unit Conversion Factors, in Terms of NIH-LH-S1[a] and NIH-FSH-S1[a]

Reference preparation	Potency (units/mg)		Assay used[c]	Reference quoted
	Vs. NIH-LH-S1	Vs. NIH-FSH-S1		
1st IRP-HMG[b]	0.00032		OAAD	Reichert (1966)
	0.009		VP	
		0.0052	S-P	
2nd IRP-HMG[b]	0.00065		OAAD	Reichert (1966)
	0.015		VP	
		0.037	S-P	
2nd hCG Int. Std. (i.e., 900 IU = 1mg NIH-LH-S1)	0.0011		OAAD, VP	Reichert (1966)
Armour 227-80	0.94		OAAD	Reichert (1966)
	1.31		VP	
Armour 264-151-X		0.55	S-P	Reichert (1966)
2nd IRP (IU)		0.038	S-P	
	0.0017		OAAD	Rosemberg (1968)
	0.019		VP	Rosemberg (1968)
2nd IRP (IU)	0.0017		OAAD	Jutisz and Tertrin-Clary (1974)

[a]The values are given in terms of the original NIH-LH-S1 and NIH-FSH-S1 reference preparations (no longer available). The potency of reference preparations subsequently prepared and distributed are compiled in the reports by Reichert and Wilhelmi (1973, 1978) or in the fliers that accompany the currently available preparations.

[b]These are widely used human menopausal gonadotropin reference preparations derived from urine. Used to define an IU for LH and FSH.

[c]Assays: OAAD, ovarian ascorbic acid depletion assay; VP, ventral prostate (weight gain) assay in hypophysectomized rats; S-P, the Steelman–Pohley hCG augmentation assay (sec section on assays).

employed for the day-to-day utilization of reference preparations. The preparations handled by the WHO International Laboratory for Biological Standards are derived as *ad hoc* preparations, donations from laboratories or institutions, etc., and generally are jointly tested in several laboratories to evaluate the expected assay response (either bioassay or radioimmunoassay) according to the type of preparation.

Although the "units" of the preparations and the "IU" of the WHO preparations are well defined in terms of that particular preparation, there is no direct cross-correlation available except reports from individual laboratories that have made the required side-by-side comparisons. Sometimes, however, such comparisons are required (for example, to compare results from different laboratories that employ different reference standards). We have accumulated some comparisons available in the literature. These comparative assay results are given in Table II for convenient reference.

It will be noted that different conversion values have been determined in several instances. This results from several factors such as the type of reference preparation vis-à-vis the type of "unknown." These differences are sometimes detected in terms of "nonparallelism" for the statistical validation of the assay. In some cases differences relate to hormone half-life differences (particularly in the *in vivo* assays). Many of these may be controlled by using the same kind of reference preparation, but that is not always convenient. A large part of the difference is, of course, due to differences in definition of unitage. The conversion factors in the table are useful for comparisons, but if the unit comparisons are critical in a study it is essential to make the appropriate comparisons

under the experimental conditions to be employed.

References

Allen, W. R., Hamilton, D. W., and Moor, R. M. (1973). *Anat. Rec.* **177,** 485–500.

Allen, W. R., and Moor, R. M. (1972). *J. Reprod. Fertil.* **29,** 313–316.

Ascoli, M., and Segaloff, D. L. (1989). *Endocrinol. Rev.* **10,** 27–44.

Atkinson, P. R., Seely, J. E., Klemcke, H. G., and Hughes, J. P. (1988). *Biochem. Biophys. Res. Commun.* **155,** 1187–1193.

Baenziger, J. U., and Green, E. D. (1988). *Biochim. Biophys. Acta* **947,** 287–306.

Bahl, O. P., and Wagh, P. V. (1986). *In* "Molecular and Cellular Aspects of Reproduction" (D. S. Dhindsa and O. P. Bahl, eds.), pp. 1–51. Plenum Press, New York.

Bambra, C. S., Lynch, S. S., Foxcroft, G. R., Robinson, G., and Amoroso, E. C. (1984). *J. Reprod. Fertil.* **71,** 227–233.

Bartosik, D., Romanoff, E. B., Watson, J. D., and Scricco, E. (1967). *Endocrinology* **81,** 186–192.

Beers, W. H., and Strickland, S. (1978). *J. Biol. Chem.* **253,** 3877–3881.

Bégeot, M., Hemming, F. J., Dubois, P. M., Combarnous, Y., Dubois, M. P., and Aubert, M. L. (1984). *Science* **226,** 566–568.

Berman, M. I., Thomas, C. G., Jr., Manjunath, P., Sairam, M. R., and Nayfeh, S. N. (1985). *Biochem. Biophys. Res. Commun.* **133,** 680–687.

Berridge, M. J. (1987). Annu. Rev. Biochem. **56,** 159–193.

Berridge, M. J., and Irvine, R. F. (1989). *Nature* **341,** 197–205.

Berson, S. A., Yalow, R. S., Bauman, A., Rothschild, M. A., and Newerly, K. (1956). *J. Clin. Invest.* **35,** 170–190.

Bewley, T. A. (1979). *Recent Prog. Hormone Res.* **35,** 155–213.

Birken, S., and Canfield, R. E. (1977). *J. Biol. Chem.* **252,** 5386–5392.

Birken, S., and Canfield, R. E. (1978). In "Structure and Function of the Gonadotropins" (K. W. McKerns, ed.), pp. 47–80. Plenum Press, New York.

Birken, S., Fetherston, J., Canfield, R., and Boime, I. (1981). *J. Biol. Chem.* **256,** 1816–1823.

Blithe, D. L., and Nisula, B. C. (1985). *Endocrinology* **117**, 2218–2228.

Bogdanov, E. M., and Nansel, D. D. (1978). In "Structure and Function of the Gonadotropins" (K. W. McKerns, ed.), pp. 415–430. Plenum Press, New York.

Boime, I. (1989). In "Glycoprotein Hormones: Structure, Synthesis, and Biologic Function" (W. W. Chin and I. Boime, eds.), pp. 111–121. Serono Symposium, Newport Beach, Calif.

Boime, I., Boothby, M., Hoshina, M., Daniels-McQueen, S., and Darnell, R. (1982). *Biol. Reprod.* **26**, 73–91.

Bolander, F. F. (1989). "Cyclic Nucleotides." Academic Press, San Diego.

Bousfield, G. R. (1981). Purification and In Vitro Biological Characterization of Equine and Chicken Gonadotropins: Demonstration of Intrinsic FSH Activity of Equine and Chicken LH in the Rat. Ph.D. dissertation, Indiana University, Bloomington, Indiana.

Bousfield, G. R., Liu, W.-K., and Ward, D. N. (1985a). *Mol. Cell. Endocrinol.* **40**, 69–77.

Bousfield, G. R., Sugino, H., and Ward, D. N. (1985b). *J. Biol. Chem.* **260**, 9531–9533.

Bousfield, G. R., and Ward, D. N. (1984). *J. Biol. Chem.* **259**, 1911–1921.

Bousfield, G. R., and Ward, D. N. (1986). 68th Annual Meeting of the Endocrine Society, Indianapolis, Indiana.

Boutin, J. M., Edery, M., Shirota, M., Jolicoeur, C., Lesueur, L., Ali, S., Gould, D., Djiane, J., and Kelly, P. A. (1989). *Mol. Endocrinol.* **3**, 1455–1461.

Boutin, J. M., Jolicoeur, C., Okamura, H., Gagnon, J., Edery, M., Shirota, M., Banville, D., Dusanter-Fourt, I., Djiane, J., and Kelly, P. A. (1988). *Cell* **53**, 69–77.

Bowers, C. Y., Friesen, H. G., Hwang, P., Guyda, H. J., and Folkers, K. (1971). *Biochem. Biophys. Res. Commun.* **45**, 1033–1041.

Bretzlaff, K. N., Elmore, R. G., and Nuti, L. C. (1989). *J. Am. Vet. Med. Assoc.* **194**, 664–668.

Burke, W. H., Licht, P., Papkoff, H., and Gallo, A. B. (1979). *Gen. Comp. Endocrinol.* **37**, 508–520.

Burzawa-Gerard, É., and Fontaine, Y.-A. (1976). *C.R. Acad. Sc. Paris.* **282**, 97–100.

Calvo, F. O., Keutmann, H. T., Bergert, E. R., and Ryan, R. J. (1986). *Biochemistry* **25**, 3938–3943.

Casey, P. J., and Gilman, A. G. (1988). *J. Biol. Chem.* **262**, 2577–2580.

Channing, C. P., Sakai, C. N., and Bahl, O. P. (1978). *Endocrinology* **103**, 341–348.

Chappel, S. C., Ulloa-Aguirre, A., and Coutifaris, C. (1983). *Endocrinol. Rev.* **4**, 179–211.

Cheng, K.-W., Glazer, A. N., and Pierce, J. G. (1973). *J. Biol. Chem.* **248**, 7930–7937.

Cheng, K. W. (1976). *Biochem. J.* **159**, 651–659.

Chin, W. W. (1985). *In* "The Pituitary Gland" (H. Imura, ed.), pp. 103–125. Raven Press, New York.

Chin, W. W. (1986). In "Molecular Cloning of Hormone Genes" (J. F. Habener, ed.), pp. 137–172. Humana Press, Clifton, New Jersey.

Chin, W. W., and Boime, J. W. (1990). "Glycoprotein Hormones: Structure, Synthesis, and Biologic Function." Plenum Press, New York.

Chin, W. W., Kronenberg, H. M., Dee, P. C., Maloof, F., and Habener, J. F. (1981). *Proc. Natl. Acad. Sci. USA* **78**, 5329–5333.

Chin, W. W., Maizel, J. V., Jr., and Habener, J. F. (1983). *Endocrinology* **112**, 482–485.

Closset, J., and Hennen, G. (1978). *Eur. J. Biochem.* **86**, 105–113.

Closset, J., Hennen, G., and Lequin, R. M. (1973). *FEBS Lett.* **29**, 97–100.

Combarnous, Y., Guillou, F., and Martinat, N. (1984). *Endocrinology* **115**, 1821–1827.

Corless, C. L., Matzuk, M. M., Ramabhadran, T. V., Krichevsky, A., and Boime, I. (1987). *J. Cell. Biol.* **104**, 1173–1181.

Croyle, M. L., Bhattacharya, A., Gordon, D. F., and Maurer, R. A. (1986). *DNA* **5**, 299–304.

Dahl, K. D., Bicsak, T. A., and Hsueh, A. J. W. (1988). *Science* **239**, 72–74.

Denamur, R., Martinet, J., and Short, R. V. (1973). *J. Reprod. Fertil.* **32**, 207–209.

Dorrington, J. H., Roller, N. F., and Fritz, I. B. (1974). *In* "Hormone Binding and Target Cell Activation in the Testis" (M. L. Dufau and A. R. Means, eds.), pp. 237–241. Plenum Press, New York.

Dufau, M. L., Rock, R., Neubaurer, A., and Catt, K. J. (1976). *J. Clin. Endocrinol. Metab.* **42**, 958–969.

Edery, M., Jolicoeur, C., Levi-Meyrueis, C., Dusanter-Fourt, I., Petridou, B., Boutin, J. M., Lesueur, L., and Kelly, P. A. (1989). *Proc. Natl. Acad. Sci. USA.* **86**, 2112–2116.

El-Deiry, S., Kaetzel, D., Kennedy, G., Nilson, J., and Puett, D. (1989). *Mol. Endocrinol.* **3**, 1523–1528.

Endo, Y., Yamashita, K., Tachibana, Y., Tojo, S., and Kobata, A. (1979). *J. Biochem.* **85**, 669–679.

Everett, J. W. (1954). *Endocrinology*, **54**, 685–690.

Fiddes, J. C., and Goodman, H. M. (1981). *J. Mol. Appl. Genet.* **1**, 3–18.

Fiddes, J. C., and Talmadge, K. (1984). *Recent. Prog. Horm. Res.* **40**, 43–74.

Garfink, J. E., Moyle, W. R., and Bahl, O. P. (1976). *Gen. Comp. Endocrinol.* **30**, 292–300.

Garnier, J. (1978). *In* "Structure and Function of the Gonadotropins" (K. W. McKerns, ed.), pp. 381–414. Plenum Press, New York.

Gilman, A. G. (1987). *Annu. Rev. Biochem.* **50**, 615–649.

Giudice, L. C., Waxdal, M. J., and Weintraub, B. D. (1979). *Proc. Natl. Acad. Sci. USA* **76**, 4798–4802.

Glenn, S. D., Liu, W. K., and Ward, D. N. (1981). *Biol. Reprod.* **25**, 1027–1033.

Godine, J. E., Chin, W. W., and Habener, J. F. (1982). *J. Biol. Chem.* **257**, 8368–8371.

Goodwin, R. G., Moncman, C. L., Rottman, F. M., and Nilson, J. H. (1983). *Nucleic Acids Res.* **11**, 6873–6882.

Gordon, D. F., Wood, W. M., and Ridgway, E. C. (1988). *DNA* **7**, 17–26.

Gordon, W. L., Bousfield, G. R., and Ward, D. N. (1989). *J. Endocrinol. Invest.* **12**, 383–392.

Gordon, W. L., and Ward, D. N. (1985). *In* "Luteinizing Hormone: Receptors and Actions" (M. Ascoli, ed.), pp. 173–197. CRC Press, Boca Raton, Florida.

Governman, J. M., Parsons, T. F., and Pierce, J. G. (1982). *J. Biol. Chem.* **257**, 15059–15064.

Graham, M. Y., Otani, T., Boime, I., Olson, M. V., Carle, G. F., and Chaplin, D. D. (1987). *Nucleic Acids Res.* **11**, 4437–4448.

Green, E. D., and Baenziger, J. U. (1988a). *J. Biol. Chem.* **263** 36–44.

Green, E. D., and Baenziger, J. U. (1988b). *J. Biol. Chem.* **263**, 25–35.

Green, E. D., Boime, I., and Baenziger, J. U. (1986). *Mol. Cell. Biochem.* **72**, 81–100.

Green, E. D., Gruenebaum, J., Bielinska, M., Baenziger, J. U., and Boime, I. (1984). *Proc. Natl. Acad. Sci. USA* **81**, 5320–5324.

Green, E. D., van Halbeek, H., Boime, I., and Baenziger, J. U. (1985). *J. Biol. Chem.* **260**, 15623–15630.

Grimek, H. J., Gorski, J., and Wentworth, B. C. (1979). *Endocrinology* **104**, 140–147.

Grotjan. H. E., Jr. (1989). *In* "Microheterogeneity of Glycoprotein Hormones" (B. A. Keel and H. E. Grotjan, Jr., eds.), pp. 23–52. CRC Press, Boca Raton, Florida.

Grotjan, H. E., Jr., and Cole, L. A. (1989). *In* "Microheterogeneity of Glycoprotein Hormones" (B. A. Keel and H. E. Grotjan Jr., eds.), pp. 219–237. CRC Press, Boca Raton, Florida.

Guidon, P. T., Whitfield, G. K., Porti, D., and Kourides, I. A. (1988). *DNA* **7**, 691–699.

Guillou, F., and Combarnous, Y. (1983). *Biochim. Biophys. Acta* **755**, 229–236.

Habener, J. F. (1987). "Molecular Cloning of Hormone Genes." Humana Press, Clifton, New Jersey.

Harris, D. C., Machin, K. J., Evin, G. M., Morgan, F. J., and Isaacs, N. W. (1989). *J. Biol. Chem.* **264**, 6706–6706.

Hertz, R. (1980). *In* "Chorionic Gonadotropin" (S. J. Segal, ed.), pp. 1–15. Plenum Press, New York.

Hirschberg, C. B., and Snider, M. D. (1987). *Annu. Rev. Biochem.* **56**, 63–87.

Hsueh, A. J. W., Adashi, E. Y., Jones, P. B. C., and Welch, T. H., Jr. (1984). *Endocrinol. Rev.* **5**, 76–127.

Hunzicker-Dunn, M., and Birnbaumer, L. (1985). *In* "Luteinizing Hormone Action and Receptors" (M. Ascoli, ed.), pp. 57–134. CRC Press, Boca Raton, Florida.

Ishii, S., and Furuya, T. (1975). *Gen. Comp. Endocrinol.* **25**, 1–8.

Jackson, R. C., and Blobel, G. (1980). *Ann. N.Y. Acad. Sci.* **343**, 391–404.

Jameson, J. L., Becker, C. B., Lindell, C. M., and Habener, J. F. (1988). *Mol. Endocrinol.* **2**, 806–815.

Jameson, J. L., Deutsch, P. J., Chattergee, K. K., and Habener, J. F. (1989). *In* "Glycoprotein Hormones: Structure, Synthesis, and Biologic Function" (W. W. Chin and I. Boime, eds.), pp. 269–277. Serono Symposium, Newport Beach, California.

Jameson, L., Chin, W. W., Hollenberg, A. N., Chang, A. S., and Habener, J. F. (1984). *J. Biol. Chem.* **259**, 15474–15480.

Jia, X.-C., and Hsueh, A. J. W. (1986). *Endocrinology* **119**, 1570–1577.

Jirgensons, B., and Ward, D. N. (1970). *Tex. Rep. Biol. Med.* **28**, 553–559.

Jutisz, M., and Tertrin-Clary, C. (1974). *Curr. Top. Exp. Endocrinol.* **2**, 195–246.

Kato, Y. (1988). *Mol. Cell. Endocrinol.* **55**, 107–112.

Keel, B. A., and Grotjan, H. E., Jr. (1989). "Microheterogeneity of glycoprotein hormones," pp. 23–52. CRC Press, Boca Raton, Florida.

Kelly, P. A., Boutin, J. M., Jolicoeur, C., Okamura, H., Shirota, M., Edery, M., Dusanter-Fourt, I., and Djiane, J. (1989). *Biol. Reprod.* **40**, 27–32.

Kessler, M. J., Mise, T., Ghai, R. D., and Bahl, O. P. (1979). *J. Biol. Chem.* **254**, 7909–7914.

Keutmann, H. T., and Williams, R. M. (1977). *J. Biol. Chem.* **252,** 5393–5397.

Keutmann, H. T., Williams, R. M., and Ryan, R. J. (1979). *Biochem. Biophys. Res. Commun.* **90,** 842–848.

Kim, K. E., Gordon, D. F., and Maurer, R. A. (1988). *DNA* **7,** 227–233.

Kornfeld, R., and Kornfeld, S. (1985). *Annu. Rev. Biochem.* **54,** 631–664.

Krupinski, J., Coussen, F., Bakalyar, H. A., Tang, W.-J., Feinstein. P. G., Orth, K., Slaughter, C., Reed, R. R., and Gilman, A. G. (1989). *Science* **244,** 1558–1564.

Landefeld, T. D., and McShan, W. H. (1974a). *Biochemistry* **13,** 1389–1393.

Landefeld, T. D., and McShan, W. H. (1974b). *J. Biol. Chem.* **249.** 3529–3531.

Lehrman, S. R., Lahm, H.-W., Meidcel, M. C., Hulmes, J. D., and Li, C. H. (1988). *Int. J. Peptide Protein Res.* **31,** 544–554.

Li, C. H., and Starman, B. (1964). *Nature* **202,** 291–292.

Liao, T.-H., and Pierce, J. G. (1970). *J. Biol. Chem.* **245,** 3275–3281.

Licht, P., Gallo, A. B., Aggarwal, B. B., Farmer, S. W., Castelino, J. B., and Papkoff, H. (1979). *J. Endocrinol.* **83,** 311–322.

Liu, L., Southers, J. L., Banks, S. M., Blithe, D. L., Wehman, R. E., Brown, J. H., Chen, H.-C., and Nisula, B. C. (1989). *Endocrinology* **124,** 175–180.

Liu, W. K., Nahm, H. S., Sweeney, C. M., Holcomb, G. N., and Ward, D. N. (1972a). *J. Biol. Chem.* **247,** 4365–4381.

Liu, W. K., Nahm, H. S., Sweeney, C. M., Lamkin, W. M., Baker, H. N., and Ward, D. N. (1972b). *J. Biol. Chem.* **247,** 4351–4364.

Liu, W. K., and Ward, D. N. (1975). *Pharmacol. Ther. B.* **1,** 545–570.

Liu, W. K., Young, J. D., and Ward, D. N. (1984). *Mol. Cell. Endocrinol.* **37,** 29–39.

Lodish, H. (1988). *J. Biol. Chem.* **263.** 2107–2110.

Loosfelt, H., Misrahi, M., Atger, M., Salesse, R., Hai-Luu Thi, M. T. V., Jolivet, A., Guiochon-Mantel, A., Sar, S., Jallal, B. Garnier, J., and Milgrom, E. (1989). *Science* **245,** 525–528.

Maghuin-Rogister G., Closset, J., and Hennen, G. (1975). *FEBS Lett.* **60,** 263–266.

Markoff, E., Sinha, Y. N., and Lewis, U. J. (1989). *In* "Microheterogeneity of Glycoprotein Hormones" (B. A. Keel and H. E. Grotjan, Jr., eds.), pp. 99–106. CRC Press, Boca Raton, Florida.

Matzuk, M. M., Keene, J. L., and Boime, I. (1989). *J. Biol. Chem.* **264,** 2409–2414.

Matzuk, M. M., Spangler, M. M., Camel M., Suganuma, N. and Boime, I. (1989). *J. Cell Biol.* **109,** 1429–1438.

McFarland, K. C., Sprengel, R. S., Phillips, H. S., Kohler, M., Rosembilt N., Nikolics, K., Segaloff, D. L., and Seeburg, P. H. (1989). *Science* **245,** 494–499.

McNeilly, A. S. (1987). *J. Endocrinol.* **115,** 1–5.

Mise, T., and Bahl, O. P. (1981). *J. Biol. Chem.* **256,** 6587–6592.

Miura, Y., Perkel, V. S., and Magner, J. A. (1988). *Endocrinology* **123,** 2207–2213.

Moore, W. T., Jr. (1978). Biological and Chemical Characterization of Pregnant Mare Serum Gonadotropin—Consideration of the Origin of the Two Activities. Ph.D. dissertation, The University of Texas Health Science Center at Houston.

Moore, W. T., Jr., and Ward, D. N. (1980). *J. Biol. Chem* **255,** 6930–6936.

Morell, A. G., Gregoriadis, G., Scheinberg, I. H., Hickman, J., and Ashwell, G. (1971). *J. Biol. Chem.* **246,** 1461–1467.

Nakamura, T., Takio, K., Eto, Y. Shibai, H., Titani, K., and Sugino, H. (1990). *Science,* **247,** 836–838.

Neer, E. J., and Clapham, D. E. (1988). *Nature* **333,** 129–134.

Nicoll, C. S. (1967). *Endocrinology* **80** 641–655.

Nicoll, C. S., and Bern, H. A. (1972). *In* "Lactogenic Hormones" (C. E. W. Wolstenholme and J. Knight, eds.), pp. 299–312. Churchill Livingstone, Edinburgh.

Nicoll, C. S., Mayer, G. L., and Russel, S. M. (1986). *Endocr. Rev.* **7,** 169–203.

Nilson, J. H., Bokar, J. A., Keri, R. A., Andersen, B., Kennedy, G., Yun, J., and Wagner, T. (1989). *In* "Glycoprotein Hormones: Structure, Synthesis, and Biologic Function" (W. W. Chin and I. Boime, eds.), pp. 259–267. Serono Symposium, Newport Beach, California.

Niswender, G. D. (1989). *In* "Glycoprotein Hormones: Structure, Synthesis, and Biologic Function" (W. W. Chin and I, Biome, eds.). pp. 421–429. Serono Symposium, Newport Beach, California.

Niswender, G. D., Roess, D. A., Sawyer, H. R., Silvia, W. J., and Barisas, B. G. (1985). *Endocrinology* **116,** 164.

O'Dowd, B. F., Hnatowich, M., Regan J. W., Leader, W. M., Caron M. G., and Lefkowitz, R. J. (1988). *J. Biol. Chem.* **263,** 15985–15992.

Okamura, H., Raguet, S., Bell, A., Gagnom, J., and Kelly, P. A. (1989). *J. Biol. Chem.* **264,** 5904–5911.

Parlow, A. F. (1961). *In* "Human Pituitary Gonadotropins" (A. Albert, ed.), pp. 300–310. Charles C. Thomas, Springfield., Illinois.

Parsons, T. F., and Pierce, J. G. (1980). *Proc. Natl. Acad. Sci. USA* **77**, 7089–7093.

Patton, P. E., Calvo, F. O., Fujimoto V. Y. Bergert, E. R., Kempers, R. D., and Ryan, R. J. (1988). *Fertil. Steril.* **49**, 620–625.

Pereira, M. E., Segaloff, D. L., Ascoli, M., and Eckstein, F. (1987). *J. Biol. Chem.* **262**, 6093–6100.

Peters, B. P., Krzesicki R. F., Hartle, R. J., Perini, F., and Ruddon, R. W. (1984). *J. Biol. Chem.* **259**, 15123–15130.

Pierce, J. G., and Parsons, T. F. (1981). *Annu. Rev. Biochem.* **50**, 465–490.

Rao, A. J., and Ramachandran, J. (1975). *Life Sci.* **17**, 441–416

Rathnam, P.,Fujiki, Y., Landefeld, T. D., and Saxena, B. B. (1978). *J. Biol. Chem.* **253**, 5355–5362.

Rebois, R. V., Fishman, P. H. (1984). *J. Biol. Chem.* **259**, 8087–8090.

Reichert, L. E., Jr. (1966). *In* "Reproduction in the Female Mammal" (G. E. Lamming and E. C. Amaroso, eds.), pp. 125–145. Butterworths, London.

Reichert, L. E., Jr., and Dattatreyamurty, B. (1989). *Biol. Reprod.* **40**, 13–26.

Reichert, L. E., Jr., Leidenberger,F., and Trowbridge, C. G. (1973). *Recent Prog. Hormone Res.* **29**, 497–526.

Reichert, L. E., Jr., and Wilhelmi, A. E. (1973). *Endocrinology* **92**, 1301–1304.

Reichert, L. E., Jr., and Wilhelmi, A. E. (1978). *Endocrinology* **102**, 982–984.

Roche, P. C., and Ryan, R. J. (1985). *In* "Luteinizing Hormone Action and Receptors" (M. Ascoli, ed.), pp. 17–56. CRC Press, Boca Raton, Florida.

Ronin, C., Papandreou, M.-J., Canonne, C., Weintraub, B. D. (1987). *Biochemistry* **26**, 5848–5853.

Rosemberg, E. (1968). In "Gondotropins" (E. Rosemberg, ed.), pp. 383–391. Geron-X, Los Altos,

Ruddon, R. W., Krzesicki, R. F., Beebe, J. S.,Loesel, L., Perini, F., and Peters, B. P. (1989). *Endocrinology* **124**, 862–869.

Ruddon, R., Krzesicki, R. F., Norton, S. E., Beebe, J. S., Peters, B. P., and Perini, F. (1987). *J. Biol. Chem.* **262**, 12533–12540.

Ryan, R. J., Keutmann, H. T., Charlesworth, M. C., McCormick, D. J., Milius, R. P., Calvo, F. O., and T. Vutyavananich, T. (1987). *Recent Prog. Horm. Res.* **43**, 383–429.

Sadighia, J. J., Kearns, W. G., Waddell, B. J., and Dimino, M. J. (1989). *Biol. Reprod.* **40**, 294–299.

Sairam, M. R. (1980). *Arch. Biochem. Biophys.* **204**, 199–206.

Sairam, M. R. (1983). *J. Biol. Chem.* **258**, 445–449.

Sairam, M. R., and Li, C. H. (1975). *Biochim. Biopys. Acta* **412**, 70–81.

Sairam, M. R., Papkoff, H., and Li, C. H. (1972a). *Arch. Biochem. Biophys.* **153**, 554–571.

Sairam, M. R., Samy, T. S. A., Papkoff, H., and Li, C. H. (1972b). *Arch. Biochem. Biophys.* **153**, 572–586.

Sairam, M. R., and Schiller, P. W. (1979). *Arch. Biochem. Biophys.* **197**, 294–301.

Schuler, L. A., Shimomura, K., Kessler, M. A., Zieler, C. G., and Bremel, R. D. (1988). *Biochemistry* **27**, 8443–8448.

Sherwood, O. D., and McShan, W. H. (1977). *In* "Reproduction in Domestic Animals," pp. 17–47. Academic Press, New York.

Shui, R. P. C., Kelly, P. A., and Friesen, H. G. (1973). *Science* **180**, 968–971.

Shome, B., and Parlow, A. F., (1973). *J. Clin. Endocrinol. Metab.* **36**, 618–621.

Shome, B., Parlow, A. F., Liu, W. K., Nahm H. S., Wen, T., and Ward, D. N. (1988). *J. Protein Chem.* **7**, 325–339.

Sibley, D. R., Benovic, J. L., Caron, M. C., and Lefkowitz, R. J. (1988). *Endocr. Rev.* **9**, 38–56.

Sinha, Y. N., and Lewis, U. J. (1986). *Biochem. Biophys. Res. Commun.* **140**, 491–497.

Smith, P. L., and Baenziger, J. U. (1988). *Science* **242**, 930–933.

Spies, H. G., Hillard, J. and Sawyer, C. H. (1968). *Endocrinology* **83**, 354–356.

Squires, E. L., and Ginther, O. J. (1975). *J. Reprod. Fertil. Suppl.* **23**, 429–433.

Steelman, S. L., and Pohley, S. L. (1953). *Endocrinology* **53**, 604–616.

Steinberger, A., and Ward, D. N. (1988). *In* "The Physiology of Reproduction" (E. Knobil and J. D. Niell, eds.), pp. 567–584. Ravan Press, New York.

Stewart, F., and Allen, W. R. (1979). *J. Reprod. Fertil. Suppl.* **27**, 431–440.

Stewart, F., Thomson, J. A., Leigh, S. E. A., and Warwick, J. M. (1987). *J. Endocrinol.* **115**, 341–346.

Strickland, T. W., and Puett, D. (1981). *Endocrinology* **109**, 1933–1942.

Strickland, T. W., Thomasen, A. R., Nilson, J. H., and Pierce, J. G. (1985). *J. Cell. Biochem.* **29**, 225–237.

Sugino, H., Takio, K., and Ward, D. N. (1989). *J. Protein Chem.* **8**, 197–219.

Talamantes, F., and Ogren, L. (1988). *In* "The Physiology of Reproduction" (E. Knobil and J. Niell, eds.). pp. 2093–2144. Raven Press, New York.

Talmadge, K., Boorstein, W. R., Vamvakopolous, N. C., Gething, M.-J., and Fiddes, J. C. (1984a). *Nucleic Acids Res.* **12,** 8414–8436.

Talmadge, K., Vamvakopolous, N. C., and Fiddes, J. C. (1984b). *Nature* **307,** 37–40.

Tanaka, T., Shiu, R. P. C., Gout, P. W., Beer, C. T., Noble, R. L., and Friesen, H. G. (1980). *J. Clin. Endocrinol. Metab.* **51,** 1058–1063.

Tashjian, A., Barowsky, N., and Jensen, D. (1971). *Biochem. Biophys. Res. Commun.* **43,** 516–523.

Tepperman, J. (1980). *In* "Metabolic and Endocrine Physiology" (J. Tepperman, ed.), pp. 29–47. Year Book Medical Publishers, Chicago.

Tsunasawa, S., Liu, W.-K., Burleigh, B. D., and Ward, D. N. (1973). *Biochim. Biophys. Acta* **492,** 340–356.

Virgin, J. B., Silver, B. J., Thomason, A. R., and Nilson, J. H. (1985). *J. Biol. Chem.* **260,** 7072–7077.

Wade, N. (1981). "The Nobel Duel." Anchor Press/ Doubleday, Garden City, New York.

Ward, D. N., and Bousfield, G. R. (1989). *In* "Glycoprotein Hormones: Structure, Synthesis, and Biologic Function" (W. W. Chin and I. Boime, eds.) pp. 81– 95. Serono Symposium, Newport Beach, California.

Ward, D. N., Bousfield, G. R., Gordon, W. L., and Sugino, H. (1989). *In* "Microheterogeneity of Glycoprotein Hormones" (B. A. Keel and H. E. Grotjan, Jr., eds.), pp. 1–21. CRC Press, Boca Raton, Florida.

Ward, D. N., Desjardins, C., Moore, W. T., Jr., and Nahm, H. S. (1979). *Int. J. Peptide Protein Res.* **13,** 62–70.

Ward, D. N., Fujino, M., and Lamkin, W. M. (1966). *Fed. Proc.* **25,** 348.

Ward, D. N., Hines, K. K., Gordon, W. L., and Bousfield, G. R. (1988). *In* "Nonsteroidal Gonadal Factors: Physiological Roles and Possibilities in Contraceptive Development" (G. D. Hodgen, Z. Rosenwaks, and J. M. Spieler, eds.), pp. 1–16. The Jones Institute Press, Norfolk, Connecticut.

Ward, D. N., Reichert. L. E., Jr., Liu, W.-K., Hahm, H. S., Hsia, J., Lamkin, W. M., and Jones, N. S. (1973). *Recent Prog. Horm. Res.* **29,** 533–554.

Watahiki, M., Tanak, M., Masuda, N., Sugisaki, K., Yamamoto, M., Yamakawa, M., Nagai, J., and Nakashima, K. (1989). *J. Biol. Chem.* **254,** 5535–5539.

Watkins, P. C., Eddy, R., Beck, A. K., Vellucci, V., Leverone, B., Tanzi, R. E., Gusella, J. F., and Shows, T. B. (1987). *DNA* **6,** 205–212.

Wondisford, F. E., Radovick, S., Moates, J. M., Usala, S. J., and Weintraub, B. D. (1988). *J. Biol. Chem.* **263,** 12538–12542.

Yalow, R. S. (1978). *Science* **200,** 1236–1245.

Ying, S.-Y. (1988). *Endocrine Rev.* **9,** 267–293.

Biochemistry and Physiology of the Gonadal Hormones

DONALD M. HENRICKS

I. Introduction

An exciting era of reproductive biology began in the late 1920s. The first papers that appeared described the presence of substances in the gonad that had effects on re-

production. Estrone was isolated from human pregnancy urine by Doisy *et al.* in 1929. From the same source estriol was isolated by Marrian in 1930. Estradiol the estrogenic hormone of most mammals was isolated from cow ovaries by MacCorquodale *et al.*

in 1935. Crystalline estrogens were isolated from testicular tissue in 1940, adrenal tissue in 1939, and placenta in 1940. This was a monumental task: for example, from 4000 kg of sows' ovaries, 12 mg of estradiol-17β was obtained. Testosterone, nominally the male sex hormone of all species, was isolated from bull testes in 1935 (David *et al*,). Progesterone was isolated from sow ovaries by Butenandt and Wesphal (1934) and Wintersteiner and Allen (1934). Between 1936 and 1942 steroids were isolated from the adrenal cortex. Paralleling these discoveries was the work of a number of talented chemists to elucidate the structure and biosynthesis of cholesterol. Their dedication brought the introduction of isotopic forms of the steroids into biology in the 1940s; which in turn brought vigor into hormone research, which had stagnated.

This chapter will discuss (albeit succinctly) the chemistry, physiological actions, mechanisms of action, and assay of the steroid hormones that integrate reproduction in domestic animals. A number of other hormones and paracrine substances (peptide in nature) have been discovered more recently, but space does not permit including them in this chapter. Indeed, the purpose of this chapter will be met if the student is stimulated to acquire an understanding of the important concepts of chemistry, molecular biology, and assay as they apply to the gonadal hormones. A number of significant references are included to encourage the student to cultivate a current understanding of the structure and function of these hormones.

II. Steroid Chemistry

As long as the known steroids were small in number, trivial names were used for the compounds. With the advent of isotopic forms of the hormones, the list of naturally occurring steroids expanded rapidly, leading to development of a systematic nomenclature. Natural and synthetic steroids now number over 1800. The trivial and systematic names of some common steroid hormones are shown in Table I. To aid in understanding this nomenclature, a brief outline of steroid chemistry is presented.

The cholesterol molecule is a system of three six-carbon cyclohexane rings (A, B, C) and a five-carbon ring (D) (Fig. 1). All steroid compounds are numbered as shown. An eight-carbon side chain and methyl groups (C-18 and C-19) are attached at C-17, -13, and -10, respectively. A drawing on the plane of the paper is not a true representation of the molecule; rather, it is laminar, because the strain induced by joining the carbons in the ring system causes the molecule to "pucker." A ring may be in either the "chair" or "boat" form. Thus the free valencies of carbon atoms in the ring are not confined to the plane of the ring. Valencies at right angles to the plane are termed axial (a), and those more or less in the plane are termed equatorial (e) valencies. By convention the angular methyl groups (C-18 and C-19) and the side chain are above the molecule's plane, which is called the β-side (indicated by a solid line), and any substituent below the plane is on the α-side (indicated by a dashed line). An attempt to represent the three dimensional structure of a steroid is shown in Fig. 1.

Six families of steroids can be classified on a structural basis and number of carbons. For each there is a stem structure that is fully saturated and devoid of functional groups. Our concern is with the progestins (C-21) (stem = pregnane), androgens

TABLE I
Nomenclature of Selected Androgens, Estrogens, and Progestins

Trivial name	Systematic name	Symbol
Androstenedione	Androst-4-ene-3, 17-dione	AD
Testosterone	17β-Hydroxy-androst-4-ene-3-one	T
Dehydroepiandrosterone	3β-Hydroxy-5-androsten-17-one	DHA
Dihydrotestosterone	17β-Hydroxy-5α-androstan-3-one	DHT
Estradiol-17β	Estra-1,3,5(10)-trien-3, 17β-diol	E_2-β
Estradiol-17α	Estra-1,3,5(10)-trien-3, 17α-diol	E_2-α
Estrone	Estra-1,3,5(10)-trien-3-ol-17-one	E_1
Estriol	Estra-1,3,5(10)-trien-3, 16α,17β-triol	E_3
Progesterone	Pregn-4-ene-3,20-dione	P_4
Pregnenolone	3β-Hydroxy-pregn-5-en-20-one	P_5
20α-Dihydroprogesterone	20α-Hydroxy-pregn-4-en-3-one	20α-OHP
20β-Dihydroprogesterone	20β-Hydroxy-pregn-4-en-3-one	20β-OHP
17α-Hydroxyprogesterone	17α-Hydroxy-pregn-4-en-3-one	17α-OHP
Pregnanediol	5β-Pregnane-3α,20α-diol	
Cholesterol	Cholest-5-en-3β-ol	C

(C-19) (stem = androstane), and estrogens (C-18) (stem = estrone). The stem term is used in constructing the formal nomenclature. The stem structures can be modified by introducing oxygen functions (OH and =O), unsaturation, and heteroatoms such as the halogens. Prefixes and suffixes are used to indicate the type of structural modification; see Table I for examples of the systematic names of steroid hormones.

As with any family of organic compounds, the steroids exhibit the property of isomerism. There are two types of isomerism: cis–trans or geometric, and optical. Space does not permit a discussion of optical isomerism other than to mention that there are asymmetric C-atoms in every steroid molecule. Cholesterol has eight of these atoms giving rise to 512 possible optical iso-

mers of cholesterol. The peculiar special arrangement of the one natural isomer dictates its biological activity.

Cis–trans or more specifically decalin isomerism must concern us if we are to understand the nomenclature of steroid hormones. Fortunately there are only three centers of asymmetry that influence much of the nomenclature of steroids; they are at the C-3, C-5, and C-17 positions.

A. Asymmetry at C-3

When the carbonyl group at C-3, which lies in the plane of the A-ring, is reduced the resulting OH group can be formed either on the β-side of the molecule or on the α-side. If placed on the same side of the C-19 methyl group, the compound becomes cis and is referred to as the 3β-ol. If placed

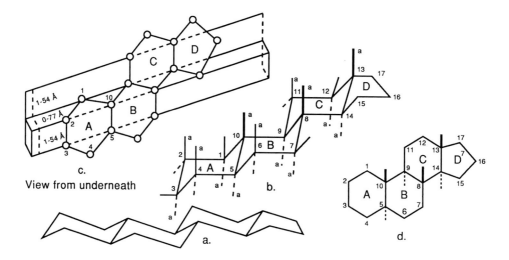

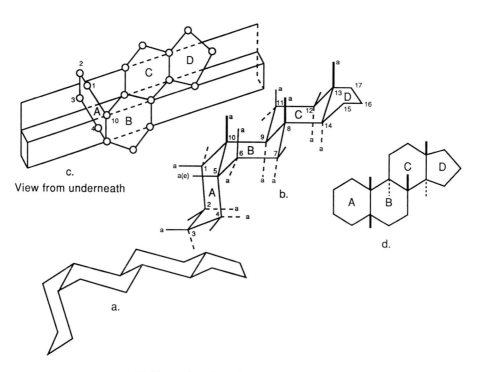

5x (Allo) series, rings A/B *trans* Shoppee, 1964).

5β (Normal) series, rings A/B *cis* (from Shoppee, 1964).

Figure 1 Cis-trans (decalin) isomerism in steroids. From Shoppee (1964).

on the opposite side (α), the compound becomes the 3α-ol.

B. Asymmetry at C-5

Another locus where asymmetric carbon atoms influence structure is the junction between each of the rings. Figure 1 illustrates the structural and spatial relationship resulting from the cis or trans A:B ring fusion in two steroids. Thus in a trans fusion of the A:B ring the α-H on the carbon-5 and the 19-methyl group are on opposite sides of the plane; in a cis fusion the β-H and the 19-methyl group are on the same side of the plane of the A:B ring. In the estrogen series with its unsaturated A-ring, cis–trans isomerism is not possible. In the metabolism of the steroids containing a Δ-4-ene-3-one structure in the A-ring (i.e., progesterone and testosterone), a family of isomers results. With the reduction of the 4-ene, two dihydro products will arise, one with A:B cis and one with A:B trans orientation.

The side chain at C-17 is the third locus where asymmetry occurs.

III. Biosynthesis of Gonadal Hormones

A. Cholesterol: Sources of Steriodogenic Cells

Cholesterol, so named by the Greeks (*chole* = bile, *stereos* = solid) is the precursor and prototype of an almost infinite number of steroids. It is an ancient compound in both the historical and evolutionary sense. It is synthesized by the prokaryotes and eukaryotes. Cholesterol is formed by *de novo* synthesis in most of the somatic cells of mammals by two-carbon addition of 18 units of acetyl-CoA (acetyl coenzyme A). Intermediates are acetoacetyl CoA and mevalonic acid. By demethylation, dehydrogenation, and phosphorylation of the isoprene unit, isopentenylpyrophosphate is formed. Thus cholesterol is related to other terpenoids such as β-carotene and vitamin K. Three isoprene units added head to tail form farnesyl pyrophosphate (C-15), and two such units combine to form squalene (C-30). Squalene cyclization then occurs via a two-step process requiring molecular O_2, NADPH, and both the soluble and microsomal fraction of a liver homogenate (Yamamato and Block, 1970). A microsomal monooxygenase, squalene epoxidase, incorporates O_2 to form squalene epoxide. The soluble fraction includes a phospholipid and a protein factor similar to the steroid carrier proteins described by Scallen *et al.* (1972). Cyclization of the squalene to lanosterol is accomplished by 2,3-oxidosqualene-sterol cyclase, with the substrate forming a chair–boat–chair–boat conformation on the microsomal enzyme's surface. Three changes are required for enzymic transformation of lanosterol into cholesterol: removal of methyl groups at C-4 and C-14, reduction of the Δ^{24} double bond of the side chain, and rearrangement of the double bond from the Δ^8 to the Δ^5 position.

Progesterone, testosterone, and estradiol-17β and closely related steroids are synthesized from cholesterol by tissues of the ovary, testes, fetal–placental unit, and adrenal gland. There are three immediate sources of cholesterol for steroidogenic cells: (1) *de novo* synthesis, (2) intracellular stores of cholesterol, and (3) blood lipoprotein cholesterol. The gonadal tissues convert acetyl-CoA to cholesterol. The rate-limiting enzyme is cytoplasmic hydroxymethyl glutaryl (HMG)-CoA reductase. All steroidogenic tissues are capable of synthesizing cholesterol *de novo*, and there is abundant literature demonstrating incorporation of [^{14}C]acetic acid into cholesterol and related products by various glands (Savard *et al.*, 1965; Preslock, 1980). Not withstanding this excellent work, current evidence makes it clear that at least for the ovary, placenta, and adrenal, uptake of lipoprotein (LP) cholesterol is the principal means of synthesis of the gonadal hormones. Whereas LP cholesterol uptake occurs in the testes, the extent of this transport is less certain. Borkowski *et al.* (1967) administered [^{14}C]cholesterol i.v. for 8 days to women and found that more than 80% of the substrate used by the corpus luteum (CL), placenta, and adrenals arises from extracellular sources. Understanding of the low-density lipoprotein (LDL) and high-density lipoprotein (HDL) has become necessary because cholesterol in the circulation is bound to these compounds, which behave as micelles (Brown and Goldstein, 1976). These complex particles consist of a lipid core covered by a protein coat of specific apoprotein. The function of the LP particle is determined not by its hydrated density, but by its apoproteins and amount and type of lipid present. Actually there are four major classes of LPs, characterized by their lipid–protein ratio, densities, electrophoretic behavior, and type of apoproteins present.

Lipoprotein cholesterol uptake is hormonally regulated in these tissues, with the exception of placenta, and coordinated with intracellular cholesterol synthesis and cholesterol ester hydrolysis to ensure a supply of free cholesterol for steroidogenesis. The LDL pathway is the primary pathway by which cells acquire cholesterol, using particles containing either apo-B or apo-E. This path is active *in vivo* in most tissues, as shown by studies in swine (Pittman *et al.*, 1979) and cows (Savion *et al.*, 1982), where the HDL pathway had no effect. Using a tissue culture of human CL cells, Carr *et al.* (1982) showed that LDL treatment enhanced progesterone synthesis. LDL binds to specific membrane receptors, which recognize the specific apoprotein. Receptor-mediated endocytosis of the cholesterol occurs followed by lysosomal degradation (see Fig. 2). A second route is the HDL-mediated pathway. It has been clearly demonstrated in rodents and less clearly in bovine granulosa and human trophoblastic cells.

With the uptake of lipoprotein cholesterol, HMG-CoA reductase activity is suppressed and acyl-CoA: cholesterol transferase is activated. This enzyme esterifies cholesterol to be stored in lipid droplets. Gonadal cells store cholesterol as long-chain fatty acid esters. This large, more stable pool is in equilibrium with free cholesterol and is maintained by the opposing action of two enzymes, acyl-coenzyme A: cholesterol acyl transferase (ACAT), and steroid ester hydrolase (cholesterol esterase). These storage enzymes and HMG-CoA reductase are under hormonal control (Strauss *et al.*, 1981). Cellular cholesterol suppresses membrane LDL receptor synthesis, resulting in lower LDL uptake. Cholesterol ester is in equilibrium with free cholesterol—the form used to synthesized steroid hormones. Indeed, it has been shown that the number of LDL receptors in the CL increases during midluteal phase (Carr *et al.*, 1982). Since the serum lipoproteins in ruminants (Puppioni, 1977) appears to be different from the species discussed above, the role of these proteins in gonadal steroidogenesis in ruminants may also differ.

The vascular anatomy of the gonad will determine which tissues will have access to

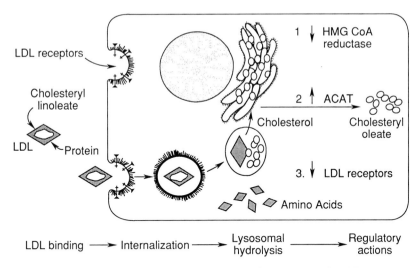

Figure 2   The low-density lipoprotein pathway in cultured human fibroblasts. HMG-CoA reductase denotes 3-hydroxy-3-methylglutaryl CoA reductase, and ACAT denotes acyl CoA: cholesterol acyltransferase. From Stryer (1981).

lipoprotein cholesterol. The CL and theca interna cells of the follicle have a highly permeable capillary bed, ensuring access to lipoproteins, while the granulosa cells are isolated by a basement mebrane. Only after luteinization do the granulosa cells obtain access to LP cholesterol. Recent experiments with cultured follicles (Strauss *et al.*, 1981) bear this out. For more on chemistry and function of lipoproteins, see Murray *et al.* (1988) and Gwynne and Strauss (1982).

B. Pregnenolone and Progesterone

A scheme for the cellular compartmentalization of steroidogenesis is shown in Fig. 3. Experiments in the late 1950s showed that cholesterol was converted to pregnenolone and isocaproic aldehyde in the mitochondria. Incubation of [^{14}C]cholesterol with bovine adrenal homogenate yielded ra-

dioactive pregnenolone as well as 20-hydroxycholesterol and 20,22-dihydroxycholesterol. In 1961 Constantopoulos and Tchen solubilized the side-chain cleavage (SCC) enzyme of adrenal mitochondria. This multienzyme complex directs the rate limiting step in gonadal steroidogenesis; see Fig. 4 for the generally accepted reaction sequence, and consult Hume and Boyd (1978) for evidence. Lieberman *et al.* (1984) suggest that stable, free hydroxylated intermediates are not involved. Rather, the cholesterol and subsequent intermediates remain bound to the SCC enzyme until pregnenolone is formed. It is released and transported from the mitochondria to the microsomes. This reaction appears to be under hormonal control and may be exerted via (1) transfer of cholesterol into mitochondria by carrier proteins, (2) availability of phospholipids important for formulation of complexes between cholesterol and the P-

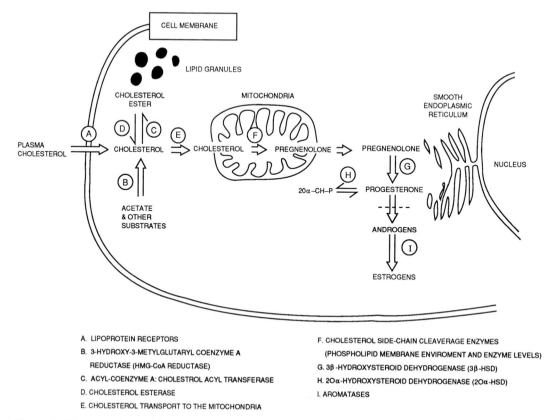

A. LIPOPROTEIN RECEPTORS

B. 3-HYDROXY-3-METYLGLUTARYL COENZYME A

    REDUCTASE (HMG-CoA REDUCTASE)

C. ACYL-COENZYME A: CHOLESTROL ACYL TRANSFERASE

D. CHOLESTEROL ESTERASE

E. CHOLESTEROL TRANSPORT TO THE MITOCHONDRIA

F. CHOLESTEROL SIDE-CHAIN CLEAVERAGE ENZYMES

    (PHOSPHOLIPID MEMBRANE ENVIROMENT AND ENZYME LEVELS)

G. 3β -HYDROXYSTEROID DEHYDROGENASE (3β-HSD)

H. 2Oα-HYDROXYSTEROID DEHYDROGENASE (2Oα-HSD)

I. AROMATASES

Figure 3 Diagram of organelles and key enzymes involved in steroid biosynthesis in ovarian cells. Letter A denotes LP receptors; B–I denote enzymes listed below diagram. Dashed line denotes deficiency of 17β-hydroxylase and 17-20-desmolase in granulosa cells. From Hsueh (1989).

450 enzyme, and (3) modulation of the P-450 enzyme concentration.

Pregnenolone converts to progesterone by action of a microsomal enzyme complex composed of Δ^5-3β-hydroxysteroid dehydrogenase and Δ^{5-4}-isomerase (Samuels *et al.*, 1951). It is the key intermediate common to all classes of steroid hormones. The enzyme complex appears to function physiologically as one entity (Hall, 1984). The reaction is irreversible and utilizes NAD^+ as an electron acceptor. Similar reactions and enzymes convert 17α-OH-P_5 and DHA to 17α-OH-P_4 and androstenedione, respectively (Fig. 4). The product, progesterone, is secreted by luteal cells or utilized as substrate for androgens and estrogens by testes and ovarian follicles, respectively. In the CL, progesterone can be catabolized to 20α- or 20β-OH-P_4 by the enzyme, 20α- or 20β-OH steroid dehydrogenase (HSD). Elevation of these enzymes may signal luteolysis. The steroid dehydrogenases are located in the smooth endoplasmic reticulum.

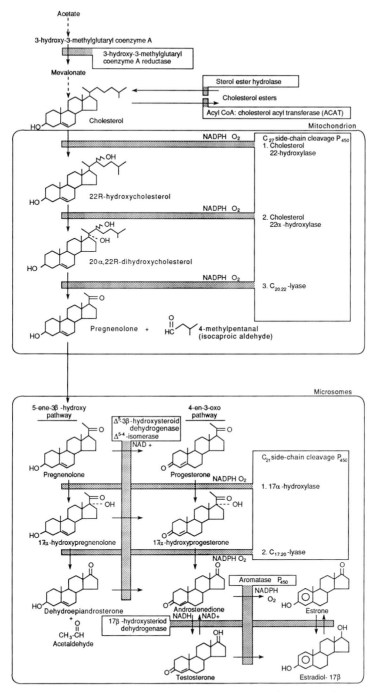

Figure 4 Biosynthetic pathways for progesterone, androgens, and estrogens in the ovary. Cholesterol may be synthesized *de novo* from acetate of derived from preformed sources. Cholesterol is metabolized sequentially by several enzyme systems, each with several catalytic functions. The enzyme systems are distributed in different organelles as indicated. From Gore-Langston and Armstrong (1988).

C. Androgens

The C-19 steroids are produced in tissues of the testes, ovary, adrenals, and placenta. The testes produce primarily androgens: testosterone, androstenedione (AD), and dehydroepiandrosterone (DHA) (Eik-Nes and Hall, 1965). Testosterone is the most potent. Its site of synthesis is the Leydig cell, which is located among the seminiferous tubules. Cholesterol stores (lipid droplets) can be seen in these cells (Christensen, 1975), as well as an extensive endoplasmic reticulum (ER), Golgi complex, and numerous lysosomes. The Sertoli cells can also synthesize and secrete some androgens.

The enzymatic reactions involved in conversion of cholesterol to testosterone are (1) hydroxylation of C-17, C-20, and C-22, (2) dehydrogenation of C-3 and C-17, (3) isomerization to move a double bond from between C-5 and C-6 (Δ^5) to C-4 and C-5 (Δ^4), and (4) C–C cleavage between C20–22 and C17–20, but not necessarily in that order. The hydroxylation and cleavage are the function of the lyase enzyme (see Fig. 4).

Throughout this discussion there have been references to a family of enzymes in the microsomes that incorporate an atom of oxygen into the steroid molecule. These enzymes are monooxygenases. They function to cleave a C–C bond (SCC enzyme or lyase enzyme) or to introduce a hydroxyl group (the hydroxylases). For an illustration of the mechanics of these enzymes' functions see Fig. 5.

Pregnenolone is metabolized to testosterone through either the Δ^4 or the Δ^5 pathway. The terminology for these pathways designates the position that the unsaturated bond maintains during the conversion of intermediates to testosterone. Whether the enzymes are separate from or are in common to both pathways is still being studied. Progesterone is the Δ^4 intermediate formed from pregnenolone by the 3β-HSD/isomerase complex. It is hydroxylated at C-17 by 17-hydroxylase to form 17-OH-P_4. The 17-hydroxylase is a monooxidase in which NADPH-cytochrome (NADPH-cyto) P-450 reductase transfers electrons from NADPH to the P-450 (Purvis et $al.$, 1973). The C-17–C-20 lyase cleaves the acetate group from C-17 of 17-OH-P_4, yielding androstenedione (AD) as product. The lyase requires NADPH with cyto P-450 functioning as oxygen acceptor and substrate binding site (Menard and Purvis, 1973). A 17β-HSD reduces the ketone on C-17 of AD to testosterone

The sequence of reactions differs in the Δ^5 pathway. Initially, pregnenolone is converted to 17OH-P_5 by 17-hydroxylase, which is cleaved by lyase to form DHA. 3β-HSD-isomerase converts DHA to AD. The 17β-HSD then converts AD to testosterone. In both pathways, the enzymes are in the microsomes of the smooth endoplasmic reticulum. There seems to be a distinct preference for utilization of the Δ^4 and Δ^5 pathways among various mammalian species, but there may be transitions from Δ^5 to Δ^4 at different levels (see Fig. 4) via the 3β-HSD/isomerase system. It seems clear that in the rat and mouse testes, the Δ^4 path predominates, whereas in the human, pig, rabbit and dog, Δ^5 is more important (see Gowers, 1984). In the special case of 16-unsaturated C-19 steroids (see Gowers and Bicknell, 1972), side-chain cleavage of the C-21 steroids occurs (mainly in boar testes) without 17-hydroxylation. Although structurally related to the androgens, these compounds have little or no androgenic activity. Because of their odor and possible phero-

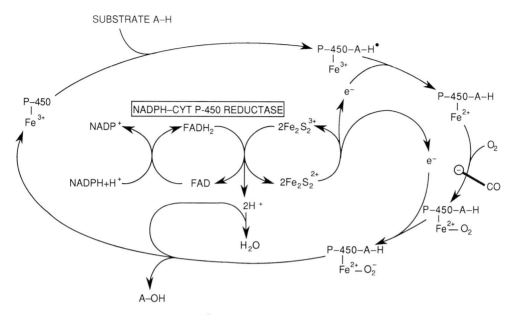

Figure 5　The typical P-450 hydroxylase cycle in microsomes. From Martin *et al.* (1985).

monal activity in pigs (Melrose *et al.*, 1971), there has been some interest in them (see review by Claus, 1979). It has been shown that the odors of both 5α-androstenenone and 5α-androst-16-en-3α-ol cause sows to take the mating stance.

1. 5A-Reductase

This enzyme is responsible for reducing the unsaturated bond that exists between C-5 and C-6 of testosterone and AD converting the H on C-5 to a 5α position, thereby forming 5α-dihydrotestosterone (5α-DHT), androsterone, and epiandrosterone. In some target tissues, namely the ventral prostate and seminal vesicles, 5α-DHT is more potent than testosterone. This conversion of testosterone to 5α-DHT occurs in numerous tissues besides the testes (adipose tissue, skin, brain, salivary glands, and lung) and is

an outstanding example of peripheral conversion of a prehormone (testosterone) to a hormone (DHT). Estrogens can arise from peripheral conversion of testosterone. Formation of DHT is high in humans and dogs, lower in rodents, and insignificant in bulls and rabbits (Gloyna and Wilson, 1969). The reductase pathway is interesting in a developmental sense. Newborn rat testes convert progesterone primarily to testosterone; at 20 days the 5α-reduced androgen is the principal metabolite, and in mature 90-day old rats, testosterone predominates again (Steinberger and Fischer, 1969).

D. Estrogens

Estradiol-17β and estrone are synthesized from the 4-ene-C_{19}-steroids testosterone and AD, respectively, by an aromatase enzyme complex in the agranular endoplasmic reticulum (see Fig. 4). Aromatization is the process of hydroxylation and loss of the C-19 methyl group and elimination of the 1 and 2 hydrogens of the A-ring. A cytochrome P-450 containing monooxidase requires 3 moles of O_2 and NADPH per mole of estrogen. The reaction is shown in Fig. 6. Most evidence supports the concept that intermediates exist in several transition states, which remain bound to the enzyme complex until the estrogen molecule is complete (Liberman et al., 1984). Ring A with its three double bonds and a phenolic 3-OH group makes the estrogen an aromatic compound. These estrogens are slightly acidic, hence extractable in $1 N$ NaOH, while the other gonadal steroids, being neutral, are extractable from serum and tissues with organic solvents.

Estrogens may arise from four glands: ovary, testes, feto–placental unit, and adrenal cortex. They may also arise from peripheral metabolism of androgens to estrogens occurring in nonglandular tissues such as adipose tissue, muscle, and brain (McDonald et al., 1971).

Early work on the pathway using organ perfusions and in vitro incubations of isotopic precursors with tissue slices showed that estrogens can arise from acetate, cholesterol, progesterone, and androgens. Total synthesis of $E_2\beta$ and E_1 from acetate by procine ovary was reported in 1953 (Werthessen, 1953). Using FSH to enchance the biosynthetic capability of human follicles, Ryan and Smith (1961) demonstrated that progesterone converted to estradiol-17β and E_1. FSH was shown to stimulate aromatase activity in dog ovaries (Hollander and Hollander, 1958); isolation of the aromatase enzyme from human placental microsomes followed (Ryan and Smith, 1965). The incu-

bation of isotopic 19-OH AD with bovine follicular fluid caused E_1 accumulation (Meyer, 1955). For a summary of the early work, see Dorfman and Unger (1965).

As shown in Fig. 4, estrogen synthesis may proceed by the Δ^5 pathway involving 17-OH-P_5 and DHA, or Δ^4 involving P_4 and 17-OH-P_4. Our understanding of the process in the ovary is complicated by the constantly changing population of cells that are involved in follicle development and CL formation and regression during the estrous cycle. See Savard (1973) for a general review of the biogenesis of the steroids in the ovary and more recent reviews by Baird (1977) and Dorrington (1977) and a review of estrogens in blood by Reed and Murray (1979).

In the follicle where there are two cell types, synthesis of estrogen occurs via the Δ^5 pathway in thecal cells and the Δ^4 pathway in granulosa cells (Ryan and Petro, 1966). The transfer of the steroids between the two cell layers and their metabolism have been studied in vivo in mares by Younglai and Short (1970). In equine follicles steroid synthesis seems to occur by the Δ^4 pathway, no evidence being found for 17-OH-P_5 or DHA. Follicular fluid of domestic species contain high concentrations of $E_2\beta$, which arises mainly from the theca interna cells. Only low concentrations of P_4 are present, suggesting that granulosa cells are relatively inactive in steroid synthesis until becoming vascularized after ovulation.

There was a controversy concerning the interrelationship between the two cell types in estrogen synthesis. No doubt some of this comes from species differences. While both thecal and granulosa cells may have all the required enzymes, most investigators now agree with the classic work of Falck (1959) that both cell types are necessary for estro-

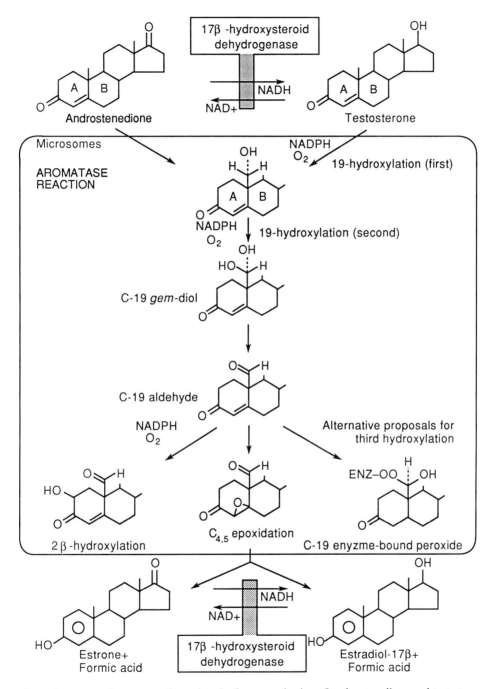

Figure 6 Proposed sequence of reactions in the aromatization of androstenedione and testosterone to estrone and estradiol-17β. From Gore-Langston Armstrong (1988).

gen synthesis. This work led to other studies by many workers that has resulted in the current "two-cell-type, two-gonadotropin" theory of steriodogenic regulation in the follicle. The early paper of Greep *et al.* (1942) using hypophysectomized immature rats provided evidence that FSH and LH act upon different cell types to promote estrogen formation. Then Hollander and Hollander (1958) demonstrated that FSH stimulated C^{14}-T conversion to E_2 by canine ovarian slices. The crucial role of FSH in estrogen synthesis in rats was established when explanted ovaries from rats produced estrogen in response to FSH but not to LH and that addition of testosterone as substrate was required. The next step was to reveal the target cell for FSH. Using rat granulosa cells in culture (Dorrington *et al.,* 1975) and the testicular Sertoli cell (Dorrington *et al.,* 1975), it was reported that treatment of these cells with FSH induced estrogen production. The granulosa cells must be supplied with an extracellular source of aromatizable androgen to synthesize estrogens as shown by studies in pigs (Evans *et al.,* 1981) and sheep (Armstrong *et al.,* 1981). C-21 steroid precursors, possibly produced by granulosa cells, can be converted to androgens by thecal cells as suggested by the studies of Luchinsky and Armstrong (1983) using porcine granulosa cells. Follicle wall preparations of the bovine, containing both theca and granulosa cells, secreted greater quantities of androstenedione than did isolated theca interna preparations (Fortune, 1986). Granulosa cells did not secrete androgens, even when provided progestin precursors. It was hypothesized that theca and granulosa cells of the bovine interact to promote androgen production in a model in which granulosa cells secrete pregnenolone, which is converted to androgens by the the-

cal cells (Fortune, 1988). Furthermore, in studies of $E_2\beta$ as a regulator of steroidogenesis in bovine follicles, the estrogen increases pregnenolone secretion by granulosa cells up to 11-fold (Fortune, 1986), suggesting that as the bovine follicle develops, its E_2 production rises and this in turn inhibits conversion of pregnenolone to progesterone in both theca and granulosa cells. This action increases availability of pregnenolone for conversion to androgens via the Δ^5 pathway in theca cells.

The "two-cell type, two-gonadotropin" theory is favored for the synthesis of estrogens by the testes. The testes of the bull (Henricks, 1988), ram (Ginther *et al.,* 1974), dog, and human (Kelch *et al.,* 1972) have been shown to produce small quantities of estrogens. Indirect evidence points to the Leydig cell as the source of estrogens, but the Sertoli cells also contributes (Dorrington and Armstrong, 1975). A luteotropin will produce a similar increase in both estrogen and androgen secretion (Longcope, 1972). In the male the major portion of the circulating E_2, however, arises from peripheral conversion of testosterone.

To complete the survey of biologically active estrogens synthesized by mammals, there are the ring-B unsaturated estrogens and catechol estrogens. The former occur during gestation in the mare. Examples are equilin and equilenin (Figure 12), which probably arise by an alternate route bypassing cholesterol in the fetal–placental unit (Bhavnani, 1988). The catechol estrogens result from hydroxylations at the C-2 or C-4 by a microsomal cytochrome *P*-450 enzyme followed by methylation to form 2- or 4-methoxy E_2 in a number of tissues. Since these estrogens are structurally similar to catecholamines, they may function in the central nervous system (CNS). Thus their hydroxylation is probably anabolic in nature and bestows a function different from that of the classical estrogens.

E. *Organization of the Steroidogenic Pathway in the Cell*

Having reviewed the steroidogenic pathways, we can turn to the subject of regulation of hormone synthesis. Regulation is exerted at many levels, one of those being the subcellular compartmentalization of the enzymes, a fact that is too often overlooked (see Figs. 3 and 4). It is clear that the pathway begins in the mitochondrion and that pregnenolone is then transferred to the microsomes to complete the process. The SCC enzyme complex is found only in the inner mitochondrial membrane. The P_5 so formed must pass out of the mitochondria into the endoplasmic reticulum before hydroxylation or dehydrogenation can occur. A number of factors (e.g., Ca^{2+}) are known to alter the permeability of the mitochondrial membrane to steroids (Koritz, 1968). The subcellular distribution of steroids and the manner in which they are transported are important in overall control of hormone synthesis.

The efforts of several workers have demonstrated product inhibition of the various transforming enzymes. For example, there are excellent examples of allostearic enzyme inhibition of the SCC enzyme by P_5 (Koritz, 1968). A summary of these studies has been given by Gowers (1984). The side-chain cleavage of 17-OH-P_4 to AD is affected by the presence of endogenous steroids. A more complete picture of steroid hormone regulation by steroid metabolites is seen in Fig. 7.

Less is known of the organization of the endoplasmic reticulum. Several problems associated with the ER have been investigated. One is the sequence of reactions. The sequence will be different depending on whether the Δ^4 or Δ^5 pathway is the preferred route. For example, rat testes use the former; pig testes prefer the latter. The choice may be dictated by the affinity of the C-21 SCC enzyme for P_5 as compared to P_4 or by the arrangement of the enzymes in the membrane. In that case it has been suggested that P_5 enters the microsome at an entry point established by a P_5-binding protein. Such a binding protein has been reported (Hall, 1984), and its location relative to two enzymes could influence the sequence of reactions.

Another problem is the location of the enzymes in the lipid bilayer. The ER has two surfaces: cytoplasmic and internal. Homogenization disrupts the ER to yield microsomes and to relate the experimental findings, such as the location of a steroidogenic enzyme, to the situation in the cell one needs to know whether they are inside or out. Figure 8 represents a suggested arrangement of the enzymes in the testicular ER. Another consideration is the influence of the microsomal environment. For example, when P_4 is incubated with adrenal microsomes, only 17α-hydroxylation products are found (Nakajin *et al.*, 1983), but testicular microsome produce a mixture of products in which C_{19} steroids predominate (Tamaoki and Shikita, 1966). However, when the 17α-hydroxylase and the hydroxylase/lyase from the two microsomal types are purified, they are indistinguishable and both produce large amounts of C-19 steroids, evidence that the microsomal environment is critical.

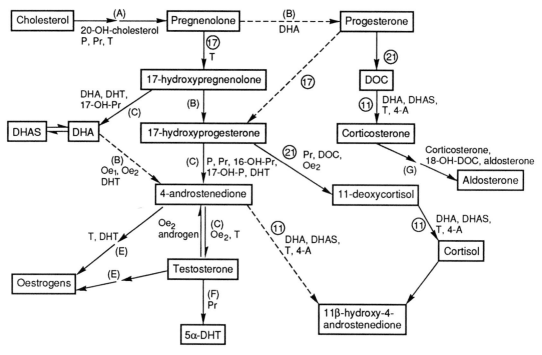

Figure 7 Inhibitory effects of some endogenous steroids on some steroid-transforming enzymes in mammalian adrenal, gonads, and prostate. This is a composite diagram to illustrate results obtained using various tissues. From Gowers (1984).

F. Regulation of Luteal Progesterone Synthesis

Many studies, beginning with those of Denamur and Mauleon (1963), support the concept that the CL of the cow, ewe, mare, and sow is dependent on anterior pituitary function both in terms of growth and secretion of P_4. Although the mechanisms involved in synthesis of P_4 are complex, the most important regulating factor is LH, being stimulatory both *in vivo* (Schomberg *et al.*, 1967) and *in vitro* (Kaltenbach *et al.*, 1967). Constant infusions of this hormone prolong CL life span in ewes (Karsch *et al.*, 1971) and enhance P_4 secretion (Kaltenbach *et al.*, 1967). Passive immunization with LH

antisera causes luteal regression (Fuller and Hansel, 1970). Similar results have been obtained with cattle by Hansel *et al.* (1973) and with horses by Ginther *et al.* (Pineda *et al.*, 1972; Ginther, 1979). Specific receptors for LH are present in ovary (Channing and Kammerman, 1974), and P_4 secretion appears to be highly correlated with number of LH receptors in cows (Rao *et al.*, 1976) and ewes (Diekman *et al.*, 1978). It appears that binding of LH and epinephrine to their respective receptors results in adenylate cyclase activation. How this process activates intracellular mechanisms that enhance luteal function and P_4 synthesis is outlined in Fig. 9. The binding of cAMP to protein kinase promotes dissociation of the enzyme's

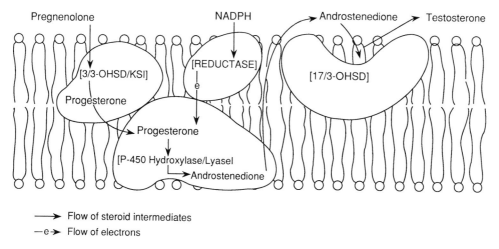

Flow of steroid intermediates
—e→ Flow of electrons

Figure 8 Suggested arrangement of enzymes in endoplasmic reticulum of Leydig cells. From Hall (1988).

catalytic subunit; the activated subunit then phosphorylates regulatory proteins, which stimulate or inhibit their activity. The consequence of such a cascade of reactions is a cellular response. The response can be terminated by dephosphorylation of the protein substrate by a phosphoprotein phosphatase. In luteal cells (or other gonadal cells) these cAMP-dependent protein kinases can affect gene expression and protein synthesis in the nucleus or in the ribosome, which may influence the steroidogenic pathway at a number of possible sites (as illustrated in Fig. 10). Thus LH may stimulate protein kinases to activate enzymes such as SCC enzyme complex and cholesterol esterase or synthesis of transport proteins, notably the steroid carrier proteins, that enhance substrate availability and removal of end products. Many of the enzymes involved in steroidogenesis are sensitive to end-product inhibition (Hochberg *et al.*, 1974). The 3β-HSD enzyme is a good example, being inhibited sixfold greater by

its product, P_4, than by P_5, its substrate (Caffrey *et al.*, 1979). Thus steroid carrier proteins may be one mechanism for enhancing steroidogenesis.

Recent studies have demonstrated that luteal tissue of the cow (Ursely and Leymarie, 1979; Koos and Hansel, 1981), sheep (Fitz *et al.*, 1982), pig (Lemon and Loir, 1977), and rabbit (Hoyer *et al.*, 1986) consists of two morphologically and functionally distinct cell types that can be separated according to size. It appears that luteal cells arise from two sources: the small luteal cells (10–20 μm) are all of thecal cell origin, and the large luteal cells found early in the cycle are of granulosa cell origin. The large cells appearing later in the cycle are of thecal origin. It has been suggested that the small cells of thecal origin differentiate into large cells as the cycle progresses (Alila and Hansel, 1984). After day 100 of pregnancy, all cells in the CL are of thecal origin. In the bovine CL, small cells are about six times more responsive to LH added *in vitro* than

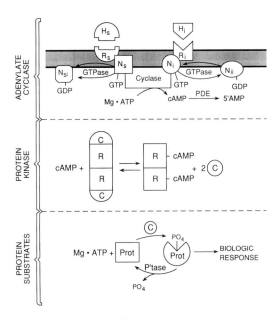

Figure 9 Schematic representation of the molecular events that occur as part of hormonal regulation of the biologic responses of target cells. *Adenylate cyclase:* Hormone, either stimulatory (Hs) or inhibitory (Hi), interacts with its cell-membrane-bound receptor (Rs and Ri), which results in activation of the appropriate regulatory components of adenylate cyclase (Ns or Ni). The dissociation of GDP from the inactive Nsi or Nii and binding of GTP activates Ns or Ni. This results in activation of the catalytic subunit (cylase), which converts ATP to cAMP. *Protein kinase:* The cAMP binds to the receptor (R) subunit of cAMP-dependent protein kinase, causing dissociation and activation of the catalytic subunit (C). *Protein substrate:* the catalytic subunit of protein kinase alters the activity of various protein substrates (Prot) via phosphorylation, leading to activation of these proteins and modification of the biologic response. Points for negative regulation of this system include activation of Ni, GTPase inactivation of Ns, phosphodiesterase (PDE) conversion of cAMP to 5'-AMP dissociation of cAMP from R, and dephosphorylation of phosphorylated protein substrate by phosphatase (Ptase). From Neswender and Nett (1988).

are those of the relatively impure preparations of the large cells (Ursley and Leymarie, 1979) and contain about 10 times more LH receptors (Fitz *et al.*, 1982). In sheep the large cells produce more progesterone than the small cells but are unresponsive to LH or manipulation of cAMP levels (Fitz *et al.*, 1982; Hoyer *et al.*, 1984). These and other findings make clear that steroidogeneis is controlled by different mechanisms in the several cell types and the products of one cell type may affect the functions of the other cell type.

The action of prostaglandin $F_2\alpha$ (PGF) on cytoplasmic granules occurs exclusively in large luteal cells (Heath *et al.*, 1983). In ovine CL, the large luteal cells contain the majority of the PGF receptors (Fitz *et al.*, 1982). A cellular mechanism for prostaglandin action has been proposed (Raymond *et al.*, 1983) in which PGF stimulates phospholipase to catalyze hydrolysis of phosphatidylinositol 4,5-bisphosphate (PIP_2), which results in rapid appearance of 1,4,5-triphosphate (IP_3) and 1,2-diacylglycerol. The latter activates protein kinase C (Nishizuka, 1986) and IP_3 mobilizes intracellular calcium (Berridge and Irvine, 1984). Since it has been shown that bovine placental cells secrete progesterone by a calcium-dependent and cAMP-independent mechanism (Shemesh *et al.*,1984), a similar mechanism may exist in one of the bovine luteal cell types. Indeed, there is evidence that progesterone synthesis in either the granulosa-derived or the theca-derived large luteal cells (or both) is controlled by the Ca^{2+}–polyphosphoinositol–C-kinase system. Synthesis in the small theca-derived cells is primarily controlled by LH and the cAMP system (Hoyer *et al.*, 1984). It has been shown that progesterone synthesis in these cells is inhib-

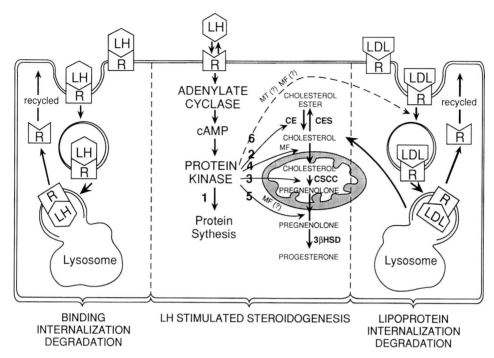

Figure 10 Schematic representation of the intracellular events involved in LH-stimulated steroido-genesis. LH binds to its receptor in the cell membrane and activates adenylate cyclase, resulting in increased intracellular levels of cAMP and activation of protein kinase (Fig. 9). The active protein kinase (1) stimulates protein synthesis; (2) activates cholesterol esterase (CE); (3) activates cholesterol side-chain cleavage complex; (4) stimulates transport of cholesterol into the mitochondrion; (5) may stimulate transport of pregnenolone out of the mitochondrion and/or (6) may stimulate the uptake of low-density lipoprotein (LDL), thus increasing cholesterol for substrate. In some species high-density lipoprotein is the preferred moiety for cholesterol uptake. See text for details of the various actions of protein kinase. The LDL is degraded in the lysosome, providing cholesterol for steroidogenesis. The LH and its receptor (R) are internalized, and the LH is degraded in lysosomes. The receptors for LH and LDL are probably recycled to the plasma membrane. From Niswender and Nett (1988).

ited in the presence of high intracellular Ca^{2+} (Hansel and Dowd, 1986).

G. *Regulation of Testicular Androgen Synthesis*

The secretion of LH is required for androgen synthesis by the testes. Hypophysec-tomy results in greatly reduced testicular androgens (Li and Evans, 1948). Utilizing two messengers, cAMP and Ca^{2+}, LH stimulates testosterone synthesis by Leydig cells. One important action locus was thought to be on the rate-limiting step in the pathway, that is, side-chain cleavage of cholesterol to pregnenolone. Since it was shown that the action of ACTH on adrenal steroidgenesis was inhibited at a step before side-chain cleavage of cholesterol, Leyding cells from rats were tested with LH (Dufau and Catt, 1975). Like ACTH, LH stimulated transport of cholesterol through the inner mitochondrial membrane (Hall *et al.*, 1979).

In Leydig cells, cAMP mediates the effect of LH on androgen synthesis and LH increases cAMP levels (Sandler and Hall, 1986). Some doubt about whether cAMP is the only mediator of LH occurs when one considers that low concentrations of LH stimulate steroid synthesis without detectable change in total cell concentrations of cAMP (Themen *et al.*, 1985). To counter this, there is growing evidence for the distribution of cAMP in the cell to vary, making it a tenable argument that rather than causing a rise in total concentrations of cAMP, LH causes a local rise available to a protein kinase. The rapid catabolism of cAMP by phosphodiesterase and the compartmentalization of cAMP as shown by the study of Dufau *et al.*, (1977) could explain how cAMP produces a localized response to one trophic hormone without stimulating every cAMP-dependent process in the cell.

Another effect of LH in the steroidogenic process in Leydig cells is the stimulation of protein synthesis. Janszen *et al.* (1977) reported that LH increased synthesis of two proteins, 21K and 33K in size. An adrenal protein 2.2K in size causes increase in side-chain cleavage of cholesterol. This protein has been found in the testes (Pederson and Brownlee, 1983).

1. Role of Calcium Ion, Phosphorylation, Phospholipids, and Protein Kinase C

Like ACTH (Birmingham *et al.*, 1953), LH is likely to require Ca^{2+} for its stimulation of steroidogenesis. The action of Ca^{2+} appears to be linked with the calcium-binding protein, calmodulin (Hall *et al.*, 1981). Other aspects of the action of Ca^{2+} on Leydig-cell androgen synthesis not linked to calmodulin are being investigated.

If cAMP is a second messenger for LH and the role of cAMP is to promote protein phosphorylation, then LH should indirectly stimulate this process in Leydig cells (Kuo and Greengard, 1969). Indeed, three such Leydig-cell proteins have been identified (Cooke *et al.*, 1977), and some correlation with the steroidogenic response to LH has been reported. Because of the importance of side-chain cleavage of cholesterol that occurs in the mitochondrion, studies need to focus on the phosphorylation of mitochondrial proteins. In one of the few studies (Inaba and Wiest, 1985), it was shown that disruption of the organelle's membrane by Ca^{2+} allows the protein kinase to stimulate side-chain cleavage of cholesterol, perhaps by facilitating access of cholesterol to the enzyme. The value of these studies in explaining the regulation of steroidgenesis will depend upon definition of how a proteins's function changes upon phosphorylation.

As a result of a new chapter in the hormone–second messenger theory, inositol triphosphate (PIPP) is now a likely candidate. It acts as a Ca^{2+} ionophore (Mitchell, 1980) to raise intracellular Ca^{2+}.

IV. Catabolism of Gonadal Hormones

In general, the process of catabolism converts the hydrophobic steroid to a more hydrophilic catabolite, which renders the molecule physiologically inactive. One way to control a hormone's activity is rapid catabolism, which is performed mainly in the liver but also in other tissues (kidney, lungs, and intestines). The catabolic reactions are mostly reductive, followed by conjugation to a steroid glucuronide or sulfate, which renders the molecule hydrophilic prior to its excretion. Attachments to glucuronic acid is by a glycosidic bond at either the C-3 or C-17. (The product is properly called a glucosiduronidate because it is a glucoside and the salt of glucuronic acid, but the trivial name "glucuronide" is more often used.) The main pathways by which the steroid hormones are catabolized are reduction of the Δ^4 double bond and the C-3 oxo group to an alcohol, reduction of the C-20 oxo group to an alcohol, and oxidation of the 17β-hydroxyl group.

A. Progesterone

An abbreviated version of the pathways is shown in Fig. 11. Reduction of ring-A by the 4-ene-5α- and 5β-reductases yield 5α- and 5β-pregnone-3,20-diones. This is followed by reduction of the C-3 oxo group by the 3α- and 3β-OHSDH to yield isomers of pregnanolones. Reduction of the C-20 oxo group forms the excretory products, six isomers of pregnanediol, the most important being 5β-pregnane-3α,20α-diol. It is the major urinary metabolite in the sow (Mayer *et al.*, 1961) and mare (Marker *et al.*, 1937). This pathway has been shown to occur in the liver of several species. Liver is the primary site of this metabolism. Some or all of this pathway can also occur in all the steroidogenic glands and the kidney, intestinal mucosa, and lungs. For example, in the ovary of domestic animals 20α- and 20β-OHSDH convert P_4 to 20α- and 20β-hydroxy-P_4, which have some progestational activity. The major route of excretion of the progestins in ruminants appears to be the feces, not the urine as in humans. For example, [^{14}C]P_4 administered to a dairy cow resulted in 16 times more ^{14}C being recovered in the feces than the urine (Williams, 1962). The fecal route seems to be prevalent for all three classes of gonadal hormones, especially in the ruminant species.

B. Androgens

The C-19 steroids, such as testosterone and AD, are inactivated by reduction in ring A followed by reduction at C-3 by means of the 4-ene-5α- and 5β-reductases and 3α- and 3β-OHSDH, respectively. The initial products are the 5α- and 5β-androstane-3,17-diones. The 3α- and 3β-OHSDH yield four 17-oxosteroids: 3α,5α androsterone, 3α,3β-actiocholanolone,, 3β,5α-epiandrosterone, and 3β,5β-epiaetiocholanolone. These along with DHA comprise the principal 11-deoxy-17-oxosteroids. It has been reported that these transformation occur not only in the liver but also in the adrenal cortex, testes, ovary, and other tissues. While not considered a catabolic action, the interconversion of testosterone and AD by oxidation and reduction is conducted by a 17-OHSDH.

C. Estrogens

Estradiol-17β (E_2β) and estrone are thought to be the primary secreted compounds. The other estrogens arise from catabolism of these two. E_2β and E_2α are interconvertible by a 17-OHSDH with estrone being an intermediate (see Fig. 12). Estriol arises from 16-hydroxylation of estrone to form 16α-OH estrone followed by reduction at C-17. E_2β can also be hydroxylated at the C-2 and C-6 to form the catechol estrogens, which are not catabolic conversions. Whereas estriol is the major end

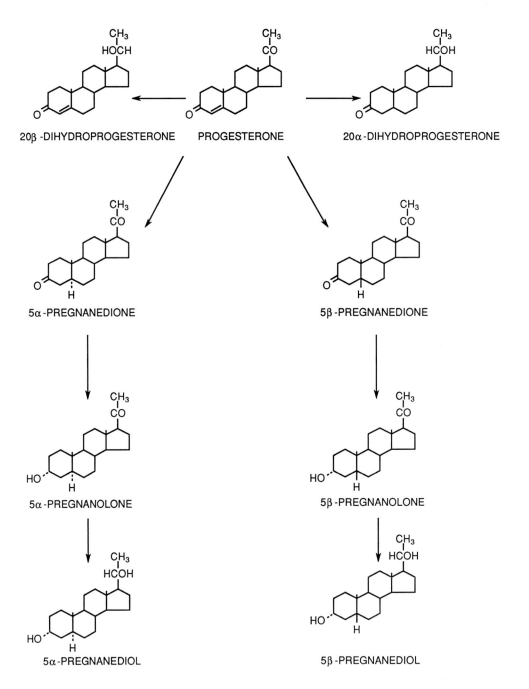

Figure 11 Catabolism of progesterone in food producing animals. From Henricks and Mayer (1977).

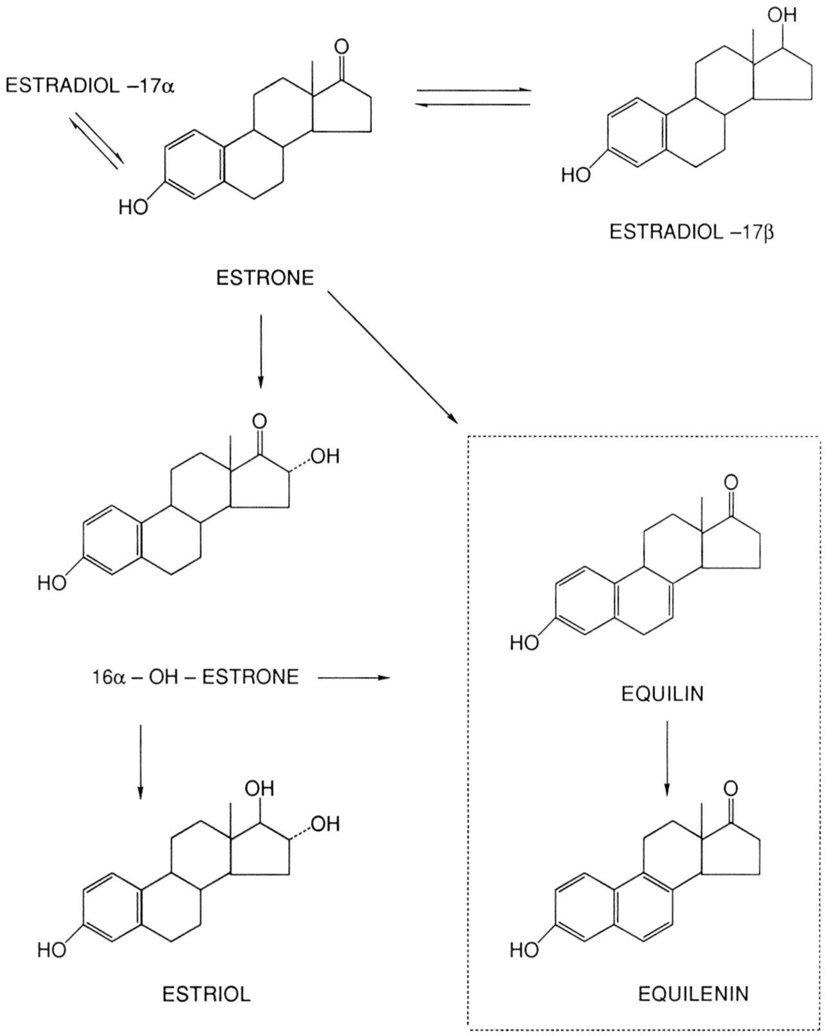

Figure 12 Interconversion and catabolism of estradiol-17β and estrone. From Henricks and Mayer (1977).

product in humans and estrone in pigs and horses and the excretory route is chiefly urine, in ruminants and rabbits the path appears to be $E_2\beta$, estrone, and $E_2\alpha$, with feces being the route of excretion (Mellin and Erb, 1966).

The liver is the major site of catabolism and conjugation of estrogens. Of the two conjugates, glucuronides or sulfates, progesterone, testosterone, and $E_2\beta$ favor the glucuronide. There are many exceptions though, depending on the steroid and the species. This type of research has been superceded by the study of the molecular biology of hormones, so that most of what can be stated with certainty only pertains to humans.

A fertile area for study is the metabolism of the secreted forms of the three gonadal hormones from the following perspectives. Which conversions and conjugations to a glucuonide or sulfate constitute inactivation (and eventual excretion) and which constitute conversion to another hormone? Certainly conversion to a sulfate could just as likely be the latter (Adlercreutz, 1970). It has been shown that the human adrenal gland secretes DHA sulfate (Baulieu, 1963). Sulfotransferases synthesize steroid sulfates in the testes, ovary, and fetal tissues as well as the liver (see review by Payne and Singer, 1979). From such studies it is clear that sulfates can take part in metabolic pathways that parallel those for free steroids. More germane to reproduction is a study (Robertson and King, 1974) in which estrone sulfate was detected in maternal blood plasma on day 16 of pregnancy, rising to a peak of 3 ng/ml between days 23 and 30, suggesting that the sulfate may be involved in implantation.

Another needed area of study in the domestic animal is extrahepatic metabolism of the gonadal hormones. The resolving power of HPLC and sensitivity of immunoassay makes such studies feasible. Extrahepatic catabolism (EHC) consists of the following steps: a steroid in the hepatic portal vein enters liver and is conjugated; the conjugated steroid is secreted in bile; hydrolysis of the conjugate occurs in the gut by bacteria, facilitating absorption from the gut; and complete reconjugation of steroid in gut wall. The steroid is absorbed from gut cells into the portal vein, which transports it to the liver. Here some of the steroid can be released into the peripheral circulation. Placement of estradiol in intestines of immature pigs results in 75% of the estrogen in the vein being estrone glucuronide or sulfate (Pohland et al., 1982; Moore et al., 1982). The conjugating enzyme is UDP-glucuronyltransferase (Rao and Brewer, 1969).

An important question in this area is: What are the effects of the EHC of the gonadal hormones on their concentration in peripheral blood? The serum concentration of a hormone is the algebraic sum of the secretion rate, catabolic rate, and EHC of the hormone. Any factor that increases rate of passage of ingesta through the gut or by sequestering the steroid in the gut, such as fiber, will decrease EHC.

It has been shown in some studies that porcine liver is a very efficient metabolizer of estrogens, and hence large differences were measured between E_2 concentrations in portal and jugular veins of mature pigs (Ruoff, 1988). Similar studies using H^3-E_2 or E_2 glucuronide injected into gut of immature pigs reported high yet similar concentrations of estrogens in the two veins, suggesting that estrogens from the hepatic portal vein were not secreted as efficiently into the bile (Moore et al., 1982; Pohland et al., 1982). In an extensive study of EHC of estrogens in pigs, Ruoff (1988) reported the concentrations of free and conjugated estrogens (E_2 and E_1) in bile were 80 times that in plasma in the prepubertal gilt. The bile concentration was 0.07 µg/ml. The concentration increased to 0.4 and 40 µg/ml, respectively, in pregnant pigs at day 30 and near term. Thus the liver has tremendous capacity to clear the blood of biologically active estrogens. Conjugated forms (E_2 and E_1 glucuronides and E_1 sulfate) go into the peripheral circulation.

V. Physiological Effects of the Gonadal Steroid Hormones

For each of the three families of sex steroids there is a complex group of hormones displaying a range of biological activities. The tissues responsive to this set of hormones are termed "target tissues" and can be divided into sexual and nonsexual. The sexual tissues are the seminal vesicles, prostate, penis, vas deferens, and epididymis (male) and the uterus, vagina, cervix, and mammary gland (female), and skin and hair growth patterns in both sexes. Nonsexual tissues include all other tissues in body, but especially muscle, bone, kidney, liver, pituitary gland, and brain. There is good evidence that these hormones have behavioral effects, both mating and social (Lunde and Hamburg, 1972). For example, castration depresses libido and androgens increase social aggressiveness.

A. Androgens

First of all, this class of steroid hormones has effects that fit in the general scheme of growth. Indeed, these compounds are noted for their anabolic activity as well as their androgenic activity. They stimulate cell hypertrophy and hyperplasia of the various organs by increasing vascularity, secretions, water and electrolyte balance, and concentrations and activities of specific enzymes. Androgens stimulate erythropoiesis via increased production of erythropoiten (Alexanian, 1969) and heme synthesis in bone marrow (Necheles and Rai, 1969).

The biological activity of testosterone, and indeed of progesterone and $E_2\beta$, is attentuated by the high affinity of specific serum binding proteins. In the case of the androgens this protein is sex-hormone-binding globulin (SHBG). This protein also binds $E_2\beta$, but at one-third the binding constant. The male hormone decreases SHBG, whereas $E_2\beta$ increases it. Thus SHBG concentration in men is one-half that in women. Since the bound fraction of a steroid hormone in plasma is not freely exchangeable with the extravascular and intracellular compartments of most organs, the biological activity of the steroid hormone is greatly affected by the plasma concentrations of the binding proteins. The liver is a notable exception in its response to the bound hormone fraction because of its unique architecture. See Partridge (1981) for a review on the significance of the binding proteins on hormone action.

A tissue protein that binds testosterone is androgen binding protein, found in the testes and epididymis. A glycoprotein, it is produced by the FSH-stimulated Sertoli cells. Its role appears to be maintenance of high concentrations of testosterone around the germ cells (Steinberger, 1975) for development of the spermatids.

1. Sexual Differentiation

The role of testosterone in this process has been reviewed by Wilson *et al.* (1981). At about 70 days gestation, the human fetal testes enlarge and actively secrete testosterone. The Wolffian ducts and urogenital sinus differentiate to become male in character. In the absence of testosterone, they become female. The testes remain quiescent during childhood until onset of puberty. Research suggests that puberty is initiated by the waning of a substance in the brain that inhibits the median basal hypothalamus (Joseph *et al.*, 1969). Removing the inhibition allows the secretion of LHRH, which causes FSH and LH secretion; LH stimulates Leydig-cell secretion of testosterone.

As mentioned in Section II,C, many accessory sex tissues contain 5α-reductase, which treats testosterone as a prehormone. The enzyme converts it to dihydrotestosterone (DHT). In the human male fetus, development of the prostate, testes, scrotum, penis, and facial hair is stimulated by DHT, whereas testosterone stimulates the seminal vesicles and epididymis as well as the anabolic facets of puberty.

What are the primary actions of the male hormones? There is great agreement that the mechanism of action is similar to that of estradiol-17β and progesterone. Testosterone and DHT diffuse into the cell and bind to a receptor, which associates with the chromatin. This assocation at a specific acceptor site stimulates the transcription of DNA to produce an mRNA, which is translated into a specific protein. Thus the androgen-directed physiological responses in a particular tissue represent events secondary to the primary action of the hormone. The mechanism of action of steroid hormones will be described in more detail for the estrogens and progesterone in the next section.

B. Progestins and Estrogens

In the interest of brevity this discussion will be confined mainly to the uterus, which is classically the primary target organ of these hormones. Here the focus will be to present the mechanism of action of a steroid hormone.

The dependence of the uterus for the sequential actions of estradiol-17β and progesterone was convincingly demonstrated early in this century (the 1920s and early 1930s). The painstaking research of Allen and Doisy (1923) showed that a substance in the ovarian follicular fluid of the sow caused vaginal cornification in the guinea pig. This finding led to the identification of estrone. Then Corner and Allen (1929) reported that extracts of porcine corpora lutea caused progestational proliferation of the rabbit uterus, which led to the identification of progesterone (Allen and Wintersteiner, 1934).

Following the isolation and identification of the chief ovarian hormones, many investigators undertook descriptive studies to define the many functions of these steroids. The studies focused on the uterus and established that a period of exposure to estrogen was necessary before progesterone was fully effective in causing endometrial proliferation. Other workers characterized the function of the ovarian hormones in the female accessory sex organs.

The early work on the mechanism of actions of these two hormones indicated that the hyperemia and vasodilation that occur in response to ovarian hormones may be a key to understanding how the hormones worked. Indeed, by studying these responses, such as mobilization of histamine, it was suggested that the primary events in the uterotropic response might be elucidated (Szego and Roberts, 1953). Others have shown that these classical effects are not the primary events, those of most importance. Rather, they occur first and probably optimize the environment for growth by making substrate, ions, and hormones available.

Taking a different approach, some workers suggested that increased metabolic activity resulted from the direct effect of estrogens on enzymes (Villee, 1962). For example, the oxidation–reduction of $E_2β$ and E_1 by E_2 dehydrogenase was involved in production of $NADP^+$ nucleotides, which then acted as transducers of the estrogenic effects (Talalay, 1961). It has been shown that while estrogens undergo oxidation–reduction reactions, this is not the uterotropic mechanism of action, nor is it in other tissues where the mechanism has been explained. It is an error to think, however, that the ovarian steroids are not responsible for elevating the activity of specific enzymes in target tissues. The classic work of Wi and Mueller (1963) demonstrated that estrogens stimulated synthesis of nucleic acids, proteins, and phospholipids in rat uterine tissue and proposed that estrogen controlled production of nucleic acid templates. Soon there were reports that an inhibitor of protein synthesis, puromycin, blocked the action of estrogens (Hamilton, 1963; Mueller *et al.*, 1961). The incorporation of tritiated (^{3}H-labeled) precursors into all classes of RNA was elevated within a few hours after estrogen administration; an RNA synthesis inhibitor, actinomycin D, would block this effect (Hamilton, 1962; Notebloom and Gorski, 1963). Protein and RNA synthesis in the chick oviduct was shown to be regulated by E_2 and progesterone (O'Malley *et al.*, 1969). The conclusion from this body of excellent early work was that the mechanism of action of the ovarian hormones probably was at the level of RNA transcription. Increasing the number of copies of specific message enchanced the concentrations of specific proteins in the cells of the uterus (O'Malley *et al.*, 1969).

1. Receptors and Hormone Action

The concept of receptors was borrowed by endocrinologists from pharmacologists. It was Ehrlich (1913) who stated "drugs don't act unless they bind." The same can be said for hormones. Muldoon (1988) wrote, "hormone action at the cellular level begins the binding of the hormone to highly specific protein receptor molecules." Hormones are elaborated by a variety of tissues and carried by a common circulatory system to all cells of body. But only those cells containing binding proteins of high affinity for the hormone have the ability to respond to the hormone. These binding proteins are termed *receptors* (R). The action of a hormone depends on its physiochemical properties, which determine its ability to cross the plasma membrane and enter cells. Steroid and thyroid hormones are lipophilic enough to diffuse across the cell membrane of all cells. Upon entering the cell, the hormone gains accessibility via an R to the nuclear replicative and proliferative machinery. Only target cells contain R in sufficient concentrations to retain the steroid long enough to gain this accessibility. From the early studies, the demonstration that ovarian steroid hormones stimulate RNA and protein synthesis led to the concept that these hormones were acting at the gene level through a receptor-mediated mechanism. Clark (1977) formalized five basic criteria for these receptors. The characteristics of specific receptor as opposed to a nonspecific one are finite binding capacity, high affinity, specificity for one hormone or a class of hormones, tissue specificity, and correlation with biological response. In other words, the biological response is a saturable phenomenon. There is a finite number of receptor sites. In Fig. 13 is a Scatchard plot for the measurement of number (*n*) of receptors and dissociation constant (K_d), which is the concentration of steroid at which 50% of the receptor sites are bound. A measure of K_d and *n* is obtained by a Scatchard analysis of data from a saturation experiment using a centrifuged fraction of the tissue's cells and a tritiated form of the

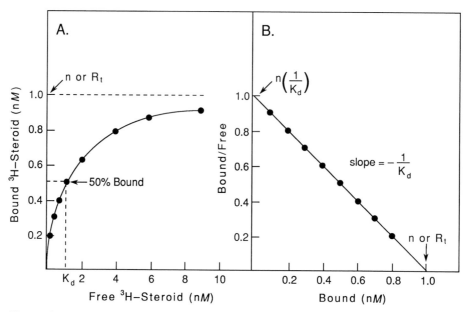

Figure 13 (a) Saturation and (b) Scatchard analysis of receptor steroid binding. Abbreviations: n or R_t, number of receptor sites; K_d, dissociation constant. From Clark and Markaverich (1988).

steroid of high specific activity. The R should have high affinity for their respective ligands (hormones), since the plasma concentrations of the ligands range between 10^{-10} and 10^{-8} M; otherwise the biological response would be weak at best. The reciprocal of K_d, termed the association or affinity constant (K_a), is obtained from the Scatchard plot, and ranges between 10^8 and 10^{10} l/mol (sometimes higher). The specificity of the receptor enables the target cell for the hormone to respond to one hormone without interference from others. Specificity of the receptor is not absolute. It has limited capacity for recognition and differentiation of various steroid structures. For example, estrogen R has some affinity for all steroids. The specificity of progesterone R appears to be less than the estrogen R (Rousseau *et al.*, 1973). If the response of

target organs results from steroid–receptor interactions, then the number of R in these tissues should be higher than that of nontarget tissues (Clark and Peck, 1979). Finally, the fifth criterion implies that the extent of response should relate to some function of R occupancy. The concepts and the complex interactions of hormone with these and other binding sites and how to resolve these interactions are discussed elsewhere (Clark and Peck, 1979).

2. Cellular Localization of Receptors

Currently there are two models dealing with cell R localization and the interaction of hormone and R after hormone entry (see Fig. 14). The nuclear translocation or two-step model of Jansen *et al.* (1968) and Gorski *et al.* (1968) was widely accepted for all steroid hormone R. This model (Fig. 14a) is

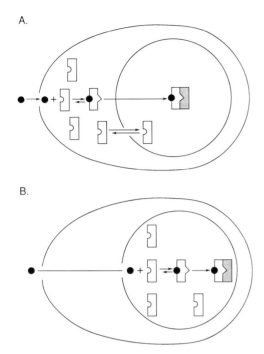

Figure 14 Models of cellular localization of steroid receptors: (a) translocation, (b) nuclear localization. From Clark and Markaverich (1988).

supported by the experiments of Munck and Foley (1976) and Pavlik *et al.* (1979). Jansen *et al.* (1968) proposed that an 8S form of R is composed of 4S subunits and binding of E_2 to the 4S transforms to a 5S R-H complex that translocates to the nucleus. Similar conclusions concerning the progesterone R were drawn by O'Malley *et al.* (1970).

The experiments of Williams and Gorski (1971) have let to an intensive effort by other workers (Martin and Sheridan, 1982; Welshons *et al.*, 1984; King and Greene 1984), and together this body of evidence supports the nuclear localization model of steroid hormone R (see Fig. 14b). These studies indicate that R reside predominately in the nucleus *in vivo* and that the cyto-

plasmic R may represent an artifact resulting from homogenization of the tissue (Sheridan, 1975). Another study (Perrot-Applanet, 1985) demonstrated nuclear localization of the progesterone R in several tissues of the rabbit and guinea pig using immunocytochemical methods and monoclonal antibodies to the progesterone R. At present it is difficult to conclude which model best fits all the experimental evidence, which at least proves the mechanism of action of hormones is still a very dynamic and exciting area in which to work. Certainly our present understanding of hormones in the sphere of molecular biology has encouraged great advances in the endocrine-based treatment of diseases such as cancers of target tissues, precocious puberty, and endometriosis, and the use of agonists to these hormones as fertility and antifertility agents.

Chemically the steroid hormone receptors appear to be acidic proteins having a molecular weight of 80–100K. The functional R consists of two subunits. Like most proteins there are functional domains: one for steroid binding and one for DNA binding, both of which possess appreciable structural homology. The interaction between R and steroid hormone causes a change in the conformational structure of the R. This change is termed "receptor activation." A conceptualization of this activation consists of a hinge region being between the two domains of the R. When the steroid binds to the R, the DNA binding region becomes exposed and the R affinity for the DNA increases. The steroid–R complex binds to specific segments of DNA, which determines which genes express themselves. The complex also serves to enchance the binding of the RNA polymerase to a newly exposed initiation site, and transcription of the adjacent gene proceeds to produce specific forms of RNA.

A subject of controversy is the presence of acceptor sites in the nucleus. There are numerous studies that show that nuclear binding of the steroid–receptor-complex is a saturable phenomenon, and other studies claim such sites don't exist. If specific binding sites do exist, they probably involve steroid–R complexes binding to both chromosomal protein and DNA. Such interactions were suggested by Schrader *et al.* (1972), who showed that the α subunit of progesterone R binds to DNA and the β subunit binds to chromatin. The binding sequence is α–β dimer to chromatin as specified by β, then α dissociating and binding to DNA. Boyd-Leinen *et al.* (1984) characterized the putative chromatin acceptor proteins as nucleoacidic proteins (NAP). The binding of NAP to the progesterone R was saturable (Spelsberg *et al.*, 1983). Similar acceptor sites have been found for the estrogen R in the avian oviduct (Kon and Spelsberg, 1982).

3. The Uterotropic Effect

To place in perspective this vast effort to understand the steroid hormone mechanism of action and gain an appreciation of how these studies at the molecular level explain a hormone's physiological functions, this short section will be presented. Estradiol-17β induces a set of uterotropic effects that have been divided into early or late responses. The latter responses culminate in cellular hypertrophy and hyperplasia, which is true growth. The early responses are certainly supportive, but not obligatory for growth. A review of these two types of responses has been written by Clark and Markaverich (1988). Early responses include hyperemia, calcium influx, histamine

release, eosinophil infiltration, increased RNA and protein precursor uptake, enhanced glucose oxidation and increased RNA and protein synthesis, and increased chromatin template activity and RNA polymerase initiation sites. There is synthesis of a protein (IP) and its mRNA (DeAngelo and Gorski, 1970).

It is now thought that these early responses are not involved in the primary mechanism of action of $E_2\beta$, as a result of a number of studies. For example, it has been observed that the estrogen catabolite estriol is a short-acting estrogen. It stimulates all of the early responses but fails to stimulate growth (Clark and Peck, 1979). It has been suggested that estriol's failure is due to its inadequate maintenance of occupancy of the R; therefore the late responses do not occur. It has been shown that sustained stimulation of RNA and protein synthesis to cause growth results from nuclear occupancy by 10–20% of the estrogen receptors for 4–6 h (Clark and Peck, 1979). Long-term nuclear occupancy is associated with elevated RNAs of all types, sustained RNA polymerase I and II activity and chromatin activity, DNA synthesis, and cellular growth.

More recently a field of work has grown showing the indirect effect of ovarian steroids on the uterine growth and function, as well as their effects on other target organs such as the vagina, mammary gland, and liver. For example, it has been shown that estrogens stimulate synthesis of polypeptide growth factors that have a paracrine function in the uterus (Sirbasku *et al.*, 1979; Ikeda and Sirbasku, 1984). There are at least four growth factors having autocrine and paracrine actions in mitogenesis and differentiation of cells in the uterus and ovary. The presence of IGF-I and -II, EGF,

TGF-β, and PDGF has been demonstrated in cells of the ovary and uterus (see Hammond *et al.*, 1988). A new chapter in reproductive biology will be written through the cooperation and imagination of biochemists, endocrinologists, and physiologists to understand the concerted actions of the paracrine peptides in reproduction and how they interact with the endocrine steroids and proteins.

VI. Assay of Steroid Hormones

Our understanding of hormones has increased in proportion to our ability to measure them in the medium being investigated. Indeed, the advances in hormone assay have been tremendous in the last 25 years and have spurred the development of modern endocrinology. The assay of hormones is an art and a science. The technology of the assay of a hormone must be perfected and then a skillful application mastered. Before putting the assay to work, however, several questions need to be asked. What is the hypothesis of the experiment? How does the assay relate to that hypothesis? In other words, does the assay measure what needs to be measured in the experiment we have designed? Unless the answers to these questions match, we will complete a flawed experiment, no matter how technically satisfying the assay.

Since the early 1900s assays have been perfected and have served their purpose in the hands of clear-thinking, skillful endocrinologists. Today we have a bewildering number of assays that can be categorized as bioassays, chemical assays, and immunoassays or hybrids of them. The mouse Leydig assay for LH is an example of a hybrid.

Assays have become more sensitive and precise, but not necessarily more accurate. Because of the hormone concentration in tissues and fluids (10^{-9} to 10^{-12} mol/liter), they had to be improved. Assays occupy the spectrum from dependence on whole animals to simple chemical compounds as the responding element in the assay.

What are the conceptual elements that distinguish assays from one another? They are structurally directed (SD) versus functionally directed (FD). The FD assays measure the biological activity of a hormone (i.e., $E_2\beta$) or more likely a class of hormones (i.e., estrogen). The SD assays measure the concentration of the compounds having structural similarity. Thus the hormone value measured by the two types of assays may differ. The objectives of these two fundamental forms of assay imply different criteria of validity that each assay must satisfy. For FD assays it is necessary that standard and unknown be functionally identical within the context of the biological system employed. For SD assays the validity depends on the fact that standard and unknown are structurally identical. Satisfying the latter criterion must be done with rigor. If perfected the assay is more sensitive and precise, so more so that current advances in hormone assay rest in the SD area. With the luxury of a cornucopia of assays for hormones, one must be critical to insure that the assay chosen fits the experimental hypothesis being tested. The FD assay still plays an irreplaceable role, especially in the pioneering phase of our understanding of a hormone and as a reference assay for establishing a new SD assay.

The same set of reliability criteria pertains to all assays. The criteria come under the broad headings of accuracy, precision, specificity, and sensitivity. Of these, the most difficult to measure is accuracy, which can be defined as the capacity to measure the true concentration (SD assay) or potency (FD assay). The "true" quantity of any compound will never be known with certainty, of course. Nevertheless, several methods are used to gauge accuracy. One is to perform a recovery experiment by spiking tissue samples with known quantities of authentic hormone, then apply the extraction method and the assay. The quantity measured is compared to that added. See Henricks *et al.* (1971) and Henricks and Torrence (1978) for examples of this approach. Another method is to use two or more assays, preferably each based on a different mechanism of measurement (i.e., immunoassay vs. chemical assay) and compare the results. One can express an index of discrimination—a concept developed by the bioassay-ist. The use of these methods can increase our confidence in the assay's capacity to measure the readily available hormone, but all of the endogenous hormone may not be extracted and therefore may not be measured. For example, if the sample is being assayed for progesterone and glucuronidase treatment is omitted, then the progesterone bound in this conjugate will not be assayed. Such a question takes on increased importance if tissue residues of the hormones are being measured. For more on the application of these assays to tissues see Henricks *et al.* (1983) and Hoffman (1978). Again, the question "What is the researcher trying to measure?" should be asked before initiating the assays.

Precision is the criterion that measures the variation of the individual measurements of the same sample. It measures the capacity to obtain the same answer each time a measurement is taken. Statistical analysis is applied to obtain mean ± SD and a coefficient of variation (CV) for both intra- and interassay variation. With high technical skill and quality of the assay components, CVs of 5–15% for immunoassays should result. How much to allow depends on the purpose of the experiment. In bioassays a different value, lambda, is used to measure variation. A λ (SD of the responses divided by slope of regression line) should be no greater than 0.3.

Specificity is the capacity to distinguish the hormone being measured from closely related compounds. It is related to accuracy. As stated earlier, the purpose of the experiment dictates how specific the assay needs to be. Certainly the assay needs to be characterized for this property. One way is to perform a cross-reactivity test with a set of compounds and their metabolites. See Niswender and Midgley (1970) for an excellent example of this type of study and how to chemically tailor a steroid immunogen to obtain a hapten radioimmunoassay (RIA) for steroid hormones. Their pioneering studies encouraged the development of steroid RIA.

Sensitivity can be defined as the least quantity that can be distinguished from zero. In statistical terms, that means the smallest quantity whose fiducial limits do not include zero. In immunoassays, sensitivity is a function of the affinity constant (K_a) of the antiserum. Sensitivity is also affected by the specific activity of the labeled hormone, the precision of the assay, and the assay blank. The assay blank is an important factor in any assay and arises from two components: reagents and tissue. It is the value obtained in the absence of the hormone. It usually acts nonspecifically and nonlinearly. In many assays, as the quantity of the hormone rises the blank is reduced. In any case the blank must be kept as low as possible and constant. It is reduced by a high K_a value, including an inert protein in the buffer system and proper choice for separating the free from the antibody-bound form of the labeled hormone.

A. Functionally Directed Assays

These assay systems necessarily fall within the class traditionally termed "bioassay," since hormonal activity, termed potency, can only be revealed in a biological environment. The classical bioassays, *in vivo* assays, were based on the ability of a class of hormones to elicit a unique biological response in test animals. The greater and more precise that response in test animals, the more useful the bioassay. A careful choice of the responding element of the assay must be made. In whole-animal assays a highly inbred strain of the species such as rat or mouse is used to obtain an acceptable precision value (λ). The protocol of the perfected assay must be rigidly applied in each assay. Whole-animal assays are less sensitive and more expensive to conduct than *in vitro* assays in which a responding tissue is incubated with the sample containing an unknown hormonal potency. Consequently, for most hormones an *in vitro* assay has been developed in which the end point is a sensitive chemical or immunoassay. Such assays are more practical than *in vivo* assays and allow for more rapid progress to occur in the field of endeavor.

An advantage of the bioassay can be the fact that the response is not limited to a hormone having one type of structure, but rather to a number of structures all having the same effect. It is essential to a valid bioassay that the log–dose response curves should have the same slope. This is equivalent to the assumption of similarity between the test and standard preparations. The test preparation must behave as though it were a dilution of the standard preparation in a diluent that is inert with respect to the response. Such preparations are said to be qualitatively similar. The essential test of validity is the demonstration of parallelism between the standard and every test preparation assayed. If the response is nonparallel, the potency value is rejected. This and other characterizations of an assay as well as potency and its confidence interval are calculated using statistical methods developed for bioassays (see Finney, 1964).

Specific bioassays for the gonadal hormones have been described in a number of excellent references; thus they will not be discussed here (see Niswender and Nett, 1977).

B. Structurally Directed Assays

To measure a gravimetric amount or concentration of a single hormonally active compound having a unique molecular structure, the saturation assay methods have come into widespread use. The original exploitation of this method came in the laboratories of Yalow and Berson to measure serum insulin (1960) and Ekins to measure serum thyroxine (1960). The fundamental principle of saturation analysis has been used in a variety of assays, the outstanding being competitive protein binding assay, radioimmunoassay, and radioreceptor assay.

The basic principle can be expressed as:

Hormone + limited reagent →
$$\text{H·reagent complex} + \text{H}$$
$$\text{(bound)} \qquad \text{(free)}$$

In an immunoassay the reagent is an antibody.

All of these methods require that the reagent, whether it is an antibody, binding protein, or receptor, is saturable and held strictly constant. If so, the distribution of hormone between bound and free fractions following reaction to equilibrium is dependent on quantity of hormone initially present. For the assay to be sensitive, the quantity of antibody must be comparable in magnitude to total amount of hormone present in the system.

C. Role of the Labeled Hormone

The use of a labeled form of the hormone that accurately reflects the distribution of the endogenous hormone provides the sensitivity and convenience inherent in saturation analysis assays. A crucial assumption is that the labeled and natural forms of the hormone behave similarly in the binding reaction. Largely because of the availability of the isotopic forms with high specific activity and a clear distinct signal and the low level of nonspecific interference coming from other biological compounds, radioimmunoassays have been exploited. The disadvantage of this approach is the high cost of isotope waste disposal, greater demand for radiation safety in the laboratory, and the intrinsic instability of γ-labeled compounds. If tritium labeling is used this is not a problem, but sensitivity is sacrificed. A greater disadvantage of RIA is the necessity for physical separation of the "free" and "bound" hormone species before measuring the radioactivity residing in each fraction. Numerous methods have been devised to accomplish precise and quick separation. To monitor the quality of this step, the amount of nonspecific binding is measured. The most used methods are charcoal absorption of free hormone such as used in the steroid RIAs, precipitation of "bound hormone" by second antibody or polyethylene glycol, and creation of a solid phase by coating the walls of plastic test tubes with the antibody.

The nonradiometric assays rely on the activity of the label remaining intact only when the hormone is attached to the free form, the activity of the bound species being suppressed as a result of steric hindrance provided by the attached antibody. Thus the distribution of the label is measured with no need for a separation step. The acceptance of these assays has depended on the development of better sensitivity, using such labels as fluorescent groups, chemiluminescent and bioluminescent groups, and the use of enzyme-linked antibodies or antigens and their substrates (ELISA).

To develop a hapten immunoassay for nonimmunogenic molecules, such as steroids, one renders them antigenic by conjugating a sufficient number of these molecules to a macromolecule such as serum albumin. The conjugation process must also be mastered to label the antigen (or antibody) with an enzyme or a fluorescent moiety. The three commonly used methods are mixed anhydride (Erlanger et al., 1957), carbodiimide (Goodfriend et al., 1964), and modified carbodiimide (Anderson et al., 1964). For a detailed discussion of this methodology as it applies to nonradiometric immunoassay, see Munro and Lasley (1988).

References

Adlercreutz, H. (1970). *J. Endocrinol.* **46,** 129.
Alexanian, R. (1969). *Blood* **33,** 564.
Alila, H., and Hansel, W. (1984). *Biol. Reprod.* **31,** 1015.
Allen, E., and Doisy, E. A. (1923). *J. Am. Med. Assoc.* **81,** 819.

Allen, W. M., and Wintersteiner, O. S. (1934). *Science* **80,** 190.

Anderson, C. W., Zimmerman, J. E., and Callahan, F. W. (1964). *J. Am. Chem. Soc.* **86,** 1839.

Armstrong, D. T., Weiss, T. J., Selstam, G., and Seamark, R. F. (1981). *J. Reprod. Fertil.* (Suppl. 30), **30,** 143.

Baird, D. T. (1977). "The Ovary," Vol. III, p. 305. Academic Press, London.

Baulieu, E. E. (1963). *J. Clin. Endocrinol. Metab.* **20,** 900.

Berridge, M. J., and Irvine, R. F. (1984). *Nature* **312,** 315.

Bhavnani, B. R. (1988). *Endocr. Rev.* **9,** 396.

Birmingham, M. K., Elliot, F. H., and Valere P. H. L. (1953). *Endocrinology* **53,** 687.

Borkowski, A. J., Levin, S., Delcroix, C., Mahler, A., and Verhas, V. (1967). *J. Clin. Invest.* **46,** 797.

Boyd-Leinen, P., Gosse, B., Rasmussen, K., Martin-Dani, G., and Spelsberg, T. C. (1984). *J. Biol. Chem.* **259,** 2411.

Brown, M. S., and Goldstein, J. L. (1976). *Science* **191,** 150.

Butenandt, A., and Wesphal, U. (1934). *Ber. Dtsch. Chem. Ces.* **67,** 1440.

Caffrey, J. L., Nett, T. M., Abel, J. H., Jr., and Niswender, G. D. (1979). *Biol. Reprod.* **20,** 279.

Carr, B. R., MacDonald, P. C., and Simpson, E. R. (1982). *Fertil. Steril.* **38,** 303.

Channing, C., and Kammerman, S. (1974). *Biol. Reprod.* **10,** 179.

Christensen, A. K. (1975). *In* "Handbook of Physiology" (R. O. Greep and E. B. Astwood, eds.), Vol. V, p. 21. American Physiological Soc., Washington, D.C.

Clark, J. H. (1977). *In* "Reproduction in Domestic Animals" (H. H. Cole and P. Cupps, eds.), p. 143. Academic Press, New York.

Clark, J. H., and Markaverich, B. M. (1988). *In* "The Physiology of Reproduction" (E. Knobil and J. Neill, eds.), p. 675. Raven Press, New York.

Clark, J. H., and Peck, E. J., Jr. (1979). "Female Sex Steroids: Recepters and Function." Springer-Verlag, Berlin.

Claus, R. (1979). *In* "Advances in Animal Physiology and Animal Nutrition, No. 10" (K. D. Gunther and M. Kirchgessner, eds.).Verlag Paul Parey, Hamburg.

Cooke, B. A., Lindh, M. L., and Janszen, F. H. A. (1977). *Biochem. J.* **168.** 43.

Constantopoulos, G., and Tchen, T. T. (1961). *J. Biol. Chem.* **236,** 65.

Corner, G. W., and Allen, W. M. (1929). *Am. J. Physiol.* **88,** 326.

David, K., Dingemanse, E., Freud, J., and Laquer, E. (1935). *Hoppe-Seyler's Z. Physiol. Chem.* **2,** 233.

DeAngelo, A. B., and Gorski, J. (1970). *Proc. Natl. Acad. Sci. USA* **66,** 693.

Diekman, M. A., O'Callaghan, P., Nett, T. M., and Niswender, G. D. (1978). *Biol. Reprod.* **19,** 999.

Denamur, R., and Mauleon, P. (1963). *C. R. Acad. Sci.* **257,** 264.

Doisy, E. A., Veler, C. D., and Thayer, S. A. (1929). *Am. J. Physiol.* **90,** 329.

Dorfman, R. I., and Ungar, F. (1965). "Metabolism of Steroid Hormones." Academic Press, New York.

Dorrington, J. H. (1977). "The Ovary," Vol. III, p. 379. Academic Press, London.

Dorrington, J. H., and Armstrong, D. T. (1975). *Proc. Natl. Acad. Sci. USA* **72,** 2677.

Dorrington, J. H., Moon, Y. S., and Armstrong, D. T. (1975). *Endocrinology* **97,** 1328.

Dufau, M. L., and Catt, K. J. (1975). *In* "Methods of Enzymology XXXIX" (J. G. Hardman and B. W. O'Malley, eds.), p. 252, Academic Press, New York.

Dufau, M. L., Tsuruhara, T., Homer, K. A., Podesta, E., and Catt, K. J. (1977). *Proc. Natl. Acad. Sci. USA* **74,** 3419.

Eik-Nes, K. B. and Hall, P. F. (1965). *Vitam. Horm.* **23,** 153.

Ekins, R. P. (1960). *Clin. Chem. Acta* **5,** 453.

Erlander, B. T., Borek, F., Beiser, S. M., and Lieberman, S. (1957). *J. Biol. Chem.* **228,** 713.

Erlich, P. (1913). *Lancet* **2,** 445.

Evans, G., Dobias, M., King, G. C., and Armstrong, D. T (1981). *Biol. Reprod.* **25,** 673.

Falck, B. (1959). *Acta. Physiol. Scand.* **47,** Suppl. 163.

Finney, D. J. (1960). *In* "An Introduction to the Theory of Experimental Design." University of Chicago Press, Chicago.

Fitz, T. A., Mayan, M. H., Sawyer, H. R., and Niswender, G. D. (1982). *Biol. Reprod.* **27,** 703.

Fitz, T. A., Hokyer, P. B., and Niswender, G. D. (1984). *Prostaglandins* **28,** 119.

Fortune, J. E. (1986). *Biol. Reprod.* **35,** 292.

Fortune, J. E. (1988). *J. Anim. Sci.* **66,** (Suppl. 2), 1.

Fuller, G. B., and Hansel, W. (1970). *J. Anim. Sci.* **31,** 99.

Ginther, O. J. (1979). "Reproductive Biology of the Mare. Basic and Applied Aspects." MacNaughton and Gunn, Ann Arbor, Michigan.

Ginther, O. J., Mapletoft, R. J., Zimmerman, N.,

Meckley, P. E., and Nuti, L. (1974). *J. Anim. Sci.* **38,** 835.

Gloyno, R. E., and Wilson, J. D. (1969). *J. Clin. Endocrinol.* **29,** 970.

Gore-Langston, R. E., and Armstrong, D. T. (1988). *In* "The Physiology of Reproduction" (E. Knobil and J. Neill, eds.), Vol. 1, pp. 331–385. Raven Press, New York.

Good friend, T. L., Levine, L., and Fosman, G. D. (1964). *Science* **144,** 1344.

Gorskei, J., Toft, D., Shyamola, G., Smith, D., and Notider, A. (1968). *Rec. Prog. Horm. Res.* **24,** 45.

Gowers, D. B. (1984). *In* "Biochemistry of Steroid Hormones" (H. L. J. Makin, ed.), pp. 170–178, 321–331. Blackwell Scientific, Oxford.

Gowers, D. B., and Bicknell, D. C. (1972). *Acta Endocrinol. (Kbh.),* **70,** 567.

Greep, R. O., Van Dyne, H. K., and Chow, B. F. (1942). *Endocrinology* **30,** 635.

Gwynne, J. T., and Strauss, J. F., III (1982). *Endocr. Rev.* **3,** 299.

Hall, P. F. (1984). *Int. Rev. Cyto.* **86,** 53.

Hall, P. F. (1988). *In* "The Physiology of Reproduction" (E. Knobil and J. Neill, eds.), Vol. 1, pp. 975–998. Raven Press, New York.

Hall, P. F., Charponnier, C., Nakamura, M., and Gobbiani, B. (1979). *J. Steroid Biochem.* **11,** 1361.

Hall, P. F., Osawa, S., and Mrotek, J. J. (1981). *Endocrinology* **109,** 1677.

Hamilton, T. H. (1962). *Science* **138,** 989.

Hamilton, T. H. (1963). *Proc. Natl. Acad. Sci. USA* **49,** 373.

Hammond, J. M., Hsu, C. J., Mondschein, J. S., and Canning, S. F. (1988). *J. Anim. Sci.* **66,** 21.

Hansel, W., Concannon, P. W., and Lukaszewska, J. R. (1973). *Biol. Reprod.* **8,** 222.

Hansel, W., and Dowd, J. P. (1986). *J. Reprod. Fertil.* **78,** 755.

Heath, E., Weinstein, P., Merritt, B., Shanks, R., and Hixon, J. (1983). *Biol. Reprod.* **29,** 977.

Henricks, D. M., and Mayer, D. T. (1977). In "Reproduction in, Domestic Animals" (H. H. Cole and P. T. Cupps, eds.), pp. 79–116. Academic Press, New York.

Henricks, D. M, and Torrence, A. K., (1978).

Henricks, D. M., Dickey, J. F., and Hill, J. R. (1971). *Endocrinology* **89,** 1350.

Henricks, D. M., Gray, S. L., and Hoover, J. L. B. (1983). *J. Anim. Sci.* **57,** 247.

Henricks, D. M., Hoover, J. L. B., Grimevely, T., and Grimes, L. W. (1988). *Hormone Metab. Res.* **20,** 494.

Hochberg, R. B., VaderHoeven, T. A., Welch, S., and Lieberman, S. (1974). *Biochemistry* **13,** 603.

Hoffman, B. (1978). *J. Assoc. Off. Anal. Chem.* **61,** 1263.

Hollander, N., and Hollander, V. P. (1958). *J. Biol. Chem.* **233,** 1097.

Hoyer, P. R., Fitz, T. A., and Niswender, G. D. (1984). *Endocrinology,* **114,** 604.

Hsueh, A. J. W. (1989). *In* "Endocrinology" (L. J. DeGroot, ed.), Vol. 3, pp. 1929–1939. W. B. Saunders, Philadelphia.

Hume, R., and Boyd, G. S. (1978). *Biochem. Soc. Trans.* **6,** 893.

Ikeda, T., and Sirbasku, D. A. (1984). *J. Biol. Chem.* **259,** 4049.

Inaba, T., and Wiest, W. G. (1985). *Endocrinology* **117,** 315.

Jansen, E. V., Suzuki, T., Kawashima, T., Stumph, W. E.,

Janszen, F. H. A., Cooke, B. A., and Van der Molen (1977). *Biochem. J.* **162,** 341.

Joseph, S. A., Knigge, K. M., and Volochin, L. (1969). *Neuroendocrinology* **4,** 42.

Jungblut, P., and DeSombre, E. R. (1968). *Proc. Natl. Acad. Sci. USA* **59,** 623.

Kaltenbach, C. C., Cook, B., Niswender, G. D., and Nalbandov, A. V. (1967). *Endocrinology* **81,** 1407.

Karsch, F. J., Roche, J. F., Noveroske, J. W., Foster, D. L., Norton, H. W., and Nalbandov, A. V. (1971). *Biol. Reprod.* **4,** 129.

Kelch, R. P., Jenner, M. R., Weinstein, S. L., Kaplan, S. L., and Grumbach, M. M. (1972). *J. Clin. Invest.* **51,** 824.

King, W. J., and Greene, G. L. (1984). *Nature* **307,** 745.

Kon, O. L., and Spelsberg, T. C. (1982). *Endocrinology* **111,** 1925.

Koos, R. D., and Hansel, W. (1981). In "Dynamics of Ovarian Function" (N. B. Schwartz and M. Hunzicker-Dunn, eds.), pp. 197–203. Raven Press, New York.

Koritz, S. B. (1968). In "Functions of the Adrenal Cortex" (K. W. McKerns, ed.) Vol. 1, p. 27. Appleton-Century-Crofts, New York.

Kuo, J. F., and Greengard, P. (1969). *Proc. Natl. Acad. Sci. USA* **64,** 1349.

Lemon, M., and Loir, M. (1977). *J. Endocrinol.* **72,** 351.

Li, C. H., and Evans, H. M. (1948). In "The Hormones" (G. Pincus and K. Thimann, eds.), Vol. 1, p. 631. Academic Press, New York.

Lieberman, S., Greenfield, N. J., and Wolfson, A. (1984). *Endocr. Rev.* **5,** 128.

Longcope, C., Widrich, W., and Sawin, C. T. (1972). *Steroids* **20,** 439.

Luchinsky, A., and Armstrong, D. T. (1983) *Can. J. Physiol. Pharmacol.* **61,** 472.

Lunde, D. T., and Hamburg, D. A. (1972). *Recent. Prog. Horm. Res.* **28,** 627.

MacCorquodale, D. W., Thayer, S. A., and Doisy, E. A. (1935). *Proc. Soc. Exp. Biol.* **32,** 1182.

Marker, R. E., Kam, O., and McGraw, R. V. (1937). *J. Am. Chem. Soc.* **59,** 616.

Marrian, G. F. (1930). *Biochem. J.* **23,** 1233.

Martin, D. W., Mayes, P. A., Rodwell, V. W., and Granner, D. K. (1985). "Harpers's Review of Biochemistry," p. 133. Appleton and Lange, Norwalk, Conn.

Martin, P. M., and Sheridan, P. J. (1982). *J. Steroid Biochem.* **16,** 215.

Mayer, D. T., Glascow, B. R., and Gawienoski, *J. Anim. Sci.* **20,** 66.

McDonald, P. C., Grodin, J. M., and Siterii, P. K. (1971). *In* "Control of Gonadal steroid Secretion" (D. T. Baird and J. A. Strong, eds.), p. 158. Edinburgh University Press, Edinburgh.

Mellin, T. N., and Erb, R. E. (1966). *Steroids* **7,** 589.

Melrose, D. R., Reed, H. C. B., and Patterson, R. L. S. (1971). *Br. Vet. J.* **129,** 497.

Menard, R. H., and Purvis, J. L. (1973). *Arch. Biochem. Biophys.* **154,** 8.

Meyer, A. S. (1955). *Biochim. Biophys. Acta* **24,** 142.

Michell, R. H. (1980). *Life Sci.* **32,** 2083.

Moore, A. B., Bottoms, G. D., Coppoc, G. L., Pohland, R. C., and Roesel, O. F. (1982). *J. Anim. Sci.* **55,** 124.

Mueller, G. C., Gorski, J., and Aizawa, Y. (1961). *Proc. Natl. Acad. Sci. USA* **47,** 164.

Muldoon, T. G., and Evans, A. C., Jr. (1988). *Arch. Intern. Med.* **148,** 961.

Munro, C. J., and Lasley, B. L. (1988). *In* "Progress in Clinical and Biological Research" (B. D. Albertson and F. P. Haseltine, eds.), Vol. 285, pp. 289–329. Alan R. Liss, New York.

Murray, R. K., Granner, D. K. Mayes, P. A., and Rodwell, V. W. (1988). "Harpers Review of Biochemistry," pp. 226–234. Appleton and Lange, Norwalk, Conn.

Nakajin, S., Shinoda, M., and Hall, P. F. (1983). *Biochem. Biophys. Res. Comm.* **111,** 512.

Necheles, T. F., and Rai, U. S. (1969). *Blood* **34,** 380.

Nishizuka, Y. (1986). *Science* **233,** 305.

Niswender, G. D., and Midgley, A. R., Jr. (1970). *In* "Immunologic Methods in Steroid Metabolism" (F. G. Peron and B. V. Caldwell, eds.) pp. 149–174. Appleton-Century-Crofts, New York.

Niswender, G. D., and Nett, T. M. (1988). *In* "The Physiology of Reproduction" (E. Knobil and J. Neill, eds.), Vol. 1, pp. 489–525. Raven Press, New York.

Notebloom, W. D., and Gorski, J. (1963). *Proc. Natl. Acad. Sci. USA.* **50,** 250.

O'Malley, B. W., McGuire, W. L., Kohler, P. O., and Koreman, S. G. (1969). *Recent. Prog. Horm. Res.* **25,** 105.

'O'Malley, B. W., Toft, O. and Sherman, M. R. (1970). *J. Biol. Chem.* **246,** 1117.

Partridge, W. M. (1981). *Endocr. Rev.* **2,** 103.

Pavlik, E. J., Rutledge, S., Eckert, R. L., and Katzenellenbogen, B. S. (1979). *Exp. Cell Res.* **123,** 177.

Payne, A. H., and Singer, S. S. (1979). *In* "Steroid Biochemistry" (R. Hobkirk, ed.), Vol. 1, p. 111. CRC Press, Boca Raton, Florida.

Pederson, R. C., and Brownlee, A. C. (1983). *Proc. Natl. Acad. Sci. USA* **80,** 1882.

Perrot-Applanet, M., Logeat, F., Groyer-Picard, M. T., and Milgron, E. (1985). *Endocrinology* **116,** 1473.

Pineda, M. H., Ginther, O. J., and McShan, W. H. (1972). *Am. J. Vet. Res.* **33,** 1767.

Pittman, R. C., Attie, A. D., Carew, T. E., and Steingerg, D. (1979). *Proc. Natl. Acad. Sci.* **76,** 5345.

Pohland, R. C., Coppoc, G. L., Bottoms, G. D., and Moore, A. B. (1982). *J. Anim. Sci.* **55,** 145.

Preslock, J. P. (1980). *J. Steroid Biochem.* **13,** 965.

Puppioni, D. L. (1977). *J. Dairy Sci.* **61,** 561.

Purvis, J. L., Canick, J. A., Latif, S. A., Rosenbaum, J. H., Hologgitas, J., and Menard, R. H. (1973). *Arch. Biochem. Biophys.* **154,** 8.

Rao, R. S., and Breuer, H. (1969). *J. Biol. Chem.* **20,** 5521.

Rao, C. V., Estergreen, V. L., Carmen, F. R., Moss, G. E., and Frandle, K. A. (1976). *Vth Int. Cong. of Endocr.* (Abstract).

Raymond, V., Leung, P. C. K. and Labrie, F. (1983). *Biochem. Biophys. Res. Commun.* **116,** 39.

Reed, M. J., and Murray, M. A. F. (1979). *In* "Hormones in Blood" (C. H. Gray and V. H. T. James, eds.), Vol. III, p. 263. Academic Press, London.

Robertson, H. A., and King, G. J. (1974). *J. Reprod. Fertil.* **40,** 133.

Rousseau, G. G., Baxter, I. D., Higgens, S. J., and Tomkins, G. M. (1973). *J. Mol. Biol.* **79,** 539.

Ruoff, W. L. (1988). Enterohepatic Circulation of Estrogens in the Domestic Pig, *Sus Scrofa*. Ph.D. Thesis, University of Illinois, Urbana.

Ryan, K. J., and Smith, O. W. (1961). *J. Biol. Chem.* **236,** 710.

Ryan, K. J., and Smith, O. W. (1965). *Recent Prog. Horm. Res.* **21,** 367.

Ryan, K. J., and Petro, Z. (1966). *J. Clin. Endocrinol. Metab.* **26,** 46.

Samuels, L. T., Helmrich, M. L., Lasater, M. B., and Reich, H. (1951). *Science* **113,** 490.

Sandler, R., and Hall, P. F. (1986). *Biochem. Biophys. Acta* **164,** 445.

Savard, K. (1973). *Biol. Reprod.* **8,** 183.

Savard, K., Marsh, J. M., and Ricee, B. F. (1965). *Recent. Prog. Horm. Res.* **21,** 285.

Savion, N., Lahertyy, R., Cohen, D., Lui, G. M., and Gospodarwitz, D. (1982). *Endocrinology* **110,** 13.

Scallen, T. J., Srikantaiah, M. W., Skirlant, H. B., and Hanbury, E. (1972). *FEBS Lett.* **25,** 227.

Schomberg, D. W., Coudert, S. P., and Short, R. V. (1967). *J. Reprod. Fertil.* **14,** 277.

Schrader, W. T., Toft, D. O., and O'Malley, B. W. (1972). *J. Biol. Chem.* **247,** 2401.

Shemesh, M., Hansel, W., and Strauss, J. F., III. (1984). *Proc. Natl. Acad. Sci. USA* **81,** 6403.

Sheridan, P. J. (1975). *Life Sci.* **17,** 497.

Shoppee, C. W. (1964). "Chemistry of the Steroids," 2nd Ed. Butterworths, London.

Sirbasku, D. A., and Benson, R. H. (1979). *Cold Spring Harbor Conf. Coll. Prolif.* **6,** 477.

Spelsberg, T. C., Littlefield, B. A., Seelke, R., Davi, G. M., Toyoda, H., Boyd-Leinen, P., Thrall, C., and Kon, O. L. (1983). *Recent. Prog. Horm. Res.* **39,** 463.

Steinberger, E., and Fisher, M. (1969). *Biol. Reprod. Supp.* **1,** 119.

Steinberger, E. (1975). *In* "Hormonal Regulation of Spermatogenesis" (F. S. French, V. Hanson, E. H. Ritzan, and S. N. Nayfel, eds.), p. 337. Plenum Press, New York.

Strauss, J. F., Schuler, L. A., Rosenblum, M. F., and Tonaka, T. (1981). *Adv. Lipid Res.* **18,** 99.

Stryer, L. (1981). "Biochemistry," 2nd Ed., p. 472. W. H. Freeman, San Francisco.

Szego, C. M., and Roberts, S. (1953). *Rec. Prog. Horm. Res.* **8,** 419.

Talalay, P. (1961). *In* "Biological Approaches to Cancer Chemotheraphy" (R. J. C. Harris, ed.). Academic Press, New York.

Tamaoki, B., and Shikita, M. (1966). *In* "Steroid Dynamics" (G. Pincus, J. Tait, and T. Nakano, eds.), p. 493. Academic Press, New York.

Themen, A. P. N., Hoogerbrugge, J. W., Rommerts, F. F. G., and Van der Molen, H. J. (1985). *Biochem. Biophys. Res. Commun.* **128,** 1164.

Ursely, J., and Leymarie, P. (1979). *J. Endocrinol.* **83,** 303.

Villee, C. A. (1962). *In* "The Molecular Control of Cellular Activity" (J. M. Allen, ed.), pp. 297–318. McGraw-Hill, New York.

Welshons, W. V., Liegerman, M. E., and Gorski, J. (1984). *Nature* **307,** 747.

Werthessen, N. T., Schwenk, E., and Baker, C. (1953). *Science* **117,** 380.

Wi, H., and Mueller, G. C. (1963). *Proc. Natl. Acad. Sci. USA* **50,** 256.

Wilson, J. D., Griffin, J. E., George, F. W., and Laskin, M. (1981). *Recent. Prog. Horm. Res.* **37,** 1.

Williams, D., and Gorski, J. (1971). *Biochem. Biophys. Res. Commun.* **45,** 258.

Williams, W. F. (1962). *J. Dairy Sci.* **45,** 1541.

Wintersteiner, O., and Allen, W. M. (1934). *J. Biol. Chem.* **107,** 321.

Yalow, R. S., and Berson, S. A. (1960). *J. Clin. Invest.* **39,** 1157.

Yamamoto, S., and Bloch, K. (1970). *In* "Natural Substances Formed Biologically from Mevalonic Acid" (T. W. Goodwin, ed.), p. 35. Academic Press, London.

Younglai, E. V., and Short, R. V. (1970). *J. Endocrinol.* **47,** 321.

CHAPTER 4

Folliculogenesis

J. C. MARIANA, D. MONNIAUX, M. A. DRIANCOURT, and P. MAULEON

I. Qualitative and Quantitative Features of Folliculogenesis

A. Formation of Primordial Follicles and Initiation of Follicular Growth

A single follicle consists of a diplotene oocyte and associated somatic cells surrounded with a basal lamina. Small follicles have a single layer of follicular cells and form the largest number in the ovary. The ovary of one ewe lamb contains about (230,000 ± 120,000) small follicles and only 250–1500 follicles that have started to grow. Most of these small follicles lie at the periphery of the ovary (Mariana and de Pol, 1986). However, some differences exist among follicles lying directly beneath albuginea and those nested deeper within the cortex of the ovary. For example, in the ovary of a 30-

day-old rabbit, the nuclei of the follicular cells close to the albuginea are flat and wrap around the oocyte almost completely.

The number of follicular cells and the labeling index of follicular cells incorporating tritiated thymidine in the S phase of the cycle are higher in follicles inside the ovary than in follicles lying in the periphery (Mariana and de Pol, 1986).

The population of small follicles is heterogeneous. This heterogeneity spreads along one gradient lying roughly from outer to inner cortex. In most mammals the first oocytes that reach the diplotene stage at the end of oogenesis lie in the central part of the cortex, whereas the peripheral cortex still contains oogonia and oocytes in the first stages of meiotic prophase. Therefore, a centripetal gradient in meiosis induction and termination occurs between the periphery and the inner part of the cortex (Mauléon, 1977).

The mesonephros contributes to the formation of the definitive gonad and to the formation of follicles in the medullary part of the ovary (ewe, Zamboni *et al.*, 1979; mouse, Upadhyay *et al.*, 1979). Epithelial celomic cells participate in the formation of follicles that lie in the outer part of the ovary. The first follicles are formed around day 70 of pregnancy in the ewe (Mauléon, 1961), day 130 of pregnancy in the cow (Erickson, 1966a), day 70 of pregnancy in the sow (Black and Erickson, 1968), and between day 25 and 30 after birth in the rabbit (Peters *et al.*, 1965). The oocyte and the proliferating follicular cells contribute to the growth of the follicle. In many animals the growth of the oocyte is biphasic in relation to the growth of the follicle (Mandl and Zuckerman, 1952). Thus in the rabbit, the oocyte reaches 80% of its maximal size when the follicle reaches only half of its preovula-

tory size, 800 μm, (Mariana and de Pol, 1986).

During later development the growth of the oocyte is slow compared to the follicle's growth. Growth of the oocyte involves the accumulation of proteins either through direct synthesis within the oocyte or by transfer from the outside. This accumulation was clearly demonstrated in the growing oocyte by Wassarman and Mrozak (1981). No data exist for the primordial follicles. During the active phase of growth of the oocyte, content of proteins is increased 100-fold and RNA content 20-fold. RNA synthesis is increased in the nucleus of the growing oocyte compared to that found in the primordial follicle. This synthesis continues during the growth of the oocyte (Moore and Lintern-Moore, 1978). In the mouse, the oocytes within the 5a-stage follicles, 100–200 cells, are about six times more active than the oocytes in the 2-stage follicles, <20 cells.

Synthetic activity decreases steadily in oocytes from follicles with an antrum, stages 6 and 7 >400 cells. The incorporation of RNA precursors can be observed *in vivo* and from collected and cultured oocytes (Bachvarova, 1981).

In vitro studies using naked oocytes from growing follicles allowed the precise measure of the synthesis and the turnover of various RNAs in the oocyte (De Leon *et al.*, 1983). It clearly appeared that most messenger and ribosomal RNAs are synthesized and accumulated in the oocyte to be reutilized later (Sternlicht and Schultz, 1981).

Follicular cells proliferate and contribute to the growth of the follicle. They multiply at different rates depending on the size of the follicle. Thus, the flat follicular cells that surround the small oocytes, about 30 μm in diameter, are rarely labeled after one injection of tritiated thymidine in the rabbit

(Mariana and de Pol, 1986) or in the rat (Hirshfield, 1989). On the other hand, follicular cells of follicles with 15–20 cuboidal follicular cells are most frequently labeled.

Proliferative activity of follicular cells increases with the size of the oocyte (Pedersen, 1970). Tentative estimates of the parameters of this proliferation (Mariana and de Pol, 1986; Hirshfield, 1989) show that the doubling times of follicular cells of follicles with a single layer of cells around the oocyte are very long, approximately 6–8 days. After a continuous infusion of tritiated thymidine in the adult cyclic rat during 1 week, 37% of follicles with four or fewer follicular cells have at least one labeled cell, and all follicles with at least 13 cells have at least one labeled cell. In the 30-day-old rabbit, where the entire small follicle population is definitively formed, identical results were obtained following repeated injections of thymidine for 48 h. All these experiments suggest either that the duration of the cellular cycle is very long, or that the proportion of proliferating cells is very low, that both occur simultaneously. The results also show that the proliferative activity of follicular cells in primordial follicles increases with the size of the oocyte, being minimum in follicles with flat cells. All studies on the growth of the oocyte and the proliferation of follicular cells suggest that the dormancy of follicles is a relative notion; one can substitute the notion of slow growth and of an active steady state (Bachvarova, 1985).

The entire population of primordial follicles can be considered as growing as soon as it is formed, with the gradient of intensity of growth increasing from the periphery to the center of the ovary. In the majority of domestic species, one can predict the following scheme for growth of primordial follicles. The smallest follicles lying in the outer cortex of the ovary grow very slowly, and few will develop during the entire life of the female. The last observable follicles in the ovaries of the oldest females are of this type. The follicles lying mostly in the inner cortex grow more actively. Thus, the gradient of growth activity can be associated with the evolution and formation of the gonad during embryonic life. The initiation of meiosis in the oocytes and colonization of the gonad by the mesonephric cells begins centrally and proceeds peripherally.

B. Qualitative Features of Follicular Growth and Atresia

The growth of a follicle is continuous from the primordial stage to the terminal stage, which ends with ovulation. It is ensured by the growth of the oocyte, follicular cell proliferation, and the enlargement of the antrum. Its growth can be stopped by degenerative changes called atresia. Biochemical and morphological changes are associated with atresia. Among the changes occuring during atresia, pycnosis within the granulosa and (or) distortion or fragmentation of the oocyte are considered as specific criteria. The theca layers seem to escape degeneration as the granulosa cells lose their ability to divide. Proliferation of theca cells (Nicosia, 1980) and steroidogenesis continue (Parshad and Guraya, 1983). However, the types and amounts of steroids produced may be changed. It is very difficult to estimate the rate of degeneration in primordial follicles, which is chiefly associated with the oocytes (Zuckerman, 1962), since the histologic structures of primordial follicles disappear very quickly.

Using a kinetic model of the follicular growth built for the mouse ovary, Faddy *et*

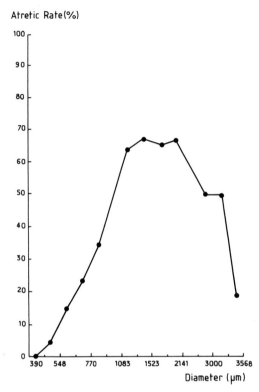

Figure 1 Rate of atresia of sheep follicles according to size in adult Timahdite breed. From Lahlou Kassi (1982).

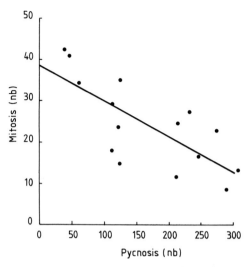

Figure 2 Relation between the number (nb) of pycnoses and the number of mitoses in rabbit follicles of diameter between 400 and 550 μm.

al. (1976) estimate that the daily rate of degeneration of the primordial growing follicles is equal to 70% of the rate of influx into the next stage of growth. The proportion of growing follicles that become atretic increases with increasing size of the follicle. Using the above model, Faddy shows that the rate of degeneration of follicles in the mouse per unit of time increases with the average size of follicles whatever age (Fig. 1).

The mechanisms of follicular atresia are far from being elucidated. Two questions, at least, can be asked. Is atresia a reversible process? A count of many follicles between 400 and 500 μm in diameter from a rabbit ovary showed that less than 5% did not have at least one pycnotic cell. Also, it is rare that a follicle containing more than 100 pycnotic cells in a single cross-section will not have mitoses or cells in the S phase labeled with tritiated thymidine. It is difficult to determine from the linear relation between the number of mitoses and the number of pycnotic cells a single point when atresia would be irreversible (Fig. 2).

A physiological role for atresia is yet to be determined. Is atresia a regulating mechanism for follicular growth? Does steroid secretion from the atretic follicles play a role in changes occuring during the estrous cycle?

C. Changes in Follicular Populations throughout Life

The types and numbers of follicles change during the life of the female, particulary during the prepubertal period. This

Table I

Average Number of Growing and Vesicular Follicles in Sheep, Cow, and Porcine Ovary at Various Times after Birth[a]

Sheep ovary (Jorio, 1987)

Age	birth	15(D)	1(M)	2(M)	3(M)	4(M)
Growing	513	476	500	209	111	114
Vesicular	139	90	211	183	81	46

Cow ovary (Erickson, 1966)

Age	1–14 (D)	22–50 (D)	60–79 (D)	90–96 (D)	113–150 (D)	180 (D)	240 (D)	12 (M)	19–24 (M)	4–6 (Y)	7–9 (Y)	10–14 (Y)	15–20 (Y)
Growing	53	93	204	137	227	220	145	248	233	211	154	152	72
Vesicular	5	10	26	32	29	35	22	27	27	28	25	20	12

Porcine ovary (Erickson, 1967)

Age	10(D)	30(D)	50(D)	70(D)	90(D)	110(D)	136(D)	300(D)
Growing	0	150	628	1228	2494	3587	2094	1237
Vesicular	0	0	0	2	44	73	378	215

[a]Days (D), months (M), years (Y).

evolution has been established in the sow (Erickson, 1966a), the cow (Erickson, 1966b), and the ewe (Jorio, 1987). In the ewe, for example, the total number of follicles with at least 2 layers of granulosa cells reaches a maximum between 15 days and 1 month of age (Table I). It decreases steadily afterward and at 4 months of age is practically reduced to four-fifths of what it was at 1 month of age.

In the ewe, cow, and sow, this precocious and large increase in the number of growing follicles depends upon the growth of a large number of small follicles shortly after birth. Simultaneously, an increase in size of the largest follicles can be observed in the ewe lamb during the 4 months following birth. In some ewe lambs 15 days old, the largest follicles may have reached the maximum size they will have in the adult ovary.

The extensive reduction in the number of growing follicles is caused by a wave of atresia that is maximum between 1 and 4 months after birth. This atresia occurs first in the oocytes and later in the granulosa. Thus, the size distribution of follicles is modified and their numbers are dramatically reduced before puberty. The evaluation of follicular populations, according to

the age, differs in individuals; some animals manifest a very precocious ovarian development, while others at the same age show an important delay in the growth of follicles. The reduction of the number of follicles observed in prepubertal atresia is also seen during seasonal anestrus in the mature ewe (Fig. 3).

During deep seasonal anestrus in the ewe, follicular growth slows down and the rate of atresia in the follicles is considerably increased. The number of growing follicles is drastically reduced by the action of these two phenomena.

The prepubertal period and the seasonal anestrus are associated with significant reductions in the secretion of LH and FSH. This reduction of gonadotropin secretion induces two situations: an increased atresia in preantral growing follicles, and a significant reduction in their number.

The population of growing follicles can be classified according to the size, the number of granulosa cells, the presence or absence of an antrum, or the number of cell layers. To these parameters may be added other functional criteria of growth, such as the number of hormonal receptors (Monniaux and de Reviers, 1989).

The distribution of follicles according to

size is very similar in the domestic species (Fig. 4). Generally, the number of follicles decreases as their size increases. It depends on the balance between their growth and their disappearance, which is associated with atresia in the classes of various sizes.

D. Follicular Dynamics

1. Estimated from Granulosa-Cell Proliferation

Important efforts to estimate the duration of growth of follicles in different spe-

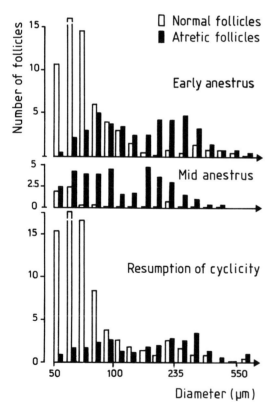

Figure 3 Follicular populations in the adult Ile-de-France sheep ovary at early anestrus, during mid-anestrus, and during the sexual season. From Oussaid (1983).

cies have been made during the last 20 years. The first method, developed by Pedersen (1970), estimated the parameters of the cell proliferation at each stage of growth. With this method, one can derive a statistical estimate of the average growth curve of one follicle. The second method enables one to estimate the inputs and the outputs of follicles per unit of time at each consecutive stage of growth (Fig. 5). With this method, one can also estimate follicular losses by atresia per unit of time at each stage of growth. This method (Faddy et al., 1976; Read et al., 1979) requires a large number of animals because the variability of the number of follicles among animals is very large. Using the first method, it was possible to estimate the average length of time required to grow one follicle in the mouse (Pedersen, 1969) and the rat (Hage et al., 1978) (Table II).

This method is difficult to apply to large domestic animals because it uses large quantities of radioactive substances. This difficulty was bypassed using so called stathmokinetc methods (Turnbull et al., 1977).

The proportion of granulosa cells blocked in metaphase after a single injection of colchicine, Colcemide, or vincristine increases with time. The rate of increase enables one to estimate the doubling times of granulosa cell numbers in each of the size classes of follicles and to estimate the times of transit of follicles in each size class (Fig. 6). One can also estimate the total duration of the follicular growth from the first stage to the preovulatory stage.

One can observe in each species studied by these methods that the growth of one follicle is, as a whole, not linear with respect to time.

The mitotic index of granulosa cells is roughly proportional to the relative increase

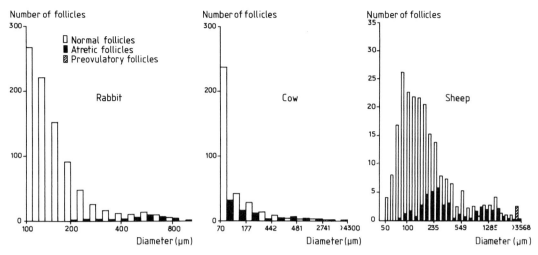

Figure 4 Distribution of follicles according to size in the rabbit cow and sheep ovary. From J. C. Mariana (1989, personal communication), Monniaux (1987), and Banoin (1988).

in the number of cells per unit of time. It increases, passes by a maximum, and decreases. In all species studied, this evolution is similar (Fig. 7). For example, the complete growth of one rabbit follicle from the one-layer stage with round cells (approximately 100 μm) to the preovulatory stage (800 μm and more) is about 100 days. Thus, if the growth of follicles smaller than 100 μm has the same characteristics with respect to time,

it takes about 100 days to completely renew all growing follicles. In the rabbit (Mariana *et al.,* 1989), extensive variability of the growth times exists for the various stages of growth of the follicles.

These results, obtained in different species, demonstrate that the growing follicle population constitutes a genuine reserve with one slow evolution with regard to the swifter phenomenon of maturation in the

Table II
Follicular Growth: Time Spent (T) in Each Class of Ovarian Follicle (C.T.) of Mouse, Rat, Sheep, Cow, and Rabbit

Immature mouse (28 days) (Pedersen, 1970)	C.T. (no. of cells) T (h)	20 — 100 ↘ 293 ↗	101 — 400 ↘ 79 ↗	401 — 600 ↘ 14 ↗	
Immature rat (28 days) (Hage, 1978)	C.T. (no. of cells) T (h)	20 — 100 ↘ 216 ↗	101 — 400 ↘ 132 ↗	401 — 600 ↘ 19 ↗	
Sheep (2 years) (Turnbull *et al.,* 1977)	C.T. (size μm) T (h)	153 — 326 ↘ 646 ↗	327 — 690 ↘ 229 ↗	691 — 2240 ↘ 84 ↗	2240 — 4930 ↘ 103 ↗
Heifer (?) (Lussier *et al.,* 1987)	C.T. (size, μm) T (h)	130 — 675 ↘ 650 ↗	675 — 3675 ↘ 182.7 ↗	3676 — 8560 ↘ 186.2 ↗	
Rabbit (20 weeks) (Mariana *et al.,* 1989)	C.T. (size, μm) T (days)	100 — 200 ↘ 83 ↗	201 — 400 ↘ 6 ↗	401 — 800 ↘ 5 ↗	801 — 950 ↘ 4 ↗

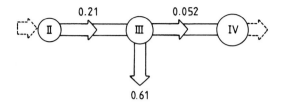

Figure 5 Explanation of the flow rates (Faddy, 1976). For example, if there are 100 follicles in one size group (III) at a particular time then follicles are moving to the next group (IV) at the rate of (0.052 × 100) per day or dying at the rate of (0.61 × 100) per day. For an individual follicle, 1/(flow rate) may be interpreted as an average time; for example, mean time to die in group III is 1/0.61 = 1.6 days, and mean time to move on to group IV is 1/0.052 = 19 days, with 0.079 = 0.052/(0.052 + 0.61) being the probability of surviving in group III and moving on alive.

last stages of growth. Indeed, through the labeling of follicles with India ink in the ewe (Smeaton and Robertson, 1971) and through the echography of the cow ovary during the estrous cycle (Pierson and Ginther, 1987a, 1987b, 1988), it has been shown that follicles larger than 4 mm are re-newed at least twice and even three times during the normal cycle (Savio *et al.*, 1988) (Fig. 8). Moreover, it is possible through luteolysis with prostaglandins to induce a succession of shortened cycles followed by one normal ovulation in the ewe without noticeably modifying the population of growing follicles with more than one layer of cells.

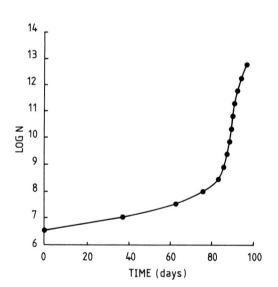

Figure 6 Average growth curve of the ovarian follicle of an adult 20-week-old rabbit. The smallest follicle taken into account has two layers of cells, and the largest one belongs to the group of large preovulatory follicles. The growth of one follicle is expressed as the logarithm of the total number of granulosa cells, log N.

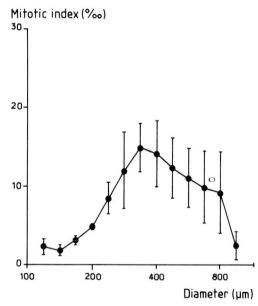

Figure 7 Pattern of mitotic index according to the size of follicles in the rabbit ovary. The smallest follicle (100 μm) taken into consideration has one to two layers of cells around the oocyte, and the largest one (800 μm) belongs to the group of preovulatory follicles.

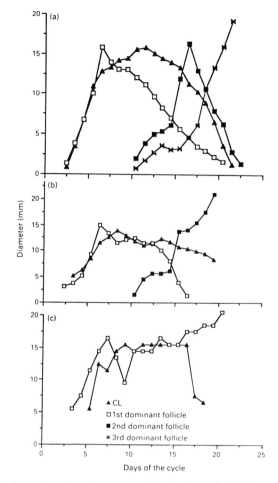

Figure 8 Growth pattern of the dominant follicle during the estrus cycle of heifers with (a) three dominant follicles, (b) two dominant follicles, and (c) only one dominant follicle. From Savio et al. (1988).

2. Estimation from Mathematical Models

The constant difficulty met for the interpretation of experiments designed to modify the population of follicles is that it is not possible to clearly separate the modification of the rate of growth of follicles from the rate of atresia. One mathematical model using a hypothesis about the renewal of growing follicles and the rate of atresia of follicles in the various stages of growth, general exponential laws (Read *et al.*, 1979; Read and Berry, 1988; Faddy *et al.*, 1976), could:

1. Fit in a satisfactory way the numbers of follicles according to size for the various ages of the female, and the changes of these numbers in various classes of size during the life of the female, with a reduced number of parameters.

2. Establish the average growth curve of one follicle at various ages, and estimate the rates of input and output of follicles with growth or atresia in the various classes by size by per unit of time during her life.

The parameters of the model, the various rates of input or output, and their variances are estimated for each time. The coefficients of variation of the values are often very large. Indeed, a large biological variation exists in animals, partly due to the nonsynchronous nature of their ovarian and follicular development (see Section I,C).

The mathematical models developed by Faddy *et al.* (1987) contain the hypothesis that the follicles behave independently, suggesting nevertheless that the follicular growth could be a "highly ordered process" during the life of the female.

E. Follicular Interactions within the Ovary

Some experimental results seem to confirm that follicles interact within the ovary. Extensive studies on large numbers of animals have shown that the numbers of follicles between classes of various sizes are often highly correlated. In the rat, cor-

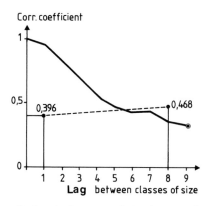

Figure 9 Level of autocorrelation between the numbers of normal adult rat follicles in various neighboring classes of size distant by lags of 1, 2, 3, . . . , 9 class interval values. The dotted line corresponds to the significant level; above it, the values of the coefficient are significant.

relations exist between the number of follicles in neighboring size classes (Fig. 9).

A significant relationship between mitotic indices exists in follicles belonging to different size classes in the same rabbit ovary. This means that under the same conditions, some does will show a high mitotic index in all sizes of the growing follicles while others will have a low index for growing follicles of all sizes. Estimates of parameters associated with the doubling time of granulosa-cell numbers are themselves correlated. These intraovarian interactions are expressed through various parameters of follicular growth and are certainly modified when the hormonal environment of the ovary is modified.

Hirshfield (1984) reported that the mitotic index of growing follicles was largely modified during the normal estrous cycle in the rat. It is maximum for all size classes during estrus, probably depending on the massive FSH release occurring at the end of

proestrus and during estrus (Schwartz, 1974).

F. Gonadotropin–Ovary Interactions

It has been known for a long time that gonadotropic hormones are essential for the growth of follicles and for the regulation of their endocrine function (for a review see Greenwald, 1974). Most results confirm that the action of gonadotropins on follicles depends on the type of follicles and the structure of the follicular population present in the ovary.

Two types of experiments have demonstrated the action of gonadotropic hormones on the follicular growth. The first type consisted of artificially inducing large and temporary variations of the levels of gonadotropins in the animal itself using hypophysectomy or hormonal supplements. The second type utilized the analysis of the follicular growth during the life, the sexual cycle, or during periods of the life in which important variations in the levels of gonadotropins occur normally at puberty, during anestrus, estrus, or pregnancy.

1. Effect of Gonadotropins on Follicular Population

We have shown (Section I,D) that the growth of small follicles with one layer of cells is very long. Hage *et al.* (1978) have shown that it takes at least 5 days in order for a small follicle with 20 cells in cross section to become a follicle containing 60 cells in cross section in the rat. This result indicates strongly that it will be difficult to measure significant modifications in the number of small follicles after a single stimulation with gonadotropins. Peters *et al.* (1973) could not highlight effects of repeated injections of exogenous chorionic gonadotropin on the number of small follicles that were already growing. A sharper analysis of the growth was made to describe modifications of the values of the parameters measured, growth of the oocytes and proliferation of follicular cells. The transcriptional activity of RNA polymerase in the nuclei of small oocytes in the 30-day-old mouse is not altered following the injection of 100 IU of eCG for 3 days before the polymerase activity was measured (Moore and Lintern-Moore, 1979). A slight increase of the labeling index of the smallest follicles (3b and 4 class of Peters classification) in the 28-day-old mouse is evident 11 h after one injection of 1.5 IU of FSH (Pedersen, 1969). The response is dependent on the age of the female; no modification of the labeling index of follicular cells is observed after one injection of FSH in rats 14 days old. Likewise, the labeling of cells is not modified after a single injection of 1 mg FSH-P in 30-day-old rabbits. However, one observes a large increase of the labeling index in small follicles with one or two layers of cells in 20-week-old rabbits 7 days after one ovary is removed and half of the remaining one is resected (Fig. 10). This increase is probably associated with FSH augmentation, which is observed when part or all the ovarian tissue is removed (Fleming *et al.*, 1984).

As mentioned in Section I,C, the population of preantral follicles is reduced in number and the size distribution is changed in the prepubertal and seasonally anestrous ewe. The pattern of secretion and the levels of gonadotropins during the periods before puberty and seasonal anestrus are well characterized (Swift and Crighton, 1980; Oussaid, 1983; Montgomery *et al.*, 1987), and it is possible to draw a parallel between the de-

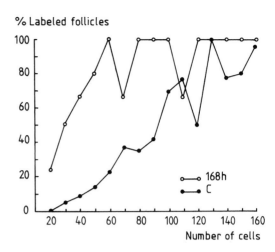

Figure 10 Proportion of labeled follicles on a fragment of ovary after removing approximately 75% of the ovarian tissue in the 20-week-old rabbit: control (c) and 168 h after resection.

cline of the follicle numbers through atresia and a slower rate of growth with the decrease in the levels of FSH. This parallel was made possible since the effects coincide with those following hypophysectomy and subsequent stimulation with exogenous hormones. These procedures significantly modified the population of growing follicles (Jones, 1961) in the mouse, rat, ewe, and cow. At the age of 15 days the sheep ovary responds to gonadotropic hormones (Fig. 11).

Dufour *et al.* (1979) showed that the population of follicles larger than 60 μm is significantly reduced following hypophysectomy. The reduction is approximately 40% in prenatal follicles, and follicles larger than 2 mm have disappeared as soon as 4 days after hypophysectomy.

Active immunization against LHRH as early as 3 days after birth entailed a significant reduction in the number of growing follicles. However, this reduction is less se-

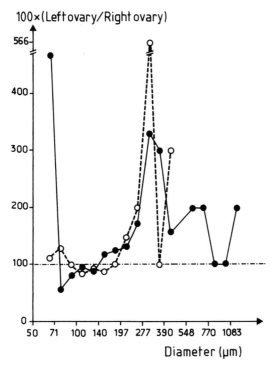

Figure 11 Relative increase of the number of follicles in a 15-day-old sheep as a consequence of an extraneous FSH stimulation (6 mg). ●–●, D′man; ○--○, Timahdite. The right ovary is collected as FSH is injected and serves as a control for the left ovary, which is collected 3 days after the right one. From Jorio (1987).

vere than that observed following hypophysectomy. Although the levels of LH after immunization can not be detected, the levels of circulating FSH are only reduced.

Practically, stimulation of the ovary in the domestic species with exogenous hormones first increases the number of eggs that are shed by an animal for embryo transfer, or, second, reactivates the ovary by loading the ovarian compartment with preovulatory follicles.

Variability of the ovulatory response fol-

lowing treatment with exogenous hormones continues to be a problem, and the structural relations associated with the dynamic growth of follicles (Section I,E) are associated with the wide range in response. Monniaux *et al.* (1984b) showed the type of response by heifers was associated with size distribution of the growing follicles in the ovary. They found that a positive significant correlation exists between the total number of growing follicles present in the ovary when treated with eCG and the number of ovulations and luteinzed follicles induced by the stimulation. Such distributions can be defined as stable or unstable, depending on whether or not they are modified following treatment with eCG.

Mariana (1979) has shown that it was necessary to double the dose of exogenous hormone (eCG) to obtain an equivalent number of ovulations in rats whose mothers had been treated with Busulfan on day 15 of pregnancy compared with the amount required in normal rats of the same age that had received 10 IU eCG. Treatment with Busulfan during pregnancy reduces the germ-cell population in the fetus and depresses the ovulatory response to exogenous gonadotropins in the offspring.

2. Gonadotropin-Induced Alterations in Cell Proliferation

The changes in the number of follicles of various class sizes following treatment with gonadotropic hormones depend on the changes in follicular growth, primarily on granulosa-cell proliferation. Therefore, the significant increase in the number of preantral follicles larger than 240 μm in the rat observed by Butcher and Kirkpatrick-Keller (1984) at estrus is associated with the significant increase in the mitotic index observed by Hirshfield (1984) in follicles of the same type in the rat during the same period of the cycle.

Pedersen (1969) has shown that as early as 21 days, the labeling index of granulosa cells in mouse follicles with less than three layers of cells increased and was maximum about 5 h after a single injection of FSH, then decreased again.

Hypophysectomy never completely suppresses the population of small follicles in the mouse, the ewe or the hamster and proliferative activity of cells is present in rats after hypophysectomy (Nakano *et al.*, 1975). Injection of FSH or eCG to hypophysectomized hamsters increases it still more (Chiras and Greenwald, 1978).

The mathematical analysis of the follicular growth in the hypophysectomized mouse (Faddy *et al.*, 1976) indicates that in follicles with less than two layers of cells, the degeneration through atresia is negligible, and the growth of follicles through these early stages is similar to that of follicles through the same stages of growth in the normal immature animal. Degeneration through atresia occurs in follicles with two or more layers of cells. The number of growing follicles is far less than the number of growing follicles that exist in the normal immature and the adult animal per unit of time. Degeneration of follicles with four or more layers of cells is major.

These results obtained in rodents indicate that the dependency of follicles on gonadotropins can be partially but never totally nonexistent. As the follicles become larger, their dependency becomes greater and gets transformed into a tight regulation. The number of preovulatory follicles is partly determined by the feedback exerted by hormones and peptides secreted by the follicles on the secretion of gonadotropins (Zeleznik and Kubik, 1986; Lacker, 1981).

II. Mechanisms Controlling Follicular Development

Differentiation of follicular cells is a multistep and multiregulated process. Many data have been obtained from rodent species, particularly the rat, and extensive reviews have been published recently (Dorrington *et al.*, 1983; Hsueh *et al.*, 1984; Erickson, 1986; Amsterdam and Rotmensch, 1987; Richards *et al.*, 1987) In this review, we will essentially refer to results obtained in domestic animals, that is, the bovine, ovine, and porcine species.

A. *Events of Granulosa and Thecal Cells Differentiation throughout Follicular Growth*

1. *Acquisition of Hormonal Receptors*

In domestic animals, the stage of follicular development when hormonal receptors appear is generally unknown, except for LH receptors in granulosa cells where data are available.

Receptors for FSH are present in granulosa cells of large antral follicles, but they probably appear at earlier stages. In the ovine and the bovine, injections of FSH stimulate the growth of preantral follicles (Monniaux *et al.*, 1984a; Jorio, 1987). In the cow, FSH binding to granulosa cells has been described in follicles larger than 5 mm in diameter; it does not change with follicular size and is only slightly reduced with advanced atresia (Merz *et al.*, 1981; Grimes *et* al., 1987; Spicer et al., 1987). In the ewe, when follicles larger than 1 mm in diameter are studied, the number of FSH receptors does not change with follicular size but decreases in late atresia (Carson et al., 1979). So, in the bovine and ovine, there is no increase in receptivity to FSH in granulosa cells during terminal follicular growth, in contrast with results obtained in rats (Monniaux and de Reviers, 1989). In the preovulatory follicle of the cow, the number of FSH receptors decreases after the LH peak (Staigmiller et al., 1982; Ireland and Roche, 1983b).

Receptors for LH have been reported in thecal cells of all large antral follicles, healthy or atretic, in the cow and ewe. The theca interna seems to be a major target tissue for LH action during the follicular phase of the estrous cycle in the ewe, since there is a marked increase in LH receptor capacity in thecal cells at this time (McNatty et al., 1986). No change in hCG binding associated with follicular size was detected in thecal cells of large follicles in the ewe and cow. However, the effect of atresia is not as clear; in the cow, hCG binding capacity is independent of follicular health (Henderson et al., 1984; McNatty et al., 1985a) but in the ewe, it decreases in atretic follicles (McNatty et al., 1986).

The acquisition of LH receptors in granulosa cells is one of the major changes reported during terminal follicular growth and differentiation. LH receptors appear in the granulosa of follicles of a definite size: for example, 3 mm diameter for ovine follicles (Carson et al., 1979) and 5 mm for porcine follicles (Lee, 1976) are critical sizes for the appearance of LH receptors. The number of LH receptors increases with follicular diameter, in correlation with estradiol production by the follicle (ewe, Carson et al.,

1979; Webb and England, 1982; sow, Lindsey and Channing, 1979; cow, Gosling et al., 1978). However, the distribution of LH receptors within a follicle is heterogeneous; granulosa cells of medium and large follicles bind 10- to 15-fold more iodinated hCG than an equivalent number of cumulus cells in porcine follicles (Channing et al., 1981). The acquisition of LH receptors in granulosa cells is indicative of the maturity of the follicle, that is, its ability to ovulate in response to the LH preovulatory discharge. During the follicular phase in the ewe, only the largest follicles bind hCG. The number of these follicles is representative of the ovulation rate, so they have been classified as "ovulatory" follicles (Webb and England, 1982). However, follicles with LH receptors on granulosa cells are found in the ovaries throughout the luteal phase, and their number is also representative of the ovulation rate (cow, Merz et al., 1981; ewe, England et al., 1981). The presence of LH receptors in granulosa is not under the control of cyclic hormonal fluctuations.

Few data are available concerning the acquisition of other hormonal receptors in domestic animals. The absence of specific receptors of LHRH has been reported in ovine, bovine, and porcine ovaries (Brown and Reeves, 1983). Prolactin receptors have been reported on granulosa cells of the sow (Rolland and Hammond, 1975), but no binding with prolactin has been found on granulosa and theca cells from the cow (Bevers et al., 1988). Epidermal growth factor (EGF) receptors (Vlodavsky et al., 1978) have been reported in bovine granulosa cells, insulin (Otani et al., 1985), and insulin-like growth factor I (IGF-I) (Maruo et al., 1988) receptors have been described in porcine granulosa cells. Recently, IGF-I receptors have been detected and quantified on granulosa cells of ovine follicles in our laboratory (Monget *et al.*, manucript in preparation). This list is not exhaustive, as more and more substances are found to have specific effects on granulosa cells.

2. Development of Steroidogenesis and Peptide Production

Analysis of differences in composition between follicular fluid and plasma is indicative of biochemical synthesis in follicular cells. Different types of products secreted by follicular cells have been identified in follicular fluid: steroids, peptides, proteoglycans (constituted of a protein core covalently bound to a mucopolysaccharide, which is heparin sulfate or chondroitin sulfate), prostaglandins, etc. In this review, we focus our attention on changes in steroidogenesis and peptidic synthesis during follicular growth.

a. Steroidogenesis Small antral follicles (1–3 mm in diameter in the ewe, <8 mm in diameter in the cow) are characterized by high levels of testosterone and low levels of progesterone and estradiol in follicular fluid (ewe, Carson *et al.*, 1981; cow, Kruip and Dieleman, 1985). When follicles enlarge, levels of estradiol (Fig. 12) and, to a lesser extent, progesterone increase dramatically, whereas levels of testosterone fall (ewe, McNatty *et al.*, 1981; Carson *et al.*, 1981; cow, Ireland and Roche, 1982; Kruip and Dieleman, 1985; Wise *et al.*, 1986). Atretic follicles are characterized by low levels of estradiol in all species; however, testosterone levels in atretic follicles are high in ewes, (Carson *et al.*, 1981) and low in the cow (Ireland and Roche, 1982), and levels of progesterone increase with atresia in cow follicles (Kruip and Dieleman, 1985).

Changes in steroid levels in follicular

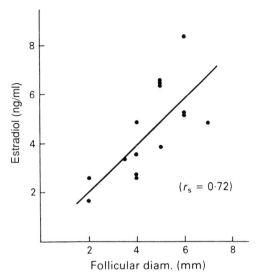

Figure 12 Relationship between estradiol-17β concentrations in follicular fluid and follicular diameter in normal sheep follicles at D13 of the cycle. From Monniaux (1987).

fluid can be correlated with changes in steroidogenic enzyme activities in thecal and granulosa cells. Granulosa cells of preantral and small antral follicles have undetectable steroidogenic enzyme activity. Granulosa cells of large antral follicles are able to synthesize progestagen compound from cholesterol substrate since they have HMG-CoA reductase, cholesterol side-chain cleavage, cytochrome (cyt) *P*-450, and ferredoxin reductase (or adrenodoxin reductase) activity (sow, Bjersing and Carstensen, 1967; Tuckey and Stevenson, 1986; Tuckey *et al.*, 1988; cow, Rodgers *et al.*, 1986, 1987a,b). Moreover, they have Δ^5-3β-hydroxysteroid dehydrogenase activity so that progesterone can be synthesized with pregnenolone as substrate (Bjersing and Carstensen, 1967). These activities are also expressed in thecal cells of antral follicles of various sizes (sow, Miyamoto *et al.*, 1985; cow, Rodgers *et al.*,

1986). Moreover, thecal cells are responsible for the synthesis of androgens in the follicle as they have 17α-hydroxylase and $C^{17,20}$ desmolase activities that are not expressed in granulosa cells. Isolated granulosa cells are unable to synthesize androgens (cow, Lacrois *et al.*, 1974; Rodgers *et al.*, 1986; ewe, Moor, 1977; sow, Tsang *et al.*, 1982, 1985; Evans *et al.*, 1981). Granulosa cells of large antral follicles express an aromatase activity in all the domestic species. Cooperation between thecal and granulosa cells for biosynthesis of estrogens by the follicles is a widespread concept; the thecal cells produce the androgens, which are used as precursors for aromatization by granulosa cells. This classic model is not totally valid in the sow and the doe (goat), where thecal cells also have aromatase activity (sow, Stoklosowa, *et al.*, 1982; Tsang and Taheri, 1987; doe, Band *et al.*, 1987). In these species, secretion of estrogens can be performed by isolated thecal cells. Aromatase activity is not detected in small antral follicles (<1 mm diameter in the ewe, <5 mm diameter in the cow); it increases with follicular size (cow, McNatty *et al.*, 1984a; ewe, Webb and England, 1982; Tsonis *et al.*, 1984a; Monniaux, 1987). In the preovulatory follicle, the aromatase activity of granulosa cells is maximal, but more than 90% of the estradiol produced is directly secreted in the bloodstream, without accumulation in the antrum (McNatty *et al.*, 1981a). After the LH preovulatory discharge, estradiol levels decrease in follicular fluid. In the sow, this decrease is the result of a rapid loss of aromatase activity in the granulosa and thecal cells while thecal production of androgens is maintained (Meinecke *et al.*, 1987). However, in the cow the loss of androgen secretion by thecal cells occurs earlier after the LH peak, and estrogen levels decrease in the follicle closely following the decrease of androgen precursors; later, aromatase activity decreases and disappears in granulosa cells while progesterone secretion increases (Dieleman *et al.*, 1983a; Dieleman and Blankenstein, 1984).

b. Peptide Production Various peptides of follicular origin can be detected in follicular fluid. Concentrations of these peptides in follicular fluid change during follicular growth, indicating that their production could be influenced by the status of differentiation of follicular cells (see Table III). Some of them have been biochemically characterized but their exact function is not always known. These peptides can be classified into four types, according to their possible function:

1. Endocrine function—inhibin, which specifically regulates FSH secretion by the hypophysis.

2. Paracrine or autocrine modulation of follicular cell differentiation during follicular growth—the growth factors IGF-I, EGF, and FGF (fibroblast growth factor), which have specific effects on multiplication and steroidogenesis by granulosa cells; FSH-RBI and LH-RBI (RBI, receptor-binding inhibitor), which prevent respectively binding of FSH and LH on their receptors; FRP (follicular regulatory protein), which has inhibitory effects on granulsa cells differentiation; and GnRH-like (GnRH, gonadotropin-releasing hormone) peptide with hypothetical actions in domestic animals.

3. Paracrine control of oocyte maturation—OMI (oocyte meiosis inhibitor), which prevents meiotic maturation of the oocyte; and AMH (anti-Müllerian hormone), whose action on meiotic maturation is still controversial.

4. Paracrine control of ovulation—oxytocin, relaxin, and plasminogen activator,

Table III
Peptides Secreted by Follicular Cells and Identified in Follicular Fluid of Domestic Animals

Peptide	Animal	Change of levels in follicular fluid during follicular growth	Origin of secretion	Reference
Inhibin (or folliculostatin)	Cow	Decrease	Granulosa cells (secretion increases with follicular size)	Henderson and Franchimont (1981)
	Sow	Decrease		Suter et al. (1986)
	Ewe	Tendency to increase		Tsonis et al. (1983)
IGF-I (insulin-like growth factor I)	Cow	Increase	Granulosa cells	Spicer et al. (1988)
	Sow	Increase		Hammond et al. (1988)
EGF-like (epidermal growth factor)	Cow	Decrease		Hsu et al. (1987)
FGF (fibroblast growth factor)	Cow		Granulosa cells	Neufeld et al. (1987)
FRP (follicular regulatory protein)	Sow	Decrease	Granulosa cells	Tonetta et al. (1988)
FSH-RBI (FSH receptor binding inhibitor)	Cow	Increase		Darga and Reichert (1978)
	Sow			Sluss and Reichert (1984)
	Ewe			Krishnan et al. (1983)
LH-RBI (LH receptor binding inhibitor)	Sow		Corpus luteum	Channing (1979)
	Ewe			Krishnan et al. (1983)
GnRH-like (gonadotropin-releasing hormone)	Cow		Granulosa cells	Aten et al. (1987)
OMI (oocyte meiosis inhibitor)	Sow	Decrease	Granulosa cells	Tsafriri and Channing (1975) Van de Wiel et al. (1983)
AMH (anti-Müllerian hormone)	Cow	Decrease	Granulosa cells	Vigier et al. (1984)
	Ewe			Bézard et al. (1988)
Oxytocin	Cow	Increase in preovulatory follicles	Granulosa cells	Wise et al. (1986)
	Goat			Walters et al. (1984)
Relaxin	Sow	Increase in preovulatory follicles	Theca and granulosa cells	Bryant-Greenwood et al. (1980) Evans et al. (1983)
Plasminogen activator	Sow	Increase in preovulatory follicles	Theca and granulosa cells	Kokolis et al. (1987)

which can act in follicular rupture at ovulation.

The precise actions of many of these peptides are still controversial and their relative importance in follicular maturation is not well understood.

3. Biochemical Events of Atresia

Atresia is the normal ultimate fate of most follicles, since only 0.1% of the follicles that begin to grow will differentiate to the preovulatory stage. Atresia is a complex and poorly understood process. It takes place in all stages of follicular growth, and the underlying mechanisms are probably different between follicles of different sizes. In preantral follicles, abnormal morphologic features are first detected in the oocytes, whereas in antral follicles the early histological changes are the appearance of pycnosis in granulosa cells concomitantly with a decrease in granulosa cell proliferation. Moreover, differences between species exist in the development of atresia; in antral follicles, degeneration of the cumulus is very late in the ewe and early in the pig (Cran *et al.*, 1983). Many factors involved in the atretic process are unknown for preantral follicles and still hypothetical for antral ones. One major difficulty in investigating atresia is that it is usually recognized only in retrospect, that is, after the follicle has undergone distinct morphological changes. Another difficulty is associated with the different criteria used to define atresia, histological, morphological (Moor *et al.*, 1978; Kruip and Dieleman, 1982), morphometric (McNatty, 1982), or biochemical.

Generally, antral atretic follicles are characterized by:

1. Loss of binding sites for gonadotropins by follicular cells. Receptor loss in granulosa cells is not an early event of atresia (cow, Grimes *et al.*, 1987; sow, Maxson *et al.*, 1985; ewe, Carson *et al.*, 1979). Loss of LH receptors by thecal cells (cow, Henderson *et al.*, 1984) or by granulosa cells in preovulatory follicles (cow, Ireland and Roche, 1982) is only detected in advanced atresia. Thus, the loss of gonadotropin binding sites would be a consequence rather than a cause of atresia. However, a loss of cellular sensitivity to gonadotropins is detected before the loss of binding sites in bovine thecal cells (McNatty *et al.*, 1985a).

2. Changes in steroidogenesis. A marked decrease of estradiol concentration in follicular fluid of atretic follicles has been reported in all domestic species. This decrease can be associated with the loss of aromatase activity in granulosa cells, which is always an early event in atresia (cow, McNatty *et al.*, 1984a; Henderson *et al.*, 1987a; sow, Max-

son *et al.*, 1985; ewe, Monniaux, 1987). Testosterone in follicular fluid of atretic follicles is elevated in ewe (Carson *et al.*, 1981), remains unchanged in sow (Maxson *et al.*, 1985), and decreases in the cow (Kruip and Dieleman, 1985). Results can be explained by the early loss of androgen secretion by thecal cells during atresia in the cow (McNatty *et al.*, 1985a) and its maintenance in the other species. In the cow, the high levels of progesterone found in the follicular fluid of atretic follicles (Kruip and Dieleman, 1985; Grimes *et al.*, 1987; Spicer *et al.*, 1987) are the consequence of the active secretion of progesterone by granulosa cells after they lose their aromatase activity (Henderson *et al.*, 1987a).

3. Changes in other biochemical syntheses: Acid and alkaline phosphatase activity and lactate dehydrogenase activity increase in atretic follicles (cow, Wise, 1987). The content of chondroitin sulfate and heparin sulfate is also increased (cow, Bushmeyer *et al.*, 1985a). In the mare, early atresia is associated with an increase in the content of prostaglandin F (Kenney *et al.*, 1979). However, some peptidic syntheses are decreased; the decrease in inhibin secretion is correlated with the lower estradiol production in atretic follicles from the ewe (Tsonis *et al.*, 1983).

The exact timing of the degenerative changes during the atretic process is not known. Atresia of antral follicles is probably caused by an inadequate hormonal environment within the follicles. A lower supply of gonadotropins can initiate atresia. In the ewe, a reduction in the size of the thecal capillary bed in atretic follicles has been described (Hay *et al.*, 1976). Moreover, during atresia, many thecal capillaries become blocked with cellular debris (O'Shea *et al.*,

1978). In the ewe, there is no reduction of thecal blood vessels adjacent to the cumulus, which remains viable during atresia, whereas in the sow, the cumulus degenerates and the underlying thecal vasculature is greatly reduced (Cran *et al.*, 1983). Whether the changes in microvasculature initiate atresia or merely accompany it remains unanswered for the present.

4. Relationships between Growth and Differentiation

During follicular development, granulosa cells replicate and the rate of proliferation changes with the size of the follicle, as reported in Section I. In all the species, the mitotic index of granulosa cells is maximal in small antral follicles, then decreases as the follicle enlarges. In the cow, as the follicle increases in size, a progressive decline in the ability of granulosa cells to clone in agar oc-

curs, and the proliferative potential of these cells as measured by colony size also decreases (Bartholomeusz *et al.*, 1988).

An inverse relationship between growth and differentiation has been described for granulosa cells:

In vivo, when the size of the follicle increases, aromatase activity of granulosa cells increases, whereas their proliferative activity decreases (ewe, see Fig. 13).

In vitro, growth and differentiation of granulosa cells are affected by the plating density (sow, May and Schomberg, 1984). In the ewe, when progesterone secretion by granulosa cells *in vitro* is measured, the cells become sensitive to FSH only when they reach confluency in culture wells (see Fig. 14).

As the antral follicle grows, differentiating cells would have a progressively re-

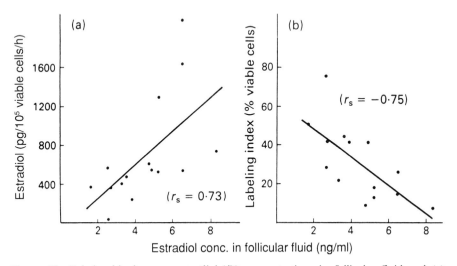

Figure 13 Relationship between estradiol-17β concentrations in follicular fluid and (a) aromatase activity of granulosa cells of sheep follicles or (b) thymidine labeling index of granulosa cells of the same follicles. Aliquo of granulosa cells were incubated with testosterone (10^{-7} *M*) for 1 h for aromatase activity determination, or with [³H]thymidine (5 μCi/ml) for 2 h for labeling index determination. From Monniaux (1987).

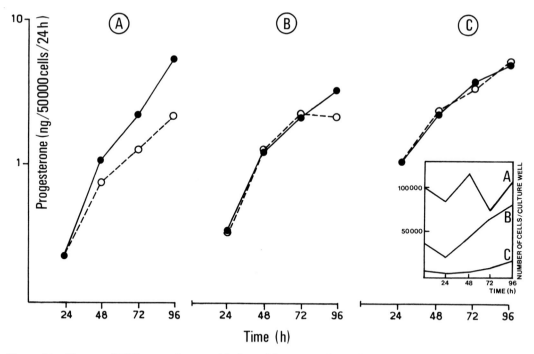

Figure 14 Changes of FSH responsiveness with time of sheep granulosa cells cultured at different seeding densities. Granulosa cells were isolated from follicles <3 mm diameter and inoculated at three densities, (A) 10^5 cells/well, (B) 4×10^4 cells/well, and (C) 7.5×10^2 cells/well, in culture wells (area = 32 mm^2/well). Cells were cultured for 4 days in F12 medium with fetal ovine serum (2%) and insulin (10 μg/ml) and with (●—●) or without (o- -o) FSH (100 ng/ml). Medium was changed every day, and progesterone determination in culture media was performed by radioimmunoassay. In inset, the changes of the number of cells per well with time is indicated. From D. Monniaux and Pisselet (unpublished data).

stricted proliferative activity; in large follicles, more and more cells leave the cell cycle of replication and enter into a differentiated state. An illustration of the changes of some of the activities of granulosa cells associated with follicular size is shown in Fig. 15.

B. Endocrine Control of Follicular Development

1. Role of Follicular Vascularization

Endocrine control of follicular development is modulated by the blood supply reaching the follicles. Important changes in follicular vascularization have been described during follicular growth. The smaller, preantral follicles have no special vascular supply of their own, lying among vessels of the stroma. A discrete follicular capillary bed, confined to the thecal layers, develops in association with the formation of the theca. In antral follicles, the thecal capillaries consist of two concentric networks of vessels in the theca interna and externa, respectively. The "inner wreath" is close to the basal lamina of the granulosa, but capillaries do not enter the granulosa of unruptured follicles. Vas-

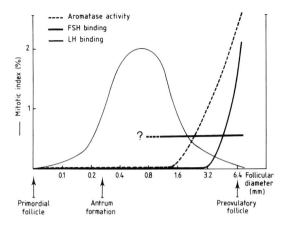

Figure 15 Schematic changes of some granulosa-cell activities during follicular development in sheep follicles.

cularization of the preovulatory follicle is particularly well developed (ewe, Moor *et al.*, 1975) and the capillary blood flow in the nonluteal tissue is high during estrus, particularly in the "ovulatory" ovary (ewe, Brown *et al.*, 1974, 1988). The marked dilatation of arterioles and enhanced permeability of capillaries and venules within the thecal vascular bed of the preovulatory follicle are the consequences of the action of the high levels of estradiol produced by this follicle (Brown and Mattner, 1984) and LH released by the hypophysis before ovulation (Niswender *et al.*, 1976). The supply of ovarian blood to the wall of the preovulatory follicle begins to decline from 12 to 16 h after the LH surge and remains low until ovulation (ewe, Murdoch *et al.*, 1983). It is tempting to postulate that the development of vascularization is an important event of the selection of the preovulatory follicle by supplying a larger amount of hormones and nutrients. A marked reduction of the microvasculature of atretic follicles has been reported (ewe, Hay *et al.*, 1976). The complete

mechanisms of angiogenesis in the ovary (Findlay, 1986) are not known, but granulosa and theca of follicles are sources of angiogenic activity (sow, Makris *et al.*, 1984).

When the hormonal control of follicular cell differentiation is studied, it must account for the nature and kinetic exchanges existing between plasma and follicular fluid. The granulosa and oocyte of the antral follicle do not have direct access to a blood supply but are bathed in a fluid that contains most serum proteins, variable amounts of gonadotropins, and high concentrations of steriods (McNatty, 1978). There is no physical barrier to the transfer of low molecular weight substances (PM < 10) from the thecal blood capillaries into the follicular antrum (ewe, Cran *et al.*, 1976; sow, cow, Szöllösi *et al.*, 1978). However, the composition of follicular fluid does differ in many important ways from that of the blood plasma: for example, in the ewe, there is a delay of approximately 3 h between the time of the LH preovulatory surge and the time when the concentrations of LH are maximal in follicular fluid. Moreover, levels are always much lower in follicular fluid (McNatty *et al.*, 1981a). Even during a preovulatory surge, the oocyte–cumulus cell complex is likely to be exposed to a much more gradual and less substantial increase in LH concentration compared with that experienced by the thecal and the granulosa cells closest to the basal lamina. In the same way, the small pulsatile discharges of LH, which occur throughout most of the estrous cycle (ewes, Baird *et al.*, 1976; Baird, 1978), are unlikely to be observed within an antral follicle.

Products of secretion of follicular cells can accumulate in the follicular fluid and/ or enter the bloodstream. Even if estradiol levels in follicular fluid of large normal follicles are particularly high, more than 90% of the estradiol produced by these follicles enters the bloodstream without first accumulating within the antrum (McNatty *et al.*, 1981a). A countercurrent transfer of ovarian steroid hormones from the ovarian vein to the ovarian artery has been found in the ewe (Walsh *et al.*, 1979), the cow (Kotwica *et al.*, 1981) and the sow (Krzymowski *et al.*, 1982). The existence of a portal subovarian vascular system would permit a local concentration of steroids in the ovarian blood, possibly acting in a paracrine way on follicular growth and differentiation. This mechanism may have an important role in regulating terminal follicular growth during the follicular phase of the cycle, when large amounts of steroids and peptides are secreted by large follicles into the bloodstream.

2. Endocrine Control of Acquisition of Hormone Receptors

As discussed above (see Section II,A,1), fluctuations in the number of FSH and LH receptors are probably directly related to the stage of development of the follicle and not under the control of cyclic hormonal fluctuations; for example, during the follicular phase in the ewe, when plasmatic LH levels increase, LH receptors in granulosa and thecal cells increase in preovulatory follicles but not in nonovulatory follicles (Webb and England, 1982). Few data are available concerning the endocrine control of acquisition of FSH and LH receptors in domestic animals. Nevertheless, in domestic animals, as in the rat, the induction of new binding sites is essentially controlled by FSH. *In vivo*, treatment with exogenous FSH increases FSH and LH receptor concentrations in follicles of chronic cystic-ovarian diseased dairy cows (Brown *et al.*, 1986). *In vitro*, FSH increases binding of FSH to porcine granulosa cells by unmasking of cryptic FSH-binding sites (Ford and La Barbera, 1988). *In vitro*, FSH induces LH receptors in porcine granulosa cells of small follicles (Channing, 1975; May *et al.*, 1980; La Barbera and Ryan, 1981a) by *de novo* synthesis (Loeken and Channing, 1985). Insulin enhances LH receptor induction by FSH on porcine granulosa cells *in vitro*, but only high concentrations are effective, suggesting that its action is not specific for insulin but that it probably acts through IGF-I receptors (Maruo *et al.*, 1988).

3. Endocrine Control of Steroidogenesis

Extensive studies performed in rodents and domestic animals have pointed out the determinant role of gonadotropins (FSH, LH) in follicular differentiation. The development of cell culture techniques has made it possible to determine how they can act on follicular cells and which substances can modulate their action. Nevertheless, if experiments *in vitro* are necessary, they are generally not sufficient to explain the mechanisms of control of follicular development.

a. Actions of Gonadotropins on Follicular Cells *in Vitro* FSH stimulates progesterone production by granulosa cells from domestic animals after a delay of approximately 48 h in culture (cow, Savion *et al.*, 1981; Skinner and Osteen, 1988; sow, Thanki and Channing, 1978; La Barbera and Ryan, 1981a; Lino *et al.*, 1985). The sensitivity of cells to FSH and the maximal level of progesterone secretion increase with the size of the follicle (ewes, D. Monniaux and Pisselet, unpublished results). In immature porcine granulosa cells, FSH, acting with estradiol, stimulates cholesterol side-chain cleavage activity (Toaff *et al.*, 1983) and also regulates the *de novo* synthesis of cholesterol (Baranao and Hammond, 1984). This increase in progesterone secretion by granulosa cells following stimulation by FSH or estradiol has also been related to an enhanced responsiveness to low-density lipoprotein (LDL) (Veldhuis *et al.*, 1984). Insulin also stimulates progesterone production by granulosa cells, probably acting through IGF-I receptors (cow, Skinner and Osteen, 1988; sow, Lino *et al.*, 1985).

To our knowledge, few papers have reported a specific effect of FSH on the induction of aromatase activity in granulosa cells of domestic animals. Chan and Tan (1986, 1988) have reported the induction of aromatase activity in granulosa cells from medium-sized follicles isolated from prepubertal gilts following treatment with FSH (1 μg/ml). The response began 24–48 h following treatment. In this model, steroids importantly modulate this induction with a stimulating effect of testosterone and DES and an inhibiting effect of 5α-DHT, pregnenolone, and progesterone. Compared to the abundant literature published on studies in the rat, the lack of results concerning the control of aromatase activity in bovine and ovine species could be indicative of a more complex mechanism of regulation.

LH stimulates progesterone production by granulosa cells from large porcine follicles (Veldhuis *et al.*, 1982), and this effect is amplified when cells are cultured with HDL and LDL (Veldhuis *et al.*, 1984; Rajkumar *et al.*, 1988). When granulosa cells are isolated from small follicles of immature gilts and cultured in a defined medium containing insulin, LH stimulates progesterone secretion and cholesterol side-chain cleavage activity

(Lino *et al.*, 1988), probably because insulin has induced LH receptors on them (May and Schomberg, 1984b).

LH acts *in vitro* on theca interna cells to stimulate androgen secretion (cow, McNatty *et al.*, 1984a). Thecal cells isolated from atretic follicles have lost their ability to secrete androstenedione following stimulation by LH, but in early atretic follicles, the cAMP response to LH is still maintained (McNatty *et al.*, 1985a).

b. Actions of Gonadotropins on Follicular Steroidogenesis and Development *in Vivo*: Present Hypothesis The actions of gonadotropins on follicular steroidogenesis and development *in vivo* have been studied by three methods in domestic animals:

By injecting gonadotropins: Generally, the products injected are not pure FSH or pure LH but eCG or pituitary extracts, which are known to have both activities. FSH action cannot be dissociated from LH action in these experiments.

By studying the temporal relationships between plasmatic levels of gonadotropins and steroids: Gonadotropins are generally measured by radioimmunoassay (RIA), and there may be some disagreement between the immunoreactivity and bioactivity of these hormones.

By studying the quantitative relationships between gonadotropin and steroid levels in follicular fluid: However, in a recent paper, Schneyer *et al.* (1988) found immunoreactive FSH detected in porcine follicular fluid was biochemically distinguishable from pituitary FSH. Despite these limits in methodology related to experiments performed *in vivo*, some hypotheses can be made concerning the actions of FSH and LH *in vivo*.

FSH increases the aromatase activity of ovine granulosa cells *in vivo,*. When ewes were treated for 24 h by injections of purified ovine FSH, aromatase activity increased in granulosa cells isolated from large (diameter > 5 mm) follicles (McNatty *et al.,* 1985b). In normal cycling ewes, irrespective of follicle size, the highest concentrations of FSH were present in follicles with high levels of estrogen, whereas the lowest levels were found in those follicles with low levels of estrogen (McNatty *et al.,* 1981b). It can be hypothesized that highly vascularized follicles are submitted to a larger amount of FSH, which accumulates in antral fluid, allowing an increase in estrogen production, by inducing and/or maintaining a high level of aromatase activity in the granulosa cells. A lower amount of FSH could initiate atresia, either because the vascularization of the follicle is not sufficiently developed, or the plasmatic levels of the hormone are too low. This mechanism might determine the selection of the preovulatory follicles in the middle of the follicular phase of the cycle, when plasmatic FSH levels decrease in response to the negative feedback of estrogens and inhibin on the hypophysis (Baird *et al.,* 1981; Miller *et al.,* 1981; Henderson *et al.,* 1986). eCG protects or rescues follicles from atresia *in vitro* (ewes, Hay *et al.,* 1979) and *in vivo* (cow, Monniaux *et al.,* 1984). Such an action is still hypothetical for FSH, but it was noted that granulosa cells, which are the more sensitive to FSH when cAMP is produced, also express a higher aromatase activity (ewes, Henderson *et al.,* 1985, 1987b).

Many *in vivo* studies have reported on the role of LH in the development of the preovulatory follicles during the follicular phase of the estrous cycle. It has been reported previously (see Section II,A,1) that the number of LH receptors increases in thecal and granulosa cells when the preovulatory follicle enlarges during the follicular phase and secretes increasing amounts of estradiol. In the preovulatory follicle, a positive relationship exists between estradiol levels in follicular fluid and the number of LH receptors in the thecal and granulosa cells (cow, Staigmiller *et al.,* 1982; ewe, Webb and England, 1982). At the end of the luteal phase, when progesterone plasmatic levels fall, a twofold increase in the frequency of episodic pulses of LH begins, each of which stimulates an increase in the secretion of estradiol (ewe, Baird *et al.,* 1981). At this time, plasmatic levels of FSH decrease and probably do not stimulate terminal follicular steroidogenesis and growth. Moreover, LH, but not FSH, stimulates testosterone and estradiol production in the autotransplanted ovary of the ewe (McCracken *et al.,* 1969).

The current hypothesis concerning the role of LH during the follicular phase is as follows: The increased frequency of episodic pulses of LH probably stimulates the secretion of androgen precursor by thecal cells, and it is the most important factor in determining the preovulatory increase in estradiol production (Baird, 1978). Nevertheless, a strict pulsatile delivery of LH may not be an absolute requirement, and follicular maturation can be achieved with widely differing patterns of LH delivery to the ovary during the preovulatory period (ewe, McNatty *et al.,* 1981c).

The effects of the preovulatory discharge of LH are particularly dramatic on follicular steroidogenesis. After a short stimulation of progesterone and testosterone secretion by the theca, the thecal production of androgens is inhibited and estrogen levels decrease in the ovulatory follicle in the cow; later, aromatase activity is inhibited while progesterone secretion increases (Fortune and Hansel, 1979; Dieleman and Blan-kenstein, 1984; Dieleman *et al.*, 1983a). In the ewe and the sow, an early loss of aroma-tase activity in granulosa cells is detected be-fore a decrease of androgen production by thecal cells (ewe, Baird *et al.*, 1981; sow, Meinecke *et al.*, 1987).

In conclusion, large follicles are depen-dent of both gonadotropins for develop-ment and steroidogenesis. Schematically, FSH would have a differentiating function in granulosa cells, by increasing steroido-genic enzyme activities and inducing LH re-ceptors. LH would stimulate aromatizable androgen production by thecal cells and sustain terminal preovulatory follicle growth. However, these actions are widely modulated by the paracine and autocrine actions of steroids and peptides acting within and also probably between follicles (see Section II,C).

4. Endocrine Control of Peptide Synthesis

The endocrine control of peptide synthe-sis has been studied in very recent investiga-tions, and only sparse data are available in domestic animals. Most results come from studies performed *in vitro,* and the state of differentiation of the cultured cells has not always been taken into account. This last point may be very important; for example, steroids plus FSH increase follicle regula-tory protein (FRP) secretion from granulosa cells of small porcine follicles while inhibit-ing FRP secretion from large follicles (To-netta *et al.*, 1988). Nevertheless, it has been reported that FSH and LH fail to stimulate inhibin production by granulosa cells from cows *in vitro*, but the secretion of inhibin would be more likely controlled by steroids (Henderson and Franchimont, 1981). FSH (Anderson *et al.*, 1985) and prolactin (Chan-ning and Evans, 1982) would stimulate OMI

production by porcine granulosa cells, but androgens would inhibit it. hCG, but not FSH, stimulates *in vitro* plasminogen activa-tor secretion by granulosa cells isolated from medium-sized porcine follicles (Shaw *et al.*, 1985). Moreover, relaxin production by theca can be stimulated by eCG and hCG *in vivo*. LH, but not FSH, stimulates relaxin production *in vitro* in preovulatory porcine follicles (Evans *et al.*, 1983). FSH and LH act differently in large follicles; FSH would stimulate the secretion of an OMI-like mate-rial that acts to maintain the cumulus–oocyte complex in an immature meiotically arrested state, whereas LH would stimulate the secretion of peptides involved in follicu-lar rupture at ovulation.

According to the important role of IGF-I in granulosa cell replication and differen-tiation (see Section II,C,3), the control of IGF-I synthesis in the follicles is of particu-lar interest. *In vitro*, FSH, LH, and estradiol stimulate immunoreactive IGF-I production by granulosa cells isolated from small folli-cles in immature gilts (Hsu and Hammond, 1987a). Growth hormone (GH), but not prolactin, also has a direct stimulating effect on IGF-I secretion, and this effect is at least partially additive to the action of estradiol plus FSH (Hsu and Hammond, 1987b). *In vivo,* eCG stimulates intrafollicular produc-tion of immunoreactive IGF-I. A synchro-nous increase of IGF-I and estradiol levels has been reported in preovulatory follicles before the LH preovulatory surge (Ham-mond *et al.*, 1988). Such regulatory systems could enhance the growth and development of ovarian follicles on a local level.

5. Endocrine Control of Cell Replication

Growth of preantral and small antral fol-licles can proceed without gonadotropic support, whereas large antral follicles are acutely dependent on gonadotropins (ewe, Dufour *et al.*, 1979; McNeilly *et al.*, 1986). However, FSH and eCG stimulate *in vivo* granulosa-cell multiplication in all types of growing follicles. For example, after eCG injection to heifers, the mitotic index doubled in preantral and early antral follicles (Monniaux *et al.*, 1984a). When thymidine incorporation was studied on granulosa cells freshly removed from follicles 2–6 mm in diameter in the ewe, incubation with FSH for 2 h increased the labeling index (Monniaux, 1987). This result suggests that an early effect of FSH on granulosa cells would be that some proliferating cells leave the G1 phase and enter into the S phase. However, when bovine granulosa cells were cultured without serum on extracellular matrix-coated dishes in medium with insulin, FGF, and HDL, they proliferated actively, but addition of FSH caused a 30% decrease in cell proliferation (Savion *et al.*, 1981). When porcine granulosa cells from small follicles from immature gilts were cultured without serum, FSH and estradiol had no effect on thymidine incorporation; however, the combination of the two caused distinct inhibition of DNA synthesis (Hammond and English, 1987). It is puzzling that FSH and estradiol stimulate granulosa-cell proliferation *in vivo* but not *in vitro*. Moreover, if FSH stimulates IGF-I secretion by granulosa cells *in vitro* (see Section II,B,4), it is unclear why IGF-I secreted *in vivo* does not stimulate granulosa cell replication. It was recently reported that FSH or estradiol induced DNA polymerase α activity in rat granulosa cells *in vivo* (Usuki and Shioda, 1986) and *in vitro* (Usuki and Shioda, 1987). *In vivo*, this action was accompanied by DNA synthesis, whereas *in vitro* there was no evidence of DNA synthesis. The authors assumed that FSH stimulated certain processes that led to

DNA synthesis but that this stimulation could not activate the processes required completely. Further investigations are necessary to elucidate the exact role of FSH in granulosa cell replication. Perhaps its action must depend on paracrine mechanisms or other cooperative relationships between follicular cells that have not been duplicated *in vitro*.

C. Intrafollicular Control of Cellular Differentiation: Paracrine and Autocrine Effects

Each follicle is a multicompartmental system in which the different compartments (thecal cells, granulosa cells, cumulus cells, oocyte, follicular fluid) may intercommunicate and interact. When a single follicular cell is considered, its products of secretion can affect another cell in two ways:

By passing through gap junctions. Small molecules with a PM < 1000, such as cAMP, calcium, prostaglandins, or polyamines, can pass between neighboring cells if gap junctions are present. In ewes, gap junctions have been described between granulosa cells of antral follicles (Hay and Moor, 1975; Cran *et al.*, 1979), between internal thecal cells (O'Shea *et al.*, 1978), and between cumulus cells and oocyte (Szöllösi, 1978; Szöllösi *et al.*, 1978).

By passing in extracellular medium. Products of secretion can accumulate in follicular fluid and then act through specific cellular receptor sites; steroid- and peptide-like growth factors may act in this way. But products secreted in extracellular medium can also bind active substances or interact with cell surfaces to inhibit specific binding of active substances. This type of action has been particularly proposed for glycosaminoglycans like chondroitin sulfate and heparin sulfate in follicular fluid (Bellin and Ax, 1984).

Thus, locally produced substances can act on follicular cell differentiation and modulate the actions of gonadotropins. The importance of such mechanisms in follicular growth is generally not well understood, and examples will be given for domestic animals only.

1. Interactions between the Oocyte and Granulosa Cells

Granulosa cells exert important actions on oocyte growth and maturation throughout follicular development, and the subject has been recently reviewed by Thibault et al. (1987). Briefly, somatic cells inhibit oocyte growth in primordial follicles by an unknown mechanism. Then, in growing follicles, a metabolic coupling by gap junctions between the oocyte and granulosa cells is a necessary prerequisite for oocyte growth. Around the period of antrum formation, the growing oocyte acquires meiotic competence, that is, becomes capable of resuming meiosis. The competent oocytes do not resume meiosis before the LH surge, but when they are cultured outside their follicles, they resume meiosis spontaneously. The current hypothesis is that granulosa cells secrete inhibitory factors, preventing meiosis resumption.

cAMP has been found to prevent meiosis resumption (ewe, Moor and Heslop, 1981), and a transfer of cAMP from cumulus cells to the oocyte via gap junctions has been proposed (sows, Racowsky, 1985). However, it is not clear that the resumption of meiosis depends on a decrease in cAMP levels after the LH preovulatory discharge. After the LH discharge, cAMP levels increase in the follicle, and the metabolic coupling existing between the oocyte and the cumulus cells decreases only after the beginning of meiotic resumption (ewe, Moor et al., 1980a; sow, Motlik et al., 1986). However, cumulus cells are devoid of LH receptors and are not directly responsive to LH. The dissociation between the granulosa cells and cumulus cells, a precocious event triggered by the preovulatory LH discharge, decreases their intracellular levels of cAMP and allows meiosis to resume.

Another inhibiting substance of peptidic nature, called OMI, has been found in porcine follicular fluid by Tsafriri and Channing (1975), but it has not been possible up to now to purify and identify it (Tsafriri, 1988). Thibault et al. (1987) hypothesized that granulosa cells synthesize a pro-OMI, which could be found in follicular fluid. This pro-OMI, after reaching cumulus cells probably via gap junctions, would be processed into a smaller molecule, which would be the true OMI or would stimulate, through cumulus-cell-specific receptors, the synthesis of the active inhibitor. The dissociation of cumulus cells from granulosa layers, triggered by the LH discharge, would allow the resumption of meiosis in the oocyte by decreasing OMI synthesis in cumulus cells.

An important role of follicular cells in cytoplasmic maturation of the oocyte has also been found. When cumulus-enclosed oocytes are matured outside their follicles, sperm nuclear decondensation does not follow the normal sequence, and further development of zygotes is impaired (cow, Thibault et al., 1975, 1976; sow, Motlik and Fulka, 1974; ewe, Moor and Trounson, 1977). Thus, the complete maturation of oocytes requires a granulosa-cell-dependent inductive phase. Granulosa cells could act by producing an adequate hormonal environ-met to the oocyte, and the ratio of estradiol to progesterone could be particularly important (ewe, Moor *et al.*, 1980b; Osborn and Moor, 1983). Recently, it was proposed that the actions of follicular cells on cyto-plasmic maturation of the oocyte could be mediated by soluble factors that cannot influence the oocyte without some direct cell–oocyte contact (sow, Mattioli *et al.*, 1988a). Follicular secretions are able to maintain a functional intercellular coupling between cumulus cells and oocyte, which is necessary for the oocyte to become penetrable by sper-matozoa and to acquire the conditions required for the formation of male pronuclei (Mattioli *et al.*, 1988b).

If many points concerning the action of follicular cells on the oocyte are still obscure, it is noticeable that nothing is known about the actions of the oocyte on follicular cells. It would be of particular importance to study the interactions between the oocyte and the somatic cells when follicles leave the pool of primordial follicles to enter a growth phase.

2. Interactions between Granulosa and Thecal Cells

a. Actions of Thecal Factors on Granulosa Cells Thecal cells secrete androgens that can act on granulosa cells in two ways: (1) by producing a substrate for aromatiza-tion and (2) by a direct androgenic action. In this second type, testosterone stimulates progesterone secretion by porcine granu-losa cells *in vitro* (Schomberg *et al.*, 1976, 1978; Haney and Schomberg, 1978) and en-hances the stimulatory effect of FSH (An-derson *et al.*, 1979). Testosterone enhances the induction of aromatase activity by FSH in porcine granulosa cells whereas 5α-DHT, a nonaromatizable androgen, has an inhibit-

ing effect (Chan and Tan, 1986). Testoster-one and 5α-DHT stimulate inhibin produc-tion by bovine granulosa cells (Henderson and Franchimont, 1981). Thus, testosterone secreted by thecal cells has overall stimula-tory effects on granulosa cells in domestic animals.

Recently, thecal cells have been found to secrete growth factors, but available data are still limited in domestic species. Thecal cells from preovulatory bovine follicles secrete transforming growth factor β (TGFβ) (Skin-ner *et al.*, 1987a). In addition to its role in neoplasia, TGFβ is a multifunctional factor, which can either inhibit or stimulate prolif-eration and differentiation of various cell types (Sporn *et al.*, 1986). With granulosa cells the functional role of TGFβ when stud-ied *in vitro* depends on the set of growth fac-tors present; for example, according to the *in vitro* conditions, TGFβ inhibits or en-hances the stimulatory action of EGF on granulosa-cell multiplication (cow, Skinner *et al.*, 1987a; sow, May *et al.*, 1988). Al-though the actions of TGFβ on granulosa cells are poorly understood, EGF actions have been more clearly delineated. In the rat, the thecal or interstitial cells produce an EGF-like substance (Skinner *et al.*, 1987b), but its cellular origin in follicles of domestic animals is still hypothetical. Nevertheless, EGF has been found in porcine follicular fluid, and its concentrations are higher in small follicles than in medium-size or large follicles (Hsu *et al.*, 1987). Furthermore, granulosa cells from small follicles have more EGF receptors than from the large, well-differentiated follicles (Buck and Schomberg, 1985, 1988). EGF has a marked mitogenic action on bovine (Gospodarowicz *et al.*, 1977) and porcine (Hammond and English, 1987; May *et al.*, 1988) granulosa cells. In addition to this clear effect on granulosa-cell replication, EGF has various effects on cellular differentiation. Inhibitory effects have been reported on progesterone secretion (Channing *et al.*, 1983) and on the induction of LH receptors by FSH in porcine granulosa cells (Mondschein *et al.*, 1981; Schomberg *et al.*, 1983). Surprisingly, EGF enhances FSH binding by porcine granulosa cells (May *et al.*, 1987). Moreover, FSH decreases the number of EGF receptors *in vitro* (May *et al.*, 1985; Buck and Schomberg, 1988) and acting at high concentrations, suppresses the mitogenic effect of EGF. These complex interactions between FSH and EGF may regulate granulosa cell functions. In small follicles, EGF levels are high, and granulosa cells are very sensitive to EGF, which may stimulate both granulosa-cell replication and FSH receptor induction. When the follicles enlarge, they are under the influence of increasing amounts of FSH by development of vascularization, and FSH, acting on strongly responding granulosa cells, decreases cellular sensitivity to EGF. In large follicles, FSH would stimulate follicular differentiation, whereas EGF would have modulating inhibitory effects. Conversely, the stimulatory effect of EGF on cellular proliferation would be low in large follicles and modulated by FSH.

b. Actions of Factors Secreted by Granulosa Cells on Thecal Cells Thecal cells have not been extensively studied, but some studies indicate that steroids and growth factors produced by granulosa cells can regulate their functions.

In domestic animals, estradiol inhibits progesterone secretion by thecal cells *in vitro* (cow, Fortune and Hansel, 1979). In the pig, progesterone and pregnenolone stimulate androgen production by thecal cells *in vitro*, probably by providing a steroid substrate.

Taken together, these data and results reported above may indicate the existence of a short-loop regulation of estradiol synthesis in the follicle. Granulosa cells could control the availability of aromatizable androgens by acting on progesterone production by thecal cells (see Fig. 16).

Granulosa cells also could act on thecal cell functions through FGF effects. Bovine granulosa cells secrete FGF, which is well known for its angiogenic effects (Gospodarowicz *et al.*, 1979). FGF can stimulate the proliferation of capillary endothelium and so contribute to the development of vascularization in thecal cells (Neufeld *et al.*, 1987).

3. Interactions within the Granulosa Layer

Factors produced by a single cell can act either on neighboring cells or on the producing cell itself. In granulosa, the existence of an autocrine action *sensu stricto*, that is, an action of the producing cell on itself, is still hypothetical, and generally it is not even known if the substance has to go out of the cell to become active. So when granulosa cells are concerned, it is cautious to talk of para/autocrine effects of a factor. As granulosa cells can communicate by gap junctions, the two different pathways previously described (see Section II,C) can exist for communication between cells.

a. Communications between Granulosa Cells through Gap Junctions When porcine granulosa cells were cocultured with mouse adrenocortical tumor cells (Y1), exposure to adrenocorticotropin hormone (ACTH) caused cAMP-dependent protein kinase dissociation in Y1 cells, which are known to be ACTH responsive. This dissociation also occurred in granulosa cells if, and only if, they contacted a responding Y1

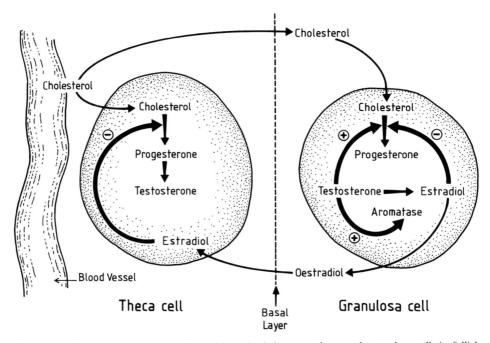

Figure 16 Short-loop regulations of steroid synthesis between theca and granulosa cells in follicles of domestic animals.

cells. When co-cultures were stimulated by FSH, granulosa cells dissociated protein kinase, as did Y1 cells in contact with granulosa cells (Murray and Fletcher, 1984). In the same way, granulosa cells did not bind LH-dissociated protein kinase if they contacted a cell that had bound hormone and dissociated enzyme (Fletcher and Greenan, 1985). Thus, receptor-mediated action can be communicated to receptorless cells, presumably by gap junctions, thereby amplifying the response to hormone. The signal transferred is likely cAMP. This mechanism could be particularly important during the LH-sustained growth of the large preovulatory follicle, in which the distribution of LH receptors is heterogeneous. Moreover, it can be understood how cumulus cells, which

are devoid of LH receptors, can respond to the preovulatory discharge of LH.

b. Para/Autocrine Regulations of Granulosa-Cell Steroidogenesis by Steroids   Progestagens and estrogens, secreted locally by granulosa cells, can act within follicles on granulosa cell steroidogenesis.

Progesterone and pregnenolone inhibit the induction of aromatase activity by FSH in granulosa cells of immature gilts *in vitro* (Chan and Tan, 1986). When cells are cultured with aminoglutethimide, which impairs progesterone synthesis, the induction of aromatase activity by FSH is enhanced (Chan and Tan, 1988). Conversely, inhibiting effects of estradiol on progesterone secretion have been reported in bovine granulosa cells of preovulatory follicles (Fortune and Hansel, 1979) and in porcine granulosa cells from small follicles, stimulated or not by FSH (Thanki and Channing, 1976). These results suggest the existence of an inverse relationship between aromatase activity and progesterone production in differentiating granulosa cells. However, stimulatory effects of estradiol on progesterone secretion have also been reported in porcine granulosa cells (Goldenberg et al., 1972), with synergistic stimulatory effects between estradiol and FSH or LH (Veldhuis et al., 1982, 1984). Veldhuis et al. (1986a, 1986b) have postulated a bipotential action of estradiol on progesterone secretion. In short-term cultures (<20 h), estradiol would inhibit progesterone secretion by blockade of pregnenolone's conversion to progesterone. In long-term cultures (>48 h), estradiol would stimulate progesterone secretion by increasing cytochrome P-450 cholesterol side-chain cleavage and 3β-hydroxysteroid dehydrogenase activities and by enhancing the utilization of extracellular lipoprotein substrate (Veldhuis and Gwynne, 1985; Rajkumar et al., 1988). Veldhuis also hypothesized (1985a, 1985b) that in vivo estradiol may act to limit the premature production of progesterone by the developing follicle ("short-term" inhibiting effect) and concomitantly would prepared granulosa cells for the later production of large quantities of progesterone after ovulation ("long-term" stimulating effect). If true, estradiol would have a determinant function in preparing the change from an estrogen-producing cell to a progesterone-producing cell.

c. Para/Autocrine Regulations of Granulosa-Cell Functions by Peptides and Other Compounds The exact function of

all the compounds secreted by granulosa cells and found in follicular fluid is not always understood, but schematically, inhibitors and stimulators of granulosa cell functions can be described.

i. Inhibitory Factors Oxytocin secreted by granulosa cells (cow, Kruip et al., 1985; Wathes et al., 1986; Jungclas and Luck, 1986) has inhibitory effects on progesterone secretion by granulosa cells of preovulatory follicles in vitro (sow, Gregoraszczuk et al., 1988). Oxytocin also inhibits androgen and estrogen production in co-cultured granulosa and thecal cells (sow, Gregorsazczuk et al., 1988).

Follicular regulatory protein (FRP), isolated from porcine follicular fluid and mainly secreted by granulosa cells of small and medium-sized porcine follicles, has various inhibitory actions on granulosa cells in rats, pigs, and primates (Ono et al., 1986). In sows, FRP inhibits aromatase activity of granulosa cells of small and medium-sized follicles (Ono et al., 1986), and acting at high concentrations inhibits progesterone secretion by decreasing 3β-hydroxysteroid dehydrogenase activity (Battin and Di Zerega, 1985). FRP also inhibits FSH induction of LH receptors (Montz et al., 1984) and FSH-responsive adenylate cyclase activity (Ujita et al., 1987) in porcine granulosa cells.

Glycosaminoglycans (GAGs) are major constituents of follicular fluid, but their functions are not well understood in follicles. The primary GAGs in follicular fluid are chondroitin sulfate and heparin sulfate. Chondroitin sulfate and heparin sulfate concentrations decrease with advancing follicular size but increase substantially in atretic follicles (sows, Ax and Ryan, 1979; cows, Bellin et al., 1983; Bellin and Ax,

1984; Bushmeyer *et al.*, 1985). GAGs could be associated with atresia by two types of inhibition on granulosa cells:

Inhibition of cellular multiplication: In CHO (Chinese hamster ovary) cells, cell-surface association of heparin sulfate is lost before mitosis occurs (Kraemer and Tobey, 1972). In bovine follicles, higher chondroitin sulfate/heparin sulfate ratios are found in healthy, growing, estradiol-active follicles (Bushmeyer *et al.*, 1985). However, the exact role of heparin sulfate is not as clear since heparin has a high affinity for FGF, which is markedly mitogenic in bovine granulosa cells (Gospodarowicz *et al.*, 1977; Savion *et al.*, 1981; Gospadorowicz, 1987). Moreover, it has been proposed that FGF synthesized by granulosa cells might be released from granulosa cells in association with heparin sulfate proteoglycans (Neufeld *et al.*, 1987). Further investigations are necessary to clarify the respective roles of FGF and heparin sulfate in mitogenesis.

Inhibition of gonadotropins and steroid actions: Addition of GAGs to cultured rat granulosa cells inhibited the availability of LH receptor and the stimulation of adenylate cyclase (Nimrod and Lindner, 1980; Salomon *et al.*, 1978). In pigs, chondroitin sulfate inhibited progesterone synthesis by granulosa cells *in vitro* (Ledwitz-Rigby *et al.*, 1984, 1987), perhaps by uncoupling the LDL receptor and restricting steroid substrate availability (Strauss *et al.*, 1985). These effects of the GAGs may be exerted via their direct association with the cell surface by topographical obstruction of receptors or interference with receptor mobility by coupling with adenylate cyclase. As GAGs are major granulosa-cell secretory products, they are prime candidates for the modula-

tion of gonadotropin and steroid effects on follicle development, as proposed by Bushymeyer *et al.* (1985).

ii. Stimulatory Factors Recent studies displayed important and various stimulating effects of the insulin-like growth factor I (IGF-I) on granulosa-cell functions, and the subject has been intensely investigated. Porcine granulosa cells secrete immunoreactive IGF-I and IGF-I-binding proteins *in vitro*, and increasing amounts of IGF-I are found in follicular fluid during terminal follicular growth (Hammond *et al.*, 1985). IGF-I production is stimulated by FSH, LH, GH, and estradiol as reported above (see Section II,B,4) and also by growth factors as EGF and TGF. TGFβ decreases the stimulating effect of EGF (Mondschein and Hammond, 1988). Specific IGF-I receptors have been identified on granulosa cells of gilts (Baranao and Hammond, 1984b) and ewes (Monget *et al.*, manuscript in preparation). Insulin, acting at concentrations 100 times higher than IGF-I, can also bind IGF-I receptors, and it is likely that many of insulin's actions reported are exerted by binding on IGF-I receptors. Estradiol enhances the sensitivity of granulosa cells to IGF-1 by increasing the number of IGF-I receptors (Veldhuis *et al.*, 1986a).

Various stimulating actions of IGF-I on porcine granulosa-cells multiplication and differentiation have been reported; these actions have always been observed with low concentrations of IGF-I:

IGF-I enhances DNA synthesis and granulosa-cell replication (sows, Baranao and Hammond, 1984b; Hammond and English, 1987; cows, Savion *et al.*, 1981).

IGF-I enhances the induction of LH receptors by FSH, but IGF-I acting alone is without effect (Maruo *et al.*, 1988).

IGF-I stimulates progesterone secretion by granulosa cells, acting alone and synergistically with FSH and estradiol (Veldhuis and Demers, 1985; Veldhuis and Furlanetto, 1985; Veldhuis *et al.*, 1986a). It also increases cytochrome P-450 cholesterol side-chain cleavage and adrenodoxin activity and synthesis (Veldhuis and Furlanetto, 1985; Veldhuis *et al.*, 1986b), increases the number of LDL receptors, and increases the rate of internalization of LDL and the amount of free and esterified cholesterol in granulosa cells (Veldhuis *et al.*, 1987).

IGF-I stimulates aromatase activity, acting alone and synergistically with FSH (Veldhuis and Demers, 1985; Maruo *et al.*, 1988).

These results suggest the existence of an autocatalytic process in granulosa cells: FSH and estradiol stimulate IGF-I secretion; estradiol enhances cellular sensitivity to IGF-I; IGF-I stimulates estradiol secretion and enhances FSH and estradiol actions. How are these mechanisms regulated to avoid a racing effect? We propose two possibilities: (1) the actions of inhibitory factors, like EGF or GAGs; (2) the modulation of IGF-I actions by IGF-I-binding proteins, which are secreted also by granulosa cells. This last eventuality has not been explored.

D. Conclusions

Regulation of follicular functions is highly complex, and it must be asked which factors are determinant and which are of secondary importance in follicular development. We suggest that during follicular growth, the nature of determinant factors for follicular cell multiplication and differ-entiation changes with the status of differentiation of the cells. We propose the existence of at least three stages in follicular maturation:

First stage: EGF-dependent stage. Preantral and early antral follicles are poorly vascularized and would be more dependent on locally produced growth factors like EGF and perhaps FGF than on endocrine factors like FSH. In these follicles, granulosa cells replicate actively and steroidogenesis is low. EGF would enhance cellular multiplication and sensitize cells to FSH.

Second stage: FSH-dependent stage. Small and medium-sized antral follicles are well vascularized and particularly sensitive to FSH actions. In these follicles, FSH would act by increasing steroidogenesis and various syntheses in follicular cells. The oocyte would be maintained in a meiotically arrested stage. Increased levels of estradiol and IGF-I would enhance the stimulatory effects of FSH autocatalytically. The existence of short-loop regulations in steroidogenesis and the action of inhibitory compounds locally produced would allow the follicles to reach a dynamic equilibrium. At this stage, granulosa cells replicate less actively, lose their sensitivity to growth factors, and become differentiated.

Third stage: LH-dependent stage. Large antral and preovulatory follicles are highly vascularized, and their final development is probably sustained by LH. However, it is modulated also by all the compounds reported earlier. The preovulatory discharge of LH would be the final signal that triggers both oocyte maturation and follicle rupture, disrupting all the settled regulatory systems.

All these regulatory mechanism are valid when a single follicle is considered. The next section of this review will describe the interactions between different follicles within an animal, the equilibrium between follicular populations, and how the ovulation rate is regulated in domestic animals.

III. Mechanisms Controlling the Development of the Ovulatory Follicles

The mechanism controlling the development of ovulatory follicles and the possible ways to alter them to produce an increased ovulation rate will be described. Most of the data presented have been obtained in ewes. However, additional data obtained in other species (cows, sows, mares, and primates) will be presented when they provide insights useful to our understanding. In contrast, the use of data obtained in hypophysectomized diethylstilbestrol (DES) treated rats will be minimized owing to the low physiological value of this model (Sadrkhanloo *et al.*, 1987).

A. *Main Events of Terminal Follicular Growth*

1. *Is Terminal Follicular Growth Restricted to the Follicular Phase?*

In animals with two or three waves of terminal follicular development during the estrous cycle (sheep, cattle), terminal follicular growth (i.e., the final growth and maturation of the ovulatory follicles) occurs outside the follicular phase (i.e., the interval between corpus luteum regression and the LH surge) as well as during it. Daily ultrasonic measurements of the population of follicles >3 mm in cattle (Pierson and Ginther,

1984; Savio *et al.*, 1987; Sirois and Fortune, 1988), ink marking of the follicles >2 mm in diameter at repeated laparotomies (Driancourt *et al.*, 1988), or histological examination of large numbers of ovaries (sheep, Brand and de Jong 1973; cattle, Rajakoski, 1960) have all demonstrated the growth of one or two large ovulatory-sized follicles during the luteal phase. In the second group (humans, Gougeon and Lefevre, 1983; pigs, Dalin, 1987), the diameter of the largest healthy follicles is markedly smaller during the luteal phase than during the follicular phase. Hence, it is likely that the mechanisms controlling terminal follicular growth work differently in sheep and cattle than in pigs and primates.

2. *Does Terminal Follicular Growth Only Occur in Cycling Animals?*

Terminal follicular growth was investigated in cattle by repeated ultrasonic measurements of the population of large follicles (R. Alberio and M. A. Driancourt, unpublished data) and in sheep by repeated laparoscopic measurements of the follicles > 2 mm (P. Chemineau and M. A. Driancourt, unpublished data). Their data showed that early in the postpartum period in cattle, and during seasonal anestrus in sheep, follicles grow to a large size. These observations have been confirmed by data obtained at dissection of the ovarian follicles of animals at these stages (anestrous sheep, Webb and Gauld, 1987; postpartum cattle, Spicer *et al.*, 1986). Furthermore, in 3-month-old lambs, histological examination has revealed that follicles >4 mm were also present (Sonjaya and Driancourt, 1989). These observations suggest that terminal follicular growth occurs at times when ovulation cannot proceed in sheep and cattle, and that the mechanisms controlling termi nal follicular growth and growth ending in ovulation are different.

3. Functional Features of the Follicles Growing during Terminal Follicular Growth

The large follicle present at the end of the luteal phase presents two distinctive features (cf. Section II): increased capabilities to produce estradiol and to bind LH on its granulosa cells. Whether the large follicles growing during the periods outside the follicular phase have similar features has been tested in two ways: (1) the presence of functional LH receptors on the granulosa cells has been measured by giving hCG and monitoring ovulation, and (2) the ability to produce estradiol has been assessed by measuring the levels of follicular fluid contents or the activity of the aromatase enzymes. Whatever the physiological condition in the ewe (prepubertal, anestrous, luteal phase), the large follicles present have functional LH receptors as demonstrated by their ability to ovulate following hCG (Driancourt et al., 1989). A similar conclusion is also valid for luteal-phase follicles in cattle (Ireland and Roche, 1984; Webb et al., 1989). In contrast, estradiol content or the follicular ability to produce estradiol is blunted during the luteal phase (cattle, Ireland and Roche, 1982, 1983), anestrus (sheep, McNatty et al., 1984b) or before puberty (Sonjaya and Driancourt, 1989). These data show that active estradiol production during terminal follicular growth is not a prerequisite for growth or acquisition of LH receptors.

4. When Can the Ovulatory Follicle Be Recognized?

Three criteria are commonly used to recognize the ovulatory follicle: (1) its size (monkeys, Clark et al., 1979); cattle, Quirk et al., 1986; Savio et al., 1987; Sirois and Fortune, 1988; sheep, Driancourt and Cahill, 1984); (2) its function as evidenced by its ability to bind LH by the theca and granulosa cells (sheep, Webb and England, 1982; monkey, Di Zerega and Hodgen, 1980a) and to produce estradiol levels high enough to distinguish between the active and inactive ovary by differences in its concentration in ovarian venous blood (monkey, Di Zerega et al., 1980; sheep, McNatty et al., 1982); and (3) by associated changes in the population of smaller follicles. These changes include (1) their decreased size when measured by ultrasound in vivo (cattle, Quirk et al., 1986; Savio et al., 1987; Sirois and Fortune, 1988), (2) the long time interval between ablation of the ovulatory follicle and ovulation (mare, Driancourt and Palmer, 1984; monkey, Goodman and Hodgen, 1979), or (3) the reduction in the superovulatory effect of exogenous gonadotropins (cattle, Pierson and Ginther, 1988; monkeys, Di Zerega and Hodgen, 1980b). Some observations indicate that the information provided by these criteria are different in mares (Driancourt et al., 1985b) and in cattle (Ireland et al., 1984; McNatty et al., 1984a). However, it is difficult to identify the ovulatory follicles in species like cattle and sheep with short follicular phases and terminal follicular growth occurring at all physiological stages of the cycle. This has been demonstrated in sheep after ink labeling of individual follicles (Driancourt and Cahill, 1984). At 0, 4, 8, 12, 24, and 48 h after a prostaglandin injection, the follicle ovulating at the following estrus was the largest in 0/7, 1/7, 1/7, 4/7, 5/7, and 5/7 ewes, respectively. Similar variability in the times at which the ovulatory follicle can be identified has been reported in cattle (McNatty et al., 1984a; Quirk et al., 1986).

In primates, identification of the ovulatory follicle is easier because it arises from the single cohort of large follicles growing during the follicular phase. It is readily identified around days 6–7 (day 0 = menses) (Gougeon and Lefevre, 1983).

5. What Are the Changes in the Population of Small Follicles Associated with the Emergence of the Ovulatory Follicle?

Changes in the population of small follicles associated with the emergence of an ovulatory follicle have been studied most easily in primates, owing to the long duration of the follicular phase. The emergence of the ovulatory follicle is associated with a decrease in the diameter of the second largest healthy follicle and a decrease in the size of the largest atretic follicle (Gougeon and Lefevre, 1983). In sheep, the size of the second largest healthy follicle after the ovulatory follicle has emerged never exceeds 2.5–2.7 mm (Brand and de Jong, 1973; Driancourt et al., 1989).

6. Where Does the Ovulatory Follicle Come From?

The minimum size a follicle with a chance to ovulate must have in sheep at luteolysis has been established by cautery experiments (Tsonis et al., 1984). At sponge removal, all follicles >4 mm, all follicles 2–4 mm in diameter, or all follicles >2 mm were electrocauterized, and the time of the LH surge was measured. Only when all follicles >2 mm in diameter were destroyed was the LH surge delayed, leading to the conclusion that all healthy follicles >2 mm at the time of corpus luteum (CL) regression have a chance to ovulate. Identification of individual follicles by ink marking at luteolysis confirmed these results. The minimum size they could have was also 2 mm (Driancourt and Cahill, 1984). This 2-mm threshold rep-

resents the size at which follicles become acutely dependent on gonadotropins as shown by their complete disappearance in hypophysectomized ewes (Dufour et al.,1979, Driancourt et al., 1987), GnRH-immunized ewes (McNeilly et al., 1986), or follicular-fluid-treated ewes (Wallace and McNeilly, 1986).

In cattle (Matton et al., 1981) and swine (Clark et al., 1979), the minimum sizes a follicle must have in order to ovulate following

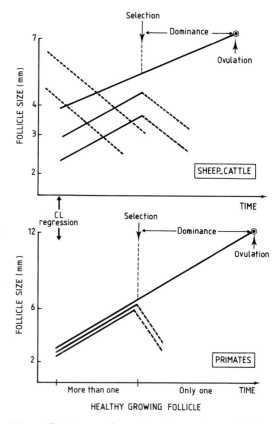

Figure 17 Pattern of growth of the ovulatory follicles between CL regression and ovulation in sheep and cattle (upper panel) and primates (lower panel). Continuous lines represent growing follicles, and dotted lines represent regressing atretic follicles.

CL regression are 3–4 and 2 mm, respectively. The schematic pattern of growth of the ovulatory follicles is presented in Fig. 17.

B. Control of Terminal Follicular Growth

Terminal follicular growth is probably regulated by both endocrine and local controls. The local controls would involve paracrine interactions between follicles and autocrine regulations within a follicle.

Terminal follicular growth cannot proceed in the absence of gonadotropins (Dufour et al., 1979; McNeilly et al., 1986), and injection of exogenous gonadotropins (eCG and hCG) is able to restore ovulation in hypophysectomized ewes (Fry et al., 1988; Driancourt et al., 1988) or GnRH-immunized ewes (Mariana et al., 1989; Driancourt et al., 1989). Thus, gonadotropins are a key component in the growth of the follicles that culminate in ovulation. The exact respective roles of LH and FSH have not been fully clarified. The close synchronization between terminal follicular growth and the increased number and decreased amplitude of LH pulses following the regression of the corpus luteum (Baird et al., 1979) had previously suggested a causal link between these two parameters. However, the pioneering studies of Picton and McNeilly (1988) in GnRH-agonist-blocked ewes have clearly demonstrated that the major parameter responsible for follicular growth and differentiation that terminates in ovulation was a synergy between FSH and basal levels of LH. Similar data have been obtained in primates (Kenigbserg et al., 1984). Indeed, FSH stimulates thymidine incorporation in small (<2 mm) follicles (Monniaux, 1987) and stimulates aromatization in large folli-

cles when administered in vivo (McNatty et al., 1986), although this effect is not apparent in vitro (Tsonis et al., 1984; Monniaux, 1987). The role of the LH pulses would then be to provide enough thecal androgens to the granulosa cells for aromatization and to stimulate antrum formation.

However, three questions still remain that prevent a full clarification of the role of FSH in follicular growth and differentiation. Primarily, for example, are there time periods of increased requirements of follicles for FSH? Such a hypothesis is supported by data reported in hypophysectomized rats (Welschen, 1973), in intact rats (Hirschfield, 1985), and in intact ewes (McNatty et al., 1985b). These last authors infused ewes with ovine FSH over the time frames −72 to −48 h, −48 to −24 h, and −24 to 0 h before prostaglandin injection. However, only the last two infusions significantly increased ovulation rate compared to the controls. Furthermore, two injections of FSH (3 mg each) given to Ile de France ewes increased ovulation rate when given at 24 and 36 h but not at 0 and 12 h after luteolysis (M. A. Driancourt, unpublished results).

Second, are the follicular requirements for FSH continuous or intermittent? Are threshold levels required? Evidence for a threshold relationship between FSH levels and terminal follicular growth has been obtained also in experiments when FSH was infused. Doubling the circulating FSH levels by such an infusion only increased ovulation rate in 50% of the ewes (Baird et al., 1985; Henderson et al., 1988). This suggests that in some ewes, thresholds may be fairly wide as big changes in FSH concentrations did not alter ovulation rate.

Third, are peripheral levels of gonadotropins an accurate measure of the levels delivered to the follicle? Carson *et al.* (1986) demonstrated that only 30% of a pulse of radioactive LH could be found at the follicle level. The existence of arteriovenous shunts within the ovarian pedicle and hilus (Mattner *et al.*, Brown and Driancourt, 1989) could explain this difference. On the other hand, it is unknown whether the mechanisms concentrating steroids in the ovarian vascular pedicle (Kotwica *et al.*, 1982) are also operative for gonadotropins.

Local regulators also play a role in controlling terminal follicular growth and differentiation, and numerous results show that they modulate the endocrine factors either positively or negatively.

Three lines of evidence demonstrate that follicular fluid contains inhibitors of follicular growth and differentiation in sheep. First, when ewes are treated with high doses of follicular fluid during the early follicular phase, follicular growth greater than 2.5 mm is prevented (Wallace and McNeilly, 1985), and ovulation is delayed for 5–6 days, despite a huge FSH surge occurring at the end of treatment (McNeilly, 1985). This time lag is much longer than the 2–3 days needed for a follicle 2 mm in diameter to reach ovulation (Driancourt and Cahill, 1984). Second, infusion of follicular fluid directly into the ovarian artery reduced LH-RH-induced estradiol secretion from the autotransplanted ovary in the ewe (Scaramuzzi *et al.*, 1986). Third, when follicular fluid is injected into hypophysectomized ewes in which follicular development is triggered by exogenous gonadotropins (eCG), the ovarian response is depressed (Cahill *et al.*, 1984).

Stimulatory effects of charcoal-treated follicular fluid have been reported on granulosa-cell division (Veldhuis *et al.*, 1987) and steroidogenesis (Ledwitz-Rigby, 1983, 1985).

Candidates for these effects should fulfil three conditions. They should be produced within the gonad, have a well-characterized activity on the gonad, and their concentration/production/response should change with increasing follicular maturation. Among the compounds that fulfil these conditions, three growth factors/ovarian compounds have a well-defined inhibitory action (EGF, FGF, and FRP), while one growth factor (IGF-I) has a stimulatory action on granulosa-cell differentiation.

At the beginning of the follicular phase of the cycle, each ewe has several healthy follicles larger than 2 mm in diameter, but the growth potential of each of them is different (Fig. 18). The heterogeneity in their growth potential is likely to be related to different contents and/or sensitivities to local regulators. Indeed, the smallest follicles are richest in EGF (Hsu *et al.*, 1987) and FRP (Tonetta *et al.*, 1988), and their response to the inhibitory effects of EGF on differentiation is maximal (Buck and Schomberg, 1988). Hence, the effect of FSH is least on the smallest follicles of the cohort of recruited follicles. The largest recruited follicle will be the first to reach 4 mm in diameter, the size at which it will start to massively secrete estradiol (Tsonis *et al.*, 1984), acquire LH receptors on the granulosa cells (Carson *et al.*, 1979), and increase its inhibin production (Tsonis *et al.*, 1983). Its massive production of negative feedback products will markedly reduce FSH secretion by the pituitary (Martin *et al.*, 1988), which will go under the threshold level necessary for growth, pushing the other recruited follicles toward atresia and blocking growth of other follicles greater than 2.5 mm in diameter. It

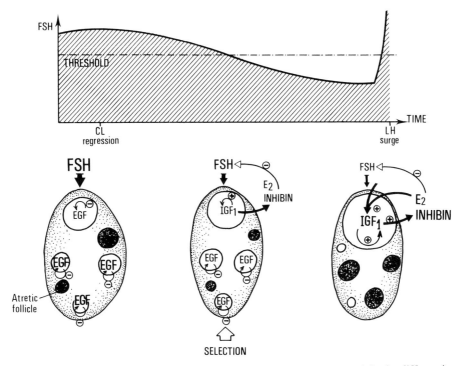

Figure 18 Interplay between FSH and growth factors (EGF, IGF₁) to explain the differentiation of a single ovulatory follicle.

has been suggested (Di Zerega *et al.*, 1983) that in primates local compounds secreted by the dominant follicle would act directly on the other follicles to inhibit their growth and differentiation. At present no evidence supports this concept in sheep. Indeed, cautery of the large dominant follicle present at day 6 of the luteal phase does not affect the ovarian response to an eCG–hCG challenge (M. A. Driancourt and R. C. Fry, unpublished). Furthermore, addition of 10% of charcoal-treated follicular fluid from large follicles to the incubation medium of intact follicles increases their estradiol production compared to 10% serum (M. A. Driancourt, unpublished).

However, the largest follicle does not suffer from the decrease in FSH concentrations as its sensitivity to FSH has steadily increased with size, as demonstrated *in vitro* in sheep (Henderson *et al.*, 1987) and *in vivo* in primates (Zeleznik and Kubik, 1986). This increased sensitivity to gonadotropins may be related to an FSH/LH-induced reduction in EGF binding (Feng *et al.*, 1987) and an increased production of IGF-I by its granulosa cells (Hammond *et al.*, 1985). Following its selection, its dominance can be explained by the following self-amplifying loop. Its high sensitivity to FSH and its high production of estradiol stimulate further IGF-I, production (Hsu and Hammond, 1987), and its receptivity to IGF-I (Adashi *et al.*, 1986). High IGF-I levels increased sensitivity to IGF-I, and the high sensitivity to FSH increase estradiol production (Adashi *et al.*, 1985), synergize to increase LH receptors (Adashi *et al.*, 1985), and maximize inhibin production (Zhiwen *et al.*, 1988).

In contrast, dominance is probably not related to either intrafollicular FSH concentrations or to the potential effects of the high levels of estradiol produced in the response to FSH, two commonly made statements. First, there is no experimental evidence supporting a higher FSH concentration in the dominant follicle (Fay and Douglas, 1987), the measurement of which is further plagued by FSH-like compounds in follicular fluid (Schneyer *et al.*, 1988). Second, there is no evidence in sheep that estradiol can have a direct positive effect on terminal follicular growth as estradiol implants delivering follicular phase levels depress follicular growth and differentiation without affecting FSH concentrations (Webb and Gauld, 1986). Furthermore, in primates, estradiol receptors are undetectable in the ovary (Hild-Petito *et al.*, 1988).

The contribution of changes in blood flow to the maintenance of the dominant follicle is unclear. Injection of labeled hCG in primates produced heavy labeling of only the dominant follicle, while all medium/large antral follicles demonstrated LH binding on their theca, suggesting a preferential delivery of LH to the dominant follicle (Zeleznik *et al.*, 1981). However, measurement of the changes in blood flow to the large follicle population (>5 mm in diameter) on days 14, 15, and 16 of the sheep estrous cycle with radioactive microspheres has not shown any significant increase associated

with the emergence of the dominant follicles at day 16 (Bruce and Moor, 1976).

C. Alterations in the Development of the Ovulatory Follicles Associated with Increased Ovulation Rate

Treatments with eCG or eCG combined with anti-eCG serum or pituitary FSH are commonly used to produce superovulation in donor cows in embryo-transfer programs. Although the way exogenous gonadotropins generate an increased ovulation rate is largely understood, its accurate control and optimum quality of ova shed have never been reached.

Two sites of action of gonadotropins to produce an increased ovulation rate have been clearly identified. First, administration of exogenous gonadotropins lowers the size at which they can trigger follicle recruitment. All healthy follicles >1.7 mm in diameter in cattle (Monniaux *et al.*, 1983) of >0.8 mm in diameter in sheep (Driancourt, 1987) are mobilized for terminal follicular growth following eCG/FSH administration. This is probably associated with the action of FSH to stimulate granulosa cell division. In contrast, antrum size is increased to a lesser extent after eCG administration. As a consequence, preovulatory follicles of eCG-stimulated cows are usually 30% smaller than preovulatory follicles in control cycles. Stimulated follicles usually contain a similar mean number of granulosa cells compared to controls, $3-5 \times 10^6$ granulosa cells per sheep preovulatory follicle. However, within-animal variation in number of granulosa cells in individual follicles is markedly increased, often reaching fourfold differences (M. A. Driancourt, unpublished data).

Whether this variability is related to the respective size of the different follicles mobilized at the time of eCG administration is unknown. The relationship between granulosa-cell number and oocyte quality of individual follicles within an animal also has not been clarified. While accelerating growth of follicles, eCG/FSH also markedly alters their steroidogenic potential. Although peripheral estradiol levels increase after administration of exogenous gonadotropins (Saumande *et al.*, 1978), the pattern of steroid production by individual follicles in treated as compared to control cycles has not been established before the LH surge. After the LH surge, estradiol production by superovulated follicles is lowered (Fortune and Hansel, 1985), as it is in the follicular fluid estradiol in hMG and clomid/hMG stimulated follicles in primates (Di Zerega *et al.*, 1983). In contrast, androgen levels in follicles from cattle or primates treated with gonadotropins were similar to those found in untreated controls.

Second, eCG/FSH administration reduces or prevents atresia in the population of follicles recruited by their administration (Dott *et al.*, 1979, McNatty *et al.*, 1982). The extent to which reversal of atresia contributes to superovulation as suggested in rats (Byskov, 1979) is unclear. Dott *et al.* (1979), using culture of intact follicles, and Driancourt *et al.* (1987), using follicles in which atresia was synchronized *in vivo* by hypophysectomy, were not able to demonstrate either a significant recovery, that is, reinitiation of growth or change to a less atretic stage in atretic follicles. In contrast, Monniaux *et al.* (1983), in their histological and functional study of the ovaries of eCG-treated cattle, suggested that reversal of atresia by eCG occurred and produced lu-

teinized follicles. It may be that only very early stages of atresia can be reversed by eCG.

The variability of the ovarian response of different cows of a similar age and in a similar environment to a given dose of exogenous gonadotropin is striking. It ranges from 0 to 72 ovulations following 1500 IU of eCG (Mariana *et al.*, 1970). A major breakthrough in understanding the underlying cause of this variability was reported by Monniaux *et al.* (1983) when they demonstrated that about 70% of the variation in eCG-induced ovulation rate in cattle was related to the number of follicles contained in the ovaries. This was confirmed later in sheep (Driancourt, 1987). However, the correlation between the response to two successive treatments is usually low (Driancourt, 1987) in sheep, and in cattle the repeatability of the number of eggs shed after successive treatments with FSH is limited (0.22; Lamberson and Lambeth, 1986). Hence, some other factors probably modulate the ovarian response to exogenous gonadotropin.

The variability of the ovulatory response may be affected by the kinetics of the disappearance of the gonadotropin injected. In women injected with hMG, the clearance of hMG was much faster in low responders than in high (Benadiva *et al.*, 1988). A negative association between the presence of large follicles on the ovarian surface and a poor response to eCG have been demonstrated in cattle (Saumande *et al.*, 1978), and administration of eCG early during the luteal phase in the presence of the large follicle usually results in a poor response (Sreenan *et al.*, 1978). Recently Grasso *et al.* (1988) confirmed this finding by monitoring follicular growth by ultrasound. In contrast, such an effect is not obvious in sheep (Driancourt and Fry, 1988). Endogenous gonadotropin levels at the time of treatment may contribute to the variability. In sheep (M. A. Driancourt and F. Castonguay, unpublished results) and in primates, no relationship exists between endogenous FSH concentrations and ovarian response. Individuals with higher FSH levels did not respond poorly (Benadiva *et al.*, 1988). However, it should be noted that when endogenous FSH levels are drastically reduced, by hypophysectomy for example, the ovarian response to exogenous gonadotropin is reduced (Bindon and Pennycuik, 1974), suggesting that drastically low levels of endogenous gonadotropin reduce the response to exogenous gonadotropins.

Administration of eCG often results in a high incidence of premature activation of the oocytes (Moor *et al.*, 1985), and now most workers prefer to use pituitary FSH preparations to maximize the quality of the ova shed. However, these preparations contain a variable amount of FSH and LH, and the profile of administration of pituitary FSH and its FSH/LH ratio may differ according to the breed treated (Chupin *et al.*, 1984).

To maximize the quantity of the ova shed, two approaches are currently being tested. In one approach, small amounts of gonadotropins have been injected in the days following ovulation to maximize the response to the superovulatory dose given later during the luteal phase. With one exception (Ware *et al.*, 1987), the results have not been rewarding (Chupin and Saumande, 1979; Lussier and Carruthers, 1987). In the other approach, treatments to standardize follicular population at the time of the injection of exogenous gonadotropin by suppressing the larger follicles through

a reduction of their gonadotropin support have been tried. This is usually achieved by active/passive immunization against GnRH through the administration of GnRH agonist/antagonists. Superovulation in these blocked animals usually requires higher doses of gonadotropins, and there is no evidence that the overall ovarian response is increased (P. Brebion and Y. Cognie, personal communication) in quantity or quality or that its variability is reduced (Mariana *et al.*, 1989; Driancourt *et al.*, 1989).

Genetic alteration of ovulation rate has been reported in several species of animals. As the information on the physiology of highly ovulatory strains of pigs—Nebraska and hyperprolific selection lines—and cattle is still scarce, this section will focus on research reported in sheep.

The growth patterns of the ovulatory follicles in most of the various strains of sheep were elucidated by ink marking of individual follicles at repeated laparotomies performed over the late luteal and follicular phases of the cycle (Fig. 19). Breed differences exist in the strategy used to generate a high ovulation rate. Booroola ewes carrying the F gene had a high ovulation rate, associated with a prolonged recruitment throughout the follicular phase and the ability of its large follicles to wait for the LH surge (Driancourt *et al.*, 1985). The high ovulation rate of Romanov ewes was associated with a large number of recruited follicles and a loss identical to that of the control animals at the time of ovulatory selection (Driancourt *et al.*, 1986). The high ovulation rate of Finn ewes was attained through reduced incidence of the follicular degeneration at the time of selection and despite no change in the number of recruited follicles (Driancourt *et al.*, 1986).

High ovulation rate also is associated with

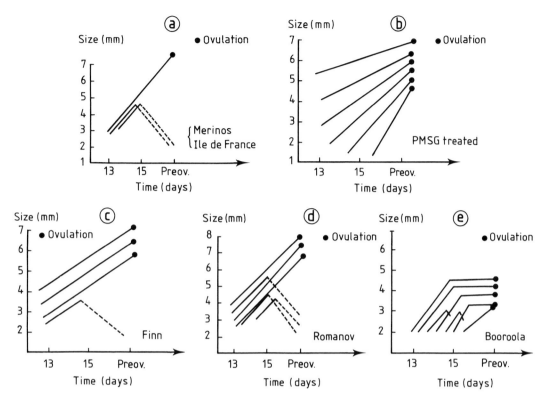

Figure 19 Pattern of growth of the ovulatory follicles between day 13 and ovulation in (a) nonprolific breeds, (c, d, e) prolific breeds (c, Finn, d, Romanov, e, Booroola), and (b) eCG-treated (PMSG-treated) Merino ewes.

alterations in the morphological and functional differentiation of ovulatory follicles (Fig. 20). These follicles are usually smaller in prolific breeds, by 20% in the Romanov and Finn breeds (Cahill *et al.*, 1979; Webb and Gauld, 1985) and by 50% in F gene carrier Booroolas (Driancourt *et al.*, 1985; McNatty *et al.*, 1986). Together with their reduced size, preovulatory follicles from Booroola and Finn ewes have a reduced number of granulosa cells, each follicle containing about one-third of the normal complement found in a preovulatory follicle from a non prolific breed (Driancourt and Fry, 1988). However, Romanov ewes produce follicles that contain as many granulosa cells as those from Ile de France ewes (Driancourt and Fry, 1988). In addition to these breed differences in the granulosa-cell number, the ovulatory follicles from all three prolific breeds produce altered amounts of estradiol and testosterone. The estradiol secretion rate of ovaries from F+ ewes containing several active follicles is identical to that of ++ ewes containing a single active follicle (McNatty *et al.*, 1985c), suggesting that estradiol production per follicle is reduced in F+ gene carriers, but that it is maintained when expressed on a per granulosa cell basis. A similar conclusion is

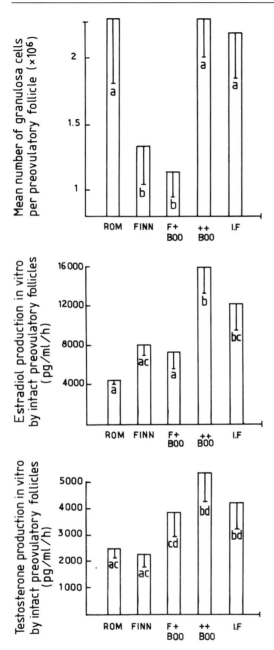

Figure 20 Alterations in (a) granulosa cell number, (b) estradiol production *in vitro*, and (c) testosterone production *in vitro* by intact large ovulatory follicles of Romanov, Finn, F+Booroola, + + Booroola, and Ile-de-France ewes. Means with a common letter don't differ significantly ($P < 0.05$)

also valid for Finn ewes (Webb *et al.*, 1989). Follicles from Romanov ewes when incubated intact *in vitro* produced 50% less estradiol than follicles from a nonprolific breed, Ile de France. These follicles from the two breeds contained identical numbers of granulosa cells, suggesting that those from the Romanov breed have a reduced ability to produce estradiol on a per cell basis.

Differences in the pattern of terminal follicular growth and in the morphological and functional differentiation of the ovulatory follicles suggest that different control mechanisms operate in the different breeds.

Several possibilities that could explain the breed differences in prolificacy have been suggested: increased follicle numbers throughout folliculogenesis, increased levels of FSH secretion, increased sensitivity to gonadotropins, and altered local regulation.

The first two have been studied extensively. Differences in the numbers of antral follicles in the ovaries are not large enough to explain the increased ovulation rate (Driancourt and Fry, 1988). Furthermore, the transit times through the different sized follicles as measured by the mitotic index and the mitotic time as calculated by Cahill and Mauleon (1980) were similar in Romanov and Ile de France ewes and in F+ and + + ewes (Driancourt *et al.*, 1985). High FSH levels during the late luteal and follicular phase have been reported in Booroola ewes carrying the F+ gene (McNatty *et al.*, 1987) and to a much more limited extent in Romanov ewes (Cahill *et al.*, 1981). Similar differences have not been reported in Finnish Landrace ewes (Webb and England, 1982; Adams *et al.*, 1987). However, the high levels of FSH found in Booroola and Romanov ewes are probably not the only reason for their high ovulation rate. Short- and long-term hypophysectomized Booroola and Romanov ewes maintained with similar levels of gonadotropins, eCG and hCG, were more prolific than their controls (Fry *et al.*, 1988; Driancourt *et al.*, 1988). The highly prolific breeds, Finnish Landrace, Booroola, and Romanov, have a different growth pattern in ovulatory follicles than ewes with normal ovulation rates that have received exogenous gonadotropin, eCG, to produce high ovulatory rates (Fig. 19).

Possible hypotheses to explain the mechanisms controlling the ovulation rate in these prolific breeds could be as follows. The high rate of blood flow to small follicles would account for the large number of growing follicles found at the end of the luteal phase of the cycle in Romanov ewes (Brown and Driancourt, 1989). Simultaneously, a dissociation occurs between the ability of FSH to stimulate granulosa-cell division and estradiol secretion. Enhanced division increases granulosa-cell numbers and total estradiol production, although the output of estradiol is decreased on a per cell basis. The high ovulation rate in Booroola ewes carrying the F+ gene as compared to those with the ++ gene is associated with an earlier shift of the granulosa cells from proliferation to differentiation. The acquisition of aromatizing ability and LH receptors occurs in smaller follicles in the prolific ewes. The respective importance of FSH and the growth factors EGF and IGF-I has not been determined. In Finnish Landrace ewes the high ovulation rate may be associated with a dissociation of the ability of FSH to reduce the rate of granulosa-cell division while maintaining estradiol production. In this breed the number of granulosa cells per preovulatory follicle is reduced and estradiol production is maintained. The multiple factors controlling the ovulation rate in this breed are unknown.

Acknowledgments

We thank Mrs. C. Merigard for the typing of the text.

References

Adams, T. E., Quirke, J. F., and Hanrahan, J. P. (1988). *J. Reprod. Fertil.* **83**, 575–584.

Adashi, E. Y., Resnick, C. E., Brodie, A. M. H., Svoboda, M. E., and Van Wyk, J. J. (1985a). *Endocrinology* **117**, 2313–2323.

Adashi, E. Y., Resnick, C. E., D'Ercole, A. J., Svoboda, M. E., and Van Wyk, J. J. (1985b). *Endocr. Rev.* **6**, 400–420.

Adashi, E. Y., Resnick, C. E., and Svoboda, M. E. (1986). *J. Biol. Chem.* **261**, 3923–3926.

Amsterdam, A., and Romentsch, S. (1987). *Endocr. Rev.* **8**, 309–337.

Anderson, L. D., Schaerf, F. W., and Channing, C. P. (1979). *Adv. Exp. Med. Biol.* **112**, 187–195.

Anderson, L. D., Stone, S. L., and Channing, C. P. (1985). *Gamete Res.* **12**, 119–130.

Aten, R. F., Ireland, J. J., Weems, C. W., and Behrman, H. R. (1987). *Endocrinology* **120**, 1727–1733.

Ax, R. L., and Ryan, R. J. (1979). *Biol. Reprod.* **20**, 1123–1132.

Bachvarova, R. (1981). *Dev. Biol.* **86**, 384–392.

Bachvarova, R. (1985). *In* "Developmental Biology" (Leon W. Browder, ed.), Vol. I, pp. 453–524.

Baird, D. T. (1978). *Biol. Reprod.* **18**, 359–364.

Baird, D. T., Swanston, I. A., and Scaramuzzi, R. J. (1976). *Endocrinology* **98**, 1490–1496.

Baird, D. T., Swanston, I. A., and McNeilly, A. S. (1981). *Biol. Reprod.* **24**, 1013–1025.

Baird, D. T., McNeilly, A. S., Wallace, J. M., and Webb, R. (1985). Proc. V Reiner de Graaf Symp., 13–17, Nijmegen.

Band, V., Kharbanda, S. M., Murugesan, K., and Farooq, A. (1987). *Biol. Reprod.* **36**, 799–806.

Banoin, M. (1988). These de doctorat de l'Université Montpellier, Montpellier, France.

Baranao, J. L. S., and Hammond, J. M. (1984a). *J. Steroid Biochem.* **20**, 1513 (Abstr).

Baranao, J. L. S., and Hammond, J. M. (1984b). *Bioch. Biophys. Res. Commun.* **124**, 484–490.

Bartholomeusz, R. K., Bertoncello, I., and Chamley, W. A. (1988). *Int. J. Cell. Cloning* **6**, 106–115.

Battin, D., and Di Zerega, G. S. (1985). *J. Clin. Endocrinol. Metab.* **60**, 116–119.

Bellin, M. E., and Ah, R. L. (1984). *Endocrinology* **114**, 428–434.

Bellin, M. E., Hinshelwood, M. M., Robinson, G. M., Ah, R. L., and Hauser, E. R. (1983). *In* "Factors regulating ovarian function" (G. S. Greenwald and P. F. Terranova, eds.), pp. 45–48. Raven Press, New York.

Benadiva, C. A., Ben Rafael, Z., Strauss, J. F., Mastroianni, L., and Flickinger, G. L. (1988). *Fertil. Steril.* **49**, 997–1001.

Bevers, M. M., Dieleman, S. J., and Kruip, T. A. M. (1988). *Anim. Reprod. Sci.* **17**, 21–32.

Bézard, J., Vigier, B., Tran, D., and Mauleon, P. (1988). *Reprod. Nutr. Develop.* **28**, 1105–1112.

Bindon, B. M., and Pennycuik, P. (1974). *J. Reprod. Fertil.* **36**, 221–224.

Bjersing, L., and Carstensen, H. (1967). *J. Reprod. Fertil.* **14**, 101–111.

Black, J. L., and Erickson, B. H. (1968). *Anat. Record* **161**, 45–55.

Brand, A., and de Jong, W. H. R. (1973). *J. Reprod. Fertil.* **33**, 431–439.

Britt, J. H., Armstrong, J. D., Cox, N. M., and Esbenshade, K. L. (1985). *J. Reprod. Fertil. Suppl.* **33**, 37–54.

Brown, B. W., and Driancourt, M. A. (1989). *J. Reprod. Fertil.* **85**, 317–323.

Brown, B. W., and Mattner, P. E. (1984). *Aust. J. Biol. Sci.* **37**, 389–392.

Brown, B. W., Hales, J. R. S., and Mattner, P. E. (1974). *Experientia* **30**, 914–915.

Brown, B. W., Cognie, Y., Chemineau, P. Poulin, N., and Salama, O. A. (1988). *J. Reprod. Fertil.* **84**, 653–658.

Brown, J. L., and Reeves, J. J. (1983). *Biol. Reprod.* **29**, 1179–1182.

Brown, J. L., Shoenemann, H. M., and Reeves, J. J. (1986). *J. Anim. Sci.* **62**, 1063–1071.

Bruce, N. W., and Moor, R. M. (1975). *J. Reprod. Fertil.* **46**, 229–239.

Bryant-Greenwood, G. D., Jeffrey, R., Ralph, M. M., and Seamark, R. F. (1980). *Biol. Reprod.* **23**, 792–800.

Buck, P. A., and Schomberg, D. W. (1985). 67th Annu. Meet. Endocr. Soc. Baltimore, 1985, p. 151 (Abstr. 602).

Buck, P. A., and Schomberg, D. W. (1988). *Endocrinology* **122**, 28–33.

Bushmeyer, S. M., Bellin, M. E., Brantmeier, S. A., Boehm, S. K., Kubajax, C. L., and Ah, R. L. (1985). *Endocrinology* **117**, 879–885.

Butcher, R. L., and Kirkpatrick-Keller, D. (1984). *Biol. Reprod.* **31**, 280–286.

Byskov, A. G. (1979). In "Ovarian Follicular Development and Function" (A. R. Midgley and W. A. Sadler, eds.), pp. 41–59. Raven Press, New York.

Cahill, L. P., and Mauleon, P. (1980). *J. Reprod. Fertil.* **58**, 321–328.

Cahill, L. P., Mariana, J. C., and Mauléon, P. (1979). *J. Reprod. Fertil.* **55**, 27–36.

Cahill, L. P., Saumande, J., Ravault, J. P., Blanc, M., Thimonier, J., Mariana, J. C., and Mauleon, P. (1981). *J. Reprod. Fertil.* **62**, 141–150.

Cahill, L. P., Clarke, I. J., Cummins, J. T., Driancourt, M. A., Carson, R. S., and Findlay, J. K. (1984). *In* "Proc. 5th Ovarian Workshop" (D. O. Toft and R. J. Ryan, eds.), pp. 35–39. Champaign, Illinois.

Carson, R. S., Findlay, J. K., Burger, H. G., and Trounson, A. O. (1979). *Biol. Reprod.* **21**, 75–87.

Carson, R. S., Findlay, J. K., Clarke, I. J., and Burger, H. G. (1981). *Biol. Reprod.* **24**, 105–113.

Carson, R. S., Salamonsen, L. A., and Findlay, J. K. (1986). *J. Reprod. Fertil.* **76**, 663–676.

Chan, W. K., and Tan, C. H. (1986). *J. Endocr.* **108**, 335–341.

Chan, W. K., and Tan, C. H. (1988). *Endocrinlogy* **122**, 2290–2298.

Channing, C. P. (1975). *Proc. Soc. Exp. Biol. Med.* **149**, 238–241.

Channing, C. P. (1979). *Adv. Exp. Med. Biol.* **112**, 327–346.

Channing, C. P., and Evans, V. W. (1982). *Endocrinology* **111**, 1746–1748.

Channing, C. P., Bae, I. H., Stone, S. L., Anderson, L. D., Edelson, S., and Fowler, S. C. (1981). *Mol. Cell. Endocrinol.* **22**, 359–370.

Channing, C. P., Garrett, R., Kroman, N., Conn, T., and Gospodarowicz, D. (1983). *In* "Factors Regulating Ovarian Function" (G. S. Greenwald and P. F. Terranova, eds.), pp. 215–220. Raven Press, New York.

Chiras, D. D., and Greenwald, G. S. (1978). *Am. J. Anat.* **152**, 307–320.

Chupin, D., and Saumande, J. (1979). *Ann. Biol. Anim. Biochim. Biophys.* **19**, 1489–1498.

Chupin, D., Combarnous, Y., and Procureur, R. (1984). *Theriogenology* **23**, 184 (Abstr.).

Clark, J. R., Dierschke, D. J., and Wolf, R. C. (1979). *Biol. Reprod.* **21,** 497–508.

Coleman, D. A., and Dailey, R. A. (1983). *Biol. Reprod.* **29,** 586–593.

Cran, D. G., Moor, R. M., and Hay, M. F. (1976). *Acta Endocrinol. (Copenh.)* **82,** 631–636.

Cran, D. G., Hay, M. F., and Moor, R. M. (1979). *Cell Tissue Res.* **202,** 439–451.

Cran, D. G., Osborn, J. C., and Rushton, D. (1983). *Reprod. Nutr. Develop.* **23,** 285–292.

Dalin, A. M. (1987). *J. Vet. Med. Assoc.* **34,** 592–601.

Darga, N. C., and Reichert, L. E. (1978). *Biol. Reprod.* **19,** 235–241.

De Leon, V., Johnson, A., Bachvarova, R. (1983). *Dev. Biol.* **98,** 400–408.

Dieleman, S. J., and Blankenstein, D. M. (1984). *J. Reprod. Fertil.* **72,** 487–494.

Dieleman, S. J., Bevers, M. M., Poortman, J., and Van Tol, H. T. M. (1983a). *J. Reprod. Fertil.* **69,** 641–649.

Dieleman, S. J., Kruip, T. A. M., Fontijne, P., De Jong, W. H. R., and Van Der Weyden, G. C. (1983b). *J. Endocrinol.* **97,** 31–42.

Di Zerega, G. S., and Hodgen, G. D. (1980a). *J. Clin. Endocrinol. Metab.* **51,** 903–910.

Di Zerega, G. S., and Hodgen, G. D. (1980b). *J. Clin. Endocrinol. Metab.* **50,** 819–829.

Di Zerega, G. S., Campeau, J. D., Nakamura, R. N., Ujita, E. L., Lobo, R., and Marrs, R. (1983a). *J. Clin. Endocrinol. Metab.* **57,** 838–844.

Di Zerega, G. S., Marut, E. L., Turner, C. K., and Hodgen, G. D. (1980). *J. Clin. Endocrinol. Metab.* **51,** 698–707.

Di Zerega, G. S., Marrs, R. P., Campeau, J. D., and Kling, O. R. (1983b). *J. Clin. Endocrinol. Metab.* **56,** 147–158.

Dorrington, J. H., McKeracher, H. L., Chan, A. K., and Gore-Langton, R. E. (1983). *J. Steroid Biochem.* **19,** 17–32.

Dott, H. M., Hay, M. F., Cran, D. G., and Moor, R. M. (1979). *J. Reprod. Fertil.* **56,** 683–689.

Driancourt, M. A. (1987). *J. Reprod. Fertil.* **80,** 207–212.

Driancourt, M. A., and Cahill, L. P. (1984). *J. Reprod. Fertil.* **71,** 205–211.

Driancourt, M. A., and Fry, R. C. (1988). *J. Anim. Sci.* **66** (suppl. 2), 9–20.

Driancourt, M. A., and Palmer, E. (1984). *Theriogenology* **21,** 591–599.

Driancourt, M. A., Cahill, L. P., and Bindon, B. M. (1985a). *J. Reprod. Fertil.* **73,** 93–107.

Driancourt, M. A., Fry, R. C., Clarke, I. J., and Cahill, L. P. (1987). *J. Reprod. Fertil.* **79,** 635–644.

Driancourt, M. A., Gauld, I. K., Terqui, M., and Webb, R. (1986). *J. Reprod. Fertil.* **78,** 565–575.

Driancourt, M. A., Gibson, W. R., and Cahill, L. P. (1985b). *Reprod. Nutr. Develop.* **25,** 1–15.

Driancourt, M. A., Philipon, P., Locatelli, A., Jacques, E., and Webb, R. (1988). *J. Reprod. Fertil.* **83,** 509–516.

Driancourt, M. A., Bodin, L., Boomarov, O., Thimonier, J., and Elsen, J. M. (1989a). *J. Anim. Sci.* (in press).

Driancourt, M. A., Castonguay, F. W., Bindon, B. M., Piper, L. R., Quirke, J. F., and Hanrahan, J. P. (1989b). *J. Anim. Sci.* (in press).

Dufour, J., Cahill, L. P., and Mauleon, P. (1979). *J. Reprod. Fertil.* **57,** 301–309.

England, B. G., Dahmer, M. K., and Webb, R. (1981). *Biol Reprod.* **24,** 1068–1075.

Erickson, B. H. (1966a). *J. Reprod. Fertil.* **10,** 97–105.

Erickson, B. H. (1966b). *J. Anim. Sci.* **25,** 800–805.

Erickson, B. H. (1967). *Int. J. Radiat. Biol.* **13,** 57–67.

Erickson, G. F. (1986). *Semin. Reproductive Endocrinol.* **4,** 233–254.

Evans, G., Wathes, D. C., King, G. J., Armstrong, D. T., and Porter, D. G. (1983). *J. Reprod. Fertil.* **69,** 677–683.

Evans, G., Dobias, M., King, G. J., and Armstrong, D. T. (1981). *Biol. Reprod.* **25,** 673–682.

Faddy, M. J. (1976). *Biometrics* **32,** 443–448.

Faddy, M. J., and Jones, M. C. (1988). *Biometrics* **44,** 587–593.

Faddy, M. J., Jones, E. C., and Edwards, R. G. (1976). *J. Exp. Zool.* **197,** 173–176.

Faddy, M. J., Telfer, E., and Gosden, R. G. (1987). *Cell. Tissue Kinet.* **20,** 551–560.

Fay, J. E., and Douglas, R. H. (1987). *J. Reprod. Fertil. Suppl.* **35,** 169–181.

Feng, P., Knecht, M., and Catt, K. (1987). *Endocrinology* **120,** 1121–1126.

Findlay, J. K. (1986). *J. Endocrinol.* **111,** 357–366.

Fleming, M. W., Rhodes, R. C., and Dailey, R. A. (1984). *Biol. Reprod.* **30,** 82–86.

Fletcher, W. H., and Greenan, J. R. T. (1985). *Endocrinology* **116,** 1660–1662.

Ford, K. A., and La Barbera, A. R. (1988). *Endocrinology* **123,** 2374–2381.

Fortune, J. E., and Hansel, W. (1979a). *Biol. Reprod.* **20** (Suppl. 1), 46 (Abstr).

Fortune, J. E., and Hansel, W. (1979b). *Endocrinology* **104**, 1834–1838.

Fortune, J. E., and Hansel, W. (1985). *Biol. Reprod.* **32**, 1069–1079.

Fry, R. C., Clarke, I. J., Cummins, J. T., Bindon, B. M., Piper, L. R., and Cahill, L, P. (1988). *J. Reprod. Fertil.* **82**, 711–715.

Goldenberg, R. L., Bridson, W. E., and Kohler, P. O. (1972). *Biochem. Biophys. Res. Commun.* **48**, 101–107.

Goodman, A. L., and Hodgen, G. D. (1979). *Endocrinology* **104**, 1304–1316.

Gosling, J. P., Morgan, P., and Sreenan, J. M. (1978). *In* "Control of Reproduction in the Cow" (J. M. Sreenan, ed.), pp. 225–236. Martinus Nijhoff, Boston.

Gospodarowicz, D. (1987). *In* "Methods in Enzymology: Peptide Growth Factors" (D. Barnes and D. Sirbaski, eds.), Vol. 147A, pp. 106–119. Academic Press, Orlando, Florida.

Gospodarowicz, D., Bialecki, H., and Thakral, T. K. (1979). *Exp. Eye Res.* **28**, 501–514.

Gospodarowicz, D., Ill, C. R., and Birdwell, C. R. (1977). *Endocrinology* **100**, 1108–1120.

Gougeon, A., and Lefevre, B. (1983). *J. Reprod. Fertil.* **69**, 497–504.

Grasso, F., Guilbault, L. A., Roy, G. L., Matton, P., and Lussier, J. G. (1989). *Theriogenology* **31**, 199 (Abstr.).

Greenwald, G. S. (1974). *In* "Handbook of Physiology," Vol. IV, Section 7, Vol IV, Ch. III, pp 293–323.

Gregoraszczuk, E., Stoklosowa, S., Tarnawska, M., and Rzasa, J. (1988). *Anim. Reprod. Sci.* **17**, 141–154.

Grimes, R. W., Matton, P., and Ireland, J. J. (1987). *Biol. Reprod.* **37**, 82–88.

Hage, A. J., Groen-Klevant, A. C., and Welschen, R. (1978). *Acta Endocrinol.* **88**, 375–382.

Halpin, D. M. G., Charlton, H. M., and Faddy, M. J. (1986). *J. Reprod. Fertil.* **78**, 119–125.

Hammond, J. M., and English, H. F. (1987). *Endocrinology* **120**, 1039–1046.

Hammond, J. M., Baranao, J. L. S., Skalers, D., Knight, A. B., Romanus, J. A., and Rechler, M. M. (1985). *Endocrinology* **117**, 2553–2555.

Hammond, J. M., Hsu, C. J., Klindt, J., Tsang, B. K., and Downey, B. R. (1988). *Biol. Reprod.* **38**, 304–308.

Haney, A. F., and Schomberg, D. W. (1978). *Biol. Reprod.* **19**, 242–248.

Hay, M. F., and Moor, R. M. (1975). *J. Reprod. Fertil.* **45**, 583–593.

Hay, M. F., Cran. D. G., and Moor, R. M. (1976). *Cell. Tiss. Res.* **169**, 515–529.

Hay, M. F., Moor, R. M., Cran, D. G., and Dott, H. M. (1979). *J. Reprod. Fertil.* **55**, 195–207.

Henderson, K. M., and Franchimont, P. (1981). *J. Reprod. Fertil.* **63**, 431–442.

Henderson, K. M., and McNatty, K. P. (1987). Proc. 4th AAAP Animal Science Congress, pp. 130–133. Hamilton, New Zealand.

Henderson, K. M., Kieboom, L. E., McNatty, K. P., Lun, S., and Heath, D. A. (1984). *Mol. Cell. Endocrinol.* **34**, 91–98.

Henderson, K. M., Kieboom, L. E., McNatty, K. P., Lun, S., and Heath, D. A. (1985). *J. Reprod. Fertil.* **75**, 111–120.

Henderson, K. M., Ellen, R. L., Savage, L. C., Ball, K., and McNatty, K. P. (1986). *Proc. N. Z. Soc. Anim. Prod.* **46**, 157–160.

Henderson, K. M., McNatty, K. P., O'Keeffe, L. E., Lun, S., Heath, D. A., and Prisk, M. D. (1987b). *J. Reprod. Fertil.* **81**, 395–402.

Henderson, K. M., McNatty, K. P., Smith, P., Gibb, M., O'Keeffe, L. E., Lun, S., Heath, D. A., and Prisk, M. D. (1987a). *J. Reprod. Fertil.* **79**, 185–193.

Henderson, K. M., Savage, L. C., Ellen, R. L., Ball, K., and McNatty, K. P. (1988). *J. Reprod. Fertil.* **84**, 187–196.

Hild-Petito, S., Stouffer, R. L., and Brenner, R. M. (1988). *Endocrinology* **123**, 2896–2905.

Hillier, S. G., Van den Bogaard, A. M. S., Reichert, L., and Van Hall, E. V. (1980). *J. Clin. Endocrinol. Metab.* **50**, 640–647.

Hirshfield, A. N. (1984). *Biol. Reprod.* **31**, 52–58.

Hirschfield, A. N. (1986). *Biol. Reprod.* **35**, 113–118.

Hirshfield, A. N. (1989). *Biol. Reprod.* (in press).

Hsu, C. J., and Hammond, J. M. (1987a). *Endocrinology* **120**, 198–207.

Hsu, C. J., and Hammond, J. M. (1987b). *Endocrinology* **121**, 1343–1348.

Hsu, C. J., Holmes, S. D., and Hammond, J. M. (1987). *Biochem. Biophys. Res. Commun.* **147**, 242–247.

Hsueh, A. J. W., Adashi, E. Y., Jones, P. B. C., and Welsh, T. H. (1984). *Endocr. Rev.* **5**, 76–127.

Ireland, J. J., and Roche, J. F. (1982). *Endocrinology* **111**, 2077–2086.

Ireland, J. J., and Roche, J. F. (1983a). *Endocrinology* **112**, 150–156.

Ireland, J. J., and Roche, J. F. (1983b). *J. Anim. Sci.* **57**, 157–167.

Ireland, J. J., Fogwell, R. L., Ohender, W. D., Ames, K., and Cowley, J. L. (1984). *J. Anim. Sci.* **59**, 764–771.

Jones, E. C. (1961). *J. Endocr.* **21**, 497–509.

Jorio, A. (1987). Thèse de doctorat de l'Universite Paris VI, Paris, France.

Jungclas, B., and Luck, M. R. (1986). *J. Endocr.* **109,** R1–R4.

Kenigsberg, D., Littman, B. A., Williams, R. F., and Hodgen, G. D. (1984). *Fertil. Steril.* **42,** 116–126.

Kenney, R. M., Condon, W., Ganjam, V. K., and Channing, C. (1979). *J. Reprod. Fertil. Suppl.* **27,** 163–171.

Kokolis, N., AleHaki-Tzivanidou, E., and Smokovitis, A. (1987). *Curr. Top. Vet. Med. Anim. Sci.* **39,** 215–220.

Kotwica, J., Williams, G. L., and Marchello, M. J. (1981). *Biol. Reprod.* **24**(Suppl. 1), 144, Abstr. 243.

Kotwica, J., Williams, G. L., and Marchello, M. J. (1982). *Biol. Reprod.* **27,** 778–789.

Kraemer, P. M., and Tobey, R. A. (1972). *J. Cell. Biol.* **55,** 713–717.

Krishnan, K. A., Vijayalakshmi, S., and Sheth, A. R. (1983). *Indian J. Exp. Biol.* **21,** 229–236.

Kruip, T. A. M., and Dieleman, S. J. (1982). *Reprod. Nutr. Develop.* **22,** 465–473.

Kruip, T. A. M., and Dieleman, S. J. (1985). *Theriogenology* **24,** 395–408.

Kruip, T. A. M., Vullings, H. G. B., Schams, D., Jonis, J., and Klarenbeek, A. (1985). *Acta Endocrinol. (Copenh.)* **109,** 537–542.

Krzymowski, T., Kotwica, J., Stefanczyk, S., Czarnocki, J., and Debek, J. (1982). *J. Reprod. Fertil.* **65,** 457–465.

La Barbera, A. R., and Ryan, R. J. (1981a). *Am. J. Physiol.* **240,** E622–E629.

La Barbera, A. R., and Ryan, R. J. (1981b). *Endocrinology* **108,** 1561–1570.

Lacker, H. M. (1981). *Biophys. J.* **35,** 433–454.

Lacroix, E., Eechaute, W., and Leusen, I. (1974). *Steroids* **23,** 337–356.

Lahlou Kassi A. (1982). Thèse de doctorat des sciences agronomiques, Rabat, France.

Lamberson, W. R., and Lambeth, V. A. (1986). *Theriogenology* **26,** 643–648.

Ledwitz-Rigby, F. (1983). *Mol. Cell. Endocrinol.* **29,** 213–223.

Ledwitz-Rigby, F., Gross, T. M., Schjeide, O. A., and Rigby, B. W. (1984). *Biol. Reprod.* **30**(Suppl. 1), 85, Abstr. 110.

Ledwitz-Rigby, F. (1985). *Int. Res. Commun. Syst. Med. Sci.* **13,** 485–487.

Ledwitz-Rigby, F., Gross, T. M., Schjeide, O. A., and Rigby, B. W. (1987). *Biol. Reprod.* **36,** 320–327.

Lee, C. Y. (1976). *Endocrinology* **99,** 42–48.

Lindsey, A. M., and Channing, C. P. (1979). *Biol. Reprod.* **20,** 473–482.

Lino, J., Baranao, S., and Hammond, J. M. (1985). *Endocrinology* **116,** 2143–2151.

Lintern-Moore, S., Peters, H., Moore, G. P. M., and Faber, M. (1974). *J. Reprod. Fertil.* **39,** 53–64.

Loeken, M. R., and Channing, C. P. (1985). *J. Reprod. Fertil.* **73,** 343–351.

Lussier, J. G., and Carruthers, T. O. (1987). *Theriogenology* **27,** 253 (Abstr.).

Lussier, J. G., Matton, P., and Dufour, J. J. (1987). *J. Reprod. Fertil.* **81,** 301–307.

Makris, A., Ryan, K. J., Yasumizu, T., Hill, C. L., and Zetter, B. R. (1984). *Endocrinology* **115,** 1672–1677.

Mandl, A. M., and Zuckerman, S. (1952). *J. Endocrinol.* **8,** 126–132.

Mariana, J. C. (1970). *Ann. Biol. Anim. Bioch. Biophys.* **10,** 575–579.

Mariana, J. C. (1979). *Ann. Biol. Anim. Bioch. Biophys.* **19**(5), 1469–1474.

Mariana, J. C., and de Pol, J. (1986). *Arch. Biol.* **97,** 139–156.

Mariana, J. C., and Hirshfield, A. N., (1989a). *Biol. Reprod.* (in press).

Mariana, J. C., and Hirshfield, A. N. (1989b). *Biol. Reprod.* (in press).

Mariana, J. C., Hulot, F., Dervin, C., Tomassone, R., and Poujardieu, B. (1989). *Arch. Biol.* **100,** 47–63.

Mariana, J. C., Mauleon, P., Benoit, M., and Chupin, D. (1970). *Ann. Biol. Anim. Bioch. Biophys* **10**(HS), 47–63.

Martin, G. B., Price, C. A., Thiery, J. C., and Webb, R. (1988). *J. Reprod. Fertil.* **82,** 319–328.

Maruo, T., Hayashi, M., Matsuo, H., Ueda, Y., Morikawa, H., and Mochizuki, M. (1988). *Acta Endocrinol. (Copenh.)* **117,** 230–240.

Mattioli, M., Galeati, G., and Seren, E. (1988a). *Gamete Res.* **20,** 177–183.

Mattioli, M., Galeati, G., Bacci, M. L., and Seren, E. (1988b). *Gamete Res.* **21,** 223–232.

Mattner, P. E., Brown, B. W., and Hales, J. R. S. (1981). *J. Reprod. Fertil.* **63,** 279–284.

Matton, P., Adelakoun, V., Couture, Y., and Dufour, J. J. (1981). *J. Anim. Sci.* **52,** 813–820.

Mauléon, P. (1961). IVth International Congress of Animal Reproduction, (The Hague, pp. 348–354.

Mauléon, P. (1977). *In* "Control of Ovulation" (D. B. Crighton, G. R. Foxcroft, N. B. Haynes, and G. E. Lamning, eds.), pp. 141–158. Butterworths, Boston.

Maxson, W. S., Haney, A. F., and Schomberg, D. W. (1985). *Biol. Reprod.* **33**, 495–501.

May, J. V., and Schomberg, D. W. (1984a). *Mol. Cell. Endocrinol.* **34**, 201–213.

May, J. V., and Schomberg, D. W. (1984b). *Endocrinology* **114**, 153–163.

May, J. V., Buck, P. A., and Schomberg, D. W. (1985). 25th Ann. Meet. Am. Soc. Cell Biol., Atlanta 1985, 380a (Abstr. 1440).

May, J. V., Buck, P. A., and Schomberg, D. W. (1987). *Endocrinology* **120**, 2413–2420.

May, J. V., Mccarty, J. K., Reichert, L. E., and Schomberg, D. W. (1980). *Endocrinology* **107**, 1041–1049.

May, J. V., Frost, J. P., and Schomberg, D. W. (1988). *Endocrinology* **123**, 168–179.

McCracken, J. A., Uno, A., Goding, J. R., Ichikawa, Y., and Baird, D. T. (1969). *J. Endocrinol.* **45**, 425–440.

McNatty, K. P. (1978). *In* "The Vertebrate Ovary" (R. Jones, ed.), pp. 215–259. Plenum Press, New York.

McNatty, K. P., Dobson, C., Gibb, M., Kieboom, L., and Thurley, D. C. (1981a). *J. Endocrinol.* **91**, 99–109.

McNatty, K. P. (1982). *In* "Follicular Maturation and Ovulation" (R. Rolland, E. V. Van Hall, S. G. Hillier, K. P. McNatty, and J. Schoemaker, eds.), pp. 1–18. Excerpta Medica, Amsterdam.

McNatty, K. P., Gibb, M., Dobson, C., and Thurley, D. C. (1981c). *J. Endocrinol.* **90**, 375–389.

McNatty, K. P., Gibb, M., Dobson, C., Ball, K., Coster, J., Heath, D. A., and Thurley, D. C., (1982). *J. Reprod. Fertil.* **65**, 111–123.

McNatty, K. P., Gibb, M., Dobson, C., Thurley, D. C., and Findlay, J. K. (1981b). *Aust. J. Biol. Sci.* **34**, 67–80.

McNatty, K. P., Heath, D. A., Henderson, K. M., Lun, S., Hurst P. R., Ellis, L. M., Montgomery, G. W., Morrison, L., and Thurley, D. C. 1984a). *J. Reprod. Fertil.* **72**, 39–53.

McNatty, K. P., Henderson, K. M., Lun, S., Heath, D. A., Ball, K., Hudson, N. L., Fannin, J., Gibb, M., Kieboom, L. E., and Smith, P. (1985c). *J. Reprod. Fertil.* **73**, 109–120.

McNatty, K. P., Hudson, N., Gibb, M., Ball, K., Henderson, K. M., Heath, D. A., Lun, S., and Kieboom, L. E. (1985b). *J. Reprod. Fertil.* **75**, 121–131.

McNatty, K. P., Hudson, N., Henderson, K. M., Lun, S., Heath, D. A., Gibb, M., Ball, K., McDiarmid, J. M., and Thurley, D. C. (1984b). *J. Reprod. Fertil.* **70**, 309–321.

McNatty, K. P., Lun, S., Heath, D. A., Kieboom, L. E., and Henderson, K. M. (1985a). *Mol. Cell. Endocrinol.* **39**, 209–215.

McNatty, K. P., O'Keeffe, L. E., Henderson, K. M., Heath, D. A., and Lun, S. (1986). *J. Reprod. Fertil.* **77**, 477–488.

McNatty, K. P., Hudon, N., Henderson, K. M., Gibb, M., Morrison, L., Ball, K., and Smith, P. (1987). *J. Reprod. Fertil.* **80**. 577–588.

McNeilly, A. S. (1985). *J. Reprod. Fertil.* **74**, 661–668.

McNeilly, A. S., Jonassen, J. A., and Fraser, H. M. (1986). *J. Reprod. Fertil.* **76**, 481–490.

Meinecke, B., Gips, H., and Meinecke-Tillmann, S. (1987). *Curr. Top. Vet. Med. Anim. Sci.* **39**, 207–213.

Merz, E. A., Hauser, E. R., and England, B. G. (1981). *J. Anim. Sci.* **52**, 1457–1468.

Miller, K. F., Nordheim, E. V., and Ginther, O. J. (1981). *Theriogenology* **16**, 669–679.

Miyamoto, H., Ishibashi, T., and Nakano, S. (1985). *Jpn. J. Zootech. Sci.* **56**, 353–360.

Mondschein, J. S., and Hammond, J. M. (1988). *Endocrinology* **123**, 463–468.

Mondschein, J. S., May, J. V., Gunn, E. B., and Schomberg, D. W. (1981). *In* "Dynamics of Ovarian Function" (N. B. Schwartz and M. Hunnzicker-Dunn, eds.), pp. 83–88. Raven Press, New York.

Monniaux, D. (1987). *J. Reprod. Fertil.* **79**, 505–515.

Monniaux, D., and De Reviers, M. M. (1989). *J. Reprod. Fertil.* **85**, 151–160.

Monniaux, D., Chupin, D., and Saumande, J. (1983). *Theriogenology* **19**, 55–81.

Monniaux, D., Mariana, J. C., and Gibson, W. R. (1984a). *J. Reprod. Fertil.* **70**, 243–253.

Monniaux, D., Mariana, J. C., Gibson, W. R., and Roux, C. (1984b). *In* "Periode-Periovulatoire" (Masson, ed.), pp. 69–84. Colloque de la société Francaise pour l'Etude de la Fertilité, Paris.

Montgomery, G. W., Martin, G. B., Blanc, M. R., and Pelletier, J. (1987). *J. Reprod. Fertil.* **80**, 271–277.

Montz, F. J., Ujita, E. L., Campeau, J. D., and Di Zerega, G. S. (1984). *Am. J. Obstet. Gynecol.* **148**, 436–441.

Moor, R. M. (1977). *J. Endocrinol.* **73**, 143–150.

Moor, R. M., and Heslop, J. P. (1981). *J. Exp. Zool.* **216**, 205–209.

Moor, R. M., and Trounson, A. O. (1977). *J. Reprod. Fertil.* **49**, 101–109.

Moor, R. M., Hay, M. F., and Seamark, R. F. (1975). *J. Reprod. Fertil.* **45**, 595–604.

Moor, R. M., Hay, M. F., Dott, H. M., and Cran, D. G. (1978). *J. Endocrinol.* **77**, 309–318.

Moor, R. M., Polge, C., and Willadsen, S. M. (1980b). *J. Embryol. Exp. Morphol.* **56**, 319–335.

Moor, R. M., Smith, M. W., and Dawson, R. M. C. (1980a). *Exp. Cell Res.* **126**, 15–29.

Moor, R. M., Osborn, J. C., and Crosby, I. M., (1985). *J. Reprod. Fertil.* **74**, 167–172.

Moore, G. P. M., and Lintern-Moore, S. (1978). *Biol. Reprod.* **18**, 865–870.

Moore, G. P. M., and Lintern-Moore, S. (1979). *Ann. Biol. Anim. Bioch. Biophys.* **19**(5), 1409–1417.

Motlik, J., and Fulka, J. (1974). *J. Reprod. Fertil.* **36**, 235–237.

Motlik, J., Fulka, J., and Fléchon, J. E. (1986). *J. Reprod. Fertil.* **76**, 31–37.

Murdoch, W. J., Nix, K. J., and Dunn, T. G. (1983). *Biol. Reprod.* **28**, 1001–1006.

Murray, S. A., and Fletcher, W. H. (1984). *J. Cell Biol.* **98**, 1710–1719.

Nakano, R., Mizuno, T., Katayama, K., and Tojo, S. (1975). *J. Reprod. Fertil.* **45**, 545–546.

Neufeld, G., Ferrara, N., Schweigerer, L., Mitchell, R., and Gospodarowicz, D. (1987). *Endocrinology* **121**, 597–603.

Nicosia, S. V. (1980). *In* "Endocrine Physiopathology of the Ovary" (R. I. Tozzini, G. Reeves, and R. L. Pineda, eds.), pp. 43–62. Elsevier Biomedical Press, New York.

Nimrod, A., and Lindner, H. R. (1980). *FEBS Lett.* **119**, 155–157.

Niswender, G. D., Reimers, T. J., Diekman, M. A., and Nett, T. M. (1976). *Biol. Reprod.* **14**, 64–81.

Ono, T., Campeau, J. D., Holmberg, E. A., Nakamura, R. M., Ujita, E. L., Devereaux, D. L., Tonetta, S. A., Devinna, R., Ugalde, M., and Di Zerega, G. S. (1986). *Am. J. Obstet. Gynecol.* **154**, 709–716.

Osborn, J. C., and Moor, R. M. (1983). *J. Steroid Biochem.* **19**, 133–137.

O'Shea, J. D., Hay, M. F., and Cran, D. G. (1978). *J. Reprod. Fertil.* **54**, 183–187.

Otani, T., Maruo, T., and Mochizuki, M. 1985. *Acta Endocrinol. (Copenh.)* **108**, 104–110.

Oussaid, B. (1983). Thèse 3ème cycle.

Parshad, V. R., and Guraya, S. S. (1983). *Proc. Indian. Acad. Sci. (Anim. Sci.)* **92**, 121.

Pedersen, T. (1969). In "Gonadotrophins and Ovarian Development" (W. R. Butt, A. C. Crooke, M. E. Ryle, and S. Livingstone, eds.). Edinburgh.

Pedersen, T. (1970). *J. Reprod. Fertil.* **21**, 81–93.

Peters, H., Levy, E., and Crone, M. (1965). *J. Exp. Zool.* **158**, 169–180.

Peters, H., Byskov, A. G., Lintern Moore, S., Faber, M., and Andersen, M. (1973). *J. Reprod. Fertil.* **35**, 139–141.

Picton, H., and McNeilly, A. S. (1988). *J. Reprod. Fertil. Abstr. Ser.* **1**, 53 (Abstr.).

Pierson, R. A., and Ginther, O. J. (1984). *Theriogenology* **21**, 495–504.

Pierson, R. A., and Ginther, O. J. (1987a). *Anim. Reprod. Sci.* **14**, 177–186.

Pierson, R. A., and Ginther, O. J. (1987b). *Anim. Reprod. Sci.* **14**, 165–176.

Pierson, R. A., and Ginther, O. J. (1988). *Anim. Reprod. Sci.* **16**, 81–95.

Pierson, R. A., *et al.* (1989). To be published.

Quirk, S. M., Hickey, G. J., and Fortune, J. E. (1986). *J. Reprod. Fertil.* **77**, 24–219.

Racowsky, C. (1985). *J. Reprod. Fertil.* **74**, 9–21.

Rajakoski, E. (1960). *Acta Endocrinol. (Copenh.) Suppl.* **52**, 1–68.

Rajkumar, K., Klingshorn, P., Chedrese, P. J., and Murphy, B. D. (1988). *Can. J. Physiol. Pharmacol.* **66**, 561–566.

Read, K. L. Q., and Berry, P. J. B. (1988). Proc. I.F.A.C.-B.M.E. Symposium on Modelling and Control in Biomedical Systems, Venice, 1988, pp. 1–6.

Read, K.L.Q., Mariana, J. C., and De Reviers, M. M. (1979). *Ann. Biol. Anim. Bioch. Biophys.* **19**(5), 1419–1433.

Richards, J. S., Jahnsen, T., Hedin, L., Lifka, J., Ratoosh, S., Durica, J. M., and Goldring, N. B. (1987). *Recent Prog. Horm. Res.* **43**, 231–276.

Richardson, S. J., Senikas, V., and Nelson, J. F. (1987). *J. Clin. Endocrinol. Metab.* **65**, 1231.

Rodgers, R. J., Rodgers, H. F, Hall, P. F., Waterman, M. R., and Simpson, E. R. (1986). *J. Reprod. Fertil.* **78**, 627–638.

Rodgers, R. J., Mason, J. I., Waterman, M. R., and Simpson, E. R. (1987a). *Mol. Endocrinol.* **1**, 172–180.

Rodgers, R. J., Waterman, M. R., and Simpson, E. R. (1987b). *Mol. Endocrinol.* **1**, 274–279.

Rolland, R., and Hammond, J. M. (1975). *Endocr. Res. Commun.* **2**, 281–298.

Sadrkhanloo, R., Hofeditz, C., and Erickson, G. F. (1987). *Endocrinology* **120**, 146–155.

Salomon, Y., Amir, Y., Azulai, R., and Amsterdam, A. (1978). *Biochim. Biophys. Acta* **544**, 262–272.

Saumande, J., Chupin, D., Mariana, J. C., Ortavant, R., and Mauleon, P. (1978). *In* "Control of Reproduction in the Cow" (J. M. Sreenan, ed.), pp. 195–225. Martinus Nijhoff, The Hague.

Savio, J. D., Keenan, L., Boland, M. P., and Roche, J. F. (1988). *J. Reprod. Fertil.* **83**, 663–671.

Savion, N., Lui, G. M., Laherty, R., and Gospodarowicz, D. (1981). *Endocrinology* **109**, 409–420.

Scaramuzzi, R. J., Downing, J. A., Campbell, B. K., and Cognie, Y. (1986). Proc. Winter Mtg. Soc. Study Fertil., Abstr. 57.

Schneyer, A. L., Reichert, L. E., Franke, M., Ryan, R. J., and Sluss, P. M. (1988). *Endocrinology* **123,** 487–491.

Schomberg, D. W., May, J. V., and Mondschein, J. S. (1983). *In* "Factors Regulating Ovarian Function" (G. S. Greenwald and P. F. Terranova, eds.), pp. 221–224. Raven Press, New York.

Schomberg, D. W., Stouffer, R. L., and Tyrey, L. (1976). *Biochem. Biophys. Res. Commun.* **68,** 77–85.

Schomberg, D. W., Williams, R. F., Tyrey, L., and Ulberg, L. C. (1978). *Endocrinology* **102,** 984–987.

Schwartz, N. B. (1974). *Biol Reprod.* **10,** 236–272.

Shaw, K. J., Campeau, J. D., Roche, P. C., and Di Zerega, G. S. (1985). *Exp. Clin. Endocrinol.* **86,** 26–34.

Sirois, J., and Fortune, J. E. (1988). *Biol. Reprod.* **39,** 308–317.

Skinner, M. K., and Osteen, K. G. (1988). *Endocrinology* **123,** 1668–1675.

Skinner, M. K., Keski-Oja, J., Osteen, K. G., and Moses, H. L. (1987a). *Endocrinology* **121,** 786–792.

Skinner, M. K., Lobb, D., and Dorrington, J. H. (1987b). *Endocrinology* **121,** 1892–1899.

Sluss, P. M., and Reichert, L. E. (1984). *Biol. Reprod.* **30,** 1091–1104.

Smeaton, T. C., and Robertson, H. A. (1971). *J. Reprod. Fertil* **25,** 243–252.

Sonjaya, H., and Driancourt, M. A. (1989). *Reprod. Fertil. Develop.* (in press).

Spicer, L. J., Convey, E. M., Leung, K., Short, R. E.,and Tucker, H. A. (1986). *J. Anim. Sci.* 62, 742–750.

Spicer, L. J., Echternkamp, S. E., Canning, S. F., and Hammond, J. M. (1988). *Biol. Reprod.* **39,** 573–580.

Spicer, L. J., Matton, P., Echternkamp, S. E., Convey, E. M., and Tucker, H. A. (1987). *Biol. Reprod.* **36,** 890–898.

Sporn, M. B., Roberts, A. B., Wakefield, L. M., and Assoian, R. K. (1986). *Science* **233,** 532–534.

Sreenan, J. M., Beehan, D., and Gosling, J. P. (1978). *In* "Control of Reproduction in the Cow" (J. M. Sreenan, eds.), pp. 144–159. Martinus Nijhoff, The Hague.

Staigmiller, R. B., England, B. G., Webb, R., Short, R. E., and Bellows, R. A. (1982). *J. Anim. Sci.* **55,** 1473–1482.

Sternlicht, A. L., and Schultz, R. M. (1981). *J. Exp. Zool.* **215,** 191–200.

Stoklosowa, S., Gregoraszczuk, E., and Channing, C. P. (1982). *Biol. Reprod.* **26,** 943–952.

Strauss, J. F., Paavola, L. G., Nestler, J. E., Soto, E. M., and Silavin, S. L. (1985). *In* "Proceedings of the Fifth Ovarian Workshop" (D. O. Toft and R. J. Ryan, eds.), pp. 275–302. Laramie, W. Y. Ovarian Workshops, Champaign, Illinois.

Suter, D. E., Bahr, J. M., Dziuk, P. J., and Schwartz, N. B. (1986). *Anim. Reprod. Sci.* **11,** 43–49.

Swift, A. D., and Crighton, A. B. (1980). *Theriogenology* **14,** 269–279.

Szöllösi, D. (1978). *Res. Reprod.* **10,** 2.

Szöllösi, D. Gerard, M., Ménézo, Y., and Thibault, C. (1978). *Ann. Biol. Anim. Bioch. Biophys.* **18,** 511–521.

Thanki, K. H., and Channing, C. P. (1976). *Endocr. Res. Commun.* **3,** 319–333.

Thanki, K. H., and Channing, C. P. (1978). *Endocrinology* **103,** 74–80.

Thibault, C., Gerard, M., and Menezo, Y. (1975). *Ann. Biol. Anim. Bioch. Biophys.* **15,** 705–714.

Thibault, C., Gerard, M., and Menezo, Y. (1976). *In* "Progress in Reproductive Biology" (P. O. Hubinont, ed.), pp. 233–240. Karger, Basel.

Thibault, C., Szöllösi, D., and Gerard, M. (1987). *Reprod. Nutr. Develop.* **27,** 865–896.

Toaff, M. E., Strauss, J. F., and Hammond, J. M. (1983). *Endocrinology* **112,** 1156–1158.

Tonetta, S. A., Yanagihara, D. L., de Vinna, R. S., and di Zerega, G. S. (1988). *Biol. Reprod.* **38,** 1001–1005.

Tsafriri, A., and Channing, C. P. (1975). *Endocrinology* **96,** 922–927.

Tsafriri, A. (1988). *In* "The Physiology of Reproduction" (E. Knobil and J. Neill, eds.), pp. 527–565. Raven Press, New-York.

Tsang, B. K., Moon, Y. S., and Armstrong, D. T. (1982). *Can. J. Physiol. Pharmacol.* **60,** 1112–1118.

Tsang B. K., Ainsworth, L., Downey, B. R., and Marcus G. J. (1985). *J. Reprod. Fertil.* **74,** 459–471.

Tsang, B. K., and Taheri, A. (1987). *Can. J. Physiol. Pharmacol.* **65,** 1951–1956.

Tsonis, C. G., Quigg, H., Lee, V. W. K., Leversha, L., Trounson, A. O., and Findlay, J. K. (1983). *J. Reprod. Fertil.* **67,** 83–90.

Tsonis, C. G., Carson, R. S., and Findlay, J. K. (1984a). *J. Reprod. Fertil.* **72,** 153–163.

Tsonis, C. G., Cahill, L. P., Carson, R. S., and Findlay, J. K. (1984b). *J. Reprod. Fertil.* **70,** 609–618.

Tuckey, R. C., and Stevenson, P. M. (1986). *Eur. J. Biochem.* **161,** 629–633.

Tuckey, R. C., Kostadinovic, Z., and Stevenson, P. M. (1988). *J. Steroid Biochem.* **31,** 201–205.

Turnbull, K. E., Braden, A. W. H., and Mattner, P. E. (1977). *Aust. J. Biol. Sci.* **30,** 229–241.

Ujita, E. L., Campeau, J. D., and Di Zerega, G. S. (1987). *Exp. Clin. Endocrinol.* **89,** 153–164.

Upadhyay, S., Luciani, J. M., and Zamboni, L. (1979). *Ann. Biol. Anim. Bioch. Biophys.* **19**(4b), 1153–1398.

Usuki, S., and Shioda, M. (1987). *Horm. Metab. Res.* **19,** 508–509.

Usuki, S., and Shioda, M. (1986). *Am. J. Obstet. Gynecol.* **155,** 447–451.

Van De Wiel, D. F. M., Bar-Ami, S., Tsafriri, A., and De Jong, F. H. (1983). *J. Reprod. Fertil.* **68,** 247–252.

Veldhuis, J. D. (1985a). *Endocrinology* **116,** 1818–1825.

Veldhuis, J. D. (1985b). *Endocrinology* **117,** 1076–1083.

Veldhuis, J. D. and Demers, L. M. (1985). *Biochem. Biophys. Res. Commun.* **130,** 234–239.

Veldhuis, J. D., and Furlanetto, R. W. (1985). *Endocrinology* **116,** 1235–1242.

Veldhuis, J. D., and Gwynne, J. T. (1985). *Endocrinology* **117,** 1321–1327.

Veldhuis, J. D., Demers, L. M., and Hammond, J. M. (1979). *Endocrinology* **105,** 1143–1152.

Veldhuis, J. D., Klase, P. A., Strauss, J. F., and Hammond, J. M. (1982). *Endocrinology* **111,** 441–446.

Veldhuis, J. D., Gwynne, J. T., Strauss, J. F., and Demers, L. M. (1984). *Endocrinology* **114,** 2312–2322.

Veldhuis, J. D., Rodgers, R. J., and Furlanetto, R. W. (1986a). *Endocrinology* **119,** 530–538.

Veldhuis, J. D., Rodgers, R. J., Dee, A., and Simpson, E. R. (1986b). *J. Biol. Chem.* **261,** 2499–2502.

Veldhuis, J. D., Nestler, J. E., and Strauss J. F., (1987). *Endocrinology* **121,** 340–346.

Vigier, B., Picard, J. Y., Tran, D., Legeai, L., and Josso, N. (1984). *Endocrinology* **114,** 1315–1320.

Vlodavsky, I., Brown, K. D., and Gospodarowicz, D. (1978). *J. Biol. Chem.* **253,** 3744–3750.

Wallace, J. M., and McNeilly, A. S. (1986). *J. Endocrinol.* **111,** 317–329.

Walsh, S. W., Yurtrzenka, C. J., and Davis, J. S. (1979). *Biol. Reprod.* **20,** 1167–1171.

Walters, D. L., Daniel, S. A. J., and Armstrong, D. T. (1984). *In* "Proc. 5th Anglo-French Meet. for Society to Study Fertility," p. 9.

Ware, C. B., Northey, D. L., and First, N. L. (1987). *Theriogenology* **27,** 292 (Abstr.).

Wassarman, P. M., and Mrozak, S. C. (1981). *Nature* **261,** 73–74.

Wathes, D. C., Swann, R. W., Porter, D. G., and Pickering, B. T. (1986). *In* "Current Topics in Neuroendocrinology" (D. Gantin and D. Pfaff, eds.), Vol. 6, pp. 129–152. Springer-Verlag, Berlin.

Webb, R., and England, B. G. (1982). *Endocrinology 110,* 873–881.

Webb, R., and Gauld, I. K. (1987). *In* "Follicular Growth and Ovulation Rate in Farm Animals" (J. F. Roche and D. O'Callaghan, eds.). Martinus Nijhoff, The Hague.

Webb, R., and Price, C. A. (1989). *J. Reprod. Fertil.* (in press).

Webb, R., and Gauld, I. K. (1985). *In* "Genetics of Reproduction in Sheep" (R. B. Land and D. W. Robinson, eds.). Butterworths, London.

Webb, R., Gauld, I. K., and Driancourt, M. A. (1989). *J. Reprod. Fertil.* (in press).

Welschen, R. (1973). *Ann. Biol. Anim. Bioch. Biophys.* **13**(HS), 195–206.

Wise, T. (1987). *J. Anim. Sci.* **64,** 1153–1169.

Wise, T., Vernon, M. W., and Maurer, R. R. (1986). *Theriogenology* **26,** 757–778.

Zamboni, L., Bezard, J., and Mauleon, P. (1979). *Ann. Biol. Anim. Bioch. Biophys.* **19**(4b), 1153–1178.

Zeleznik, A. J., and Kubik, C. J. (1986). *Endocrinology* **119,** 2025–2032.

Zeleznik, A. J., Schuler, H. M., and Reichert, L. E. (1981). *Endocrinology* **109,** 356–362.

Zhiwen, Z., Carson, R. S., Herington, A. C., Lee, V. W. K., and Burger, H. G. (1987). *Endocrinology* **120,** 1633–1637.

Zuckerman, S. (1962). *In* "The Ovary" (S. Zuckerman, ed.), Vol. I, pp. 251–253. Academic Press, New York.

Spermatogenesis

LARRY JOHNSON

Reproduction in Domestic Animals, Fourth Edition
Copyright © 1991 by Academic Press, Inc.

I. Introduction

A. History

Castration, a classical means to study reproductive biology, was introduced by Aristotle as early as 300 B.C. However, since spermatozoa and related germ cells are microscopic, their observation was not possible until Von Leeuwenhoek invented the microscope. He reported his findings on spermatozoa in 1679. Other major contributors followed. Von Kolliker (1841) discovered that spermatozoa develop from specialized cells located in seminiferous tubules of the testis (Fig. 1). Franz von Leydig in 1850 and E. Sertoli in 1865 described spermatogenic support cells. Von LaVallette (1876) classified germ cells morphologically, and work of Von Ebner (1871) and Benda (1887) led to comprehension of the orderly spermatogenic process. These findings led to the concept of the spermatogenic cycle (Regaud, 1901; Von Ebner, 1902). These classical

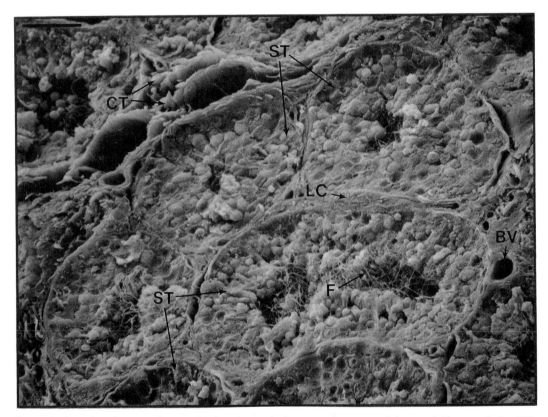

Figure 1 Scanning electron micrograph of equine testicular parenchyma containing seminiferous tubules (ST), blood vessels (BV), Leydig cells (LC), and connective tissue (CT). The majority of the testis is composed of seminiferous tubules, some of which have flagella (F) of developing spermatids extending into the lumen. Bar length equals 100 μm. From Johnson *et al.* (1978b).

studies set the basis for our present understanding of spermatogenesis (see review by Steinberger and Steinberger, 1975).

B. Definition of Spermatogenesis

Spermatogenesis is the lengthy, chronological process whereby a few stem-cell spermatogonia, lining the base of seminiferous tubules, divide by mitosis to maintain their own stem-cell numbers and to cyclically produce primary spermatocytes that, in turn, undergo meiosis to produce haploid spermatids, which differentiate into spermatozoa released into the tubular lumen (Fig. 2). Hence, spermatogenesis is the division and differentiation process by which spermatozoa are produced in seminiferous tubules of the testis. Seminiferous tubules constitute the major component of the testis (Fig.1).

For most species, the duration of spermatogenesis can be fairly equally divided into

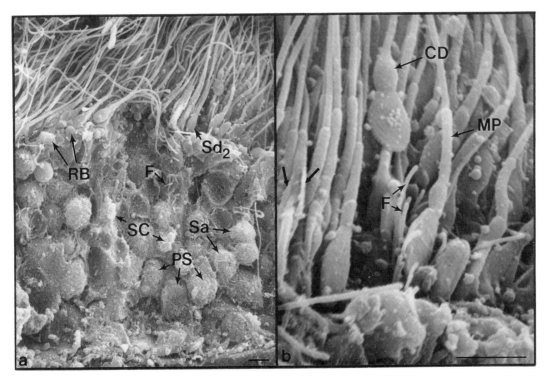

Figure 2 Scanning electron micrographs of seminiferous epithelium in stage VIII, the spermatogenic stage of spermiation, revealing (a) the entire epithelial height and (b) an enlarged luminal view. (a) Based on size and location, pachytene primary spermatocytes (PS), Sa spermatids with spherical nuclei (Sa), Sd_2 spermatids (Sd_2) whose entire length is exposed in the lumen, Sertoli cells (SC), and residual bodies (RB) can be identified. (b) The middle pieces (MP) of Sd_2 spermatids are enlarged with mitochondria and attached to residual cytoplasm by cytoplasmic stalks (arrows). The released spermatozoon has the characteristic cytoplasmic droplet (CD) of testicular spermatozoa. Developing flagella (F) from (a) Sa spermatids below the luminal surface project toward or (b) into the lumen. Bar length equals 5 μm. From Johnson *et al.* (1978b).

three phases: spermatocytogenesis (mitosis), meiosis, and spermiogenesis. In spermatocytogenesis, stem-cell spermatogonia divide by mitosis to produce other stem cells in order to continue the lineage throughout the adult life of the male and divide cyclically by mitosis to produce committed spermatogonia, which immediately proliferate to produce primary spermatocytes. During meiosis, the exchanged genetic material between homologous chromosomes of primary spermatocytes is followed by a reduction division producing haploid spermatids. Spermiogenesis is the differentiation of spherical spermatids into mature spermatids, which are released at the luminal free surface as spermatozoa.

II. Spermatozoon, Product of Spermatogenesis

A description of the main product of spermatogenesis, the spermatozoon, may enhance the understanding of the series of complex events that are necessary during spermatogenesis to create this remarkably differentiated cell.

The spermatozoon is composed of a head and tail (Fig. 3), and the general structures for domestic species are similar to those described for bulls (Saacke and Almquist, 1964) or stallions (Johnson et al., 1980). Horse spermatozoa are different in that a high percentage of tails are abaxial in that they do not attach to the middle of the head. The head contains the nucleus whose DNA represents the male genetic material (male genome) to be delivered to the egg and an acrosome, which contains the hydrolytic enzymes necessary for penetration of the egg vestments during fertilization. The tail includes the neck region, where it is attached to the implantation fossa of the head, the middle piece, which contains mitochondria, the principal piece, which contains the fibrous sheath, and the end piece. The plasma membrane encloses both the head and tail. From the apical region of the head, the plasma membrane covers the underlying acrosome, whose own membrane encloses various segments: the apical ridge, principal, and equatorial segments of the acrosome. The size and shape of the apical ridge vary widely among domestic species (Garner and Hafez, 1980). The acrosome

Figure 3 Micrographs of equine spermatozoa. Views include those produced by (a, c–i) transmission electron, (b) scanning electron, and (j) phase-contrast microscopy. Attaching lines between (a) and (b) or (b) and (c) and between (d–i) and (j) reveal corresponding regions of the spermatozoon. (a) The plasma membrane (PM) encloses the entire head and tail. The head is composed of the nucleus (N), overlying acrosome (A), and postacrosomal region (PR). The acrosome can be divided into the apical segment (ASA), principal segment (PSA), and equatorial segment (ESA). The inner acrosomal membrane (IAM) is in juxtaposition with the nuclear membrane (NM). The outer acrosomal membrane (OAM) fuses with the plasma membrane to discharge the acrosomal contents during the acrosome reaction. (j) The tail is composed of the middle piece (MP), principal piece (PP), and end piece (EP). (c) The tail attaches at the implantation fossa (IF), and it contains mitochondria (M) in the middle piece and the fibrous sheath (FS) in the principal piece. (d–i) The distal centriole is continuous with the outer nine doublet microtubules (DI) of the axoneme, which also has the characteristic central pair (CP) of microtubules. The nine dense fibers (DF) parallel the axoneme and extend to different lengths of the principal piece. In the end piece, the axonemal doublets become disorganized and ultimately separate into 20 single microtubules (SM), but still are enclosed in the plasma membrane. (b, c) The cytoplasmic droplet (CD), located at the proximal end of the middle piece of these spermatozoa from the equine efferent ducts and caput epididymidis, is characteristic of spermatozoa that have not yet matured in the epididymis. Bar length equals (a) 0.5 μm, (b) 0.85 μm, (c) 0.75 μm, (d–i) 0.24 μm, or (j) 1.28 μm. From Johnson et al. (1978a, 1980).

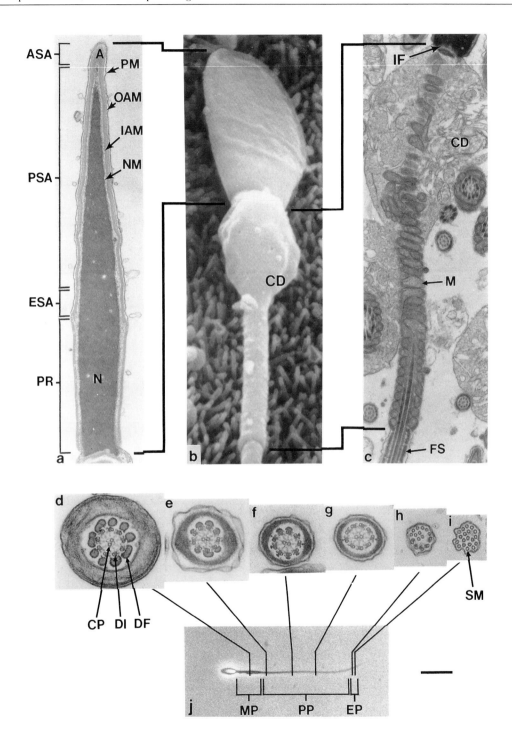

contains hydrolytic enzymes necessary for penetration of the layers of the egg. The postacrosomal region is the portion of the head not covered by the acrosome. Attached to the base of the head is the neck region of the tail. The proximal centriole remains attached at the implantation fossa in the neck, but the distal centriole has given rise to the axoneme (nine microtubular doublets and a central pair of microtubules) during tail development. The axoneme is surrounded by nine dense fibers, which arise from densities attached to the nine doublets (Fawcett and Phillips, 1970; Irons and Clermont, 1982b). While biochemical studies have been conducted on the outer dense fibers, their functional role remains uncertain (Olson and Sammons, 1980). Fawcett (1975a) proposed that they function as passive elastic structures. A passive elastic function is supported by biochemical studies; however, the studies could not rule out function in axoneme-coupled active motility nor a restriction of motility during the capacitated state or other elevated phases of spermatozoan movement (Olson and Sammons, 1980). The annulus marks the end of the middle piece and the beginning of the principal piece. This latter piece is lengthy and contains a fibrous sheath with rib-like structures to facilitate bending during movement and dense fibers that extend to different lengths within it. The fibrous sheath is proteinaceous and is composed of two longitudinal columns bridged by attaching ribs (Irons and Clermont, 1982a). The axoneme continues through the middle piece, principal piece, and end piece. The end piece is composed strictly of the axoneme or single microtubules and plasma membrane.

There are species differences in size and shape of spermatozoa (Garner and Hafez, 1980). Bull spermatozoa, like boar and ram, have flattened heads (Saacke and Almquist, 1964). The equine spermatozoon (Johnson et al., 1980) is similar in size and shape to that of humans (Johnson, 1982) and other primates. In mammals, spermatozoa are not mature when released from the seminiferous tubules. Each contains a cytoplasmic droplet on its middle piece (Figs. 2 and 3) (Johnson et al., 1978a, 1978b) and must undergo maturation in the epididymis to gain progessive motility, structural stability, and fertilizing ability (Johnson et al., 1980; Amann and Schanbacher, 1983; Amann, 1988).

III. Germ Cells in the Three Phases of Spermatogenesis

A. Spermatogonia— Spermatocytogenesis

Spermatocytogenesis is the first phase of spermatogenesis and involves spermatogonia. Spermatogonia arise postnatally from gonocytes and are located at the base of seminiferous tubules in adults. Stem-cell spermatogonia divide by mitosis to produce other stem cells to continue their lineage throughout the male's adult life and to repopulate the testis following damage of the more advanced germ cells or seasonal regression in some species. Repopulation of the germ cells following testicular damage may arise from a distinct reserve population of spermatogonia that does not routinely enter into cyclic divisions; however, this has been questioned (Huckins, 1971, 1978). In addition to stem-cell renewal, cyclic mitotic activity of spermatogonia produces committed spermatogonia, which immediately enter into proliferation and/or differentiation toward the production of primary spermatocytes. Hence, spermatocytogenesis functions to (1) continue the lineage of stem cells by the production of uncommitted spermatogonia and (2) produce committed spermatogonia, which result in production of spermatozoa after the end of duration of spermatogenesis for that species.

The three types of spermatogonia originally described include the dust-like (Regaud, 1901) A spermatogonia (Allen, 1918), intermediate spermatogonia (Clearmont and Leblond, 1953), and the crust-like (Regaud, 1901) or B spermatogonia (Allen, 1918). The A spermatogonia have relatively large oval or spherical nuclei. Differences in appearance of the types of spermatogonia are due to the changes in chromosomes, which are longer in A spermatogonia and shortened in B spermatogonia (Knudsen, 1958; Ortavant *et al.*, 1977).

In the horse (Fig. 4), five different sub-

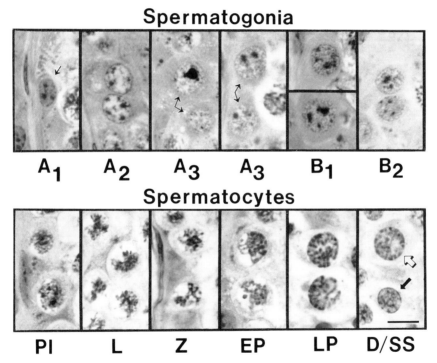

Figure 4 Composite of bright-field micrographs of 5-μm methacrylate sections of spermatogonia and spermatocytes. Subtypes of spermatogonia are classified by nuclear profiles, which include the most primitive cells with a small, oval or flattened nucleus (A_1), a light center (A_2), a large single nucleolus (A_3), two or three nucleoli (A_3), or large nucleoli plus fragmented nucleoli (B_1), or with a small spherical nucleus (B_2). Arrows designate the cells of interest when other cells are in view. Spermatocytes include preleptotene (Pl), leptotene (L), zygotene (Z), early pachytene (EP), late pachytene (LP), diplotene (D; open arrow) primary spermatocytes, and secondary spermatocytes (SS; closed arrow). Bar length equals 10 μm.

types of spermatogonia are characterized by either (1) small flattened nuclei (A_1); (2) large nuclei with apparently empty centers composed of euchromatin (A_2); (3) the largest nuclei with either one large nucleolus or large fragments of nucleoli (A_3); (4) large oval to spherical nuclei with large chromatin flakes (intermediate or B_1); or (5) small spherical nuclei with small chromatic flakes, as does the conventional B_2 spermatogonia, whose division yields preleptotene primary spermatocytes (Johnson, 1990). The number of spermatogonial divisions in various mammals, (3–5), cocks, and fish have been summarized and have been shown to range from 1 to 14 (Courot *et al.*, 1970).

The mechanism by which the number of A spermatogonia is seasonally modulated in horses is unclear. The spermatogonial number is dependent upon (1) the number of stem cells per testis; (2) the scheme of stem cell renewal; (3) the number of cell divisions from the stem cell to the primary spermatocyte (Hochereau-de-Reviers, 1981); and (4) the amount of degeneration of specific subtypes of spermatogonia (Huckins, 1978). Only the interval at which newly committed spermatogonia enter (i.e., 12.2-day cycle length) has been determined for horses (Swierstra *et al.*, 1974; Amann, 1981). The number of spermatogonial subtypes has not been confirmed for horses, but it is considered to be constant at 4–6 subtypes for a given species (Hochereau-de-Reviers, 1981) and is not likely to be involved in seasonal changes in daily sperm production.

Seasonal modulation of A spermatogonia in horses may result from proliferation of renewing stem cells. Reserve stem cell spermatogonia are the youngest form of germ cells and may be dormant in testes active in spermatogenesis (Clermont and Bustos-

Obregon, 1968). These are the cells which carry on the lineage throughout the life of adult males. Seasonal variation in the number of renewing stem cells has been found in other seasonal breeders such as rams and red deer stags (Hochereau-de-Reviers, 1981).

The numbers of each subtype of equine spermatogonia per 100 Sertoli cells in each of eight spermatogenic stages (to be described later) reveal cyclic differences (Fig. 5) (Johnson, 1989a). These cyclic differences reflect mitotic activity and degeneration of the different subtypes of spermatogonia in different stages of the spermatogenic cycle. The yield of the division of A_3 to B_1 or B_1 to B_2 was low at approximately one instead of the theoretical two cells resulting from each division. In stallions, spermatocytogenesis produces an overabundant number of spermatogonia in the breeding season, and germ-cell degen-

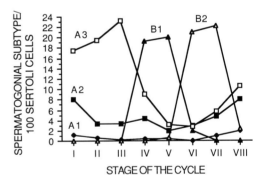

Figure 5 Number of each subtype of spermatogonia per 100 Sertoli cells in each of the eight stages of the spermatogenic cycle in equine testes obtained in the breeding season. Cyclic differences in numbers of cells are noted for each cell type; however, the greatest stage-dependent differences occur for the more advanced spermatogonia, namely, A_3, B_1, and B_2. (See Fig. 4 for the description of the different spermatogonial subtypes.) From Johnson (1989a).

eration is an important mechanism to reduce its yield (Johnson, 1985). Seasonal differences in spermatogonial number [which largely control seasonal differences in daily sperm production (Johnson, 1985)] are mostly caused by an increased yield of spermatogonial division early in spermatocytogenesis in the breeding season (Johnson, 1990).

Based on data quantifying A spermatogonia and their mitotic activity in different stages of the spermatogenic cycle (Johnson, 1985, 1987), Amann and Johnson proposed a pattern for the process of spermatogonial renewal in the stallion (Fig. 6). Degeneration of germ cells occurs both in the nonbreeding (early in spermatocytogenesis) and in the breeding season (late in spermatocytogenesis when twice as many spermatogonia are present; Johnson, 1989a). A similar pattern of stem-cell renewal has been proposed in classical studies of the bull (Hochereau-de Reviers, 1981), boar (Frankenhuis et al., 1980), and ram (Ortavant, 1958). However, since several factors must be considered (cell division, germ-cell degeneration, differential, etc.), interpretation of these data is difficult, and alternative proposed models are possible (Amann and Schanbacher, 1983). Even in the rat, whose spermatocytogenic process has been studied extensively, the stem cell has been attributed to every type of spermatogonia (A_1, A_2, A_3, A_4, or A_5) other than B spermatogonia (Huckins, 1971). Spermatocytogenesis plays a pivotal role in regulation of spermatogenesis; however, its details remain relatively obscure, especially in nonrodent species.

B. Spermatocytes—Meiosis

Meiosis is the process by which genetic material is exchanged between homologous

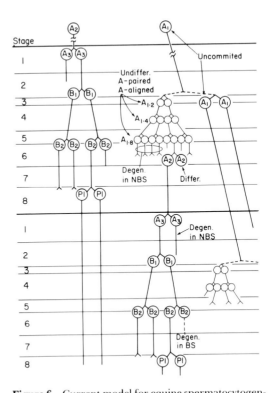

Figure 6 Current model for equine spermatocytogenesis. Stem-cell renewal includes formation of two new isolated and uncommitted A_1 spermatogonia as a major function of spermatocytogenesis. Commitment of A_1 spermatogonia toward differentiation by forming a pair of undifferentiated $A_{1.2}$ spermatogonia that divide to form a chain of committed but undifferentiated $A_{1.4}$ and then $A_{1.8}$ spermatogonia is presented as the second function of spermatocytogenesis. Spermatogenesis begins with formation of a pair of committed $A_{1.2}$ spermatogonia. Commitment of another A_1 spermatogonium to form a pair of $A_{1.2}$ spermatogonia occurs only in stage III of the cycle of the seminiferous epithelium. Although degeneration of spermatogonia can occur at any point in spermatocytogenesis, degeneration during the breeding season involves mostly B_2 spermatogonia, whereas in the nonbreeding season it involves $A_{1.8}$ and A_3 spermatogonia. Degeneration of the B_2 spermatogonia during their mitosis in the breeding season results from an overkill (overpopulation) of A spermatogonia whose progeny are too numerous to be supported. From Amann (1989a).

chromosomes and haploid spermatids are produced. Meiosis occurs only in germ cells. In males these cells are spermatocytes (Fig. 4). Following the mitotic division of B_2 spermatogonia, preleptotene primary spermatocytes result. While these cells appear very similar to B_2 spermatogonia, they immediately begin meiosis by active DNA synthesis for 14–28 h, depending upon the species (Courot *et al.*, 1970). This is considerably longer than the DNA synthesis of spermatogonia. Incorporation of [^{3}H]thymidine during DNA synthesis of preleptotene primary spermatocytes facilitates the timing of the kinetics of spermatogenesis when radioactively labeled cells are followed by autoradiography throughout spermatogenesis (Amann, 1970b; Swierstra *et al.*, 1974).

Spermatocytes (like spermatogonia, spermatids, or residual bodies) are attached to one another by intercellular bridges (Fig. 7). These bridges, between cells in the same developmental step, may function to facilitate (1) synchronous development or degeneration of similar germ cells, (2) the production of committed spermatogonia (Huckins, 1978), (3) the differentiation of haploid spermatids, which now have only one sex chromosome, and/or (4) phagocytosis and digestion of residual bodies left behind at spermiation.

Chromatin clumps located near the nuclear envelope in B_2 spermatogonia and preleptotene primary spermatocytes disperse to produce the fine chromatin filaments of leptotene primary spermatocytes as these primary spermatocytes undergo meiotic development (Fig. 4). During the preleptotene phase, homologous chromosomes duplicate themselves (Courot *et al.*, 1970). Often the chromatin filaments are not evenly distributed, such that one side of the nucleus appears light in leptotene phase. The zygotene phase marks the beginning of the exchange of genetic material through the intimate pairing of homologous

Figure 7 High-voltage electron micrographs of equine Sertoli and germ cells embedded in Epon 812 and sectioned at 1.0 μm. The basal (a) and apical (b) regions of the Sertoli cells are characterized. (a) The close association between Sertoli cells and germ cells representing different development steps is evident. Laminar processes (LP) extend from the main body of a Sertoli cell and can be seen between germ cells. One forms a rim of Sertoli cell cytoplasm from the main body of the Sertoli cell to the site of an intercellular bridge (IB) between two primary spermatocytes (PS). Long, slender mitochondria (arrows) oriented with the long axis of the Sertoli cell characterize the main body cytoplasm, while numerous dense lipid and lipofuscin granules were found in the basal cytoplasm. Mitochondria (M) of germ cells are contrasted with long, slender ones in Sertoli cells. Rough endoplasmic reticulum (RER) can be seen in the cytoplasm of germ cells. Portions of Sertoli-cell nuclei (N) depict the highly indented nature of the nuclear surface. An A spermatogonium (AS) shares the intercellular space (IS) with a Sertoli cell. While Golgi-phase spermatids (Sa) have not yet developed an acrosomic vesicle, the developing flagellum (F) from the centrioles (Ct) can be seen within the cytoplasm of one such spermatid. (b) Apical region of a Sertoli cell containing long, slender mitochondria (arrows) and four embedded Sd_1 spermatids with elongated nuclei are shown in this profile. A portion of the nucleus and three mitochondria of an Sa spermatid (Sa) are present. By following Sertoli cell cytoplasm marked by intercellular space (IS) between Sd_1 spermatids and the Sertoli cell, it is evident that Sertoli-cell cytoplasm is located between each spermatid. These Sd_1 spermatids are deeply embedded in the apex of the Sertoli cell via tubular indentations in the Sertoli cell plasma membrane. The annulus (An), manchette (Mn), and mitochondria (M) of the Sd_1 spermatids are indicated. The flagellar canal (FC), seen as the space between the developing flagellum and surrounding manchette, is produced by an infolding of the plasma membrane from the surface of the spermatid cell body. The manchette is situated to provide structural support to the flagellar canal. Bar length is (a) 2 μm (b) 1 μm. From Johnson (1986a).

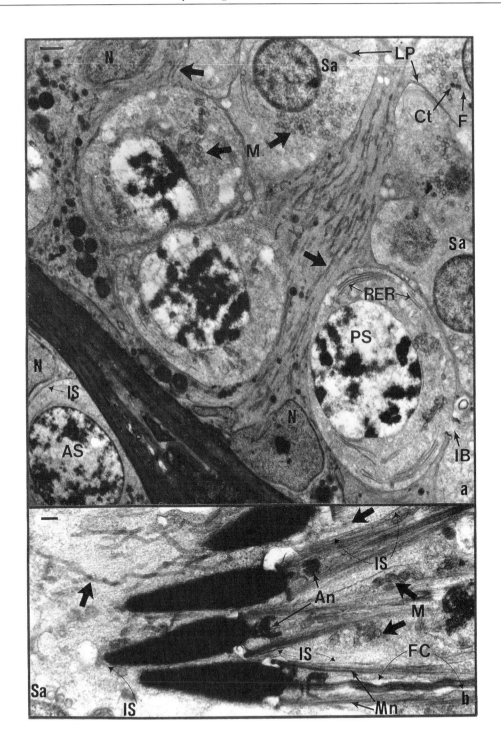

chromosomes via the synaptonemal complex, forming bivalents or tetrads of chromatids (duplicated paired homologous chromsomes). The synaptonemal complex is the site of genetic exchange in paired chromosomes. The sex chromosomes are separated into the sex vesicle (Burgos *et al.*, 1970). The size of chromosomes increases between the leptotene and zygotene phases. In pachytene phase the chromosomes appear largest as each chromosome divides longitudinally into two chromatids and coiling of the tetrads continues. The pachytene phase is the longest phase, with exchange of genetic material persisting. Following the pachytene phase, chiasmata form between homologous chromosomes in diplotene phase. Diakinesis is the short-lived, final phase of meiotic prophase in which the synaptonemal complex disappears and chromosomes become contracted prior to the first meiotic division (Knudsen, 1958; Courot *et al.*, 1970; Ortavant *et al.*, 1977).

During the first meiotic division, primary spermatocytes rapidly undergo nuclear-membrane dissolutions, metaphase (tetrad-alignment), anaphase (separation of dyads of sister chromatids), and telophase (complete separation of dyads). Secondary spermatocytes result from this division (Fig. 4) and contain a haploid number of duplicated chromosomes (i.e., XX or YY). Hence, alignment of duplicate paired homologous chromosomes (tetrad) in meiosis differs from mitosis in which unpaired but duplicated chromosomes line up at metaphase. Secondary spermatocytes have spherical nuclei with chromatin flakes of varying sizes. These cells are in a short interphase with no DNA synthesis prior to the second meiotic division, which produces spermatids with a haploid number of chromosomes. Secondary spermatocytes can easily be located adja-

cent to cells in metaphase or anaphase of the meiotic divisions. Hence, one duplication of chromosomes followed by two divisions (one to separate paired homologous chromosomes and the second to separate sister chromatids) results in the production of haploid numbers of chromosomes in spermatids.

In addition to DNA synthesis during early phases of meiosis, these spermatocytes are active in RNA synthesis: RNA synthesis is highest during the mid-pachytene phase and the end of meiotic prophase (Loir, 1972a). Recent studies have revealed that mRNA specific for structural proteins increases during prophase of meiosis (Slaughter *et al.*, 1989).

C. Spermatids—Spermiogenesis

Spermiogenesis is the morphologic differentiation of spermatids into spermatozoa (Fig. 8). Spermatids, produced by the second meiotic division, differentiate from spherical cells with spherical nuclei to cells that have (1) a streamlined head containing penetrative enzymes and a condensed nucleus carrying the male genome and (2) a tail necessary for cellular motility.

Nucleotide synthesis occurs during the early period of spermiogenesis. Since spermiogenesis does not involve cell division, DNA synthesis is minimal and probably restricted to DNA repair. Chromatin proteins (histone) change in arginine and cystine content, and the number of disulfide bonds changes during and after elongation (Loir, 1972b; Monesi, 1965). RNA synthesis (measured by tritiated uridine incorporation) is high in early spermatids of the ram (Loir, 1972a) and thought to be responsible for the production of specific proteins and structures unique to spermatids. These

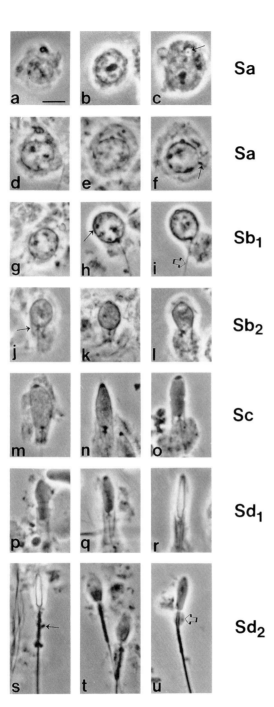

Sa

Sa

Sb₁

Sb₂

Sc

Sd₁

Sd₂

Figure 8 Phase-contrast micrographs of equine spermatids in testicular homogenates revealing different steps of development during spermiogenesis. (a–c) Nuclei of the newly formed spermatids are spherical and contain coarse and fine granules not unlike the nucleoplasm of secondary spermatocytes. (a–f) Nuclei of early spermatids, like nuclei of all spermatids with spherical nuclei, are the smallest germinal cell nuclei. Early spermatids may be characterized by a large, single, centrally located nucleolus and the lack of an acrosomic structure. (c) However, later, the acrosomic vesicle with its granule (arrow) appears in the cytoplasm. (d–f) Spherical nuclei of more developed spermatids show further signs of acrosomic development with (d) the spherical acrosomic vesicle and granule present or (e) the flattened acrosomic vesicle attached to the nucleus. The formation and extension of the acrosomic cap over the nucleus and (f) the first distinct appearance of the flagellum (arrow) attached to the side of the nucleus adjacent to the developing acrosome characterize later development of Sa spermatids or early Sb₁ spermatids. (g–i) The latest spermatids with spherical nuclei are characterized by the acrosomic cap (arrow) and developing flagellum (open arrow). Nuclei of these Sb₁ spermatids are essentially spherical and are similar in size to earlier spermatids. (j–l) Nuclei of spermatids undergoing elongation are characterized by the first appearance of the manchette (arrow), which extends from the middle of the nucleus, not covered by the developing acrosome, to a short distance down the flagellum. The developing flagellum now has a distinct annulus just below its connection with the nucleus. These are Sb₂ spermatids. (m–o) Nuclei of Sc spermatids develop further by elongation without distinct chromatin condensation. The manchette extends further down the developing flagellum, and the dense acrosomic granule is still concentrated at the anterior tip of the nucleus. (p–r) The early Sd₁ spermatid has the manchette, which now encloses only about a fourth of the nucleus and extends down the developing flagellum. The manchette is removed during Sd₁ spermatid development. The nucleus has undergone condensation of its chromatin and has taken on the shape of a typical equine spermatozoon. (s–u) Sd₂ spermatids have undergone further chromatin condensation and flagellar development. The annulus has migrated to its most distal location, and mitochondria (arrow) have surrounded the middle piece from the nucleus to the annulus. (u) The cytoplasmic droplet (open arrow) located at the proximal position, next to where the flagellum attaches to the head, can be seen on flagella of the most advanced Sd₂ spermatids. Sa spermatids are classified as Golgi phase, Sb₁ as cap phase, Sb₂ and Sc as acrosome phase, and Sd₁ and Sd₂ as maturation phase. Bar length is 5 μm. From Johnson and Thompson (1983).

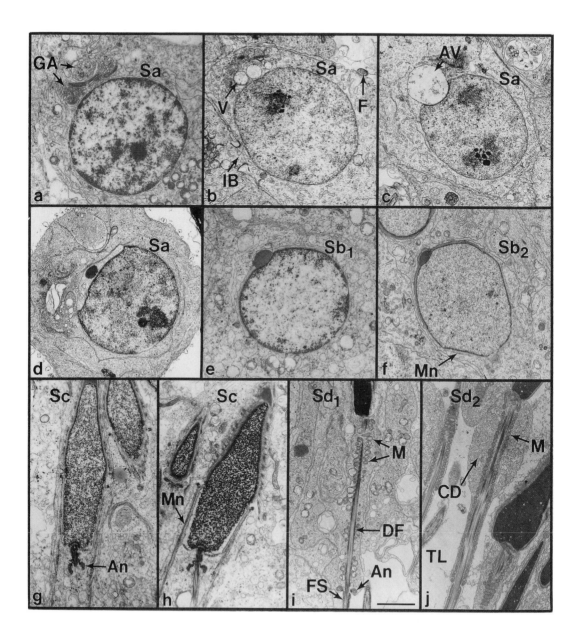

structures include the manchette, which may be involved in nuclear elongation, as will be discussed later. Transcription of haploid cells has been illustrated by the synthesis of poly $(A)^+$ RNAs by round spermatids (D'Agostino *et al.*, 1978; Erickson *et al.*, 1980; Kleene *et al.*, 1983). Alternatively, slow metabolism of heterogenous nuclear RNA (HnRNA) in germ cells may provide a mechanism whereby HnRNA can be made prior to spermiogenesis but still be active as a messenger during late spermiogenesis (Soderstrom and Parvinen, 1976).

Development of spermatids has been divided into four phases (Golgi, cap, acrosomal, and maturation phases), based largely upon the development of the acrosome in the different spermatid types (Fig. 8). Compared to domestic species, these phases are more pronounced in rodents, the animals in which they were first described (Leblond and Clermont, 1952). However, early phases of development and series of events in domestic species are similar to those in rodents.

The acrosome is an enzyme-containing vesicle over the nucleus of the spermatozoon (Fig. 3). The Golgi apparatus of a newly formed spermatid gives rise to the acrosome (Fig. 9), as it does for other enzyme-containing membrane systems such as lysosomes. The Golgi phase spermatid has a prominent Golgi apparatus that produces membrane-bound enzymes on its mature face; these small vesicles fuse to form the acrosomic vesicle adjacent to the nucleus. The acrosomic vesicle makes contact with the nuclear envelope and exerts a certain amount of pressure, as noted by an indentation of the nuclear envelope. This pressure may be positive toward the nuclear envelope by the ever-increasing size of the acrosomic vesicle as small vesicles from the Golgi continue to be added. The centrioles migrate to a region near the nuclear envelope where the distal centriole gives rise to the developing axoneme inside the cytoplasm (Fig. 7) before its growth is projected away from the cell body toward the lumen (Fig. 2). The nuclear envelope again is indented—this time

Figure 9 Transmission electron micrographs of equine spermatids representing different steps of development during spermiogenesis. (a–d) The Sa spermatid (Sa) is characterized by a large Golgi apparatus (GA), that produces vesicles (V) that fuse to form larger vesicles and ultimately to form the acrosomic vesicle (AV). The acrosomic vesicle indents the nuclear envelope and then (d) flattens over the nucleus. (b) Developing flagella (F) from neighboring Sa spermatids extend away from the spermatid cell body into the extracellular space. Intercellular bridges (IB) connect adjacent spermatids. (e) In the Sb_1 spermatids (Sb_1), the acrosome has formed a head cap over the nucleus. (f) In the Sb_2 spermatid (Sb_2), the appearance of the manchette (Mn) coincides with the beginning of nuclear elongation. (g, h) The Sc spermatid (Sc) has a distinct manchette (Mn), an elongating and condensing nucleus, an attached flagellum with a distinct annulus (An), and a well-defined acrosome over the anterior portion of the nucleus. (i, j) The Sd_1 and Sd_2 spermatids (Sd_1, Sd_2, respectively) have further condensation of their nuclei. (b) The flagellum (F) begins in the early Sa spermatid and (g, h) appears as a growing axoneme in Sb_1, Sb_2 (not shown), and Sc spermatids. (i) The late Sd_1 spermatid is characterized by dense fibers (DF), a completed fibrous sheath (FS), and mitochondrial (M) migration around the flagellum. Prior to mitochondrial migration in the Sd_1 spermatid, the manchette is removed and the annulus (An) migrates to its permanent position at the distal end of the middle piece. Note the abaxial (not center of the head) attachment of the flagellum. (j) The late Sd_2 spermatid is mostly extended into the tubular lumen (TL) and has complete migration of mitochondria (M) around the middle piece. Excess cytoplasm of the spermatid remains in the proximal region of the middle piece as a cytoplasmic droplet (CD). Bar length equals 2 μm.

by the attaching flagellum at the implantation fossa of the nucleus (Fig. 3). The latter nuclear indentation potentially may be caused by a positive pressure from a flagellating axoneme, which also could explain how the plasma membrane (via the flagellar canal; Figs. 7 and 10) is drawn near the nucleus. The flagellar canal is a tubular infolding of the spermatid plasma membrane, from the surface of the cell body to the annulus located just below the attachment of the flagellum to the nucleus, through which the flagellum extends toward or into the lumen (Fig. 2). The flagellar canal provides a mechanism by which new growth of the flagellum can be directed away from the spermatid cell body to allow subsequent flagellar growth. Also, the flagellar canal may potentially facilitate minute flagellation: to assure nuclear membrane contact during flagellar attachment to the nucleus, and to direct the need for ATP (energy) to the middle piece to draw mitochondria into this region where they will be attached in the spermatozoon immediately after the annulus migrates to its final position. It is known that cells with a high energy use have a large number of mitochondria. Flagellation (vibrating) of maturation-phase spermatids can be observed in slides made from fresh, diced seminiferous tubules. Alternatively, the flagellar canal may function to prevent mitochondria from attaching to the axoneme or outer dense fibers during development of these structures. This is consistent with the possibility that mitochondria (which are motile in cells) may end up in the middle piece simply because they stick to the outer dense fibers during their random movement after annulus migration (David Phillips, personal communication).

When the acrosomic vesicle flattens and begins to spread, it forms a cap over the nucleus (Figs. 8 and 9). The spermatid is now in the cap phase. The Golgi apparatus becomes less conspicuous as it moves to the caudal end (away from the nucleus) of the cell. Flagellar development becomes more obvious as the flagellum projects further from the surface of the cell body of the spermatid. The nucleus is still spherical.

As the cap extends further over the nucleus (over two-thirds the length in the boar, bull, ram, and horse; Ortavant et al., 1977) and as the nucleus begins its elongation, the spermatid is in the acrosome phase. The manchette, a transient organelle found only in spermatids, is composed of microtubules attached by linking arms and arranged in a sheath, which forms around the lower half (caudal region) of the elongating nucleus (Fig. 7–9), and extends down the developing flagellum (Fig. 10). Acrosome-phase spermatids are embedded in deep recesses within the Sertoli-cell apex (Figs. 7 and 10). Nuclear elongation begins when the manchette appears at the caudal region of the spermatid where the flagellum is attached (Figs. 8j, 9f). Previously, it was thought that the manchette disappears (Burgos et al., 1970) after elongation is complete. In the horse, some components of the manchette remain in the residual body that is to be left behind at spermiation, where it degenerates (Goodrowe and Heath, 1984). Also in my laboratory, components of the manchette (a large group of microtubules attached with linking arms) have been observed in the residual bodies of the horse testis. In the horse as in other species, the manchette is not normally found in the spermatozoon itself.

In the maturation phase, the final phase of spermatid development, the manchette migrates caudally where it may provide a

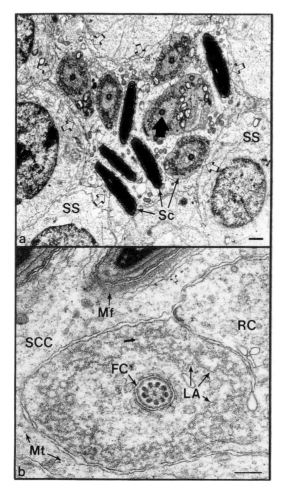

Figure 10 Transmission electron micrographs of cross sections of (a) the apex of a Sertoli cell and 10 embedded Sc spermatids and (b) the developing flagellum of an Sd_1 spermatid. (a) The plasma membrane of the Sertoli cell (large open arrows), surrounded by 8–10 secondary spermatocytes (SS), can be traced around the entire cluster of Sc spermatids (Sc). The axoneme with the dense fibers (closed arrow) can be seen within a cross section of the developing flagellum. (b) An enlargement of Sd_1 spermatids embedded in a Sertoli cell reveals the linking arms (LA) that attach adjacent microtubules of the manchette. Unattached microtubules (small closed arrow) also are seen. The residual cytoplasm (RC) is already being sequestered to one side but is still attached to the elongated body of the spermatid. Microtubules attached together with linking arms also are found in the residual cytoplasm. In the center, the axoneme is surrounded by nine dense fibers and the plasma membrane of the flagellum. The flagellar canal (FC) is seen as the space between the plasma membrane surrounding the flagellum and that of the spermatid cytoplasm. Sertoli-cell cytoplasm (SCC) has microtubules (Mt) and a high density of microfilaments (Mf) plus endoplasmic reticulum (open arrow) near the head of Sd_1 spermatids. Bar length equals (a) 1 μm, (b) 0.3 μm. (a) From Johnson (1986a).

shaft that supports the flagellar canal and freedom of movement during flagellar development (Figs. 7–9). Indeed, the dissolution of the manchette corresponds to the migration of the annulus to its permanent location at the junction of the middle and principal pieces of the spermatozoon and the shortening/disappearance of the flagellar canal (Fawcett, 1986). Quickly, mitochondria move in around the flagellum in the middle piece region following annulus migration. This process appears to occur quickly, as it is difficult to find a spermatid whose annulus has migrated but mitochondria have not surrounded its middle piece. Also indicative of the rate of mitochondrial migration is the observation that some spermatids within a group embedded in the same Sertoli cell will exhibit initiation of this process while others have it completed (see Holstein and Roosen-Runge, 1981, for an example).

The mechanism of production of the nine dense fibers and fibrous sheath of the spermatozoon (Fig. 3) is unclear. Dense fibers may arise from the columns of the capitellum (system of nine cross-striated columns that fuse to the centrioles) at the articular fossa (Burgos *et al.*, 1970). Others provided evidence that the dense fibers arise from lateral densities attached to the nine doublets of the axoneme (Irons and Clermont, 1982b) and that they grow along the length of the tail simultaneously, in contrast to the growth direction of the axoneme (Fawcett and Phillips, 1970). The fibrous sheath (two columns attached by ribs that extend down the principal piece between the plasma membrane and outer dense fibers) formation may involve a spindle-shaped body, the transient tubular complex found in the middle piece region of the developing human flagellum (Holstein and

Roosen-Runge, 1981). In the rat, the fibrous sheath starts with the formation of minuscule longitudinal columns found at the distal end of the principal piece and is completed by the formation of columns and ribs at the proximal end of the middle piece (Irons and Clermont, 1982a).

A large portion of cytoplasm found in the spermatid is left behind, creating the residual body to be phagocytized and digested by Sertoli cells. The developing spermatid is attached to its residual body by cytoplasmic stalks (Fig. 2). The neck region of a spermatid is attached to this excess cytoplasm, remaining as a cytoplasmic droplet, located on the middle piece of the luminally released spermatozoon. The release of spermatids as spermatozoa is known as spermiation, the counterpart to ovulation.

IV. Kinetics of Spermatogenesis

A. *Stages of the Spermatogenic Cycle*

The spermatogenic cycle (Fig. 11) is superimposed on the three phases of spermatogenesis (spermatocytogenesis, meiosis, and spermiogenesis). The spermatogenic cycle (cycle of the seminiferous tubular epithelium; Leblond and Clermont, 1952) is "a series of changes in a given [region] of seminiferous epithelium between two appearances of the same developmental stages." If we use spermiation as a reference point, the cycle would be all the events that occur between two consecutive occasions of spermatozoan release from a given region of the seminiferous epithelium.

To understand the cycle and development of germ cells throughout spermatogensis, it may be useful to compare spermatogenesis with a 4-year college model (Fig.

12) (Johnson, 1989b). Using graduation as the reference point in the college model, the cycle would be all the events between two consecutive graduations. In both processes, the cycle length (time between two consecutive releases) is dictated by the frequency of cells or classes entering the process and is less than the duration of the entire process. While spermatozoa are released each 12.2 days in horses, the duration of spermatogenesis is 57 days (Swierstra *et al.*, 1974). While graduation occurs yearly, college takes 4 years to complete (Fig. 12). Hence, multiple generations of germ cells (spermatogonia, spermatocytes, and spermatids) or classes (freshmen, sophomores, juniors, and seniors) must occur simultaneously. The amount of time between consecutive generations or classes is one cycle length. Furthermore, germ-cell degeneration in spermatogenesis and dropouts in college reduce the product yield (number of cells spermiated and number of students graduated, respectively).

In both cases (Fig. 12), once the process starts, it continues at a defined rate such that cells (Golgi phase spermatids) or students (incoming freshmen) at a given developmental step are almost always associated with other cells (maturation-phase spermatids) or students (incoming sophomores, juniors, or seniors) at respective developmental steps. Indeed, any combination of germ cells or classes that differ in developmental age by a multiple of the cycle length plus a constant will constitute a stage of the cycle. A stage in the cycle is defined as associations of four or five generations of germ cells formed in specific, chronological developmental steps, whose developmental age differs by a multiple of the cycle length plus a constant (Johnson, 1989b).

Differences between our college model and spermatogenesis may help in further understanding of the dynamic process of spermatogenesis. Cell division magnifies the yield in spermatogenesis; however, no multiplying component occurs in college. Each committed spermatogonium entering spermatogenesis has a potential yield of 64 spermatozoa. In contrast, less than one student graduates for each student entering college. In spermatogenesis, the frequency of product release is greater than in our college model. Resulting from college largely beginning only once a year in the fall, graduation for most students occurs only once a year, usually in late spring. In spermatogenesis, the production of newly committed spermatogonia is not synchronized among tubules and not simultaneous along the length of the same tubule. Therefore, somewhere in the testis, newly committed spermatogonia enter the process of spermatogenesis each second. This creates a continual release of spermatozoa into the lumen of seminiferous tubules. If college started each second instead of once in the fall, graduation would occur each second instead of once in the late spring. While this continual release seems wasteful, this assures that at least some male gametes are always available, even after exhaustive sexual activity, whenever female gametes are available for fertilization.

Stages of the cycle represent man-made divisions of naturally occuring and continuously changing cellular associations in a given region of the seminiferous tubules (Fig. 11). Two types of classification of stages have been described and are based on (1) the morphological changes of germ cells (Roosen-Runge and Giesel, 1950) or (2) the acrosomic development of spermatids in each stage (Leblond and Clermont, 1952). Man-made divisions include 14 stages in the rat (Perey *et al.*, 1961) and bull (Berndtson

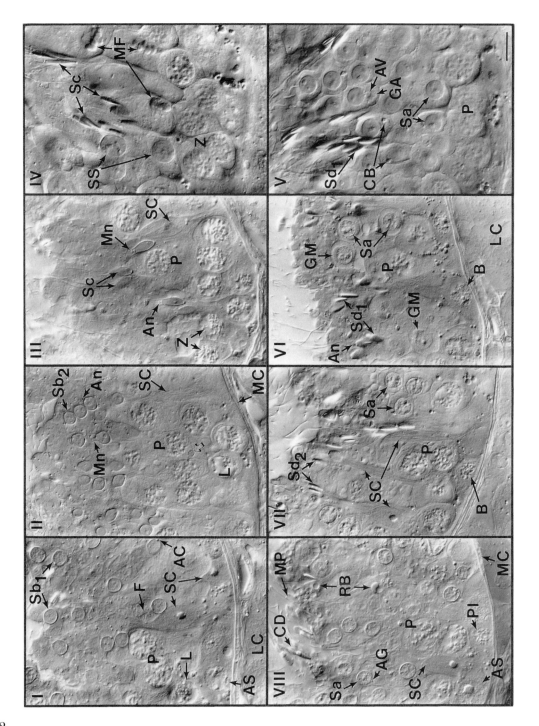

Figure 11 A clockwise arrangement of eight stages of the cycle of the equine seminiferous epithelium observed by Nomarski optics in unstained, 20-μm Epon histologic sections. Both nuclear and cytoplasmic details are revealed in Leydig cells (LC), Sertoli cells (SC), various types of germ cells, and myoid cells (MC). A spermatogonia (AS) of different types and Sertoli cells are found in all stages of the cycle. Classification of spermatid development was based on that described for humans (Clermont, 1963) and used in horses (Johnson, 1985). Briefly, the Sa spermatid is the earliest form as it contains a spherical nucleus and a large Golgi with either no acrosomic vesicle, a developing acrosomic vesicle, or an acrosomal cap. The Sb_1 spermatid has a spherical nucleus but also has an attached flagellum and distinct acrosome covering half of the nuclear surface. The Sb_2 spermatid has begun nuclear elongation with the appearance of the manchette. The Sc spermatid has a distinct manchette and a more elongated nucleus than the Sb_2 spermatid. The Sd_1 spermatid is undergoing final maturation with the removal of the manchette and migration of mitochondria around the tail. The Sd_2 spermatid is the final form, and a large portion of the cell, including the head, projects into the tubular lumen in Stage VIII. Stage numbers are marked in corners. Stage I is characterized by prelepto-tene or leptotene (L) and pachytene (P) primary spermatocytes as well as Sb_1 spermatids (Sb_1). These spermatids are characterized by spherical nuclei, attached acrosomal caps (AC), and developing flagella (F). Stage II is characterized by leptotene (L) and pachytene (P) primary sperm-atocytes, and Sb_2 spermatids (Sb_2). The labeled pachytene primary spermatocyte is displaying its large spherical Golgi apparatus (open arrow). The Sb_2 spermatid has an elongating nucleus, a manchette (Mn) and an annulus (An). Stage III has zygotene (Z) and pachytene (P) primary spermatocytes, and bundles of Sc spermatids (Sc). The Sc spermatid has an elongating and condensing nucleus, and a distinct manchette (Mn) and annulus (An). Stage IV is characterized by the presence of zygotene primary spermatocytes (Z), diplotene primary spermatocytes or secondary spermatocytes (SS), meiotic figures (MF), and Sc spermatids (Sc). Stage V is composed of pachytene primary spermatocytes (P), newly formed Sa (Sa), and Sd_1 (Sd_1) spermatids. The Sa spermatid has a distinct Golgi apparatus (GA), developing acrosomic vesicle (AV), and chromatoid body (CB), and Sd_1 spermatids form bundles that are deeply embedded in the seminiferous epithelium. State VI has type B spermatogonia (B), pachytene primary spermatocytes (P), Sa spermatids (Sa) with their grouped mitochondria (GM), and Sd_1 spermatids (Sd_1). The Sd_1 spermatid has an annulus (An) and a less distinct manchette. Stage VII is characterized by type B spermatogonia (B), pachytene primary spermatocytes (P), Sa spermatids (Sa), and Sd_2 spermatids (Sd_2). Sd_2 spermatids, the most advanced form, are migrating toward the tubular lumen in this stage. Stage VIII has newly formed preleptotene (Pl) and pachytene (P) primary spermatocytes, late Sa spermatids (Sa) with their acrosomic granule (AG) and newly attached developing flagellum, and Sd_2 spermatids lining the lumen with their migrated annulus, enlarged middle pieces (MP), and attached cytoplasmic droplet (CD). Residual bodies (RB) left behind by spermiation of spermatids can be seen near the luminal surface and in transit toward the base within Sertoli cell cytoplasm. Bar length equals 10 μm. Figure and legend from Johnson *et al.* (1990b).

193

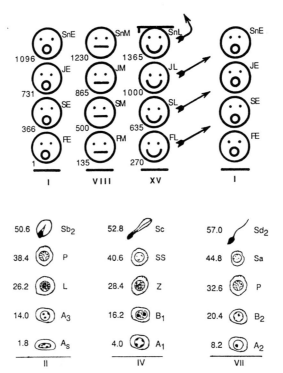

Figure 12 Comparison of the kinetics of spermatogenesis in horses with timing of student development in 4 years of college. To facilitate comprehension of the orderly process of spermatogenesis, 4 years of college (top) and spermatogenesis (bottom) have been divided into classes or generations (rows) and stages (columns). In the college model, 1 year of college has been divided into 15 stages (three shown here), and the placement of the same students in the first stage of the second year is indicated by the arrows. Three of the eight stages in spermatogenesis in horses are illustrated. Numbers indicate the developmental age (in days) of A or B spermatogonia (A, B), leptotene, zygotene, or pachytene primary spermatocytes (L, Z, P), secondary spermatocytes (SS), and spermatids of various types (Sa, Sb$_2$, Sc, Sd$_2$). In the college model, the numbers represent days of completion as freshmen (FE, FM, or FL), sophomores (SE, SM, or SL), juniors (JE, JM, or JL), or seniors (SnE, SnM, or SnL) progress through the early (FE, SE, JE, or SnE) middle (FM, SM, JM, or SnM), or late (FL, SL, JL, or SnL) period or stage of that year (respectively). College and spermatogenesis are similar in that (1) time between consecutive releases is less than the duration of the entire process, (2) multiple classes or generations of cells occur simultaneously, (3) the amount of time between classes or generations is one cycle length, and (4) the process continues at a defined rate such that students or cells in specific developmental steps are always associated with other students or cells at respective developmental steps. Since the early period of the year occurs at the same time for each class, incoming freshmen are in the same stage of the cycle as are incoming sophomores, juniors, and seniors. Hence, any combination of students or cell types that differ in developmental age by a multiple of the cycle length plus a constant will constitute a stage of the cycle. The cycle refers to the maturation of students within one academic year between two consecutive graduations or maturation of cells between two spermiations. Modified from Johnson (1989b).

and Desjardins, 1974); eight stages in the bull (Hochereau, 1963), ram (Ortavant, 1959), boar (Swierstra, 1968), and horse (Swierstra *et al.*, 1974); and six stages in humans (Clermont, 1963). The larger number of stages described for the same species is made possible by identifying different steps in acrosomal development of spermatids (Leblond and Clermont, 1952; Berndtson and Desjardins, 1974). It is easy to see that differences in spermatid development (Figs. 8 and 9) would facilitate classification of the spermatogenic cycle into several stages. Classification of stages of the spermatogenic cycle also has been based on the presence and location of specific germ cells (i.e., B spermatogonia, one or two generations of primary spermatocytes or spermatids, or secondary spermatocytes) and on changes associated with nuclear elongation, location, and release of spermatids (Roosen-Runge and Giesel, 1950). Though this has been done for the bull, ram, boar, and horse using conventional histology, it is presented here for the horse using Nomarski (differential interference contrast) optics of tissues embedded in 20-μm Epon sections (Fig. 11). This different approach reveals details of the equine spermatogenic cycle. Details regarding differences among stages have been included in the legend.

B. Cycle Length and Duration of Spermatogenesis

As mentioned earlier, the length of a cycle is the amount of time between consecutive releases of spermatozoa (spermiations) and is dictated by the frequency of newly committed spermatogonia entering the process. Hence, the cycle length is not a man-made division of the duration of sper-

matogenesis. The cycle lengths in days for various species are 7.2 for prairie voles, 8.6 for the boar, 8.7 for the hamster, 8.9 for the mouse, 10.4 for the ram, 13.5 for the bull, 13.6 for the beagle dog, 9.5 for the rhesus monkey, 12.2 for the horse, 12.9 for the rat, and 16 for humans (Amann *et al.*, 1976a; Amann, 1981, 1986; Clermont, 1972).

The duration of spermatogenesis can only be approximated for any species due to the difficulty of determining the point at which committed A spermatogonia enter the process. In the absence of direct determination, it has been considered to be about 4.5 times the cycle length for a given species (Amann, 1986). This corresponds to four or five generations of germ cells found in each stage of the spermatogenic cycle. The duration of spermatogenesis can be approximated at 74 days in humans, 61 days in bulls and dogs, 60 days in rats, 57 days in stallions, 47 days in Ile de France rams, and 39 days in boars (Amann, 1986). The cycle length and duration of spermatogenesis for several species, including fish and the cock, have been summarized elsewhere (Courot *et al.*, 1970). Since the duration of spermatogenesis does not change even after hypophysectomy (Harvey, 1962), it is considered to be constant for a species (Courot *et al.*, 1970). Recent studies have revealed that spermatogenic stages in the testis can be synchronized to three or four stages by the testicular injection of retinol in vitamin A-deficient rats (Griswold *et al.*, 1988). Synchronization apparently occurs by blocking the development of preleptotene primary spermatocytes by vitamin A deficiency and continuing their development when retinol is injected. Hence, the duration of spermatogenesis must be prolonged during this process.

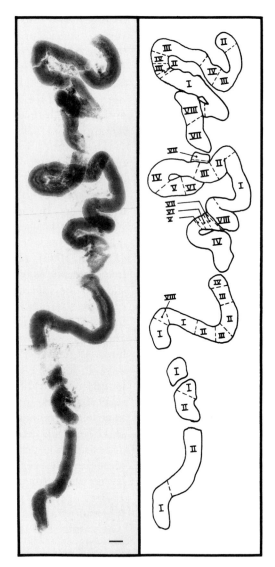

Figure 13 Spermatogenic wave revealed by an enzymatically isolated equine seminiferous tubule, fixed in glutaraldehyde followed by osmium tetroxide, infiltrated with Epon, mounted *in toto* in Epon, and observed by bright-field microscopy. The corresponding drawing reveals the consecutive stages along the length of the tubule as determined by observation with Nomarski optics. While modulations (reversal of order) occur, adjacent stages are in consecutive order. Later stages (more developed; higher numbers) are observed in the same direction of two sets of all stages I–VIII. Bar length equals 200 μm. From Johnson *et al.* (1990b).

C. Frequencies Of Stages

The frequency of each stage of the spermatogenic cycle is similar for members of the same species, but differs among species. The frequency of a spermatogenic stage is expressed as a percentage and is determined by dividing the number of tubular sections in a given stage by the total number of all tubular sections in any stage observed. This frequency multiplied by the cycle length represents the duration of a given stage and the time a given cell type stays in that given stage of the cycle. Using differences in stage frequencies, Ortavant *et al.* (1977) noted that the duration of different aspects of spermiogenesis varies among species. Changes in stage frequencies within a species may be used to identify spermatogenic toxins (Hess *et al.*, 1988), but the cause and importance of frequent changes are unclear.

D. Spermatogenic Wave

The spermatogenic wave is the spatial, sequential order of stages along the length of seminiferous tubules at any given time (Fig.13). The origin of the wave is unknown, but it results from synchronized but not simultaneous division of stem-cell spermatogonia in adjacent tubular segments along the length of the seminiferous tubule. The spermatogenic wave may function as a mechanism to (1) assure a constant release of spermatozoa, (2) reduce competition for hormones and metabolites used in a given stage, (3) reduce tubular congestion that could be produced if spermiation occurred simultaneously along the entire length of the tubule, (4) assure a constant flow of seminiferous tubular fluid to maintain the vehicle for spermatozoan transport and hormones needed by the epididymal epithelium, and (5) facilitate maturation of spermatozoa in the epididymis by a constant flow of spermatozoa and fluid from the testis. As mentioned earlier, a constant release of spermatozoa is essential to maximize the opportunity for fertilization when female gametes only are available intermittently (cyclicly).

While changes in the spermatogenic cycle are seen over time in a given region of the tubule, the spermatogenic wave refers to the distribution of consecutive stages along the length of the seminiferous tubules at any given instance in time (Fig.13). In 1901, Regaud expanded Von Ebner's (1871) concept of the "wave of the seminiferous epithelium" by stating that "the wave is in space what the cycle is in time."

The spermatogenic wave has been described for the mouse, bull, guinea pig, rabbit, ram, boar, dog, cat (Benda, 1887; Curtis, 1918), rat (Perey *et al.*, 1961), horse (Johnson *et al.*, 1990b), marsupials (Furst, 1887), and human (Schulze, 1982; Schulze and Rehder, 1984). Recently, it was found that enzymatic dispersion and Nomarski optics facilitated staging of unstained equine seminiferous tubules embedded in Epon and mounted *in toto* (Fig.13). As in other species, the spermatogenic wave in the horse consists of stages in consecutive order with modulations (reversal of the order of consecutive stages) in the wave. In the horse, the stage order almost always returns once again by reversing its order beginning with the last stage of the modulation (Johnson *et al.*, 1990b).

In the unit-segment theory of the wave evolution (Perey *et al.*, 1961), smaller unit segments make up longer tubular segments occupied by a given stage. Hence, the number of unit segments in the same stage would determine the length a tubular segment occupying the same stage (Amann, 1981; Johnson *et al.*, 1990b). The number of unit segments of the same stage occupying a tubular segment would depend on the number of synchronized simultaneous spermatogonial divisions to produce committed spermatogonia. In the case of humans, only a very small number of unit segments (which do not entirely extend around the tubular cross section) are in the same stage. This results in the spermatogenic wave being seen only if one considers the spiral course of the unit segments at different stages along the length of the tubule (Schulze, 1982; Schulze and Rehder, 1984).

The spermatogenic wave has been exploited to determine stage-specific changes in the seminiferous epithelium (Parvinen, 1982). Both FSH and androgens, essential hormones for spermatogenesis, seem to have different preferential stages in which they act. Also, there are stage-dependent differences in the production of androgen-binding protein, maximal secretion of plasminogen activator, and of meiosis-inducing substance (Parvinen, 1982). These kinds of studies were made possible by staging live, dispersed rat or mouse seminiferous tubules by transillumination (Parvinen and Vanha-Perttula, 1972; Parvinen and Hecht, 1981). Seminiferous tubules in domestic species are tightly bound in the testicular parenchyma and are not easily isolated for these types of studies. Recently, transillumination has been successfully employed to stage enzymatically dispersed seminiferous tubules in the horse (Fig.14) (Johnson and Hardy, 1989; Johnson *et al.*, 1990c). This procedure should prove fruitful to improve our understanding of spermatogenesis in domestic animals.

V. Evaluation of Spermatogenesis

A. Ejaculate

Spermatogenesis in paired testes is reflected in daily sperm output, the total number of spermatozoa found in an ejaculate (Amann, 1970b; Johnson, 1986b). Daily sperm output closely relates to daily sperm production (prediction of the number of spermatozoa produced per day) in several species: boars (Swierstra, 1968), rabbits (Amann, 1970a), bulls (Amann *et al.*, 1976a), horses (Gebaurer *et al.*, 1974), dogs (Olar *et al.*, 1983), and even humans (Johnson, 1982).

The number of spermatozoa in an ejaculate is influenced not only by the number of spermatozoa produced by the testes but also by the ejaculation frequency (Johnson, 1982). Sufficient numbers of ejaculates are needed in sexually rested males before the extragonadal (epididymal) spermatozoan reserves are stabilized. It is only after these reserves have been stabilized that daily sperm output in the ejaculate will accurately reflect daily sperm production (Amann, 1970b; 1981; Johnson, 1986b).

B. Testis

The circumference of the scrotum in bulls and rams or total scrotal width in horses and pigs is highly correlated with testicular weight and spermatozoan production (Willett and Ohms, 1957; Foote, 1978; Coulter, 1980). This measurement is easily made and is valuable in an andrological examination for breeding soundness (Amann and Schanbacher, 1983).

Spermatogenesis viewed in histologic sections has been evaluated qualitatively by the

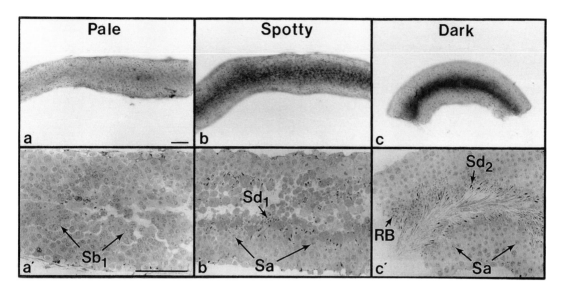

Figure 14 Enzymatically isolated equine seminiferous tubules staged (a, b, c) by transillumination and (a', b', c') confirmed by bright-field histologic observation of the same tubules. (a, a') Homogeneously pale tubules containing Sb_1 spermatids with spherical nuclei (Sb_1) as the most advanced cell type are characteristic of Stage I of the spermatogenic cycle. (b, b') Spotty tubules by transillumination contain Sa (Sa) and Sd_1 (Sd_1) spermatids and are composed of mostly stages V and VI. (c, c') Dark tubules by transillumination contain Sa (Sa) and Sd_2 (Sd_2) spermatids found in stage VIII. The entire lengths of the Sd_2 spermatids extend into the lumen while residual bodies (RB) from the Sd_2 spermatids line the lumen. Bar lengths equal 100 μm. From Johnson *et al.* (1990c).

general appearance of tubules and quantitatively by counting different germ-cell types (Amann, 1970b, 1981; Berndtson, 1977; Johnson, 1986b). When using the general appearance, it must be considered that species differ in the duration and, hence, the frequency of stages that contain maturation-phase spermatids or spermiation. Given the life span and theoretical yield of a specific germ cell, an estimate of daily sperm production can be obtained from the number of germ cells of that type in the testis (Fig. 15). The duration of spermatogenic stages in which a cell type occurs equals that cells's life span. The theoretical yield is equal to 2^n where n is the number of cell divisions be-

tween that cell type and spermatids (Johnson, 1989b).

Daily sperm production per gram of decapsulated testis is useful in species comparisons as it is a measure of spermatogenic efficiency. This efficiency is lower in humans (4–6 × 10^6/g) than all other species evaluated (Fig. 15; Johnson, 1986b). The values (× 10^6/g) for other species are estimated at 23 in rhesus monkeys, 25 in rabbits, 20–24 in rats, 23 in boars, 16–19 in stallions, 21 in rams, 24 in hamsters, 16 in dogs, and 12 in bulls (Amann *et al.*, 1976a; Lebovitz and Johnson, 1983; Amann, 1986; Johnson *et al.*, 1984a, 1986; Johnson, 1989b).

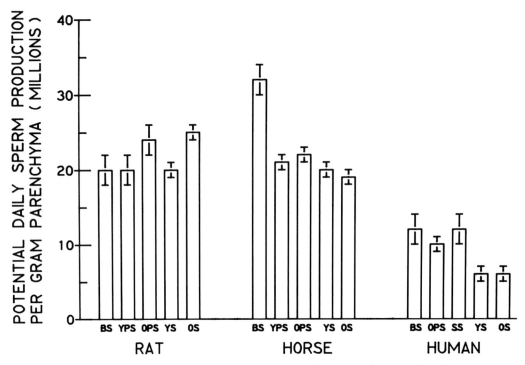

Figure 15 Potential daily sperm production per gram parenchyma at different developmental steps in spermatogenesis of rats, horses, and humans. Cell types on which potential daily sperm production per gram or daily sperm production per gram was based included B spermatogonia (BS), preleptotene plus leptotene (young primary) spermatocytes (YPS), pachytene plus diplotene (old primary) spermatocytes (OPS), secondary spermatocytes (SS), Golgi- and cap-phase (young) spermatids (YS), and maturation-phase (old) spermatids (OS). Adult (400 g) rats experienced no significant loss during these different steps in spermatogenesis (Lebovitz and Johnson, 1983; Johnson et al., 1984a). Adult horses had early losses in spermatogenesis (end of spermatocyte formation) with no subsequent losses (Johnson, 1985). Younger adult humans had significant losses during the second meiotic division. From Johnson (1986b).

The lower efficiency of spermatogenesis in humans compared to that in rats or domestic animals (Fig. 15) results from (1) a lower germ-cell density and (2) a longer duration of spermatogenesis (and longer cycle length; Johnson, 1986b). Likewise, the spermatogenic efficiency is lower in bulls and dogs, who also have a longer duration of spermatogenesis, approximately 60 days. Rats overcome this negative component of a longer duration (61 days) by having a high density of germ cells and a higher percentage of the testis occupied by seminiferous epithelium. The percentage of the human testis (62%) occupied by seminiferous tubles is lower than that for rats (83%) and for horses (72%) (Johnson, 1986b). Likewise, the seminiferous tubules occupy 77% of the testis in boars, 76% in bulls, and 84% in dogs (Swierstra *et al.*, 1974).

VI. Germ-Cell Degeneration during Spermatogenesis

A. *Germ Cells That Degenerate*

Germ-cell degeneration during specific steps of development has been estimated by (1) comparing the ratio of germ-cell numbers before and after a given developmental step or (2) comparing daily sperm production per gram parenchyma based on germ cell types in different developmental steps (Johnson, 1986b). The latter method has revealed species differences in the steps of germ-cell degeneration during spermatogenesis (Fig. 15).

Germ-cell degeneration occurs throughout spermatogenesis; however, the greatest amount and impact occurs during spermatocytogenesis and meiosis. Degeneration rate during spermatocytogenesis has been determined to be 25% in mice (Clermont, 1962), 11% in Sherman rats, and 75% in adult Sprague-Dawley rats (Huckins, 1978). In rams, long-day illumination of the nonbreeding season causes increased spermatogonial degeneration (Ortavant, 1958). Likewise in the stallion, the number of A spermatogonial (that is doubled in the equine breeding season) is dampened with an increased degeneration of B spermatogonia (Fig. 16; Johnson, 1985). Recent studies (Johnson, 1989a) have shown that the twofold increase in the number of A spermatogonia in the breeding season results from an increased yield of early spermatogonia (yield of A_2 spermatogonia or less degeneration of A_3 spermatogonia; see Fig. 4).

Meiotic divisions account for 13% of germ cell loss in mice, 2% in Sprague-Dawley rats, 27% in Sherman rats, 25% in rabbits, 25% in horses during the breeding season, and <5% during the nonbreeding season of horses (Johnson, 1989b). Fewer spermatids were found in long-day illumination of the nonbreeding season of rams (Ortavant, 1958). In humans, 30–40% spermatocyte degeneration occurs during the meiotic divisions (Johnson *et al.*, 1983; Barr *et al.*, 1971).

Degeneration of spermatids has been reported in Sherman rats (Clermont, 1962), mice (Oakberg, 1956), and horses in the nonbreeding season (Johnson, 1985). Neither horses (Johnson, 1985), adult (>400 g) Sprague-Dawley rats (Johnson *et al.*, 1983, 1984a), humans (Johnson *et al.*, 1981; Fig. 15), nor bulls (Amann, 1962), suffered significant degeneration of spermatids during spermiogenesis.

B. *Etiology of Germ-Cell Degeneration*

Degeneration of germ cells may result from chromosomal abnormalities, which may serve to limit the number of spermatogonial progeny (B spermatogonia or primary spermatocytes). Germ-cell degeneration may be the result of a mechanism to eliminate cells containing abnormal chromosomes (Oakberg, 1956; Clermont, 1962). However, simple selection to eliminate chromosomal abnormalities cannot explain the fact that only certain types of spermatogonia usually undergo degeneration and the magnitude of degeneration is relatively constant for given types of A spermatogonia (Huckins, 1978). Alternatively, the number of cells that can be sustained by the available Sertoli-cell population may determine the amount of degeneration (Huckins, 1978; Johnson, 1986b). The number of Sertoli cells in humans is positively correlated to

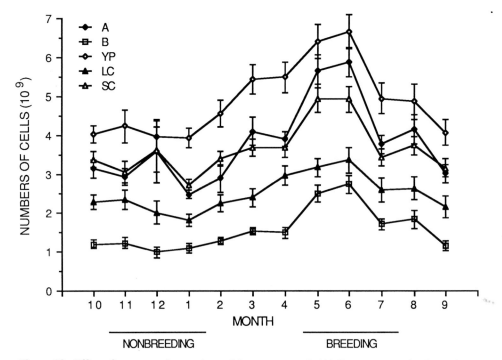

Figure 16 Effect of season on the numbers of A spermatogonia (A), B spermatogonia (B), prelepto-tene and leptotene plus zygotene primary spermatocytes (YP), Leydig cells (LC), and Sertoli cells (SC) per testis in 13–17 adult (4–20 years old) stallions each month. Numbers of A and B spermato-gonia in the breeding season were twice their values in the nonbreeding season. However, monthly differences in numbers of cells were seen in each cell type. There was no obvious trend for increases in a given cell type to precede increases in another cell type as the breeding season approached. However, A spermatogonia increased at a higher rate than did other cell types. From Johnson and Tatum (1989).

daily sperm production in humans (John-son *et al.*, 1984b). In stallions, if a large num-ber of B spermatogonia did not degenerate in the breeding season of stallions when the number of A spermatogonia is doubled, each Sertoli cell would have to accommo-date 44 instead of the typical 28 germ cells (Table I) (Johnson, 1986a). This is probably physically impossible. Table I illustrates the seasonal differences in the number of germ cells that are supported by available Sertoli cells in the stallion.

While germ-cell degeneration plays a piv-otal role in quantitative and qualitative sper-matogenesis, its mechanism and approaches to its prevention remain to be discovered (Johnson, 1989b).

VII. Sertoli Cells

A. *Role in Spermatogenesis*

Sertoli cells, the somatic component of the seminiferous epithelium, are critical to germ-cell development during spermato-

Table I

Seasonal Variation in Number of Sertoli Cells per Testis and Number of Germ Cells Accommodated by a Single Sertoli Cell in Stage VIII Tubules Based on 28 Horses per Season[a]

Item	Season		Significance
	Nonbreeding	Breeding	
Number of Sertoli cells per testis[b] (10^9)	2.6 ± 0.2	3.6 ± 0.2	$p < 0.01$
Germ-cell type	1.1 ± 0.1	1.5 ± 0.1	
Type A spermatogonia			$p < 0.01$
Preleptotene and leptotene plus zygotene primary spermatocytes	2.9 ± 0.4	2.9 ± 0.2	NS
Pachytene plus diplotene primary spermatocytes	2.6 ± 0.2	3.0 ± 0.2	NS
Spermatids with round nuclei	8.1 ± 0.8	10.9 ± 0.8	$p < 0.05$
Spermatids with elongated nuclei nuclei	8.0 ± 0.8	10.2 ± 0.7	$p < 0.05$
All germ-cell types combined	22.8 ± 2.1	28.5 ± 1.7	$p < 0.05$

[a]Means ± SEM. From Johnson (1986a).
[b]Based on homogenates of fixed testes (Johnson and Thompson, 1983).

genesis. Numbers of Sertoli cells are important as they have been related to the spermatogonial number, spermatid number, or spermatozoan production rates in rams (Hochereau-de Reviers and Courot, 1978), bulls (Berndtson et al., 1987), horses (Johnson 1986a), rats (Orth et al., 1988; Berndtson, 1989), and humans (Johnson et al., 1984b). The proposed functions of Sertoli cells include (1) structural support and nutrition of germ cells, (2) spermiation of mature spermatids, (3) movement of young germ cells, (4) phagocytosis of degenerating germ cells and residual bodies left by released spermatozoa, (5) secretion of luminal fluid and proteins, (6) formation of the blood–testis barrier, and (7) cell-to-cell communication (Dym and Madhwa Raj, 1977). Probably all of these are important. Sertoli cells also secrete lactate, a necessary energy source for germ cells (Jutte et al., 1982,

1983), and mitogenic polypeptides (Feig et al., 1980; Tres et al., 1986; Buch et al., 1988; Bellve and Zheng, 1989). This latter possibly functions to stimulate germ-cell or Sertoli-cell proliferation. Co-culture of germ cells with Sertoli cells stimulated RNA and DNA synthesis by the germ cells (Rivarola et al., 1985).

The precise role of Sertoli cells in spermatogenesis remains unknown. Sertoli cells provide structural support to developing germ cells based on their intimate contact (Figs. 7 and 10) (Fawcett, 1975b) and communication with germ cells. Tight junctional complexes attach adjacent Sertoli-cell processes and constitute the blood–testis barrier, which restricts the flux of serum components to spermatocytes and spermatids (Setchell and Waites, 1975). The unique environment in which spermatocytes develop and spermatids differentiate into spermatozoa is created by the Sertoli cells and their junctional complexes (Waites, 1977).

Sertoli-cell junctional complexes form between adjacent Sertoli cells and separate the seminiferous epithelium into basal and adluminal compartments (Fawcett, 1975b). Spermatocytogenesis occurs in the basal compartment where preleptotene primary spermatocytes are formed. These preleptotene primary spermatocytes migrate through the blood–testis barrier to the adluminal compartment, where meiosis and spermiogenesis occur. The barrier is maintained by forming new junctional complexes below the preleptotene primary spermatocytes prior to the dissolution of junctional complexes above the preleptotene primary spermatocytes. Plasminogen activator protein, secreted by Sertoli cells during the stage of the cycle when spermatocytes cross the barrier (Parvinen, 1982; Hettle et al., 1988), may be involved in local modification of the Sertoli-cell plasma membrane in this process.

The blood–testis barrier isolates spermatocytes and spermatids from serum born components and constitutes an immunologic isolation for these germ cells. It probably (1) prevents loss of specific concentrations of androgen binding protein, inhibin, and enzyme inhibitors within the luminal compartment of the tubules (Garner and Hafez, 1980) and (2) provides an environment unique for the development of germ cells. The blood–testis barrier does not appear to have an effect on the incorporation of radioactive amino acids by germ cells inside the barrier (Dadoune et al., 1981).

Sertoli cells, like other nurse cells in the body, function to provide material, stimulate growth, and remove waste. Sertoli cells phagocytize degenerating germ cells and the residual bodies at the end of spermia-

tion. Indeed, the lipid content of Sertoli cells, resulting from phagocytosis of residual bodies, varies with the stage of the cycle (Lacy and Roy; 1960; Brokelmann, 1963; Kerr and deKretser, 1975). The transit of residual bodies from the apex of the seminiferous epithelium to the base of Sertoli cells, easily seen in rats, can also be seen in the appropriate spermatogenic stage of domestic species with the use of Nomarski optics and 20-μm (thick) histologic sections (Fig. 10; horse, Stage VIII).

B. Pubertal Development of and Effect of Age on Sertoli-Cell Numbers

Numbers of Sertoli cells in young animals increase to adult values, stabilize, and then tend to decline with advancing age. Holstein bulls at 16, 20, 24, 28, and 32 weeks of age have 2×10^6, 202×10^6, 3520×10^6, 7927×10^6, and 8862×10^6 Sertoli cells per testis, respectively (Curtis and Amann, 1981). Hence, Sertoli-cell numbers per testis after completion of Sertoli-cell formation in Holstein bulls (7×10^9 to 9×10^9) are similar to the values reported for Normandy bulls (6×10^9 to 8×10^9; Attal and Courot, 1963). In stallion testes obtained in the breeding season (summer), the numbers of Sertoli cells are 1.2×10^9 by 1–2 years of age, 2.8×10^9 by 2–3 years of age, 3.6×10^9 by 4–5 years of age, and stabilized at approximately 3.5×10^6 until 20 years of age (Johnson and Thompson, 1983). The numbers of Sertoli cells per horse reach adult levels at 3 years of age (Johnson et al., 1990d). In testes obtained in the nonbreeding season (winter), the number of Sertoli cells increased with age up to 4–5 years of age and was decreased in 13- to 20-year olds. However, the number of Sertoli cells was significantly lower in the winter than in the summer (Johnson and Thompson, 1983; Johnson and Nguyen, 1986; Johnson, 1986a; Johnson and Tatum, 1989). It was reported (Jones and Berndtson, 1986) that the number of Sertoli cells decreases with age in horses to 20 years of age and that the ratio of germ cells per Sertoli cell increases with age. An age-related loss of Sertoli cells does appear to be appropriate for other species and perhaps all species if aged individuals are compared. In humans, the ratio of germ cells to Sertoli cell was not influenced by age; however, Sertoli-cell numbers declined with age paralleling the loss in spermatogenesis (Johnson et al., 1984b). Hence, it appears that the loss of numbers of Sertoli cells with age was not necessarily accompanied by loss of function of individual Sertoli cells. The influence of Sertoli-cell loss on the blood–testis barrier is unknown (Johnson, 1986b). Given that Sertoli cells are lost yearly in the horse (Johnson, 1986a), the horse may be a useful animal model to study their loss and its effect on the completeness of the blood–testis barrier.

C. Effect of Season on Sertoli-Cell Numbers in Adults

Sertoli cells generally are assumed to have a stable population size in adults (Lino, 1971; Steinberger and Steinberger, 1977; Hochereau-de Reviers and Courot, 1978). This assumption was supported by the failure to observe mitotic figures in rat Sertoli cells after 15 days of age (Clermont and Perey, 1957) and by the failure of adult rat Sertoli cells to display mitotic figures or incorporate tritiated thymidine in culture (Steinberger and Steinberger, 1971, 1977; Orth, 1982; Orth et al., 1988).

Conflicting data exist. Although the numbers of animals were small and findings were not statistically significant, the number of Sertoli cells in the adult red deer was 36% greater in the breeding season than in the nonbreeding season (Hochereau-de Reviers and Lincoln, 1978). In a large number of adult stallions, the number of Sertoli cells was greater in the breeding season (summer) than in the Winter (Johnson and Thompson, 1983). In a subsequent experiment involving 186 stallions (Johnson and Nguyen, 1986), it was found that adult stallions had more Sertoli cells during the breeding season than the nonbreeding season, and intermediate absolute values were found between seasons (Fig. 16). Photoperiod drives seasonal changes in serum concentrations of sex hormones in the stallion (Clay et al., 1988) and probably is responsible for seasonal changes in numbers of testicular cells including the Sertoli cells (Johnson and Tatum, 1989).

While the source of additional Sertoli cells in the breeding season is unknown, mitotic activity has not been evaluated in the horse (Johnson and Nguyen, 1986). Sertoli cells in vitro can maintain their ability to produce androgen binding protein while retaining their ability to undergo DNA synthesis and cell division by mitosis (Kierszenbaum and Tres, 1981).

In the rat, Sertoli-cell proliferation (measured by incorporation of [^{3}H]thymidine) occurs during prenatal and postnatal periods to day 15 of age (Steinberger and Steinberger, 1971; Orth, 1982). Sertoli-cell division is influenced by FSH (Griswold et al., 1977; Orth et al., 1984), opiate-like peptides of testicular origin (Orth, 1986), and insulin-like growth factor (GF$_1$) (Borland et al., 1984; Smith et al., 1987). Sertoli-cell uptake of [^{3}H]thymidine can be enhanced by hemicastration on the day of birth (Orth et al., 1984). Bulls (Curtis and Amann, 1981) and horses (Johnson and Thompson, 1983) differ from rats in that the numbers of Sertoli cells increase over a period of weeks or months during prepubertal development compared to 15–25 days in prenatal and postnatal rats (Orth, 1982). Hence, due to the extremely short proliferation period in rats, the mechanism(s) that control Sertoli-cell proliferation may be different. However, recent studies have shown that the number of Sertoli cells in rats is significantly larger at 60 days of age than at 15–30 days (Johnson et al., 1990e).

Interstital growth (from within as opposed to from the ends) along the length of the tubule, possibly by mitosis, appears to be the logical means by which Sertoli-cell augmentation occurs during the breeding season of stallions (Johnson and Nguyen, 1986). While the length of tubules increases significantly from 1.9 to 2.9 km during the breeding season, there is no seasonal difference in tubular diameter, size of individual Sertoli-cell nuclei, or number of Sertoli cells with nucleoli per cross section of tubules (Johnson and Thompson, 1983). Furthermore, a positive correlation between the total number of Sertoli cells per testis and the total tubular length was reported for rats, rams, and bulls (Hochereau-de Reviers and Courot, 1978).

Regardless of the source or fate of Sertoli cells in adults, seasonal variation in Sertoli-cell numbers in stallions (Johnson and Thompson, 1983; Johnson, 1986a) increase in Sertoli-cell number at day 60 in rats (Johnson et al., 1990e), and age-related loss of Sertoli cells in humans (Johnson et al., 1984b) are clear examples of the instability of the Sertoli-cell population in adults and challenge the present understanding of the dynamics of the seminiferous epithelium. Obvious changes in numbers of Sertoli cells

in stallions may be an indication of the turnover in Sertoli cells in adults of other species that do not experience a net change in Sertoli-cell number. Indeed, the age-related reduction in Sertoli-cell numbers in humans may be net loss in the turnover of a larger percentage of cells. The effect of new Sertoli cells entering the seminiferous epithelium during the approaching breeding season or cell loss at the end of the breeding season in horses and in aged men on the completeness of the blood–testis barrier is unknown.

VIII. Role of Other Nongerminal Testicular Cells in Spermatogenesis

A. Myoid Cells

Myoid cells (peritubular cells) constitute the cellular component of the boundary tissue around the seminiferous epithelium (Fig. 11). Myoid cells, like Sertoli cells and Leydig cells, are influenced by testicular growth factors (Bellve and Zheng, 1989). In addition to myoid cells, the boundary tissue contains a relatively thick (visible at the light-microscopic level) basal lamina and collagen elastic fibrils. In domestic species, myoid cells form a single layer, but in humans three to five layers of these flattened cells surround the seminiferous epithelium (Johnson et al., 1986, 1988). These cells contain fine cytoplasmic filaments believed to be actin and are thought to be involved in rhythmical contractions of seminiferous tubules. Myoid cells maintain tubular integrity and may assist in propulsive movements of spermatozoa and fluid from the seminiferous tubules (Fawcett, 1986).

In many instances of male infertility, the

boundary tissue layer of tubules becomes greatly thickened (DeKretser *et al.*, 1975; Salomon and Hedinger, 1982). The myoid cell layer becomes thickened with age in humans, resulting from reduced total seminiferous tubular length without loss in the total boundary tissue volume (Johnson *et al.*, 1986). Seasonally reduced length of seminiferous tubules in horses is not associated with a thickened tubular boundary tissue (Johnson and Neaves, 1981).

Myoid-cell secretions alter Sertoli cell function as measured by secretion of transferrin and androgen-binding protein (Skinner and Fritz, 1986). However, myoid cells can inhibit Sertoli-cell secretion of plasminogen activator protein (Hettle *et al.*, 1988). Indeed, a possible explanation for the low efficiency of spermatogenesis in men versus domestic and laboratory species (Johnson, 1986b) may be the negative influence on production of plasminogen activator protein by the high number of myoid cells next to the seminiferous epithelium in humans. This could trap preleptotene primary spermatocytes outside the blood–testis barrier and reduce spermatogenic efficiency in this species.

B. Leydig Cells

Leydig cells support spermatogenesis by the production of high concentrations of testosterone in close proximity to seminiferous tubules. The role of testosterone will be discussed later. Equine Leydig cells (Fig.17) have an abundant smooth endoplasmic reticulum (SER) where cholesterol is synthesized and pregnenolone is rapidly metabolized to testosterone. The volume density of Leydig-cell SER (percentage SER/testis) among several species almost totally explained the variation in *in vitro* production

of testosterone among species (Zirkin *et al.*, 1980), and significant correlations between SER volume per testis and serum concentrations of testosterone or intratesticular testosterone content were found in horses (Johnson and Thompson, 1987). Leydig cells have many mitochondria, which house the rate-limiting step of testostereone production, namely, cholesterol side-chain cleavage (Christensen, 1975).

Leydig cells are functional in the early embryo to establish the male sex ducts, regress in the late embryonic and early neonatal period (see Fig. 18, below), and then become functional again before puberty (Hooker, 1970). Leydig cells are also active during the neonatal period in boars (Van Straaten and Wensing, 1978).

Leydig-cell development corresponds to changes in darkness of testicular parenchyma and tubular development in horses. Fetal testes are laden with plump Leydig cells surrounding space cords of seminiferous tubules (Fig. 18). This gives the parenchyma a dark appearance when viewed grossly. The parenchyma becomes lighter as debris from fetal Leydig cells is removed and seminiferous tubules develop during prepubertal development. In 15–20 g testes from 1–1.5 year-old-horses, the parenchyma has light centers and darker peripheral regions. Dark regions are characterized by small undeveloped tubules with mostly Sertoli cells and gonocytes (Fig. 18b), but seminiferous tubules in the light regions are larger, with forming or completed lumena, and more advanced germ cells (Fig. 18c-d). Dark regions are characterized by an interstitium in which degenerating fetal interstitial cells are being removed, large pigment cells (macrophages) are abundant, and Leydig cells cannot be recognized (Bouin and Ancel, 1905; Nishikawa and Horie, 1955;

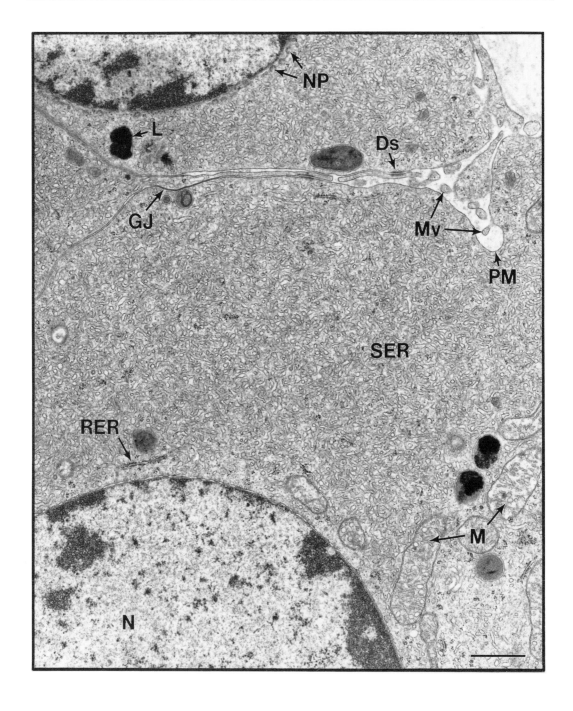

Johnson *et al.*, 1990c). Hence, local development of seminiferous tubules characterizes the horse testis.

Leydig-cell development appears first in the left, then the right testis. The left testis weighs more than the right 80% of the time in a large series of horses aged 1–54 months (Nishikawa and Horie, 1955). The fact that the left testis leads the right in development, and the center leads the peripheral region in development is consistent with a local control of Leydig-cell development (or removal of fetal interstitial cells) and local control in the establishment of spermatogenesis (Johnson *et al.*, 1990d).

Establishment of the Leydig-cell population corresponds temporally with changes in testicular size and parenchymal darkness. Leydig-cell numbers increase with age in horses from 1.4×10^9 at 2–3 years of age to 4.7×10^9 at 13–20 years of age (Johnson and Neaves, 1981). This corresponds to an increase in Leydig-cell volume per testis from 6 ml at 2–3 years of age to 32 ml in 13- to 20-year olds, an increase in testicular size from 117 to 213 g, and a darkening of the parenchyma. Indeed, a parenchymal pigmentation score was shown to be indicative of a horse's age from 2–20 years of age (Johnson and Neaves, 1981).

The Leydig-cell population size, like that of germ cells or Sertoli cells, cycles yearly in the horse (Fig. 16) (Johnson and Thompson, 1986; Johnson and Tatum, 1989). Seasonal changes in Leydig-cell number are responsible for seasonal changes in the volume of smooth endoplasmic reticulm in Leydig cells and intratesticular testosterone content in stallions (Johnson and Thompson, 1987). New Leydig cells are recruited to the existing population by mitosis in adult hamsters, regardless of whether the testes are recrudescing or active (Johnson *et al.*, 1987). Reduction in the rate of Leydig cell recruitment could explain the decline in Leydig cells in aging humans (Neaves *et al.*, 1984, 1987). Daily sperm production is positively correlated with the number of Leydig cells ($r = 0.76$) in the horse (Johnson and Thompson, 1986) and the amount of Leydig-cell smooth endoplasmic reticulum in humans (Johnson *et al.*, 1990a).

IX. Establishment of Spermatogenesis

Establishment of spermatogenesis takes a long period of time in domestic animals and even longer in humans. The process starts before birth, when primordial germ cells of the early developing embryo migrate into the undifferentiated fetal gonads from the yolk sac (Everett, 1945; Mintz, 1959). These primordial germ cells undergo several divisions in the gonads to produce gonocytes within the fetal male sex cords (Fig. 18). Sex cords are composed of the light-staining

Figure 17 Transmission electron micrograph of Leydig cells in the stallion during the breeding season, but also characteristic of the nonbreeding season. Cytoplasm contains an abundance of branched tubular profiles of smooth endoplasmic reticulum (SER). Polyribosomes and profiles of rough endoplasmic reticulum (RER) are present but not in abundance. Mitochondria (M) with their characteristic tubular cristae usually form clusters but are largely surrounded by SER. Secondary lysosomes (L) in various degrees of development into lipofuscin granules are presented. The nucleus (N) is largely euchromatic with heterochromatin clumps near the nuclear envelope where numerous nuclear pores (NP) are found. At the free surface of cells, the plasma membrane (PM) forms microvilli (Mv). At surfaces where adjacent Leydig cells touch, gap junctions (GJ) and desmosomes (Ds) are found. Bar = 1 μm. From Johnson and Thompson (1987).

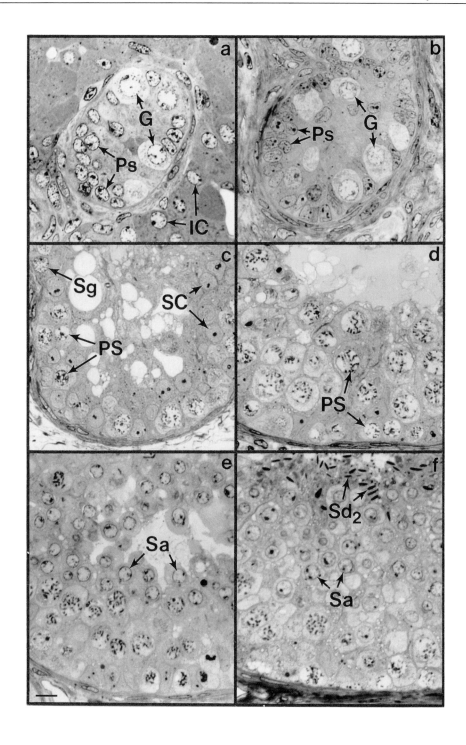

gonocytes and the darkly staining indifferent supporting cells that differentiate into Sertoli cells. Gonocytes persist in the male sex cords until just before puberty, at which time they divide and differentiate to produce spermatogonia. In the lamb, spermatogonia proliferate to produce spermatocytes by 100 days and spermatids by 120–125 days with complete spermatogenesis by 140–150 days (Ortavant et al., 1977). In the calf, spermatogonial proliferation occurs during 16–20 weeks after birth, primary spermatocytes are present by 24 weeks, spermatids by 28 weeks, and there is completion of spermatogenesis by 32 weeks (Curtis and Amann, 1981).

Figure 18 illustrates the events in establishing of spermatogenesis in the stallion. Gonocytes were present in the sex cords during early and late gestation. By 12 months of age spermatogonia and primary spermatocytes may be present, followed by round spermatids at about 16 months and mature spermatids at 36 months. However, there is considerable variation in the timing of the appearance of specific germ cells among stallions and among tubules within a stallion. In horses, as in other species (Clermont and Perey, 1957; Attal and Courot, 1963; Curtis and Amann, 1981), large numbers of degenerating germ cells can be found during the establishment of spermatogenesis.

Once spermatogenesis is complete, the efficiency of yield is increased, then stabilized at the adult level. Increased spermatozoan production rates per testis above the level at puberty is due to increased testicular size. At 28 weeks, 50% of bulls have a daily sperm production over 0.5×10^6 per gram, for a mean 0.60×10^6/g parenchyma. At 32 weeks, the efficiency of spermatogenesis measured by daily sperm production/g was 4.0×10^6/g (Curtis and Amann, 1981) and increased through the first year when it reached a mature level of 12×10^6 (Amann and Almquist, 1976). In stallion testes obtained in the summer, daily sperm production/g parenchyma was 6×10^6 at 2–3 years, 18×10^6 at 4–5 years, and 20×10^6 for 6–20 years of age (Johnson and Thompson, 1983).

Establishment of spermtogenesis apparently occurs at random throughout the testis in most species (Courot et al., 1970). The horse is an exception in that development of seminiferous tubules starts at the center of the testis and is follwed by development of peripheral regions (Bouin, 1904). A gross section through a developing testis from a

Figure 18 Steps in development of spermatogenesis in the horse from (a) 270 days of gestation through (b, c) 10–12 months, and (d, e) 16 months to (f) 36 months of age. (a) Fetal male sex cords, scarce among a high density of robust interstitial cells (IC), are composed of darker-staining support cells or pre-Sertoli cells (Ps) and lighter-staining gonocytes (G). (b) The seminiferous tubules at 10–12 months are similar to the fetal cords containing gonocytes (G); however, some nuclear maturation can be seen in pre-Sertoli cells (Ps), with a larger number of pre-Sertoli cell nuclei located in the center of the tubule. (c) Other 10- to 12-month old colts exhibit considerable maturation of Sertoli cells (SC) whose apical regions extend toward the forming lumen. Gonocytes have matured into spermatogonia (Sg) and some primary spermatocytes (PS) are present. (d) By 16 months, two generations of primary spermatocytes (PS) are present, and the lumen is completely formed. (e) Also at 16 months, some seminiferous tubules contain Sa spermatids with spherical nuclei (Sa) as the most advanced cell type. (f) By 36 months, spermatogenesis has been completed, as is evident by numerous Sa (Sa) and Sd_2 (Sd_2) spermatids. Variation in the most advanced cell type present in the seminiferous epithelium exists among tubules of the same testis and among horses. Bar length equals 10 μm.

1- or 2-year-old horse has light testicular parenchyma with expansion of seminiferous
tubules in the center and dark on the peripheral regions where tubules are less developed. As indicated (see Section VIII.B),
a local control mechanism for the initiation

of spermatogenesis is implied by central followed by peripheral development and by
the left testis developing before the right
(Johnson *et al.*, 1990d).

In stallions, as in other domestic animals,
the maximum efficiency of spermatogenesis

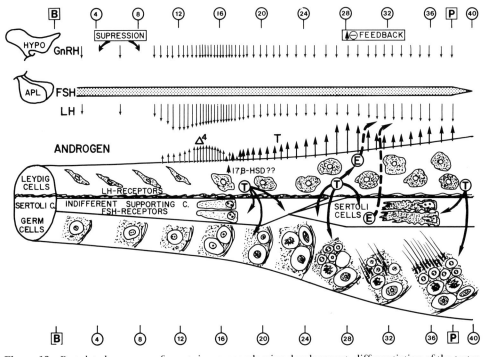

Figure 19 Postulated sequence of events in neuroendocrine development, differentiation of the testes,
and the initiation of spermatogenesis between birth (B) and puberty (P) in well-fed Holstein bulls. Age
is shown in weeks. Starting at about 12 weeks, a sequence of events induces (a) initiation of pulsatile
discharge of LH, possibly resulting from a diminished suppression of GnRH discharge from the hypothalamus or from an increased capacity of the anterior pituitary to respond to GnRH; (b) LH-induced
differentiation of Leydig cells with increased secretion of androstenedione in response to LH stimulation; (c) further differentiation of Leydig cells resulting in LH-stimulated secretion of testosterone; (d)
testosterone-induced differentiation of indifferent supporting cells to Sertoli cells concomitant with
testosterone induced differentiation of gonocytes to prespermatogonia and A-spermatogonia; (e) an
increased sensitivity of the hypothalamic–pituitary complex to the negative feedback action of gonadal
steroids; (f) a somewhat diminished frequency and amplitude of LH discharges; (g) cessation of Sertoli
cell formation followed by formation of junctional complexes between adjacent Sertoli cells and establishment of the blood–testis barrier; (h) sequential formation of primary spermatocytes, spermatids and
spermatozoa; and (i) continued increases in the efficiency of spermatogenesis until sufficient sperm are
produced to provide the first ejaculum around 37 weeks. Figure and legend from Amann and Schanbacher (1983).

is obtained prior to maximum spermatozoan production per testis. Based on seminal characteristics, puberty occurs at 1–2.2 years (Cornwell, 1972) or 1.1–1.9 years (Naden *et al.*, 1990) in Quarter Horse colts. Based on testicular development (Nishikawa, 1959), puberty occurs between 1.4 and 1.8 years in Anglo-Norman stallions. By 2.5 years of age, equine testes from lightweight breeds obtain adult levels for the percentage of seminiferous tubules with elongated spermatids (Sd_1 or Sd_2 spermatids; Figs. 8–9) or residual bodies (Fig. 2) left behind by released spermatozoa (Johnson *et al.*, 1990d). Adult levels of daily sperm production per gram parenchyma are obtained by 3 years of age in the stallion, but testicular parenchymal weight and daily sperm production increases to year 4. The percentage of the testis occupied by seminiferous tubules was similar at about 72% for horses aged 2–3, 4–5, and 13–20 years when testicular weight increased from 117 to 213 g per testis and the volume of seminiferous tubules/testis increased from 72 to 127 ml (Johnson and Neaves, 1981). Increase in seminiferous tubular volume was due to a slight increase in the diameter and an increased length from 2.0 km to 2.8 km over the same time period. A similar pattern of testicular growth was seen in the bull where testicular weight increased from 117 to 350 g without alterating the percentage of testis occupied by seminiferous tubules (Amann, 1962; Attal and Courot, 1963; Swierstra, 1966; Humphrey and Ladds, 1975; Curtis and Amann, 1981).

The coordination of the development of spermatogenesis with the establishment of Sertioli-cell and Leydig-cell populations and hormonal influences in the bull has been summarized (Fig. 19) by Amann and Schanbacher (1983). The completion of this coor-

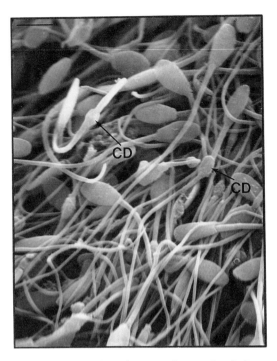

Figure 20 Scanning electron micrograph of ejaculated equine spermatozoa. Most of the spermatozoa in the ejaculate are normal. However, the cytoplasmic droplets on some spermatozoa are still attached to the proximal end (CD) and at the distal end (CD′) of the middle piece of the spermatozoan tail. The transition between the middle piece and principal piece is less distinct in ejaculated spermatozoa compared to testicular (Fig. 2) or efferent ductual (Fig. 3) spermatozoa. Bar length equals 5 μm. From Johnson *et al.* (1978a).

dinated effort in domestic species results in the production of a large number of ejaculated spermatozoa (Fig. 20).

X. Control of Spermatogenesis

A. Hormonal Influences

The rate at which a spermatozoon develops (duration of spermatogenesis or spermatogenic cycle length) is not influenced by hormones in normal or in hypophysecto-mized males (Harvey, 1962; Courot et al., 1979). However, the amount of germ-cell degeneration and spermatozoan product rates are influenced by hormones (Cler-mont, 1972; Steinberger and Steinberger, 1975; Setchell, 1978; Courot et al., 1979; Hochereau-de Reviers et al., 1987; Courot, 1988; Amann, 1989b).

The initiation of spermatogenesis is thought to depend largely upon concentra-tions of testosterone and follicle-stimulating hormone (FSH). High concentrations of tes-tosterone, produced following luteinizing hormone (LH) stimulation of Leydig cells in juxtaposition with seminiferous tubules, are required to maintain spermatogenesis in the adult. Seminiferous tubules in the horse are bathed in testosterone at 70 mg/g testicular parenchyma (Johnson and Thompson, 1987), compared to a serum concentration of 300 pg/ml (Johnson and Thompson, 1983). FSH stimulates Sertoli cells directly to stimulate their functions in support of germ-cell development.

Secretion of both FSH and LH, by the an-terior pituitary, is under positive control by gonadotropin-releasing hormone from the hypothalamus and under negative control by both testosterone and estradiol from the testis and locally produced estradiol (Chris-tensen, 1975). Leydig cells are regulated to produce testosterone by prolactin, which acts synergistically with LH and other hor-mones including a paracrine substance se-creted by seminiferous tubules at specific stages of the spermatogenic cycle (Amann, 1989b).

It has been reasoned (Amann, 1989b) that the production of LH and FSH appar-ently is differentially controlled by the inter-play of feedback from Sertoli cells via in-

hibin and activin and by the influence of the ratio of estradiol and testosterone on the go-nadotropin secretion of the anterior pitu-itary. The ratio of estradiol to testosterone is controlled by direct secretion and clear-ance of testosterone and by the conversion of testosterone to estradiol in target tissues.

Hormonal control of spermatogenesis varies among species (Courot, 1988). In the ram, FSH is necessary (Courot et al., 1979; Hochereu-de Reviers et al., 1987) and LH may also have a direct role in the regulation of spermatogonial proliferation (Courot, 1988). In contrast, spermatogenesis can be maintained by testosterone alone in the rat. In the horse, photoperiod-driven (Clay et al., 1988) serum concentrations of FSH, LH, testosterone, and prolactin were highest in the summer when testicular weight, in-tratesticular testosterone levels, numbers of Sertoli cells or Leydig cells, number of sper-matogonia, and daily sperm production are highest (Fig. 16) (Amann, 1989a; Johnson, 1986a, 1989a; Johnson and Nguyen, 1986; Johnson and Tatum, 1989; Johnson and Thompson, 1983, 1986, 1987; Thompson and Johnson, 1987; Thompson et al., 1986).

B. Other Influences

Elevated temperature or adversely cold conditions are detrimental to spermatogen-esis. Artificially increasing scrotal tempera-ture to 41°C for 3 h in rams and boars causes rapid destruction of pachytene spermatocytes (Mazzari et al., 1968; Waites and Ortavant, 1968). A transient effect on spermatogonial divisions occurs with in-creased numbers of cells in prophase and metaphase (Waites and Ortavant, 1968). While elevated scrotal temperature also caused loss of round spermatids in the boar (Mazzarri et al., 1968), pachytene spermatocytes in meiotic prophase appear to be the most thermosensitive (Waites and Ortavant, 1968). High ambient temperature reduces spermatozoan production in boars (Wettemann *et al.*, 1976). Other factors, such as radiation, drugs, vitamin A deficiency, limited diet, and toxic chemicals, are detrimental to spermatogenesis.

Testes from rams (Ortavant, 1958; Ortavant *et al.*, 1977; Lincoln and Short, 1980) and stallions (Amann, 1989a; Johnson and Thompson, 1983) undergo regressive changes without total cessation of spermatogenesis. In the ram, seasonal regressed testes have fewer spermatogonia and reduced spermatozoan production (Hochereau-de Reviers, 1981). In stallions (Fig. 16), seasonal-regressed testes have reduced weight, reduced number of Leydig and Sertoli cells, fewer spermatogonia, reduced hormone concentrations, and reduced daily sperm production (Johnson, 1985, 1986a; Johnson and Neaves, 1981; Johnson and Nguyen, 1986; Johnson and Tatum, 1989; Johnson and Thompson, 1983, 1986, 1987; Amann, 1989a).

Age reduces numbers of germ cells or daily sperm production in rabbits (Ewing *et al.*, 1972), mice (Gosden *et al.*, 1982), and rats (Johnson and Neaves, 1983). In horses, daily sperm production reaches its peak at about 4–5 years of age and remains at this level until 20 years of age (Johnson and Neaves, 1981; Johnson and Thompson, 1983). What happens beyond 20 years remains unknown. In Holstein bulls located at artificial insemination organizations, about 46% of bulls selected for both genetic and reproductive desirability are removed for reproductive problems (Kratz *et al.*, 1983).

The fact that some males retain their fertility even with the advancement of age is indicated by fertility in bulls 19 years of age (Bishop, 1970) and a successful paternity in a 94-year-old man (Seymour *et al.*, 1935).

XI. Concluson

Spermatogenesis is a long process by which spermatozoa, found in the ejaculate (Fig. 20), are produced. Spermatogenesis involves mitosis to increase its yield, meiosis to reduce chromosome number, and an unsurpassed example of cell differentiation in the production of the self-propelled, penetrative-enzyme-containing, male-genome delivery system, namely, the spermatozoon.

Acknowledgments

The author is grateful to Dr. Janice Grumbles for assistance in organizing references and typing the original manuscript, to Ms. Judy Gloyna for typing the intermediate and final copies, to Mr. Vince Hardy for producing most of the plates, and to Drs. R. P. Amann and R. G. Saacke for comments on the chapter.

This chapter was supported in part by NIH grants AG02260 and HD16773.

References

Allen, E. (1918). *J. Morphol.* **31,** 33.
Amann, R. P. (1962). *Am. J. Anat.* **110,** 69.
Amann, R. P. (1970a). *Fertil. Steril* **21,** 662.
Amann, R. P. (1970b). *In* "The Testis" (A. D. Johnson, W. R. Gomes, and N. L. Van Demark, eds.), Vol. I., p. 433. Academic Press, New York.
Amann, R. P. (1981). *J. Androl.* **2,** 37.
Amann, R. P.(1986). *Environ. Health Perspect.* **70,** 149.
Amann, R.P. (1988). Proc. Int. 11th Cong. Animal Reproduction and Artificial Insemination, p. 320, Dublin, Ireland.
Amann, R.P. (1989a). *In* "Management of the Stallion for Maximum Reproductive Efficiency. II" (B. W. Pickett, R. P. Amann, A. O. McKinnon, E. L. Squires, and J. L. Voss, eds.), p. 27. Colorado State University, Ft. Collins, Colorado.

Amann, R. P. (1989b). *J. Am. Coll. Toxicol.* **8,** 457.

Amann, R. P., and Almquist, J. O. (1976). Proc. VI Tech. Conf. Artificial Insemination and Reproduction, Columbia, Ohio, p. 1.

Amann, R. P., and Schanbacher, B. D. (1983). *J. Amin. Sci.* **57**(Suppl.2),380.

Amann, R. P., Johnson, L., Thompson, D. L., Jr., and Pickett, B. W. (1976a). *Biol. Reprod.* **15,** 586.

Amann, R. P., Kavanaugh, J. F., Griel, L. C., Jr., and Voglmayr, J. K. (1976b). *J. Dairy Sci.* **57,** 93.

Attal, J., and Courot, M. (1963). *Ann. Biol. Anim. Biochem. Biophys.* **3,** 219.

Barr, A. B., Moore, D. J., and Paulsen, C. A. (1971). *J. Reprod. Fertil.* **25,** 75.

Bellve, A. R., and Zheng, W. (1989). *J. Reprod. Fertil.* **85,** 771.

Benda, C. (1887). *Arch. Microscop. Anat.* **30,** 49.

Berndtson, W. E. (1989). *Biol. Reprod.* **40** (Suppl. 1),80.

Berndtson, W. E. (1977). *J. Anim. Sci.* **44,** 818.

Berndtson, W. E., and Desjardins, C. (1974). *Am. J. Anat.* **140,** 167.

Berndtson, W. E., Igboeli, G., and Pickett, B. W. (1987). *J. Anim. Sci.* **64,** 241.

Bishop, M. W. H. (1970). *J. Reprod. Fertil. Suppl.* **12,** 65.

Borland, K., Mita, M., Oppenheimer, C. L., Blinderman, L. A., Massague, J., Hall, P. F., and Czech, M. P. (1984). *Endocrinology* **114,** 240.

Bouin, P., and Ancel, P. (1905). *Arch. Zool. Exp. Gen.* **3,** 391.

Bouin, P. (1904). *C. R. Soc. Biol.* **11,** 658.

Brokelmann, J. Z. (1963). *Z. Zellforch Mikroscop. Anat.* **59,** 820.

Buch, J. P., Lamb, D. J., Lipshultz, L. I., and Smith, R. G. (1988). *Gert. Steril.* **49,** 658.

Burgos, M. H., Vitale-Calpe, R. and Aoki, A. (1970). "The Testis" (A. D. Johnson, W. R. Gomes, and N. L. Van Demark, eds.), Vol. I, p. 551. Academic Press, New York.

Christensen, A. K. (1975). *In* "Handbook of Physiology" (R. O. Greep and E. B. Astwood, eds.), Section 7, Vol. V, p. 57. American Physiological Society, Washington, D.C.

Clay, C. M., Squires, E. L., Amann, R. P. and Nett, T.M. (1988).*J. Anim. Sci.* **66,** 1246.

Clermont, Y. (1962). *Am. J. Anat.* **111,** 111.

Clermont, Y. (1963). *Am. J. Anat.* **112,** 35.

Clermont, Y. (1972). *Physiol. Rev.* **52,** 198.

Clermont, Y., and Bustos-Obregon, E. (1968). *Am. J. Anat.* **122,** 237.

Clermont, Y., and LeBlond, C. P. (1953). *Am. J. Anat.* **93,** 475.

Clermont, Y., and Perey, B. (1957). *Am. J. Anat.* **100,** 241.

Cornwell, J.C. (1972). M.S. Thesis, Louisiana State University.

Coulter, G. H. (1980). Proc. 7th Tech. Conf. Nat. Assoc. Animal Breeders, Columbia, Ohio, p. 106.

Courot, M. (1988) Proc. Inter. 11th Cong. Animal Reproduction and Artificial Insemination **5,** 312 Dublin, Ireland.

Courot, M., Hochereau-de Reviers, M. T., Monte-Kuntez, A., Locatelli, A., Pisselet, C., Blanc, M. R., and Dacheux, J.L. (1979) *J. Reprod. Fertil. Suppl.* **26,** 165.

Courot, M., Hochereau-de Reviers, M. T. and Ortavant, R. (1970). *In* "The Testis" (A. D. Johnson, W. R. Gomes, and N. L. Van Demark, eds.), Vol. I, p. 339. Academic Press, New York.

Curtis, G. M. (1918). *Am. J. Anat.* **24,** 339.

Curtis, S. K., and Amann, R. P. (1981). *J. Anim. Sci.* **53,** 1645.

Dadoune, J. P., Fain-Maurel, M. A., Alfonsi, M. F., and Katsanis, G. (1981). *Biol. Reprod.* **24,** 153.

D'Agostino, A., Geremia, R., and Monesi, V. (1978). *Cell Differen.* **7,** 175.

DeKretser, D. M., Kerr, J. B., and Paulsen, C. A. (1975). *Biol. Reprod.* **12,** 317.

Dym, M., and Madhwa Raj, J. G. (1977). *Biol. Reprod.* **17,** 676.

Erickson, R. P., Erickson, J. M., Betlach, C. J., and Meistrich, M. L. (1980). *J. Exp. Zool.* **214,** 13.

Everett, N. B. (1945). *Biol. Rev.* **20,** 45.

Ewing, L. L., Johnson, B. H., Desjardins, C., and Clegg, R. F. (1982). *Proc. Soc. Exp. Biol. Med.* **140,** 907.

Fawcett, D. W. (1975a). *Devel. Biol.* **44,** 394.

Fawcett, D. W. (1975b). *In* "Handbook of Physiology" (R. O. Greep and E. B. Astwood, eds.), Section 7, Vol. V, p. 21. American Physiological Society, Washington, D.C.

Fawcett, D. W. (1986). *In* "Bloom and Fawcett: A Textbook of Histology," 11th Ed., p. 820. W. B. Saunders Co., Philadelphia.

Fawcett, D. W., and Phillips, D. M. (1970). *In* "Comparative Spermatology" (B. Bacetti, ed.), Proc. Int. Symp. Rome and Siena, p. 2. Academic Press, New York.

Feig, L. A., Bellve, A. R., Erickson, N. H., and Klagsbrun, M. (1980). *Proc. Natl. Acad. Sci. USA* **77,** 4774.

Foote, R. H. (1978). *J. Anim. Sci.* **47**(Suppl.2), 1.

Frankenhuis, M. T., de Roooij, D. G., and Kramer, M. F. (1980). Proc. 9th Int. Congr. Animal Reproduction and Artificial Insemination **5,** 17.

Furst, C. (1887). *Arch. Mikroskop. Anat. Entwicklungsmech.* **30,** 336.

Garner, D. L. and Hafez, E. S. E. (1980). *In* "Reproduction in Farm Animals" (E. S. E. Hafez, ed.), 4th Ed., p. 167. Lea and Febiger, Philadelphia.

Gebauer, M. R., Pickett, B. W., and Swierstra, E. E. (1974). *J. Anim. Sci.* **39,** 732.

Goodrowe, K. L., and Heath, E. (1984). *Anat. Rec.* **209,** 177.

Gosden, R. G., Richardson, D. W., and Davidson, D. W. (1982). *J. Reprod. Fertil.* **64,** 127.

Griswold, M. D., Morales, C., and Sylvester, S. R. (1988). *Oxford Rev. Reprod. Biol.* **10,** 124.

Griswold, M. D., Solari, A., Tung, P., and Fritz, I. B. (1977). *Mol. Cell. Endocrinol.* **7,** 151.

Harvey, S. C. (1962). M.Sc. Thesis, McGill University, Montreal, Canada.

Hess, R. A., Linder, R. E., Strader, L. F. and Perreault, S. D. (1988). *J. Androl.* **9,** 327.

Hettle, J. A., Balekjian, E., Tung, P. S., and Fritz, I. B. (1988). *Biol. Reprod.* **38,** 359.

Hochereau, M. T. (1963). *Ann. Biol. Anim. Biochem. Biophys.* **3,** 93.

Hochereau-de Reviers, M. T. (1981). *In* "Reproductive Processes and Contraception" (K. W. McKerns, ed.), p. 307. Plenum Press, New York.

Hochereau-de Reviers, M. T., and Courot, M. (1978). *Ann. Biol. Anim. Biochem. Biophys.* **18,** 573.

Hochereau-de Reviers, M. T., and Lincoln, G. A. (1978). *J. Reprod. Fertil.* **54,** 209.

Hochereau-de Reviers, M. T., Monet-Kumtz, C., and Courot, M. (1987). *J. Reprod. Fertil. Suppl.* **34,** 101.

Holstein, A. F., and Roosen-Runge, E. C. (1981). "Atlas of Human Spermatogenesis," p.140 Grosse Verlag, Berlin.

Hooker, C. W. (1970). *In* "The Testis" (A. D. Johnson, W. R. Gomes, and N. L. Van Demark, eds.), Vol. I, p. 483. Academic Press, New York.

Huckins, C. (1971). *Anat. Rec.* **169,** 533.

Huckins, C. (1978). *Anat. Rec.* **190,** 905.

Humphrey, J. D., and Ladds, P. W. (1975). *Res. Vet. Sci.* **19,** 135.

Irons, M. J., and Clermont, Y. (1982a). *Am. J. Anat.* **165,** 121.

Irons, M. J., and Clermont, Y. (1982b). *Anat. Re.* **202,** 463.

Johnson, L. (1982). *Fertil. Steril.* **37,** 811.

Johnson, L. (1985). *Biol. Reprod.* **32,** 1181.

Johnson, L. (1986a). *Anat. Rec.* **214,** 231.

Johnson, L. (1986b). *J. Androl.* **7,** 331.

Johnson, L. (1987). Proc. Ann. Mtg. Am. Assoc. Vet. Anat., Madison, Wisconsin, p. 28.

Johnson, L. (1989a). *Biol. Reprod.* **40**(Suppl.1), 79.

Johnson, L. (1989b). Proc. of the 1988 Serono International Symposium on Gamete Physiology, Newport Beach, California.

Johnson, L. (1990). *Biol. Reprod.* (in press).

Johnson, L., and Hardy, V. B. (1989). 14th Ann. Mtg. Am. Soc. Androl, New Orleans, Louisiana. *J. Androl.* **10,** 41.

Johnson, L., and Neaves, W. B. (1981). *Biol. Reprod.* **24,** 703.

Johnson, L., and Neaves, W. B. (1983). *J. Androl.* **4,** 162.

Johnson, L., and Nguyen, H. B. (1986). *J. Reprod. Fertil.* **76,** 311.

Johnson, L., and Tatum, M. E. (1989). *Biol. Reprod.* **40,** 994.

Johnson, L., and Thompson, D. L., Jr. (1983). *Biol. Reprod.* **29,** 777.

Johnson, L., and Thompson, D. L., Jr. (1986). *Biol. Reprod.* **35,** 971.

Johnson, L., and Thompson, D. L., Jr. (1987). *J. Reprod. Fertil.* **81,** 227.

Johnson, L., Abdo, J. G., Petty, C. S., and Neaves, W. B. (1988). *Fertil. Steril.* **49,** 1045.

Johnson, L., Amann, R. P., and Pickett, B. W. (1978a). *Am. J. Vet. Res.* **39,** 1428.

Johnson, L., Amann, R. P., and Pickett, B. W. (1978b). *Fertil. Steril.* **29,** 208.

Johnson, L., Amann, R. P., and Pickett, B. W. (1980). *Am. J. Vet. Res.* **41,** 1190.

Johnson, L., Grumbles, J. S., Chastain, S., and Goss, H. F., Jr., and Petty, C. S. (1990a). *J. Androl.* **11,** 155.

Johnson, L., Hardy, V. B., and Martin, M. T. (1990b). *Anat. Rec.* **227** (in press).

Johnson, L., Kattan-Said, A. F., Hardy, V. B., and Scrutchfield, W. L. (1990c). *J. Reprod. Fert.* **89** (in press).

Johnson, L., Lebovitz, R. M., and Samson, W. K. (1984a). *Anat. Rec.* **209,** 501.

Johnson, L., Matt, K. S., Bartke, A., Nguyen, H. B., and Le, H. T. (1987). *Biol. Reprod.* **37,** 727.

Johnson, L., Petty, C. S., and Neaves, W. B. (1981). *Biol. Reprod.* **25,** 217.

Johnson, L., Petty, C. S., and Neaves, W. B. (1983). *Biol. Reprod.* **29,** 207.

Johnson, L., Petty, C. S., and Neaves, W. B. (1986). *J. Androl.* **7,** 316.

Johnson, L., Varner, D. D., and Thompson, D. L., Jr., (1990d). *J. Reprod. Fert.* (submitted).

Johnson, L., Vaughan, S. C., and Ventura, M. S. F. (1990e) *J. Androl.* **11,** 45.

Johnson, L., Zane, R. S., Petty, C. S., and Neaves, W. B. (1984b). *Biol. Reprod.* **31,** 785.

Jones, L. S., and Berndtson, W. E. (1986). *Biol. Reprod.* **35,** 138.

Jutte, H. P. N., Jansen, R., Grootegoed, J. A., Rommerts, F. F., Clausen, O. P. E., and Van der Molen, H. J. (1982). *J. Reprod. Fertil.* **65,** 431.

Jutte, H. P. N., Jansen, R., Grootegoed, J. A., Rommerts, F. F., Clausen, O. P. E., and Van der Molen, H. J. (1983). *J. Reprod. Fertil.* **68,** 219.

Kerr, J. B., and deKretser, D. M. (1975). *J. Reprod. Fertil.* **43,** 1.

Kierszenbaum, A. L., and Tres, L. L. (1981). *In* "Bioregulators of Reproduction" (G. Jagiello and H. J. Vogel, eds.), p. 207. Academic Press, New York.

Kleene, K. C., Distel, R. J., and Hecht, N. B. (1983). *Develop. Biol.* **98,** 455.

Knudsen, O. (1958). *Int. J. Fertil.* **3,** 389.

Kratz, J. L., Wilcox, C. J., Martin, F. G., and Becker, R. B. (1983). *J. Dairy Sci.* **66,** 642.

Lacy, D., and Roy, J. (1960). *Microsc. Soc.* **79,** 209.

Leblond, C. P., and Clermont, Y. (1952). *Am. J. Anat.* **90,** 167.

Lebovitz, R. M., and Johnson, L. (1983). *Bioelectromagnetics* **4,** 107.

Lincoln, G. A., and Short, R. V. (1980). *Recent Prog. Horm. Res.* **36,** 1.

Lino, B. F. (1971). *Anat. Rec.* **170,** 413.

Loir, M. (1972a). *Ann. Biol. Anim. Biochem. Biophys.* **12,** 203.

Loir, M. (1972b). *Ann. Biol. Anim. Biochem. Biophys.* **12,** 411.

Mazzari, G., du Mensil du Buisson, F., and Ortavant, R. (1968). Proc. VI Congr. Reprod. Anim. Artificial Insemination, Vol. I, p. 305. Paris.

Mintz, B. (1959). *Arch. Anat. Microscop. Morphol. Exp.* **48** (bis), 155.

Monesi, V. (1965). *Exp. Cell Res.* **39,** 197.

Naden, J., Amann, R. P., and Squires, E. L. (1990). *J. Reprod. Fert.* **88,** 167.

Neaves, W. B., Johnson, L., and Petty, C. S. (1987). *Biol. Reprod.* **36,** 301.

Neaves, W. B., Johnson, L., Porter, J. C., Parker, C. R., Jr., and Petty, C. S. (1984). *J. Clin. Endocrinol. Metab.* **59,** 756.

Nishikawa, Y., *Japan Racing Assoc. Tokyo,* p. 2 (1959).

Nishikawa, Y., and Horie, T. (1955). *Nat. Inst. Agric. Sci. Jpn. Bull. Ser. G. Anim. Husb.* **10,** 229.

Oakberg, E. F. (1956). *Am. J. Anat.* **99,** 391.

Olar, T. T., Amann, R. P., and Pickett, B. W. (1983). *Biol. Reprod.* **29,** 1114.

Olson, G. E., and Sammons, D. W. (1980). *Biol. Reprod.* **22,** 319.

Ortavant, R. (1958). D.Sci. Thesis, University of Paris, France.

Ortavant, R. (1959). *Ann. Zootech.* **8,** 183, 271.

Ortavant, R., Courot, M., and Hochereau-de Reviers, M. T. (1977). *In* "Reproduction in Domestic Animals" (H. H. Cole and P. T. Cupps, eds.), 3rd Ed., p. 203. Academic Press, New York.

Orth, J. M. (1982). *Anat. Rec.* **203,** 485.

Orth, J. M. (1986). *Endocrinology* **119,** 1876.

Orth, J. M., Gunsalus, G. L., and Lamperti, A. A. (1988). *Endocrinology* **122,** 787.

Orth, J. M., Higginbotham, C. A., and Salisbury, R. L. (1984). *Biol. Reprod.* **30,** 263.

Parvinen, M. (1982). *Endocr. Rev.* **3,** 404.

Parvinen, M., and Vanha-Perttula. T. (1972). *Anat. Rec.* **174,** 435.

Parvinen, M., and Hecht, N. B. (1981). *Histochemistry* **71,** 567.

Perey, B., Clermont, Y., and Leblond, C. P. (1961). *Am. J. Anat.* **108,** 47.

Regaud, C. (1901). *Arch. Anat. Microscop. Morphol. Exp.* **4,** 101, 231.

Rivarola, M. A., Sanchez, P., and Saez, J. M. (1985). *Endocrinology* **117,** 1796.

Roosen-Runge, E. C., and Giesel, L. O. (1950). *Am. J. Anat.* **87,** 1.

Saacke, R. G., and Almquist, J. D. (1964). *Am. J. Anat.* **115,** 163.

Salomon, F., and Hedinger, C. E. (1982). *Lab. Invest.* **47,** 543.

Schulze, W. (1982). *Andrologia* **14,** 200.

Schulze, W., and Rehder, U. (1984). *Cell. Tissue Res.* **237,** 395.

Sertoli, E. (1865). *Morgagni* **7,** 31.

Setchell, B. P. (1978). "The Mammalian Testis." Cornell University Press, Ithaca, New York.

Setchell, B. P., and Waites, G. M.. H. (1975). *In* "Male Reproductive System" (R. O. Greep and D. W. Hamilton, eds.), p.143. American Physiological Society, Bethesda, Maryland.

Seymour, F. I., Duffy, C., and Koerner, A. (1935). *J. Am. Med. Assoc.* **105,** 1423.

Skinner, M. K., and Fritz, I. B. (1986). *Mol. Cell. Endocrinol.* **44,** 85.

Slaughter, G. R., Meistrich, M. L., and Means, A. R. (1989). *Biol. Reprod.* **40:**395.

Smith, E. P., Soboda, M. E., Van Wyk, J. J., Kierszenbaum, A. L., and Tres, L. L. (1987). *Endocrinology* **120,** 186.

Soderstrom, K. O., and Parvinen, M. (1976). *Mol. Cell. Endocrinol.* **5,** 181.

Steinberger, A., and Steinberger, E. (1971). *Biol. Reprod.* **4,** 84.

Steinberger, A., and Steinberger, E. (1975). *In* "Handbook of Physiology" (R. O. Greep and E. B. Astwood, eds), Section 7, Vol. V. p. 1. American Physiological Society, Washington, D.C.

Steinberger, A., and Steinberger, E. (1977). *In* "The Testis" (A. D. Johnson and W. R. Gomes, eds.), Vol. IV, p. 371. Academic Press, New York.

Swierstra, E. E. (1966). *Can. J. Anim. Sci.* **46,** 107.

Swierstra, E. E. (1968). *Anat. Rec.* **161,** 1.

Swierstra, E. E., Gebauer, M. R., and Pickett, B. W. (1974). *J. Reprod. Fertil.* **40,** 113.

Thompson, D. L., Jr., and Johnson, L. (1987). *Domestic Anim. Endocrinol.* **4,** 17.

Thompson, D. L., Jr., Johnson, L., St. George, R. L., and Garza, F., Jr. (1986). *J. Anim. Sci.* **63,** 854.

Tres, L. L., Smith, E. P., Van Wyk, J. J., and Kierszenbaum, A. L. (1986). *Exp. Cell Res.* **162,** 33.

Van Straaten, H. W. M., and Wensing, C. J. G. (1978). *Biol. Reprod.* **18,** 86.

Von Ebner, V. (1871). "Untersuchunger aus dem Institute fur Physiologie und Histologie in Graz," Vol. 2, p. 200.

Von Ebner, V. (1902). *In* "Kolliker's Handbuch der Gewebelehre des Menschen," Leipzig, Vol. III.

Von Kolliger, R. A. (1841). *In* "Beitrage zur Kenntnis der Geschlechtverhaltnisse und der Samenflussigkeit Wirbelloser Tiere," Berlin.

Von LaVallette, St. G. (1876). *Arch. Mikroscop. Anat.* **12:**797.

Von Leeuwenhoek. (1679). *Phil. Trans. R. Soc. Ser. B.* **12,** 1040.

Von Leydig, F. (1850). "Von Leydig, F. (1850). *Wiss. Zool.* **2,** 211.

Waites, G. M. H. (1977). *In* "The Testis" (A. D. Johnson and W. R. Gomes, eds.), Vol. IV, p. 91. Academic Press, New York.

Waites, G. M. H., and Ortavant, R. (1968). *Ann. Biol. Anim. Biochem. Biophys.* **8,** 323.

Wettemann, R. P., Wells, M. E., Omtvedt, I. T., Pope, C. E., and Turman, E. J. (1976). *J. Anim. Sci.* **42,** 664.

Willett, E. L., and Ohms, J. I. (1957). *J. Dairy Sci.* **40,** 1559.

Zirkin, B. R., Ewing, L. L., Kromann, N., and Cochran, R. C. (1980). *Endocrinology* **106,** 1867.

CHAPTER 6

Male Reproductive Organs and Semen

B. P. SETCHELL

I. Introduction

The reproductive organs in male domestic mammals comprise the testes, epididymides, accessory organs, and penis (Fig. 1). The testes produce the spermatozoa and the male sex hormones, the most important of which is probably testosterone. The spermatozoa pass from the testis into the epididymis where they acquire the capacity for motility and for fertilizing ova; mature spermatozoa are stored in the tail of the epididymis. During sexual excitement, the penis becomes erect, and at ejaculation, some

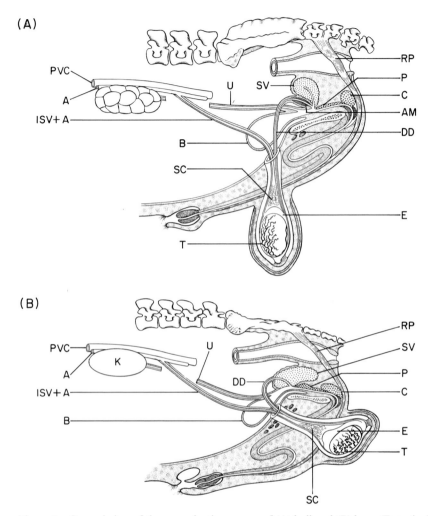

Figure 1 General view of the reproductive organs of (A) bull and (B) boar. T, testis; E, epididymis; SC, spermatic cord; DD, ductus deferens; ISV + A, internal spermatic vein and artery; A, aorta; PVC, posterior vena cava; AM, ampulla; SV, seminal vesicles; P, prostate; C, Cowper's gland; RP, retractor penis muscle; B, bladder; U, ureter, shown cut, not to obscure the internal spermatic vein and artery, which pass ventromedially; K, kidney. Drawings after Ackerknecht (1943) by J. R. Fuller.

of the fluid and spermatozoa in the epididymis passes along the ductus deferens to the urethra, and mixes with the secretions from the various accessory glands to constitute the ejaculated semen.

II. The Testes

All domestic mammals have paired scrotal testes. As a percentage of body weight, they are larger in rams (each testis weighs about

300 g, 0.4% of body weight) than in bulls (about 400 g, 0.05% of body weight), boars (about 500 g, 0.16% of bodyweight) or stallions (250 g, 0.06% of body weight). These values all fall reasonably close to the line relating log testes mass to log body mass, as calculated by Kenagy and Trombulak (1986), and deviations may be related to the pattern of sexual behavior of the particular species (Harcourt *et al.*, 1981).

Each testis is enclosed by a tough capsule, which consists mostly of fibrous tissue, but in some species also contains contractile cells, which may be involved in maintaining fluid pressure inside the testis or in moving fluid and sperm into the epididymis; the larger blood vessels, both arteries and veins, run in or just under the capsule. The parenchyma of the testis consists of the seminiferous tubules separated by interstitial tissue, which includes the smaller blood and lymphatic vessels and nerves, as well as the hormone-producing Leydig cells. In the human testis, the parenchyma is divided into clearly defined lobules, but this subdivision is much less clear-cut in the domestic mammals. In the center of the testes of rams, bulls, and boars, there is a fibrous mediastinum, containing the channels of the rete testis through which the spermatozoa and the fluid in which they are suspended leave the testis, and coils at the ends of the centripetal arteries that run in from the surface of the testis (see Setchell, 1970a; Setchell and Brooks, 1988; Hees *et al.*, 1989). In the stallion, the channels of the rete testis are arranged around a "central" vein, which lies near the dorsal edge of the testis (Amann *et al.*, 1977).

A. *The Scrotum and Descent of the Testes*

Although the testes of all domestic mammals are in a well-developed scrotum, they do not originate there. In the fetus, the gonads first become apparent when the promordial germ cells migrate to a ridge on one side of each fetal kidney, the mesonephros. At first, it is not possible to distinguish a testis from an ovary, but after a time, the germ cells in males become enclosed by the progenitors of Sertoli cells to form cords and their divisions are arrested, whereas in the female the germ cells are not enclosed by somatic cells and they undergo meiosis prenatally (McLaren, 1988). By the time the sex of the gonad is apparent, the mesonephros is degenerating and the testis appropriates its duct to convert it into the epididymis and ductus deferens, while the final kidney, the metanephros, moves cranially to lie anterior to the gonad (Gier and Marion, 1970). At the caudal end of the testis, a structure known as the gubernaculum develops, and grows out through the inguinal canal beyond the abdominal wall. The gubernaculum then enlarges, and in the process, pulls the testis across the abdominal cavity and into the inguinal canal. Then regression of the gubernaculum causes the testis to enter the scrotal cavity where the gubernaculum had been (Fig. 2). The outgrowth of the gubernaculum appears to be independent of androgens, as it continues in decapitated fetuses in which the Leydig cells are regressed, and cannot be stimulated by treatment with androgens; it does depend on something coming from the testis, as its development stops in castrated fetuses. On the other hand, regression of the gubernaculum is androgen dependent (Wensing and Colenbrander, 1988). In most domestic mammals, descent is complete by birth, but in dogs it is completed postnatally. The testis of the fetal horse reaches its maximal prepubertal size when the fetus is only about 50 cm long, and then regresses until

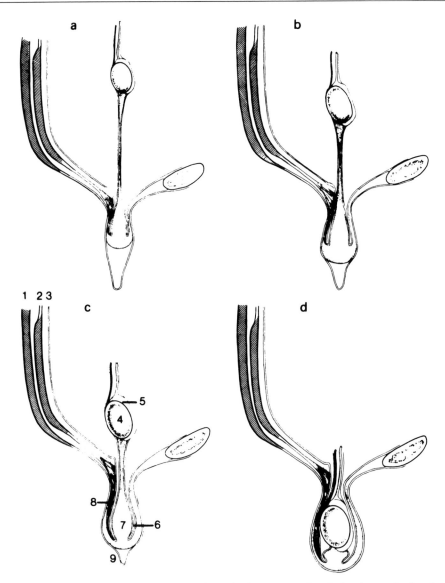

Figure 2 Schematic drawing of four phases in the process of testicular descent in the pig fetus: (a) 60 days p.c., (b) 70 days p.c., (c) 80 days p.c., (d) term, 114 days p.c. 1, External oblique muscle; 2, internal oblique muscle; 3, parietal peritoneum; 4, testis; 5, epididymis; 6, cavity of vaginal process; 7, gubenaculum; 8, cremaster muscle; 9, external spermatic fascia. Note how the extra-abdominal part of the gubenaculum enlarges much more than the intra-abdominal part, while its total length hardly changes. This growth of the extra-abdominal part carries it beyond the external inguinal ring into the region that becomes the scrotum, and exerts traction on the testis to pull it towards the deep inguinal ring. Then as the gubenaculum regresses, the testis is drawn into its final position in the scrotum. Reproduced by permission from Wensing and Colenbrander (1986).

puberty (Cole *et al.*, 1933), so that the testis may be more obviously scrotal in a newborn foal than in one a little older.

The scrotal sac in which the testis is located consists of an outpocketing of peritoneum, to which the testis and epididymis are attached by a mesorchium. The two sacs are enclosed in a smooth muscle coat, the tunica dartos, which is responsible for the variation in the position of the testes according to temperature, holding them close to the abdominal wall in cold conditions and allowing them to hang further away if it is hot. The skin covering the two peritoneal sacs is thinner than elsewhere in the body, and less well covered with hair or wool. In rams, it has an abundant population of sweat glands, in contrast to other skin, and these can discharge synchronously if the environmental temperature is raised, to keep the scrotal subcutaneous tissue at a temperature several degrees cooler than the air. This cooling is transferred to the testis itself by a countercurrent exchange of heat in the spermatic cord (see next section). The scrotal skin is also well endowed with temperature receptors that provide information to the brain, which can override input from other sources, so that if the scrotum alone is heated, a ram can reduce its body temperature by panting or vasodilatation to levels that would normally provoke shivering (see Waites, 1970; Setchell, 1978; Setchell and Brooks, 1988).

If the testis fails to descend normally, as often happens in pigs and horses, but less often in cattle and sheep, that testis is described as "cryptorchid" and is incapable of producing spermatozoa, so the if both testis are affected, the animal is sterile; however, the cryptorchid testis still produces some androgenic hormones, so that the animal (called a "rig" or "ridgeling" in the horse world) still exhibits male behavior (Cox, 1988). The abnormality in the testis appears to be due to its failure to reach the cooler environment of the scrotum, as normal spermatogenesis can be induced in a cryptorchid testis if it is cooled to a normal scrotal temperature (Frankenhuis and Wensing, 1979). Similarly, if the temperature of a scrotal testis is raised to body temperature or above, by (1) exposing the whole animal to a suitably hot environment, (2) immersing just the scrotum and its contents in warm water or air, or (3) surrounding just the scrotum with insulation, the animal can be made temporarily or permanently sterile. This is dependent on the temperature and the duration of exposure; delays of up to 14 days have been reported (see Waites, 1970; Setchell, 1978). Certain cells in the testis are particularly susceptible to heat, and this causes the delay, as other testicular cells and the immature spermatozoa in the epididymis appear to be more resistant.

B. Vascular Supply and the Spermatic Cord

As the testis descends from its point of origin into the scrotum, its artery does not change its origin, but elongates as the testis migrates. In addition, the artery in its course outside the inguinal canal becomes convoluted so that several meters of artery are coiled up into a structure about 10 cm long. The venous drainage runs mostly on the surface of the testes in bulls, rams, and boars; there is a substantial central vein in horses. When these veins reach the dorsal pole of the testis, they divide again to form a plexus of small veins known as the pampiniform plexus, which surrounds the coiled artery to form the spermatic cord. This cord can be quite a substantial stucture, weighing more than 15 g in rams even when empty of blood. In cross section, the small veins invest the coils of the artery very closely. Several large lymphatic vessels usually lie on the outside of the cord (see Setchell, 1978).

No one yet knows why this structure has evolved or what advantage it confers on the animal. However, several functions of the cord have been demonstrated (Fig. 3). It acts as a very efficient countercurrent heat exchanger, cooling the arterial blood from body temperature to that of the subcutaneous scrotal tissue before it reaches the testis, while the venous blood, which leaves the testis at the lower temperature, is warmed up so that it reaches the inguinal canal at body temperature. Similar systems are found in the limbs of animals and birds living in cold environments, but in these circumstances the heat exchangers probably act to limit heat loss through the extremities (see Waites, 1970; Setchell, 1978). In the head of sheep, there is also a system for cooling the arterial blood going to the brain by venous blood returning from the nasal passages (Baker and Hayward, 1968). It is not clear whether the spermatic cord has evolved to stop heat loss through the scrotum or to keep the testis cool in hot weather, but it certainly has the latter effect, and as already mentioned, if the temperature of the testes is raised to body temperature, the animal will become sterile.

The spermatic cord also acts to eliminate the pulse from the arterial blood, while reducing the mean pressure only slightly. No one has been able to reintroduce pulsatile flow to the testis, so we do not know if this is important. There is also evidence for the passage in the cord of diffusible substances such as water and inert gases from the artery to the veins and vice versa. If this occurs

with a substance being produced or taken up by the testis, it would lead to increased or decreased concentrations respectively, in the tissue, but no convincing evidence has yet been produced that this occurs to an appreciable extent. Testosterone, which would be an obvious candidate, is transferred from vein to artery but in only very small amounts, so that while the concentration in the artery reaching the testis is greater than in general arterial blood (Fig. 3), it is still well below that in testicular interstitial fluid or venous blood. The higher concentrations in arterial blood may be important for other androgen-dependent tissues such as the epididymis (see Free, 1977; Setchell, 1978).

Recently, interest has been reawakened in arterio–venous anastomoses in the spermatic cord. The first experiments to demonstrate their presence were undertaken by de Graaf in the seventeenth century (see Jocelyn and Setchell, 1972; Setchell, 1984a). Only recently convincing evidence of their existence and function has been obtained (Noordhuizen-Stassen et al., 1985, 1988). A substantial fraction of the blood entering the artery at the top of the cord may bypass the testis through the anastomoses, producing a lower concentration of testosterone in venous blood above the cord than in venous blood from the surface of the testis (see Fig. 3). However, the full significance and control of the anastomoses are still not known.

Even when the artery reaches the testis, it does not divide immediately into branches to supply the tissue, but runs down the caudal edge and round the distal pole of the testis before branching. The branches continue on the surface for some distance before turning and running into the interior near the mediastinum. There they form tight coils before forming ramifying

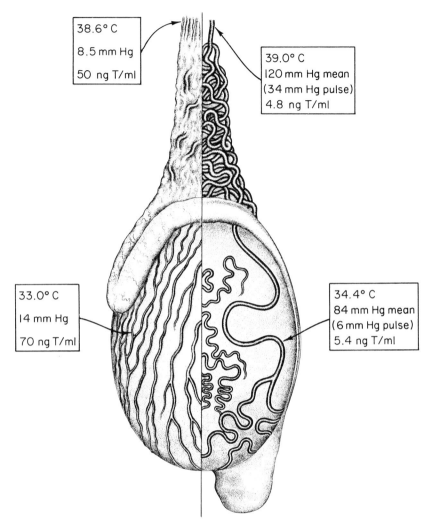

38.6° C

8.5 mm Hg

50 ng T/ml

39.0° C
120 mm Hg mean
(34 mm Hg pulse)
4.8 ng T/ml

33.0° C

14 mm Hg

70 ng T/ml

34.4° C
84 mm Hg mean
(6 mm Hg pulse)
5.4 ng T/ml

Figure 3 Diagram of the blood vessels of a ram testis, with arteries on the right side and veins on the left. The figures in the boxes give the temperature, the mean blood pressure (and pulse pressure), and the concentration of testosterone in blood at the point shown. Note the convolution of the artery in the spermatic cord and its sinuous course on the surface of the testis. (Original drawing by J. R. Fuller.)

branches, which are directed toward the surface of the testis (see Setchell, 1970a, 1978; Hees *et al.*, 1989). The function of this curious anatomical arrangement is unknown.

C. Interstitial Tissue

The interstitial tissue lies between the seminiferous tubules and consists of the Leydig cells, blood and lymph vessels, and nerves, with a variable population of other cells, macrophages, and mast cells being common in some species. In rams, the interstitial tissue comprises about 15% of the testis, and the Leydig cells comprise about 7.5% of the interstitial tissue, that is, about 1% of the testis; the walls and lumina of the blood and lymph vessels comprise about 37% of the interstitial tissue, that is, about 6% of the testis. A ram testis weighing about 250 g, contains about 8×10^9 Leydig cells, each about 350 μm^3 in volume (Monet-Kuntz et al., 1988). The values for bulls are similar, and in both these species the Leydig cells are found in clumps, often associated with a small blood vessel (Fig. 4). In horses and especially in pigs, the interstitial tissue makes up a much greater proportion of the testis (28% and 37%, respectively), and the larger Leydig cells comprise the majority of this part of the organ (see Christensen, 1975; Johnson and Neaves, 1981). The Leydig cells in rams and bulls are thus appreciably smaller, and those in stallions and boars are larger than those in rats (Mori and Christensen, 1980) and hamsters (Sinha Hikim et al., 1988).

Leydig cells in mammals are in general relatively large, polyhedral, epithelioid cells, surrounded by a typical plasma membrane, which is usually covered with microvilli; each cell has a single, spherical or ovoid, eccentrically located nucleus, with one to three large nucleoli. Granules of chromatin are distributed around the nuclear membrane, making it appear unusually thick. The cytoplasm (particularly in boars among the domestic animals) is usually abundant and contains much smooth endoplasmic reticulum. The cytoplasm contains moderate numbers of average-sized mitochondria, with mostly lamellar cristae, and a well-developed Golgi complex. The Leydig cells of some species

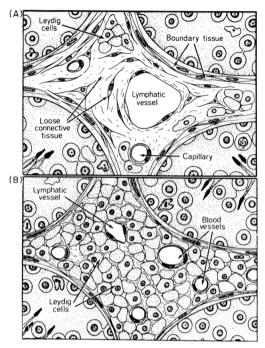

Figure 4 A diagram of the interstitial tissue of (A) ram and (B) boar testis. In the ram the interstitial tissue contains a central lymph vessel, blood capillaries, and clumps of Leydig cells in a loose connective tissue. Some of the Leydig cells are grouped around capillaries, others are not. In the boar, most of the interstitial space is filled by the numerous Leydig cells with small lymph vessels and blood capillaries in among them. Reproduced by permission of the publishers from Fawcett (1973).

contain large numbers of lipid droplets, however, not in rams, bulls, boars, or stallions (see Fawcett et al., 1973; Christensen, 1975). The appearance of the Leydig cells shows considerable variation in such seasonal breeders as the ram, and—particularly—the red stag (Hochereau-de Reviers and Lincoln, 1978).

The identification of the Leydig cells as the source of secreted steroids is based on the much higher production (when compared with seminiferous tubules) of the hormone *in vitro* by isolated Leydig cells or separated interstitial tissue. These studies have been performed primarily in rodents in which the tubules are easy to separate from the interstitial tissue (see Maddocks and Setchell, 1988a). Leydig cells from young pigs have also been cultured and secreted steroids in considerable amounts (Benahmed *et al.*, 1982).

The Leydig cells use cholesterol as the starting material for the synthesis of the steroid hormones. This is either synthesized from acetate or taken up from the blood, probably in the form of lipoproteins (see Setchell, 1978; Hall, 1988). The mechanism of secretion of the steroids into the bloodstream is still largely unknown (see Maddocks and Setchell, 1988a).

Analyses of interstitial extracellular fluid collected from the testes (Maddocks and Setchell, 1988a) and from lymph collected from a vessel above the spermatic cord (see Setchell, 1970a, 1982b, 1986) suggest that these fluids are very similar in composition to testicular venous blood plasma. Notable exceptions are the conjugated steroids, which are present in spermatic-cord lymph from the testes of pigs and horses in concentrations many times higher than in testicular venous blood (Setchell and Cox, 1982; Setchell *et al.*, 1983). Almost 80% of the dehydroepiandrosterone sulfate produced by pig testis leaves the testis via the lymph; pig testes produce about three times as much of this steroid as testosterone. The endocrinological significance of these conjugated steroids is still not certain. In vasectomized rams and pigs (Ball and Setchell, 1983), lymph from a vessel in the spermatic cord contains appreciable numbers of spermatozoa, these would no doubt contribute to the immunological reaction seen in many male animals against their own spermatozoa fol-lowing vasectomy (Tung, 1980). The testicular interstitial fluid also contains appreciable concentrations of many peptides active elsewhere in the body, including substances like oxytocin, vasopressin, neurophysin, prostaglandins, renin, β-endorphin, and other pro-obiomelanocortin (POMC) derived peptides (see Maddocks and Setchell, 1988a).

It has been known for a number of years that tissue grafts survive better in the interstitial tissue of the testis of rodents than in many other tissues; this is not so in rams (Maddocks and Setchell, 1988b), and may be related to the presence in rat intertitial fluid of interleukin-1 (Gustafsson *et al.*, 1988) and an immunosuppressive peptide (Pollanen *et al.*, 1988). The latter is not found in ram testes (Pollanen *et al.*, 1989). Recently, interest has increased in possible paracrine effects of seminiferous tubules on adjacent Leydig cells in rats (Bergh, 1985); several possible paracrine peptides have been suggested (Sharpe, 1984, 1986; Sharpe and Cooper, 1987). Whether similar peptides operate in the testes of domestic mammals is not yet known. In ram testes made aspermatogenic by a brief exposure to heat, the Leydig cells were larger and fewer in number (Setchell *et al*, 1989), suggesting that similar paracrine mechanisms might be operating.

The Leydig cells respond to stimulation with luteinizing hormone (LH) by releasing increased amounts of testosterone. This increased is not associated with a change in testicular blood flow (Laurie and Setchell, 1978), and the testis appears to respond maximally to a normal-sized peak of LH. If a supraphysiological peak of LH is produced by an increased dose of luteinizing hormone releasing hormone (LHRH), there is no increase in the size of the testosterone peak (D'Occhio and Setchell, 1984). A similar but more prolonged effect on testosterone is seen in rams treated with human chorionic gonadotropin (hCG), which is chemically very similar to LH, but is cleared more slowly from the circulation. One injection of hCG produces a rise in testosterone that peaks at about 3 h, then falls off again to rise again to a second peak between 48 and 72 h after injection (Sundby and Torjesen, 1978; Setchell, 1984b). In rats, treatment with hCG results in a rise in vascular permeability to albumin and lymph flow (Setchell and Sharpe, 1981) and an accumulation of fluid in the interstitial tissue (see Sharpe, 1983). These responses are probably mediated by leukocytes (Bergh *et al.*, 1986), but there is no change in flow rate of testicular lymph after hCG injection in either rams or boars (unpublished results).

D. Seminiferous Tubules

The majority of the parenchyma of the testis (~85% in rams and bulls, ~75% in stallions, and ~65% in boars) consists of seminiferous tubules in which the spermatozoa are formed. These tubules are not penetrated by any blood or lymph vessels or nerves; they are basically convoluted, two-ended, hollow fluid-filled cylinders, the two ends of which open into the rete testis through straight tubules (or tubuli recti). Their diameter is between 200 and 250 μm, so that an average ram testis contains nearly 3000 m of tubules. The tubules are surrounded by a specialized peritubular tissue, which consists of several layers of noncellular material in domestic mammals, and of myoid cells, which are contractile and probably cause peristaltic contractions of the tubules to assist in the transport of spermatozoa (see Setchell and Brooks, 1988).

Inside the tubules are various germ cells—diploid spermatogonia dividing mitotically, tetraploid primary spermatocytes undergoing the long prophase of meiosis, short-lived secondary spermatocytes between the first and second meiotic divisions, and haploid spermatids and developing spermatozoa awaiting release into the lumen. Also inside the tubules are the somatic Sertoli cells, which envelop the germ cells (Fig. 5) and form the main component of the blood–testis barrier (see next section). They also secrete the fluid that fills the lumina of the tubules, and a number of specific proteins, which are found in tubular fluid exclusively or in much higher concentrations than elsewhere in the body.

The ionic composition of tubular fluid has been determined in samples removed by micropuncture from the seminiferous tubules of rats and hamsters. It can also be calculated by comparing the analyses of fluid separated from one testis whose efferent ducts had been ligated 24 h earlier and the contralateral unligated testis of the same animal. Concentrations of potassium and inositol are much higher than in serum and sodium and chloride are lower. There is much less protein, and virtually no glucose, probably because the glucose entering the tubules is converted by the Sertoli cells into lactate or pyruvate for utilization by germ cells (see Setchell *et al.*, 1978; Setchell and Brooks, 1988). Total fluid secretion can be measured by the weight gain of the testis up to 24 h after ligation of the efferent ducts (Setchell, 1970b; Setchell and Brooks, 1988), although this procedure is more easily carried out in rats than in any of the domestic mammals. The tubules empty into the rete testis, and it has been possible to collect rete-testis fluid and testicular spermatozoa in rams, bulls, and boars over the course of days or

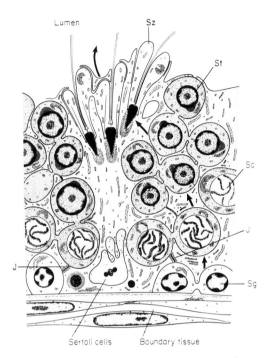

Figure 5 Diagram of the arrangement of the germinal cells and the Sertoli cells, in the seminiferous tubule. The diploid spermatogonia (Sg) are confined between the Sertoli cell cytoplasm and the boundary tissue of the tubule. The spermatocytes (Sc) and early spermatids (St) are sandwiched between pairs of Sertoli cells, on the luminal side of the specialized junctions (J) between adjacent Sertoli cells. The late spermatids (Sz) are embedded in the luminal surface of the Sertoli cell cytoplasm, until they are finally extruded into the lumen and leave the testis. Reproduced by permission of the publishers from Fawcett (1974).

even weeks from an implanted catheter. Rete-testis fluid (RTF) is a dilute suspension of spermatozoa (about 10^8/ml) in a fluid that is quite different from blood plasma or testicular lymph; it has some similarities to and some differences from tubular fluid in the rat, in which it is possible to collect both fluids. The main features of rete-testis fluid are its high concentration of potassium, glutamate, and inositol, and low concentrations of glucose and protein, especially immunoglobulin (see Setchell et al., 1969, 1978; Setchell, 1970a, 1978; Setchell and Brooks, 1988). A number of blood proteins cannot be detected in RTF, but it contains others that either are not present or are present at much lower concentrations in blood plasma. In many species, but not in the boar, these proteins include an androgen-binding protein (ABP), which is secreted by the Sertoli cells and is similar but not identical to the sex hormone-binding globulin secreted by the liver in man and sheep, but not in the boar or adult rat (see Jegou et al., 1979; Setchell and Brooks, 1988). There is also a specific transferrin in RTF secreted by the Sertoli cells, and several enzymes are present in high concentrations. It also contains a nonglycosylated peptide growth factor that stimulates the division in culture of mouse fibroblasts. A glycosylated protein, called clusterin, which causes the agglutination of several types of cells in culture, makes up about 20% of the total protein. Ram RTF is a potent source of inhibin, which reduces follicle-stimulating hormone (FSH) secretion by the pituitary, but may also have paracrine effects in the testis. Mullerian-inhibiting factor can be found in RTF from adult boars, although its primary action appears to be to cause regression of the female ducts in fetal life (see Setchell and Brooks, 1988).

Steroids in general are present in RTF in lower concentrations than in venous blood; conjugated steroids are present in much lower concentrations. The high concentration of ABP in RTF from some species ensures that most of the testosterone and dihydrotestosterone (DHT) is bound and these steroids may be transported in this form to the epididymis (see Setchell and Brooks, 1988).

E. Blood–Testis Barrier

The large differences in composition between tubular and rete-testis fluid versus blood plasma and interstitial fluid could not be maintained unless some movement of substances across the walls of the seminiferous tubules is restricted. With the aid of radioactive and other tracers, it was found that large water-soluble molecules like proteins are virtually excluded and lipid-soluble molecules and water enter readily; other substances like ions enter slowly, and some compounds like 3-O-methylglucose are brought into the tubules by saturable, specific carriers (see Setchell *et al.*, 1969; Setchell and Waites, 1975; Setchell, 1980b, 1986; Setchell and Brooks, 1988). Although proteins such as albumin appear to cross the walls of the blood vessels in the testes without much restriction, some dyes stain the interstitial tissue less readily in adult animals (Kormano, 1967), suggesting some "restriction" at the endothelium. These endothelial cells contain high levels of the enzyme, γ-glutamyl transaminase, that is often associated with transport of amino acids (Niemi and Setchell, 1986). Some evidence suggests that a saturable transport system for leucine and phenylalanine exists in these cells (Bustamante *et al.*, 1982).

Another way of demonstrating the existence of the barrier is to compare the volume of the interstitial tissue, as determined by morphometry with the space of distribution (calculated by dividing the counts/μg testis by counts/μl plasma) after the intravenous administration of a suitable radiolabeled marker such as Cr-EDTA, which is excluded from the tubular fluid. In adult rats, the Cr-EDTA space was found to be about 130 μl/g, whereas the interstitial tissue comprises only about 10% of the testis (100 μl/g) (Setchell and Sharpe, 1981; Setchell *et al.*, 1988), but this was probably an overestimate, and the correct value for the space in the parenchyma, excluding the capsule, is about 60 μl/g (Setchell, 1990). In animals in which the barrier is broken down or in immature animals, the Cr-EDTA/interstitial tissue volume ratio can be double adult values (Setchell, 1986; Setchell *et al.*, 1988).

The anatomical site for the barrier to large water-soluble molecules was located in the tubular wall in studies using dyes (Kormano, 1967), and the ultrastructural basis was subsequently identified at the electron-microscopic level in the specialized junctions between pairs of adjacent Sertoli cells (Fig. 5). In rodents, the myoid cell layer also imposes some restriction of entry. The Sertoli cell junctions divide the seminiferous epithelium into a basal and an adluminal compartment (Dym and Fawcett, 1970; see also Bellve, 1979), with the spermatogonia and the preleptotene spermatocytes in the basal compartment and the pachytene and later spermatocytes and the spermatids in the adluminal compartment. The leptotene spermatocytes pass through the barrier when new junctions form outside them; then the old junctions on the luminal side open up (Dym and Cavicchia, 1977; Russell, 1977). It is still not clear whether the multiple layers of noncellular material around the tubules in the ram, bull, and boar have any effect. The Sertoli cell junctions would only be significant for substances that enter the tubules by a paracellular route, such as large water-soluble molecules. Lipophilic substance would cross the cytoplasm of the Sertoli cells, and the process involved in their movement are likely to be quite different. For example, testosterone enters tubular and rete-testis fluid much more quickly than dihydrotestosterone (DHT), although the latter is in fact slightly more lipid soluble. The entry of testosterone *in vivo* can be reduced if the concentration of nonradioactive testosterone is raised (see Setchell, 1980b), suggesting that some sort of saturable carrier system might be involved. A similar effect is not seen with isolated tubles *in vitro*, so it has proved very difficult to gather information on the nature and location of this mechanism.

The specialized junctions form in rats between the ages of 16 and 18 days (Vitale-Calpe *et al.*, 1973), but the functional barrier to Cr-EDTA develops much more gradually and only reaches fully adult proportions at about 45 days of age (Setchell *et al.*, 1988). The barrier breaks down when the testis loses its turgidity about 36 h after efferent duct ligation. At the same time the EDTA space rises well above the measured interstitial tissue and the concentrations of ABP in blood rise. Before the barrier breaks down, the secreted fluid is retained within the tubules and rete, leading to distention and turgidity of the testis (see Setchell, 1986; Gunsalus *et al.*, 1980). Treatment with cadmium salts causes an increase in permeability of the tubles to rubidium, as well as a very large increase in vascular permeability to albumin (Setchell and Waites, 1970). Many other treatments of the testis such as heating or irradiation are without effect on the barrier although spermatogenesis can be seriously disrupted (see Setchell and Brooks, 1988). The barrier is sufficiently resistant to withstand exposure of the tubules to hypertonic solutions, which produce shrinkage of the cells in the basal compartment but not those inside the junctions (Gilula *et al.*, 1976). Using this technique, it can be shown that a barrier is absent from the testes of some mice with certain genetic abnormalities (such as testicular feminization and sex reversal) but not from others (Fritz *et al.*, 1983).

The significance of the barrier is difficult to define, but presumably it is involved in creating conditions appropriate for meiosis, as it is found in many, but not all, classes of invertebrate animals, as well as in all vertebrates so far studied (see Setchell and Pilsworth, 1989). It is also important in preventing the entry of antibodies into the tubules and the segregation of most of the immunologically foreign germ cells in the tubules (Tung, 1980). More recent evidence (Yule *et al.*, 1988) suggests that some autoantigenic germ cells lie outside the barrier. This latter observation emphasizes the importance of specific immunosuppressive mechanisms in the testes (see Pollanen *et al.*, 1989). Endocrinologically, the barrier blocks the entry of peptide hormones into the tubules. Therefore, they must act on the basal surfaces of the Sertoli cells, where the receptors for FSH are indeed found (Orth and Christensen, 1977). Similarly, any proteins or peptides secreted by the Sertoli cells pass into the luminal fluid and the intersitial extracellular fluid, presumably in proportion to the surface area of the cell directed toward the adluminal and basal compartments, respectively. They only estimates of this compartmentalization have been made in the rat, where the adluminal surface is about 20 times greater than the basal surface (Weber *et al.*, 1983). However, the secretion of ABP, inhibin, and transferrin appears to be in the ratio of ~3 : 1 toward the luminal fluid, suggesting that secretion may be polarized to some extent as well (see Setchell and Brooks, 1988). Polarized secretion *in vitro* has been demonstrated by Sertoli cells in a two-chamber culture system (Hadley *et al.*, 1987).

F. Testicular Spermatozoa

Throughout spermatogenesis, the germ cells develop as clones of cells linked by cytoplasmic bridges and in close contact with the Sertoli cells (see Chapter 5). At the time of spermiation, the link joining the spermatozoon to the residual cytoplasm is broken; the latter is retained by the Sertoli cell, and each spermatozoon begins an independent separate life. A remnant of the cytoplasm persists as the kinoplasmic or cytoplasmic droplet; this moves down the midpiece to the junction with the tail proper. In the bull and ram, the droplet is not usually shed before ejaculation, as it is in rodents (see Amann, 1987). The testicular spermatozoa are immotile, and while they do acquire some capacity for movement if stored *in vitro* under appropriate conditions, the motility reached is never comparable with that of ejaculated or caudal epididymal sperm. Testicular sperm are also incapable of fertilizing ova, either *in vitro* or following concentration and insemination (see Setchell *et al.*, 1969; Voglmayr, 1975; Voglmayr *et al.*, 1978).

The composition of testicular sperm, particularly with respect to lipid, is quite different from caudal epididymal sperm. Testicular sperm have an appreciably higher content of phospholipid than epididymal sperm. The fatty acids in the phospholipids of testicular sperm are rich in palmitic acid, whereas caudal epididymal sperm contain more unsaturated fatty acids. When incubated in conventional buffered saline media resembling serum, testicular sperm convert a high proportion of glucose to carbon dioxide, amino and carboxylic acids, and inositol, as well as some unidentified perchloric acid-extractable compounds. Caudal epididymal and ejaculated sperm convert most glucose to lactate, even under aerobic conditions (see Setchell *et al.*, 1969; Voglmayr, 1975; Dacheux *et al.*, 1979). If the medium is made to resemble rete testis fluid and is buffered with bicarbonate, the addition of inositol, glutamate, ascorbate (substances present in high concentrations in rete testis fluid) or of glucose has very little effect on oxygen uptake. However, the same spermatozoa incubated in native rete testis fluid have a higher oxygen uptake than those incubated in saline (Evans and Setchell, 1978). It is also worth noting that testicular sperm metabolize comparatively little glucose via the pentose phosphate cycle (similar to ejaculated sperm) but in contrast to testis tissue, which has an active pentose cycle metabolism (see Setchell *et al*, 1969).

III. The Epididymis

A. Structure

When the spermatozoa leave the testis, they pass from the rete testis through the efferent ducts (vasa efferentia) into the epididymis, a compact fibrous organ closely attached to the posterior or superior border of the testis. Four to 14 efferent ducts join the single epididymal duct, which is about 80 m long in rams. The epididymis is usually divided anatomically into three parts, caput or head, corpus or body, and cauda or tail, according to the gross appearance. The corpus is the narrow part between the two enlarged ends, the caput and cauda. However, another system of subdivision has been proposed by Glover and Nicander (1971), into initial, middle, and terminal segments, according to the histology and function of the epithelium. In this system fluid concentration and first stage of sperm maturation occur in the initial segment; maturation is completed in the middle segment, and the sperm are stored in the terminal segment. The boundaries between the divisions in the two systems do not necessarily correspond, especially in different species, and both are useful. The initial segment has been further subdivided histologically into an initial segment proper and a distal initial segment. In the initial segment proper, the cells are tall and have prominent stereocilia that virtually fill the lumen. It contains very few sperm (Jones and Clulow, 1987). The middle and terminal segments are lined with two types of cells, principal and basal. The former are the more common and are tall cells extending the entire depth of the epithelium, whereas the basal cells are located near the basement membrane. The height of the epithelium decreases progressively from the initial segment onward and is quite low in the cauda. Conversely, the muscle layer around the duct, which is involved in ejaculation, becomes more obvious as the cauda is approached (see Hamilton, 1975; Amann, 1987; Setchell and Brooks, 1988).

The blood supply to the head and body of the epididymis leaves the testicular artery in the spermatic cord, whereas the tail is supplied by the deferential artery. Important differences exist in the microvascular anatomy of the different regions, and the capillaries of the initial segment are fenestrated, unlike those in the testis and elsewhere in the epididymis. The epididymis also has an abundant adrenergic and cholinergic innervation, particularly in those areas with well-developed smooth muscle. Likewise, the ductus deferens has an extensive nerve supply, so that this tissue has become a widely used system for some neuropharmacological studies (see Setchell and Brooks, 1988).

B. Luminal Contents

The rete testis fluid that enters the epididymis contains about 1×10^8 spermatozoa/ml; this is equivalent to a spermatocrit (the volume of the fluid occupied by cells after centrifugation at 3000 g for 15 min in a hematocrit centrifuge) of about 1%. In the fluid in the lumen of the cauda of the epididymis, the spermatocrit is about 40%, because almost all the fluid leaving the testis is reabsorbed in the initial segment (Fig. 6). The absorption of fluid is probably linked to the transport of sodium chloride, and consequently, the potassium concentration in the fluid (which is already higher than blood plasma, see Section II,D) rises (Fig. 6), as does the potassium to sodium ratio. There is virtually no glucose in the epididymal fluid, but the inositol concentration, which is already higher in RTF than blood, also rises in the first part of the epididymis; in rodents, more inositol is secreted by the cells lining the epididymal duct, so that in the cauda, concentrations of up to 80 mM have been recorded. The situation is quite different in the ram, bull, and boar, in which the highest inositol levels are reached in the caput, and the concentrations in the ductus deferens are about the same as in RTF (see Hinton *et al.,* 1980; Amann, 1987; Setchell and Brooks, 1988). Another substance that is present in high concentrations in epididymal luminal fluid is carnitine, which is taken up by the cells from blood and then secreted into the lumen (see Setchell and Brooks, 1988). As with inositol, the concentrations of carnitine found in the

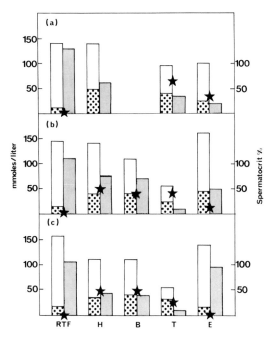

Figure 6 Graph of the sodium (clear columns), potassium (large dots), and chloride (small dots) concentrations in the seminal plasma and the spermatocrit (stars) in ejaculated semen (E) and in semen from the rete testis (RTF), head (H), body (B), and tail (T) of the epididymis of (a) rams, (b) bulls, and (c) boars. The epididymal samples were from slaughterhouse material, the rest from conscious animals. Based on data from Crabo (1966), Mann (1964), Mann and Lutwak-Mann (1981), White (1973), Setchell (1970a), and Pholpramool *et al.* (1985).

ram and boar are not as high as in the rat (Hinton *et al.*, 1979), and in the bull, it appears to be much lower (see Brooks, 1990). Glycerylphosphorylcholine (GPC) is found in higher concentrations in the caudal epididymal fluid of the domestic mammals than in rodents; this compound is synthesized in the epididymis, probably from lecithin associated with blood lipoproteins (see Setchell and Brooks, 1988). In rodents, it has been shown that these three compounds represent a large fraction of the osmotic

pressure and are secreted in different regions of the epididymis (see Setchell and Hinton, 1981). The epididymal fluid contains high concentrations of a number of amino acids, especially glutamic acid, glutamine, asparagine, taurine, and hypotaurine; the luminal fluid is significantly acidified as it passes through the caput (see Setchell and Brooks, 1990). A number of androgen-dependent specific proteins are secreted into the fluid (see Robaire and Hermo, 1988; Brooks, 1990, and these have been implicated in the functional maturation of the spermatozoa (see Section III,C).

The epididymis has significant 5α-reductase activity. Consequently, most of the testosterone entering in the rete testis fluid or in the blood is converted to dihydrotestosterone (see Robaire and Hermo, 1988; Setchell and Brooks, 1988).

C. *Spermatozoal Maturation*

The motility of epididymal spermatozoa has been assessed in both undiluted epididymal fluid and in that diluted in saline media. There is clear evidence that the potential for motility after dilution develops quite sharply (Fig. 7) in bulls, rams, and boars as the sperm pass through the corpus (Acott and Carr, 1984; Dacheux and Paquignon, 1980; Fournier-Delpech *et al.*, 1979). The situation is less clear for sperm in native epididymal fluid, and the results obtained

Figure 7 (Top) Development of the ability of spermatozoa from various regions (A, head; B, body; C, tail) of the epididymis to fertilize ova in several species. The figures give the fertility rate and the dark area indicates where fertile spermatozoa are present. Reproduced with permission from Dacheux and Paquignon (1980). (Bottom) Summary of the changes in fertility, motility, metabolism, and lipid synthesis in boar spermatozoa as they move through the epididymis. Reproduced with permission from Dacheux and Paquignon (1980).

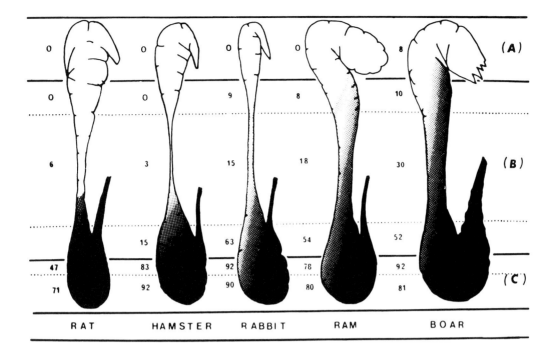

	RAT	HAMSTER	RABBIT	RAM	BOAR	
	0	0	0	0	8	(A)
	0	0	9	8	10	
	6	3	15	18	30	(B)
		15	63	54	52	
	47	83	92	78	92	(C)
	71	92	90	80	81	

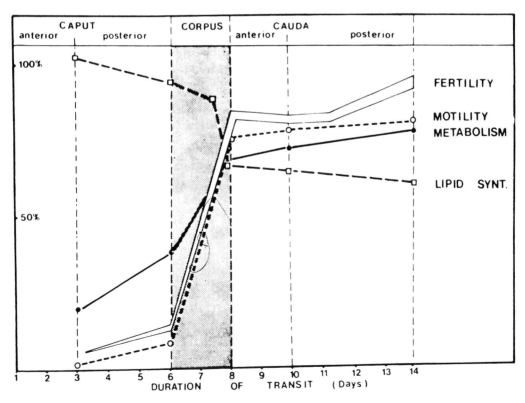

depend on the species and the technique used. When bull sperm were studied on glass slides under a cover-slip, they remained immotile (Carr and Acott, 1984); when they were examined in capillary tubes or as droplets suspended in paraffin oil, they showed appreciable motility (Cascieri *et al.*, 1976; Morton *et al.*, 1978; Pholpramoool *et al.*, 1985). It is not clear whether the glass surfaces are influencing motility (see Stephens *et al.*, 1981) or whether contact with the oil changes some constituent of the fluid (e.g, pH). Boar sperm, like rodent sperm, do not show any motility in their own fluid (Dacheux and Paquignon, 1980). The development of motility is associated with a rise in the concentration of cAMP in the sperm, and motility can be induced in immotile sperm by phosphodiesterase inhibitors and a "forward motility protein" that is found in caudal epididymal fluid and semen (Hoskins *et al.*, 1979, but cf. Stephens *et al.*, 1981) but not in semen from vasectomized animals (Brandt *et al.*, 1978). However, caudal epididymal fluid from bulls also contains a "quiescence factor" (Carr and Acott, 1984), which interacts with pH to keep the spermatozoa from moving in the epididymis. This factor appears to be quite different from rat "immobilin", which appears to act by virtue of its high viscoelasticity (Usselman and Cone, 1983).

At the same time (or slightly after) the sperm become motile, they also become capable of fertilizing ova (Holtz and Smidt, 1976; Fournier-Delpech *et al.*, 1979; Dacheux and Paquignon, 1980). However, although sperm from the distal portion of the corpus in the ram are capable of fertilizing eggs (Fig. 7), the embryos that result have a significantly increased chance of dying *in utero* (Fournier-Delpech *et al.*, 1979).

During passage through the epididymis, sperm also undergo changes in fine structure and composition involving both lipids and proteins, metabolism, and surface properties (see Setchell *et al.*, 1969; Bedford, 1975; Evans and Setchell, 1979; Dacheux and Paquignon, 1980; Amann, 1987; Hammerstedt and Parks, 1987; Robaire and Hermo, 1988). However, it is difficult to associate any of these specific changes with the two functional changes in motility and fertilization ability. The most promising seem to be the absorption by the spermatozoa of some of the proteins secreted by the epididymis.

D. Passage of Spermatozoa through the Epididymis and Fate of Unejaculated Spermatozoa

The time required for spermatozoa to pass through the epididymis can be determined by injecting radioactive thymidine into the testes and following the movement of the labeled sperm through the duct. A time of about 13 days is required in the ram, about 11 days in the bull, and 9–14 days in the boar. There is a slight decrease, usually no more than 20%, in the time required to traverse the epididymis in animals that are ejaculating frequently. The movement of the spermatozoa is not affected by ligation of the efferent ducts (see Bedford, 1975), which blocks fluid leaving the testis, but spontaneous and catecholamine-induced contractions of the smooth muscle surrounding the epididymal duct are probably important, particularly in the lower part of the epididymis (Knight, 1974). Subsequent studies in rats have extended our understanding of the control of these contractions

(Laitinen and Talo, 1981; Pholpramool and Triphrom, 1984).

The fate of spermatozoa that are not ejaculated has been the subject of some controversy. In some species, including the ram, spermatozoa are regularly voided in the urine by sexually inactive males (Lino and Braden, 1972). In other species, spermiophagy, that is, the destruction and uptake

of sperm by the lining of the epididymis, occurs. It may be stimulated by chemical or mechanical manipulation of the epididymis or ductus deferens, and luminal macrophages may be involved (see Robaire and Hermo, 1988). Following vasectomy, appreciable numbers of intact spermatozoa can be found in spermatic cord lymph in rams and boars (Ball and Setchell, 1983), presumably following leakage from the spermatocele that usually forms at the site of ligation (see Bedford, 1975).

E. Control of Epididymal Function

Several secretory functions of the epididymis and the structure of the epithelial cells are dependent on androgens, primarily supplied by the bloodstream (see Brooks, 1981, 1983; Robaire and Hermo, 1988). However, the initial segment may be affected by substances in rete testis fluid, which can stimulate secretion and influence the ultrastructure of cells (see Robaire and Hermo, 1988; Jones *et al.*, 1987, 1989). Androgens also appear to have the unexpected effect of slowing transit of spermatozoa through the epididymis (Sujarit and Pholpramool, 1985).

IV. Accessory Glands

A. Anatomy

The primary accessory sex glands in the male are the ampullae, seminal vesicles, prostate, and Cowper's glands (also known as the bulbo-urethral glands, Fig. 8), all of which produce secretions that contribute to ejaculated semen. There are also urethral

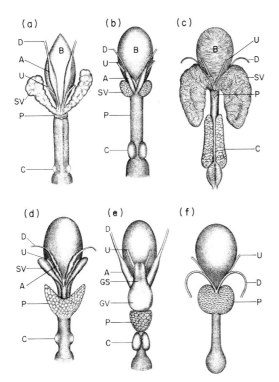

Figure 8 Diagrammatic view of the dorsal aspect of the accessory glands of (a) bull, (b) ram, (c) boar, (d) stallion, (e) rabbit, and (f) dog. SV, seminal vesicle; P, prostate, in the ram inside the urethral muscle; C, Cowper's glands; A, ampulla of the ductus deferens (D); U, ureter; B, bladder; GS, glandula seminalis; GV, glandula vesicularis. (a), (c), and (d) after diagrams in Eckstein and Zuckerman (1956) (f) after diagrams in Ackerknecht, (1943) and (b) and (e) from fresh specimens by J. R. Fuller.

glands (Littre's glands) and preputial glands in some species, but little is known of their function or the composition of their secretions.

The ampullae are enlargements at the urethral end of each ductus deferens; they are particularly well developed in stallions but absent in boars. They are lined by an epithelium, which forms numerous, irregular-branching folds, between which are branched outpocketings penetrating deep into the muscular covering. These glands are lined by a single layer of columnar cells.

The seminal vesicles are situated on either side of the neck of the bladder, and were so named because they were thought to act as a store for "semen" from the testis. This is no longer thought to be true, although in humans the duct of the seminal vesicle joins the ductus deferens to form an ejaculatory duct, and consequently sperm are sometimes found in the vesicles post mortem. An ejaculatory duct is not present in any of the domestic mammals, and the ductus deferens and the duct of the seminal vesicle open separately into the urethra. Dogs have no seminal vesicles. In bulls and especially in boars, they are large and contribute a large fraction of the volume of the ejaculate; in rams, they are relatively small. In stallions, they are bag-shaped glands, with a folded lining; in rams, bulls, and boars, they are compact lobed glands, with a system of ramifying secretory ducts, lined by a pseudo-stratified epithelium.

The prostate lies caudal to the seminal vesicles, close to the junction of the bladder and urethra. In rams and bulls, the prostate is inconspicuous, as it is disseminate and does not protrude outside the muscular covering of the urethra. In horses and pigs, it has a disseminate portion, but also a discrete lobe outside the urethral wall. In dogs as in humans, the prostate is the principal accessory gland, a relatively large prominent organ, completely surrounding the urethra.

The Cowper's glands lie near the pelvic part of the urethra, into which they empty through a short duct. In bulls, rams, and stallions, these glands are comparatively small, and are absent in dogs; in boars, they are large cylindrical organs, producing a viscous secretion that is important in the process of "gelation" of the semen following ejaculation. The glands are lined with mucous-like cells (see Eckstein and Zuckerman, 1956; Mann, 1964; Mann and Lutwak-Mann, 1981; Setchell and Brooks, 1988).

B. Secretions of the Accessory Glands

In most species, the secretions of the various accessory glands make up the majority of the volume of the ejaculated semen. These secretions contain a bewildering array of chemical constituents, many with still unknown functions. Furthermore, a substance such as fructose, which is found in the semen of many, but not all, mammals originates from different glands in different species. However, in any given species, a specific constituent can often indicate the functional activity of a specific gland. A summary of the available information on the origins of the consituents of seminal plasma, based primarily on the work of Mann and colleagues, is given in Table I. Fructose is formed in the tissue from blood glucose via sorbitol, and all other constituents appear to be formed in the glands concerned, although the chemistry of their formation is not well understood (see Mann and Lutwak-Mann, 1981).

Table I
The Origins of Characteristic Substances Found in Semen[a,b]

	Bull	Ram	Goat	Boar	Stallion	Dog	Rabbit
Seminal vesicle	FCi(e−)	FCPr	F	fCIE	C	Gland absent	FC
Prostate				c		P(c−,f−)	Fc
Cowper's gland				S(c−,i−)		Gland absent	c
Ampulla	fc		f	Gland absent	Ei(F−,c−)		FC

[a]Based on data from Mann (32, 33).

[b]F, fructose; C, citric acid; I, inositol; E, ergothioneine; Pr, prostaglandins; P, proteolytic enzymes; S, sialoproteins. Capital letters indicate high concentrations, lower case letters indicate moderate to low concentrations, lower case with minus sign in parenthesis indicates substance virtually absent.

C. Control of Accessory Gland Function

All the accessory glands respond to androgens, in terms of both structure and function, and the rat prostate has been a popular tissue for elucidating the molecular mechanism of action of androgens (see Mainwaring, 1977; Mainwaring et al., 1988; Coffey, 1988). Fructose secretion, as judged by concentration in semen or in the seminal vesicles, is a useful index of androgenic activity, and has been extensively used for this purpose (see Mann, 1964). It is probably more useful in demonstrating subnormal androgen secretion than increases above normal (see Setchell et al., 1965).

V. Erection and Ejaculation

For deposition of the semen in the female, it is necessary for the male to exhibit normal sexual behavior, including erection of the penis. More general aspects of sexual behavior and its endocrinological control are dealt with elsewhere (see Chapter 1). Erection of the penis is clearly under nervous control, as it does not occur in men with certain spinal lesions and can be produced by local nerve stimulation. However, the actual mechanics of erection involve dilatation of the arterial supply to the penis, and possibly a restriction in venous outflow, although total blood flow through the penis appears to be increased during erection. Erection can be produced in humans by injection into the corpus cavernosum of compounds that block adrenergic transmission, but there is evidence that VIP (vasoactive intestinal peptide) is the neurotransmitter normally involved (see Brindley, 1985; Benson, 1988). The vascular changes have been studied in anesthetized dogs, in which erection was induced by stimulation of the pelvic nerve (see Carati et al., 1987, 1988). However, in these studies the pressures inside the corpus cavernosum of the penis never exceeded arterial blood pressure, whereas during erection in conscious animals, pressures greatly in excess of arterial pressure have been observed (Beckett et al., 1972, 1973; Purohit and Beckett, 1976). This suggests that contraction of the muscles around the bulb of the penis may also be involved in normal erection; animals in which the ischiocavernosus muscle is anesthetized cannot copulate (Purohit and Beckett, 1976).

The mechanism of erection may also be slightly different in animals such as the stallion, dog, and human with a vascular-type penis and those such as the bull, ram, and boar with a fibro-elastic penis, which forms a sigmoid flexure when flaccid. In the latter type of penis, there is little increase in length or diameter during erection, simply a straightening out of the flexure, and muscular involvement may be greater; another function of some of the muscles may be to reintroduce the flexure when sexual excitement has passed (Watson, 1964, see Fig 1). In the dog and stallion, there may also be a delayed engorgement of the bulbus glandis, which does not normally occur until after intromission (Purohit and Beckett, 1976).

Ejaculation involves a nervously mediated contraction of the muscles of the epididymis, ductus deferens, accessory glands, and closure of the neck of the bladder. The contractions of these muscles appear to be tonic, not peristaltic, although rhythmic contractions of the urethral muscle and the muscles around the bulb of the penis may assist in the expulsion of semen. In humans, retrograde ejaculation into the bladder is a recognized abnormality (see Brindley, 1985). Not all parts of the male tract contract at the same time, with the result that the ejaculate can be divided into so-called "split ejaculates." The boar and the stallion are particularly suited to this procedure, as ejaculation in these species is very much prolonged, compared with the bull and ram. In the stallion, pre-sperm, sperm-rich, and post-sperm fractions can be collected, and a later fraction that is emitted after the stallion dismounts is not at all representative of the entire ejaculate (see Mann and Lutwak-Mann, 1981).

VI. Semen

A. Spermatozoa

The structure of a typical ejaculated mammalian spermatozoon is illustrated in Fig. 9. The head consists mainly of highly condensed chromatin, combined with strongly basic small proteins called protamines. The anterior half of the head is covered by the acrosome, which in domestic mammals, in contrast to some rodents, is a reasonably inconspicuous structure. The acrosome consists of an inverted sac of membrane containing a specific lipoprotein complex, including a number enzymes such as hyaluronidase and acrosin. The former breaks down mucopolysaccharides, and is probably involved in dispersing the cumulus oophorus. The latter is a protease that is involved in the penetration of the spermatozoa through the zona pellucida of the egg. The postacrosomal region of the surface of the nucleus is important because it is this area that attaches and fuses to the egg. The caudal face of the head is indented to allow attachment of the midpiece and tail, which have a central core consisting of nine plus two microtubules, an arrangement common to many cilia in a wide variety of organisms. Outside each of the nine microtubules in spermatozoa is a "dense outer fiber" once believed to have some contractile properties, but now more likely thought to contribute to the passive elastic properties of the tail. Each of the nine microtubules is in fact a doublet, consisting of two subunits, one of which (A) is a complete cylinder, the other (B) an incomplete cylinder applied closely to one side of the A subunit. The A subunits have along their other surface a series of dynein arms that contain ATPase activity. During movement of the tail, the dynein

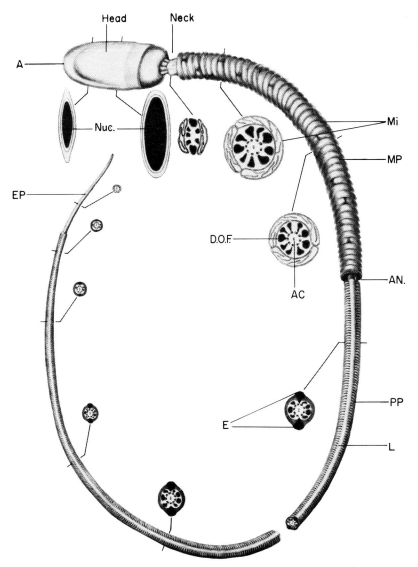

Figure 9 Diagram of an ejaculated mammalian spermatozoon with cross sections at various levels, as indicated. A, acrosome; Nuc, nucleus; Mi, mitochondria, which are helically arranged around the midpiece (MP); AC, axonemal complex of two single central and nine pairs of peripheral microtubules; DOF, nine dense outer fibers, one of which corresponds to each of the nine peripheral pairs of microtubules; AN, annulus marking the end of the midpiece and mitochondria, and the beginning of the principal piece (PP). In the first part of the principal piece, dense outer fibers 3 and 8 terminate and are replaced by inward extensions (E) of the longitudinal columns (L) of the fibrous sheath. The fibrous sheath stops at the junction of the principal piece and the endpiece (EP). Reproduced by permission of the publishers from Fawcett, (1975).

arms undergo a cycle of extension, mechanochemical transduction followed by rigor, and then a return to the flattened form. The interation between these dynein arms and the neighboring doublet can produce a sliding movement, which is probably the basis for the movement of the tail (see Fawcett, 1977; Satir, 1979; Gibbons, 1979, 1981). Around the dense outer fibers in the midpiece is a helix of mitochondria, and in the principal piece beyond the annulus where the mitochondria end, the dense outer fibers are surrounded by a fibrous sheath with two thickened longitudinal columns. The end piece beyond the end of the dense outer fibers consists of the nine-plus-two microtubules covered only by the plasma membrane (see Fawcett, 1977; Setchell, 1982a). The tail of a bull sperm beats at about 10/s, and the sperm can move forward at a rate of about 100 μm/s in saline and rather less in cervical mucous (Bishop, 1962).

The nucleus contains a haploid set of chromosomes, and as half the sperm have an X chromosome and the other half a Y chromosome, much attention has been directed toward ways of separating the male-producing from the female-producing sperm. However, no method has yet been reported that gives unequivocal, repeatable results (see Mohri *et al.*, 1987). The mature spermatozoa contain very little cytoplasm, and therefore it is not suprising that they contain very little RNA. They do contain some lipids that may act as energy reserves as well as important membrane-constituents. Peroxidation of these lipids may exert important toxic effects on the sperm during storage. The metabolism of spermatozoa has been studied in great detail. Bull and ram sperm break down glucose or fructose rapidly to lactic acid under anaerobic

conditions, while boar and stallion sperm do so at a much slower rate; sperm from all species oxidize sugars to carbon dioxide under aerobic conditions. A number of other substrates can also be utilized, including lactic, acetic, and pyruvic acids, glycerol, and sorbitol. Anaerobic fructolysis of bull and ram sperm is closely correlated with motility, but the relationship to fertility is less clear. Motilty is also highly correlated with intracellular levels of cyclic AMP, which is readily lost from sperm during storage or following damage (see Mann and Lutwak-Mann, 1981).

B. Seminal Plasma

Seminal plasma is the fluid fraction of the semen after removal of the sperm by centrifugation or filtration. As indicated in previously Section IV, seminal plasma contains a number of substances not usually found in animals and others in much higher concentrations than found elsewhere in the body. Many of these substances are usually associated with plants rather than animals. Typical concentrations found in different species are given in Table II. It should be emphasized that the values determined will depend on the rate of testosterone secretion by the testis, and will vary in different animals of the same species, in different samples of semen from the same animal, and in successive fractions of the single ejaculate. For example, in stallions the pre-sperm fraction contains only low concentrations of citric acid and ergothioneine; most of the latter is found in the sperm-rich fraction, while most of the citric acid is found in the post-sperm fraction. If electroejaculation is used, the proportion derived from the various glands depends on the placement of the electrode, and even wider variations can be

Table II
Some Details of the Composition of Semen of the Domestic Animals[a]

	Bull	Ram	Goat	Boar	Stallion	Dog	Rabbit
Semen							
Volume (ml)	2–10	0.5–2	0.5–2.5	150–500	20–300	2–15	0.4–6
Dry weight (%)	9.5	14.8		4.6	4.3	3.8	
Sperm concentration ($\times 10^{-6}$/ml)	300–2000	2000–5000	1000–5000	25–350	30–800	60–300	50–350
Spermatocrit (%)	10	33		2	3		
pH	6.48–6.99	5.9–7.3		6.85–7.9	6.2–7.8	6.1–7.0	6.59–7.5
Specific gravity	1.035				1.013	1.011	
Seminal plasma							
Protein (gm/100 ml)	3–8					2.1	
Fructose (mg/100 ml)	120–540	150–600		20–40	<1	<1	40–150
Sorbitol (mg/100 ml)	10–136	26–120		6–18	20–60	<1	80
Citric acid (mg/100 ml)	357–1000	137		36–325	8–53		
Ascorbic acid (mg/100 ml)	8.7	5					
Inositol (mg/100 ml)	25–46	10–15		380–610	19–47		
Ergothioneine (mg/100 ml)	Trace		Absent	6–30	3.5–13.7		
Glutamic acid (mg/100 ml)	35–41	76					
Glycerylphosphorycholine (mg/100 ml)	110–500	1600–2000	1400–1600(WS)[b]	110–240	40–110(WS)[b]	180(WS)[b]	215–370(WS)[b]
Sodium (mmol/liter)	117	78		122	114	114	
Potassium (mmol/liter)	44	23		16	26	8.1	
Calcium (mmol/liter)	9.3	1.9			6.5	0.35	
Magnesium (mmol/liter)	3.4	2.4			3.8	0.25	
Chloride (mmol/liter)	49	18		96		152	
Bicarbonate (mmol/liter)	7.1(WS)[b]	7.1(WS)[b]				2.9(WS)[b]	
α-Mannosidase (units/ml)	400	50					
β-N-Acetylglucosaminidase (units/ml)	15,000	16,000					

[a] Based on data from Mann, 1964; Mann and Lutwak-Mann, 1981.
[b] WS, whole semen.

found than in samples collected with an artificial vagina (see Mann and Lutwak-Mann, 1981).

References

Ackerknecht, E. (1943). In "Ellenberger-Baum Handbuch der vergleichenden Anatomie der Haustiere" (O. Zietschman, E. Ellenberger, and H. Gran, eds.) Springer-Verlag, Berlin.

Acott, T. S., and Carr, D. W. (1984). *Biol. Reprod.* **30,** 926–935.

Amann, R. P., Johnson, L., and Pickett, B. W. (1977). *Am. J. Vet. Res.* **38,** 1571–1579.

Amann, R. P. (1987). *J. Reprod. Fertil. Suppl.* **34,** 115–131.

Baker, M. A., and Hayward, J. N. (1968). *J. Physiol.* (London) **198,** 561–579.

Ball, R. Y. and Setchell, B. P. (1983). *J. Reprod. Fertil.* **68**, 145–153.

Barenton, B., Blanc, M. R., Caraty, A., Hochereau-de Reviers, M. T., Perreau, C., and Saumande, J. (1982). *Molec. Cell. Endocrinol.* **28**, 13–25.

Beckett, S. D., Hudson, R. S., Walker, D. F., Vachon, R. I., and Reynolds, T. M. (1972). *Biol. Reprod.* **7**, 359–364.

Beckett, S. D., Hudson, R. S., Walker, D. F., Reynolds, T. M., and Vachon, R. I. (1973). *Am. J. Physiol.* **225**, 1072–1075.

Bedford, J. M. (1975). *In* "Handbook of Physiology," Section 7, Endocrinology, Vol. V, "Male Reproductive System," (D. W. Hamilton and R. O. Greep, eds.), pp. 303–317. American Physiological Society, Washington, D.C.

Bellve, A. R. (1979). *Oxford Rev. Reprod. Biol.* **1**, 159–261.

Benahmed, M., Bernier, M., Ducharme, J. R., and Saez, J. M. (1982). *Mol. Cell. Endocrinol.* **28**, 705–716.

Benson, G. S. (1988). *In* "Physiology of Reproduction" (E. Knobil and J. D. Neill, eds.), pp. 1121–1390. Neill, Raven Press, New York.

Bergh, A. (1985). *Int. J. Androl.* **8**, 80–85.

Bergh, A., Widmark, A., Damber, J.-E., and Cajander, S. (1986). *Endocrinology* **119**, 586–589.

Bishop, D. W. (1962). *Physiol. Rev.* **42**, 1–59.

Brandt, H., Acott, T. S., Johnson, D. J., and Hoskins, D. D. (1978). *Biol. Reprod.* **19**, 830–835.

Brindley, G. S. (1985). *In* "Genito-urinary Surgery" (H. N. Whitfield and W. F. Hendry, eds.), pp. 1083–1094. Churchill Livingstone, Edinburgh.

Brooks, D. E. (1981). *Physiol. Rev.* **61**, 515–555.

Brooks, D. E. (1983). *Aust. J. Biol. Sci.* **36**, 205–221.

Brooks, D. E. (1990). *In* "Marshall's Physiology of Reproduction," Vol. II (G. E. Lamming, eds.). Churchill Livingstone, Edinburgh.

Bustamante, J. C., Jarvis, L. G., and Setchell, B. P. (1982). *J. Physiol.* **330**, 62–63.

Carati, C. J., Creed, K. E., and Keogh, E. J. (1987). *J. Physiol.* **384**, 525–538.

Carati, C. J., Creed, K. E., and Keogh, E. J. (1988). *J. Physiol.* **400**, 75–88.

Carr, D. W., and Acott, T. S. (1984). *Biol. Reprod.* **30**, 913–925.

Cascieri, M., Amann, R. P., and Hammerstedt, R. H. (1976). *J. Biol. Chem.* **251**, 787–793.

Christensen, A. K. (1975). *In* "Handbook of Physiology," Section 7, "Endocrinology," Vol. V, "Male Reproductive System" (D. W. Hamilton and R. O.

Greep, eds.), pp. 57–94. American Physiology Society, Washington, D.C.

Coffey, D. S. (1988). *In* "Physiology of Reproduction" (E. Knobil and J. D. Neill, eds.), pp. 1081–1119. Raven Press, New York.

Cole, H. H., Hart, G. H., Lyons, W. R., and Catchpole, H. R. (1933). *Anat. Rec.* **56**, 275–293.

Cox, J. E. (1988). *Anim. Reprod. Sci.* **18**, 43–50.

Crabo, B. (1968). *Acta Vet. Scand.* **6** (Suppl. 5), 1.

Dacheux, J. L., O'Shea, T., and Paquignon, M. (1979). *J. Reprod. Fertil.* **55**, 287–296.

Dacheux, J. L., and Paquignon, M. (1980). *Reprod. Nutr. Dev.* **20**, 1085–1099.

D'Occhio, M. J., and Setchell, B. P. (1984). *J. Endocrinol.* **103**, 371–376.

Dym, M., and Cavicchia, J. C. (1977). *Biol. Reprod.* **17**, 390–403.

Dym, M., and Fawcett, D. W. (1970). *Biol. Reprod.* **3**, 308–326.

Eckstein, P., and Zuckerman, S. (1956). *In* "Marshall's Physiology of Reproduction," Vol. 1A (A. S. Parkes, ed.), pp. 43–155. Longmans Green, London.

Evans, R. W., and Setchell, B. P. (1978). *J. Reprod. Fertil.* **52**, 15–20.

Evans, R. W., and Setchell, B. P. (1979). *J. Reprod. Fertil.* **57**, 189–196.

Fawcett, D. W. (1977). *In* "Frontiers in Reproduction and Fertility Control" (R. O. Greep and M. A. Koblinsky, eds.), pp. 353–378. MIT Press, Cambridge, Massachusetts.

Fawcett, D. W. (1973). *Adv. Biol. Sci.* **10**, 83.

Fawcett, D. W. (1974). *In* "Male Fertility and Sterility" (R. E. Mancini and L. Martini, eds.), pp. 13–36. Academic Press, London.

Fawcett, D. W. (1975). *Devel. Biol.* **44**, 394.

Fawcett, D. W., Neaves, W. B., and Flores, M. N. (1973). *Biol. Reprod.* **9**, 500–532.

Fournier-Delpech, S., Colas, G., Courot, M., Ortavant, R., and Brice, G. (1979). *Ann. Biol. Anim. Biochem. Biophys.* **19**, 597–605.

Frankenhuis, M. T., and Wensing, C. J. G. (1979). *Fertil. Steril.* **31**, 428–433.

Free, M. J. (1977). *In* "The Testis," Vol. IV (A. D. Johnson and W. R. Gomes, eds.), pp. 39–90. Academic Press, New York.

Fritz, I. B., Lyon, M. E., and Setchell, B. P. (1983). *J. Reprod. Fertil.* **67**, 359–363.

Gier, H. T., and Marion, G. B. (1970). *In* "The Testis," Vol. I (A. D. Johnson, W. R. Gomes, and N. L. Vandemark, eds.), pp. 1–45. Academic Press, New York.

Gibbons, B. H. (1979). *In* "The Spermatozoon," (D. W. Fawcett and J. M. Bedford, eds.), pp. 91–97. Urban and Schwarzenburg, Baltimore.

Gibbons, I. R. (1981). *J. Cell Biol.* **91,** 107s–124s.

Gilula, N. B., Fawcett, D. W., and Aoki, A. (1976). *Develop. Biol.* **50,** 142–168.

Glover, T. D., and Nicander, L. (1971). *J. Reprod. Fertil. Suppl.* **13,** 39–50.

Gunsalus, G. L., Musto, N. A., and Bardin, C. W. (1980). *In* "Testicular Development, Structure and Function" (E. Steinberger and A. Steinberger, eds.), pp. 291–297. Raven Press, New York.

Gustafsson, K., Soder, O., Pollanen, P., and Ritzen, E. M. (1988). *J. Reprod. Immunol.* **14,** 139–150.

Hadley, M. A., Djakiew, D., Byers, S., and Dym, M. (1987). *Endocrinology* **120,** 1097–1103.

Hall, P. F. (1988). *In* "Physiology of Reproduction" (E. Knobil and J. D. Neill, eds.), pp. 975–998. Raven Press, New York.

Hamilton, D. W. (1975). *In* "Handbook of Physiology, Section 7, Endocrinology, Male Reproductive System," (D. W. Hamilton and R. O. Greep, eds.), pp. 259–301. American Physiological Society, Washington D.C.

Hammerstedt, R. H. and Parks, J. E. (1987). *J. Reprod. Fertil. Suppl.* **34,** 133–149.

Harcourt, A. H., Harvey, P. H., Larson, S. G., and Short, R. V. (1981). *Nature* **293,** 55–57.

Hees, H., Wrobel, K.-H., Kohler, T., Abou Elmagd, A., and Hees, I. (1989). *Cell Tissue Res.* **255,** 29–39.

Hinton, B. T., Snoswell, A. M., and Setchell, B. P. (1979). *J. Reprod. Fertil.* **56,** 105–111.

Hinton, B. T., White, R. W., and Setchell, B. P. (1980). *J. Reprod. Fertil.* **58,** 395–399.

Hochereau-de Reviers, M.-T., and Lincoln, G. A. (1978). *J. Reprod. Fertil.* **54,** 209–213.

Holtz, W., and Smidt, D. (1976). *J. Reprod. Fertil.* **46,** 227–229.

Hoskins, D. D., Johnson, D., Brandt, H., and Acott, T. S. (1979). *In* "The Spermatozoon" (D. W. Fawcett and J. M. Bedford, eds.), pp. 43–53. Urban & Schwarzenberg, Baltimore.

Jegou, B., Dacheux, J. L., Terqui, M., Garnier, D. H., Colas, G., and Courot, M. (1979). *J. Reprod. Fertil.* **57,** 311–318.

Jocelyn, H. D., and Setchell, B. P. (1972). *J. Reprod. Fertil. Suppl.* **17,** 1–76.

Johnson, L., and Neaves, W. B. (1981). *Biol. Reprod.* **24,** 703–712.

Jones, R. C., and Clulow, J. (1987). *In* "Proceedings 1st Asian & Oceanian Physiological Society, Bangkok" (C. Pholpramool and R. Sudsuang, eds.), pp. 229–240. The Physiological Society (Thailand).

Jones, R. C., Clulow, J., Stone, G. M., and Setchell, B. P. (1987). *In* "New Horizons in Sperm Cell Research" (H. Mohri, ed.), pp. 63–74. Japan Science Press, Tokyo.

Jones, R. C., Sujarit, S., Lin, M., Setchell, B. P., and Chaturaparich, G. (1989). *In* "Proceedings IVth International Congress on Andrology, Florence, May 14–18," pp. 359–380. Monduzzi Editore, Bologna.

Kenagy, G. J., and Trombulak, S. C. (1986). *J. Mammal.* **67,** 1–22.

Knight, T. W. (1974). *J. Reprod. Fertil.* **40,** 19–29.

Kormano, M. (1967). *Histochemie* **9,** 327–338.

Laitinen, L., and Talo, A. (1981). *J. Reprod. Fertil.* **63,** 205–209.

Laurie, M. S., and Setchell, B. P. (1978). *J. Physiol.* **287,** 10P.

Lino, B. F., and Braden, A. W. H. (1972). *Aust. J. Biol. Sci.* **25,** 351–358.

Maddocks, S., and Setchell, B. P. (1988a). *Oxford Rev. Reprod. Biol.* **10,** 53–123.

Maddocks, S., and Setchell, B. P. (1988b). *Immunol. Cell Biol.* **66,** 1–8.

Mainwaring, W. I. P. (1977). "The Mechanism of Action of Androgens." Monographs on Endocrinology, 10. Springer-Verlag, New York.

Mainwaring, W. I. P., Haining, S. A., and Harper, B. (1988). *In* "Hormones and Their Actions, Part 1" (B. A. Cooke, R. J. B. King, and H. J. van der Molen, eds.), pp. 169–195. Elsevier, Amsterdam.

Mann, T. (1964). "Biochemistry of Semen and of the Male Reproductive Tract." Methuen, London.

Mann, T., and Lutwak-Mann, C. (1981). "Male Reproductive Function and Semen." Springer-Verlag, New York.

McLaren, A. (1988). *Oxford Rev. Reprod. Biol.* **10,** 162–179.

Monet-Kuntz, C., Hochereau-de Reviers, M.-T., Pisselet, C., Peneau, C., Fontaine, I., and Schanbacher, B. D. (1988). *J. Androl.* **9,** 278–283.

Mori, H. and Christensen, A. K. (1980). *J. Cell Biol.* **84,** 340–354.

Mohri, H., Oshio, S., Kodayasi, T., and Iizuka, R. (1987). *In* "New Horizons in Sperm Cell Research" (H. Mohri, ed.), pp. 469–481. Japan Science Society Press, Tokyo.

Morton, B. E., Sagadraca, R., and Fraser, C. (1978). *Fertil. Steril.* **29,** 695–698.

Niemi, M., and Setchell, B. P. (1986). *Biol. Reprod.* **35,** 385–391.

Noordhuizen-Stassen, E. N., Charbon, G. A., de Jong, F. H., and Wensing, C. J. G. (1985). *J. Reprod. Fertil.* **75,** 193–201.

Noordhuizen-Stassen, E. N., de Jong, F. H., MacDonald, A. A., Schamhardt, H. C., and Wensing, G. C. J. (1988). *Int. J. Androl.* **11,** 493–505.

Orth, J., and Christensen, A. K. (1977). *Endocrinology* **101,** 262–278.

Pholpramool, C., and Triphrom, N. (1984). *J. Reprod. Fertil.* **71,** 181–188.

Pholpramool, C., Zupp, J. L., and Setchell, B. P. (1985). *J. Reprod. Fertil.* **75,** 413–420.

Pollanen, P., Soder, O., and Uksila, J. (1988). *J. Reprod. Immunol.* **14,** 125–138.

Pollanen, P., Soder, O., Punnonen, J., Kaipia, A., Kangasniemi, M., Sainio-Pollanen, S., Zalewski, P., Forbes, I., Uksila, J., and Setchell, B. P. (1989). Proceedings of IVth International Congress of Andrology, Florence, May 14–18, 1989, pp. 177–182. Monduzzi Editore, Bologna.

Purohit, R. C., and Beckett, S. D. (1976). *Am. J. Physiol.* **231,** 1343–1348.

Robaire, B., and Hermo, L. (1988). *In* "Physiology of Reproduction" (E. Knobil and J. D. Neill eds.), pp. 999–1080. Raven Press, New York.

Russell, L. (1977). *Am. J. Anat.* **148,** 313–328.

Satir, P. (1979). *In* "The Spermatozoon" (D. W. Fawcett and J. M. Bedford, eds.), pp. 81–90. Urban and Schwarzenburg, Baltimore.

Setchell, B. P. (1970a). *In* "The Testis," Vol. I (A. D. Johnson, W. R. Gomes, and N. L. Vandemark, eds.), pp. 101–239. Academic Press, New York.

Setchell, B. P. (1970b). *J. Reprod. Fertil.* **23,** 79–85.

Setchell, B. P. (1978). "The Mammalian Testis." Elek Books, London.

Setchell, B. P. (1980a). *In* "Animal Models in Human Reproduction" (M. Serio and L. Martini, eds.), pp. 135–147. Raven Press, New York.

Setchell, B. P. (1980b). *J. Androl.* **1,** 3–10.

Setchell, B. P. (1982a). *In* "Reproduction in Mammals" (C. R. Austin and R. V. Short, eds.), pp. 63–101. Cambridge University Press, Cambridge.

Setchell, B. P.(1982b). *Comp. Biochem. Physiol.* **73A,** 201–205.

Setchell, B. P. (1984a). "Male Reproduction." Benchmark Papers in Human Physiology Series, Number 17. Van Nostrand Reinhold New York.

Setchell, B. P. (1984b). *In* "Reproduction in Sheep" (D. R. Lindsay and D. T. Pearce, eds.), pp. 62–72. Australian Academy of Science, Canberra.

Setchell, B. P. (1986). *Aust. J. Biol. Sci.* **39,** 193–207.

Setchell, B. P. (1990). *Reprod. Fertil. Dev.* **2,** 291–309.

Setchell, B. P., and Brooks, D. E. (1988). *In* "Physiology of Reproduction" (E. Knobil and J. D. Neill, eds.), pp. 753–836. Raven Press, New York.

Setchell, B. P., and Cox, J. E. (1982). *J. Reprod. Fertil. Suppl.* **32,** 123–127.

Setchell, B. P., and Pilsworth, L. M. (1989). *In* "The Testis," Second Edition (H. Burger and D. de Kretser, eds.) pp. 1–66. Raven Press, New York.

Setchell, B. P., Davies, R. V., Gladwell, R. T., Hinton, B. T., Main, S. J., Pilsworth, L., and Waites, G. M. H. (1978). *Ann. Biol. Anim. Biochim. Biophys.* **18,** 623–632.

Setchell, B. P., and Hinton, B. T. (1981). *Prog. Reprod. Biol.* **8,** 58–66.

Setchell, B. P., Laurie, M. S., Flint, A. P. F., and Heap, R. B. (1983). *J. Endocrinol.* **96,** 127–136.

Setchell, B. P., Locatelli, A., Perreau, C., Pisselet, C., Fontaine, I., Kuntz, C., Saumande, J. and Hochereau-de Reviers, M.-T. (1989). *Proc. Aust. Soc. Reprod. Biol.* **21,** 7.

Setchell, B. P., Pollanen, P., and Zupp, J. L. (1988). *Int. J. Androl.* **11,** 225–233.

Setchell, B. P., Scott, T. W., Voglmayr, J. K., and Waites, G. M. H. (1969). *Biol. Reprod. Suppl.* **1,** 40–66.

Setchell, B. P., and Sharpe, R. M. (1981). *J. Endocrinol.* **91,** 245–254.

Setchell, B. P., and Waites, G. M. H. (1970). *J. Endocrinol.* **47,** 81–86.

Setchell, B. P., and Waites, G. M. H. (1975). *In* "Handbook of Physiology, Section 7, Endocrinology, Vol. V, Male Reproductive System" (D. W. Hamilton and R. O. Greep, eds.), pp. 143–172. American Physiological Society, Washington D.C.

Setchell, B. P., Waites, G. M. H., and Lindner, H. R. (1965). *J. Reprod. Fertil.* **9,** 149–162.

Sharpe, R. M. (1983). *Q. J. Exp. Physiol.* **68,** 265–287.

Sharpe, R. M. (1984). *Biol. Reprod.* **30,** 29–49.

Sharpe, R. M. (1986). *Clin. Endocrinol. Metab.* **15,** 185–207.

Sharpe, R. M., and Cooper, I. (1986). *J. Endocrinol.* **113,** 89–96.

Sinha Hikim, A. P., Bartke, A., and Russell, L. D. (1988). *Biol. Reprod.* **39,** 1225–1237.

Stephens, D. T., Acott, T. S., and Hoskins, D. D. (1981). *Biol. Reprod.* **25,** 945–949.

Sujarit, S., and Pholpramool, C. (1985). *J. Reprod. Fertil.* **74,** 497–502.

Sundby, A., and Torjesen, P. A. (1975). *Acta Endocrinol.* **99,** 787–792.

Tung, K. S. K. (1980). *In* "Immunological Aspects of Infertility and Fertility Regulation" (D. H. Dhindsa and G. F. B. Schumacher, eds.), pp. 33–91. Elsevier North Holland, New York.

Usselman, M. C., and Cone, R. A. (1983). *Biol. Reprod.* **29,** 1241–1253.

Vitale-Calpe, R., Fawcett, D. W., and Dym, M. (1973). *Anat. Rec.* **176,** 333–344.

Voglmayr, J. K. (1975). *In* "Handbook of Physiology, Section 7, Endocrinology, Vol. V, Male Reproductive System" (D. W. Hamilton and R. O. Greep, eds.), pp. 437–451. American Physiological Society, Washington, D.C.

Voglmayr, J. K., White, I. G., and Parks, R. P. (1978). *Theriogenology* **10,** 313–321.

Waites, G. M. H. (1970). *In* "The Testis," Vol. I (A. D. Johnson, W. R. Gomes, and N. L. Demark, eds.) pp. 241–279. Academic Press, New York.

Watson, J. W. (1964). *Nature* **204,** 95–96.

Weber, J. E., Russell, L. D., Wong, V., and Peterson, R. N. (1983). *Am. J. Anat.* **167,** 163–179.

Wensing, C. J. G., and Colenbrander, B. (1986). *Oxford Rev. Reprod. Biol.* **8,** 130–164.

White, I. G. (1973). *J. Reprod. Fertil. Suppl.* **18,** 225.

Yule, T. D., Montoya, G. D., Russell, L. D., Williams, T. M., and Tung, K. S. K. (1988). *J. Immunol.* **141,** 1161–1167.

CHAPTER 7

Artificial Insemination

D. L. GARNER

I. Introduction

The process of artificial insemination (AI), the introduction of semen into the female tract without contact between the male and female, is a common practice throughout the world. This technology is the first application of genetic engineering to the live-stock industry. Moreover, it has served as a model of the application of basic laboratory research to production agriculture. The major advantages of AI are the potential genetic gains, disease control, safety, and cost effectiveness. The AI process also provides flexibility, but involves several sequential steps including collection of semen, its processing and storage, the introduction of po-tentially fertile spermatozoa into the female tract at the appropriate time, and finally, rigorous evaluation of the success of the en-tire process. The physiological aspects of the AI process for several domestic species of livestock are discussed briefly in this chap-ter. Much additional, in-depth information, including reviews and monographs, is avail-able on the subject (Parkes, 1960; Salisbury and VanDemark, 1961; Maule, 1962; Mann, 1964; Foote, 1969, 1974, 1980; Ha-fez, 1970, 1974, 1986; Pickett *et al.*, 1974; Saacke, 1974; Gomes, 1977; Salisbury *et al.*, 1978; Faulkner and Pineda, 1980; Mann and Lutwak-Mann, 1981; Evans and Max-well, 1987).

II. Semen Production and Collection

The success of AI depends on collection of relatively large numbers of potentially fer-tile spermatozoa from genetically superior sires. The semen that is used for AI is gen-erally collected from most domestic farm animals using an artificial vagina (AV). This collection process is used routinely for dairy bulls, many beef bulls, stallions, rams, and goats (Faulkner and Pineda, 1980). The AV has not proved to be entirely suitable for some types of farm animals. Semen is col-lected more effectively from some domestic animals, such as swine, dogs, and poultry by manual manipulation of the reproductive organs (Foote, 1969, 1974, 1980; Gomes, 1977). For some males it is, however, neces-sary to collect semen using electroejacula-tion for a variety of reasons, including the fact that some animals are not readily trained, as well as animal behavior prob-lems, injuries, health limitations, facilities,

etc. (Ball, 1976; Faulkner and Pineda, 1980; Foote, 1980).

A. General Principles

Collection of semen requires appropriate facilities to prevent injury to the animals and their handlers. The area to be used for collection should be free of distractions and relatively quiet. The collection process re-quires specialized equipment and proce-dures to maximize the physiological respon-siveness of the animal in producing semen. Maintenance of libido is very important be-cause cooperation of the animal is necessary for most semen collection techniques. The collection area should not be used for pro-curement of blood samples or other stress-ful procedures.

1. Artificial Vagina

The AV is designed to simulate the vagi-nal orifice of the female, and its use requires cooperation of the male. The first AV was developed for the dog by Amantea (Saacke *et al.*, 1982). Similar AVs for the bull, stal-lion, and ram were designed in the Soviet Union (Sorensen, 1938; Saacke *et al.*, 1982). The most commonly used AV for the bull is the Danish model, which comprises two major layers, an outer casing and an inner latex sleeve (Fig. 1). The space between these two layers forms an enclosed cavity, which is filled with water maintained at 42–46°C. A latex cone, which directs the ejacu-late into a test tube for collection, is attached to the anterior end of the AV casing. The collection tube is normally insulated to pro-tect the semen from temperature changes (Faulkner and Pineda, 1980). The AV method is highly effective for those animal's possessing penile receptors that are highly temperature sensitive (Foote, 1969, 1974).

Equine AV   Bovine AV   Ovine AV

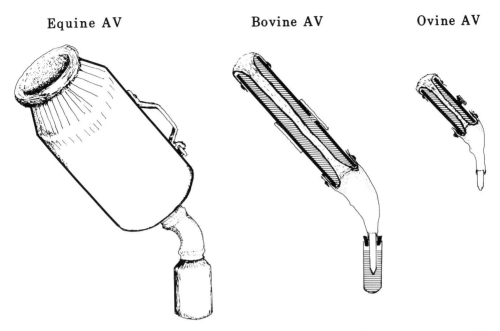

Figure 1 Artificial vaginas (AVs) for the stallion, bull, and ram. The equine AV, Japanese model, has an outer aluminum shell with a latex inner liner (Nishikawa, 1959, 1962, 1964; Hafez, 1986). The bovine and ovine AVs are sectioned to show the inner chamber (cross lines) for maintaining water at 42–46°C. The bovine AV also has a outer tube surrounding the collection vial (cross lines) in which water is maintained at near body temperature to prevent cold shock.

For collection using an AV, a mount or "teaser" animal is normally required to stimulate copulatory behavior. Generally, an estrous or hormonally treated female is used, but some males are readily trained to mount a dummy (phantom), a castrated male, or another male animal (Gomes, 1977).

The AV that is used for the ram is similar to the Danish model for the bull, but is smaller (Fig. 1) (Salamon, 1976; Faulkner and Pineda, 1980). Most rams are easily trained and a ewe or dummy can be used as a mount during collection (Aamdal, 1960). Several variations of the AV have been developed for the horse, including the Missouri (Lambert and McKenzie, 1940), Japanese (Fig. 1) (Nishikawa, 1959, 1962, 1964;

Komarek *et al.*, 1965), and Colorado models (Pickett and Back, 1973). A filter is frequently used on the Japanese and Colorado models to allow collection of the fluid portion of the ejaculate while retaining the viscous, gel portion (Komarek *et al.*, 1965; Pickett and Back, 1973; Faulkner and Pineda, 1980). Estrous or nymphomaniac mares are good teaser animals. Special care must be exercised when collecting semen from stallions. The mare should be restrained with breeding hobbles, her anal-genital region washed, and her tail wrapped with cloth to insure collection of an uncontaminated ejaculate (Faulkner and Pineda, 1980; Tischner, 1988). Suitable protective clothing including a helmet is recommended for the person collecting semen from a horse (Pickett and Back, 1973).

2. Manual Manipulation

The technique of manual manipulation is effective for the males of species in which pressure, rather than temperature, is the predominant sensory stimulus for which the penile receptors are the most sensitive. Efficient collection of semen from swine is normally accomplished using the "gloved hand" technique (Norman *et al.*, 1962; King and Macpherson, 1973). The hand, while encased in an examination or surgical glove, readily adapts to the cork-screw shape of the boar's penis and can provide the constant pressure necessary during the stimulatory and ejaculatory phases of the collection process. The manual approach also is the most effective method for collecting semen from the dog (Faulkner and Pineda, 1980). It is possible to obtain semen from bulls by massaging the ampullae and vesicular glands via the rectum, but the quality of semen obtained by this method is relatively poor (Salisbury *et al.*, 1978). Epididymal semen can be readily obtained from recently killed or castrated males by retrograde flushing via the vas deferens (Czarnetzky and Henle, 1938; Lessley and Garner, 1983) or by stripping the spermatozoa from the cauda epididymides (Jones, 1973).

3. Cloacal Manipulation

Collection of semen from avian species requires skilled manipulation of the cloaca of the bird. The cloacal orifice is inverted and the fingers are used to stimulate, through the abdominal wall, the seminal ducts to release semen (Lake, 1962). Relatively large amounts of highly concentrated semen can be collected from turkey toms

and roosters using this method (Lake, 1962).

4. Electroejaculation

Semen can be collected from uncooperative males, or from animals in which health problems limit physical abilities, with the aid of electrical stimulation. A probe with electrodes on its surface is placed within the rectum adjacent to the nerve trunks leading to the reproductive organs (Ball, 1976). Carefully controlled low-level, intermittent electrical pulses can elicit a neural stimulus response resulting in erection and ultimately ejaculation (Ball, 1976). This method, which is somewhat stressful to the animal, is commonly used for semen collection during veterinary breeding soundness evaluations of bulls (Faulkner and Pineda, 1980; Hudson, 1972; Ball, 1976). Rams are routinely collected using a hand-held, battery-powered electroejaculator (Kimberling et al., 1986). This is especially important for semen collection under field conditions. Semen quality has been considered to be equivalent to that obtained with the AV with some notable exceptions. Ram spermatozoa may be more susceptible to cold shock and less likely to freeze properly (Quinn et al., 1968). Furthermore, unless caution is exercised during electrostimulation, samples tend to be more dilute than when collected with an AV. It must be noted that proper restraint devices are necessary to prevent injury to the animals and their handlers during semen collection using electroejaculation (Curtiss, 1988).

Similar electrical stimulation procedures have been adapted for use in certain avian species (Betzen, 1985). Such procedures could be extremely useful for collecting semen from endangered species of birds.

B. *Efficiency of Collection*

Both the quantity and quality of semen that can be collected with an AV can be enhanced by "teasing" the male prior to collection. Such erotic stimuli appear to maximize the ejaculatory processes. For the bull, this teasing procedure involves allowing one or more "false mounts" and in some instances a waiting period before the animal is allowed to ejaculate (Hafs *et al.*, 1962). The number of motile spermatozoa that can be collected from a bull can be increased by 50% by proper sexual stimulation (Fig. 2) (Hafs *et al.*, 1962; Salisbury and VanDemark, 1961; Salisbury *et al.*, 1978). Both the number of spermatozoa and the percentage of motile spermatozoa in the ejaculate of boars have been increased by restraint (Cassou, 1968). Similar sexual preparation procedures did not effectively increase sperm

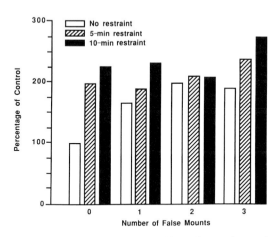

Figure 2 Illustration of the effect of restraint and false mounts on the number of spermatozoa per ejaculate from bulls that were unrestrained or restrained for 5 or 10 min. The degree of teasing is reflected by the number of false mounts. Adapted from Hafs *et al.* (1962).

numbers in stallion ejaculates, but did increase seminal volume, resulting in a lower sperm concentration (Pickett and Back, 1973). Certainly, sexual stimulation is effective for increasing the number of spermatozoa per ejaculate for males of some domestic species, but does not appear to be effective for males of all species (Gomes, 1977).

III. Semen Evaluation

Semen composition is highly variable both within and between male animals. Reliable estimations of semen quality require careful examination of a minimum of three ejaculates collected at intervals of a few days (Hudson, 1972). Although such an approach is not always practical, the resultant data have a much greater likelihood of being representative of the seminal quality of that particular animal. Immediately upon collection, semen should be protected from light and placed in an environment where temperature can be maintained at 30–37°C until the sample can be examined and further processed. It should be noted that any semen evaluation should be joined with a detailed clinical examination of the animal (Hudson, 1972; Uwland, 1984).

A. *Classical Techniques*

The initial evaluation of semen should occur immediately after collection and should include examination of the general appearance of the sample. The ejaculate should be free of contaminants such as hair, feces, blood, urine, etc. In addition, clear, watery samples usually contain an inadequate number of spermatozoa and the presence of clumps or a lack of uniformity are indicative of infective processes such as seminal vesiculitis (Roberts, 1971) and epididymitis (Kimberling et al., 1986). The seminal characteristics of domestic animals are provided in Table I.

The volume of the ejaculate can be readily measured when the sample has been collected directly into a graduated vial. Semen volume can vary greatly depending on a number of factors, including the season of the year, the method of collection, and the sexual preparation of the animal.

The estimation of sperm numbers includes determining the total spermatozoa per ejaculate and the concentration of spermatozoa per unit volume of the ejaculate. The sperm concentration of a sample can be determined using a variety of techniques including hemocytometer counts, turbidimetric measurements, and electronic counting systems (Foote, 1969, 1974, 1980; Hafez,

1986). All sperm counting techniques require careful calibration to ensure reasonable accuracy of the measurements (Foote, 1974, 1978).

The most unique feature of spermatozoa is their ability to move. Classical motility evaluations include estimates of both the percentage of progressively motile spermatozoa and the rate of progressive motility. The accuracy of motility evaluations requires careful control of the temperature of the semen during preparation and evaluation. Although the motility of a semen sample from some farm animals, especially the ram, can be estimated by the swirling motion at low-power magnification, the most widely accepted method for assessment of sperm motion is the track motility of individual spermatozoa using a microscope equipped with phase-contrast optics and a heated stage (Elliott, 1967; O'Conner *et al.*,

Table I
Seminal Characteristics of Domestic Animals[a]

Domestic animal	Seminal characteristics					
	Volume (ml)	Sperm concentration ($\times 10^9$/ml)	Total sperm ($\times 10^9$)	Motile sperm (%)	Normal sperm (%)	Ejaculation collection/ week (no.)
Bovine, dairy	6	1.2	7	70	89	4
Bovine, beef	4	1.0	4	65	80	4
Ovine	1	3.0	3	75	90	20
Caprine	0.8	2.4	2	80	70	20
Porcine	225[b]	0.2	45	60	60	3
Equine	60	0.15	9	70	70	3
Canine	5	0.3	1.5	85	80	3
Chicken	0.5	3.5	1.8	85	90	3
Turkey	0.5	7.0	3.5	80	90	3

[a]Adapted from Gomes (1977).
[b]Gel-free volume.

1982; Amann, 1988). The tracks of individual spermatozoa are conveniently measured from timed photomicrographs of the sample using still exposures on 35-mm film (Elliott, 1967). These measurements tend to be repeatable and precise under controlled laboratory conditions (Elliott, 1967; Amann, 1988). The development of video recording technology may simplify this approach, but the motility estimates obtained do not appear to be any more precise than those obtained by 35-mm still photography (Amann, 1988). Microscopic estimates of sperm motility tend to vary greatly between different laboratories. Some of this variability is due to the method of estimation and the individual evaluators. Sometimes low motility estimates reflect improper handling rather than poor semen quality. Thus, sperm motility estimates under field conditions are highly variable and in many cases unreliable. For this reason evaluations of semen quality, as part of breeding soundness examinations, tend to emphasize sperm morphology assessments rather than motility (Hudson, 1972; Morrow, 1980). The morphological features of spermatozoa from domestic animals are depicted in Fig. 3. An ejaculate normally contains some morphologically abnormal spermatozoa (Adams, 1962; Foote, 1968; Faulkner and Pineda, 1980). The normalcy of the ejaculate, however, depends on the percentage of spermatozoa that possess primary and secondary abnormalities. An increasingly important morphological assessment has been estimation of the normalcy of the acrosome because of its obvious role in fertilization (Saacke, 1972a, 1972b, 1978). Other seminal characteristics such as fructolysis, glutamic-oxalic transaminase, pH, and proacrosin content have been used to estimate seminal quality, but few have received widespread application for evaluation of semen from domestic animals (Foote, 1968; Gomes, 1977; Garner, 1984b).

B. Sperm Penetration and Fusion Tests

The first major biological fluid that spermatozoa are exposed to upon insemination is vaginal fluid, cervical mucus, or uterine secretions, depending on the site of insemination. When cervically inseminated, spermatozoa move from seminal plasma or extender into the more viscous cervical mucus, only slight changes in the pattern of motility are observed (Hafez, 1986). The ability of spermatozoa to penetrate cervical mucus has been considered to be a potentially important characteristic because this attribute might be useful in predicting fertilizing ability (Lorton and First, 1977). The cervical mucus penetration test has been used as an *in vitro* laboratory test for semen analysis for a number of species, including humans (Kremer, 1965; Kremer and Jager, 1976), but has not received widespread use and does not appear to provide any additional information to that obtained in routine motility assessments (Lorton and First, 1977).

Another so-called penetration test, the zona-free hamster oocyte test, is more correctly a gamete fusion test. This test involves a number of steps including *in vitro* capacitation, superovulation of hamsters, removal of the zona pellucida from the recovered oocytes, incubation of the prepared gametes, and careful examination for sperm penetration (Thadani, 1980; Wright, 1981; Amann and Seidel, 1982; Martin *et al.*, 1982; Brandriff *et al.*, 1986). It has been suggested as a suitable method for *in vitro* testing of fertilizing capacity, especially if

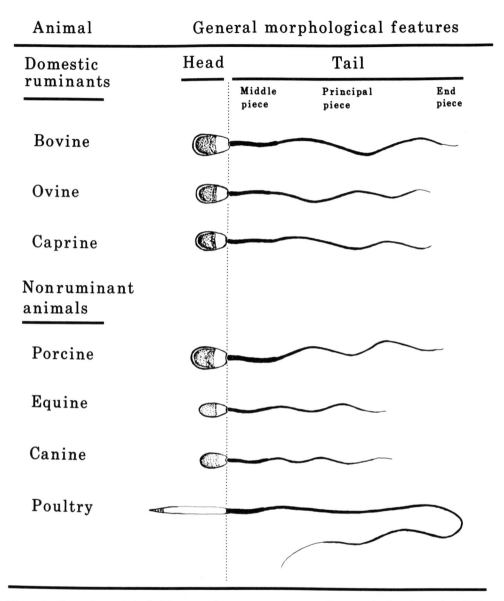

Figure 3 General morphological features of spermatozoa from ruminants and other domestic animals showing the relative size of the gametes.

bovine oocytes are used to test sperm penetration (Amann, 1984). Such tests have been used to assess the chromosome complement of individual spermatozoa (Martin *et al.*, 1982; Brandriff *et al.*, 1986). Certain problems such as variability and the fact that it is not physiologically correct without the major penetration barrier, the zona pellucida, are noteworthy. The principle restrictions to more widespread use in assessing the spermatozoa of domestic animals, however, appear to be the labor intensity, time commitments, and cost of the procedure.

C. Automated Sperm Evaluation Technologies

Computer-assisted sperm analysis systems have been adapted for use on semen from domestic livestock (Amann, 1988). Although some information can be obtained on morphological parameters, the predominant feature measured by these systems is sperm motility (Katz *et al.*, 1985; Katz and Davis, 1987; Amann, 1988). The computer-assisted systems have enabled identification of several patterns of sperm motility within semen samples in both epididymal (Stephens and Hoskins, 1986) and ejaculated semen (O'Conner *et al.*, 1982; Budworth *et al.*, 1987, 1988; Katz and Davis, 1987; Amann, 1988). These systems provide simultaneous measurement of several motility parameters including the percentage of motile and progressively motile spermatozoa, curvilinear and straight-line velocities, linearity of movement, amplitude of lateral head displacement, and beat cross frequency, in addition to the track motility of individual spermatozoa (Katz and Davis, 1987; Amann, 1988). No doubt computer-assisted systems of evaluation will become an increasing important method of assessing

seminal quality. The attractiveness of these automated systems is in the rapidity and repeatability of the measurements. Their precision, at least at this stage of development, does not appear to be better than that obtained using photographic track motility estimates (Amann, 1988). Some major questions remain as to the general applicability of computer-assisted motility analyses to the semen of several domestic species (Amann, 1988).

Flow cytometry offers a rapid and highly precise method of analyzing spermatozoa (Gledhill *et al.*, 1976; Pinkel *et al.*, 1983; Garner, 1984a, 1984b; Garner *et al.*, 1986; Ballachey *et al.*, 1987, 1988). The precision is such that it is capable of resolving the difference in DNA content between X- and Y-chromosome-bearing spermatozoa (Pinkel *et al.*, 1983; Garner *et al.*, 1983; Garner, 1984a, 1984b; Johnson *et al.*, 1987). Viable rabbit spermatozoa have been flow sorted into the X- and Y-chromosome-bearing sperm populations using a modified flow cytometer to detect differences in DNA content (Johnson *et al.*, 1989). The sex ratio of the live progeny obtained using these sorted sperm were altered as predicted by the ratio of X- to Y-spermatozoa. This is a major accomplishment and is a significant step toward development of a practical means for sexing the semen of domestic animals. The fluorescence emitted from stained viable spermatozoa also can be rapidly quantitated using a flow cytometer (Evenson *et al.*, 1982; Garner *et al.*, 1986; Ericsson *et al.*, 1989). Flow cytometry, because of its flexibility in data acquisition, offers a means of simultaneously characterizing a variety of sperm characteristics (Ericsson *et al.*, 1989). It also allows one to measure a relatively large number of individual spermatozoa within a semen sample, thereby reducing measurement errors (Garner, 1984b; Garner *et al.*, 1986).

Automated morphometric analyses can readily identify very subtle differences in sperm head shape. (Moruzzi *et al.*, 1985). Such image analyses systems have only recently become efficient enough to be of practical significance for analyzing semen samples from domestic livestock.

D. Heterospermic Insemination/ Competitive Fertilization

The mixing of spermatozoa from two males prior to insemination provides a means of directly comparing the fertilizing ability of gametes from the two different males in the same female tract. This competitive procedure, which is termed heterospermic insemination, provides a comparison of the difference in fertilizing ability that is 170–340 times more powerful than comparison of the fertility using homospermic inseminations (Beatty *et al.*, 1969). The source of the spermatozoon that fertilizes each ovum must be precisely determined either by genetic markers or by labeling the spermatozoa (Bedford and Overstreet, 1972; Saacke *et al.*, 1980a, 1980b). Another means of making an approximation as to which spermatozoa fertilized an oocyte is by determining the source of the accessory spermatozoa embedded in the zona pellucida of the zygote (Saacke *et al.*, 1988). This provides a highly effective means of comparing fertility among different males.

IV. Laboratory Processing of Semen

Ejaculated spermatozoa are highly dynamic cells and do not survive well unless they are diluted with a medium (Salisbury *et al.*, 1978; Mann and Lutwak-Mann, 1981). These extending media must maintain the proper osmotic pressure and electrolyte balance, provide nutrients as an energy source, protect against rapid cooling, provide buffering capacity against metabolic acidosis, and inhibit microbial growth (Foote, 1980).

A. Dilution of Semen and Extenders

The dilution of semen in appropriate media not only provides an environment suitable for extended storage of the spermatozoa, but also dilutes the sperm concentration to a level appropriate for making multiple inseminations (Picket and Berndtson, 1974; Graham, 1978; Salisbury *et al.*, 1978; Foote, 1980). Nearly all semen extenders contain egg yolk, milk, or some combination of the two (Salisbury *et al.*, 1978; Foote, 1980; Amann, 1984; Corteel and Paquignon, 1984). A variety of other biological materials have been tried as extender components, but none have received wide application (Salisbury *et al.*, 1978; Foote, 1980).

Egg-yolk-containing media is commonly diluted with isotonic sodium citrate or an organic buffer and is sometimes combined with heated milk or skimmed milk. Milk must be heated (92–95°C for 10 min) to destroy lactenin, a spermacide (Flipse *et al.*, 1954). In addition to 2.9% sodium citrate, various complex buffering agents have been employed in extenders including sodium phosphate, Tris [tris(hydroxymethyl) aminomethane), TES [N-tris(hydroxymethyl) methyl-2-aminomethanesulfonic acid], MES [2-(N-morpholino)ethanesulfonic acid], HEPES (N-2-hydroxyethylpiperazine-N'-2-ethanesulfonic acid), PIPES [piperizine-N,N'-bis(2-ethanesulfonic acid),

MOPS [2-(*N*-morpholino)propane sulfonic acid], and BES [*N*,*N*-bis(2-hydroxyethyl)-2-aminoethane sulfonic acid] (Flipse *et al.*, 1954; Salisbury *et al.*, 1978). The use of these organic zwitterion buffers is increasing (Salisbury *et al.*, 1978; Amann, 1984; Corteel and Paquignon, 1984).

Several variations of these formulas are used for dilution of boar semen (Pursel and Johnson, 1975a, 1975b; Paquignon and Courot, 1976; Corteel and Paquignon, 1984). Those used principally for storage of liquid boar semen are the Guelph or Kiev extenders (Haeger and Mackle, 1971). Several other diluents have been used successfully to store boar spermatozoa, including an egg yolk–glycerol–glucose extender (Polge *et al.*, 1970), the INRA–ITP medium (Paquignon and Courot, 1976), and the Westendorf medium (Westendorf *et al.*, 1975). It should be noted that boar spermatozoa, unlike bull and ram gametes, are particularly sensitive to dilution in egg-yolk-containing media. The egg yolk appears to increase acrosomal degeneration (Watson and Plummer, 1985). Other components, such as the synthetic detergent Orvus Ex Paste (OEP), have been used to overcome this sensitivity and appear to enhance preservation of the fertilizing capacity of boar spermatozoa (Pursel and Johnson, 1975b; Pursel *et al.*, 1978). The ability of goat spermatozoa to undergo storage is greatly enhanced by removal of seminal plasma due to its interaction with extender components (Corteel, 1974). The same three-way interaction between the spermatozoa, seminal plasma, and various extender components affects preservation of boar gametes (Pursel and Johnson, 1975b). Removal of the seminal plasma gel fraction from boar and stallion semen appears to expedite processing and storage procedures (Corteel, 1974;

Pursel and Johnson, 1975b; Amann, 1984).

B. Microbial Contamination and Antimicrobial Agents

Most AI programs require at least some short-term storage of spermatozoa. Because the media used for storage of spermatozoa are particularly supportive of microbial growth, it is necessary to include antimicrobial agents to prevent massive proliferation of the microorganisms present in ejaculated semen (Gunsalus *et al.*, 1941; Foote and Salisbury, 1948; Almquist, 1949a, 1949b). The inclusion of antimicrobial agents is necessary whether the spermatozoa are stored short term in a liquid medium or cryopreserved (Salisbury *et al.*, 1978; Shin *et al.*, 1987; Lorton *et al.*, 1988a, 1988b). The level of contamination is influenced greatly by the collection process (Riek *et. al.*, 1980). Certain precautions, such as clipping preputual hair and allowing only one intromission into an AV, can minimize the microbial contamination (Amann, 1984). The various species of microorganisms found in semen include bacilli, diptherioids, micrococci, coliform organisms, streptococci, pseudomonas, actinomyces, proteus, yeasts, mycoplasma, and viral agents (Foote and Berndtson, 1976; Foote, 1978). Some organisms can survive freezing and, if present in frozen-thawed spermatozoa can infect inseminated females (Getty and Ellis, 1967). Other infectious microorganisms that can be transmitted via contaminated semen include *Brucella abortus*, *Campylobacter fetus*, *Trichomonas fetus*, *Leptospira pomona*, *Mycobacterium tuberculosis*, and *Mycobacterium paratuberculosis* (Foote and Berndtson, 1976; Foote, 1978). The organism responsible for most of the epididymitis problem in rams,

Brucella ovis, is transmitted through semen (Kimberling *et al.*, 1986). The combination of penicillin, dihydrostreptomycin, and poylmixin B has been widely used by the commercial AI industry (Sullivan *et al.*, 1966). Some problems with resistant organisms have been noted (Amann, 1984). Recently, a new combination of antibiotics comprised of gentamicin, tylosin, and Linco-Spectin has been used successfully to control certain resistant microorganisms in cryopreserved bull semen (Shin *et al.*, 1987; Lorton *et al.*, 1988a, 1988b; Ericsson *et al.*, 1989).

V. Preservation of Spermatozoa

The spermatozoa to be used for insemination must be not only alive but also fertile if AI is to be successful. Hence, preservation of spermatozoa is a process in which not only the viability of the cells but also their fertilizing capacity must be maintained during storage. Sperm motility, which has been used as a general measure of sperm viability, is important but is not a guarantee of fertility (Salisbury *et al.*, 1978). The development of extenders using egg yolk provided the basis for most of the media used for preservation of spermatozoa (Salisbury *et al.*, 1978). Several factors must be taken into account when preserving spermatozoa. These include temperature, energy source, osmotic pressure, electrolyte balance, pH and buffering capacity, microbial control, and an appropriate dilution rate (Salisbury *et al.*, 1978).

A. Short-Term Liquid Storage of Semen

Fertile spermatozoa can be preserved for a short period of time without freezing. The gametes of many species can be stored in a liquid media which is maintained either at ambient temperature or in a refrigerator at 4–5°C. There are definite advantages to being able to store gametes at ambient temperature, especially where refrigeration facilities may not be readily available. The daily fluctuations encountered in storage temperature can present serious problems in maintaining fertile spermatozoa. Ambient-temperature storage requires prevention of microbial contamination and effective prevention of microbial proliferation without impairment of gamete viability. For successful short-term liquid storage of semen it is necessary either to extend the time that spermatozoa are able to maintain high metabolic rates or to slow the process down. The first effective ambient-temperature storage medium was based on the fact that carbon dioxide markedly reduced the rate of sperm metabolism (Salisbury *et al.*, 1978). This storage medium, which was termed illini variable temperature (IVT) extender, was saturated with carbon dioxide by bubbling the pure gas through an egg yolk–citrate extender. Other ambient temperature extenders, such as CUE (Cornell University extender), utilize similar metabolic inhibition principles (Foote *et al.*, 1960). Another ambient temperature storage medium called Caprogen extender was extensively field tested in New Zealand and adopted for routine use at a storage temperature of 18–24°C (New Zealand Dairy Board, 48th Farm Production Report, 1971–1972; Shannon, 1972). The Caprogen extender is a modified citrate-buffered egg yolk medium containing caproic acid, glucose, glycine, sulfacetamide, penicillin, and streptomycin and is saturated with nitrogen gas (Shannon, 1972). The fertility of bovine spermatozoa can be maintained for about 3 days using the common ambient-temperature storage media (Foote *et al.*, 1960).

Most systems for preserving spermatozoa for more than 3 days require some type of cooling and refrigeration. In this context, it is important to note that ejaculated spermatozoa are highly sensitive to a rapid decline in temperature. This phenomenon, termed "cold shock," results in an impairment of membrane integrity that is demonstrated by the leakage of potassium, ATP, lipoproteins, enzymes, and other important intracellular constituents from cold-shocked gametes. The plasma membrane, acrosome, and mitochondria are very sensitive to cold shock (Watson and Plummer, 1985). It has been demonstrated that certain substances, including lecithin and phospholipids, could protect spermatozoa from the effects of rapid cooling (Mayer and Lasley, 1945; Salisbury *et al.*, 1978). Adequate levels of these substances were found to be present in egg yolk and milk (Salisbury *et al.*, 1978). The rate of cooling is an important factor, but not all species respond in the same manner to cold shock. For instance, boar spermatozoa are quite sensitive to cooling below 15°C (Polge, 1956; Watson and Plummer, 1985), whereas spermatozoa from other domestic animals such as the fowl are relatively resistant to cold shock below this temperature (Watson, 1981).

B. Cryopreservation Techniques

The obvious limitations of fluid storage of semen, whether it be at ambient or refrigeration temperature, have been overcome by cryopreservation. The discovery by Polge and colleagues (Polge *et al.*, 1949) of the cryoprotectant properties of glycerol emphasized the advantages of preserving semen at ultra-low temperatures. Spermatozoa of many species can be stored at liquid nitrogen temperatures (-196°C) for indefinite periods and, after thawing, retain relatively high rates of fertility (Salisbury *et al.*, 1978; Foote, 1982, 1988; Graham *et al.*, 1984. Even under the best conditions, approximately 40–50% of motile cells in a sample succumb during the freezing and thawing processes currently used for bovine spermatozoa. Spermatozoa are damaged during these processes due to (1) internal ice crystal formation causing structural alterations, (2) increases in solute concentrations to toxic levels due to the withdrawal of pure water from both the intracellular milieu and the suspending media by freezing, and (3) a combination of these two physical phenomena (Mazur, 1963; Salisbury *et al.*, 1978). It also is noteworthy that spermatozoa from all species do not undergo cryopreservation in a manner that is compatible with routine AI procedures. For instance, ram spermatozoa can be cryopreserved successfully, but fertility is acceptable only if insemination of the thawed spermatozoa is intrauterine rather than vaginal or cervical (Evans and Maxwell, 1987).

Glycerol, the commonly used cryoprotectant for bull spermatozoa, continues to be the most satisfactory agent, even though several other potential cryoprotectants have been examined including ethylene glycol, propylene glycol, dimethyl sulfoxide (DMSO), and a number of saccharides (Salisbury *et al.*, 1978; Graham *et al.*, 1984). Evaluations of post-thaw sperm motility indicated that glycerol should be added slowly or at least stepwise to extended semen to attain the highest percentage of progressively motile spermatozoa (Choong and Wales, 1962). The observed enhancement in post-thaw motility, however, does not appear to be reflected in higher fertility levels when comparative samples are glycerolated quickly or slowly over a 2-h period (Hunter, 1968). The cryoprotective properties of glycerol have not been fully delineated, but it appears that several factors are involved (Salisbury et al., 1978). The beneficial aspects of glycerol include binding of water, thereby decreasing the freezing point of solutions, a salt-buffering action, and an apparent alteration in ice crystal formation (Mazur, 1963; Graham et al., 1984). It is likely that other factors are involved. One of the advancements in attaining successful cryopreservation of fertile boar spermatozoa was a reduction of the glycerol content of the media and the inclusion of the synthetic detergent OEP in the extender (Pursel and Johnson, 1975b; Westendorf et al., 1975). The OEP which is an amino–sodium lauryl sulfate, increased the postthaw percentage of both normal acrosomes and motile spermatozoa (Graham et al., 1971; Westendorf et al., 1975; Pursel et al., 1978) and enhanced the preservation of fertilizing capacity (Pursel et al., 1978). The precise action of OEP is unknown, but it appears to act by altering the composition of the egg yolk rather than the spermatozoal membranes (Pursel et al., 1978). Thawing appears to be the most sensitive step for cryopreserved boar spermatozoa (Courtens and Paquignon, 1985). Several thawing solutions, such as the Beltsville thawing solution (BTS; Pursel and Johnson, 1975a), have been developed to improve spermatozoal survival and conception rate (Paquignon, 1985).

The techniques that were successful for preserving bovine spermatozoa have not been adequate when applied to cryopreservation of stallion spermatozoa (Amann, 1984; Tischner, 1988). Recent developments of other complex media have been encouraging but have not yielded consistent fertility results. Many of the reports on the fertility of cryopreserved stallion spermatozoa are difficult, if not impossible, to interpret due to the lack of information on sperm numbers, inseminations per cycle, and the number of cycles during which the mares were inseminated (Amann, 1984; Tischner, 1988). Interpretation of data is a major problem in species where estrus is extended and multiple inseminations per cycle are sometimes required. In addition, a large variation appears to exist in the freezability of spermatozoa from different stallions (Loomis et al., 1984; Christanelli et al., 1984). A report comparing the fertility of fresh and cryopreserved spermatozoa from three stallions suggests that the fertility rate of frozen semen (56%) can approach that obtained for fresh samples (65%) (Christanelli et al., 1984). Alterations in the cryopreservation techniques may yield more acceptable fertility rates. Salazar-Valencia (1983) achieved a pregnancy rate of 66% in 204 mares with spermatozoa cryopreserved using a modified extender. Using another extender modification, Christanelli et al. (1984) successfully froze semen from 24 of 37 stallions. Briefly, the semen was extended in a medium containing 25% seminal plasma, 5% egg yolk, 2.5% glycerol, and 0.4% OEP, along with lactose, fructose, glucose, ethylenediamine tetraacetic acid (EDTA), sodium citrate, and sodium bicarbonate. After thawing, at least 50% of the spermatozoa were motile in 90% of the samples that underwent successful cryopreservation.

It appears to be necessary to remove the seminal plasma from goat semen before successful cryopreservation can be accomplished (Corteel, 1974). The washing of ejaculated caprine spermatozoa by centrifugation and resuspension before dilution in a milk–glucose–glycerol extender greatly enhances their survival (Corteel, 1974). The secretions of the bulbouretheral glands and seminal vesicles appear to inhibit both sperm motility and survival in milk-based extenders (Corteel, 1974).

There is very limited information on the fertility of cryopreserved dog spermatozoa (Amann, 1984). Although some degree of success has been reported in freezing canine semen (Seager and Fletcher, 1973), even the more recent data remain equivocal (Amann, 1984). Approximately half as many bitches became pregnant when inseminated with cryopreserved spermatozoa as compared to bitches inseminated with fresh spermatozoa from the same three males (Amann, 1984). It must be noted that the frozen samples contained only one-fourth the number of progressively motile spermatozoa.

The proper thawing of cryopreserved spermatozoa is as important as the freezing process if acceptable post-thaw recoveries of sperm motility and fertility are to be achieved (Mazur, 1963). The rate at which cryopreserved spermatozoa should be thawed depends on a number of factors. These include the size and shape of the container (ampule, straw, or pellet), the composition of the container (glass, plastic, etc.), the thawing medium (water, air), the temperatures used and the type of thawing procedure employed (Pickett, 1971; Salisbury et al., 1978).

Semen that has been processed and stored in glass ampules is best thawed by placing the cryopreserved samples in an ice bath (about 5°C) (Pickett, 1971). This slow thawing procedure, taking 8–10 min, appears to result in recovery of the highest percentage of progressively motile spermatozoa (Miller and VanDemark, 1954). Some

reports have indicated that thawing bovine semen in ampules in water maintained at 38–40°C was superior to that obtained with ice water (Blackshaw, 1955; Bruemmer et al., 1963). The warmer temperature, about 35°C, has been found to be the most practical method of thawing semen stored in 0.5-ml French straws (Cassou, 1972; Jondet, 1972), even though the optimum thawing procedure appears to be in water maintained at 75°C for 12 s (Aamdal and Andersen, 1968). Most of the recommended thawing procedures are consistent with the hypothesis that the faster the freezing process, the more rapidly the semen should be thawed (Salisbury et al., 1978). Thus, a small cylindrical container such as the straw, which freezes more rapidly due to a greater surface area, should be thawed rapidly.

Although some boar semen is frozen using the Hulsenberg straw (Westendorf et al., 1975), much is cryopreserved using the Beltsville pellet method, requiring a specialized thawing procedure (Pursel and Johnson, 1975b). The pelleted boar semen is thawed by rapidly dumping the frozen pellets into a heated thawing solution. This thawing procedure requires careful consideration of both temperature and thawing medium to achieve a rate of thawing that will result in acceptable post-thaw recoveries of the spermatozoa (Pursel and Park, 1985). It was found that superior post-thaw motilities were obtained by thawing pelleted boar semen in a solution heated to 50°C (Richter and Liedicke, 1972). This procedure apparently minimized the time that the spermatozoa had to spend at temperatures detrimental to their survival (Pursel and Park, 1985). Pursel and Johnson (1976) found that the length of time spermatozoa were subjected to these critical temperature ranges was decreased if the pelleted semen was poured into a dry styrofoam box for 3 min prior to placing the pellets in the 50°C thawing solution. Although the pelleted system is widely used for cryopreservation of boar spermatozoa, the application of maxi-straws for preservation of boar semen should increase once the optimum cooling rates, glycerol concentrations, and thawing procedures have been established (Corteel and Paquignon, 1984; Pursel and Park, 1985).

C. Semen Additives

Various additives have been incorporated into semen extenders in the hopes that they might enhance sperm preservation and fertility (Maule, 1962). The acrosomal enzyme hyaluronidase, which was one of the first additives tried, yielded inconsistent results (Maule, 1962). Other enzymes, including α- and β-amylase and β-glucuronidase, have resulted in enhancement of various laboratory tests but have not been shown to be effective in increasing fertility when rigorously tested (Kirton et al., 1968; Hafs et al., 1971; Salisbury et al., 1978). Recently, the addition of catalase to cryopreserved bull semen was found to be without benefit (Gaffer, 1988). Several other agents including a vitamin analog, thiamine propyl disulfide; a tranquilizer, cholorpromazine; and various hormonal preparations have been added to semen with limited success that has not often been repeatable (Salisbury et al., 1978).

Certain protective agents such as the chelator EDTA have been effectively utilized to enhance sperm survival and are constituents of many commonly used extenders (Salisbury et al., 1978; Paquignon, 1985). One of the most noteworthy additives has been the synthetic detergent, OEP. Several antioxidants, including α-tocopherol acetate, di-mercaprol sodium sulfonate, butyloxytoluol, and butyloxyanisol have been added to boar semen with reported increases in motility, survival, and fertilizing capacity (Kononov and Narizhnyi, 1982).

VI. Insemination Procedures

The success of artificial insemination depends on placing sufficient numbers of viable spermatozoa in the proper site at the optimum time relative to ovulation. Thus, all of the preceding preparation procedures such as collection, handling, and processing are fruitless unless the final phase of the process is properly carried out. This makes the condition and fertility status of the females, the detection of estrus, and the timing of insemination extremely important factors to any artificial insemination program. Several in-depth reviews on insemination procedures for the various domestic animals are available (Maule, 1962; Perry, 1968; Hafez, 1970, 1974; Salamon, 1976; Salisbury et al., 1978; Faulkner and Pineda, 1980; Evans and Maxwell, 1987).

Although semen is deposited in the vagina during natural mating in cattle, and there is no doubt that the first method used for artificial insemination was vaginal, it has been largely replaced by other more efficient methods. This is because a large number of spermatozoa are required to achieve reasonable fertility levels when semen is deposited vaginally (Salisbury et al., 1978). The requirement for substantial sperm numbers can be reduced by depositing the semen directly in the cervix with the aid of a speculum. This system, however, has been essentially replaced by the rectovaginal technique because the conception rates are about 10% higher with the latter method (Salisbury et *al.,* 1978). With this technique the inseminator uses a gloved hand to locate and manipulate the cervix through the rectal wall. The insemination tube is inserted into the vagina and guided into and through the cervical rings with the gloved hand to aid deposition of semen in the anterior cervix or vagina. Achievement of the optimum conception rates attainable using this method requires practice to reach the skill level required for repeatable success. This technique is more difficult to learn but has several advantages. These advantages are the higher conception rate and the need for less equipment, much of which is now disposable. It is also important to point out that manipulation of the cervix tends to simulate natural mating, which may be an additional benefit (VanDemark and Moeller, 1951; Salisbury *et al.,* 1978).

Because of the anatomical limitation of size, it is not possible, to effectively use the rectovaginal technique for sheep and goats. Instead, a speculum is used to aid deposition of semen in the vagina or, more preferably, in the cervix (Evans and Maxwell, 1987). A speculum is not normally used during vaginal insemination; instead introduction of the inseminating pipette is aided by the free hand and by gently twisting the pipette as it enters the vulvar orifice. The semen is simply deposited in the anterior portion of the vagina in hopes that sufficient numbers of spermatozoa will migrate into the cervix and reach the site of fertilization. The hindquarters of the ewe are normally elevated to facilitate the insemination process (Evans and Maxwell, 1987). Deposition of semen in the cervix or uterus is certainly the site of choice, at least for fresh semen (Gomes, 1977; Evans and Maxwell, 1987). Millions of sheep are routinely inseminated in many parts of the world, especially East-

ern Europe, using freshly collected, undiluted ram semen (Corteel and Paquignon, 1984). Short-term preservation of ram semen is becoming more common. Ram semen has been diluted and used effectively for a period of up to 12 h after collection (Barillet, 1983). Conception rates of 60% or greater have been attained using estrus-synchronized ewes (Barillet, 1983; Corteel and Paquignon, 1984). Acceptable fertility levels are attainable with cryopreserved ram spermatozoa only if intrauterine insemination is used (Evans and Maxwell, 1987). Thus, the cryopreservation process alters the ability of spermatozoa to ascend the female tract (Hafez, 1986; Evans and Maxwell, 1987). The problem can be bypassed by placing the thawed cryopreserved spermatozoa directly in the uterine lumen by aid of a laproscope (Evans and Maxwell, 1987). A summary of currently used insemination procedures for domestic ruminants is given in Table II.

The general procedure for artificially inseminating swine uses a spiral-tipped or cuffed inseminating catheter that is connected to a polyethylene bottle with a length of flexible tubing (Hafez, 1986). The semen is slowly expressed into the uterus by squeezing the bottle. This approach is necessary due to the volume of semen normally used for insemination of pigs.

The vaginocervical technique is the most common approach used to artificially inseminate the mare (Foote, 1980; Loomis *et al.,* 1984). Other insemination techniques have been used occasionally, including a rectocervical approach, deposition of semen into the uterus with the aid of a speculum, or direct placement of a gelatin capsule containing the semen in the uterus (Hafez, 1986). During the more commonly used vaginocervical insemination procedure, the index finger is

Table II
Summary of Insemination Procedures for Ruminant Animals

Domestic animal	Method and site	Insemination time	Estrus detection	Sperm numbers	Insemination volume
Bovine	Rectovaginal	Onset estrus + 9 h	Cycle history, observation, Kamar, marker bull	$10-15 \times 10^6$	0.25–1.0 ml
	Cervix & uterus (1st)				
Ovine	Speculum & pipette	Onset estrus + 10–20 h	Cycle history, vasectomized marker ram	$>50 \times 10^6$	0.05–0.2 ml
	Cervix & uterus (frozen)			Frozen 200×10^6	
Caprine	Speculum & pipette	Onset estrus + 12–36 h	Cycle history, vasectomized marker buck	$>50 \times 10^6$	0.1–0.2 ml
	Cervix & uterus (frozen)				

used to guide the catheter into the cervical orifice (Foote, 1980).

Insemination of the bitch is normally accomplished by depositing the semen in the anterior region of the vagina (Seager *et al.*, 1975; Seager and Platz, 1977a, 1977b; Faulkner and Pineda, 1980). This is necessary because it is almost impossible to catheterize the cervix due to the presence of a pseudocervix and the oblique angle of the cervical orifice (Faulkner and Pineda, 1980). A summary of insemination procedures for nonruminant domestic animals is provided in Table III.

A. Estrus and Insemination Time

One of the most important steps in AI is the introduction of fertile spermatozoa into the female tract at the optimum time necessary for fertilization to occur. The expected lifetimes of the male gametes and the oocyte are such that to achieve optimum fertilization capacitated spermatozoa should be present when the egg enters the ampulla of the oviduct. Thus, insemination must take place prior to the projected time of ovulation, and detection of estrus becomes imperative for the timing of insemination (Foote, 1980; Hafez, 1986). The cycling cow, sow, and caprine doe exhibit estrus at 21-day intervals, whereas estrus in the ewe occurs at 16- to 17-day intervals (Foote, 1980; Evans and Maxwell, 1987). A mare will cycle at about 24-day intervals during the spring months, but the length of the estrus tends to vary (Faulkner and Pineda, 1980). The duration of estrus shortens from about 7 to 5 days as the breeding season progresses (Ginther *et al.*, 1972). Both sheep and goats are seasonal and generally do not cycle during the spring (Foote, 1980; Evans and Maxwell, 1987). The bitch has an estrous cycle of approximately 6 months (Faulkner and

Table III
Summary of Insemination Procedures for Nonruminant Domestic Animals

Domestic animal	Method and site	Insemination time	Estrus detection	Sperm number	Insemination volume
Equine	Vaginal; cervix and uterus	Every 2nd day starting on Day 2 of estrus	Cycle history; observe heat signs	1.0×10^9	20–40 ml
Porcine	Pipette; cervix and uterus	Onset estrus + 10–30 hr	Cycle history; observe heat signs; boar	5.0×10^9	50 ml
Canine	Speculum and pipette; anterior vagina	Onset of estrus + 1 and 2 day	Cycle history; vasectomized dog	$>50 \times 10^6$	0.5–1.0 ml
Turkey	Pipette; venting, midvagina	Every 3 wk	N/A	3.5×10^6	0.1–0.5 ml
Chicken	Pipette; venting, midvagina	Weekly, afternoon	N/A	1.0×10^8	0.1–0.5 ml

Pineda, 1980). It should be noted that careful records of the estrous cycles of the female population to be inseminated can be a immense help in determining the proper time for insemination.

Estrus in the cow is identified by restlessness, bawling, and attempting to mount other animals or standing for other animals to mount (Britt, 1977; Hafez, 1986). The cow, which normally ovulates 8–10 h after the end of an 18- to 24-h estrous period, should be inseminated toward the end of estrus. Estrus is detected by observing the behavior changes in the cow. A variety of aids are available for detection of estrus (Faulkner and Pineda, 1980; Foote, 1980). One accepted practice is the "AM/PM rule," where cows noted in heat in the morning are inseminated that afternoon and those identified in the afternoon are inseminated the next morning. This approach works well with dairy cattle, but the close scrutiny required is not easily adapted to beef cattle operations.

The ewe exhibits few behavioral characteristics identifying estrus and ovulates toward the end of a 24- to 42-h estrus. Thus, ewes should be inseminated 8–10 h after the onset of estrus. This is because ovulation would normally occur within 25–30 h after estrus begins (Evans and Maxwell, 1987). Sheep AI programs can be enhanced greatly by using a vasectomized ram to detect estrus or by implementing an estrus synchronization program. Seasonal differences in sheep breeds can make this process even more difficult (Hafez, 1986). The caprine doe exhibits a marked seasonality in her estrous cycle (Perry, 1968; Sherman, 1984). The doe tends to ovulate some 20–36 h after the beginning of estrus and should be inseminated 12–18 h after the onset (Foote, 1980; Evans and Maxwell, 1987). Estrus in the doe is difficult to detect in that the behavioral signs are minimally displayed.

Two to three days before the onset of estrus, the sow's vulva swells and becomes reddened (Faulkner and Pineda, 1980; Hafez, 1986). Detection of estrus can be augmented by the use of vasectomized boars, because the sow responds well to the presence of males (Faulkner and Pineda, 1980). The sow exhibits estrus for about 3 days and should be inseminated 30–36 h after the onset of estrus, since ovulation normally occurs 40–45 h after the begining of the estrous period (Dziuk, 1970). Insemination during the latter part of the second day results in the highest conception rate (Pursel and Johnson, 1975a).

The mare exhibits marked behavioral changes at the onset of estrus (Faulkner and Pineda, 1980). The process of detecting estrus, however, should be done only by trained personnel using a routine program of teasing. The behavior of most estrus mares is such that trained observers readily identify changes that are characteristic of estrus including evertion of the vulvar lips ("winking") and extrusion of clear copious fluid from the vagina (Faulkner and Pineda, 1980). Such responses are readily evoked by a teaser stallion when the mare is in estrus. The ideal teaser is a small gentle stallion with a high sex drive. The time of ovulation tends to vary to the extent that the time of insemination can be determined by periodic examination of the ovaries by palpation. Without the aid of palpation the mare should be inseminated at least on the begining of the fourth day of estrus (Faulkner and Pineda, 1980). If handling permits, the mare should be inseminated every second day beginning with the second day of estrus

until she stops responding to teasing (Foote, 1980). The time of ovulation is somewhat irregular, but estrus normally ceases a day or two following ovulation (Faulkner and Pineda, 1980).

The first identifiable sign of an impending estrus in the bitch is the bloody vaginal discharge that occurs at proestrus (Faulkner and Pineda, 1980). The first reliable sign of estrus, however, is acceptance of the male for mating. The most reliable method of AI for the bitch is to use a vasectomized male to detect estrus and to inseminate on the first and third days of estrus (Faulkner and Pineda, 1980).

Many avian species can be readily inseminated with fresh undiluted (neat) semen with achievement of reasonable fertility levels (Lake, 1962), provided the insemination is accomplished within 30 min. Insemination is accomplished with the aid of an assistant to evert the cloaca to reveal the oviduct. A syringe is used to deposit semen at a depth in the oviduct of approximately 3 cm. The chicken should be inseminated at weekly intervals, whereas the turkey requires insemination at 2- to 3-week intervals (Lake, 1962; Faulkner and Pineda, 1980; Hafez, 1986). Hens should be inseminated in late afternoon to avoid the presence of a hard-shelled egg in the uterus.

B. Anatomical Structure, Sperm Numbers, and Transport

Successful fertilization requires spermatozoa to reach the ampullar region of the oviduct at the proper time relative to ovulation. The mechanisms by which spermatozoa move from the site of insemination to the site of fertilization are complex and interrelated. Successful insemination practices must take into consideration the potential barriers imposed by the structures of the female tract. Although uterine deposition is favored for most species, the available practical options may be severely limited (Faulkner and Pineda, 1980; Hafez, 1986; Evans and Maxwell, 1987). For instance, it is almost impossible to deposit semen in the uterus of maiden ewes or bitches, whereas uterine deposition is a routine for cows, mares, and sows. The number of spermatozoa required for an acceptable level of fertility is influenced by the site of semen deposition. For any given species, far fewer spermatozoa are required for most uterine inseminations than for situations where the semen must be deposited in the vagina.

The common methods of insemination, the site of sperm deposition, the number of spermatozoa required, and the normal insemination volume used for most common domestic livestock are given in Table I.

C. Estrous-Cycle–Synchronized Females

Synchronization of the estrous cycle can greatly simplify AI for animals that are not routinely observed on a daily basis. Two principal pharmacological methods have been successful for synchronizing estrus in domestic livestock. These successful methods use either progestational agents or prostaglandins (Day, 1984; Roche and Ireland, 1984). Although effective hormonal products for cattle have been commercially available since the 1960s, widespread application of these products has not occurred. Significant progress was made by development of sustained-release implants containing synthetic progestogen. These commercially available implants were used with a single injection of estradiol valerate and acceptable conception rates were achieved (Wiltbank *et al.*, 1971). The field was advanced greatly by

discovery of the luteolytic properties of the prostaglandin $F_{2\alpha}$ ($PGF_{2\alpha}$) and its subsequent chemical synthesis (Hafs *et al.*, 1974). Two injections of $PGF_{2\alpha}$ at an interval of 11 days has been used effectively to synchronize groups of cows (Shultz, 1987). The animals, however, must be cycling for the luteolytic action of $PGF_{2\alpha}$ to be effective (King *et al.*, 1982). This basic approach can be augmented somewhat by the use of luteinizing hormone or gonadotropin-releasing hormone (Sequin, 1984). Although synchronization agents can be cost-effective when used as management tools to simplify insemination procedures, it must be noted that groups of pregnancy-synchronized animals, if not properly planned for, can become a major problem at calving time.

Just before the onset of the breeding season ewes can be at least partially synchronized by the introduction of sterile rams into groups of ewes that had been isolated from males for a few weeks (Hafez, 1986; Evans and Maxwell, 1987). This synchronization approach also has been effective with goats. (Evans and Maxwell, 1987). Ewes can be synchronized pharmacologically using a combination of sustained-release progestogen-containing intravaginal pessaries (sponges) (Robinson, 1964) and timed injections of gonadotropins to induce ovulations (Gordon, 1975; Evans and Maxwell, 1987). It should be noted that this synchronization regimen is more effective with cycling ewes than during the anestrus period (Hafez, 1986). Both cycling ewes and caprine does respond well to injections of $PGF_{2\alpha}$ (Evans and Maxwell, 1987). The injections must be given at intervals of 10 to 14 days because the corpus luteum is sensitive to $PGF_{2\alpha}$ only between days 5 and 14 for the ewe and days 6 and 17 for the doe (Haresign, 1978; Evans and Maxwell, 1987) Prostaglandin treatments are not effective in the anestrus ewe (Day, 1984).

The prostaglandins have not proved to be effective for regulating the estrous cycle of the pig (Diehl and Day, 1974). In swine, however, estrus can be synchronized effectively using the progestational agent, altrenogest (Day, 1984). This synthetic steroid can be used to synchronize estrus to a 4-day period in approximately 95% of gilts that have exhibited a previous estrus, by feeding 15–20 mg of altrenogest daily for 18 days (Day, 1984). Altrenogest also has been used to synchronize estrus in mares during the breeding season (Christanelli et al., 1984).

VII. Fertility Proficiency

The success of any AI program is maintaining a satisfactory fertility level. If a large portion of the females within a population fails to conceive, or if the conceptus is lost during gestation, or if neonatal death occurs, then the economic and genetic advantages of AI are lost. The loss of milk production in individual dairy animals that fail to breed also constitutes an economic loss. The effect of a number of factors must be considered in evaluating an AI program. First, it must be economically sound. This depends on the reproductive efficiency of the herd or flock. Several measures of reproductive efficiency are useful, including services per conception, calving rate, and nonreturn rate (Salisbury et al., 1978). These measures were developed mainly for cattle, but can be applied, for the most part, to other domestic animals with the exception of the pig, where litter size is extremely important. In addition, the effect of season

must be considered because most domestic animals are influenced, at least to some degree, by seasonal changes (Foote, 1988).

A. Fertility Estimates

Pregnancy in the cow is an all-or-none situation, and variability of the fertility potential of individual females within a population makes estimation of the success of AI or the fertility of a particular semen sample very difficult. The exclusion of sterile females when calculating services per conception is especially important (Salisbury et al., 1978). The average number of services per conception is approximately 1.5 for dairy cattle (Foote and Hall, 1954). This measure has not been widely used because it can be affected markedly by the number of services permitted for unidentified sterile individuals (Salisbury et al., 1978).

The detection of pregnancy by palpation is a common means of evaluating reproductive efficiency in dairy cattle (Sequin, 1981) and to some degree in beef cattle (Faulkner and Pineda, 1980). This approach is becoming common but the economical and logistical difficulties can be limiting. Calving interval is used also as a measure of reproductive efficiency (Salisbury et al., 1978). This herd characteristic is an extremely important economic factor and can be a useful means of monitoring reproductive efficiency (Salisbury et al., 1978).

Pregnancy can be detected in the ewe by assaying for pregnancy-associated antigens and hormones (Hafez, 1986) or, more commonly, by ultrasound to detect the presence of developing fetuses (Taverne and de Bois, 1984; Langford et al., 1984; Hafez, 1986; Arnesson et al.,1988). The increasing use of ultrasound should make early diagnosis more practical. It has been used effectively for monitoring pregnancy in mares (Ginther, 1984).

The only valid estimate of fertility is production of live offspring. This measure of reproductive efficiency is commonly used in large beef herds and is expressed as calving rate (Salisbury *et al.*, 1978). A consideration of the number of sterile cows in the herd should be made. An approximation of 3% has been used.

The most common measure of the success of AI is nonreturn rate (Salisbury *et al.*, 1978) This estimate of reproductive efficiency is the percentage of animals that are inseminated during a given period, usually 1 month, and that have not been recorded as having been bred again within a specified number of days, usually 60–90 days. The ranges in days may vary (30–60, 60–90, 150–180) depending on the system used by the organization. The assumption that if a cow does not return for service she is pregnant is not always valid. Animals that have been sold or died are recorded as having conceived to the insemination. Such data are only useful for estimating the reproductive efficiency of large populations. The limitations make the use of nonreturn information for estimating the fertility of individual bulls, breeds, inseminators, or semen-producing organizations a questionable approach (Salisbury *et al.*, 1978; Foote, 1988). Follow-up measures of successful conception such as verification of pregnancy by palpation can be useful if precise measures are needed. A summary of the fertility proficiency of domestic livestock is provided in Table IV.

B. Field Practice Factors

Several factors relative to the success of AI in the field must be considered. Early in the development of AI in cattle, most of the inseminations were carried out by trained technicians (Foote, 1980). This, however, has changed, and an increasing number of

Table IV
Fertility Proficiency of Domestic Livestock[a]

Animal	Pregnancy rates			Fertility estimate, nonreturn rates (%)	Sterility estimate, barren animals
	Natural mating (%)	After one insemination, fluid semen (%)	After one insemination, frozen semen (%)		
Dairy cattle	60–75	50–65	50–65	70–85	8–9
Beef cattle	70	60	—	—	15–20
Sheep	80–90	60–80	3–71	65	6–8
Goats	70–80	42–69	27–67	—	8–9
Swine	85–95	65–90	62–64	90–95	6–7
Horses	40–75	50–65	38–46	—	25–60

[a]Adapted from Amann (1984), Rasbech (1984), and Evans and Maxwell (1987).

inseminations is done by the owner or herdsman. The obvious result is a lowering of overall conception rate as more moderately trained individuals carry out the insemination process.

References

Aamdal, J. (1960). *In* "The Artificial Insemination of Farm Animals" (E. J. Perry, ed.), 3rd ed., p. 264. Rutgers University Press, New Brunswick, New Jersey.

Aamdal, J., and Andersen, K. (1968). *Proc. 6th Intern. Congr. Anim. Reprod. Artifical Insemination, Paris* **2,** 973–976.

Adams, C. E. (1962). *In* "The Semen of Animals and Artificial Insemination" (J. P. Maule, ed.), p. 318. Commonwealth Agricultural Bureau, Farnham Royal, England.

Almquist, J. O. (1949a). *J. Dairy Sci.* **32,** 722.

Almquist, J. O. (1949b). *J. Dairy Sci.* **32,** 950.

Amann, R. P. (1984). *Proc. 10th Intern. Congr. Anim. Reprod. Artificial Insemination, Urbana,* **IV,** II28–II36,

Amann, R. P. (1988). *Proc. 12th Tech. Conf. Artificial Insemination Reprod., NAAB,* Milwaukee, pp. 38–44.

Amann, R. P., and Seidel, G. E. (eds). (1982). "Prospects for Sexing Mammalian Sperm." Colorado University Press, Boulder.

Arnesson, P., Blomgren, E., Rosman, E., and Gustafsson, H. (1988). *Proc. 11th Intern. Congr. Anim. Reprod. Artificial Insemination, Dublin* **3,** 226.

Ball, L. (1976). *In* "Applied Electronics for Veterinary Medicine and Animal Physiology" (W. R. Klemm, ed.), pp. 394–441. Charles C. Thomas, Springfield, Illinois.

Ballachey, B. E., Evenson, D. P., and Saacke, R. G. (1988). *J. Androl.* **9,** 109.

Ballachey, B. E., Hohenboken, W. D., and Evenson, D. P. (1987). *Biol. Reprod.* **36,** 915.

Barillet, F. (1983). "Insémination Artificielle et Améliorstions Génétique: Bilan et Perspectives Critiques," Toulose-Auzeville, France.

Beatty, R. A., Bennet, G. H., Hall, J. G., Hancock, J. L., and Stewart, D. L. (1969). *J. Reprod. Fertil.* **19,** 491.

Bedford, J. B., and Overstreet, J. W. (1972). *J. Reprod. Fertil.* **31,** 407.

Betzen, K. M. (1985). M. S. Thesis, "Techniques for Electrical Semen Collection from Birds," Oklahoma State University, Stillwater.

Blackshaw, A. W. (1955). *Australian Vet. J.* **31,** 238.

Brandriff, B. F., Gordon, L. A., Haendel, S., Singe, S., Moore, D. H., II, and Gledhill, B. L. (1986). *Fertl. Steril.* **46,** 678.

Britt, J. H. (1977). *J. Dairy Sci.* **60,** 1345.

Bruemmer, J. H., Eddy, R. W., and Duryea, W. J. (1963). *J. Cellular Comp. Physiol.* **62,** 113.

Budworth P. R., Amann, R. P., and Chapman, P. L. (1988). *J. Androl.* **9,** 41–54.

Budworth P. R., Amann, R. P., and Hammerstedt, R. H. (1987). *J. Dairy Sci.* **70,** 1927–1936.

Cassou, R. (1968). *Proc. 6th Intern. Congr. Anim. Reprod. Artificial Insemination, Paris* **II,** 1009.

Cassou, R. (1972). *Proc. 7th Intern. Congr. Anim. Reprod. Artificial Insemination, Munich, Vol. II,* pp. 1421–1425.

Choong, C. H., and Wales, R. G. (1962). *Aust. J. Biol. Sci.* **15,** 543.

Christanelli, J. J., Squires, E. L., Amann, R. P., and Pickett, B. W. (1984). *Theriogenlogy* **22,** 39.

Corteel, J. M. (1974). *Ann. Biol. Anim. Biophys.* **14,** 741.

Corteel, J. M., and Paquignon, M. (1984). *Proc. 10th Intern. Congr. Anim. Reprod. Artificial Insemination, Urbana,* **VI,** II20–II27.

Courtens, J. L., and Paquignon, M. (1985). *In* "Deep Freezing of Boar Semen" (L. A. Johnson and K. Larsson, eds.), pp. 61–87. Swedish University of Agricultural Sciences, Uppsala.

Curtiss, S. E. (ed.) (1988). "Guide for the Care and Use of Agricultural Animals in Agricultural Research and Teaching," 1st ed. Consortium, Champaign, Illinois.

Czarnetzky, E. J., and Henle, W. (1938). *Proc. Soc. Exp. Biol. Med.* **38,** 63.

Day, B. N. (1984). *Proc. 10th Intern. Congr. Anim. Reprod. Artificial Insemination, Urbana, Illinois* **VI,** IV1–IV8.

Diehl, J. R., and Day, B. N. (1974). *J. Anim. Sci.* **39,** 392.

Dziuk, P. (1970). *J. Reprod. Fertil.* **22,** 277.

Elliott, F. I. (1967). *Proc. 20th Ann. Conv. Natl. Assoc. Artificial Breeders, Hershey, Pennsylvania,* p. 109.

Ericsson, S. A., Garner, D. L., Redelman, D., and Ahmad, K. (1989). *Gamete Res.* **22** 355.

Evans, G., and Maxwell, W. M. C. (1987). "*Salamon's Artificial Insemination of Sheep and Goats.*" Butterworths, Sydney.

Evenson, D. P., Darzynkiewicz, Z., and Melamed, M. R. (1982). *J. Histochem Cytochem.* **30,** 279.

Faulkner, L. C., and Pineda, M. H. (1980). *In* "Veterinary Endocrinology and Reproduction" (L. E. McDonald, ed.), p. 330–366. Lea and Febiger, Philadelphia.

Flipse, R. J., Patton, S., and Almquist, J. O. (1954). *J. Dairy Sci.* **32,** 1205.

Foote, R. H. (1968). *In* "Current Veterinary Therapy" (R. W. Kirk, ed.), p. 686. Saunders, Philadelphia.

Foote, R. H. (1969). *In* "Reproduction in Domestic Animals" (H. H. Cole and P. T. Cupps, ed.), 2nd ed., p. 313. Academic Press, New York.

Foote, R. H. (1974). *In* "Reproduction in Farm Animals" (E. S. E. Hafez, ed.), 3rd ed., p. 409. Lea and Febiger, Philadelphia.

Foote, R. H. (1978). *Proc. 7th Tech. Conf. Artificial Insemination Reprod., NAAB,* Madison, pp. 55–61.

Foote, R. H. (1980). *In* "Reproduction in Farm Animals" (E. S. E. Hafez, ed.), Lea and Febiger, Philadelphia.

Foote, R. H. (1982). *J Androl.* **3,** 85.

Foote, R. H. (1988). *Proc. 11th Intern. Congr. Anim. Reprod. Artificial Insemination, Dublin* **V,** 126.

Foote, R. H., and Berndtson, W. E. (1976). *Proc. 6th Tech. Conf. Artificial Insemination and Reproduction., NAAB,* Milwaukee, pp. 23–30.

Foote, R. H., and Hall, A. C. (1954). *J. Dairy Sci.* **37,** 673.

Foote, R. H., Hauser, E. R., and Casida, L. E. (1960). *J. Anim. Sci.* **19,** 238.

Foote, R. H., and Salisbury, G. W. (1948). *J. Dairy Sci.* **31,** 763.

Gaffer, T. (1988). *Proc. 11th Intern. Congr. Anim. Reprod. Artificial Insemination, Dublin,* **3,** 46.

Garner, D. L. (1984a). *Proc. 10th Tech. Conf. Artificial Insemination Reprod., NAAB,* Milwaukee, pp. 87–92.

Garner, D. L. (1984b). *Proc. 10th Intern. Congr. Anim. Reprod. Artificial Insemination, Urbana, Illinois* **IV,** X9-X15.

Garner, D. L., Gledhill, B. L., Pinkel, D., Lake, S., Stephenson, D., Van Dilla, M. A., and Johnson, L. A. (1983) *Biol. Reprod.* **28,** 312.

Garner, D. L., Pinkel, D., Johnson, L. A., and Pace. M. M. (1986). *Biol. Reprod.* **34** 127.

Getty, S. M., and Ellis, D. J. (1967). *J. Am. Vet. Med. Assoc.* **151,** 1688.

Ginther, O. J. (1984). *Theriogenology* **21,** 633-644.

Ginther, O. J., Whitmore, H. L., and Squires, E. L. (1972). *Am. J. Vet. Res.* **33,** 1935.

Gledhill, B. L., Lake, S., Steinmetz, L. L., Gray, J. W., Crawford, J. R., Dean, P. N., and Van Dilla, M. A. (1976). *J. Cell. Physiol.* **87,** 367.

Gomes, W. R. (1977). *In* "Reproduction in Domestic Animals" (H. H. Cole and P. T. Cupps, ed.), 3rd ed., p. 257. Academic Press, New York.

Gordon, I. (1975). *Anna. Biol. Anim. Biochim. Biophys.* **15,** 303.

Graham, E. F. (1978). *In* "The Integrity of Frozen Spermatozoa," Proc. Conf. Nat. Acad. Sci. p. 4. National Academy of Science, Washington, D. C.

Graham E. F., Rajamannan, A. H. J., Schemehl, M. K. L., Maki-Larila, M., and Bower, R. S. (1971). *Artif. Insem. Dig.* **19,** 12.

Graham, E. F., Schmehl, M. LO., and Deyo, R. C. M. (1984). *Proc. 10th Tech Conf. Artificial Insemination Reprod., NAAB,* Milwaukee, pp. 4–29.

Gunsalus, I. C., Salisbury, G. W., and Willet, E. L. (1941). *J. Dairy Sci.* **24,** 911.

Haeger, O., and Mackle, N. (1971). *Dtsch. Tuierarztl. Wochenschr.* **27,** 345–397.

Hafez, E. S. E. (ed.). (1970). "Reproduction and Breeding Techniques for Labortary Animals." Lea and Febiger, Philadelphia.

Hafez, E. S. E. (ed.). (1974). "Reproduction in Farm Animals," 3rd ed. Lea and Febiger, Philadelphia.

Hafez, E. S. E. (ed.). (1986). "Reproduction in Farm Animals," 5th ed. Lea and Febiger, Philadelphia.

Hafs, H. D., Boyd, L. J., Cameron, S., Johnson, W. L., and Hunter, A. G. (1971). *J. Dairy Sci.* **54,** 420.

Hafs, H. D., Knisely, R. C., and Desjardins, C. (1962). *J. Dairy Sci.* **45,** 788.

Hafs, H. D., Louis, T. M., and Noden, P. A. (1974). *J. Anim. Sci.* **38,** Suppl.2, 10.

Haresign, W. (1978). *In.* "Control of Ovulation" (D. B. Crighton, G. R. Foxcroft, N. B. Haynes, and G. E. Lamming, eds.). Butterworths, London.

Hudson, R. S. (1972). Symp. Reprod. in Cattle and Horses, Am. Vet. Soc. for Study of Breeding Soundness, American College of Theriogenology, Mimeo, Hastings.

Hunter, W. U. (1968). *Proc. 6th Intern. Congr. Anim. Reprod. Artificial Insemination, Paris* **II,** 1053-1056.

Johnson, L. A., Flook, J. P., and Look, M. V. (1987). *Gamete Res.* **17,** 203.

Johnson, L. A., Flook, J. P., and Hawk, H. W. (1989). *Biol. Reprod.* **41,** 199.

Jondet, R. (1972). *Proc. 7th Intern. Congr. Anim. Reprod. Artificial Insemination, Munich* **II,** 1371–1382.

Jones, R. C. (1973). *Nature (Lond.)* **243,** 5401.

Katz, D. F., and Davis, R. O. (1987). *J. Androl.* **8,** 170.

Katz, D. F., Davis, R. O., Delandmeter, B. A., and Overstreet, J. W. (1985). *Comp. Methods Prog. Biomed.* **21,** 173.

Kimberling, C. V., Arnold, K. S., Schweitzer, D. J., Jones, R. L., VonByern, H., and Lucas, M. (1986). *J. Am. Vet. Med. Assoc.* **189,** 73.

King, G. J., and Macpherson, J. W. (1973). *J. Anim. Sci.* **36,** 563.

King, M. E., Kiracofe, G. H., Stevenson, J. S., and Schalles, R. R. (1982). *Theriogenology* **18,** 191.

Kirton, K. T., Boyd, L. J., and Hafs, H. D. (1968). *J. Dairy Sci.* **51,** 1426.

Komarek, R. J., Pickett, B. W., Gibson, E. W., and Lanz, R. N. (1965). *J. Reprod. Fertil.* **10,** 337.

Kononov, V. P., and Narizhnyi, A. G. (1982). *Vestnik. Sel' Shokhozyaistvennoi Navki* **5** 123.

Kremer, J. (1965). *Int. J. Fertil.* **10,** 209.

Kramer, J., and Jager, S. (1976). *Fertil. Steril.* **27,** 335.

Lake, P. E. (1962). *In* "The Semen of Animals and Artificial Insemination" (J. P. Maule, ed.), p. 338. Commonwealth Agricultural Bureau, Farnham Royal, England.

Lambert, W. V., and McKenzie, F. F. (1940). Artificial Insemination in Livestock Breeding. USDA Circular, pp. 567.

Langford, G. A., Shrestha, J. N. B., Fiser, P. S., Ainsworthy, L., Heaney, P. P., and Marcus, G. J. (1984). *Theriogenology* **21,** 691.

Lessley, B. A., and Garner, D. L. (1983). *Gamete Res.* **7,** 49.

Loomis, P. R., Amann, R. P., Squires, E. L., and Pickett, B. W. (1984). *J. Anim. Sci.* **56,** 687.

Lorton, S. P., and First, N. L. (1977). *Fertil. Steril.* **28,** 1295-1300.

Lorton, S. P., Sullivan, J. J., Bean, B., Kaprotl, M., Kellgren, M., and Marshall, C. (1988a). *Theriogenology* **29,** 593.

Lorton, S. P., Sullivan, J. J., Bean, B., Kaprotl, M., Kellgren, H., and Marshall, G. (1988b). *Theriogenology* **29,** 609.

Mann, T. (1964). "The Biochemistry of Semen and the Male Reproductive Tract." Menhuen, London.

Mann, T., and Lutwak-Mann, C. (1981). "Male Reproductive Function and Semen." Springer-Verlag, New York.

Martin, R. H., Liu, C. C., Balkan, W., and Burns, K. (1982). *Am. J. Hum. Genet.* **34,** 454.

Maule, J. P. (ed.). (1962). "The Semen of Animals and Artificial Insemination." Commonwealth Agricultural Bureaux, Farnham Royal, England.

Mayer, D. T., and Lasley, J. R. (1945). *J. Anim. Sci.* **4,** 261–269.

Mazur, P. (1963). *J. Gen. Physiol.* **47,** 347.

Miller, W. N., and Van Demark, N. L. (1954). *J. Dairy Sci.* **37,** 45.

Morrow, D. A. (1980). "Current Therapy in Theriogenology: Diagnosis, Treatment and Presentation of Reproductive Disorders in Ovine," p. 330. W. B. Saunders, Philadelphia.

Moruzzi, J. F., Wyrobek, A. J., Mayall, B. H., Gordon, L. A., and Gledhill, B. L. (1985). Proc. Intern. Conf., Analytical Cytology XI, Hilton Head, Hawaii, Abstr. 458.

New Zealand Dairy Board, 48th Farm Production Report. (1971–1972).

Nishikawa, Y. (1959). "Studies on Reproduction in Horses." Japan Racing Association, Toyko.

Nishikawa, Y. (1962). *Kyoto Univ. English Bull.* **2,** 43.

Nishikawa, Y. (1964). *Proc. 5th Intern. Congr. Anim. Reprod. Artificial Insemination, Trento* **7,** 162.

Norman, C., Goldberg, E., and Porterfield, I. D. (1962). *Exp. Cell Res.* **28,** 69.

O'Conner, M. T., Amann, R. P., and Saacke, R. G. (1982). *J. Anim. Sci.* **53,** 1368.

Paquignon, M. (1985). *In* "Deep Freezing of Boar Semen" (L. A. Johnson, and K. Larsson, eds.), pp. 129–145. Swedish University Agricultural Sciences, Uppsala.

Paquignon, M., and Courot, M. (1976). *Proc. 8th Intern. Congr. Anim. Reprod. Artificial Insemination, Kracow,* pp. 1041–1044.

Parkes, A. S. (ed.). (1960). "Marshall's Physiology of Reproduction," Vol. 1, Part 2. Longmans, Green, New York.

Perry, E. J. (ed.). (1968). "The Artificial Insemination of Farm Animals," 4th ed. Rutgers University Press, New Brunswick, New Jersey.

Pickett, G. W. (1971). *Artif. Insem. Dig. 19,* 8.

Pickett, B. W., and Back, D. G. (1973). Information Series No. 2-1, Animal Reproduction Lab., Colorado State University, Ft. Collins.

Pickett, B. W., Back, D. G., Burwash, L. D., and Voss, J. L. (1974). *Proc. 5th Tech. Conf. Artificial Insemination Reprod., NAAB,* Chicago, p. 47.

Pickett, B. W., and Berndtson, W. E. (1974). *J. Dairy Sci.* **57,** 1287.

Pinkel, D., Lake, S., Gledhill, B. L., Stephenson, D., and Van Dilla, M. A. (1983). *Cytometry* **3,** 1.

Polge, C. (1956). *Vet. Rec.* **68,** 62.

Polge, C., Smith, A. U., and Parkes, A. S. (1949). *Nature* **164,** 666.

Polge, C., Salamon, S., and Wilmut, I. (1970). *Vet. Rec.* **87,** 424.

Pursel, V. G., and Johnson, L. A. (1975a). *J. Anim. Sci.* **40,** 99.

Pursel, V. G., and Johnson, L. A. (1975b). *J. Anim. Sci.* **41,** 374.

Pursel, V. G., and Johnson, L. A. (1976). *J. Anim. Sci.* **42,** 927.

Pursel, V. G., Schulman, L. L., and Johnson, L. A. (1978). *J. Anim. Sci.* **47,** 190.

Pursel, V. G., and Park, C. S. (1985). *In* "Deep Freezing Boar Semen" (L. A. Johnson and K. Larsson, eds.), pp. 147–166. Swedish University of Agriculture, Uppsala.

Quinn, P. J., Salamon, S., and White, I. G. (1968). *Aust. J. Agric. Sci.* **19,** 119.

Rasbech, N. O. (1984). *In* "The Male in Farm Animal Reproduction" (M. Courot, ed.), pp. 2–23. Martinus Nijhoff, Boston.

Richter, L., and Liedicke, A. (1972). *Proc. 7th Intern. Congr. Anim. Reprod. Artificial Insemination, Munich* **2,** 1617.

Riek, P. M., Pickett, B. W., and Creighton, K. A. (1980). *Proc. 8th Tech. Conf. Artificial Insemination Reprod., NAAB,* pp. 67–70.

Roberts, S. J. (1971). "Veterinary Obstetrics and Genital Diseases," 2nd ed., S. J. Roberts, Ithaca, New York.

Robinson, T. J. (1964). *Proc. Aust. Soc. Anim. Prod.* **5,** 47.

Roche, J. R., and Ireland, J. J. (1984). *Proc. 10th Intern. Congr. Anim. Reprod. Artificial Insemination, Urbana* **IV,** IV9–IV17.

Saacke, R. G. (1972a). *Proc. 4th Tech. Conf. Artificial Insemination Reprod., NAAB,* Chicago, p. 17.

Saacke, R. G. (1972b). *Proc. 4th Tech. Conf. Artificial Insemination Reprod., NAAB,* Chicago, p. 22.

Saacke, R. G. (1978). *Proc. 7th Tech. Conf. Artificial Insemination Reprod., NAAB,* Madison, p. 3.

Saacke, R. G. (1974). *Proc. 8th Conf. Artificial Insemination in Beef Cattle, NAAB,* Denver, p. 11.

Saacke, R. G., Marshall, C. E., Venson, W. E., O'Conner, M. L., Chandler, J. E., Mullins, J., Amann, R. P., Wallace, R. A., Vincel, W. N., and Kellgren, H. C. (1980a). *Proc. 9th Int. Congr. Anim. Reprod. Artificial Insemination, Madrid* **5,** 75.

Saacke, R. G., Nebel, R. I., Karabinus, D. S., Bame, J. H., and Mullins, J. (1988). *Proc. 12th Tech. Conf. Artificial Insemination Reprod., NAAB,* Milwaukee, pp. 7–14.

Saacke, R. G., Nebel, R. I., Pennington, D., and Miller, R. (1982). *Proc. 9th Tech. Conf. Artificial Insemination Reprod., NAAB,* Milwaukee, pp. 63–67.

Saacke, R. G., Venson, W. E., O'Conner, M. L., Chandler, J. E., Mullins, J., Amann, R. P., Marshall, C. E., Wallace, R. A., Vincel, W. N., and Kellgren, H. C. (1980b). *Proc. 8th Tech. Conf. Artificial Insemination Reprod., NAAB,* pp. 71–78.

Salazar-Valencia, F. (1983). *Theriogenology* **19,** 146.

Salamon, S. (1976). "Artificial Insemination of Sheep." University of Sidney, New South Wales, Australia.

Salisbury, G. W., and VanDemark, N. L. (1961). "Physiology of Reproduction and Artificial Insemination of Cattle." W. H. Freeman, San Francisco.

Salisbury, G. W., VanDemark, N. L., and Lodge, J. R. (1978). "Physiology of Reproduction and Artificial Insemination of Cattle," 2nd ed. W. H. Freeman, San Francisco.

Seager, S. W. J., and Fletcher, W. S. (1973). *Vet. Rec.* **92,** 6.

Seager, S. W. J., and Platz, C. C. (1977a). *Vet. Clin. North Am.* **7,** 757.

Seager, S. W. J., and Platz, C. C. (1977b). *Vet. Clin. North Am.* **7,** 765.

Seager, S. W. J., Platz, C. C., and Fletcher, W. S. (1975). *J. Reprod. Fertil.* **45,** 189.

Sequin, B. E. (1981). *Comp. C. E. Prac. Vet.* 3, S445.

Sequin, B. E. (1984). *Proc. 10th Intern. Congr. Anim. Reprod. Artificial Insemination, Urbana* **1V,** IV25–IV30.

Shannon, P. (1972). *Proc. 7th Intern. Congr. Anim. Reprod. Artificial Insemination, Munich* **11,** 1441.

Sherman, D. M. (1984). *Dairy Goat J.* **62,** 75.

Shin, S. J., Lein, D. H., Patten, V. H., and Ruhnke, H. L. (1987). *Theriogenology* **29,** 557.

Shultz, R. H. (1987). Synchronization of Estrus, *In* "Cow Manual," pp. 104–113. Society for Theriogenology, Hastings.

Sorensen, E. (1938). Den. Kgl. Veterinaer-Og Landbohojskole Aarlskrift, pp. 83–146. Copenhagen.

Stephens. D. T., and Hoskins, D. D. (1986). *Biol. Reprod.* **34,** (Suppl. 1), 189.

Sullivan, J. J., Elliott, F. I., Barlett, D. E., Murphy, D. M., and Kuzdas, D. C. (1966). *J. Dairy Sci.* **49,** 1569.

Taverne, M. A. M., and de Bois, C. J. W. (1984). *Proc. 10th Int. Congr. Anim. Reprod. Artificial Insemination* **II,** 113d.

Thadani, V. M. (1980). *J. Exp. Zool.* **212,** 435.

Tischner, M. (1988). *Proc. 11th Int. Congr. Anim. Reprod. Artificial Insemination, Dublin* **5,** 355–362.

Uwland, J. (1984). *In* "The Male in Farm Animal Reproduction" (M. Courot, ed.), pp. 269–289. Martinus Nijhoff, Boston.

VanDemark, N. L., and Moeller, A. N. (1951). *Am. J. Physiol.* **165,** 674.

Watson, P. F. (1981). *In* "Effects of Low Temperature on Biological Membrane" (G. J. Morris and A. Clark, eds), pp. 189–218. Academic Press, London.

Watson, P. F., and Plummer, J. M. (1985). *In* "Deep Freezing Boar Semen" (L. A. Johnson and K. Lars-son, eds.), pp. 113–127. Swedish University of Agricultural Science, Uppsala.

Wiltbank, J. N., Sturges, V. C., and Wideman, D. (1971). *J. Anim. Sci.* **33,** 600–606.

Westendorf, A., Richter, L., and Treu, H. (1975). *Dtsch. Tierarztl. Wochenschr.* **82,** 261–267.

Wright, R. W., Jr. (1981). *J. Anim. Sci.* **53** (Suppl. 1), 495.

CHAPTER 8

Fertilization, Early Development, and Embryo Transfer

GARY B. ANDERSON

I. Introduction

Much of what is known about fertilization and development of the early embryo has been learned from studying invertebrates and lower vertebrates. Greater availability and relative size of ova from nonmammalian species often facilitate studies of developmental processes. Over the past 25 years, refinements in superovulation procedures, embryo culture systems, and equipment for micromanipulation have allowed extensive study of developmental processes in mammalian embryos. For reasons of economy and convenience, the laboratory mouse has been the species of choice for study of early development; hypotheses concerning fertilization and embryo development in farm animals have been formed chiefly using results obtained from research with mice. The commercial embryo transfer industry provided the impetus for study of farm animal embryos, identifying both similarities and differences between embryos of laboratory and large animal species. This chapter deals with fertilization and early development in farm animals. When available, information pertaining to large domestic animals is given; in other instances cautious extrapolation from laboratory animals to farm animal species is presented.

II. The Oocyte at Ovulation

A. *Morphology*

The cytoplasm of a freshly ovulated oocyte is surrounded by the vitelline membrane, the zona pellucida, and the cumulus oophorus. The vitelline membrane is the cell or plasma membrane of the occyte. The zona pellucida is a noncellular mucoprotein structure that immediately surrounds the vitelline membrane. It is formed during follicular development by gradual accumulation of material, presumably from the growing oocyte, between the oocyte and surrounding follicular cells. While the oocyte is in the follicle, the zona pellucida is penetrated by follicular cell processes that form gap junctions with oocyte microvilli, maintaining what is theorized to be biochemical communication between the oocyte and follicular cells (Anderson and Albertini, 1976; Gilula *et al.*, 1978). These cellular processes are retracted and the gap junctions disrupted shortly before ovulation. The zona pellucida probably functions both as mechanical protection and as a regulator of the immediate chemical environment of the delicate oocyte. Survival of early embryos is reduced when the zona pellucida is damaged or removed, perhaps even more so in large domesticated species than in the mouse (Bronson and McLaren, 1970; Moore *et al.*, 1968; Trounson and Moore, 1974). The zona pellucida also contains receptor molecules important for binding spermatozoa, and in some species it provides a barrier to fertilization by more than one spermatozoan. If zona-free embryos contact one another *in vitro*, they tend to aggregate into a single, large embryo; the zona pellucida may also prevent such aggregation from occurring *in vivo*. The cumulus oophorus consists of layers of loosely packed granulosa cells. The granulosa cells closest to the zona pellucida are more tightly packed, forming a layer called the corona radiata. A viscous matrix high in hyaluronic acid holds the granulosa cells of the cumulus oophorus together. These cells cling to the oocyte for several hours after ovulation, offering an additional barrier to be negotiated by the fertilizing spermatozoan. The cumulus oophorus is lost soon after ovulation from ova of the sheep, cow, and pig.

B. Stage of Maturation

Meiosis is initiated in the fetal ovary but interrupted at the diplotene stage of prophase I (Donahue, 1972). In most mammals, meiosis I is resumed after the ovulatory surge of gonadotropin. The chromatin then condenses, the germinal vesicle breaks down, and the first polar body is extruded. Meiosis II is halted at metaphase. It is at this stage, metaphase II, that oocytes of most mammals are ovulated. Notable exceptions are oocytes of the dog and fox, which are thought to be ovulated prior to completion of meiosis I. Sperm penetration of primary oocytes may occur in these species; however, the stimulus for induction of maturation is unknown. In early reports the horse also was included among species that ovulate primary rather than secondary oocytes; recent morphological examinations have failed to substantiate this contention, however (Enders *et al.*, 1987).

When full-sized oocytes from essentially any mammalian species examined to date are removed from their follicles and placed in suitable culture medium, they spontaneously resume meiosis and develop to metaphase II, similar to what occurs after the ovulatory surge of gonadotropin. This process is referred to as *in vitro* maturation of the oocyte. Some component of the follicle or follicular fluid that prevents oocyte maturation *in vivo* has been theorized. Culture of immature oocytes in follicular fluid or in coculture with granulosa cells can inhibit *in vitro* maturation (Sirard and First, 1988;

Wassarman, 1988). Hypoxanthine, a low-molecular-weight component of follicular fluid, has been proposed to contribute to maintenance of the murine oocyte in meiotic arrest (Eppig *et al.*, 1985; Eppig and Schroeder, 1986). A number of other compounds have been shown to inhibit *in vitro* maturation, including membrane-permeable analogs of cAMP, inhibitors of cyclic nucleotide phosphodiesterase, and a selection of agents that increase intracellular cAMP. Some maturation inhibitors are effective only in the presence of cumulus cells. Furthermore, maintenance of meiotic arrest *in vitro* with these inhibitors results in a rapid decline in postfertilization developmental competency (Downs *et al.*, 1986). It appears that a complex system is employed by the intact follicle to maintain the oocyte in meiotic arrest. Until recently, oocytes matured *in vitro* were considered to have low fertilizability (Cross and Brinster, 1970). Recent improvements in culture systems have yielded procedures by which *in vitro*-matured oocytes can be fertilized and will develop to term (Cheng *et al.*, 1986; Lambert *et al.*, 1986; Goto *et al.*, 1988; Leibfried-Rutledge *et al.*, 1989a) Some of these procedures employ culture media that contain sera and hormones, and some involve coculture of the oocyte with granulosa cells. A review of meiotic arrest and oocyte maturation is available (Leibfried-Rutledge *et al.*, 1989b).

III. Gamete and Embryo Transport

Ovulation results in release of one or more ova by rupture of the Graafian follicle(s). In domestic mammals, survival time of gametes is short and deposition of spermatozoa in the female tract must be closely synchronized with ovulation. Freshly ovulated oocytes are picked up by the infundibulum and directed down the oviduct. Sperm cells are ejaculated into either the vagina (cow, sheep, goat, rabbit, and primates) or cervix and uterus (horse, dog, pig, and rodents). Since fertilization occurs in the ampulla of the oviduct, transport of oocytes down the female tract to the site of fertilization must be coincident with transport of sperm cells up the tract from their site of deposition. The female reproductive tract employs carefully controlled mechanisms whereby ova and spermatozoa are transported in opposite directions at the same time.

A. Transport of Spermatozoa to the Site of Fertilization

At mating or insemination, spermatozoa are deposited in the vagina, cervix, or uterus of the female and must reach the ampulla of the oviduct in order for fertilization to occur. Some authors have ascribed a role to sperm motility in normal transport, while others consider sperm transport totally a function of the female tract (Freund, 1973). Numerous physiological factors affect the rate of transport, including volume of ejaculate, site of deposition and anatomy of the female tract. Table I depicts some of these variables. These values were obtained by a number of different workers using a variety of techniques, and variations in the estimates exist. Regardless of species, however, relatively few sperm cells reach the ampulla; the majority are lost from the female tract by retrograde flow through the vagina and/or phagocytosis by white blood cells. It is also true that spermatozoa reach the site of fertilization in the oviduct relatively soon, often within minutes after deposition in the female tract. It has been proposed that the

Table I
Characteristics of the Ejaculate and Sperm Transport in Various Species

Species	Volume of ejaculate (ml)	Site of deposition	Interval from ejaculation to appearance of sperm in oviducts	Number of sperm reaching site of fertilization
Mouse	>0.1	Uterus	15 minutes	<100
Hamster	>0.1	Uterus	2–60 minutes	Few
Rat	0.1	Uterus	15–30 minutes	5–100
Guinea pig	0.15	Uterus	15 minutes	25–50
Rabbit	1.0	Vagina	Several minutes	250–500
Cat	0.1–0.3	Vagina and cervix	No data	40–120
Dog	10.0	Uterus	Several minutes	5–100
Sheep	1.0	Vagina	6 minutes	600–700
Cow	4.0	Vagina	2–13 minutes	Few
Pig	250	Cervix and uterus	15–30 minutes	1000
Man	3.5	Vagina	5–30 minutes	Few

[a] Adapted from Blandau (1973) and Harper (1988).

first spermatozoa to reach the oviduct do not participate in fertilization since they are nonmotile, presumably having been damaged by rapid transport through the female tract (Overstreet and Cooper, 1979). Even nonmotile sperm cells can undergo rapid transport; thus, it appears to be a function of the female tract. It has been further proposed that sperm transport occurs in two phases in many species, a rapid phase and a prolonged or sustained phase. After rapid transport, sperm cells are continuously transported from storage reservoirs in the female tract to the site of fertilization. The role of rapid sperm transport is unknown, but prolonged transport is thought to maintain a supply of fertile sperm cells at the site of fertilization. After a time spermatozoa in the ampulla are transported out the oviduct into the body cavity and replaced by new sperm cells from reservoirs further down the tract.

In the horse and pig, a large volume of semen is ejaculated and at least part of this volume is propelled through the cervix into the uterus. In rodents, a small volume is ejaculated, but ejaculation is directly into the uterus. In these animals, the cervix of the female provides no barrier to transport of spermatozoa. In the cow, ewe, rabbit, goat, and woman, sperm cells are deposited into the vagina and must traverse the cervix before reaching the uterus and oviduct. In these females the cervix may serve as a sperm reservoir; sperm migrate into the cervix, are stored in crypts, and then slowly released into the uterus over a period of several to many hours. It has been theorized that by virtue of this constant release of spermatozoa from the cervix, and phagocytosis of sperm cells within the tract, a population of viable sperm cells is maintained at the site of fertilization (Morton and Glover,

1974). There appears to be continual movement of spermatozoa anteriorly from the cervix in sheep and cattle because sperm numbers in the uterus and oviducts increase gradually between 1 and 24 h after mating or insemination into the vagina (Quinlivan and Robinson, 1969; Dobrowolski and Hafez, 1970).

The mechanism(s) by which spermatozoa are transported through the cervix still is subject to debate. One theory proposed to explain rapid transport through the cervix is that contractions of the vagina and uterus during intromission and orgasm in the female result in drawing of seminal fluid into the cervical canal. This theory has been used to explain sperm transport through the cervix of the human female but perhaps is less applicable to species such as the cow, sheep, and goat, where intromission and ejaculation take place very quickly. In the ewe, dead sperm cells enter the cervix only in small numbers (Mattner and Braden, 1969). Furthermore, motile sperm cells are capable of penetrating and migrating through cervical mucus (Hunter *et al.*, 1980), a process that may be facilitated by seminal plasma. These observations suggest that motility is important for transport across the cervix. Passage of sperm cells through the cervix in the rabbit is enhanced by a second coital stimulus with a vasectomized male (Bedford, 1971), suggesting involvement of either a neurohumoral response or a substance within the seminal plasma that stimulates muscular activity.

Transport of spermatozoa through the uterus to the uterotubal junction is rapid and primarily the function of uterine muscular contractions. Oxytocin released during mating in the cow and sheep increases uterine activity and may facilitate sperm transport. Prostaglandins in seminal plasma may have a similar effect.

Analysis of mechanisms for transport of spermatozoa through the uterotubal junction is complex due to anatomical and histological differences observed among species. The uterotubal junction selects against dead spermatozoa and spermatozoa of a foreign species in the rat (Blandau, 1973) but not the pig (First *et al.*, 1968).

Like the cervix in some species, the isthmus of the oviduct is thought to serve as a sperm reservoir, a site where spermatozoa accumulate for later transport to the site of fertilization. It has been observed in the cow (Hunter and Wilmut, 1982, 1984), sheep (Hunter *et al.*, 1980, 1982; Hunter and Nichol, 1983), pig (Hunter, 1975, 1981, 1984), and rabbit (Overstreet *et al.*, 1978; Overstreet, 1982) that spermatozoa accumulate in the isthmus for a number of hours after mating or insemination and move up the oviduct close to the time of ovulation. Spermatozoa have been described as being relatively quiescent in the isthmus but active when they move into the ampulla (Overstreet and Cooper, 1979; Cummins, 1982). Transport of sperm cells through the isthmus is accomplished primarily by muscular contractions. In some nonmammalian systems, two types of cilia beat in opposite directions and result in transport of sperm cells up and ova down the female tract (Parker, 1931). In mammals, cilia usually beat toward the uterus and are ineffective in sperm transport up the tract. A mechanism involving both cilia and muscular contractions has been described in the sow, however (Blandau and Gaddum-Rosse, 1974). Fluids that flow toward the ovary may also be involved.

Attempts have been made to administer various compounds to domestic animals at insemination in an effort to enhance sperm transport and fertility. This subject has been reviewed by Hawk (1983).

B. Transport of the Oocyte to the Site of Fertilization

The mechanism by which freshly ovulated oocytes are transported from the ruptured follicle into the oviduct depends upon a number of factors, including the anatomic configuration of the fimbriae (fringe-like projections on the infundibulum) and their relationship to the surface of the ovary at the time of ovulation; the manner by which the oocyte and surrounding granulosa cells are expressed from the follicle; and the physical characteristics of the antral fluids and matrix of the cumulus oophorus. At ovulation in the rabbit, guinea pig, cat, ungulates, and primates, the fimbriae fill with blood, become contractile, and surround the ovaries. The freshly ovulated oocytes in their associated cumulus mass come into contact with ciliated cells lining the fimbriae, which help to direct the oocytes into the oviduct. In the mouse, rat, and hamster, the ovary is enclosed completely by a thin periovarial sac into which a relatively small opening of the oviduct projects. The oviduct makes only superficial contact with the ovaries, and oocytes are shed into the fluid-filled periovarial sac. Contractions of the mesovarium result in movement of the ovary, fluid, and oocytes within the periovarial sac. When the oocytes come into contact with ciliated cells at the ostium, they are swept into the oviduct.

Oocytes usually are transported rapidly from the infundibulum to the site of fertilization in the ampulla of the oviduct. The action of cilia is thought to be the primary mechanism for transport through the ampulla, but peristaltic and segmental muscular contractions may also be involved. Movement of fluid through the ampulla may also provide a means for oocyte transport in some species, but usually fluid flow is in the opposite direction. In polytocous animals, the mass of cumulus cells and ova often are transported as a single entity until the granulosa cells are lost. After fertilization embryos remain above the isthmoampullar junction for a variable number of hours, depending on the species.

C. Transport of the Embryo through the Oviduct

In most mammals, transport of ova through the oviduct to the uterus takes 3–4 days (Table II). Rabbit ova may reach the junction of the ampulla and isthmus as quickly as 6 min after ovulation, but more than half still are above this junction 24 h later. The remainder of the time in the ovi-duct is spent in slow transport through the isthmus (Bodkhe and Harper, 1973). Mouse ova are transported rapidly through the ampulla and then detained at the isthmoampullar junction for 24 h. Transport through the isthmus is again rapid with a 30-h delay at the uterotubal junction.

In the pig (Oxenreider and Day, 1965) and cow (Aref and Hafez, 1973), ova spend most of the time in the oviduct above the isthmoampullar junction and then are transported rapidly through the isthmus into the uterus. Whether the isthmoampullar junction acts to retain embryos in the ampulla in all mammals and how spermatozoa are allowed to enter the ampulla while ovum transport is inhibited are unknown. In some species, low doses of exogenous estrogen cause a phenomenon known as "tube locking" where ova are retained at the isthmoampullar junction; larger doses of estrogen hasten transport through the isthmus and into the uterus. Hawk (1988) concluded

Table II
Rates of Development of Various Species[a]

Species	One-cell stage (hours)	Morula stage (h)	Blastocyst stage (days)	Entry into uterus (days)	Gestation (days)
Mouse	0–24	68–80	3–4	3	21
Rat	0–24	72–80	3–4	3–4	22
Rabbit	0–14	48–68	3–4	3–4	30
Cat	0–24	72–96	5–6	4–8	60
Pig	0–15	72–96	5–6	2–4	115
Goat	0–30	120–140	6	3–4	147
Sheep	0–38	96	6–7	3–4	150
Human	0–24	96	5–8	3	270
Cow	0–27	144	7	3–4	284
Horse	0–24	98	6	4–5	340

[a]Times estimated from ovulation. Adapted from Brinster (1974), Davis and Hesseldahl (1971), and Harper (1988).

that embryo transport is accelerated in su-perovulated versus single-ovulating cows. In the mare, the developing embryo is trans-ported into the uterus, bypassing unfertil-ized ova from previous estrous cycles that are retained in the oviduct (VanNiekerk and Gerneke, 1966).

IV. Fertilization

A. Sperm Capacitation

Austin (1951) and Chang (1951) inde-pendently reported that freshly ejaculated rat and rabbit spermatozoa are incapable of penetrating an ovum. Only after spending a period of time in the female tract did they acquire fertilizing ability, a process called ca-pacitation. The original observations have been extended to include spermatozoa of essentially all mammals examined, includ-ing domesticated species. While the need for incubation of spermatozoa in the female tract prior to fertilization has been estab-lished, the time required varies among spe-cies. Mouse spermatozoa require less than 1 h while rabbit spermatozoa require at least 5–6 h to attain full fertilizing potential. Ovine and porcine spermatozoa require $1\frac{1}{2}$ and 3–6 h, respectively (Austin, 1974).

Capacitation involves removal or alter-ation of substances on the sperm plasma membrane. Chang (1957) observed that ca-pacitated rabbit spermatozoa lose their abil-ity to fertilize when incubated in rabbit, bull, or human seminal plasma. These "decapaci-tated" spermatozoa regained their fertiliz-ing ability after further exposure to the fe-male tract. These observations were interpreted to mean that a decapacitation factor exists in seminal plasma, coats the sperm cell, and must be removed before the

spermatozoan gains fertilizing ability. It was subsequently learned that epididymal sper-matozoa also require capacitation in order to fertilize ova, despite not having been ex-posed to seminal plasma. It appears that substances removed from the sperm cell during capacitation are adsorbed on or inte-grated into the sperm plasma membrane during epididymal maturation and/or con-tact with seminal plasma. A number of met-abolic and other changes have been re-ported to be associated with capacitation (reviewed by Fraser, 1984; Yanagimachi, 1988), including a change in the pattern of motility referred to as hyperactivation. Spermatozoa that have undergone hyperac-tivation exhibit vigorous motility thought to be important for migration through the am-pulla of the oviduct and through the outer investments of the ovum at fertilization. Hyperactivated spermatozoa are relatively short-lived.

Capacitation can occur in either the uterus or oviduct, but spermatozoa are most readily capacitated if exposed to both envi-ronments. In general, estrogens mediate changes in the female tract that have a stim-ulatory effect on capacitation, while proges-terone can inhibit capacitation. Mating and capacitation normally occur while the fe-male is under the influence of estrogen. Ca-pacitation in the uterus seems to be more re-sponsive to hormones than is capacitation in the oviduct.

Capacitation also involves changes in the plasma membrane covering the anterior portion of the sperm head—changes that allow a phenomenon known as the acro-some reaction to occur. The acrosome reac-tion involves the progressive breakdown and fusion of the plasma membrane and outer acrosomal membrane of the sperm cell (Fig. 1). Vesicle formation allows leakage of enzymes from the acrosome with ultimate loss of the plasma membrane–acrosome complex, exposing the inner acrosomal membrane. The release of acrosomal enzymes is thought to be important in penetration of the outer investments of the ovum by the fertilizing spermatozoan. It has been proposed that capacitated spermatozoa begin to undergo the acrosome reaction, with consequent liberation of enzymes, in close proximity to or at initial contact with the cumulus–ovum mass. In this scenario, enzymes are released gradually to dissolve the matrix between the granulosa cells, thereby facilitating passage of the spermatozoan toward the zona pellucida. Complete loss of the outer acrosomal membrane occurs before the sperm cell penetrates the zona pellucida. Some investigators have further proposed that loss of the acrosome exposes a perforatorium, which acts as a mechanical tool to assist in penetration of the zona pellucida. The sequence for initiation and completion of the acrosome reaction described in Fig. 1, though seeming to make sense physiologically, is not completely supported by experimental observations. In the golden hamster, for example, spermatozoa can reach the zona pellucida with acrosomes still intact (Cherr *et al.*, 1986). In this species, both acrosome-intact and acrosome-reacted spermatozoa can bind to the zona pellucida (Suarez *et al.*, 1984), a prelude to penetration of the zona. In the guinea pig, only acrosome-reacted spermatozoa firmly bind to the zona pellucida (Huang *et al.*, 1981), while acrosome-reacted murine spermatozoa will bind but may be unable to penetrate the zona pellucida. In the pig, even uncapacitated spermatozoa can bind to the zona pellucida. Results of experiments carried out *in vitro* suggest that components of both the cumulus mass and zona pellucida are ca-

pable of inducing the acrosome reaction. In a number of species, including the mouse (Florman and Storey, 1982), golden hamster (Cherr *et al.*, 1986), rabbit (O'Rand and Fisher, 1987), human (Cross *et al.*, 1988), and pig (Berger *et al.*, 1989b), solubilized zona pellucida will stimulate the acrosome reaction *in vitro*. Indeed, various compounds have been tested and shown to be capable of inducing loss of the acrosome *in vitro*. In some cases, however, loss of the acrosome has not been associated with attainment of capacity to fertilize an ovum (Fraser, 1984). Degenerative changes in the sperm cell that result in loss of the acrosome are sometimes referred to as a false acrosome reaction. Until these conflicting observations can be fully explained, descriptions of fertilization in mammals will continue to be modified or at least recognized to vary somewhat among species.

Some researchers consider the acrosome reaction to be part of the capacitation process while others consider it a process that occurs subsequent to capacitation (i.e., changes in the spermatozoan during capacitation allow the acrosome reaction to occur). Regardless of the interpretation, it is clear that the acrosome reaction occurs widely throughout the animal kingdom and, in mammals, is preceded by changes in the spermatozoan that occur in the female tract.

B. Sperm Penetration

In most mammals, the sperm cell that successfully fertilizes the oocyte must be capable of traversing the cumulus oophorus, corona radiata, and zona pellucida before it can cross the plasma or vitelline membrane of the ovum. Hyaluronidase and acrosin are acrosomal enzymes that may contribute to penetration of the ovum by the sperm cell;

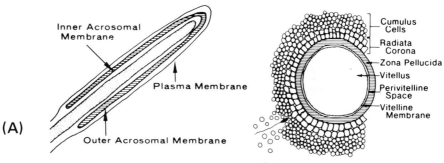

(A)

Inner Acrosomal Membrane

Plasma Membrane

Outer Acrosomal Membrane

Capacitated Sperm

Cumulus Cells

Radiata Corona

Zona Pellucida

Vitellus

Perivitelline Space

Vitelline Membrane

Penetrating Cumulus Oophorus

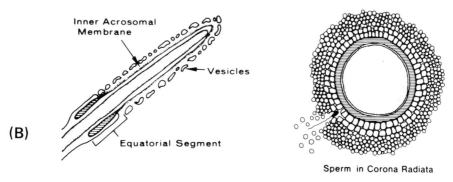

(B)

Inner Acrosomal Membrane

Vesicles

Equatorial Segment

Membrane Vesiculation

Sperm in Corona Radiata

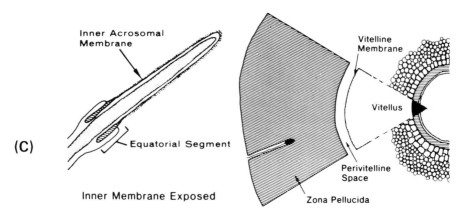

(C)

Inner Acrosomal Membrane

Equatorial Segment

Inner Membrane Exposed

Vitelline Membrane

Vitellus

Perivitelline Space

Zona Pellucida

Sperm Penetrating Zona

they represent two of over 20 enzymes reported to be of acrosomal origin (Yanagimachi, 1988). Hyaluronidase is thought to be released during the acrosome reaction and to be important in dissolving the hyaluronic acid complex matrix of the cumulus oophorus as the spermatozoan passes through. Some reevaluation of this mechanism may be in order for species in which spermatozoa can reach the zona pellucida prior to having undergone the acrosome reaction; moreover, rooster spermatozoa do not contain hyaluronidase but are capable of penetrating the cumulus oophorus of hamster ova. It may be that leakage of acrosomal enzymes occurs in some species without other obvious changes associated with the acrosome reaction, which is not in conflict with the generally accepted role of hyaluronidase in penetration of the cumulus oophorus. A second acrosomal enzyme, acrosin, is sometimes referred to as a zona lysin because it is thought to assist the sperm cell in penetration of the zona pellucida. Acrosin is stored in an inactive form (proacrosin) in the acrosome and may be activated as a result of the acrosome reaction. Sperm motility is thought to be essential during this phase of fertilization, since the whipping motion of the sperm tail facilitates passage through the outer investments of the ovum. In an interesting experiment in the rabbit, it was demonstrated that acrosome-reacted spermatozoa removed from the perivitelline space of recently penetrated ova are capable of penetrating a second ovum (Kuzan et al., 1984).

Experimental evidence suggests that the zona pellucida exhibits a species-specific receptor, making spermatozoa from foreign species unable to attach and penetrate. In some species the sperm receptor has been identified and at least partially characterized. Three different families of glycoproteins have been identified in zonae pellucidae of the mouse, pig, and hamster (Greve and Wassarman, 1985; Hedrick and Wardrip, 1986, 1987; Ahuja and Bolwell, 1983). In the mouse they are distinguished as ZP1, ZP2, and ZP3 (i.e., zona proteins 1, 2, and 3). ZP1 has been identified as the primary zona sperm receptor in this species; its activity is lost when the O-linked saccharide moiety is removed. The peptide moiety of ZP3 has been proposed as responsible for induction of the acrosome reaction. In the pig, one of two 55-K glycoproteins and a 90-K glycoprotein can block sperm binding to the zona pellucida in an in vitro assay, suggesting they may be involved with binding to the zona pellucida in vivo; the carbohydrate moiety was shown to be important for binding activity (Berger et al., 1989a). The 90-K and a 55-K porcine zona glycoprotein also have been shown to be capable of stimulating the acrosome reaction in porcine spermatozoa (Berger et al., 1989b); again, activity was lost when the carbohydrate moiety was removed.

If the zona pellucida is removed from a rat or mouse ovum, neither hamster nor guinea pig spermatozoa are able to penetrate the vitelline membrane. When the zona pellucida is removed from a hamster ovum,

Figure 1 Sperm acrosome reaction and penetration of outer investments of ovulated oocyte. (A) Capacitated spermatozoon (left) penetrates the cumulus oophorus (right). (B) Acrosome reaction of spermatozoon (left) as it penetrates corona radiata (right). (C) Acrosome-reacted spermatozoon (left) penetrates zona pellucida. Not all experimental data support this scheme for the acrosome reaction and penetration of the oocyte; see text for details. Reproduced from McRorie and Williams (1974).

capacitated spermatozoa of other species can penetrate into the ooplasm (Yanagimachi, 1972). This finding indicates that species specificity for fertilization in the hamster resides in the zona pellucida and led to development of an *in vitro* assay for fertility of spermatozoa from particular males. The zona-free hamster egg assay has been used as a clinical test to assess male infertility and as a research tool to study the fertilization process. Among large animals, it has been used to test spermatozoa from the bull, stallion (Brackett *et al.*, 1982b), and boar (Berger and Horton, 1988).

Incorporation of the sperm cell by the ooplasm is considered to be an active process on the part of the vitelline membrane. When the spermatozoan comes in contact with the vitelline membrane, it establishes contact at the equatorial segment immediately posterior to the exposed inner acrosomal membrane. The plasma membrane covering the equatorial segment fuses with the vitelline membrane, and the ooplasm around the area of contact elevates to actively surround the sperm head. The vitelline membrane fuses around the sperm head, incorporating it into the ooplasm. The thick lamina resulting from fusion of the sperm plasma membrane and the vitelline membrane disappears, releasing the sperm nucleus into the cytoplasm of the ovum. During the early stages of this incorporation process, the sperm cell including the tail is contained in the perivitelline space. Except in a few species, such as the Chinese hamster and field vole, the entire length of the sperm tail is incorporated into the egg by progressive envelopment and fusion of the vitelline membrane over the spermatozoan (Yanagimachi, 1973). As a result of sperm–ovum membrane fusion, the sperm plasma membrane becomes incorporated into the ovum plasma membrane; sperm antigens have been observed to disperse gradually from the point of sperm–ovum fusion (Gaunt, 1983). There has been some question whether or not spermatozoa contribute cytoplasmic factors, such as mitochondria, to the developing embryo. Evidence in the mouse indicates that mitochondrial inheritance is strictly maternal (Ferris *et al.*, 1982). It is generally accepted that nonnuclear sperm elements degenerate in the ovum.

C. Consequences of Fertilization

Penetration of the egg by the sperm cell results in a series of events leading to normal development of the fertilized ovum. In most mammals, erection of a barrier to prevent fertilization by more than one sperm cell will occur. This barrier includes the zona reaction and the vitelline block. Penetration of the ovum also stimulates resumption and completion of meiosis II with extrusion of the second polar body. The male and female pronuclei are formed, and maternal and paternal chromosomes combine to form the diploid zygote. Fertilization also results in activation of the ovum to initiate cleavage and enhanced biochemical activity.

1. Erection of a Barrier to Polyspermy

Polyspermic fertilization, when it does occur, can result in a polyploid individual that is not viable. It is important that a mechanism exists to limit fertilization of the ovum to only one sperm cell. The chances of polyspermy occurring are reduced by the limited number of spermatozoa at the site of fertilization. Deposition of excessive numbers of spermatozoa directly into the oviduct increases polyspermic fertilization (Hunter, 1973). Polyspermy can also occur with fertilization *in vitro* where sperm numbers in proximity to the ovum are considerably greater than *in vivo*. The ovum itself displays a mechanism for preventing penetration by more than one spermatozoan. The block to polyspermy in mammalian ova operates at two levels in most species: the zona pellucida (zona reaction) and the vitelline membrane (vitelline block). Attachment of the fertilizing sperm cell to the vitelline membrane results in exocytosis of ooplasmic structures called cortical granules. The membrane surrounding the cortical granules fuses with the vitelline membrane, and contents of the granules are released into the perivitelline space. Release of the cortical granule contents results in some change in the character of the zona pellucida (zona reaction), which makes it impenetrable to other spermatozoa. Results from experiments with mouse and hamster ova suggest that the cortical granules release a trypsin-like protease that hydrolyzes the sperm receptor on the zona pellucida (Gwatkin *et al.*, 1973). This theory is supported by observations in the hamster that spermatozoa do not attach to the zona pellucida after sperm-induced rupture of the cortical granules (Barros and Yanagimachi, 1971). Many sperm cells can be found on the zona pellucida of the fertilized porcine ovum. The change that occurs at the vitelline membrane that renders it unresponsive to sperm attachment or prevents sperm attachment (vitelline block) is poorly understood. It has not been proven conclusively that the vitelline block is due to contents of the cortical granules or that it is electrical in nature as in some invertebrates.

Most available information regarding barriers to polyspermy has been obtained from laboratory animals. Research with these and other animals has shown that dif-

ferences exist among species. In some (sheep, dog, and hamster) the zona reaction is relatively quick and effective and few sperm cells other than the fertilizing spermatozoan are found in the perivitelline space. In the mouse and rat, extra sperm cells in the perivitelline space are more common, which suggests that the zona reaction is slower in these species. The rabbit has an ineffective zona reaction and many spermatozoa pass through the zona pellucida; however, an effective vitelline block allows only one spermatozoan to enter the ooplasm and effect fertilization.

2. Completion of Meiosis II

At the time of sperm penetration in most domestic mammals except the dog, the ovum is at metaphase of the second meiotic division. The chromosomes are arranged mostly in pairs and aligned on the equator of the meiotic spindle. Migration of the chromosomes to opposite poles of the spindle and cytokinesis result in formation of the haploid ovum and second polar body. The first polar body can be distinguished from the second polar body by its larger size and morphological features. The first polar body usually contains cortical granules because it is extruded prior to sperm penetration. In addition, chromosomes of the first polar body, like those of the oocyte entering meiosis II, contain no nuclear envelope. Chromosomes of the second polar body are enclosed in a nuclear membrane, having undergone a process similar to formation of the female pronucleus at the end of meiosis II.

3. Pronucleus Formation, Syngamy, and Activation of Cleavage

The chromosomes remaining in the fertilized ovum at the end of the second meiotic division are surrounded by a nuclear membrane. This haploid structure is referred to as the female pronucleus.

Upon entry into the ooplasm, the sperm head swells. The sperm tail and sperm head elements are separated from the nucleus and degenerate. The sperm nuclear membrane degenerates, exposing paternal chromatin directly to the ooplasm. Decondensation of paternal chromatin is mediated by removal of sperm-specific nuclear proteins, protamines (in contrast to histones, which are nuclear proteins found in somatic cells), which are thought to permit condensation of chromatin and repression of DNA activities. A new nuclear membrane is acquired, producing the male pronucleus. There is evidence that DNA damage is repaired in the male pronucleus before paternal and maternal chromosomes combine (Brandriff and Pedersen, 1981). An ooplasmic factor that controls development of the male pronucleus, male pronucleus growth factor, has been theorized (Thibault and Gerard, 1973). Oocytes that mature *in vitro* under suboptimal conditions often fail to support development of the male pronucleus, perhaps due to a lack of an appropriate cytoplasmic factor(s).

The male and female pronuclei move into close proximity to one another in the center of the ovum. Chromatin duplication takes place in both pronuclei, and, as prophase of the first cleavage division is approached, chromosomes condense and pronuclear envelopes break down. In the mouse (Zamboni, 1972), rabbit (Longo and Anderson, 1969) and perhaps other mammals, pronuclei do not fuse prior to pronuclear membrane breakdown to form a single diploid nucleus (a process called syngamy) as is common in invertebrates. At the end of prophase, two distinct sets of condensed chromosomes can be observed in the zygote. During metaphase the two groups of chromosomes move together and arrange themselves on the equatorial plate. As mitosis continues, the sister chromatids separate and migrate in opposite directions. With the formation of nuclear membranes and deepening of the cleavage furrow, two diploid blastomeres are formed, marking the end of the first cleavage division.

D. Aging of Gametes

In most species mating shortly precedes ovulation. The coincidence of mating and ovulation ensures that fertile sperm cells will be at the site of fertilization when the oocyte is released from the follicle. In species where sperm capacitation is necessary or prolonged, spermatozoa must be in the female tract for several hours in order to fertilize the oocyte. Timing is important since the fertile lives of gametes are short (Table III). Most mammalian ova should be fertilized within a few hours of ovulation and few can be fertilized later than 12 h after ovulation; the dog is a possible exception (Cole, 1975). In domestic animals, the fertile life of spermatozoa in the female tract is also short. This fragility of the gametes is particularly important to livestock producers utilizing artificial insemination. Accurate detection of estrus or ovulation is imperative to proper timing of insemination. Failure of fertilization to occur while the gametes are "fresh" can affect not only events of fertilization but also subsequent embryonic development. It has been shown that optimum fertility in the pig and sheep is achieved if insemination occurs 12 h prior to ovulation (Dzuik, 1970). This schedule allows time for adequate numbers of capacitated sperm

Table III
Survival Time and Fertility of Gametes in the Female Tract

Species	Retention of sperm motility (h)	Retention of fertility (h)	
		Egg[a]	Sperm
Human	60–144	6–24	28–48
Rabbit	43–50	6–8	30–36
Cow	15–56	8–12	28–50
Sheep	48	16–24	30–48
Guinea pig	41	20	21–22
Mouse	13	6–15	6–12
Rat	17	8–12	14
Ferret	126	30	48–126
Horse	144	6–8	72–120
Pig	50	8–10	24–48

[a]Values are approximate and based on estimated time from ovulation.
Adapted from Hamner (1973) and Dukelow and Riegle (1974).

cells to reach the site of fertilization by the time of ovulation and prior to decreased fertility of the spermatozoa.

Aging of ova can result in decreased fertilizability and an increase in abnormal embryos (Salisbury and Hart, 1970). Senescence appears to effect failure of the block to polyspermy and deterioration of the female genome and/or its division apparatus. In porcine and murine ova, aging is expressed as a deterioration of the ovum's defenses against polyspermy, leading to penetration by more than one spermatozoan. The pig also shows digyny in aged ova, the formation of two female pronuclei resulting from retention of the second polar body or fragmentation of the female pronucleus. In aged mouse ova, on the other hand, fertilization occurs, but the female pronucleus fails to develop normally. If fertilization is delayed in the rabbit, ova undergo a disruption of the meiotic apparatus as evidenced by alterations in the orientation and structure of the meiotic spindle and wandering of chromosomes (Longo, 1974) as well as digyny and hypodiploidy (Salisbury and Hart, 1970).

Aging of spermatozoa in the female tract also has a detrimental effect upon fertility, although probably little effect upon embryonic development. Table III gives the approximate intervals from ejaculation that spermatozoa remain motile and fertile in the female tract. The application of artificial insemination to domestic livestock has made the effect of *in vitro* aging of spermatozoa important. When sperm cells are held at 4–5°C prior to insemination, fertility is decreased and embryonic mortality increased with increasing storage periods (Salisbury and Hart, 1970). Storage at −196°C dramatically delays, but does not prevent, gamete senescence. It has been suggested that these results are due to time-dependent changes in the genetic information contributed by spermatozoa.

E. In Vitro *Fertilization*

Procedures for *in vitro* fertilization of oocytes from laboratory animal species have been available for two decades and those for human oocytes are now widely used. Progress on *in vitro* fertilization schemes for use in livestock has been slow. Birth of the first calf from an *in vitro* fertilized oocyte was reported in 1982 (Brackett *et al.*, 1982a), but duplication of these results by other labs came only after several years and modification of procedures. Recent advances have led to improved procedures for *in vitro* fertilization of not only bovine oocytes but ovine and porcine oocytes as well (Cheng *et al.*, 1986). Among significant breakthroughs were development of procedures for *in vitro* maturation of oocytes (Ball *et al.*, 1984; Hensleigh and Hunter, 1985) and *in vitro* sperm capacitation, the latter made possible by incubation in high-ionic-strength medium (Brackett *et al.*, 1982a; Lambert *et al.*, 1986), heparin (Parrish *et al.*, 1985, 1986, 1988), elevated pH (Cheng *et al.*, 1986), and with long incubation periods (Wheeler and Seidel, 1987). Although results from *in vitro* fertilization in livestock species have improved dramatically in the past few years, they still may be somewhat unpredictable. Variables that are difficult to control, such as those due to male differences, can affect results (Iritani *et al.*, 1988; Eyestone and First, 1989). Reviews of *in vitro* fertilization procedures for use in farm animals are available from First and Parrish (1987) and Leibfried-Rutledge *et al.* (1989a,b).

Several new procedures involving micromanipulation are under development for assisting fertilization *in vitro*. Zona drilling involves weakening discrete areas of the zona pellucida by application of acidified medium; it has been proposed as a means to facilitate passage of the sperm cell through the zona pellucida in cases of low sperm numbers (Gordon and Talansky, 1986; Depypere *et al.*, 1988). Direct microinjection of a spermatozoan into an oocyte to effect fertilization has been accomplished in both the mouse (Markert, 1983; Mann, 1988) and the rabbit (Yang *et al.*, 1988).

F. Androgenesis, Gynogenesis, and Parthenogenesis

Parthenogenesis, development without fertilization, is observed in a variety of vertebrates, including birds, reptiles, fish, and amphibians. Since development takes place with contributions from only the female genome, maternal and paternal genomes in these species appear to be functionally equivalent. Unfertilized mammalian oocytes sometimes spontaneously undergo parthenogenetic cleavage and can be readily activated experimentally. In mice, preimplantation and early postimplantation development often appears normal, but in no case has development to term been observed. Markert and Petters (1977) and Hoppe and Illmensee (1977) described procedures for production of homozygous, uniparental embryos by removal of a single pronucleus from a pronuclear-stage zygote and diploidizing the remaining pronucleus. The resulting embryos developed to the blastocyst stage, regardless of whether the embryonic genome originated from the maternal or paternal pronucleus, but did not develop to term (Modlinski, 1980; Markert, 1982). Homozygous diploid embryos were shown to be capable of contributing to normal development of a chimera when combined with a normally fertilized embryo, however (Anderegg and Markert, 1986; Otani *et al.*, 1987). Mann and Lovell-Badge (1984) obtained viable young when a pronucleus from a fertilized ovum was transferred to the cytoplasm of an enucleated parthenogenetic ovum, but development failed with the reciprocal transplantation, demonstrating that failure of development in parthenotes has a nuclear rather than cytoplasmic basis. Neither diandric (developing from two male pronuclei, androgenesis) nor digynic (developing from two female pronuclei, gynogenesis) mouse embryos survive to term, even when the two female pronuclei are from different embryos and different strains (McGrath and Solter, 1984a,b; Surani *et al.*, 1984). Normal development was observed only when a male and a female pronucleus were combined, suggesting complementary roles of maternal and paternal genomes. It has been proposed that some form of selective change takes place during gametogenesis that distinguishes maternal from paternal genome, a process referred to as imprinting of the genome (Surani *et al.*, 1986, 1987). Further evidence for imprinting of the genome is differential expression of maternally and paternally derived alleles (Cattanach and Kirk, 1985). The exact mechanism for imprinting of the genome is not understood at this time, but it accounts for the need for both maternal and paternal genomes and failure of development of embryos derived from parthenogenesis, gynogenesis, and androgenesis.

V. Embryo Development

A. *Cleavage*

The fertilized mammalian ovum begins its development as one of the largest cells of the body and with a high ratio of cytoplasmic to nuclear volume. Ferilization activates the ovum to initiate mitotic divisions, the first of which results in formation of a two-cell embryo. Each cell, or blastomere, is essentially the same size and is half the size of the original single-celled zygote. Unlike mitosis in other cells where daughter cells go through a growth phase prior to the next division, both blastomeres of the two-cell embryo divide again, yielding a four-cell embryo. Each blastomere is now one-quarter the size of the original zygote. The four blastomeres divide to form an eight-cell embryo, then a 16-cell embryo and so on (Fig. 2). Cell divisions are nearly, but not exactly, synchronous, and uneven numbers of blastomeres are seen. For example, a three-cell embryo is formed when one blastomere of a two-cell embryo divides slightly before the other; a normal three-cell embryo is easily distinguished by one large and two small blastomeres. There is evidence that the developmental fate of early-dividing blastomeres is different from that of the late-dividing blastomeres at the two-cell (Kelly *et al.*, 1978; Surani and Barton, 1984), four-cell, and eight-cell stages (Spindle, 1982). Early-dividing blastomeres are more likely to contribute to the inner cell mass of the blastocyst than are late-dividing cells, which are more likely to contribute to trophectoderm. This effect may be due to spatial configuration within the developing embryo assumed by the early- and late-dividing blastomeres (i.e., the former, which are smaller, may be more likely to be pushed to the interior of the developing mass of cells, an arrangement known to favor development into inner cell mass).

Throughout cleavage, cells become progressively smaller with no net increase in size of the embryo. In essence, the cytoplasm

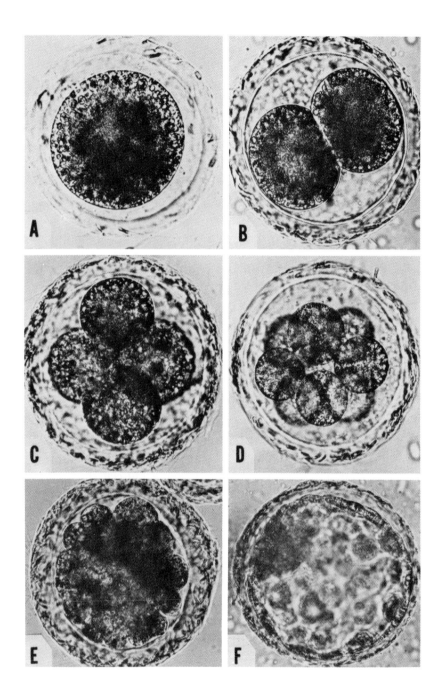

of the zygote is partitioned into smaller packages until a solid ball of cells called a morula is formed. The number of cells present at the morula stage is species dependent. Most species form a morula at approximately 16–32 cells; with continued mitotic divisions the number of cells continues to increase and cell size continues to decrease. This change in the ratio of cytoplasmic to nuclear volume is thought to be important in regulation of development, particularly formation of the blastocyst (Brinster, 1974; Smith and McLaren, 1977). While the size of individual blastomeres is reduced until it reaches that of normal adult body cells, the embryo as a whole does not increase in size and may actually decrease in total mass during the first several days of development (Brinster, 1967).

Other major morphological events that take place during cleavage include the process of compaction and development of cell polarity. Compaction is marked by establishment of gap junctions for direct communication between blastomeres and incipient tight junctions between blastomeres on the outside of the embryo, which will be important in formation of trophectoderm and the blastocyst. Individual blastomeres become more difficult to distinguish and dissociate from one another as a result of junctional differentiation. Compaction occurs at the eight-cell stage in the mouse but not until the 16-cell and 32-cell stage, respectively, in the sheep and cow (Prather and First, 1988). The outer cells become polarized as they develop a higher density of microvilli on their outer, free surface compared to surfaces apposed to other blastomeres (Ducibella and Anderson, 1975); inner cells remain apolar (Handyside, 1980; Ziomek and Johnson, 1980). Polar cells tend to contribute to formation of trophectoderm and apolar cells to inner cell mass of the blastocyst (Johnson et al., 1981; Johnson and Maro, 1986).

Rates of cleavage vary among species (See Table II). Laboratory species generally have shorter gestation periods and slightly greater rates of cleavage than large domestic species, yet development during the first several days after fertilizaton is remarkably similar for most mammals studied. Passage from the oviduct into the uterus is coincidental with blastocyst formation in several laboratory species, which led to speculation regarding involvement of the uterus in initiation of blastocyst formation. Considerable interest was generated in uterine secretory product(s) called blastokinin or uteroglobin, which were described as having stimulatory effects on embryo development (Krishnan and Daniel, 1967; Beier, 1968). The more recent demonstration that embryos of several species can be cultured in vitro indicated that probably no single uterine factor controls development. On the other hand,

Figure 2 Pig embryos collected at various intervals from ovulation demonstrating development from fertilization to the blastocyst stage, ×400. Sperm cells adherring to the zona pellucida are conspicuous. (A) One-cell zygote. (B) Two-cell embryo recovered from the oviduct 18 h after ovulation. Note that each blastomere is half the size of the original one-cell zygote. (C, D) Four-cell (recovered from oviduct) and eight-cell (recovered from upper uterine horn) embryos collected 41 and 55 h after ovulation, respectively. (E) Morula containing 16–32 cells recovered from upper uterine horn 73 h after ovulation. Note that the size of the morula is the same as that of the one-cell zygote. (F) Blastocyst recovered from upper uterine horn 117 h after ovulation. The fluid-filled blastocoele is evident. Expansion of the blastocyst has resulted in thinning of the zona pellucida. Reproduced from Hunter (1974).

treatments that hasten or delay passage of the embryo from the oviduct to the uterus and asynchronous embryo transfers result in decreased embryo viability, indicating that the maternal environment does affect early development.

B. Blastocyst Formation

After the morula stage, a cavity called the blastocoele forms in the embryo, producing a fluid-filled structure called the blastocyst (Fig. 2). The blastocoele forms from fluid transport across the outer cells of the compacted morula. Fluid accumulation occurs at the 16-cell stage in the pig and hamster, the 32-cell stage of the mouse, 64-cell stage in sheep, and 128-cell stage in the rabbit (Pedersen, 1988). The blastocyst stage marks the first overt sign of allocation of cells to different types in the developing embryo. The outer single layer of cells is called trophectoderm and will contribute to the outer layers of the placenta. Inside the trophectoderm is a group of cells called the inner cell mass (ICM), from which the embryo proper and portions of the placenta will form. Trophectoderm overlying the ICM is designated polar trophectoderm, while the remainder is mural trophectoderm. In the mouse, of the 32 cells that comprise the blastocyst, only three or four form ICM whereas the remainder become trophectoderm (Gardner, 1972). In non-mammalian species, cytoplasmic differences in the unfertilized ovum are reflected in cells of the embryo and affect differentiation; there is evidence that in mammals, formation of the ICM in blastocysts is due to location or position of cells in regard to one another, rather than inherent cytoplasmic differences. According to this theory, cells that are forced to the interior of the morula

differentiate into ICM, while the remainder become trophectoderm.

The forces responsible for blastocyst formation and accumulation of fluid between cells of the morula are thought to be due to action of a Na^+/K^+ pump across the outer cells of the compacted morula (Cross and Brinster, 1969; Benos and Biggers, 1981). According to this theory, the Na^+/K^+ pump transports ions across the membranes into the center of the compacted morula. Osmotic action then causes transport of water into the interior of the morula, resulting in formation of a blastocoele. Once formed, it continues to enlarge, resulting in an increase in size of the blastocyst (expanded blastocyst). Magnitude and rate of expansion are species-dependent. Blastocysts of the primate, rabbit, dog, cat, and most livestock species increase to many times their initial size during blastocyst expansion. The volume of the rabbit blastocyst increases 4000-fold between days 4 and 7 of development (Brinster, 1974). Blastocysts of most laboratory rodents, including the mouse, rat, and guinea pig, show only modest expansion; the diameter of the implanting blastocyst is similar to that of the ovulated oocyte.

Formation of the blastocyst prepares the embryo for gastrulation, formation of the three primary germ layers (ectoderm, mesoderm and endoderm) from which the various tissues and organs will develop. Blastocyst expansion may also function to "turn off" transport mechanisms of the uterus. When the blastocyst reaches a critical size, an inhibition of propulsion may result that helps orient the embryo in the uterus prior to implantation; this may be particularly important in species where implantation of the embryo takes place at the blastocyst stage. Most blastocysts "hatch" or escape from the zona pellucida at the expanded blastocyst stage. Expansion and the resultant thinning of the zona pellucina may play a role in the hatching process. Uterine production of a zona lysin to aid in hatching has been proposed (McLaren, 1970), but blastocysts will also hatch *in vitro*.

C. Loss of Totipotency

Allocation of cells to form different types (ICM and trophectoderm) is observed at the blastocyst stage. Prior to blastocyst formation, biochemical and developmental differences identify prospective cells of these two types (Section V,A). Individual blastomeres of the early mammalian embryo are totipotent, or contain the full range of developmental capabilities necessary to form a complete individual. When all blastomeres except one were destroyed in two-, four- and eight-cell sheep and rabbit embryos, the remaining blastomere developed into a normal young after transfer to a suitable recipient (Moore, 1973). Viable young have been produced from single blastomeres taken from four- and eight-cell embryos of rabbits (Moore *et al.,* 1968), sheep (Willadsen, 1981), cows, pigs, and horses (Willadsen, 1982). Genetically identical triplet and quadruplet lambs have been produced with single blastomeres from a four-cell embryo and pairs of blastomeres from an eight-cell embryo, respectively (Willadsen, 1981). There is evidence that totipotency is maintained in a number of species to at least the 16-cell stage (Rossant and Vijh, 1980; Ziomek *et al.,* 1982; Papaioannou and Ebert, 1986). In the mouse, outer cells of the morula and ICM cells are totipotent (Rossant and Vijh, 1980).

Early research with murine embryos indicated that single blastomeres from two-cell embryos will develop into viable young, but development of blastomeres from four- and eight-cell embryos is limited (Tarkowski and Wroblewska, 1967; Fiser and Macpherson, 1976; Rossant, 1976). This lack of development may be due to a low number of cells at blastocyst formation, rather than to lack of developmental capacity, however. The problem may be peculiar to the mouse, which forms the blastocyst after fewer cleavage divisions than other species. Demonstration of totipotency of single blastomeres often requires novel procedures to overcome the reduced amount of cytoplasm in a single blastomere compared to that of the intact embryo. Willadsen and Fehilly (1983) produced five genetically identical lambs from a single eight-cell embryo by combining a blastomere from an eight-cell embryo with a blastomere from a four-cell embryo. The former contributed to formation of ICM (and, thus, the lamb) and the latter contributed to trophectoderm (and placental trophoblast).

D. Activation of the Embryonic Genome

In invertebrate and amphibian embryos, proteins and RNA required for early development are of maternal origin and provided in the ovum. In mammals, products of the embryonic genome are essential for orderly developmental events, which can be disrupted by inhibitors of embryonic protein and RNA synthesis. Development beyond the two-cell stage appears to require mRNA synthesis. Paternal markers (i.e., markers where maternal and paternal forms can be distinguished) have been used to identify the time period during development when products of the embryonic genome are first detected; expression of an allele passed to the embryo from the paternal genome is evidence that the embryonic genome has been activated. Isozymes of β-glucuronidase (Wudl and Chapman, 1976) and paternal isoform of β_2-microglobulin (Sawicki *et al.*, 1982) can be detected in the two-cell mouse embryo. Modest mRNA synthesis may occur even in the one-cell fertilized mouse ovum, but its functional significance is not clear (Clegg and Piko, 1983). Recent results from experiments with ovine and bovine embryos indicate that the embryonic genome in these species may not be activated until the eight- to 16-cell stage (Camous *et al.*, 1986; Crosby *et al.*, 1988).

These observations should not be interpreted to mean that there is no maternal control or contribution to early mammalian developmemt, only that the mammalian embryo appears to exert far greater influence on its development via its own genome than does the nonmammalian embryo. Reviews on embryonic genome expression during preimplantation development are available (Johnson, 1981; Schultz, 1986), as are general reviews of preimplantation development (Pedersen, 1988; Prather and First, 1988).

VI. *In Vitro* Manipulation

A. *Embryo Culture*

Reviews on embryo culture are available (Bavister, 1987; Kane, 1987), including those particularly focused on embryos from domestic livestock (Wright and Bondioli, 1981; Anderson, 1983, 1985; Wright and O'Fallon, 1987). The history of embryo culture systems has been described by Biggers (1987). The ability to culture mammalian embryos outside the maternal environment has been important in extending knowledge of early development to include mammals.

Culture systems that will support development to the blastocyst stage are available for a number of laboratory and livestock species. Requirements for culture of embryos vary with species, and in some cases with genetic strain within a species. Questions remain regarding how closely *in vitro* development mimics development in the female tract, however. Controlled studies of *in vitro* development have indicated that, although it proceeds through the appropriate stages, development usually is slower *in vitro* than *in vivo* (Bowman and McLaren, 1970; Tervit and Rowson, 1974; Binkerd and Anderson, 1979; Kreitmann and Hodgen, 1981). Even more disturbing are reports that cultured embryos are less viable than those that developed to a comparable stage *in vivo*, despite their normal morphology (Bowman and McLaren, 1970; Binkerd and Anderson, 1979; Hahn and Schneider, 1982). Renard *et al.* (1980) reported higher embryonic losses between days 21 and 60 of pregnancy in bovine embryos that were cultured for 24 h than is usually observed with noncultured embryos. Extensively manipulated embryos (e.g., isolated blastomeres) often will develop more successfully in the oviducts of a host female, even of another species, than *in vitro* (Willadsen, 1982).

Mammalian embryos benefit from coculture with other cells, including such diverse cell types as uterine or testicular fibroblasts (Kuzan and Wright, 1982), endometrial cell monolayers (Allen and Wright, 1984), and oviductal epithelium (Eyestone *et al.*, 1987). The contribution by "feeder cells" to *in vitro* embryonic development is unknown; in some cases it is sufficient to culture embryos in medium that has been previously exposed to the other cell type ("conditioned" medium), suggesting that the feeder cells secrete stimulatory growth factors into the medium. Other experiments have shown that contact with the feeder layer is important for maximal benefit. Beneficial effects have even been observed from culturing embryos with other embryonic cells. Trophoblastic vesicles can be produced from elongating blastocysts (e.g., approximately day 13 to 15 in cattle), which are cut into sections and cultured for several days until they form spheres. Improved development *in vitro* has been observed with coculture of bovine and ovine embryos with trophoblastic vesicles (Camous *et al.*, 1984; Heyman *et al.*, 1987b). Interestingly, transfer of trophoblastic vesicles with bovine frozen–thawed blastocysts and bisected embryos has been reported to improve rates of pregnancy in recipient females (Heyman and Menezo, 1987; Heyman *et al.*, 1987a), presumably due to their antiluteolytic action in the pregnant female (Heyman *et al.*, 1984).

Stage of development at which culture is initiated is an important variable affecting the success with which embryos can be grown *in vitro*. In general, early cleavage-stage embryos cannot be cultured as successfully as later stage embryos. It has been proposed that early mammalian embryos must pass through certain "blocks" (i.e., stages of development where mammalian embryos are arrested *in vitro*). For example, embryos from some strains of mice will cleave *in vitro* to the two-cell stage but no further. The *in vitro* block for bovine embryos is said to occur at the eight-cell stage. These blocks to embryo development are not absolute, since some embryos will culture successfully from early cleavage stages to the blastocyst stage (Wright *et al.*, 1976; Peters *et al.*, 1977). Failure of early embryos to develop *in vitro* has become an important issue with the advent of procedures that involve manipulation of very early embryos (e.g., *in vitro* fertilization,

gene transfer). Usually one-cell zygotes are cocultured with other cells that promote development *in vitro*.

B. *Cryopreservation of Embryos*

When the mammalian embryo is cooled to temperatures below 20°C, mitosis ceases and development is arrested. Holding embryos at low temperatures is an effective method for maintaining viability during storage *in vitro*. Embryos of most laboratory and livestock species will survive refrigeration temperatures (approximately 4°C) for one to several days (BonDurant *et al.*, 1982; Hughes and Anderson, 1982; Lindner *et al.*, 1983; Carnevale *et al.*, 1987; Sertich *et al.*, 1988). Porcine embryos are an exception; they usually do not survive refrigeration for even short periods (Wilmut, 1972, 1986), although occasional successful results have been reported (Nagashima *et al.*, 1988).

For long-term storage *in vitro*, mammalian embryos can be frozen to −196°C in liquid nitrogen and held indefinitely. The first report of successful freezing and thawing of mammalian embryos was published in 1972 (Whittingham *et al.*, 1972). A year later freezing procedures were successfully applied to the bovine embryo (Wilmut and Rowson, 1973), followed by embryos of many other species. Successful freezing and thawing of mammalian embryos require addition of a cryoprotectant such as glycerol or dimethyl sulfoxide. Often a nonpermeating, extracellular compound such as sucrose is also added to reduce osmotic stress during dilution of the cryoprotectant, which often is toxic to embryos at room temperature. Embryonic death during storage at −196°C is considered to be negligible; frozen–thawed embryos are damaged or die during the freezing or thawing processes, not while at $-196°C$. Unique characteristics of embryos, including their size, low surface-to-volume ratio, and low permeability to water at low temperatures, initially required unusually slow freezing and thawing rates compared to those used for other animal cells. Progress now is being made with rapid freezing by direct plunging of room-temperature embryos into liquid nitrogen.

A number of modifications in freezing and thawing procedures are under study with the aim of simplification for practical on-the-farm use. For example, a procedure has been described whereby an embryo is transferred to a recipient female directly from the plastic straw in which it was frozen and thawed; with this technique the critical step of diluting cryoprotectant from the embryo is carried out in the straw (Leibo, 1984; Massip and VanDerZwalmen, 1984; Massip et al., 1987). Another method allows rapid cooling and thawing by using high concentrations of cryoprotectants, which supercool to very low temperatures and increase in viscosity before solidifying without formation of ice (vitrification) (Rall and Fahy, 1985; Massip et al., 1987; Rall, 1987). Under optimal conditions, only a slight reduction of viability is observed with transfer of frozen–thawed embryos to recipient females compared to viability after transfer of nonfrozen embryos; survival is affected by many variables, however.

C. Blastomere Isolation and Embryo Splitting

The mammalian embryo will continue normal development when a portion of its cells is removed or destroyed, a capability widely used to study retention of totipotency by individual blastomeres or groups of embryonic cells. The ability of a cluster of cells removed from an embryo to develop into a complete animal can also be used to produce genetically identical multiplets from a single embryo. Both morulae and blastocysts can be split, usually with only a slight reduction in viability of each "half-embryo" compared to that of the original intact embryo (cattle: Ozil et al., 1982; Williams et al., 1984; Baker and Shea, 1985; Takeda et al., 1986; Arave et al., 1987; sheep: Gatica et al., 1984; Shelton and Szell, 1988; pigs: Ash et al., 1988). When blastocysts are split, it is important that the inner cell mass be bisected; morulae can be split in any plane. It has been estimated that approximately 15% of embryonic cells are lost during the splitting procedure (Chesna et al., 1987; Skrzyszowska and Smorag, 1987). Often each half-embryo is placed inside a zona pellucida, although this step appears to be unnecessary (Warfield et al., 1986). Efforts to split morulae into quarters result in substantial reduction in viability (Lehn-Jensen and Willadsen, 1983; Voelkel et al., 1986), apparently due to excessive reduction in the amount of cytoplasm available to each quarter-embryo. Half-embryos can be frozen–thawed, but viability is reduced compared to that of nonfrozen split embryos, regardless of whether freezing and thawing are carried out before or after splitting (Lehn-Jensen and Willadsen, 1983; Niemann et al., 1986; Chesna et al., 1987; Takeda et al., 1987). Embryo splitting usually is carried out under high magnification and with the aid of micromanipulators, but simplified procedures also have been reported (Williams and Moore, 1988). An alternative approach to producing identical animals from a single embryo is separation of early cleavage-stage embryos into their component blastomeres; successful production of young from isolated blastomeres was discussed earlier (Section V,C).

D. Production of Chimeras

The ability of the mammalian embryo to regulate its developmental processes is exhibited by experimental chimeras, composite animals with different cell populations derived from more than one fertilized ovum. Reviews on mammalian chimeras and their uses in research are available from McLaren (1976), LeDouarin and McLaren (1984), and Anderson (1987b).

Mammalian chimeras usually are produced from preimplantation-stage embryos by aggregation of two or more embryos (aggregation chimera) or injection of cells into a blastocyst (injection chimera). Blastocysts that develop from aggregated embryos are larger than normal, but regulation of size occurs during gestation so that chimeras have normal birth weights and normal morphology. The laboratory mouse has been the species of choice for most research with experimental chimeras, but chimeras also have been produced in the rat (Weinberg *et al.*, 1985), rabbit (Babinet and Bordenave, 1980), sheep (Fehilly *et al.*, 1984a; Butler *et al.*, 1987), and cow (Brem *et al.*, 1984). Viable chimeras can also be produced by combining embryonic cells from different species, including two species of mouse (*Mus musculus* and *Mus caroli:* Rossant and Frels, 1980), *Bos taurus* and *Bos indicus* cattle (Summers *et al.*, 1983), and the sheep and goat (*Ovis aries* and *Capra hircus:* Fehilly *et al.*, 1984b; Polzin *et al.*, 1987). Interspecies chimeric embryos where trophoblast develops from cells of the same species as the recipient species have resulted in successful pregnancies between species that will not otherwise carry such pregnancies to term

(Rossant *et al.*, 1983; Fehilly *et al.*, 1984b; Meinecke-Tillman and Meinecke, 1984; Polzin *et al.*, 1987).

Experimental chimeras have been useful models for study of early mammalian development. For example, research with chimeras has led to the conclusion that only a few cells of the blastocyst contribute to formation of the embryo proper; all other cells of the ICM contribute to extraembryonic structures. Mintz (1970) proposed that only three cells in the blastocyst form the embryo. This hypothesis was based on the observation that in about 25% of animals produced by aggregation of two embryos, only one component cell line is expressed. Markert and Petters (1978) demonstrated that at least three cells are allocated to produce the embryo when they produced a triply chimeric mouse by aggregation of three embryos. Their results set a lower but not an upper limit on the number of cells that originally contribute to the embryo. They subsequently reported the production of an aggregation chimera that expressed four different genotypes (Petters and Markert, 1980), which may indicate that four, not three, cells of the blastocyst contribute to the embryo. It is possible, however, that allocation of cells in an embryo that is four times normal size is different from that in a single embryo.

Experimental chimeras have also been used in research aimed at understanding control of normal development. For example, while parthenogenetically activated mammalian embryos are incapable of developing to term (Section IV,F), they can contribute to tissues of chimeras with normally fertilized embryonic cells (Surani *et al.*, 1977; Stevens, 1978). Such chimeras produced germ cells from both the normal and parthenogenetic lines (Stevens, 1978), demonstrating that parthenogenetically activated ova retain the ability to contribute to both somatic and germ cells. An even more dramatic example in chimeras of the ability of normal embryonic cells to direct development of cells otherwise incapable of developing to term is the reversal of malignancy of embryonal carcinoma cells, which are stem cells derived from a teratocarcinoma. When injected into blastocysts (Mintz and Illmensee, 1975; Stewart and Mintz, 1981) or aggregated with embryos (Stewart, 1982), embryonal carcinoma cells have the ability to contribute to development of normal tissues and organs of a chimeric animal.

Furthermore, colonization of the germ-cell line can occur so that the resulting chimera produces gametes that contain the haploid genotype of the embryonal carcinoma cell line. Pluripotent cells called embryonic stem cells have also been isolated from murine ICM (Martin, 1981). Embryonic stem cells can be maintained in culture, be frozen and thawed, incorporate foreign DNA, and still contribute to development of normal tissues including germ cells of a chimera.

E. Semen and Embryo Sexing

Efforts to control the sex ratio at birth by separation of X- and Y-bearing spermatozoa have been largely unsuccessful, despite abundant claims to the contrary (reviewed by Amann and Seidel, 1982; Batzofin, 1987; Gledhill, 1988; Seidel, 1988; Amann, 1989). Available data often are selected and/or reflect small sample sizes; nonetheless, reports of positive results deserve careful examination.

The ultimate test of a semen-sexing procedure is a consistent and predictable shift in sex ratio after insemination of a large number of females with treated semen. Such breeding trials are expensive and time-consuming and usually are preceded by laboratory tests aimed at detecting enrichment of semen samples with either X-bearing or Y-bearing spermatozoa. Among such laboratory tests are detection of a fluorescent spot (F-body) on the Y chromosome after staining with quinacrine mustard (Ogawa et al., 1988); flow cytometry, which separates spermatozoa based on differences in DNA content (Pinkel et al., 1985); fusion of spermatozoa with zona-free hamster ova and analysis of metaphase chromosome spreads (Brandriff et al., 1986); and use of Y-specific DNA probes.

Procedures for separation of X- and Y-bearing spermatozoa are based on detecting differences resulting from dissimilarities in the X and Y chromosomes. The procedure for which positive results are most frequently reported involves allowing spermatozoa to swim through an albumin gradient (Beernink and Ericsson, 1982). Samples collected from beneath the gradient have been reported to be enriched for Y-bearing spermatozoa and resulted in 73–76% males when sexed human semen was used for insemination [cited as personal communications by Batzofin (1987) and Amann (1989)]. Results with semen from domestic animals have failed to provide shifts in the sex ratio, however (Beal et al., 1984). Flow cytometric techniques have been used in domestic animals to separate spermatozoa into two populations based on differences in the amount of DNA (Garner et al., 1983); the X chromosome in domestic animals contains slightly more DNA than does the Y chromosome. The procedure requires sophisticated equipment and usually renders treated spermatozoa infertile. Recent modifications in procedures have resulted in fertile sperm cells after separation by flow cytometry and yielded encouraging results in regard to sex of offspring produced from sexed semen (Morrell *et al.*, 1988; Johnson *et al.*, 1989). Other procedures for which positive results have been reported but for which limited data are available include separation of X- and Y-bearing spermatozoa by antibodies to H-Y antigen (Zavos, 1983; Bradley and Heslop, 1988), laminar-flow fractionation (Sarkar *et al.*, 1984), electrophoresis (Mohri *et al.*, 1986), density-gradient centrifugation (Iizuka *et al.*, 1987), and Sephadex gel filtration (Adimoelja, 1987).

An alternative approach to affecting the sex ratio at birth is identification of embryonic sex in conjunction with embryo transfer procedures; with this approach only embryos of the desired sex are transferred to recipients and allowed to develop to term (reviewed by Anderson, 1987a). Cytological procedures have been used whereby metaphase sex chromosomes prepared from a biopsy of an embryo are examined (Betteridge *et al.*, 1981; King, 1984; Picard *et al.*, 1985). Although accurate, the low proportion of embryos from which a metaphase chromosomal spread can be prepared limits widespread use. Differences in X-linked enzyme activity in embryos prior to X-chromosome inactivation can also be used to distinguish male from female embryos (Williams, 1986; Monk and Handyside, 1988). Detection of embryonic male-specific antigen (usually referred to as H-Y antigen) by indirect immunofluorescence has resulted in approximately 85% accuracy of sexing embryos in the mouse (White *et al.*, 1983), cow (White *et al.*, 1987a), pig (White *et al.*, 1987b), and sheep (White *et al.*, 1987c). Assessment of fluorescence is highly subjective, and results can be affected by embryo quality, stage of development, and other

variables. The newest approach to embryo sexing involves use of a DNA probe for Y-specific repetitive sequences (Bondioli *et al.*, 1989). A biopsy of an embryo is removed by micromanipulation and subjected to DNA analysis. A high degree of accuracy in identifying male versus female embryos is possible.

F. Nuclear Transplantation

Nuclear transplantation procedures are useful in studying the ability of nuclei from more or less differentiated tissue to support normal development in cytoplasm at another stage of development; another use is to produce groups of genetically identical offspring (clones). Procedures for removal of diploid nuclei from embryos and transplantation to oocytes have been available for use in amphibians for many years (Briggs and King, 1952; reviewed by McKinnell, 1981). Repeatable procedures for successful nuclear transplantation in mammals have been developed only recently. Current procedures involve fusion of blastomeres or karyoplasts (nuclei surrounded by a small amount of cytoplasm and cell membrane) with enucleated oocytes, zygotes, or blastomeres (cytoplasts). Developmental stages of the karyoplast and cytoplast and nuclear/cytoplasmic interactions are important variables in the success of nuclear transplantation procedures (Howlett *et al.*, 1987; Smith *et al.*, 1988). Mouse karyoplasts from four- and eight-cell blastomeres have only limited potential to support further development when fused to enucleated zygotes (McGrath and Solter, 1984); when fused with enucleated two-cell blastomeres, however, they are capable of development into blastocysts *in vitro* (Robl *et al.*, 1986) and to term *in vivo* (Tsunoda *et al.*, 1987).

Cytoplasts derived from mature oocytes may be better able to reprogram a later-stage nucleus to support normal development after transplantation than is a zygote or two-cell embryo (Robl *et al.*, 1987; Prather *et al.*, 1987; Robl and Stice, 1989). Nuclei from eight-cell mouse blastomeres do not appear to support normal development after transplantation (McGrath and Solter, 1986; Robl *et al.*, 1986). Nuclei from domestic species may retain the capability of being reprogrammed and supporting development in oocytes later in development than do mouse nuclei. Nuclei from bovine and ovine eight- and 16-cell blastomeres (Willadsen, 1986; Prather *et al.*, 1987) and from bovine morulae (Marx, 1988) and ovine ICM (Smith and Wilmut, 1989) will support development to term after transplantation to enucleated oocytes. The latter observations do not support the hypothesis that nuclear totipotency in mammals is lost at the time of embryonic genome activation (Surani *et al.*, 1987).

G. Gene Transfer

Perhaps no area of biological research has developed as rapidly in recent years as molecular genetics. In addition to creating an understanding of gene control and action, molecular techniques can be applied to modifying physiological processes (e.g., accelerated growth; Palmiter *et al.*, 1982) and correcting simply inherited genetic defects. Current gene-transfer procedures entail manipulation of the pronuclear stage zygote. Foreign DNA is most frequently inserted into ova by microinjection, retrovirus-mediated incorporation, or indirectly via pluripotent embryonic stem cells. Recent developments include introduction of DNA by injection of dissected chromosome frag-

ments (Richa and Lo, 1989) and using sperm cells as vectors (Lavitrano *et al.*, 1989). The laboratory mouse embryo has been used for most gene-transfer experiments (Pedersen, 1988); frequencies of incorporation and expression of foreign DNA by mouse embryos are substantially higher at this time than can be achieved with domestic animal embryos (Brinster *et al.*, 1985; Hammer *et al.*, 1985; Ward *et al.*, 1986; Biery *et al.*, 1988).

For comprehensive discussions of gene-transfer procedures and their applications to livestock production, the reader is referred to recent reviews on the subject (Petters, 1986; Land and Wilmut, 1987; Renard and Babinet, 1987; Simons and Land, 1987; Strojek and Wagner, 1988). The reader should be aware, however, that the field is advancing so rapidly that even up-to-date literature reviews on the subject soon become outdated.

VII. Embryo Transfer: Applications and Procedures

Transfer of embryos from one female to another has been a useful research tool when it is desirable to separate fetal and maternal genetic effects. Commercialization of bovine embryo transfer in the early 1970s was exceedingly important in stimulating basic and applied research in embryo development. Early embryo transfer and associated procedures usually included superovulation of the donor female and techniques for collection and transfer of the embryo. As new procedures for embryo manipulation were developed, they often were included under the general term embryo transfer. Today, "embryo transfer" is used to include a number of the associated procedures discussed under previous sections (Section VI, A–G). The subject of embryo transfer has been reviewed extensively (e.g., Seidel, 1981; Seidel and Seidel, 1982; Kuzan and Seidel, 1986), and a complete discussion will not be repeated here. Interested readers are referred to January issues of the scientific journal *Theriogenology*, in which the proceedings of the annual meeting of the International Embryo Transfer Society are published. Symposium papers and abstracts on a large number of topics related to embryo transfer are presented.

An important use of embryo transfer in farm animals is production of offspring from valuable females. The value of the donor female may be based on either her economic value or her genetic value; the latter is often the reason given for entering a particular female in an embryo transfer program, but the former usually provides the underlying justification. Under appropriate circumstances embryo transfer can augment conventional genetic selection, but estimated genetic progress usually is considered to be more modest than can be achieved with artificial insemination of semen from proven sires (Bradford and Kennedy, 1980; VanVleck, 1982). Embryo transfer procedures are useful under some circumstances to produce offspring from females able to produce fertile gametes but not able to carry a pregnancy to term. Interspecies embryo transfer may provide an avenue for propagation of endangered species (Anderson, 1988). An application important especially to the swine industry is to reduce the risk involved with introduction of new animals into a closed herd, as is required with replacement of herd sires. The risk of new disease organisms being introduced is thought to be lessened by bringing in embryos that develop to term in uteri of females already in the herd, a contention supported by results of research on disease transmission by embryos (Singh, 1987). International exchange of germ plasm also is facilitated by embryo transfer, although restrictions based on possible disease transmission may curtail such activity.

Primarily nonsurgical embryo collection and transfer procedures are used in cattle and horses; in sheep, goats, and pigs embryos usually are collected and transferred via laparotomy. Research continues on development of nonsurgical and laparoscopic procedures for these species as well. One of the more problematic aspects of the entire embryo-transfer procedure is variable response by the donor to superovulatory treatment; despite much attention, and great progress in other associated procedures, superovulatory response remains unpredictable. One can only surmise how this and other problems that limit usefulness of embryo transfer in farm animals will be solved. It should be remembered, however, that many of today's routine procedures were considered impractical or impossible only a decade or two ago. Embryo transfer and its associated procedures hold exciting and perhaps even unknown applications for the future.

References

Adimoelja, F. X. A. (1987). "New Horizons in Sperm Cell Research" (H. Mohri, ed), p. 491. Japanese Scientific Society Press, Tokyo.

Ahuja, K. K., and Bolwell, C. P. (1983). *J. Reprod. Fertil.* **69,** 49.

Allen, R. L., and Wright, R. W., Jr. (1984). *J. Anim. Sci.* **59,** 1657.

Amann, R. P. (1989). *Theriogenology* **31,** 49.

Amann, R. P., and Seidel, G. E., Jr. (1982). "Prospects for Sexing Mammalian Sperm," Colorado Associated University Press, Boulder, Colorado.

Anderegg, C., and Markert, C. L. (1986). *Proc. Natl. Acad. Sci. USA.* **83,** 6509.

Anderson, E., and Albertini, E. (1976). *J. Cell Biol.* **71,** 680.

Anderson, G. B. (1983). *Adv. Vet. Sci. Comp. Med.* **27,** 129.

Anderson, G. B. (1985). *J. Anim. Sci.* **61** (Suppl. 3), 1.

Anderson, G. B. (1987a). *Theriogenology* **27,** 81.

Anderson, G. B. (1987b). *J. Reprod. Fertil. Suppl.* **34,** 251.

Anderson, G. B. (1988). *Biol. Reprod.* **38,** 1.

Arave, C. W., Bunch, T. D., Mickelsen, C. H., and Warnick, K. (1987). *Theriogenology* **28,** 372.

Aref, I., and Hafez, E. S. E. (1973). *Obstet. Gynecol. Surv.* **28,** 679.

Ash, K., Anderson, G. B., BonDurant, R. H., Pashen, R. L., Parker, K. M., and Berger, T. (1988). *Theriogenology* **29,** 214.

Austin, C. R. (1951). *Aust. J. Sci. Res.* **4,** 581.

Austin, C. R. (1974). *Proc. Roy. Soc. Med.* **67,** 925.

Babinet, C., and Bordenave, G. R. (1980). *J. Embryol. Exp. Morphol.* **60,** 429

Baker, R. D., and Shea, B. F. (1985). *Theriogenology* **23,** 3.

Ball, G. D., Leibfried, M. L., Lenz, R. W., Ax, R. L., Bavister, B. D., and First, N. L. (1984). *Biol. Reprod.* **28,** 717.

Barros, C., and Yanagimachi, R. (1971). *Nature* **233,** 268.

Batzofin, J. H. (1987). *Urol. Clin. North Am.* **14,** 609.

Bavister, B. D. (1987). "The Mammalian Preimplantation Embryo." Plenum Press, New York.

Beal, W. E., White, L. M., and Garner, D. L. (1984). *J. Anim. Sci.* **58,** 1432.

Bedford, J. M. (1971). *J. Reprod. Fertil.* **25,** 211.

Beernink, R. J., and Ericsson, R. J. (1982). *Fertil. Steril.* **38,** 493.

Beier, H. M. (1968). *Biochim. Biophys. Acta* **160,** 289.

Benos, D. J., and Biggers, J. D. (1981). *In* "Fertilization and Embryonic Development in Vitro" (L. Mastroianni and J. Biggers, eds.), p. 283. Plenum Press, New York.

Berger, T., Davis, A., Wardrip, N. S., and Hedrick, J. L. (1989a). *J. Reprod. Fertil.* **86,** 559.

Berger, T., Turner, K. O., Meizel, S., and Hedrick, J. L. (1989b). *Biol. Reprod.* **40,** 525.

Berger, T., and Horton, M. B. (1988). *Gamete Res.* **19,** 101.

Betteridge, K. J., Hare, W. C. D., and Singh, E. L. (1981). *In* "New Technologies in Animals Breeding" (B. G. Brackett, G. E. Seidel, Jr., S. M. Seidel, eds.), p. 109. Academic Press, New York.

Biery, K. A., Bondioli, K. R., and DeMayo, F. J. (1988). *Theriogenology* **29,** 224 (abstr).

Biggers, J. D. (1987). *In* "The Mammalian Preimplantation Embryo" (B. D. Bavister, ed.), p. 1. Plenum Press, New York.

Binkerd, P. E., and Anderson, G. B. (1979). *Gamete Res.* **2,** 65.

Blandau, R. J. (1973). *In* "Handbook of Physiology" (R. O. Greep, ed.), Vol. II, Sect. 7, p. 151. Williams & Wilkins, Baltimore, Maryland.

Blandau, R. J., and Gaddum-Rosse, P. (1974). *Fertil. Steril.* **25,** 61.

Bodkhe, R. R., and Harper, M. J. K. (1973). *In* "Regulation of Mammalian Reproduction" (S. J. Segal, R. Crozier, P. A. Corfman, and P. G. Condliffe, eds.), p. 364. Charles C. Thomas, Springfield, Illinois.

Bondioli, K. R., Ellis, S. B., Pryor, J. H., Williams, M. W., and Harpold, M. M. (1989). *Theriogenology* **31,** 95.

BonDurant, R. H., Anderson, G. B., Boland, M. P., Cupps, P. T., and Hughes, M. A. (1982). *Theriogenology* **17,** 223.

Bowman, P., and McLaren, A. (1970). *J. Embryol. Exp. Morphol.* **25,** 203.

Brackett, B. G., Bousquet, D., Boice, M. L., Donawick, W. J., Evans, J. F., and Dressel, M. A. (1982a). *Biol. Reprod.* **27,** 147.

Brackett, B. G., Cofone, M. A., Boice, W. J., and Bousquet, D. (1982b). *Gamete Res.* **5,** 217.

Bradford, G. E., and Kennedy, B. W. (1980). *Theriogenology* **13,** 13.

Bradley, M. P., and Heslop, B. F. (1988). *J. Dairy Sci.,* **71,** *Suppl 1* (Abstr P318).

Brandriff, B. F., Gordon, L. A., Haendel, S., Sionger, S., Moore, D. H., and Gledhill, B. L. (1986). *Fertil. Steril.* **46** 678.

Brandriff, B., and Pedersen, R. A. (1981). *Science* **211,** 1431.

Brem, B., Tenhumberg, H., and Kraublich, H. (1984). *Theriogenology* **22,** 609.

Briggs, R., and King, T. J. (1952). *Proc. Natl. Acad. Sci. USA* **38,** 455.

Brinster, R. L. (1967). *J. Reprod. Fertil.* **13,** 413.

Brinster, R. L. (1974). *J. Anim. Sci.* **38,** 1003.

Brinster, R. L., Chen, H. Y., Trumbauer, M. E., Yagle, M. K., and Palmiter, R. D. (1985). *Proc Natl. Acad. Sci. USA 82,* 4438.

Bronson, R. A., and McLaren, A. (1970). *J. Reprod. Fertil.* **22,** 129.

Butler, J. E., Anderson, G. B., BonDurant, R. H., Pashen, R. L., and Penedo, M. C. T. (1987). *J. Anim. Sci.* **65,** 317.

Camous, S., Heyman, Y., Meziou, W., and Menezo, Y. (1984). *J. Reprod. Fertil.* **72,** 479.

Camous, S., Kopechy, V., and Flechon, J-E. (1986). *Biol. Cell* **58,** 195.

Carnevale, E. M., Squires, E. L., and McKinnon, A. O. (1987). *J. Anim. Sci.* **65,** 1775.

Cattanach, B. M., and Kirk, M. (1985). *Nature* **315,** 496.

Chang, M. C. (1951). *Nature* **168,** 997.

Chang, M. C. (1957). *Nature* **179,** 258.

Cheng, W. T. K., Moor, R. M., and Polge, C. (1986). *Theriogenology* **26,** 146 (abstr).

Cherr, G. N., Lambert, H., Meizel, S., and Katz, D. F. (1986). *Dev. Biol.* **114,** 119.

Chesna, P., Heyman, Y., Chupin, D., Procureur, R., and Menezo, Y. (1987). *Theriogenology* **27,** 221 (abstr).

Clegg, K. B., and Piko, L. (1983). *J. Embryol. Exp. Morphol.* **74,** 169.

Cole, H. H. (1975). *Biol. Reprod.* **12,** 194.

Crosby, I. M., Gandolfi, F., and Moor, R. M. (1988). *J. Reprod. Fert.* **82,** 769.

Cross, M. H., and Brinster, R. L. (1969). *Exp. Cell Res.* **58,** 125.

Cross, N. L., Morales, P., Overstreet, J. W., and Hanson, F. W. (1988). *Biol. Reprod.* **38,** 235.

Cross, P. C., and Brinster, R. L. (1970). *Biol. Reprod.* **3,** 298.

Cummins, J. M. (1982). *Gamete Res.* **6,** 53.

Davis, J., and Hesseldahl, H. (1971). *In* "The Biology of the Blastocyst" (R. J. Blandau, ed.), p. 27. Univ. of Chicago Press, Chicago, Illinois.

Depypere, H. T., McLaughlin, K. J., Seamark, R. F., Warnes, G. M., and Matthews, C. D. (1988). *J. Reprod. Fertil.* **84,** 205.

Dobrowolski, W., and Hafez, E. S. E. (1970). *J. Anim. Sci.* **31,** 940.

Donahue, R. P. (1972). *In* "Oogenesis" (J. D. Biggers and A. W. Scheutz, eds.), p. 413. University Park Press, Baltimore, Maryland.

Downs, S. M., Schroeder, A. C., and Eppig, J. J. (1986). *Gamete Res.* **15,** 305.

Ducibella, T., and Anderson, E. (1975). *Dev. Biol.* **47,** 45.

Dukelow, W. R., and Riegle, G. D. (1974). *In* "Oviduct and Its Functions" (A. D. Johnson and C. W. Foley, eds.), p. 193. Academic Press, New York.

Dzuik, P. (1970). *J. Reprod. Fertil.* **22,** 277.

Enders, A. C., Liu, I. K. M., Bowers, J., Lantz, K. C., Schlafke, S., and Suarez, S. (1987). *Biol. Reprod.* **37,** 453.

Eppig, J., Ward-Bailey, P., and Coleman, D. (1985). *Biol. Reprod.* **33,** 1041.

Eppig, J. J., and Schroeder, A. C. (1986). *Theriogenology* **25,** 97.

Eyestone, W. H., Vignieri, J., and First, N. L. (1987). *Theriogenology* **27,** 228 (abstr).

Eyestone, W. H., and First, N. L. (1989). *Theriogenology* **31,** 191 (abstr).

Fehilly, C. B., Willadsen, S. M., and Tucker, E. M. (1984a). *J. Reprod. Fertil.* **70,** 347.

Fehilly, C. B., Willadsen, S. M., and Tucker, E. M. (1984b). *Nature* **307,** 634.

Ferris, S. D., Sage, R. D., and Wilson, A. C. (1982). *Nature* **295,** 163.

First, N. L., Short, R. E., Peters, J. B., and Stratman, F. W. (1968). *J. Anim. Sci.* **27,** 1037.

First, N. L., and Parrish, J. J. (1987). *J. Reprod. Fertil. Suppl.* **34,** 151.

Fiser, P. S., and Macpherson, J. W. (1976). *Can. J. Anim. Sci.* **56,** 33.

Florman, H. M., and Storey, B. T. (1982). *Dev. Biol.* **91,** 121.

Fraser, L. R. (1984). *In* "Oxford Reviews of Reproductive Biology, Vol. 6" (J. R. Clarke, ed.), p. 174. Clarendon Press, Oxford.

Freund, M. (1973). *In* "Regulation of Mammalian Reproduction" (S. J. Segal, R. Crozier, P. A., Corfman, and P. G. Condliffe, eds.), p. 352. Charles C. Thomas, Springfield, Illinois.

Gardner, R. L. (1972). *J. Embryol. Exp. Morphol.* **28,** 279.

Garner, D. L., Gledhill, B. L., Pinkel, D., Lake, S., Stephenson, D., VanDilla, M. A., and Johnson, L. A. (1983). *Biol. Reprod.* **28,** 312.

Gatica, R., Boland, M. P., Crosby, T. F., and Gordon, I. (1984). *Theriogenology* **21,** 555.

Gaunt, S. J. (1983). *J. Embryol. Exp. Morphol.* **75,** 277.

Gilula, N., Epstein, M., and Beers, W. (1978). *J. Cell Biol.* **78,** 58.

Gledhill, B. L. (1988). *Gamete Res.* **20,** 377.

Gordon, J. W., and Talansky, B. E. (1986). *J. Exp. Zool.* **239,** 347.

Goto, K., Kajihara, Y., Kosaka, S., Koba, M., Nakanishi, Y., and Ogawa, K. (1988). *J. Reprod. Fertil.* **83,** 753.

Greve, J. M., and Wassarman, P. M. (1985). *J. Mol. Biol.* **181,** 253.

Gwatkin, R. B. L., Williams, D. T., Hartman, J. F., and Kniazuk, M. (1973). *J. Reprod. Fertil.* **32**, 259.

Hahn, J., and Schneider, U. (1982). *Exp. Biol. Med.* **7**, 170.

Hammer, R. E., Pursel, V. G., Rexroad, C. E., Jr., Wall, R. J., Bolt, D. J., Ebert, K. M., Palmiter, R. D., and Brinster, R. L. (1985). *Nature* **315**, 680.

Hamner, C. E. (1973). *In* "Regulation of Mammalian Reproduction" (S. J. Segal, R. Crozier, P. A., Corfman, and P. G. Condliffe, eds.), p. 203. Charles C. Thomas, Springfield, Illinois.

Handyside, A. H., (1980). *J. Embryol. Exp. Morphol.* **60**, 99.

Harper, M. J. K. (1988). *In* "The Physiology of Reproduction" (E. Knobil and J. Neill, eds.), p. 103. Raven Press, New York.

Hawk, H. W. (1983). *J. Dairy Sci.* **66**, 2645.

Hawk, H. W. (1988). *Theriogenology* **29**, 125.

Hedrick, J. L., and Wardrip, N. J. (1986). *Anal. Biochem.* **157**, 63.

Hedrick, J. L., and Wardrip, N. J. (1987). *Dev. Biol.* **121**, 278.

Hensleigh, H. C., and Hunter, A. G. (1985). *J. Dairy Sci.* **68**, 1456.

Heyman, Y., Camous, S., Fevre, J., Meziou, W., and Martal, J. (1984). *J. Reprod. Fertil.* **70**, 533.

Heyman, Y., Chesne, P., Chupin, D., and Menezo, Y. (1987a). *Theriogenology* **27**, 477.

Heyman, Y., Menezo, Y., Chesne, P., Camous, S., and Garnier, V. (1987b). *Theriogenology* **27**, 59.

Heyman, Y., and Menezo, Y. (1987). *In* "The Mammalian Preimplantation Embryo" (B. D. Bavister, ed.), p. 175. Plenum Press, New York.

Hoppe, P. C., and Illmensee, K. (1977). *Proc. Natl. Acad. Sci. USA* **74**, 5657.

Howlett, S. K., Barton, S. C., and Surani, M. A. (1987). *Development* **101**, 915.

Huang, T. T. F., Fleming, A. D., and Yanagimachi, R. (1981). *J. Exp. Zool.* **217**, 286.

Hughes, M. A., and Anderson, G. B. (1982). *Theriogenology* **18**, 257.

Hunter, R. H. F. (1973). *J. Exp. Zool.* **183**, 57.

Hunter, R. H. F. (1974). *Anat. Rec.* **178**, 169.

Hunter, R. H. F. (1975). *Br. Vet. J.* **131**, 681.

Hunter, R. H. F. (1981). *J. Reprod. Fertil.* **63**, 109.

Hunter, R. H. F. (1984). *J. Reprod. Fertil.* **72**, 203.

Hunter, R. H. F., Nichol, R., and Crabtree, S. M. (1980). *Reprod. Nutr. Develop.* **20**, 1869.

Hunter, R. H. F., Barwise, L., and King, R. (1982). *Br. Ver. J.* **138**, 225.

Hunter, R. H. F., and Nichol, R. (1983). *J. Exp. Zool.* **228**, 121.

Hunter, R. H. F., and Wilmut, I. (1982). *Anim. Reprod. Sci.* **5**, 167.

Hunter, R. H. F., and Wilmut, I. (1984). *Reprod. Nutr. Develop.* **24**, 597.

Iizuka, R., Kaneko, S., Aoki R., and Kobayashi, T. (1987). *Hum. Reprod.* **2**, 573.

Iritani, A., Utsumi, K., Miyake, M., Hosoi, Y., and Saeki, K. (1988). *Annals NY Acad. Sci.* **541**, 583.

Johnson, L. A., Flook, J. P., and Hawk, H. W. (1989). *Biol. Reprod. Suppl.* **1**, 162 (Abstr).

Johnson, M. H. (1981). *Biol. Rev.* **56**, 463.

Johnson, M. H., Pratt, H. M. P., and Handyside, A. H. (1981). *In* "Cellular and Molecular Aspects of Implantation" (S. R. Glasser and D. W. Bullock, eds.), p. 55. Plenum, New York.

Johnson, M. H., and Maro, B. (1986). *In* "Experimental Approaches to Mammalian Embryonic Development" (J. Rossant and R. A. Pedersen, eds.), p. 35. Cambridge University Press, Cambridge.

Kane, M. T. (1987). *Theriogenology* **27**, 49.

Kelly, S. J., Mulnard, J. G., and Graham, C. F. (1978). *J. Embryol. Exp. Morphol.* **48**, 37.

King, W. A. (1984). *Theriogenology* **21**, 7.

Kreitmann, O., and Hodgen, G. D. (1981). *J. Am. Med. Assoc.* **246**, 627.

Krishnan, R. S., and Daniel, J. C., Jr. (1967). *Science* **158**, 490.

Kuzan, F., Flemming, A. D., and Seidel, G. (1984). *Fertil. Steril.* **41**, 766.

Kuzan, F. B., and Seidel, G. E., Jr. (1986). *In* "Manipulation of Mammalian Development" (R. B. L. Gwatkin, ed.), p. 249. Plenum Press, New York.

Kuzan, F. B., and Wright, R. W., Jr. (1982). *J. Anim. Sci.* **54**, 811.

Lambert, R. D., Sirard, M. A., Bernard, C., Beland, R., Rioux, J. E., Lelerc, P., Menard, D. P., and Bedoya, M. (1986). *Theriogenology* **25**, 117.

Land, R. B., and Wilmut, I. (1987). *Theriogenology* **27**, 169.

Lavitrano, M., Camaioni, A., Fazio, V. M., Dolci, S., Farace, M. G., and Spadafora, C. (1989). *Cell* **57**, 717.

LeDouarin, N., and McLaren, A. (1984). "Chimeras in Developmental Biology." Academic Press, London.

Lehn-Jensen, H., and Willadsen, S. M. (1983). *Theriogenology* **19**, 49.

Leibfried-Rutledge, M. L., Critser, E. S., Parrish, J. J., and First, N. L. (1989a). *Theriogenology* **31**, 61.

Leibfried-Rutledge, M. L., Florman, H. M., and First, N. L. (1989b). *In* "The Molecular Biology of Fertilization" (H. Schatten and G. Schatten, eds.), p. 259. Academic Press, San Diego, California.

Leibo, S. P. (1984). *Theriogenology* **21**, 767.

Lindner, G. M., Anderson, G. B., BonDurant, R. H., and Cupps, P. T. (1983). *Theriogenology* **20**, 311.

Longo, F. J. (1974). *Biol. Reprod.* **11**, 22.

Longo, F. J., and Anderson, E. (1969). *J. Ultrastruct. Res.* **29**, 86.

Mann, J. R. (1988). *Biol. Reprod.* **38**, 1077.

Mann, J. R., and Lovell-Badge, R. H. (1984). *Nature* **310**, 66.

Markert, C. L. (1982). *J. Hered.* **73**, 390.

Markert, C. L. (1983). *J. Exp. Zool.* **228**, 195.

Markert, C. L., and Petters, R. M. (1977). *J. Exp. Zool.* **201**, 295.

Markert, C. L., and Petters, R. M (1978). *Science* **202**, 56.

Martin, G. R. (1981). *Proc. Natn. Acad. Sci. USA* **78**, 7634.

Marx, J. L. (1988). *Science* **239**, 463.

Massip, A., VanDerZwalmen, P., and Ectors, F. (1987). *Theriogenology* **27**, 69.

Massip, A., and VanDerZwalmen, P. (1984). *Vet. Rec.* **115**, 327.

Mattner, P. E., and Braden, A. W. H. (1969). *Aust. J. Biol. Sci.* **22**, 1069.

McGrath, J., and Solter, D. (1984a). *Cell* **37**, 179.

McGrath, J., and Solter, D. (1984b). *Science* **226**, 1317.

McGrath, J., and Solter, D. (1986). *J. Embryol. Exp. Morphol. Suppl.* **97**, 277.

McKinnell, R. G. (1981). *In* "New Technologies in Animal Breeding" (B. G. Brackett, G. E. Seidel and S. M. Seidel, eds.), p. 163. Academic Press, New York.

McLaren, A. (1970). *J. Embryol. Exp. Morphol.* **23**, 1.

McLaren, A. (1976). "Mammalian Chimaeras." Cambridge University Press Cambridge.

McRorie, R. A., and Williams, W. L. (1974). *Annu. Rev. Biochem.* **43**, 777.

Meinecke-Tillman, S., and Meinecke, B. (1984). *Nature* **307**, 637.

Mintz, B. (1970). *Symp. Int. Soc. Cell Biol.* **9**, 15.

Mintz, B., and Illmensee, K. (1975). *Proc. Natl. Acad. Sci. USA* **75**, 3585.

Modlinski, J. A. (1980). *J. Embryol. Exp. Morphol.* **60**, 153.

Mohri, H., Oshio, S., and Kaneko, A. (1986). *In* "Progress in Developmental Biology, Part A" p. 179. A. R. Liss, New York.

Monk, M., and Handyside, A. H. (1988). *J. Reprod. Fertil.* **82**, 365.

Moore, N. W. (1973). *J. Reprod. Fert., Suppl.* **18**, 111.

Moore, N. W., Adams, C. E., and Rowson, L. E. A. (1968). *J. Reprod. Fertil.* **17**, 527.

Morrell, J. M., Keeler, K. D., Noakes, D. E., Mackenzie, N. M., and Dresser, D. W. (1988). *Vet. Rec.* **122**, 322.

Morton, D. B., and Glover, T. D. (1974). *J. Reprod. Fertil.* **38**, 131.

Nagashima, H., Kato, Y., Yamakawa, H., and Ogawa, S. (1988). *Theriogenology* **29**, 280.

Niemann, H., Brem, G., Sacher, B., Smidt, D., and Drausslich, H. (1986). *Theriogenology* **25**, 519.

O'Rand, M. G., and Fisher, S. J. (1987). *Dev. Biol.* **119**, 551.

Ogawa, S., Yamakawa, H., Yamanoi, J., Nishida, S., Kano, Y., Takeshima, T., Tauchi, K., and Nagashima, H. (1988). *Theriogenology* **29**, 1083.

Otani, H., Yokoyama, M, Nozawa-Kimura, S., Tanaka, O., and Katsuki, M. (1987). *Develop. Growth Differ.* **29**, 373.

Overstreet, J. W. (1982). *In* "Mechanism and Control of Animal Fertilization" (J. F. Hartman, ed.), p. 499. Academic Press, New York.

Overstreet, J. W., Cooper, G. W., and Katz, D. F. (1978). *Biol. Reprod.* **19**, 115.

Overstreet, J. W., and Cooper G. W. (1979). *J. Reprod. Fertil.* **55**, 53.

Oxenreider, S. L., and Day, B. N. (1965). *J. Anim. Sci.* **24**, 413.

Ozil, J.-P., Heyman, Y., and Renard, J.-P. (1982). *Vet. Rec.* **110**, 126.

Palmiter, R. D., Brinster, R. L., Hammer, R. E., Trumbauer, M. E., Gosenfield, M. B., Birnberg, N. C., and Evans, R. M. (1982). *Nature* **300**, 611.

Papaioannou, V. E., and Ebert, K. M. (1986). *In* "Experimental Approaches to Mammalian Embryonic Development" (J. Rossant and R. A. Pedersen, eds.), p. 67. Cambridge University Press, Cambridge.

Parker, G. H. (1931). *Phil. Trans. R. Soc.* **219**, 381.

Parrish, J. J., Susko-Parrish, J. L., and First, N. L. (1985). *Theriogenology* **24**, 537.

Parrish, J. J., Susko-Parrish, J., Leibfried-Rutledge, M. L., Critser, E. S., Eyestone, W. H., and First, N. L. (1986). *Theriogenology* **25**, 591.

Parrish, J. J., Susko-Parrish, J., Winer, M. A., and First, N. L. (1988). *Biol. Reprod.* **38**, 1171.

Pedersen, R. A. (1988). *In* "The Physiology of Reproduction" (E. Knobil and J. Neill, eds.), p. 187. Raven Press, New York.

Peters, D. F., Anderson, G. B., and Cupps, P. T. (1977). *J. Anim. Sci.* **45**, 350.

Petters, R. M. (1986). *J. Anim. Sci.* **62**, 1759.

Petters, R. M., and Markert, C. L. (1980). *J. Hered.* **71**, 70.

Picard, L., King, W. A., and Betteridge, K. J. (1985). *Vet. Rec.* **117,** 603.

Pinkel, D., Garner, D. L., Gledhill, B. L., Lake, S., Stephenson, D., and Johnson, L. A. (1985). *J. Anim. Sci.* **60,** 1303.

Polzin, V. J., Anderson, D. L., Anderson, G. B., BonDurant, R. H., Butler, J. E., Pashen, R. L., Penedo, M. C. T., and Rowe, J. D. (1987). *J. Anim. Sci.* **65,** 325.

Prather, R. S., Barnes, R. L., Sims, M. M., Robl, J. M., Eyestone, W. H., and First, N. L. (1987). *Biol. Reprod.* **37,** 859.

Prather, R. S., and First, N. L. (1988). *J. Anim, Sci.* **66,** 2626.

Quinlivan, T. D., and Robinson, T. J. (1969). *J. Reprod. Fertil.* **19,** 73.

Rall, W. F. (1987). *Cryobiology* **24,** 387.

Rall, W. F., and Fahy, G. M. (1985). *Nature* **313,** 573.

Renard, J.-P., Heyman, Y., and Ozil, J.-P. (1980) *Vet. Rec.* **107,** 152.

Renard, J.-P., and Babinet, C. (1987). *Theriogenology* **27,** 181.

Richa, J., and Lo, C. W. (1989). *Science* **245,** 175.

Robl, J. M., Gilligan, B., Crister, E. S., and First, N. L. (1986). *Biol. Reprod.* **34,** 733.

Robl, J. M., Prather, R., Barnes, R., Eyestone, W., Northey, D., Gilligan, B., and First, N. L. (1987). *J. Anim. Sci.* **64,** 642.

Robl, J. M., and Stice, S. L. (1989). *Theriogenology* **31,** 75.

Rossant, J. (1976). *J. Embryol. Exp. Morphol.* **36,** 283.

Rossant, J., Croy, B. A., Clark, D. A., and Chapman, V. M. (1983). *J. Exp. Zool.* **228,** 223.

Rossant, J., and Frels, W. I. (1980). *Science* **208,** 419.

Rossant, J., and Vijh, K. M. (1980). *Dev. Biol.* **76,** 475.

Salisbury, G. W., and Hart, R. G. (1970). *Biol. Reprod. Suppl.* **2,** 1.

Sarkar, S., Jolly, D. J., Friedmann, T., and Jones, O. W. (1984). *Differentiation* **27,** 120.

Sawicki, J. A., Magnuson, T., and Epstein, C. J. (1982). *Nature* **294,** 450.

Schultz, B. A. (1986). *In* "Experimental Approaches to Mammalian Embryonic Development" (J. Rossant and R. A. Pedersen, eds.), p. 239. Cambridge University Press, Cambridge.

Seidel, G. E. (1981). *Science* **211,** 351.

Seidel, G. E., Jr. (1988). *Proc. 11th Int. Congr. Anim. Reprod.* **5,** 136.

Seidel, G. E., Jr., and Seidel, S. M. (1982). *In* "New Technologies in Animal Breeding" (B. G. Brackett, G. E. Seidel, Jr., and S. M. Seidel, eds.), p. 41. Academic Press, New York.

Sertich, P. L., Love, L. B., Hodgson, M. R., and Kenney, R. M. (1988). *Theriogenology* **30,** 947.

Shelton, J. N., and Szell, A. (1988). *Theriogenology* **30,** 855.

Simons, J. P., and Land, R. B. (1987). *J. Reprod. Fertil. Suppl.* **34,** 237.

Singh, E. L. (1987). *Theriogenology* **27,** 9.

Sirard, M. A., and First, N. L. (1988). *Biol. Reprod.* **39,** 229.

Skrzyszowska, M., and Smorag, Z. (1987). *Theriogenology* **27,** 276 (Abstr.).

Smith, L. C., Wilmut, I., and Hunter, R. H. F. (1988). *J. Reprod. Fertil.* **84,** 619.

Smith, L. C., and Wilmut, I. (1989). *Biol. Reprod.* **40,** 1027.

Smith, R., and McLaren, A. (1977). *J. Embryol. Exp. Morphol.* **41,** 79.

Spindle, A. I. (1982). *J. Exp. Zool.* **219,** 361.

Stevens, L. C. (1978). *Nature* **276,** 266.

Stewart, L. C. (1982). *J. Embryol. Exp. Morphol..* **67,** 167.

Stewart, T. A., and Mintz, B. (1981). *Proc. Natl. Acad. Sci. USA* **78,** 6314.

Strojek, R. M., and Wagner, T. W. (1988). *Genet. Eng. Principles Methods* **10,** 221.

Suarez, S. S., Katz, D. F., and Meizel, S. (1984). *Gamete Res.* **10,** 253.

Summers, P. M., Shelton, J. N., and Bell, D. (1983). *Anim. Reprod. Sci.* **6,** 91.

Surani, M. A. H., Barton, S. C., and Kaufman, M. H. (1977). *Nature* **270,** 601.

Surani, M. A. H., Barton, S. C., and Norris, M. L. (1984). *Nature* **308,** 548.

Surani, M. A. H., Barton, S. C., and Norris, M. L. (1986). *Cell* **45,** 127.

Surani, M. A. H., Barton, S. C., and Norris, M. L. (1987). *Biol. Reprod.* **36,** 1.

Surani, M. A. H., and Barton, S. C. (1984). *Dev. Biol.* **102,** 335.

Takeda, T., Hallowell, S. V., McGauley, A. D., and Hasler, J. F. (1986). *Theriogenology* **25,** 204 (abstr).

Takeda, T., Henderson, W. B., and Hasler, J. F. (1987). *Theriogenology* **27,** 285 (Abstr.).

Tarkowski, A. K., and Wroblewska, J. (1967). *J. Embryol. Exp. Morphol.* **18,** 155.

Tervit, H. R., and Rowson, L. E. A. (1974). *J. Reprod. Fertil.* **38,** 177.

Thibault, C., and Gerard, M. (1973). *Ann. Biol. Anim. Biochim. Biophys.* **13,** 145.

Trounson, A. O., and Moore, N. W. (1974). *J. Reprod. Fertil.* **41**, 97.

Tsunoda, Y., Yasui, T., Shioda, Y., Nakamura, K., Uchida, T., and Sugie, T. (1987). *J. Exp. Zool.* **242**, 147.

VanNiekerk, C. H., and Gerneke, W. H. (1966). *Onderstepoort J. Vet. Res.* **33**, 195.

VanVleck, L. D. (1982). *In* "New Technologies in Animal Breeding" (B. G. Brackett, G. E. Seidel, Jr., and S. M. Seidel, eds.), p. 222. Academic Press, New York.

Voelkel, S. A., Rorie, R. W., McFarland, C. W., and Godke, R. A. (1986). *Theriogenology* **25**, 207 (Abstr.).

Ward, K. A., Franklin, I. R., Murray, J. D., Nancarrow, C. D., Raphael, K. A., Rigby, N. W., Byrne, C. R., Wilson, B. W., and Hunt, C. L. (1986). *Proc. 3rd Wld. Congr. on Genetics Applied to Livestock Production*, p. 6, Lincoln, Nebraska.

Warfield, S. J., Seidel, G. E., Jr., and Elsden, R. P. (1986). *Theriogenology* **25**, 212 (Abstr).

Wassarman, P. M. (1988). *In* "The Physiology of Reproduction" (E. Knobil and J. Neill, eds.), p. 69. Raven Press, New York.

Weinberg, W. C., Howard, J. C., and Iannaccone, P. M. (1985). *Science* **227**, 524.

Wheeler, M. B., and Seidel, G. E., Jr. (1987). *Gamete Res.* **18**, 237.

White, K. L., Lindner, G. M., Anderson, G. B., and BonDurant, R. H. (1983). *Theriogenology* **19**, 701.

White, K. L., Anderson, G. B., and BonDurant, R. H. (1987a). *Biol. Reprod.* **37**, 867.

White, K. L., Anderson, G. B., Berger, T. J., BonDurant, R. H., and Pashen, R. L. (1987b). *Gamete Res.* **17**, 107.

White, K. L., Anderson, G. B., Pashen, R. L., and BonDurant, R. H. (1987c). *J. Reprod. Immunol.* **10**, 27.

Whittingham, D. G., Leibo, S. P., and Mazur, P. (1972). *Science* **178**, 411.

Willadsen, S. M. (1981). *J. Embryol. Exp. Morphol.* **64**, 167.

Willadsen, S. M. (1982). *In* "Mammalian Egg Transfer" (C. E. Adams, ed.), p. 185. CRC Press, Boca Raton, Florida.

Willadsen, S. M. (1986). *Nature* **320**, 63.

Willadsen, S. M., and Fehilly, C. B. (1983). *In* "Fertilization of the Human Egg in Vitro—Biological Basis and Clinical Applications," p. 353. Springer-Verlag, Berlin.

Williams, T. J. (1986). *Theriogenology* **25**, 733.

Williams, T. J., Elsden, R. P., and Seidel, G. E., Jr. (1984). *Theriogenology* **22**, 521.

Williams, T. J., and Moore, L. (1988). *Theriogenology* **29**, 477.

Wilmut, I. (1972). *J. Reprod. Fertil.* **31**, 513.

Wilmut, I. (1986). *In* "Manipulation of Mammalian Development" (R. B. L. Gwatkin, ed.), p. 217. Plenum Press, New York.

Wilmut, I., and Rowson, L. E. A. (1973). *J. Reprod. Fertil.* **33**, 352.

Wright, R. W., Jr., Anderson, G. B., Cupps, P. T., and Drost, M. (1976). *J. Anim. Sci.* **43**, 170.

Wright, R. W., Jr., and Bondioli, K. R. (1981). *J. Anim. Sci.* **53**, 702.

Wright, R. W., Jr., and O'Fallon, J. V. (1987). *In* "The Mammalian Preimplantation Embryo" (B. D. Bavister, ed.), p. 251. Plenum Press, New York.

Wudl, L., and Chapman, V. (1976). *Dev. Biol.* **48**, 104.

Yanagimachi, R. (1972). *J. Reprod. Fertil.* **28**, 477.

Yanagimachi, R. (1973). *In* "The Regulation of Mammalian Reproduction" (S. J. Segal, R. Crozier, P. A. Corfman, and P. G. Condliffe, eds.), p. 215. Charles C. Thomas, Springfield, Illinois.

Yanagimachi, R. (1988). *In* "The Physiology of Reproduction" (E. Knobil and J. Neill, eds.), p. 135. Raven Press, New York.

Yang, X., Chen, J., and Foote, R. H. (1988). *J. Reprod. Fertil. Abstr. Ser.* **1**, 13.

Zamboni, L. (1972). *In* "Biology of Mammalian Fertilization and Implantation" (K. S. Moghissi and E. S. E. Hafez, eds.), p. 213. Charles C. Thomas, Springfield, Illinois.

Zavos, P. M. (1983). *Theriogenology* **20**, 235.

Ziomek, C. A., and Johnson, M. H. (1980). *Cell* **21**, 935.

Ziomek, C. A., Johnson, M. H., and Handyside, A. H. (1982). *J. Exp. Zool.* **221**, 345.

The Development of the Conceptus and Its Relationship to the Uterus

P. F. FLOOD

I. Introduction

The central theme of this chapter is the morphological relationship between the developing conceptus and the uterus. Therefore it will largely be concerned with the placenta and fetal membranes, but to put these structures in their appropriate context I must also describe a variety of associated structures and events. These will include the uterus itself, the morphology of the preattachment embryo—with the rapid and astonishing changes that it undergoes—and some of the immunological problems posed by pregnancy.

The tubal stages of pregnancy are described in Chapter 8 and the endocrine changes that parallel the events considered here are primarily dealt with in Chapter 10, though I occasionally draw attention to features particularly closely related to maternal recognition of pregnancy.

A few essential definitions are appropriate at this stage. The "conceptus" is simply the products of conception and includes those parts of the placenta of embryonic origin and the embryo or fetus itself. The words "embryo" and "conceptus" are synonymous between fertilization and the formation of the extraembryonic membranes, but after that stage "embryo" applies only to the structures that give rise directly to the free-living adult. At a rather arbitrary stage of development the embryo becomes known as the fetus. This is usually when a cursory examination of the developing organism proper, as opposed to the extraembryonic structures, reveals its species and sex. The exact time varies somewhat from one species to another but is between 35 and 45 days gestation in the domestic animals.

Our knowledge of placentation, even in the species dealt with here, has now become so extensive that it is impossible to give a comprehensive account in this chapter. Hence my object is to provide a serviceable summary for the student, the clinician, and the general research worker. In so doing, my choice of material and the level of detail have inevitably been colored by my own interests and the scientific preoccupations of recent years. Those who require more detailed information are referred to the original publications and in particular to the classic review by Amoroso (1952), the magnificent compendium by Mossman (1987), the late Professor Amoroso's 80th birthday commemorative volume (*J. Reprod. Fertil. Suppl.* **31,** 1982), and Steven's (1975a) collection of essays. Those seeking a well-illustrated account of the essentials are referred to Latshaw's (1987) excellent

Figure 1 The uterus and cervix in the common domestic species. The uterine body has been opened from its dorsal side and the dorsal half of the cervix has been removed to reveal a dorsal section. Top left, sow: note that the ovary is contained in a loose and capacious ovarian bursa and that the uterine horns are very long and irregularly convoluted. The body of the uterus is partially divided by the velum uteri and the cervix is exceedingly tortuous. Top right, cow: The uterus of the cow is essentially similar to that of the pig except that the horns are shorter and the cervix is simpler, consisting of a series of fibrous rings (not all of these are shown). Lower left, bitch: The ovary is completely enclosed within the bursa and the horns are elongated. The inset shows the cervix in median section; the dotted line indicates the plane of the section depicted in the main drawing. Lower right, mare: The ovarian bursa is so shallow that it cannot enclose the large ovary. The uterine horns are similar in length to the undivided uterine body. They are shown curved backward for convenience though they are not permanently retained in this position as they are in cows. The cervix is very simple.

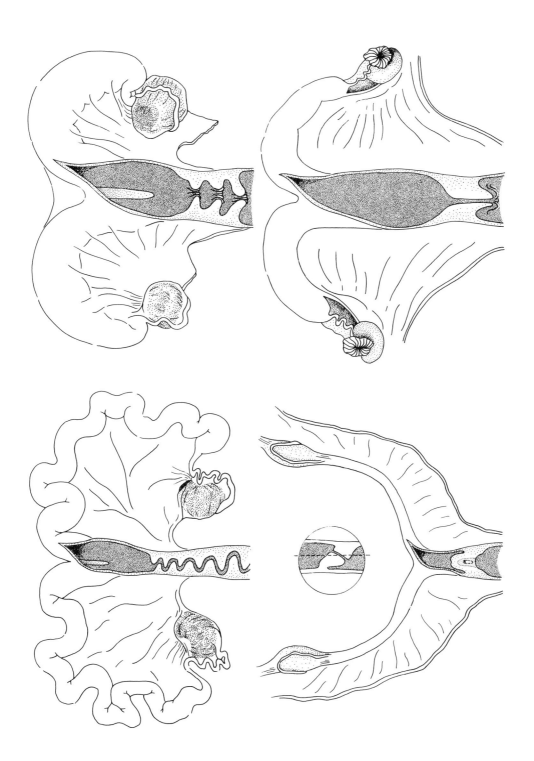

synopsis and Björkman's (1981) fine introduction to placental histology.

II. Anatomy of the Uterus

A. Macroscopic Features

1. The Body, Horns, and Cervix

Detailed anatomical descriptions of the uteri of the domestic animals are readily available (Dyce *et al.*, 1987; Schummer *et al.*, 1979) and only the principal features will be mentioned here. In all of them the uterus is Y-shaped. The arms of the Y, the uterine horns, are relatively long and end close to the ovaries. The body of the uterus, the stem of the Y, is usually short and ends at the cervix (Fig. 1).

In the dog and cat the long uterine horns are irregularly coiled in the dorsal abdomen. During pregnancy the middle parts of the horns sink to the abdominal floor, leaving the ovarian and cervical extremities in a comparatively dorsal position. The body of the uterus is particularly short and is not normally involved in placentation. The cervix lies within the abdomen; it is short and simple and does not protrude into the vagina. The uterine opening of the cervical canal is dorsal to the vaginal opening (Fig. 1).

In ruminants, the exterior of the uterus is deceptive because the fused caudal thirds of the uterine horns give the impression that quite a large uterine body is present. In reality an internal septum, velum uteri, extends caudally from the point of external fusion and separates the lumina of the two horns almost to the cervix, leaving only a very short undivided body. The shape of the uterine horns is characteristic; they taper gradually from their middles to the

ovarian ends and are curled laterally, ventrally, and caudally so they have the appearance of two symmetrical, loosely coiled snail shells with the ovary at the apex. This is possible because the ovaries have migrated caudally in the cloven-hooved species, perhaps as a result of the profound rearrangement of the abdomen consequent upon the massive development of the digestive tract. The ovaries no longer lie just caudal to the kidneys but are found close to the lateral walls of the abdominal cavity, at the pelvic inlet. The cervix consists of a series of more or less well-defined fibrous rings and protrudes somewhat into the vagina, forming a distinct dorsal fornix.

The arrangement of the uterus and ovaries of the pig is similar to that of the ruminants but the uterine horns are enormously elongated in association with litter bearing. They may reach 2 m in length during diestrus, though they are shorter during estrus (Rigby, 1968). The horns end fairly abruptly at their ovarian extremities, but they are curled in much the same way as those of ruminants. However, because of their great length, the curl is thrown in further irregular convolutions. The cervix is elongated and has a characteristic series of interlocking fibrous protrusions that arise from the lateral walls of the canal. The lumen of the cervix is extremely narrow at the uterine end but becomes progressively larger and more distensible in a caudal direction and eventually blends imperceptibly with the vagina.

The uterus of the horse is unique among the domestic species in that the length of the horns and the length of the body are similar. They are also rather similar in diameter, and the ovarian ends of the horns end bluntly. The cervix is large and remarkably distensible. It protrudes into the vagina and

has well-developed dorsal and ventral for-
nices.

2. *The Broad Ligaments and Blood Supply*

For convenience, the peritoneal folds attaching the reproductive tract to the body
wall, the mesovarium, the mesosalpinx, and
the mesometrium, can be referred to collectively as the broad ligaments. They carry the
uterine nerves and vasculature and contain
remarkably large amounts of smooth muscle, especially in ruminants and the pig. The
position of the broad ligaments and the arrangement of their vessels differs between
species and has important physiological consequences.

In the carnivores, in which the arrangement is the simplest, the broad ligaments
are attached to the dorsal wall of the abdomen and the uterus is supplied by two pairs
of arteries: the caudally directed uterine
branches of the ovarian arteries and the cranially directed uterine arteries that arise
from the pudendal arteries. The veins are
satellites of the arteries.

The broad ligament is also attached dor-
sally in the horse, but in this species there is
an additional pair of arteries that lies caudal
to the ovarian and cranial to the pudendal
arteries. These vessels, the uterine arteries,
are the largest supplying the uterus and
arise close to the bifurcation of the aorta.
Again, the veins follow a similar course to
the arteries, but it is the ovarian veins that
receive most of the uterine blood.

In ruminants and the pig the uterus is
also supplied by three pairs of vessels much
as in the horse, but in this case the pattern
of the veins is very different from that of
the arteries. While most blood reaches the
uterus through the uterine arteries, the vein
accompanying this vessel is small or nonexistent and the majority of blood leaving the

uterus does so via the large utero-ovarian
veins. These are satellites of the ovarian arteries and the two vessels lie in close apposition, the artery forming elaborate convolutions on the surface of the much larger vein
(Del Campo and Ginther, 1973) (Fig. 2).

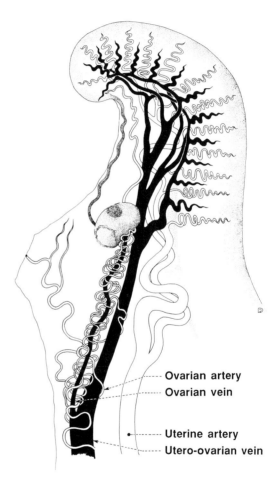

Ovarian artery
Ovarian vein

Uterine artery
Utero-ovarian vein

Figure 2 The vasculature of the left uterine horn of
sheep. Note the large size of the uterine artery and the
utero-ovarian vein, and the curious relationship between the ovarian vein and the tortuous ovarian artery.
The ovarian artery also ramifies extensively on the surface of the utero-ovarian vein, but only a small part of
this structure could be shown in this drawing.

B. Microscopic Anatomy

Naturally the outside of the uterus is covered by peritoneum everywhere but along the line of mesometrial attachment; beneath it lies the myometrium, consisting of outer longitudinal and inner circular layers of smooth muscle. The endometrium has a connective tissue matrix containing a rich vascular meshwork and the uterine glands; its inner surface is covered by the luminal or uterine epithelium. The connective tissue usually contains many macrophages, lymphocytes, and eosinophils. The uterine glands branch to a varying degree and normally have a columnar epithelium. The uterine epithelium is also usually columnar and delimits a lumen that most commonly appears as an irregular slit when seen in transverse section, though each species shows it own subtle variations. The lumen extends from near the site of mesometrial attachment toward the antimesometrial side of the uterus. If the uterus is sectioned longitudinally in a plane at right angles to the plane of the broad ligament, the lumen is usually seen to follow a zigzag course that is much longer than the uterus itself.

In most species the endometrium appears to be uniform throughout the uterus but, as Fabricius illustrated in 1604 (Adelmann, 1942), in ruminants it is differentiated into two distinct types, the caruncles and the intercaruncular area. The caruncles are pallid, raised, elliptical areas devoid of uterine glands; they appear early in development and are readily recognizable in the uterus of the bovine fetus where they form four longitudinal rows in each horn (Atkinson *et al.*, 1984). During pregnancy, the caruncles become attached to corresponding specialized areas of the allantochorion, the cotyledons, and together they form placen-

tal units known as placentomes. The reported average number of caruncles in the uteri of sheep ranges from 100 to 150 depending on breed (Alexander, 1964), and there is a similar though equally variable number in cattle (Atkinson *et al.*, 1984). These species are said to be polycotyledonary in contrast to the oligocotyledonary cervids, which have as few as six (Hamilton *et al.*, 1960).

III. Anatomy of the Conceptus

A. The Blastocyst

Division of the fertilized egg leads to the formation of the early morula, a fairly regular group of spherical cells held together by the zona pellucida. At about the eight-cell stage the blastomeres change in shape and become packed more closely together. This change is referred to as *compaction* and is a consequence of the formation of intercellular tight junctions. Fluid begins to accumulate in the irregular intercellular interstices. These coalesce to form a central cavity, the blastocoel. Simultaneously the embryonic cells differentiate into the outer trophoblast, destined to form the all-important interface between mother and fetus, and the inner cell mass, which will give rise to the embryo proper (Biggers *et al.*, 1988; Betteridge and Fléchon, 1988). In the next critical step the cells of the inner cell mass become organized into a regular sheet of columnar epithelium, which is elliptical in shape. This structure, the embryonic disc, becomes incorporated into the wall of the embryo, which now has the form of a hollow sphere.

The blastocyst or morula arrives in the uterus still encased in the zona pellucida. The zona is shed or "hatched" as a consequence of enzyme action (Menino and Williams, 1987) and embryonic movements (Fléchon and Renard, 1978). From then on the embryo is either directly exposed to the uterine enviroment or, as in horses, is further protected by a secondary acellular membrance (Bonnet, 1889). The times of hatching are shown in Table I.

B. Formation of the Primary Germ Layers

The trophoblast and the embryonic disc are collectively known as the ectoderm, which is the first of the primary germ layers. No sooner is the embryonic disc established than a second layer is formed by delamination of cells from its inner surface. These endoderm cells form another complete sphere within the first. Shortly thereafter a third and final layer is formed between the first two by a wave of cellular emigration from the embryonic disc. This is the meso-derm (Fig. 3). The parts of the primary layers lying beyond the confines of the embryonic disc, the extraembryonic parts, give rise to the extraembryonic membranes and hence to the embryonic components of the placenta.

C. Formation of the Extraembryonic Membranes

In the domestic species the extraembryonic membranes are formed by the simple folding process seen in the eggs of birds and reptiles. In rodents and higher primates membranogenesis has become so foreshortened that some of the steps are hard to equate with the ancient folding process, but this need not concern us further here. The allantois, that vascular outpouching of the hind gut so essential to mammalian placentation, is likewise unmodified in the domestic mammals in that it possesses a large fluid-filled cavity like that of birds and reptiles.

Table I
Duration of Tubal Transport and Time of Hatching in Domestic Species

| Species | Time after ovulation | | Source |
	Entry to uterus (h)	Hatching (days)	
Horses	120–144	7–8[a]	Betteridge et al. (1982) Flood et al. (1983)
Cattle	72–84	9–10	Betteridge and Fléchon (1988) Fléchon and Renard (1978)
Sheep	66–72	7–8	Holst (1974)
Pigs	46–48	6	Hunter (1974)
Dogs	132–156	11–12	Holst and Phemister (1971) Harvey et al. (1989)
Cats	120–160	11–12	Denker et al. (1978a, 1978b) Harper (1982) Leiser (1979, 1982)

[a]The zona pellucida is replaced by the capsule.

A

B

C

D

E

F

G

H

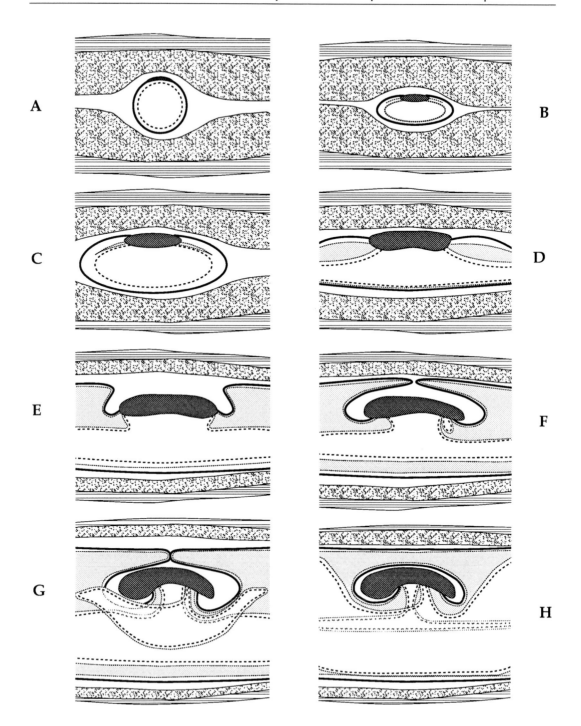

The allantoic cavity is absent or rudimentary in many rodents and primates, but the allantoic mesoderm is highly developed and continues to vascularize the definitive placenta.

Membranogenesis by folding can best be understood by following the series of diagrams shown in Fig. 3. The first vital event is the development of a narrow cavity separating the mesoderm into an inner and outer layer. The outer layer is the somatic mesoderm and the inner layer is the splanchnic mesoderm; the intervening cleft is the extraembryonic coelom or exocoel. These structures take their names from the fate of their homologs within the embryo. Thus derivatives of the splanchnic mesoderm are mainly associated with the viscera, while the somatic mesoderm gives rise to much of the connective tissue of the body wall. In the extraembryonic region the somatic mesoderm fuses with the overlying ectoderm to form the somatopleure. Likewise, the splanchnic mesoderm fuses with the underlying endoderm to form the splanchnopleure. Folding of the somatopleure and splanchnopleure results in the formation of the familiar amnion, chorion, and yolk sac.

The amniotic folds arise like an elliptical wave in the somatopleure surrounding the developing embryo, but instead of spreading outward the wave arches inward and eventually completely covers the embryo. It can be seen from Fig. 3F that this subdivides the somatopleure into inner and outer layers. The inner layer, the amnion, has ectoderm on its inner surface and mesoderm on its outer surface. The outer layer, the chorion, has ectoderm outside and mesoderm inside; the outer ectoderm is usually known as the trophoblast. The space between the mesodermal layers, the extraembryonic coelom, is now a large cavity.

Further folds form in the splanchnopleure and push inward beneath the embryo to subdivide the cavity that was once the blastocoel into the primitive gut and the yolk sac (Fig. 3E). The yolk-sac wall has extraembryonic endoderm on its surface (the surface that absorbs the yolk of telolecithal eggs) and splanchnic mesoderm on the outside. The splanchnic mesoderm of the yolk sac is the first site of hemopoiesis in the developing conceptus. In some mammalian species, especially rodents, the yolk sac is an important and elaborate structure throughout pregnancy and has an important absorptive function, but in the domestic species it is small and transient. It seems to have an absorbtive function during very early pregnancy in horses and carnivores.

The processes described above take place

Figure 3 Formation of the extraembryonic membranes by folding. The events depicted occur from the second to the fourth weeks of pregnancy. The diagrams do not represent any particular species but they apply best to the elongated conceptus of artiodactyls. The process is essentially similar in horses and carnivores but the shapes of the structures are slightly different. Myometrium = horizontal hatching. Endometrium = irregular stipple. Embryo proper = dark stipple. Extraembryonic coelom = light stipple. Ectoderm = thick black line. Mesoderm = thin dotted line. Endoderm = thick dashed line. Uterine lumen, amnion, yolk sac, allantois = white. (A) The blastocyst. (B) The extraembryonic mesoderm has just appeared. (C) The mesoderm has split into inner splanchnic and outer somatic layers. (D) The extraembryonic coelom has enlarged but the mesoderm has not yet fused with the ectoderm and endoderm. (E) The mesoderm has fused to the ectoderm and endoderm. (F) The amniotic folds are almost complete and the allantoic bud has appeared. (G) The allantois is almost as large as the yolk sac. (H) At the bottom center the allantois has fused with the chorion to form the first small piece of the allantochorion.

between 10 and 20 days gestation in the domestic animals. Toward the end of this period an event occurs that is essential to the formation of the definitive placenta in all the eutherian mammals, the apperance of the allantois. This is a highly vascular outpouching of the hindgut (Fig. 3F) that grows with such astonishing speed that within about a week of its formation, it comes to occupy all of the extraembryonic coelom. More significant still, it fuses with the chorion and in so doing brings an abundant vasculature to an otherwise avascular structure.

To understand the apparent complexity of the formation of the extraembryonic membranes and the placenta it is important to realize that angiogenesis is strictly confined to the splanchnic mesoderm and is never seen in the somatic mesoderm, much less in the ectoderm or endoderm. Formation of a placenta depends on the availability of a rich embryonic vascular bed, and in mammals this is provided by the splanchnic mesoderm of either the yolk sac or the allantois. The intimately fused allantois and chorion is known as the allantochorion, and it is this structure, in association with the endometrial tissues, that forms the major functional component of the placenta in all higher mammals.

IV. Movement of the Early Embryo

A. Entering the Uterus and Migration within It

The time of arrival of the embryo in the uterus is shown in Table I.

Early observations implied that there were marked differences between species in the behavior of the embryo in the uterus. In the domestic ruminants [though not in some African bovids (Buechner, 1961)] the embryo is usually found in the uterine horn on the same side as the corpus luteum (Scanlon, 1972), suggesting that it has only a limited tendency to migrate and rarely passes through the uterine body into the contralateral horn. This is particularly noticeable in cattle, where single fetuses are almost invariably found on the same side as the corpus luteum and where, if two ovulations occur from one ovary, one embryo passes through the uterine body in only about 10% of cases. Sheep embryos show a slightly greater tendency to migrate, and 8% of single embryos are found on the opposite side from the corpus luteum. In the case of double ovulations from one ovary, migration occurs on about 90% of occasions.

In the horse there is no correlation between the side of ovulation and the side of final attachment (Butterfield and Matthews, 1979), but there is a strong tendency for embryos to attach on opposite sides in successive years (Allen and Newcombe, 1981; Pascoe, 1982). Transuterine migration is also normal in the pig, and when embryos from black pigs are transferred to the tip of one uterine horn and embryos from white pigs are introduced into the other, they are found to be distributed almost at random when the uterus is opened close to term (Dziuk et al., 1964). Porcine embryos are fairly evenly distributed throughout the uterus by Day 11, even after unilateral ovariectomy that obliges the entire litter to begin their migration from the tip of one uterine horn (Dhindsa et al., 1967). Little further movement occurs after Day 12 (Polge and Dzuik, 1970). These observations demonstrate that the embryo readily passes through the uterine body in pigs and horses.

Recently, transrectal ultrasonography has elegantly confirmed that the horse conceptus moves rapidly throughout the uterus from the time it can first be observed until Day 15, movement being maximal between Days 11 and 14 (Ginther, 1986). Some insight into the speed of this process, which is apparently a consequence of myometrial activity, can be gained from the fact that an embryo was observed to pass from the caudal part of the left horn to its tip, back to the body, to the tip of the right horn, and back to the caudal part of the right horn during a single 2-h observation period. There is evidence that embryonic estrogens play a role in the migration in pig embryos (Pope *et al.*, 1982), and equine embryos are known to produce large amounts of estrogens (Flood and Marrable, 1975; Flood *et al.*, 1979; Zavy *et al.*, 1979). The conceptus has usually ceased to migrate by Day 16 and adopts a position at the bend in the uterus at the caudal end of either the right or left horn (Ginther, 1986). Here it occupies a chamber at the antimesometrial side of the uterine lumen (Enders and Liu, 1991).

The small spherical embryos of cats (Markee and Hinsey, 1933) and dogs (Holst and Phemister, 1971) also migrate freely from one uterine horn to another, apparently to achieve an equable distribution.

B. Orientation of the Blastocyst and Its Position in the Uterine Lumen

Before attachment begins, the conceptus develops a specific orientation in relation to the uterus (Wimsatt, 1975), and this orientation is constant within the well-defined mammalian taxonomic groups. In the Carnivora, Artiodactyla, and Perissodactyla (which include all the common domestic species) the early conceptus is arranged so that the yolk sac is found on the mesometrial side of the uterine lumen. The opposite occurs in rodents. It is difficult to determine the orientation of the early embryos of large species, but because the embryonic disc normally lies opposite the yolk sac it is likely that the embryonic disc lies on the antimesometrial side of the conceptus in all the domestic animals, as it apparently does in the horse (Enders and Liu, 1991).

The embryo also adopts a characteristic position within the uterus. In pigs the filamentous blastocyst is always found lying at the extreme mesometrial angle of the uterine lumen (Perry and Rowlands, 1962), and the situation is probably similar in other cloven-hooved species. Once fixed in position, the globular horse conceptus occupies a large chamber on the antimesometrial side of the uterine lumen (Enders and Liu, 1991). The position of the carnivore blastocyst is unknown before expansion begins, but once the conceptus reaches a diameter of a few millimeters it inevitably tends to occupy the center of the lumen and attachment rapidly involves the entire circumference of the uterine cavity.

C. Spacing and Embryo Location

In the horse the conceptus always settles down at the caudal end of one or other of the uterine horns (Ginther, 1986). In ruminants the early embryo is usually found in the middle of one of the uterine horns (Leiser, 1975; Lee *et al.*, 1977), and a bulge in the filamentous blastocyst is first seen ultrasonically in the middle of the uterine horn in cattle (Curran *et al.*, 1986).

In the litter-bearing species the fetuses are normally fairly evenly distributed along the uterine horns. How this even spacing is achieved is a matter for conjecture. Initially it seems likely that the embryos are distributed more or less at random, but once an embryo has settled down in the uterus it may create changes in its vicinity that are inconducive to the attachment of other embryos (Mossman, 1987). Alternatively, a settled embryo may initiate centrifugal waves of uterine contraction that discourage encroachment by other embryos (Dziuk, 1985). Differential uterine growth in the neighborhood of each conceptus might further contribute to regular spacing (Knight *et al.*, 1977), and this effect is probably enhanced by the death of some embryos that are too close to their neighbors. The death of malpositioned embryos is certainly frequent in the pig (Flood, 1974b). The filamentous nature of the blastocysts, to be described below, may also play a role in the even spacing of embryos.

V. Attachment and Placental Ontogeny

A. *Description of the Placenta*

Readers may notice that I have assiduously avoided the use of the word "implantation" in this account. The reason is simple. "Implantation" graphically describes the events that occur during early pregnancy in higher primates and rodents. In these species the blastocyst becomes rapidly embedded in the uterine wall and is in contact with vascular and connective tissue on all sides; in practical terms, it no longer lies within the lumen of the uterus. In the domestic species the conceptus remains within the uterine lumen, and the surrounding allantochorion acquires a close relationship with the endometrium over some or all of its surface. The details of the relationship differ between

species and change with time, but the blastocyst never becomes embedded in the uterine wall in the manner suggested by "implantation." For this reason, I like to follow Mossman's (1987) usage and refer to the establishment of the placenta in the domestic species as "attachment" and reserve the word "implantation" for circumstances in which it is more apt.

At this stage it is necessary to say something about the interhemal membrane, that vital layer that separates the fetal and maternal blood. At first there are six readily recognizable cellular layers separating the two blood streams. These are the maternal endothelium, the maternal connective tissue, the uterine epithelium, the trophoblast—the outermost cell layer of the chorion—the fetal connective tissue, and the fetal endothelium. To facilitate diffusion, the thickness of the interhemal membrane is progressively reduced during pregnancy and in some cases whole cell layers are lost. As a result the trophoblast may come to lie in contact with uterine epithelium, the maternal connective tissue, the maternal endothelium, or maternal blood. The placentae thus formed are referred to as epitheliochorial, syndesmochorial, endotheliochorial, and hemochorial, respectively. The hemochroial placentae, characteristic of higher primates and rodents, are not found in the domestic species, and the term syndesmochorial can only be applied to the placentae of certain ruminants with careful qualifications to be discussed later. This chapter is therefore chiefly concerned with epitheliochorial and endotheliochorial placentae.

Placentae may also be described according to the shape of the main area of haemotrophic exchange. Where this is disc shaped the placenta is "discoid," where it forms an equatorial zone around the placenta it is "zonary," and where the entire chorion is more or less equally involved, the placenta is "diffuse" (Fig. 4). Where the principal areas of exchange are fairly small, discrete and scattered, the placenta is said to be cotyledonary.

The events occuring at birth provide another means of placental classification. In deciduate placentae, maternal tissue is lost at parturition; in nondeciduate placentae it is not. Placentae that are contradeciduate are completely or partially retained in the uterus after birth and resorbed.

Despite the seemingly endless variety of placental structure, there are some consistent features. The trophoblast always forms a complete envelope around the conceptus (With the exception of the necrotic tips of artiodactyls and the inverted yolk sac of rodents), and the fetal endothelium is never lost. In general the fetal connective tissue is

retained, though it may be eliminated over small areas. It is axiomatic that in the normal placenta the fetal and maternal bloodstreams never intermingle.

While it is customary to discuss the initial establishment phases of placentation in isolation from a description of the definitive placenta, as our knowledge of both the early and late stages of the feto-maternal relationship increases, it becomes ever more clear that there is no definitive state of placental structure because it changes continuously to meet the changing circumstances of the fetus and its mother. I will therefore treat the pig, the horse, the ruminants and the domestic carnivores separately, dealing with events from attachment to parturition in sequence. There is probably no ideal order in which to discuss these species, but the order used here is one of increasing complexity of at least some placenta features.

B. The Pig

1. Blastocyst Expansion

Between 11 and 13 days gestation the blastocysts undergo an astonishing transformation. On Day 10 they are ellipsoidal structures 4–5 mm in length; by Day 13 many of them exceed 1 m in length but have decreased markedly in diameter (Marrable, 1971; Geisert et al., 1982b). In the majority of pigs, the exact time of expansion of individual embryos is variable, and spherical and very elongated blastocysts can be found in the same uterus; however, early development of the prolific Meishan breed of pig is highly synchronized, which perhaps accounts for their reduced level of embryonic mortality (Bazer et al., 1988). The fully expanded blastocysts can be seen as a fine glistening thread lying in the mesometrial angle

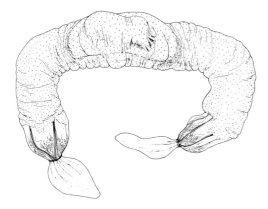

Figure 4 Pig placenta at about 50 days gestation. Note the fetus and the amniotic cavity extending a little beyond it in either direction. The remainder of the sac is filled with allantoic fluid. The extremities have undergone necrosis to form the necrotic tips but the old blood vessels are still visible as faint lines. Paraplacentae are present with functional elongated vessels. The small dots on the placental part of the allantochorion are areolae. From Ashdown and Marrable (1967).

of the uterine lumen. Oddly, these delicate structures arrange themselves end to end in the uterus and rarely overlap (Perry and Rowlands, 1962; Anderson, 1978): it would be tempting to believe that the filamentous blastocysts played a part in embryo spacing were it not for the inconvenient observation that the embryonic disc is inconsistently located along the filament. The primary function of the filamentous blastocyst is almost certainly related to maternal recognition of pregnancy, since if the number of embryos is reduced to less than four (Polge *et al.*, 1966) or blastocysts are excluded from even a fairly small part of the uterus (Dhindsa and Dziuk, 1968) pregnancy fails as a consequence of premature luteolysis.

The sudden transformation of the pig conceptus from the ellipsoid to the filamentous occurs at a time when its DNA content is increasing exponentially. However, the amount of DNA turns out to be unrelated to the shape of the blastocyst, and it is therefore clear that the change in form is brought about by cellular reorganization and not proliferation (Geisert *et al.*, 1982b). Indeed, it is hard to imagine that cell multiplication alone could account for the observed maximum rate of elongation of 30–45 mm/h.

2. Early Endometrial Responses

The earliest morphological difference between the pregnant and nonpregnant uterus is seen on Day 11 when the apical surfaces of the cells of the uterine epithelium become flattened where they impinge on the blastocyst (Geisert *et al.*, 1982a; King *et al.*, 1982). This coincides with an increase in uterine blood flow that is dependent on the presence of blastocysts (Ford and Christenson, 1979).

On Day 13 the endometrial surface acquires a remarkable green fluorescence in the region of the blastocyst (Keys and King 1988), and numerous small epithelial protrusions develop that interlock with matching depressions in the trophoblast (Dantzer, 1985). At the same time the subepithelial capillaries of the uterus show ultrastructural evidence of increased permeability in the region of the blastocyst. By Day 15 the uterine epithelium adjacent to the conceptus is rich in 17β-hydroxysteroid dehydrogenase, but this enzyme is not found in other parts of the epithelium or in the non-pregnant uterus (Flood, 1974a). By Day 16 of pregnancy the number of intraepithelial lymphocytes in the uterine lining is much lower than it is at an equivalent stage of the estrous cycle (King, 1988).

During the period of expansion, the blastocysts exert a luteotropic effect (van der Meulen *et al.*, 1988) and show a wide range of synthetic actions. They produce large amounts of estrogens (Perry *et al.*, 1973; Gadsby *et al.*, 1980; King and Ackerley, 1985; Bate and King, 1988; van der Meulen *et al.*, 1989), an interferon (Cross and Roberts, 1989), a major basic protein (Baumbach *et al.*, 1988) and prostaglandins (Lewis, 1989).

3. The Attachment Process

The trophoblast and the uterine epithelium become closely apposed to one another as soon as the elongation of the blastocyst is complete, and it seems likely that from then on movement is severely restricted by the endometrial protrusions and their trophoblastic caps, demonstrated by Dantzer (1985). King *et al.*, (1982) note that on Day 16, the distal ends of the conceptus float free from the wall of the opened uterus with gentle flushing but discrete areas around the middle are already firmly adherent. The microvilli of the two epithelial surfaces first begin to interdigitate with one another at this time or a little earlier, especially over the domes (Dantzer, 1985), which probably accounts for the adhesion observed. The microvillar junction (Fig. 5) thus formed is fully established by Day 18 (Crombie, 1970).

From Day 18 to Day 28 the highly vascular allantois grows at an astonishing rate, changing from a small bud hardly visible to the unaided eye into a cylindrical sac with rounded ends that is about 30 cm in length and 5 cm in diameter. By the end of this period the allantois has fused to the overlying chorion throughout and the microvillar junction has expanded to involve the entire trophoblast, with the exception of those areas opposite the openings of the uterine glands, and the very ends of the chorionic sacs. It is common to state that the attach-

ment between the trophoblast and the uterine epithelium is not particularly secure, but this is not the case during life or within a few minutes of death; then it requires great care, as emphasized by Dantzer (1984), to manually separate the trophoblast from the endometrium, and it may be impossible to do so without leaving trophoblastic remnants adhering to the endometrium. However, the least autolysis allows the two layers to slide apart easily.

Adherence between the two surfaces is made even more secure by the development of an elaborate system of folds that primarily serves to increase the area of the placental exchange (Fig. 6). Microscopic ridges are first evident at about Day 20, and by midgestation primary and secondary folds are visible macroscopically and carry primary

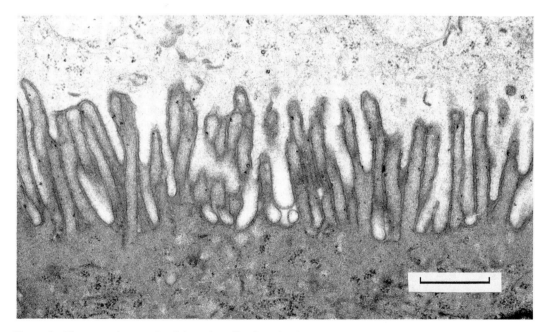

Figure 5 Electron micrograph of the microvillar junction between the uterine epithelium (bottom) and the trophoblast (top) of the pig, Day 31. Scale bar = 0.05 μm.

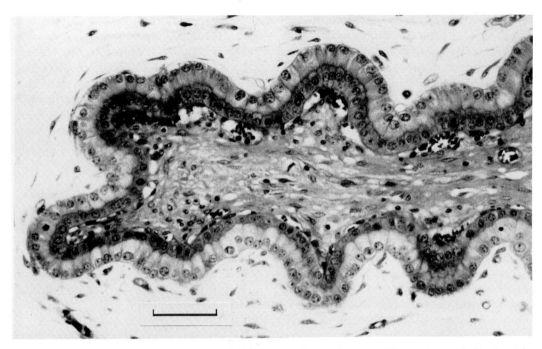

Figure 6 A fold of the porcine placenta, day 36. The maternal tissue is central. The uterine epithelium and the trophoblast are firmly adherent and there are abundant maternal subepithelial capillaries. H & E, Helly's fluid; scale bar = 100 μm.

and secondary microscopic ridges. These run circumferentially around the chorionic sac with the exception of the secondary ridges, which are perpendicular to the primary ones (Dantzer, 1984). The free edges of the ridges are thicker than the rest, so the maternal and fetal components are held together by the same principle as the zipper.

As soon as the ridges are well established it becomes clear that the placenta is far from uniform and that there is a degree of local specialization. Over the crests of the chorionic ridges both the maternal and fetal epithelia are low and by mid-gestation the capillaries on both sides deeply indent their respective epithelia; as term approaches this process has been carried so far that the in-

terhemal membrane is reduced to 2 μm or less. The reduction is brought about not only by the epithelial changes but by virtual elimination of the connective tissue. The basal laminae of the trophoblast and the fetal capillaries fuse and on the maternal side the corresponding basal laminae are brought into close apposition (Friess et al., 1980). It is therefore clear that the chorionic ridges are specialized for the transport of highly diffusible, light molecules.

In the intervening troughs, the epithelia are higher, the capillaries do not invade the epithelium, and the trophoblast often contains distinctive cytoplasmic droplets. These areas are assumed to transport higher-molecular-weight compounds (Dantzer et al.,

1981). Toward the end of gestation the epithelial height is reduced throughout the placenta; this is consistent with the increased need for respiratory exchange and the maturing metabolic capacity of the fetus. The microcirculation of the porcine placenta has been elegantly displayed by scanning electron microscopy of corrosion casts by Leiser and Dantzer (1988), who conclude that both

cross-current and countercurrent exchange systems exist.

4 Areolae

The trophoblast opposite the openings of the uterine glands is specialized for the absorption of glandular secretion; it forms an inverted cup over the gland mouth, which acts as a reservoir for the secretion (Fig. 7).

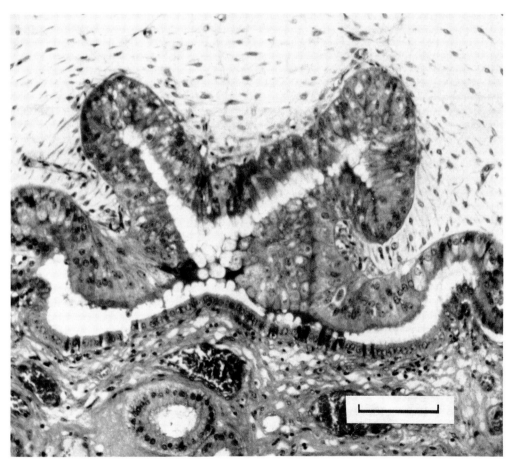

Figure 7 A porcine areola in section, day 36. The gland-filled endometrium is at the bottom. The cavity of the areola is central and contains precipitated uterine milk, which has pulled away from the surrounding epithelia during fixation. The trophoblast of the areola is very deep and at higher magnification reveals numerous absorptive vacuoles. H & E, Helly's fluid; scale bar = 100 μm.

The trophoblast lining the cup is an absorptive columnar epithelium with cytoplasm that is rich in endocytotic vacuoles (Friess *et al.*, 1981). Because the uterine glands are elaborately branched structures in the pig, their mouths, and therefore the areolae, are fairly infrequent (Fig. 4) but quite large when compared with those of other species. The areolae absorb uteroferrin secreted by the uterine glands and therefore have a specific role in iron transport (Raub *et al.*, 1985) as well as a general histotrophic function.

In addition to these so-called regular areolae there are a number of much larger, irregular pale areas known as irregular areolae (Amoroso, 1952a); these apparently receive secretion from a number of glands and have a modified uterine epithelium and an unspecialized trophoblast, the opposite of the condition in the regular areolae (Perry, 1981).

5 Fusion between the Amnion and Chorion

The amnion and chorion are extensively fused in the pig throughout the fetal period (Marrable, 1971) and form the so-called amniochorion. The only parts of the amnion that are not involved in amniochorion formation are the poles and a segment of one surface that lies adjacent to the remaining part of the allantoic cavity. The amniochorion in no less vascular than the allantochorion, owing to the persistence of a layer of vascular allantoic mesoderm between the amnion and chorion.

6 Necrotic Tips

The developing allantois does not normally invade the entire length of the filamentous blastocyst, but the delicate, undilated, blastocyst tips soon degenerate. Thus the entire surface of the chorion is supplied by a vigorous allantoic vasculature at about 28 days gestation. Remarkably, by 31 days, the ends of the allantochorion have become necrotic; the trophoblast and the allantoic endoderm have disappeared, leaving only a dehydrated connective tissue envelope to retain the allantoic fluid within. Blood flow ceases in these extremities, but degenerating hemoglobin leaves brownish tracts that indicate where vessels once existed (see Fig 4). These degenerate ends of the chorionic sac are usually known as necrotic tips. Their formation coincides with gonadal sex differentiation, and it is normally assumed that they prevent the formation of vascular anastomoses between adjacent allantochorionic circulations and the widespread occurrence of freemartins in pigs. The mechanism is not invariably effective (Crombie, 1972). Necrosis of the allantochorionic extremities is preceded by the differentiation of palid, relatively ischemic zones in the apposed endometrium (Flood, 1973). The ends of most chorionic sacs undergo a series of necroses, each of which increases the size of the necrotic tip. The necrotic tips of adjacent conceptuses are normally invaginated into one another in a way that makes them hard to separate. These invaginations also involve the living, but poorly vascularized, paraplacentae that form between the placenta proper and the necrotic tips (Ashdown and Marrable, 1967). In the latter half of pregnancy, the pallid endometrial zones separating the conceptuses are invaded and eventually obliterated by the highly vascular placenta zones, reflecting a need for increasing exchange area by the growing fetus. The adjacent placentae then become adherent and give the superficial impression of a continuous tube, but vascular interconnections do not form at this late stage (Ashdown and Marrable, 1970).

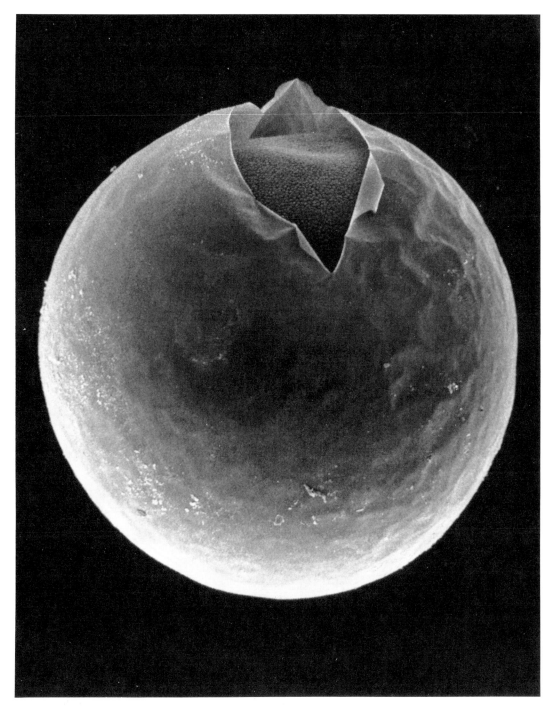

Figure 8 A scanning electon micrograph of a Day 10.5 equine embryo with the capsule opened to reveal the trophoblast within. Diameter = 1 mm. Kindly supplied by M. Guillomot and K. J. Betteridge.

C. The Horse

1. The Capsule of the Equine Embryo

The horse embryo is surrounded by an acellular membrane (Fig. 8) (Bonnet, 1889; Betteridge, 1989) that is about 4 μm thick and composed of a glycoprotein distinct from that of the zona pellucida (Bousquet *et al.*, 1987). The membrane or capsule is first detectable as an electron-dense deposit on the inner surface of the zona pellucida on day 6 (Flood *et al.*, 1982) (Fig. 9). The remnants of the true zona pellucida are shed from its outer surface by day 7 or 8, and by the end of the second week of pregnancy the capsule has become a strong, resilient, glistening envelope (Marrable and Flood, 1975). Its function is obscure. It may facilitate the movement of the embryo during its intrauterine ramblings and afford physical support. It may also provide a degree of protection from uterine microorganisms (Schlafer *et al.*, 1987) and potentially hostile maternal leucocytes and antibodies. The capsule has disappeared by the end of the fourth week after mating (Betteridge,

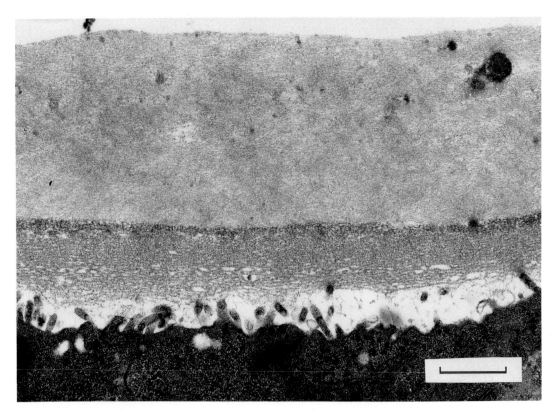

Figure 9 The zona pellucida of the horse with the new capsule forming on its inner surface. Note the distinct difference in texture and staining properties of the two embryonic coverings. The outer surface of the trophoblast cells can be seen at the bottom. Age 6.25 ± 0.25 days. Embryo diameter, 183 μm. Scale bar = 1μm. Prepared in collaboration with K. J. Betteridge.

1989), coincident with the sudden resumption of growth by the conceptus (Ginther, 1986). It is at this time that the conceptus first comes into direct contact with maternal tissues.

2. The Shape of the Conceptus

Since the equine conceptus remains more or less globular until the end of the fifth week and is roughly ellipsoidal for some time thereafter (Marrable and Flood, 1975), the interrelationships of the yolk sac and allantois are quite different from those of artiodactyls. Though the essentials were documented by Bonnet (1889), it was not until much later that beautifully detailed accounts appeared (van Nierkerk, 1965; van Nierkerk and Allen,, 1975; Ginther, 1979), some spurred by the advent of transrectal ultrasonography (Ginther, 1986). Briefly, the embryonic pole of the conceptus, with the associated exocoel, comes to lie at the antimesometrial side of the uterus at about the time that embryonic migration ceases. The forming amnion then lifts the embryo proper a little away from the antimesometrial wall of the uterus. During the fourth week of pregnancy the allantois grows at an astonishing rate and almost completely obliterates the exocoel; with the accumulation of allantoic fluid the embryo is raised further. The process is continued, as the

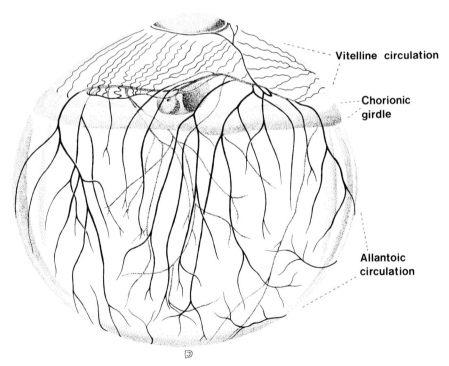

Figure 10 The equine conceptus at 34 days gestation. The mesometrial aspect is uppermost. Because some vessels are superimposed the vitelline and allantoic circulations occasionally appear to interconnect, but this is not the case.

growing allantois gradually overwhelms the yolk sac and the embryo eventually comes to lie close to the mesometrial side of the uterus (Fig. 10). Only with the formation and elongation of the umbilical cord will the fetus—as it may now be called—descend once more.

In the mare, invasion of the body and nonpregnant horns of the uterus by the conceptus is slower than it is in ruminants; the fetal membranes enter the uterine body by day 56 and the nonpregnant horn by Day 77 (Douglas and Ginther, 1975; Marrable and Flood, 1975). It would appear that placental development progresses fairly rapidly in the newly colonized areas of endometrium because the external surface of the allantochorion appears remarkably uniform.

One unusual feature of the fetal membranes of the horse, clearly recognized by Ruini in 1589 (Steven, 1982), is the complete separation of the allantochorion and amnion; they are extensively fused in the artiodactyla. As a consequence, the umbilical cord is distinctly divided into allantoic and amniotic parts.

3. Differentiation of the Trophoblast

Ewart (1897) described two types of trophoblast in the horse, the placental trophoblast covering the majority of the conceptus and the specialized band that encircles the vitelline pole of the conceptus and is known as the chorionic girdle (Fig. 10). These two cell types behave quite differently.

a. The Chorionic Girdle The chorionic girdle can be seen by the unaided eye during fourth week of gestation and reaches its full development by Day 35 (Allen *et al.*, 1973). At that time it is a belt surrounding

the conceptus close to the junction of the territories served by the allantoic and vitelline circulations. Its opalescent appearance contrasts sharply with the transparency of the remainder of the chorion. By Day 36, the trophoblast cells of the girdle, which are binucleate at this stage (F. B. P. Wooding, personal communication), have emigrated from the chorion, penetrated and then destroyed the adjacent uterine epithlium, and established themselves in the endometrial stroma (Allen *et al.*, 1973) (Fig.11). There they hypertrophy and eventually produce distinctive pale plaques about 2 cm in diameter, the endometrial cups. These form a discontinuous ring in the dilated, caudal end of the uterine horn in which the conceptus first settled. The established trophoblast cells of the endometrial cups are epithelioid and very large (Fig. 12); they are the source of equine chorionic gonadotropin. The endometrial cups persist for a variable period but have usually disappeared by Day 130 (Clegg *et al.*, 1954). The mechanisms responsible for their demise are discussed later.

b. The Placental Trophoblast The conceptus can be rolled out of the uterus without very obvious signs of adhesion until about 35 days of pregnancy, although the vascularized chorion and the endometrium are closely associated from 25 days onward and the underlying capillaries are dilated (Enders and Liu, 1991). By Day 40 of gestation some adhesion between maternal and embryonic epithelia is evident (Ginther, 1979), the chorionic surface has developed small primary villi (van Niekerk, 1965), and the opposing microvilli of the trophoblast and uterine epithelia have begun to interdigitate (Kurnosov, 1973). At Day 60, the

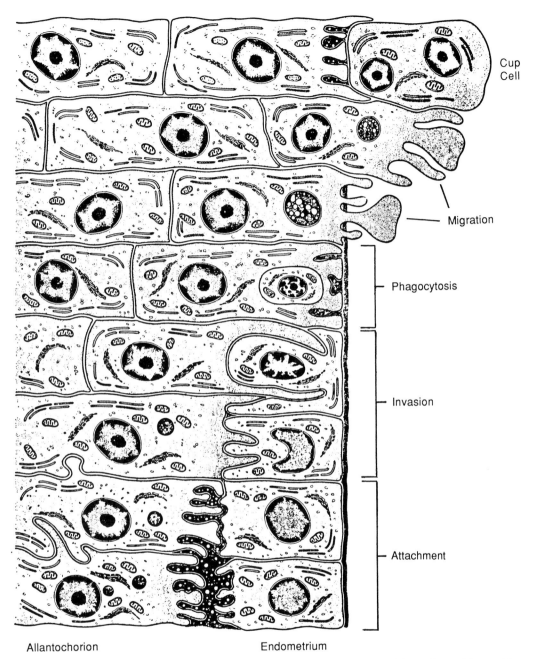

Cup
Cell

Migration

Phagocytosis

Invasion

Attachment

Allantochorion   Endometrium

Figure 11. This diagram shows the principle events leading to the formation of the endometrial cups. First (bottom), the girdle cells and the endometrium come into appostion. They then adhere to each other and the girdle cells invade and phagocytize the cells of the uterine epithelium. The migrant cells then disrupt the basement membrane and colonize the endometrial stroma before differentiating further. From Moor *et al.* (1975).

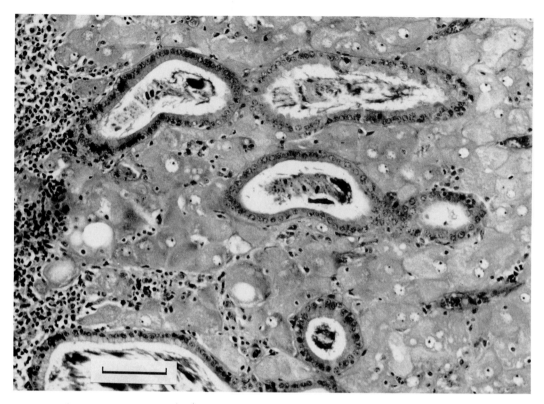

Figure 12 Endometrial cup, Day 56. Equine chorionic gonadotrophin is secreted by the large cells with large open nuclei seen packed between the endometrial glands. The endometrial surface is to the right and a wall of lymphocytes lies to the left between the endometrial cup cells and the normal endometrium. H & E, Helly's fluid; scale bar = 100 μm.

trophoblast and the maternal epithelium are closely attached to one another by a carpet of interlocking microvilli and folded in a fairly simple way (Samuel *et al.*, 1974). No generalized erosion of maternal tissues occurs, so the mare's placenta is said to be epitheliochorial.

At first glance, the equine placenta appears to be entirely uniform, with the exception of those areas that lie adjacent to the cervix and the endometrial cups and do not take part in normal placental exchange; they are pale and poorly vascularized. However, on close examination of the surface of the fully developed allantochorion, it is clear that the chorionic villi have secondary and even tertiary branches arranged in little tufts or microcotyledons (Turner, 1876). Each microcotyledon lies in a complementary and equally elaborate invagination of the endometrium. The endometrial glands, which are generally simple tubes, are numerous and open between the microcotyledons.

4. Microcotyledons

The primary villi forming the base of each microcotyledon first appear in a rudimentary form along the course of the developing allantoic vessels, and by 40 days gestation they cover most of the surface of the chorion (van Niekerk, 1965). By 100 days the primary villi have developed branches and, eventually, the fusion of adjacent groups of villi results in the formation of the thousands of globular microcotyledons seen throughout the latter half of pregnancy (Fig.13).

The thickness of the interhemal membrane is progressively reduced during the life of the microcotyledon. This is achieved by the formation of ever-deepening grooves in the inner surface of the trophoblast for the accommodation of the fetal capillaries, and progressive thinning of the uterine epithelium. While this results in almost complete elimination of fetal connective tissue from the interhemal membrane, all the

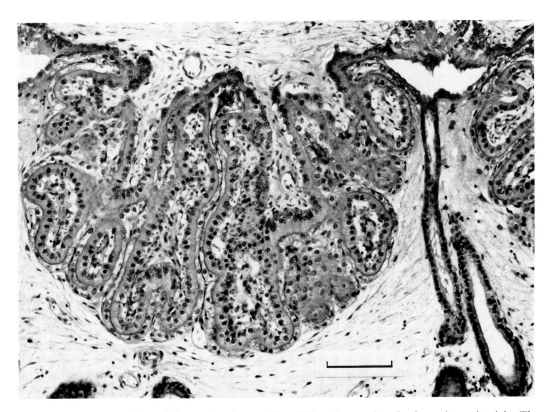

Figure 13 A microcotyledon of the equine placenta (Day 122) with a uterine gland opening to its right. The maternal tissues are at the bottom of the photomicrograph. Part of another microcotyledon is also visible. The relatively tall trophoblast cells are firmly attached to the low cuboidal cells of the maternal epithelium. Tissue shrinkage has opened the lumen of the arcade into which the gland opens. H & E, Helly's fluid; scale bar = 100 μm.

other layers remain, though their combined thickness is reduced to about one-third of the original (Samuel *et al.*, 1976). The oxygen-rich, uterine arteries enter the fetal aspect of the microcotyledons, and the uterine veins leave from the maternal side (Steven and Samuel, 1975) This may facilitate a countercurrent exchange system and explain why the oxygen tension in the umbilical vein can exceed that in the uterine vein under experimental conditions (Comline and Silver, 1970).

5. Histotrophic nutrition

While it is apparent that the microcotyledons are specialized for the rapid interchange of small molecules, it is equally clear that the areas between the microcotyledons are involved in histotrophic nutrition of the fetus. It is here that the endometrial glands open (Fig.13) and the adjacent trophoblast shows clear evidence of endocytosis (Samuel *et al.*, 1977). The space receiving the glandular secretion is not divided into circumscribed areolae as in the other species dealt with here; instead it forms a continuous meshwork of channels that surround the microcotyledons and extend throughout the placenta. These structures, which are of course homologous with the areolae of other species, are referred to as arcades (Ginther, 1979). The uterine glands are normally in groups of four and are active throughout pregnancy (Samuel *et al.*, 1977).

D. The Ruminants

The earliest stages of placentation are similar in ruminants and the pig, but later the situation is complicated by those two peculiarly ruminant features, the caruncles and the placental binucleate cells.

1. Blastocyst Expansion and Attachment

Like the porcine blastocyst, the blastocyst of ruminants undergoes a remarkable, though slightly less extreme, elongation before attachment (Fig.14). This begins on about Day 11 in sheep (Rowson and Moor, 1966; Bindon, 1971; Guillomot *et al.*, 1981) and a day later in cattle (Betteridge *et al.*, 1980;). In sheep the rate of increase in length is estimated to reach 1 cm/h and the filiform conceptus begins to invade the horn on the side opposite the ovulating ovary by Day 13. The growth of the bovine embryo was investigated by Leiser (1975), whose results are summarized in Fig. 15. His findings have been confirmed by ultrasonography, which reveals that the conceptus usually occupies the entire length of the ipsilateral horn by Day 17 and reaches the tip of the contralateral horn by Day 20 (Kastelic *et al.*, 1988).

In ruminants the first intimate connection between the embryonic and maternal tissues is provided by numerous, minute papillae that penetrate the mouths of the uterine glands (Guillomot *et al.*, 1981). They form by Day 13 in sheep (Wooding *et al.*, 1982) and Day 18 in cattle (Guillomot and Guay, 1982), and they dissappear in both species around Day 20. The papillae apparently form first close to the embryonic disc and possibly serve to immobilize this region.

In sheep the cells of the trophoblast and the uterine epithelium come into close apposition on Day 14 (Guillomot *et al.*, 1981; Wooding, 1984) and there is some adhesion on Day 15 (Boshier, 1969). The uterine epithelium responds to the presence of the filamentous blastocyst by forming steroid-metabolizing enzymes on day 12 (Flood and Ghazi, 1981) and alkaline and acid phospha-

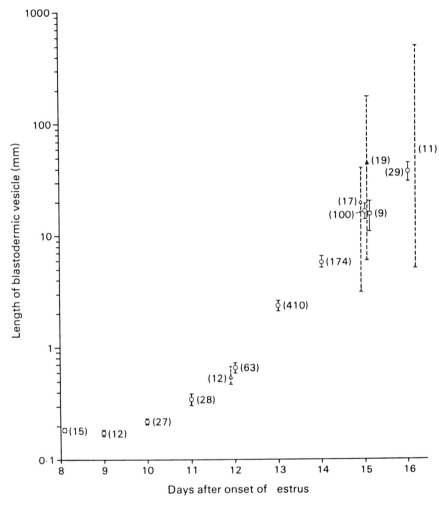

Figure 14 Collated data from several sources on the growth of embryos in cattle. Mean lengths are indicated on a logarithmic scale (solid error bars = SEM, broken error bars = range, *n* in parentheses). The points shown by ○ are from superovulated animals, and the others are from natural ovulations. From Betteridge *et al.* (1980).

tases at least as early as Day 14 (Boshier, 1969). On Day 14 too, the centers of the caruncles become depressed and some of their epithelial cells develop characteristic cytoplasmic protrusions (Guillomot *et al.*, 1981); on Day 15 the caruncular capillaries show increased permeability (Boshier, 1970). Between Days 16 and 18 there is increasing interpenetration of the uterine microvilli and the cytoplasmic projections of the

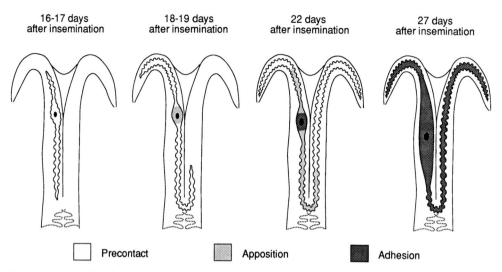

Figure 15 Attachment in cattle. In the precontact stage the maternal and embryonic epithelia are aligned with one another. In the apposition stage their apical cell membranes are in contact. Adhesion is characterized by interdigitation of embryonic and maternal microvilli. Redrawn from Leiser (1975).

trophoblast cells, leading to effective attachment (Guillomot *et al.*, 1981).

In pregnant cattle the maternal and embryonic epithelia become closely aligned in the region of the embryo on Day 16 (Leiser, 1975) and the intercaruncular epithelial cells retain the apical protrusions that they would otherwise lose at this time (Guillomot and Guay, 1982). By Day 18 the intercaruncular epithelium is reduced in height (King *et al.*, 1981). Two days later, extensive though rather tenuous adhesion between the trophoblast and the uterine epithelium is seen in the uterine horn ipsilateral to the corpus luteum. Early in the fourth week the maternal and trophoblastic microvilli begin to interdigitate, giving a significant degree of adhesion (Leiser, 1975; King *et al.*, 1980; Wathes and Wooding, 1981). Attachment is always more advanced close to the embryo and less so at the extremities of the blastocyst (Fig.15), but at first events in the carun-

cular and intercaruncular areas are similar and more or less synchronous (King *et al.*, 1980).

2. Development of the Placentomes

In ruminants the placenta is essentially diffuse until the fourth week of pregnancy (King *et al.*, 1980), but by Day 30 in sheep (Boshier, 1969) and Day 36 in cattle (King *et al.*, 1979) those areas of the allantochorion attached to the caruncles have developed short trophoblastic villi that occupy corresponding crypts in the caruncular tissue. The villi become vascularized by allantoic capillaries and rapidly increase in length. Their growth and that of the associated uterine tissue leads to the formation of the placentomes that are so characteristic of the placenta of ruminants. These vary in shape somewhat from species to species; the caruncles of sheep are concave while those of cattle are convex, and those of goats are intermediate in profile (Fig. 16). Caruncles are normally round or elliptical in outline, but the ellipse may become very much elongated toward the end of gestation in the pregnant horn.

The villi lying in their maternal crypts form the principal area of hemotrophic exchange in ruminants. However, there is evidence of histotrophic activity even in the placentomes because degeneration of the margins of the maternal septa (that lie in the depths of the recesses between the chorionic villi; Fig. 16) leads to the release of blood and tissue debris. This is absorbed by the adjacent trophoblast cells, which are consequently rich in iron (Myagkaya and De Bruijn, 1982; Myagkaya et al., 1984). These histotrophic "arcades" first appear in the sheep between 70 and 80 days, reaching a peak of activity at Day 125 (Myagkaya and Vreeling-Sindelarova, 1976). The arcades seem to be homologous with the areas in the depths of the chorionic folds in the pig, which are adapted for transport of high-molecular-weight substances.

Those who have studied the histology of the placentome have never doubted that the fetal and maternal basement membranes were separated by two layers of cells, but whether the layer adjacent to the maternal basement membrane was of embryonic or maternal origin has been the subject of debate throughout this century. There is probably no simple answer. To appreciate this it is necessary to understand the origin and fate of the binucleate giant cells.

3. The Binucleate Giant Cells

Naturally the binucleate giant cells are large and have two nuclei; they are also roughly elliptical and contain characteristic glycoconjugate granules.

a. Morphology in Sheep In the sheep, binucleate cells appear as soon as elongation of the blastocyst is complete at Day 14 and increase rapidly in numbers until they constitute 15–20% of the trophoblast cells on Day 18. At first they seem to be most abundant in the trophoblast that is opposed to the caruncles (Wooding, 1984). About one-fifth of the binucleate cell population—always fully granulated cells—are found bridging the junction between the embryonic and maternal tissues from Day 16 onward; here they interrupt both the tight junction barrier and the microvillar junction (Wooding, 1980; Wooding et al., 1980; Morgan and Wooding, 1983). This is particularly noteworthy in view of the long-standing contention that the uterine epithium is either augmented or replaced by migrant binucleate cells (Assheton, 1906; Wimsatt, 1951; Amoroso, 1952a): a contention supported by the similarity between the granules in the binucleate cells and those found in syncytial masses that begin to replace the originally columnar uterine epithelium about 2 days after the binuclear cells first appear (Lawn et al., 1969). Not only can this similarity be seen by conventional histology (Wimsatt, 1951), it has also been demonstrated by electron microscopy (Lawn et al., 1969; Guillomot et al., 1981), phosphotungstic acid staining (Wooding, 1980), immunocytochemisty (Lee et al., 1985, 1986), and lectin binding (Munson et al., 1989).

Using electron microscopy with phosphotungstic staining, Wooding (1984) has now demonstrated a series of intermediate stages that strongly indicate that ovine binucleate cells are formed continuously in the trophoblast throughout gestation, and they then migrate through the microvillar junction before fusing with the cells on the maternal

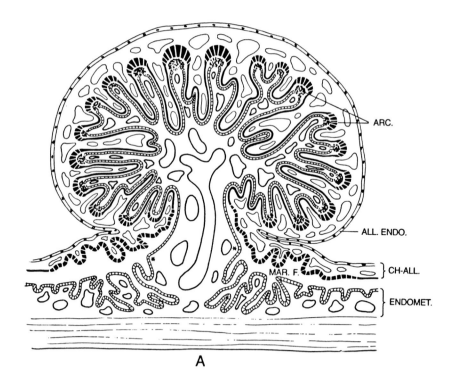

ARC.

ALL. ENDO.

CH-ALL.

MAR. F.

ENDOMET.

A

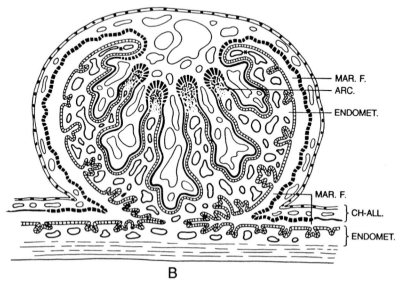

MAR. F.
ARC.
ENDOMET.

MAR. F.

CH-ALL.

ENDOMET.

B

side to form the syncytium (Fig. 17). In the early stages of this process hybrid trinucleate cells formed from a binucleate cell and a uterine epithelial cell are common, but by Day 30, 90% of the layer replacing the uterine epithelium consists of syncytial plaques containing about 20 nuclei. Thymidine labeling of binucleate cells has provided additional support for the view that they make an important contribution to the syncytium that comes to line the maternal crypts (Wooding, 1982).

During chorionic attachment, some of the uterine epithelial cells die and are absorbed by the trophoblast, and occasionally limited areas of the maternal epithelium are completely denuded, causing transient exposure of the basement membrane to the trophoblast. In the region of the placentomes the syncytium adjacent to the maternal basement membrane persists until term (Wooding *et al.*, 1980).

b. Morphology in Cattle   In cattle binucleate cells constitute around 10% of the trophoblast cells on Days 18 and 20 and subsequently about 20% until close to term; some 20% of these cells are in the process of migrating. They first appear in the uterine epithelium on Day 19, where they apparently fuse with the uterine epithlial cells to form multinucleate giant cells that make up about 50% of the maternal epithelial surface by Day 24. Events thus far are very similar to those in sheep, but in cattle mitotic activity in the residual maternal cells leads to almost total reconstitution of the uterine epithe-

(a) Sheep and goat

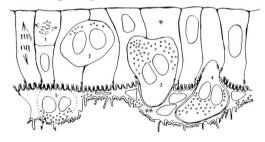

(b) Cow and deer

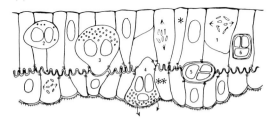

Figure 17   Migration of binucleate cells (stages 1–4 and 1–5) in (a) sheep and goats at day 24 and (b) cows and deer at Day 28. Mitoses are frequent in the trophoblast. In sheep and goats the syncytium (**) apposed to the maternal basement membrane increases in area by augmentation from migrating binucleate cells. In the cow and deer mitoses also occur in the uterine epithelium but binucleate cell migration continues to produce transient trinucleate cells throughout gestation. By means of this process binucleate cell granules are transported across the microvillar border. Kindly supplied by F. B. P. Wooding.

lium in a way that does not occur in sheep, and from 40 days of pregnancy onward giant cells are absent, though a small number of minisyncytia can always be found (Wooding and Wathes, 1980; Wathes and Wooding, 1981; Wooding, 1983). In the

Figure 16   Axial sections showing the shape of the placentomes of (A) cattle and (B) sheep. The trophoblast is shown in solid black. The uterine epithelium and the lining of the endometrial crypts is shown as open cells. ARC = histotrophic arcades, ALL. ENDO. = allantoic endoderm, CH-ALL. = allantochorion, ENDOMET. = endometrium, MAR. F. = marginal folds (an area of marked histotrophic activity). From Mossman (1987).

peripartum period the frequency of binucleate cells is reduced to about a quarter of that seen earlier (Williams *et al.*, 1987).

c. Function It would appear that the function of the binucleate cells is to facilitate the transfer of complex molecules across the embryonic and maternal tight junction seals and the microvillar junction, into the maternal circulation (Wooding, 1982). Exocytosis of binucleate cell granules from the uterine epithelium into the maternal connective tissue has been observed (Wooding, 1984), and tracer materials introduced into the uterus readily follow this route (Guillomot *et al.*, 1986). There is a widespread (Martal *et al.*, 1977; Watkins and Reddy, 1980; Wooding, 1981; Morgan *et al.*, 1989) though not universal (Carneige *et al.*, 1982) view that the granules contain placental lactogen. The cells apparently also contain pregnancy-specific protein B (Sasser *et al.*, 1989) and they are important sites of steroidogenesis (Reimers *et al.*, 1985; Goss and Williams, 1988). The binucleate cells seen in any single histological section almost always vary in their staining characteristics, but it is not clear whether these differences are cuased by maturation or whether more than one binucleate cell population exists (Lee *et al.*, 1986). Binucleate cell migration would seem to be at least partially under the control of the fetus (Wooding *et al.*, 1986).

d. Classification of Ruminant Placentae Appropriate classification and histological description of the placentomes of ruminants is a problem because none of the terms in general use embrace the idea that part of the interhemal membrane might be composed of hybrid cells derived from both embryo and fetus. There is little doubt that the intercotyledonary areas to be described

shortly and the established placentomes of cattle can reasonably be referred to as epitheliochorial: to describe the placentomes of sheep and the early attachment phases of bovine placenta in the same way, or to refer to them as syndesmochorial, is to neglect a fascinating phenomenon.

4. The Intercotyledonary Areas

There is a tendency to dismiss the intercaruncular area of the ruminant placenta as an inert structure of little significance, but its appearance shows that this is far from true. Initial attachment of the embryonic and maternal epithelia is similar in the caruncular and intercaruncular regions in both cattle (King *et al.*, 1981) and sheep (Wooding, 1984). Later, in sheep, migration of the binucleate cells results in extensive syncytium formation in the intercaruncular epithelium, but sufficient of the original epithelial cells remain to permit reestablishment of the normal uterine epithelium by about 40 days; this, of course, is in sharp contrast to the situation in the caruncles (Wooding, 1984; Mercer *et al.*, 1989). In cattle, as in the placentomes, no syncytia are formed in the intercaruncular epithelium, but abundant giant cells are present during the fourth week of pregnancy, though these rapidly decline in numbers (King *et al.*, 1981).

As in other species, there are specialized areas for the absorption of uterine glandular secretion. Areolae, like those of the pig, are present in both sheep (Wimsatt, 1950) and cattle (Björkman, 1954) until mid-gestation and though less obvious, they are more numerous. In addition, where the uterine glands are particularly well developed around the caruncles, the adjacent folds of the allantochorion (the marginal folds) show vigorous endocytosis (Mossman, 1987).

During the latter half of pregnancy, adhesion between the fetal and maternal intercaruncular epithelia is lost in cattle (Björkman, 1954) and there is no microvillar junction at term in sheep (Perry et al., 1975) though it is retained at least until Day 120 (Wooding, 1980).

5. Fusion between the Amnion and Chorion

The amnion and chorion fuse to form an amniochorion in ruminants just as they do in pigs. However, the area of fusion is reduced late in gestation in cattle (Authur, 1959), which may allow some calves to be delivered while still partially enclosed within the amnion.

6. Changes Late in Gestation

At term the number of binucleate giant cells in the placenta decreases to about 25% of its original level in both cattle (Williams et al., 1987) and sheep (Wooding, 1982). Small foci of necrosis are common in bovine placentomes at the end of pregnancy, and separation of the placentomes apparently begins at their edges. Though it has been stated that the bovine placenta is syndesmochorial at term (Grunert et al., 1976; Williams et al., 1987), the ultrastructural studies that would be necessary to unequivocally show this do not seem to have been performed. If there is any loss of the maternal epithelium late in gestation it is very limited, and the epthelium of the caruncular crypts remains largely intact in both cattle (Björkman and Sollén, 1960, 1961; Holm et al., 1964) and sheep (Perry et al., 1975). Many crypts contain whole trophoblastic villi after delivery of the afterbirth in cattle (Björkman and Sollén, 1960). In sheep the cotyledonary trophoblast is normally retained within the crypts and subsequently resorbed; the line of separation is at the level of the trophoblast basement membrane, which is shed with the placenta (Perry et al., 1975; Steven, 1975b). Probably the placentae of ruminants are all contradeciduate in varying degrees.

E. The Carnivores

1. Attachment

Attachment in carnivores begins at the embryonic pole of the conceptus—though the area immediately adjacent to the embryonic disc is not involved at first—and spreads rapidly to involve the whole equatorial zone, hence the zonary placenta. The details, as we know them are as follows.

a. The Cat The early feline blastocyst arrives in the uterus about 6 days after mating, encased in the zona pellucida. By Day 12 the blastocyst has grown into an ellipsoidal structure some 1.5×2 mm, and it is at this stage that dissolution of the embryo covering occurs, beginning at the abembryonic pole. This permits alignment of the maternal and embryonic epithelia. By the next day the blastocyst has become orientated so that the embryonic pole is antimesometrial and junctional complexes have formed between the embryonic and maternal epithelia. On Day 14, erosion of the maternal epithelium occurs and the attachment sites are marked by local uterine swellings (Denker et al., 1978a, 1978b; Leiser, 1979, 1982).

b. The Dog The timing of events in early pregnancy in the dog presents a problem because of the confusion surrounding the time of ovulation and the time taken for the ovulated primary oocyte to undergo its maturation divisions. If it is assumed that the day of last acceptance of the male usually occurs 2 or 3 days after the last day of vaginal cornification (Holst and Phemister, 1971), that loss of cornification occurs about 1 day after ovulation, and that ovulation occurs 1 or 2 days after the luteinizing hormone (LH) peak, then the chronologies of Holst and Phemister (1971), Barrau *et al.* (1975), and Harvey *et al.* (1989) are in reasonable agreement.

On this basis it appears that canine embryos enter the uterus as early blastocysts about 6 days after ovulation; they become dispersed throughout the cornua during the next 2 or 3 days and take up fixed positions in the uterus around Day 9. On Day 11 or 12 the blastocyst covering is lost and local endometrial edema causes distinct uterine swellings. A day later, the newly differentiated syncytiotrophoblast begins to displace the uterine epithelium and invade the mouths of the uterine glands. The timing of events in the dog and cat therefore seems to be similar, given the delay between mating and ovulation in the cat.

2. *The Placenta*

The general structure of the placentae of carnivores is remarkably consistent and that seen in the dog and cat is typical. Three parts are clearly distinguishable at first inspection: First is the highly vascular placental labyrinth which is the principal site of hemotrophic nutrition; this forms a broad zone around the middle of the conceptus in the domestic species but may have somewhat different shapes in other carnivores. Second are the darkly pigmented zones at the placental margins; these are often known as marginal hematomas, though they are better called hemophagous organs; they are areas where histotrophic nutrition predominates. Third are the poles of the chorionic sacs; these are more or less transparent and relatively poorly vascularized but may be involved in the absorbtion of glandular secretion.

As in other eutherians, the definitive carnivore placenta is perfused by the allantoic vasculature, but in addition there is a well-developed and persistent yolk sac or vitelline circulation. The highly vascular yolk sac retains its contact with the mesometrial side of the chorion until it is ousted, during the fourth week of pregnancy, by the expanding extraembryonic coelom, which is in its turn obliterated by the allantois (Evans, 1979). During this early stage it probably has an absorptive function and from 14 to 45 days it is an important site of erythropoiesis in the cat (Tiedemann, 1977). Its function after this stage is doubtful but it remains highly vascular; it has been suggested that it acts as an extracorporial liver (Lee *et al.*, 1983; Tiedemann, 1976).

a. The Placenta Labyrinth Once the uterine epithelium has been breached, the trophoblast rapidly advances, destroying the superficial parts of the glands and much of the related connective tissue as it does so. The syncytium becomes closely applied to the endometrial capillaries, and the maternal connective tissue elements that remain become the maternal giant cells (Barrau *et al.*, 1975; Malassiné, 1974). After the initial period of invasion the advancing trophoblast becomes cellular once more and is highly phagocytic; the syncytium is confined to the areas of exchange adjacent to the maternal vessels. The glandular layer is progressively engulfed during pregnancy and at term has all but disappeared. While this is happening the superficial and very deep parts of the glands retain a fairly normal appearance but the middle sections become enormously dilated and are only separated from each other by thin, but highly vascular, layers of maternal connective tissue. This "spongy layer" (Amoroso, 1952a) is one of the most striking histological features of the carnivore placenta. While it is usual to describe the establishment of the carnivore placenta as a process of invasion as I have done here, it should be remembered that the final structure is much thicker than the original uterine wall; growth of the maternal tissues must therefore play a crucial part.

By mid-gestation the placenta consists of a series of alternating maternal and fetal laminae arranged in a plane perpendicular to the long axis of the uterus (Kehrer, 1973) (Fig. 18). They are very regular in the cat but less so in the dog. The fetal laminae consist of capillaries, minimal connective tissue, discontinuous cytotrophoblast, and a continuous layer of syncytiotrophoblast (Anderson, 1969; Wynn and Corbett, 1969; Malassiné, 1974).

The trophoblast is separated from the maternal capillaries by a rather irregular

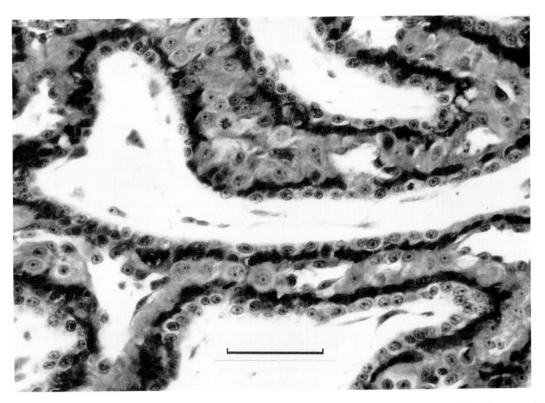

Figure 18 The established placental labyrinth of the cat during the fifth week of pregnancy. The pale tracts of diffuse tissue are fetal connective tissue. This is surrounded by a moderately stained layer of cellular trophoblast. The syncytiotrophoblast lies outside the cellular layer and is intensely stained. The syncytium is in direct contact with either the maternal endothelium or the decidual cells. H & E, Helly's fluid; scale bar = 50 μm.

but generally fairly thick interstitial membrane ("la substance inerte interstitielle" of Malassiné). This universal feature of the placenta of carnivores forms an almost complete barrier between the fetal and maternal tissues but it is penetrated by small cytoplasmic protrusions of the syncytium in cats (Malassiné, 1974). The microvillar junction that is such a prominent feature of ungulate placentae is absent in carnivores.

The maternal laminae consist of capillaries with a remarkably thickened endothelium and interspersed "giant cells." The latter are obvious histologically in the cat but can only be recognized with certainty in electron micrographs of the dog placenta (Anderson, 1969; Barrau et al., 1975). The function of the giant cells is obscure, but Malassiné (1974) notes that they show cytological evidence of synthetic activity.

Thus the interhemal membrane of carnivores consists of the fetal endothelium with its normal basement membrane, the syncytiotrophoblast, and the maternal endothelium with its thickened basement membrane. As term approaches the membrane is progressively thinned by indentation of the syncytium by the fetal capillaries and attenuation of the syncytium and the maternal endothelium. As a result the two circulations are separated by less than 2 μm in the cat (Leiser, 1982) and dog (Wynn and Björkman, 1968; Anderson, 1969) in late pregnancy. The two placental capillary beds of the cat are arranged in a cross-current exchange system (Leiser and Kohler, 1984).

Following parturition, the placentae separate from the remainder of the uterus, leaving only the deep remnants of the endometrial glands. Complete regeneration follows though the positions of the placentae are marked by so-called "placental scars," which are in reality brownish areas rich in hemosiderin-laden macrophages.

b. The Hemophagous Organs Early in the third week of pregnancy the endometrial folds flanking either end of the labyrinthine zones begin to degenerate at their free edges and as a result blood and epithelial debris are released into the uterine lumen (Barrau et al., 1975; Leiser and Enders, 1980b). The cause of the endometrial degeneration is not clear, but even in the hemophagous organ there are areas where the trophoblast and uterine epithelium are at least temporarily in direct contact so the cytolytic properties of the trophoblast may come into play (Leiser and Enders, 1980a). Most of the trophoblast of the hemophagous organ shows vigorous endocytosis of the luminal contents (Malassiné, 1982). In dogs the rate of release of endometrial debris, perhaps with an admixture of glandular secretion, would seem to exceed the rate of absorption so there is considerable accumulation of stagnant blood between the chorionic folds; this degenerates, releasing bile pigments that give this part of the placenta its characteristic green color. No such accumulation occurs in cats and the placental margins are dull red or brown. The hemophagous organs are the principle source of iron for the fetus.

c. The Polar Zones The allantochorion at the poles of the chorionic sacs is apposed to the mouths of seemingly active uterine glands and absorbs their secretion. Small areolae like those of artiodactyls have been recognized in the dog (Amoroso, 1952b) but not in the cat.

VI. Fetal Fluids

The formation of the blastocoel and its subsequent partition and supplementation by other fluid filled cavities has already been mentioned. It remains to comment on the two major fluid cavities of the established fetal membranes, the amnion and the allantois.

A. Amniotic Fluid

A minute amount of uterine luminal fluid may be trapped within the closing amnion, but this is rapidly augmented by a mildly hypotonic fluid that can only be derived from either the embryonic or amniotic ectoderm (McCance and Dickerson, 1957). This is because prior to the breakdown of the cloacal membrane toward the end of the fourth week of pregnancy (Marrable, 1971), the products of renal secretion are unable to enter the amnion and must pass into the allantois via the urachus. After this, urine is free to pass into the amnion until the development of the external genitalia has progressed far enough to effectively occlude the urogenital sinus once again. Significant quantities of urine are then denied access to the amnion until the urethra has matured sufficiently to allow further flow. This apparently occurs during mid-gestation in the sheep (Alexander and Nixon, 1961). Late in pregnancy, a further stage of maturation is postulated in cattle (Arthur, 1965) in which increasing tone in the urethral sphincter redirects flow toward the allantois yet again, but the composition of amniotic fluid suggests that some urine continues to enter the amnion until term (Reeves et al., 1972).

With increasing maturity, saliva and pulmonary secretions also enter the amnion, which possibly accounts for the increased viscosity and improved lubricant properties of the fluid (Alexander and Nixon, 1961, Arthur, 1965). Amniotic fluid is swallowed, and the fetus makes respiratory movements that cause fluid to be moved in and out of the trachea (Dawes et al., 1972). Fetal feces are only found within the amnion as a consequence of serious fetal distress late in gestation.

In very general terms, the volume of amniotic fluid rises slowly at first, then increases rather rapidly until mid-gestation, remaining fairly constant thereafter (Fig. 19). The details differ somewhat between species, and the exact volume of fluid is rarely the same, even within a litter (Arthur, 1969; Bongso and Basrur, 1976; Marrable, 1971; Ginther, 1979).

B. Allantoic Fluid

The volume of allantoic fluid rises very rapidly during early pregnancy in all the domestic species and though the mesonephric kidneys appear to be very active at this stage (Marrable, 1971), their size alone suggests that they cannot be responsible for all the fluid accummulated (McCance and Dickerson, 1957). This fact and the active ultrastructural appearance of the allantoic endoderm (Tiedemann, 1979a; 1979b; 1982; Steven et al., 1982) indicate that this membrane transmits most of the fluid in the embryonic period, but raised allantoic urea concentrations from about Day 40 onward (McCance and Dickerson, 1957) suggest that the fetal kidneys then make a significant contribution. In the last half of gestation, the volume of allantoic fluid increases in cattle (Arthur, 1957; Bongso and Basrur, 1976), is very variable in sheep (Arthur,

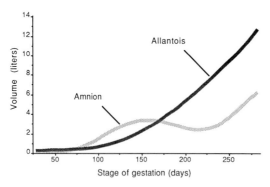

Figure 19 Changes in placental fluid volume in cattle. Accumulation of allantoic fluid is very rapid before day 50 but the total volumes are too small to show clearly on this scale. In mid-gestation the volume of the amnion exceeds or equals that of the allantois but in the last third of pregnancy the allantois predominates once more. Polynomial regression lines fitted to the combined data of Arthur (1959, 1965) and Bongso and Basrur (1976).

1. Hippomanes

A hippomanes is a strange concretion that is invariably present in the allantoic fluid of mares and occasionally of other species. At term it looks like a small, well-used cake of brown soap but its consistency is more rubbery. Hippomanes first begin to appear in mares around Day 60 and grow slowly by annular accretion thereafter, reaching 4–8 cm in length by term (King, 1967). They mainly consist of a mucoprotein matrix infiltrated with precipitated calcium phosphate (Dickerson *et al.*, 1967). Whether it is advantagous to the fetus to sequester minerals in this way is doubtful, but it offers a less fanciful explanation than many of those advanced since the time of Aristotle (King, 1967).

1956), but declines in horses (Ginther, 1979) and pigs (Marrable, 1971).

The turnover of water in allantoic and amniotic fluids is probably very rapid since fetal sheep produce more than a liter of urine daily in late pregnancy (Wlodek *et al.*, 1988) and tritiated water studies indicate that pig allantoic fluid is exchanged every 1 to 2 h (Stanier, 1965). This perhaps accounts for the great variability of fetal fluid volumes. Urea, on the other hand, does not seem to leave the allantoic fluid very rapidly (Stanier, 1965), and creatinine concentrations in the allantoic fluid of cattle slowly rise during pregnancy (Thomsen and Edelfors, 1976; Baetz *et al.*, 1976).

The fetal fluids contain a wide variety of metabolic products, electrolytes (Baetz *et al.*, 1976; McCance and Dickerson, 1957), and proteins (Godkin *et al.*, 1988; Lambert *et al.*, 1987; Newton *et al.*, 1989).

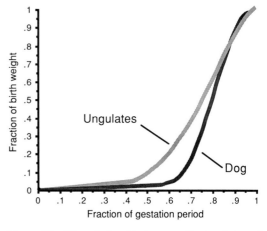

Figure 20 The relationship between fetal weight and stage of gestation. The curve for ungulates is shown as a single line because the data points for cattle, sheep, horses, and pigs were inextricably intermingled. The data on dogs, formed a distinctly separate line. Slightly modified polynomial regression lines fitted to the combined data of Marrable and Ashdown (1967), Ginther (1979), Green (1946), and Evans (1979).

VII. Growth of the Conceptus

Data on the growth of the conceptus have been reviewed thoroughly elsewhere (Evans and Sack, 1973; Ginther, 1979; Evans, 1979; Marrable, 1971) and the reader is referred to these sources. Figure 20 is a composite diagram indicating the proportion of birth weight acquired at different stages of gestation; it is derived from data on pigs, cattle, sheep, horses, and dogs. Evidently the first half of pregnancy makes minimal nutritional demands on the mother, but requirements for growth increase rapidly toward term, only to rise yet further during lactation.

VIII. Immunological Aspects of Pregnancy

In preparing this necessarily skeletal account of what is to many an extraordinarily fascinating but confusing subject, I have drawn heavily on accounts by Antczak and Allen (1989), Hansen *et al.*, (1989), Simpson (1984), Wegmann (1983), Billington and Bell (1983), Head and Billingham (1984), and Koch (1985). Readers are referred to these works where no specific source is indicated.

A. Is There a Problem?

Since the idea was first clearly articulated by Medawar (1953), it has been conventional to regard the placenta as an intrauterine semi-allograft, which, by analogy with almost all other transplanted tissues, should undergo acute immunological rejection. The reason this does not occur is a perennial biological conundrum. Graft rejection is a consequence of genetic differences between the donor and the recipient involving the major histocompatibility complex (MHC), a highly polymorphic group of co-dominant, antigenic, cell-surface glycoproteins that are crucial to the ability of the immune system to distinguish "self" from "nonself." MHC antigens are found on almost all cells. The analogy between the placenta and a tissue graft is very apt in the case of the those much-studied species, man and mouse, because in their hemochorial placentae circulating maternal blood comes into direct and continuous contact with allogeneic fetal tissue. In the domestic species the analogy may be less appropriate because, with one notable exception, the fetal tissues never come into direct contact with the bloodstream of the mother. I will return to this point later.

B. Can the Problem Be Defined Further?

Many hypotheses have been advanced to explain the immunological recalcitrance of the placenta and usually, to quote Cohen (1984), "an *ad hoc* array of not-quite-individually-adequate mechanisms is proposed." There is no obvious reason to suppose that any single mechanism is responsible for the immunological protection of the conceptus: the extraordinary diversity of placental structure, embodying differences between species and between the parts and stages of a single placenta, would suggest that the predominant mechanism ought to differ in different circumstances.

These are some important, very general considerations in the argument:

1. Pregnant females successfully immunized against paternal MHC antigens carry

normal fetuses to term without obvious placental damage.

2. Pregnant females may or may not spontaneously form antibodies against paternal MHC antigens. In either case they carry normal fetuses to term. A proportion of females are therefore immunologically cognisant of the presence of the conceptus.

3. Nonplacental tissues grafted into the endometrium undergo rejection.

4. Trophoblast cells grafted beneath the renal capsule are rejected.

5. Some paternal MHC antigens are exposed to maternal blood in species with hemochorial placentae, but they may not be present throughout the trophoblast and may not be there all of the time.

6. In species with epitheliochorial placentae like horses (Oriol et al., 1989) and pigs (Whyte and Robson, 1984), the trophoblast possesses unique antigenic determinants not found in other tissues of the body.

7. The myometrium is rich in lymphatics but these are very sparse or absent in the endometrium.

It may be deduced from the foregoing that the afferent arm of the maternal immune response is frequently, though not always, compromised; that the efferent arm is effectively blocked; but that the uterus has no very special priviliges regarding immunological rejection. Further, the attractive hypothesis that the trophoblast, a tissue differentiated before any other, lacks paternal antigens, must be abandoned, at least in its simplest form. I should add at this stage that there is no evidence of significant, generalized immunological depression in pregnant females; in any case such a situation would seem to have little adaptive merit.

C. The Importance of the Trophoblast

The critical importance of the trophoblast is demonstrated in hybrid pregnancies or in experiments in which embryos of one species are transferred to the uterus of a closely related one. This frequently seems to impose an intolerable burden on the antigenic hospitality of the recipient uterus and the embryos die, at least in part for immunological reasons. The sheep × goat hybrid is a good example (Dent et al., 1971a, 1971b). If, on the other hand, embryos are artificially constructed having the inner cell mass of one species and the trophoblast of another, they can be carried to term in the uterus of the species from which the trophoblast is derived (mice: Rossant et al., 1983; sheep: Roth et al., 1989).

Local mechanisms at the feto-maternal interface are therefore very important and may take a variety of forms. First, endocrine agents produced by the conceptus are present in specially high concentrations at this site and may have immunosuppressive actions. Progesterone is the most important agent in this class and, as well as having general immunosuppressive properties, it discourages leukocytes from entering the uterine lumen (Staples et al., 1983). Other potentially active placental products include corticosteroids, prostaglandin E, early pregnancy factor (Koch, 1985), interferons (Roberts, 1989; Cross and Roberts, 1989), and a variety of glycoproteins and "megasuppressin" (Hansen et al., 1989).

Second, there is evidence that exposed antigens are obscured by maternal "blocking" antibodies that discourage any further immunological response. Formation of blocking antibodies is facilitated if the antigens are presented to the responding organism, not via the lymphatic system but by the bloodstream; this seems to be the case in the placenta. In addition, potentially damaging antigen–antibody complexes that are formed on the surface of the trophoblast may be internalized by the trophoblast cells and broken down, rendering them innocuous (Singh *et al.*, 1983).

Third, suppressor lymphocytes may accumulate adjacent to antigenic sites and inhibit an aggresssive response.

D. The Anomalous Situation of Horses

Horses provide us with a curious natural experiment because they possess two easily distinguished kinds of trophoblast, one of which is invasive and the other not. The endometrial cup cells, the invasive trophoblast, are surrounded by a dense layer of leucocytes almost as soon as they enter the uterus. This is followed by the appearance of paternally directed cytotoxic antibodies in the maternal blood. No such antibodies are present in maiden mares. Sixty to 80 days later the endometrial cups are invaded by leukocytes, become necrotic, and are sloughed from the endometrial surface.

The leukocytic response is more severe and sloughing is hastened in mares carrying mules (Antczak and Allen, 1989), but the endometrial cups apparently survive much longer than normal in mares carrying foals sired by their own co-twin (Spincemaille *et al.*, 1975). It therefore appears that the death of the cup cells has an immunological basis and that they represent the only allogeneic placental tissue known to perish in this way. Meanwhile the placental trophoblast is entirely unaffected. The invading endometrial cup cells are unusually rich in MHC antigens though they seem to lose

them later, but MHC antigens are absent from the noninvasive trophoblast. Thus the placental trophoblast may owe its survival either to lack of antigens or to the protection of the uninterrupted uterine epithelium.

E. Is the Placenta Really an Allograft in the Domestic Species?

Tissue transplantation immunology is normally based on the presumption of a confluence of host and graft connective tissue and vasculature. This never occurs in the placenta and, with the exception of the endometrial cups, is not even approached in the epitheliochorial and endotheliochorial placentae of the domestic species. Further, among these species it is only mares, with their invasive trophoblast, that respond systemically to paternal antigens during midpregnancy; in pigs (Linklater, 1968) and cows (Newman and Hines, 1980) such responses are inconsistent and do not occur until the placenta is beginning to degenerate, just before or at parturition. Perhaps the blood–uterine lumen barrier (McRae, 1988), which is the analogue of the blood–brain barrier, may help the mother to ignore the antigenic contents of her uterus. The allograft analogy therefore seems tenuous in the domestic animals and has been questioned (Enders, 1983; Koch, 1985).

But people's fascination with hybrids and chimeras of their own creation, and their desire to facilitate the reproduction of endangered species through embryo transfer, presents even the epitheliochorial placental barrier with a problem it cannot reliably solve (Anderson, 1988). It is therefore clear that we have only begun to understand the intricacies of the physiological and immunological conversation going on between the conceptus, the endometrium, and the mother, and its relationship to the slightly better known languages of inflammatory and immunological interactions.

Acknowledgments

I am most grateful to my wife for amiably suffering my asocial working habits during the last months and to my colleagues William Latshaw and Susan Tedesco for reviewing the manuscript. Figures 1, 2, 3, 10, and 15 were kindly drawn by Juliane Deubner.

References

Adelmann, H. B. (1942). "The Embryological Treates of Hieronymus Fabricius." Cornell University Press, Ithaca, New York.

Alexander D. P., and Nixon, D. A. (1961). *Br. Med. Bull.* **17**, 112–117.

Alexander, G. (1964). *J. Reprod. Fertil.* **7**, 289–305.

Allen, W. E., and Newcombe, J. R. (1981). *Equine Vet. J.* **13**, 51–52.

Allen, W. R., Hamilton, D. W., and Moor, R. M. (1973). *Anat. Rec.* **177**, 485–502.

Amoroso, E. C. (1952a) *In* "Marshall's Physiology of Reproduction" (A. S. Parkes, ed.), 3rd ed., Vol. II, pp. 127–311. Longmans, Green, New York.

Amoroso, E. C. (1952b). *J. Anat.*, **86**, 481–482.

Anderson, G. B. (1988). *Biol. Reprod.* **38**, 1–15.

Anderson, J. W. (1969). *Anat. Rec.* **165**, 15–36.

Anderson, L. L. (1978). *Anat. Rec.* **190**, 143–154.

Antczak, D. F., and Allen, W. R. (1989). *J. Reprod. Fertil. Suppl.* **37**, 69–78.

Arthur, G. H. (1956). *J. Comp. Pathol.* **66**, 345–353.

Arthur, G. H. (1957). *Brit. Vet. J.* **113**, 17–28.

Arthur, G. H. (1959). *Vet. Rec.* **71**, 345.

Arthur, G. H. (1965). *Vet. Rec.* **77**, 623–624.

Arthur, G. H. (1969). *J. Reprod. Fertil. Suppl.* **9**, 45–52.

Ashdown, R. R., and Marrable, A. W. (1967). *J. Anat.* **101**, 269–275.

Ashdown, R. R., and Marrable, A. W. (1970). *Res. Vet. Sci.* **11**, 227–231.

Assheton, R. (1906). *Phil. Trans. R. Soc. B.* **198**, 143–225.

Atkinson, B. A., King, G. J., and Amoroso, E. C. (1984). *Biol. Reprod.* **30**, 763–774.

Baetz, A. L., Hubert, W. T., and Graham, C. K. (1976). *Am. J. Vet. Res.* **37**, 1047–1052.

Barrau, M. D., Abel, J. H., Torbit, C. A., and Tietz, W. J. (1975). *Am. J. Anat.* **143**, 115–130.

Bate, L. A., and King, G. J. (1988). *J. Reprod. Fertil.* **84**, 163–169.

Baumbach, G. A., Climer, A. H., Bartley, N. G., Kattesh, H. G., and Godkin, J. G. (1988). *Biol. Reprod.* **39**, 1171–1182.

Bazer, F. W., Thatcher, W. W., Martinat-Botte, F., and Terqui, M. (1988). *J. Reprod. Fertil.* **84**, 37–42.

Betteridge, K. J. (1989). *Equine Vet. J. Suppl.* **8**, 92–100

Betteridge, K. J., Eaglesome, M. D., Mitchell, D., Flood, P. F., and Beriault, R. (1982). *J. Anat.*, **135**, 191–209.

Betteridge, K. J., Eaglesome, M. D., Randall, G. C. B., and Mitchell, D. (1980). *J. Reprod. Fertil.* **59**, 205–216.

Betteridge, K. J., and Fléchon, J.-E. (1988). *Theriogenology* **29**, 155–187.

Biggers, J. D., Bell, J. E., and Benos, D. J. (1988). *Am. J. Physiol.* **255**, C419–C432.

Billington, W. D., and Bell, S. C. (1983). *In* "Reproductive Immunology 1983" (S. Isojima and W. D. Billington, eds.), pp. 147–155. Elsevier Science, Amsterdam.

Bindon, B. M. (1971). *Aust. J. Biol. Sci.* **24**, 131–147.

Björkman, N., and Sollén, P. (1960). *Acta Vet. scand.* **1**, 347–362.

Björkman, N., and Sollén, P. (1961). *Acta Vet. scand.* **2**, 157–177.

Björkman, N. H. (1954). *Acta Anat. Suppl.* **21**, 1–91.

Björkman, N. (1981). In "Placentation" (H.-D. Dellman & E. M. Brown, eds.), "Textbook of Veterinary Histology, pp. 337–355. Lea and Febiger, Philadelphia.

Bongso, T. A., and Basrur, P. K. (1976). *Can. Vet. J.*, **17**, 38–41.

Bonnet, R. (1889). *Verh. Anat. Ges.* **3**, 17–38.

Boshier, D. P. (1969). *J. Reprod. Fertil.* **19**, 51–61.

Boshier, D. P. (1970). *J. Reprod. Fertil.* **22**, 595–596.

Bousquet, D., Guillomot, M., and Betteridge, K. J. (1987). *Gamete Res.* **16**, 121–132.

Buechner, H. K. (1961). *Nature* **190**, 738–739.

Butterfield, R. M., and Matthews, R. G. (1979). *J. Reprod. Fertil. Suppl.* **27**, 447–452.

Carnegie, J. A., Chan, J. S. D., McCully, M. E., Robertson, H. A., and Friesen, H. G. (1982). *J. Reprod. Fertil.* **66**, 9–16.

Clegg, M. T., Boda, J. M., and Cole, H. H. (1954). *Endocrinology* **54**, 448–463.

Comline, R. S., and Silver, M. (1970). *J. Physiol., Lond.* **208**, 587–608.

Crombie, P. R. (1970). *J. Physiol. Lond.* **210**, 101P–102P.

Crombie, P. R. (1972). *J. Reprod. Fertil.* **29**, 127–129.

Cross, J. C., and Roberts, R. M. (1989). *J. Reprod. Fertil.* **40**, 1109–1118.

Curran, S., Pierson, R. A., and Ginther, O. J. (1986). *J. Am. Vet. Med. Assoc.* **189**, 1289–1294.

Dantzer, V. (1984). *Acta Anat.* **118**, 96–106.

Dantzer, V. (1985). *Anat. Embryol.* **172**, 281–293.

Dantzer, V., Björkmann, N., and Hasselager, E. (1981). *Placenta* **2**, 19–28.

Dawes, G. S., Fox, H. E., Leduc, B. M., Liggins, G. C., and Richards, R. T. (1972). *J. Physiol.* **220**, 119–143.

Del Campo, C. H., and Ginther, O. J. (1973). *Am. J. Vet. Res.* **34**, 305–316

Denker, H.-W., Eng., L. A., and Hamner, C. E. (1978a). *Anat. Embryol.* **154**, 39–54.

Denker, H.-W., Eng, L. A., Mootz, U., and Hamner, C. E. (1978b). *Anat. Anz.* **144**, 457–468.

Dent, J., McGovern, P. T., and Hancock, J. L. (1971a). *Nature* **231**, 116–117.

Dent, J., McGovern, P. T., and Hancock, J. L. (1971b). *J. Anat.* **109**, 361–363.

Dhindsa, D. S., and Dziuk, P. J. (1968). *J. Anim. Sci.* **27**, 668–672.

Dhindsa, D. S., Dziuk, P. J., and Norton, H. W. (1967). *Anat. Rec.* **159**, 325–330.

Dickerson, J. W. T., Southpate, D. A. T., and King, J. M. (1967). *J. Anat.* **101**, 285–293.

Douglas, R. H., and Ginther, O. J. (1975). *J. Reprod. Fertil. Suppl.* **23**, 503–505.

Dyce, K. M., Sack, W. O., and Wensing, C. J. G. (1987). "Textbook of Veterinary Anatomy." W. B. Saunders, Philadelphia.

Dziuk, P. (1985). *J. Reprod. Fertil. Suppl.* **33**, 57–63.

Dziuk, P. J., Polge, C., and Rowson, L. E. A. (1964). *J. Anim. Sci.* **23**, 37–42.

Enders, A. C. (1983). *In* "Immunology of Reproduction" (T. G. Wegmann and T. J. Gill, eds.), pp. 163–177. Oxford University Press, New York.

Enders, A. C., and Liu, I. K. L. (1991). *J. Reprod. Fertil. Suppl.* 44 (in press).

Evans, H. E. (1979). *In* "Anatomy of the Dog" H. E. Evans and G. C. Christensen, pp. 13–77. W. B. Saunders, Philadelphia.

Evans, H. E., and Sack, W. O. (1973). *Anat. Histol. Embryol.* **2**, 11–45.

Ewart, J. C. (1897). "A Critical Period in the Development of the Horse." Adam and Charles Black, London.

Fléchon, J.-E., and Renard, J. P. (1978). *J. Reprod. Fertil.* **53**, 9–12.

Flood, P. F. (1973). *J. Reprod. Fertil.* **32**, 539–543.

Flood, P. F. (1974a). *J. Endocrinol.* **63**, 413–414.

Flood, P. F. (1974b). *Res. Vet. Sci.* **17**, 102–105.

Flood, P. F., Betteridge, K. J., and Diocee, M. S. (1982). *J. Reprod. Fertil. Suppl.* **32**, 319–327.

Flood, P. F., Betteridge, K. J., and Irvine, D. S. (1979). *J. Reprod. Fertil. Suppl.* **27**, 413–420.

Flood, P. F., and Ghazi, R. (1981). *J. Reprod. Fertil.* **61**, 47–52.

Flood, P. F., and Marrable, A. W. (1975). *J. Reprod. Fertil. Suppl.* **23**, 569–573.

Ford, S. P., and Christenson, R. K. (1979). *Biol. Reprod.* **21**, 617–624.

Friess, A. E., Sinowatz, F., Skolek-Winnisch, R., and Träutner, W. (1980). *Anat. Embryol.* **158**, 179–191.

Friess, A. E., Sinowatz, F., Skolek-Winnisch, R., and Träutner, W. (1981). *Anat. Embryol.* **163**, 43–53.

Gadsby, J. E., Heap, R. B., and Burton, R. D. (1980). *J. Reprod. Fertil.* **60**, 409–417.

Geisert, R. D., Renegar, H. R., Thatcher, W. W., Roberts, R. M., and Baser, F. W. (1982a). *Biol. Reprod.* **27**, 925–939.

Geisert, R. D., Brookbank, J. W., Roberts, R. M., and Bazer, F. W. (1982b). *Biol. Reprod.* **27**, 941–955.

Ginther, O. J. (1979). "Reproductive Biology of the Mare." O. J. Ginther, Cross Plaines, Wisconsin.

Ginther, O. J. (1986). "Ultrasonic Imaging and Reproductive Events in the Mare." Equiservices, Cross Plains, Wisconsin.

Godkin, J. D., Lifsey, B. J., and Baumbach, G. A. (1988). *Biol. Reprod.* **39**, 195–204.

Goss, T. S., and Williams, W. F. (1988). *J. Reprod. Fertil.* **83**, 565–573.

Green, W. W. (1946). *Am. J. Vet. Res.* **7**, 395–402.

Grunert, E., Schulz, L.-Cl., and Ahlers, D. (1976). *Ann. Rech. Vétér.* **7**, 135–138.

Guillomot, M., Betteridge, K. J., Harvey, D., and Goff, A. K. (1986). *J. Reprod. Fertil.* **78**, 27–36.

Guillomot, M., Fléchon, J., and Wintenberger-Torres, S. (1981). *Placenta* **2**, 169–182.

Guillomot, M., and Guay, P. (1982). *Anat. Rec.*, **204**, 315–322.

Hamilton, W. J., Harrison, R. J., and Young, B. A. (1960). *J. Anat.* **94**, 1–33.

Hansen, P. J., Stephenson, B. G., Low, B. G., and Newton, G. R. (1989). *J. Reprod. Fertil. Suppl.* **37**, 55–61.

Harper, M. J. K. (1982). *In* "Reproduction in Mammals: Germ Cells and Fertilization" (C. R. Austin and R. V. Short eds.), pp. 102–127. Cambridge University Press, Cambridge.

Harvey, M. J., Boyd, J. S., Ferguson, J. M., and Renton, J. P. (1989). *J. Reprod. Fertil. Abstr. Ser.* **3**, 34.

Head, J. R., and Billingham, R. E. (1984). *In* "Immunological Aspects of Reproduction in Mammals" (D. B. Crighton, ed.), pp. 133–152. Butterworths, London.

Holm, L. W., Salvatore, C., and Zeek-Minning, P. (1964). *Am. J. Obstet. Gynecol.* **88**, 479–483.

Holst, P. A., and Phemister, R. D. (1971). *Biol. Reprod.* **5**, 194–206.

Holst, P. J. (1974). *J. Reprod. Fertil.* **36**, 427–428.

Hunter, R. H. F. (1974). *Anat. Rec.* **178**, 169–186.

Kastelic, J. P., Curran, S., Pierson, R. A., and Ginther, O. J. (1988). *Theriogenology* **29**, 39–54.

Kehrer, A. (1973). *Z. Anat. Entwickl.-Gesch.* **143**, 25–42.

Keys, J. L., and King, G. J. (1988). *Biol. Reprod.* **39**, 473–487.

King, G. J. (1988). *J. Reprod. Immunol.* **14**, 41–46.

King, G. J., and Ackerley, C. A. (1985). *J. Reprod. Fertil.* **73**, 361–367.

King, G. J., Atkinson, B. A., and Robertson, H. A. (1979). *J. Reprod. Fertil.* **55**, 173–180.

King, G. J., Atkinson, B. A., and Robertson, H. A. (1980). *J. Reprod. Fertil.* **59**, 95–100.

King, G. J., Atkinson, B. A., and Robertson, H. A. (1981). *J. Reprod. Fertil.* **61**, 469–474.

King, G. J., Atkinson, B. A., and Robertson, H. A. (1982). *J. Reprod. Fertil. Suppl.* **31**, 17–30.

King, J. M. (1967). *J. Anat.* **101**, 277–284.

Knight, J. W., Bazer, F. W., Thatcher, W. W., Franke, D. E., and Wallace, H. D. (1977). *J. Anim. Sci.* **44**, 620–637.

Koch, E. (1985). *J. Reprod. Fertil. Suppl.* **33**, 65–81.

Kurnosov, K. M. (1973). *Dokl. Biol. Sci.* **210**, 229–233.

Lambert, P. P., Lammens-Verslijipe, M. J., Sennesael, J., Herremans, N., and Delsaux, B. (1987). *Biol. Reprod.*, **37**, 887–899.

Latshaw, W. K. (1987). "Veterinary Developmental Anatomy: A Clinically Oriented Approach." B. C. Decker, Toronto.

Lawn, A. M., Chiquoine, A. D., and Amoroso, E. C. (1969). *J. Anat.* **105**, 557–578.

Lee, C. S., Gogolin-Ewens, K., White, T. R., and Brandon, M. R. (1985). *J. Anat.* **140**, 565–576.

Lee, C. S., Wooding, F. B. P, and Brandon, M. R. (1986). *Placenta* **7**, 495–504.

Lee, S. Y., Anderson, J. W., Scott, G. L., and Mossman, H. W. (1983). *Am. J. Anat.* **166**, 313–327.

Lee, S. Y., Mossman, H. W., Mossman, A. S., and del Pino, G. (1977). *Am. J. Anat.* **150**, 631–640.

Leiser, R. (1975). *Anat. Histol. Embryol.* **4**, 63–86.

Leiser, R. (1979). *Anat. Histol. Embryol.* **8**, 79–96.

Leiser, R. (1982). *Biblio. Anat.* **22**, 93–107.

Leiser, R., and Dantzer, V. (1988). *Anat. Embryol.* **177**, 409–418.

Leiser, R., and Enders, A. C. (1980a). *Acta Anat.* **106**, 293–311.

Leiser, R., and Enders, A. C. (1980b). *Acta Anat.* **106**, 312–326.

Leiser, R., and Kohler, T. (1984). *Anat. Embryol.* **170**, 209–216.

Lewis, G. S. (1989). *J. Reprod. Fertil. Suppl.* **37**, 261–267.

Linklater, K. A. (1968). *Vet. Rec.* **83**, 203–204.

Malassiné, A. (1974). *Anat. Embryol.* **146**, 1–20.

Malassiné, A. (1982). *Biblio. Anat.* **22**, 108–116.

Markee, J. E., and Hinsey, J. C. (1933). *Proc. Soc. Exp. Biol. Med*, **31**, 267–270.

Marrable, A. W. (1971). "The Embryonic Pig: A Chronological Account." Pitman, London.

Marrable, A. W., and Ashdown, R. R. (1967) *J. Agric. Sci. Camb.*, 69, 443–447.

Marrable, A. W. and Flood, P. F. (1975). *J. Reprod. Fert., Suppl.* **23**, 499–502.

Martal, J., Djiane, J., and Dubois, M. P. (1977). *Cell Tissue Res.* **184**, 427–433.

McCance, R. A., and Dickerson, J. W. T. (1957). *J. Embryol. Exp. Morphol.* **5**, 43–50.

McRae, A. C. (1988). *J. Reprod. Fertil.* **82**, 857–873.

Medawar, P. B. (1953). *Symp. Soc. Exp. Biol.* **7**, 320–338.

Menino, A. R., and Williams, J. S. (1987). *Biol. Reprod.* **36**, 1289–1295.

Mercer, W. R., Gogolin-Ewens, K. J., Lee, C. S., and Brandon, M. R. (1989). *Placenta* **10**, 71–82.

Moor, R. M., Allen, W. R., and Hamilton, D. W. (1975). *J. Reprod. Fertil. Suppl.* **23**, 391–396.

Morgan, G., and Wooding, F. B. P. (1983). *J. Ultrastruct. Res.* **83**, 148–160.

Morgan, G., Wooding, F. B. P., Beckers, J. F., and Friesen, H. G. (1989). *J. Reprod. Fertil.* **86**, 745–752.

Mossman, H. W. (1987). "Vertebrate Fetal Membranes." Macmillan, London.

Munson, L., Kao, J. J., and Schlafer, D. H. (1989). *J. Reprod. Fertil.* **87**, 509–517.

Myagkaya, G., and Vreeling-Sindelarova, H. (1976). *Acta Anat.* **95,** 234–238.

Myagkaya, G. L., and De Bruijn, W. C. (1982). *Biblio. anat.* **22,** 117–122.

Myagkaya, G. L., Schornagel, K., Van Veen, H., and Everts, V. (1984). *Placenta* **5,** 551–558.

Newman, M. J., and Hines, H. C. (1980). *J. Reprod. Fertil.* **60,** 237–241.

Newton, G. R., Hansen, P. J., Bazer, F. W., Leslie, M. V., Stephenson, D. C., and Low, B. G. (1989). *Biol. Reprod.* **40,** 417–424.

Oriol, J. G., Poleman, J. C., Antczak, D. F., and Allen, W. R. (1989). *Equine Vet. J. Suppl.* **8,** 14–18.

Pascoe, R. R. (1982). *J. Reprod. Fertil. Suppl.* **32,** 441–446.

Perry, J. S. (1981). *J. Reprod. Fertil.* **62,** 321–335.

Perry, J. S., and Rowlands. I. W. (1962). *J. Reprod. Fertil.* **4,** 175–188.

Perry, J. S., Heap, R. B., and Ackland, N. (1975). *J. Anat.,* **120,** 561–570.

Perry, J. S., Heap, R. B., and Amoroso, E. C. (1973). *Nature Lond,* **245,** 44–47.

Perry, J. S. and Rowlands, I. W. (1962) *J. Reprod. Fertil.* **4,** 175–188.

Polge, C., and Dzuik, P. J. (1970). *J. Anim. Sci.* **31,** 565–566.

Polge, C., Rowson, L. E. A., and Chang, M. C. (1966). *J. Reprod. Fertil.* **12,** 395–397.

Pope, W. F., Maurer, R. R., and Stormshak, F. (1982). *Biol. Reprod.* **27,** 575–579.

Raub, T. J., Bazer, F. W., and Roberts, R. M. (1985). *Anat. Embryol.* **171,** 253–258.

Reeves, J. T., Daoud, F. S., Gentry, M., and Eastin, C. (1972). *Am. J. Vet. Res.* **33,** 2158–2167.

Reimers, T. J., Ullman, M. B., and Hansel, W. (1985). *Biol. Reprod.* **33,** 1227–1236.

Rigby, J. P. (1968). *Res. Vet. Sci.* **9,** 551–556.

Roberts, M. R. (1989). *Biol. Reprod.* **40,** 449–452.

Rossant, J., Croy, D. A., and Chapman, V. M. (1983). *J. Exp. Zool.* **228,** 223–333.

Roth, T. L., Anderson, G. B., Bon Durant, R. H., and Pashen, R. L. (1989). *Biol. Reprod.* **41,** 675–682.

Rowson, L. E. A., and Moor, R. M. (1966). *J. Anat.* **100,** 777–785.

Samuel, C. A., Allen, W. R., and Steven, D. H. (1974). *J. Reprod. Fertil.* **41,** 441–445.

Samuel, C. A., Allen, W. R., and Steven, D. H. (1976). *J. Reprod. Fertil.* **48,** 257–264.

Samuel, C. A., Allen, W. R., and Steven, D. H. (1977). *J. Reprod. Fertil.* **51,** 433–437.

Sasser, R. G., Crock, J., and Ruder-Montgomery, C. A. (1989). *J. Reprod. Fertil. Suppl.* **37,** 109–113.

Scanlon, P. F. (1972). *J. Anim. Sci.* **34,** 791–794.

Schlafer, D. H., Dougherty, E. P., and Woods, G. L. (1987). *J. Reprod. Fertil. Suppl.* **35,** 695.

Schummer, A., Nickel, R., and Sack, W. O. (1979). "The Viscera of the Domestic Animals." Paul Parey, Berlin.

Simpson, E. (1984). *In* "Immunological Aspects of Reproduction in Mammals" (D. B. Crighton, ed.), pp. 1–11. Butterworths, London.

Singh, B., Raghupathy, R., Anderson, D. J., and Wegmann, T. G. (1983). *In* "Immunology of Reproduction" (T. G. Wegmann and T. J. Gill, eds.), pp. 229–250. Oxford University Press, New York.

Spincemaille, J., Bouters, R., Vanderplassche, M., and Bonte, P. (1975). *J. Reprod. Fertil. Suppl.* **23,** 415–418.

Stanier, M. W. (1965). *J. Physiol.* **178,** 127–140.

Staples, L. D., Heap, R. B., Wooding, F. B. P., and King, G. J. (1983). *Placenta* **4,** 339–350.

Steven, D. H. (1975a). "Comparative Placentation: Essays in Structure and Function." Academic Press, London.

Steven, D. H. (1975b). *J. Exp. Physiol.* **60,** 37–44.

Steven, D. H. (1982). *J. Reprod. Fertil. Suppl.* **31,** 41–55.

Steven, D. H., and Samuel, C. A. (1975). *J. Reprod. Fertil.* **23,** 579–582.

Steven, D. H., Burton, G. J., and Keeley, V. L. (1982). *Biblio. Anat.* **22,** 128–133.

Thomsen, J. L., and Edelfors, S. (1976). *J. Dairy Sci.,* **59,** 288–292.

Tiedemann, K. (1976). *Cell Tissue Res.,* **173,** 109–127.

Tiedemann, K. (1977). *Cell Tissue Res.,* **183,** 71–89.

Tiedemann, K. (1979a). *Anat. Embryol.* **156,** 53–72.

Tiedemann, K. (1979b). *Anat. Embryol.* **158,** 75–94.

Tiedemann, K. (1982). *Anat. Embryol.* **163,** 403–416.

Turner, W. (1876). "Lectures on the Comparative Anatomy of the Placenta." A. and C. Black, Edinburgh.

van der Meulen, J., Helmond, F. A., and Ouden-Aarden, C. P. T. (1988). *J. Reprod. Fertil.* **84,** 157–162.

van der Meulen, J., te Kronnie, G., van Deursen, R., and Geelen, J. (1989). *J. Reprod. Fertil.* **87,** 783–788.

van Niekerk, C. H., and Allen, W. R. (1975). *J. Reprod. Fertil. Suppl.* **23,** 495–498.

van Niekerk, C. H. (1965). *J. S. Afr. Vet. Med. Assoc.* **36,** 483–488.

Wathes, D. C., and Wooding, F. B. P. (1981). *Am. J. Anat.* **159,** 285–306.

Watkins, W. B., and Reddy, S. (1980). *J. Reprod. Fertil.* **58,** 411–414.

Wegmann, T. G. (1983). *In* "Reproductive Immunology 1983." (S., Isojima and W. D. Billington, eds.), pp. 111–117. Elsevier Science, Amsterdam.

Whyte, A., and Robson, T. (1984). *Placenta* **5,** 533–540.

Williams, W. F., Margolis, M. J., Manspeaker, J., Douglas, L. W., and Davidson, J. P. (1987). *Theriogenology* **28,** 213–224.

Wimsatt, W. A. (1950). *Am. J. Anat.* **87,** 391–457.

Wimsatt, W. A. (1951). *Am. J. Anat.* **89,** 233–282.

Wimsatt, W. A. (1975). *Biol. Reprod.* **12,** 1–40.

Wlodek, M. E., Challis, J. R. G., and Patrick, J. (1988). *J. Dev. Physiol.* **10,** 309–319.

Wooding, F. B. P. (1980). *Biol. Reprod.* **22,** 357–365.

Wooding, F. B. P. (1981). *J. Reprod. Fertil.* **62,** 15–19.

Wooding, F. B. P. (1982). *J. Reprod. Fertil. Suppl.* **31,** 31–39.

Wooding, F. B. P. (1983). *Placenta* **4,** 527–540.

Wooding, F. B. P. (1984). *Am. J. Anat.* **170,** 233–250.

Wooding, F. B. P., Chambers, S. G., Perry, J. S., George, M., and Heap, R. B. (1980). *Anat. Embryol.* **158,** 361–370.

Wooding, F. B. P., Flint, A. P. F., Heap, R. B., Morgan, E., Buttle, H. L., and Young, I. R. (1986). *J. Reprod. Fertil.* **76,** 499–512.

Wooding, F. B. P., Staples, L. D., and Peacock, M. A. (1982). *J. Anat.* **134,** 507–516.

Wooding, F. B. P., and Wathes, D. C. (1980). *J. Reprod. Fertil.* **59,** 425–430.

Wynn, R. M., and Björkman, N. (1968). *Am. J. Obstet. Gynecol.* **102,** 34–43.

Wynn, R. M., and Corbett, J. R. (1969). *Am. J. Obstet. Gynecol.* **103,** 878–887.

Zavy, M. T., Mayer, R., Vernon, M. W., Bazer, F. W., and Sharp, D. C. (1979). *J. Reprod. Fertil. Suppl.* **27,** 403–411.

Hormonal Mechanisms in Pregnancy and Parturition

HUBERT R. CATCHPOLE

I. Introduction

Pregnancy is defined as the interval between the implantation of a fertilized ovum in the uterus and the expulsion from it of the fetus and its associated membranes. While the end stage is fairly precise, the initial events are not. Implantation may be relatively prompt after the fertilized ovum reaches the uterus, as in rodents, carnivores, and humans, or quite prolonged as in the equids when 4 weeks may elapse before the first tentative junction is established. In sheep and pigs, this period is about 2 weeks, and in the cow nearly 3. Independently of actual junction, which is hard to define even histologically, some progestational events in the uterus may precede, and be independent of, the arrival or even the presence of the ovum. It is most convenient in most cases to date pregnancy from the time of insemination, although those animals with delayed fertilization pose a dilemma.

Descriptions of individual species in later chapters will usually include pregnancy as one phase in the whole reproductive cycle. This chapter will attempt a general and comparative approach. Pregnancy represents the central event or events in the life cycle of the female mammal, and both single and multiple terms of pregnancy and lactation are relatively prolonged in actual time. In all species, the breeding period is delimited at the beginning by the processes of growth, maturation, and sexual development, and at the end by less well-defined aging processes signaled in humans and other primates by a definite menopause and

in all species by dwindling fertility associated with aging. Gestation spans are given in still valuable compilations of Kenneth (1947) and Asdell (1964), and some of them relevant to the present chapter are given (Table I) with additional pertinent data. A rough correspondence exists between animal weight and duration of pregnancy, which occupies 3 or 4 weeks in laboratory rodents and rabbits, and 2–10 months in the common domestic forms, such as dogs (2 months), pigs (4 months), sheep (5 months), cows and horses (9 months and 10 months, respectively), and elephants (24 months). Both hamsters (15 days) and guinea pigs ($2\frac{1}{2}$ months) appear exceptional, the latter in the relatively advanced stage of development of the newborn pups. The order

Table I
Length of Gestation, Litter Size, and Result of Oophorectomy during Pregnancy[a]

Species	Litter size	Length of gestation (days)	Aborts if oophorectomized before given day of gestation
Cat	4	63	50
Cow	1–2	277–290	Term
Dog	Multiple, mean 7.0	61	Term
Guinea pig	2–4	68	38
Goat	1–3	145–151	Term
Horse	1	330–345	150,200
Monkey (Macaca mulatta)	1	168	25
Man	1	280 ± 9.2	30–60
Mouse	Multiple, mean 6.0	19–20	Term
Pig (domestic)	8–12	112–115	Term
Rabbit (domestic)	Multiple, mean 8.0	31	Term
Rat	6–9	22	Term
Sheep (domestic)	1–2	144–152	55

[a]Data compiled from many sources (Asdell, 1964; Kenneth, 1947). See also text.

Marsupalia, which is based in part on the feature of birth of the young in an embryonic stage, forms a major exception. Following a gestation period ranging from 8 to 40 days, embryos transfer to mammary-gland teats within or in the absence of a mammary pouch structure. In this order, placentation is usually lacking, gestation short, and lactation prolonged. Nor is the gestation length at all related to maternal size. These fascinating animals, which have substituted a prolonged and elaborate lactation period for uterine pregnancy, will not be further considered in this chapter. However, reference may be made to a spirited defense of the order (Renfree, 1981) against the suggestion that these are second-class mammals.

Problems of deriving scaling relations between neonatal size or gestation period and maternal size in mammals were examined (Martin and McLaren, 1987). It was pointed out that artificial breeding in the dog has produced a wide range of adult body sizes across breeds with no significant variation in gestation length.

Pregnancy is marked by progressive adaptation of the uterus to the growing conceptus which may represent single or multiple fetuses. The new organisms achieve a total bulk at term that is a significantly large fraction of the maternal body weight: up to 33% in mice, 15% in the sows, and 10% in thoroughbred mares. Extrapolated from a small sample of birth announcements, the figure in humans is estimated to be about 7%, the same as that for the Yale Colony macaques. All of this considerable weight represents the passage of water and dissolved oxygen and nutrients across the "placental barrier," a term which regarded in this sense is a considerable misnomer. As the uterus expands to accommodate the fe-

tuses with their accompanying fluid-filled sacs and attachment membranes, it is natural to suppose that there would be some stretching and thinning of the uterine wall. However, stretching is an excellent stimulus for contraction, which definitely does not occur during most of pregnancy. The process must be regarded rather as one of accommodation, mediated in part by an increase in myometrial cell size (up to 10-fold in volume) and in part by an increase in amount, and very likely in molecular organization, of the extracellular matrix of collagen and glycosaminoglycans. Cell size increase is accompanied by a large increase in the actin and myosin content of the uterus. Following parturition, the uterus involutes, a rather rapid but complex process involving loss in water and of the chemical constituents added during pregnancy. The final state of the uterus, in size and weight, is usually at a higher baseline than previous to a given pregnancy.

During pregnancy the mammary glands of the mother undergo growth and development to the point of prompt lactation immediately postpartum, with the production of colostrum followed by true milk.

All aspects of pregnancy are mediated by hormones. These are supplied initially by the maternal endocrine organs. Subsequently, and often sooner than later, the developing fetus and the unique organ of pregnancy, the placenta, begin to make various hormonal contributions. The use of the term feto–placental unit refers to possible synergistic actions of fetus and placenta in achieving an overall successful gestation.

Important topics that concern the earlier stages of pregnancy are the recognition by the maternal ovary of the presence of a fetus in the uterus (how does the mother know she is pregnant?), and the problem of the fetal allograft (why is the fetus not rejected since half of its genes are foreign?). Both of these are discussed in Chapter 9.

Questions that have occasioned much work are the relative quiescence of the uterus for the greater part of pregnancy and the reactivation of the contractile process at parturition; and that of preventing luteolysis over the critical bridging period between cycle and fertilization (antiluteolysis) and ensuring the active life of the corpus luteum of pregnancy (luteotropic actions) (Procknor *et al.*, 1986). In some species a luteolytic process may also be present at parturition.

Technical procedures introduced since the last review include the addition of enzyme immunoassay to the radioimmunoassay (RIA) revolution, increasing the possibilities of easy and accurate field determinations of hormone levels, which are a standby of endocrinological research (Stanley *et al.*, 1986; Chang and Estergreen, 1983).

In addition to reviews that cover the exact topic of the present one (Bazer and First, 1983), a number of active areas have elicited specialized coverage as follows: maternal recognition of pregnancy (Basu, 1985; Findlay, 1981; Hearn, 1986), endocrine control of uterine blood flow (Ford, 1982; Resnick, 1983), pregnancy diagnosis in domestic ruminants (Sasser and Ruder, 1987). and maternal and fetal endocrinology in mares (Pashen, 1984).

No brief summary can do justice to the review of Stewart *et al.* (1988) on the contribution of recombinant DNA techniques to reproductive biology. While much of this work refers to laboratory species (including *Drosophila*), the answers now appearing show the importance of these methods to the solution of problems in the domestic species, and part of the reported work bears

directly on them. Some pertinent references are to chemistry and biology of the neurophysins (Breslow, 1979), structure and function of the chorionic gonadotropins and lactogens (Boime *et al.*, 1978; Fiddes *et al.*, 1979), progesterone (Heap and Flint, 1979), estrogens (Knowler and Beaumont, 1985), and steroid receptor-regulated transcription (Yamamoto, 1985).

II. Maternal Endocrine Patterns

A. *Estrogen and Progesterone Patterns in Blood during Pregnancy and Parturition*

The constant presence of the ovarian hormones, estrogen and progesterone, throughout pregnancy makes their temporal patterns of considerable interest since it is upon these that the actions of many other hormones are superimposed. Curves were presented previously (Catchpole, 1977) for the cow, ewe, sow, goat (doe), mare, bitch, and rabbit (doe) with some apologies for their relative incompleteness. All values were obtained by methods of RIA. While many values have been reported over the past decade, problems of devising adequate curves from the literature have not entirely vanished: for example, many results are reported for quite restricted periods of pregnancy. New values are not felt to have changed overall impressions to an important degree except in the case of pig values, which are revised (Adair *et al.*, 1982; Anderson *et al.*, 1982; King and Rajamahendran, 1988; Robertson and King, 1974). These data are presented as profiles so that interspecies comparisons are easier to visualize (Figs. 1 and 2). In each case equal abscissae

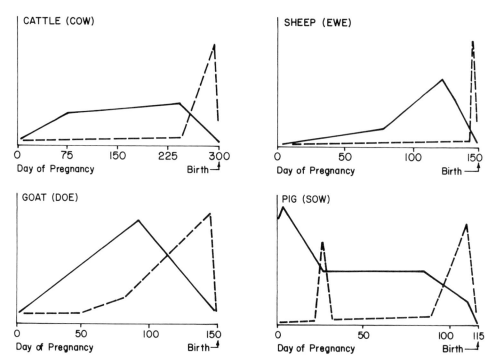

Figure 1 Pregnancy blood levels of progesterone (solid line) and estrogen (dotted line) in cows, sheep, goats, and pigs. Drawn from data in Catchpole (1977), which gives original references; see also text.

represent the whole span of pregnancy. A profile is also added for the laboratory rat and/or mouse, based on rather few available data (Barkley *et al.*, 1977; Pepe and Rothchild, 1974). Also, absolute values are omitted but can mostly be obtained from the previous detailed charts.

B. The Maternal Ovaries in Pregnancy

In the presence of a fertilized and implanting ovum, changes are initiated in the ovary that largely define its character for the remainder of pregnancy. Corpora lutea, which normally undergo regression as a result of luteolytic actions, become stabilized as corpora of gestation. These bodies tend

to be larger than those of the estrous cycle, although individual cells are not greatly altered. These are typical steroid-secreting cells as defined histochemically and ultrastructurally.

Involution of the corpus luteum (corpora lutea) destined to become the active gland(s) of pregnancy is halted by antiluteolytic influences beginning soon after the blastocyst enters the uterus. The presence of agents that could be responsible for this signal to the mother has been described in fluids or extracts derived from the conceptus in the mare (Sharp *et al.*, 1989; Zavy *et al.*, 1984a), and ewe (Findlay, 1978; Bazer *et al.*, 1986). These agents are believed to suppress the secretion of endometrial prostaglandin

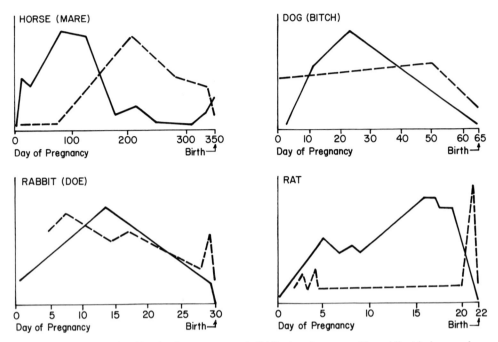

Figure 2   Pregnancy blood levels of progesterone (solid line) and estrogen (dotted line) in horses, dogs, rabbits, and rats. For rats, estrogen values are for uterine blood. Source as in Fig. 1.

PGF-2, considered to be a major luteolytic factor (Auletta and Flint, 1988). In the mare, restricted mobility of the conceptus in the uterus caused abortion, possibly since these inhibitory substances were no longer being distributed (McDowell *et al.*, 1988).

Luteotropic agents are present in the same sources, as described in bovines (Hansel *et al.*, 1989; Hickey *et al.*, 1989; O'Neill *et al.*, 1989). Conceptus products acting directly or indirectly via the uterus on the maternal ovary may include proteins, glycoproteins, and steroids. They act to ensure the longevity of the corpus luteum and are therefore agents of maternal recognition. In the pig, this function is subserved by estrogen, acting as an antiluteolysin (Bazer *et al.*, 1986). In the rabbit, estrogen acts as a luteo-

tropin together with a luteotropic placental hormone (Gadsby and Keyes, 1984).

The growth of follicles, which would normally signal the onset of a new estrous cycle, does not generally occur, and these continue to be suppressed throughout pregnancy in domestic animals with the exception of mares (p. 371).

Ovariectomy at any point terminates pregnancy in the cow, goat, pig, rabbit, rat, and mouse, by resorption or abortion of embryos or fetuses (Table I). Pregnancy can be maintained by the prompt and continuous administration of progesterone with or without estrogen to term or until such a time that other sources of these hormones may become available. Hormones are given empirically to achieve the normal expected pregnancy blood levels, as emphasized in the mare (Shideler *et al.*, 1982). In humans the ovary is dispensable after day 40, and in the ewe after day 55. In the mare, ovariectomy after day 50 gave a survival rate of 11 out of 20 pregnancies; when ovariectomy was delayed until 140 and 210 days, pregnancies persisted (Holtan *et al.*, 1979). These outcomes are related to the amount of extraovarian progesterone available. In support of this interpretation, ovariectomy of pregnant sheep at 110, 122, and 135 days changed blood levels of progesterone very little (Fylling, 1970).

1. Role of Estrogens

The primary estrogen secreted by the ovary and possibly the placenta of all mammals is estradiol-17. Following estrus and fertilization, blood levels tend to fall from mid-cycle levels but estrogen continues to be secreted, and it is upon estrogen-primed tissues that progesterone, which is being secreted in increasing amounts, now acts. During pregnancy the estrogen profiles show a general tendency to increase, with peaking toward mid to later pregnancy in the mare and bitch, and high peaks terminally around the time of parturition in rat, mouse, cow, ewe, and sow, with the rabbit showing a variable elevation throughout (Figs. 1 and 2). Rodents apparently show early variable peaks, while the pig shows a pronounced peak between days 23 and 30, falling to baseline until the final rise at term. It is difficult to apportion pregnancy estrogens between ovarian and extraovarian sources.

Placentas would appear to be a source of estrogen synthesis in all species during pregnancy, but postpartum estrus in some species shows that the ovary may also be prepared to secrete increasing amounts to-ward the end of pregnancy, and could be largely responsible for the late estrogen peaks.

Estrogenic effects on the reproductive tract include multiplication of uterine epithelial cells and simple glandular extension; hypertrophy of uterine smooth muscle cells and synthesis of the contraction-related proteins actin and myosin; synthesis of cellular DNA and RNA and their related protein synthetic activity; glycogen deposition in smooth muscle cells of the uterus and blood vessels; and synthesis of collagen and glycosaminoglycans of the extracellular matrix. Estrogens produce water uptake in the sex skin of primates, in which there is an elaboration of extracellular macromolecules, and dye uptake in the uteri of estrogen-treated rodents. It seems more useful to focus on the specific action of estrogen on connective tissue components rather than on concepts of vascular permeability to explain both these effects (Catchpole, 1973).

Estrogen appears to act as a primer or prerequisite for the action of other hormones. Examples are the preparation of the endometrium for the action of progesterone, of the mammary gland for the actions of progesterone and prolactin, and of the symphysis pubis for that of relaxin. Its preliminary actions could represent the synthesis of binding sites for these subsequently acting hormones.

Besides the more specific endocrine actions listed above, many of which serve to explain uterine growth and adaptation to the conceptus at the morphological level, estrogens exert effects on uterine metabolism in order to meet the increasing demands of fetal growth.

Estrogen is excreted in the urine of domestic animals in increasing amounts as pregnancy advances. Interest in the urinary estrogens dwindled with the possibility of assaying for minute amounts present in the blood, and they are perhaps of significance now only in the metabolic transformations undergone by active metabolic hormones either in the liver or the placenta to produce biologically less active derivatives such as estrone or estriol. In the equids, a series of biologically inactive, highly unsaturated steroids related to the estrogens appears in the urine of pregnancy in some abundance.

The rise in estrogen over the span of pregnancy shows considerable agreement in all species recorded, although the actual time spans of pregnancy range from 20 to 340 days. This agreement points to common functions, of which maintenance of the maternal uterus would appear to be the most important. The reason for increasing amounts of estrogens could be related to the necessity of increasing the number of binding sites (e.g., for progesterone) in an expanding uterus. The relation of estrogens to progesterone binding sites was explored in a previous volume (Clark et al., 1977).

2. Role of Progesterone

The action of progesterone on an estrogen primed uterus is to produce a progestational proliferation, represented histologically by an increasing tortuosity and sacculation of the endometrial glands, which become inserted deep in the mucosa. This adaptation appears to be necessary for the reception and attachment of the fertilized ovum. An important function of progesterone from the inception of pregnancy is to produce "progesterone block" (Csapo, 1956), an idea that has withstood the test of time. Myometrial contractions are inhibited, and in order to initiate parturition this block must be removed either by a luteolytic agent directed to the ovarian corpus luteum or to

some mechanism acting at the level of the feto-placenta. The actual mechanism of myometrial suppression has tapped every fashionable hypothesis. Current ideas favor oxytocin and prostaglandins acting at the level of ionic channels of the myometrium (Popescu et al., 1984).

Progesterone titers in the blood rise during pregnancy to reach peaks at somewhat different times for different species. For goat, rabbit, and bitch this is at or just before mid-pregnancy; for cow, ewe, and rat definitely after mid-pregnancy, although the rat has an earlier and smaller peak; for the mare prior to mid-pregnancy; and for the pig a wide plateau covering almost the span of pregnancy (Anderson et al., 1982). Nevertheless, for all species progesterone is constantly present, from before implantation until the day of parturition. Progesterone is usually accepted as "the" hormone of pregnancy, to which the proviso should be added "in the presence of an estrogen environment" until proven otherwise. Part of the variability in progesterone profiles may be due to the differing times at which placental production of the hormone is triggered in.

Two experiments showed specifically the role of progesterone. In the mouse, a monoclonal progesterone antibody successfully blocked pregnancy in 17 of 17 animals treated 32–65 h postcoitus (Wright et al., 1982). Removal of progesterone from the blood of 10-day pregnant rats by binding to an entrapped antiprogesterone antiserum lowered progesterone levels from 50 ng/ml to less than 2 ng/ml within 36 h, with eventual fetal resorption (Cheesman et al., 1982).

There is a peculiar disparity in absolute amounts of progesterone present during peak values between the smaller laboratory animals (rats, 130 ng/ml; guinea pigs, 265 ng/ml) and those recorded for domestic species (range 14–35 ng/ml), which is unexplained. Also, pregnancy values for progesterone appear to be comparable to, or only slightly higher than, those recorded during an estrous cycle. Progesterone represents a large proportion of all secreted, hormonally active steroids during pregnancy.

The corpus luteum is traditionally regarded as composed of cells from two sources: large cells derived from granulosa cells and smaller cells from the theca interna. Both secrete progesterone, as demonstrated by experiments in which the two cell populations were separated from the corpus luteum from cows during late pregnancy (Weber et al., 1987).

Progesterone-binding proteins in blood were measured in guinea pigs and other hystricomorphs with peak values between days 20–25 and days 50 to term, the patterns being parallel to progesterone concentrations. This was suggested to be a progesterone-conserving mechanism in animals with a prolonged gestation (Heap et al., 1981).

The additional corpora lutea crop produced in the mare at about the time of appearance of placental gonadotropins in this species (i.e., at 6–8 weeks) appear to be hormonally active along with the existing corpus luteum of pregnancy. These bodies all share an histological and functional demise after day 160–180, when presumably, from the evidence of castration, the placenta is fully able to supply sufficient hormone to ensure continuance of pregnancy.

A further effect of progesterone is on the mammary gland, which during pregnancy develops sufficiently to achieve lactational status at parturition. The action of progesterone, somewhat analogous to that on the uterus, is to develop to its fullest extent the acinar structure, on a ductal system prepared by estrogens, so that prompt lactation can occur under the influences of prolactin and suckling.

3. Role of Relaxin

Relaxin is a peptide hormone synthesized in the corpus luteum of pregnancy and released into the blood prior to parturition. It was originally isolated from the ovaries of pigs by Hisaw in 1928. Much experimental work was done on guinea pigs in which the pubic symphysis exhibits marked softening and wide extension at term to accommodate delivery of the large fetuses characteristic of this species. The peptide sequence of relaxin was inferred from the encoding genomic clone (Hudson et al., 1983). The biologically active hormone in all species consists of two linked (A and B) polypeptide chains. For long considered a chemical oddity in its ovarian milieu, the synthesis of relaxin shifts to the feto-placental unit during pregnancy.

Blood levels tend to rise to variable peaks in the latter half of pregnancy in the cat (Addiego, 1987), rabbit, pig, and cow, although the net contribution from ovary and placenta is not known. Its classical effect was on the fibrocartilagenous ligament of the guinea pig pubic symphysis, which was transformed into a malleable fibrous tissue with concomitant effects on the extracellular matrix. Mobilization of the sacroileac joints in the cow, sheep, and probably humans are related effects, all tending to a widening of the birth canal at parturition. In the pubic symphysis pretreatment with estrogens strongly potentiates the action of relaxin, which may point to one function of the high prepartum estrogen peaks found in some species. Relaxin also acts on the cervix and uterus when given alone or with estrogens and on the mammary gland to suppress lactation. The biological actions of relaxin in pigs and beef cattle were reviewed (Anderson *et al.*, 1982). Theories involving possible relaxin effects in early pregnancy on the uterus involve a suppression of oxytocin release and a direct effect on the smooth muscle of the myometrium (Summerlee *et al.*, 1984). There is no obvious change in the secretion of relaxin at the onset of pregnancy to suggest that its action in pacifying the uterus is a sudden pharmacological event. It appears, rather, to be a gradual change occurring in the milieu of falling estrogen and increasing progesterone over the period in which the conceptus is becoming established. While an action directly on smooth muscle is possible, involvement with the uterine connective tissue matrix does not seem to be ruled out. Possible actions of relaxin, including paracrine effects, were reviewed (Bryant-Greenwood, 1985), and a critical look was taken at proposed therapeutic uses (Sarma, 1983).

In the mare, development of a sensitive homologous equine relaxin RIA test showed the first appearance of relaxin at day 75, a peak at 175–200 days followed by a fall at 225–250 days, and a rise to a higher peak at the day of parturition. The variation could reflect lower corpus luteum activity at midpregnancy and a placental contribution thereafter. All values found were very significantly increased compared with the use of a porcine hormone in the assay test (Stewart *et al.*, 1982).

In the dog (beagles and Labradors), immunoreactive relaxin appeared at day 21–28, rising to a peak at 2–3 weeks prior to whelping (4–5 ng/ml) and declining toward term. Significant levels were found for 4–9 weeks postpartum and during lactation. No source was suggested for the relaxin secreted during pregnancy, but secretion in any case shifts back to the ovary postpartum (Steinetz *et al.*, 1987).

C. The Hypophysis of Pregnancy

1. Anterior Pituitary

Hypophysectomy of many common species before mid-pregnancy leads to fetal death and resorption, and after this point varying degrees of failure are encountered. Rabbits and ferrets are totally intolerant; pregnant gilts operated upon at 70–90 days aborted within 60 h, and goats operated at 38–120 days aborted in 3–9 days. Ewes hypophysectomized at 50–83 days went to term, but aborted if operated upon earlier. Animals operated upon as early as 10 days, however, could be maintained by a combination of pituitary prolactin and luteinizing hormone (LH), both recognized as luteotropins. These experiences suggest pituitary activity in maintaining the supply of progesterone, especially in earlier stages of pregnancy. More recent experiments with goats showed that these animals could be maintained after hypophysectomy at 40 days of pregnancy by the use of luteinizing hormone primarily, but prolactin is synergistic with it and placental lactogen can substitute for prolactin, thus defining a luteotropic complex of pregnant goats (Buttle, 1983).

In later pregnancy, when large conceptuses are present, pituitary removal may have grave metabolic effects related to the thyroid and adrenals, which are difficult to correct (unless they are being supplied, in part, endogenously). It should be remembered that nonpregnant rodents survive after hypophysectomy only under laboratory conditions of diet and temperature control and that they are intolerant of stress, related to the above deficits. However, the rule is supported that pregnant animals tolerate hypophysectomy in inverse proportion to their dependence on ovarian estrogen and progesterone.

Studies of the pituitary either histologically or from the standpoint of hormonal content during pregnancy have always seemed rather disappointing because of the hormone content versus hormone output dilemma. Better insights seem to be promised by measurement of messenger RNAs, as well as of hormone content (Wise *et al.*, 1985). Luteinizing hormone and messengers for its alpha and beta subunits, for prolactin, and for growth hormone were measured in pregnant ewes at days 50 and 140 of gestation and at several days postpartum. LH declined between 50 and 140 days, was low 2 days postpartum, and returned to cycling levels 22 days postpartum. This decline was paralleled by a decrease in both subunit messengers, which also began to increase postpartum and were maximal at 13 days postpartum, thus preceding LH by several days. Growth hormone messenger did not change and prolactin changed only slightly, so that the gestation effects were cell specific (basophils or gonadotropes versus acidophils or somatotropes/lactotropes). Failure of prolactin to increase terminally is a little disappointing, but the older reports of a fall in pituitary gonadotropins after mid-pregnancy seem somehow vindicated. However, vastly more needs to be done along the above lines to determine the true state of pituitary activity during pregnancy.

In a previous review of this topic (Catchpole, 1977), minimal quantities of gonadotropic factors of pituitary origin were found to be present in pregnancy in the species under consideration. Reports were sparse and remain so. Ovulation is generally preceded by the familiar surge in LH, but thereafter LH values settle to low but detectable levels in pregnancy in the cow, sow, sheep, and dog. The constant presence of progesterone suggests that LH may be acting as a luteotropin. However, pituitary prolactin is also present in the ewe, goat, and cow, shows terminal increases in the ewe, and is also suggested as a luteotropic agent (Kelly *et al.*, 1974).

FSH and LH values were measured in the sera of four horses and four donkeys entering pregnancy and continued when possible to 100 days. LH values were not obtainable after the appearance of equine chorionic gonadotropin (eCG) at day 35–40 due to cross-reaction. LH showed a peak after ovulation but thereafter the titers were flat at baseline until the arrival of eCG. FSH values fluctuated markedly, but irregularly for the first 100 days in all animals observed, with pronounced surges at times. Since follicular activity in the mare is present before the appearance of eCG, follicular growth was attributed to pituitary FSH rather than to the more spectacular chorionic hormone, which is able to elicit follicular stimulation in a variety of test animals (although possibly not in the mare herself). Transformation of these follicles to corpora lutea is still considered to be the function of eCG (Urwin, 1983).

Plasma prolactin was measured in pony mares during parturition and early lactation, showing a marked rise in the final week of pregnancy with values remaining high but variable in early lactation, declining to basal levels 1–2 months postpartum (Worthy *et al.*, 1986).

Prolactin, choriomammotropin (placental lactogen, PL), and progesterone were determined in pregnant sheep and give one of the few comparisons of pituitary and placental hormone outputs available. Prolactin increased from early pregnancy to term (60 to 170 ng/ml) and PL from a low value initially to an 8-fold increase over prolactin at term (16 to 1350 ng/ml). Progesterone showed the expected rise from baseline values of 4.8 ng/ml to 24 ng/ml at day 135. Also, PL titers fell precipitously postpartum, while prolactin tended to increase at least to day 18 of lactation before falling to baseline values at 95–135 days (Vernon *et al.*, 1981). In the same publication, somatotropin was also found to increase from 2.0 to 8.5 ng/ml from early pregnancy to term, with a decided increase to 17 ng/ml during lactation, the only figures found in the literature.

2. Posterior Pituitary-Hypothalamus

Discovered by Dale in 1909, oxytocin is the clinical agent of choice to precipitate human labor, and is involved in physiological neural processes associated with suckling or milking and with actual milk ejection from the mammary gland. Isolated from the posterior pituitary, and now established as having its source in neurons of hypothalamic nuclei, its isolation from ovine corpora lutea came as a distinct surprise (Wathes and Swann, 1982). It is present also in human and bovine corpora lutea. Its role at parturition is considered later. Actions of oxytocin in early pregnancy were suggested (Fairclough *et al.*, 1984; Homeida and Cooke, 1983). Oxytocin is considered to be normally luteolytic in ruminants, acting by the release of uterine prostaglandin (Flint and Sheldrick, 1982). The embryo blocks this reaction. Adaptations suggested at this time include a fall in uterine oxytocin receptors, and in oxytocin synthesis and release, by the corpus luteum of the ewe (Flint *et al.*, 1989), thereby avoiding uterine stimulation and contributing to its quiescence.

D. Uterine Secretions

In the pregnant uterus, uterine secretions proper arise in the uterine glands. The isolation of proteins from uterine washings and an investigation of their possible functions has revived an interest in the ancient term "embryotrophe." William Harvey already had the idea of a nutritive juice, and spoke of it as "an albuminous fluid" capable of sustaining the embryo (Marshall, 1910). Nineteenth-century analyses of material from the uteri of ruminants showed the presence of large amounts of protein and fat. An adequate review (Roberts and Bazer, 1988) emphasizes the relative importance of such secretions in animals with a noninvasive type of placentation (e.g., pig and horse) and in the earlier stages at least of cattle and sheep pregnancies. In the pig, secretory activity of the endometrium increases after day 30, is maximal between days 60 and 75, and declines with the approach of term. Progesterone is identified as the major stimulus. Proteins isolated include uteroferrin, which is an iron transporter with acid phosphatase activity, related basic polypeptides with phosphatase properties, and lysozyme, an antibacterial agent. Functions are readily available for all of these in the feto-maternal economy.

E. The Mammary Glands of Pregnancy

Mammary gland development in an animal entering its first pregnancy, like the uterus, reflects its hormonal history. The doe rabbit, which does not ovulate spontaneously, shows simple ductal growth and rudimentary alveoli characteristic of estrogen stimulation. The bitch, that has experienced a single, unmated cycle with ensuing pseudopregnancy will show extensive alveolar proliferation to actual lactation, as a result of estrogen plus progesterone plus prolactin actions. Domestic species will usually show some intermediate variant. In any event, the corpus luteum of pregnancy, presumably rescued from luteolysis, persisting and acting on a gland already primed with estrogen, ensures mammary growth from the initiation of pregnancy. This type of stimulus will continue with the help of extraovarian sources of estrogens and progesterone.

The experimental induction of lactation in virgin heifers was done under the aegis of H. H. Cole over 50 years ago, using a crude pituitary lactogenic extract prepared by another pioneer, W. R. Lyons. Now fully characterized chemically and as to its gene structure (Stewart *et al.*, 1988), prolactin is responsible for the prolonged lactation of the postpartum. During pregnancy, its actions are presumably augmented by placental lactogen. Both hormones are considered to be luteotropic in most species. Lactation represents the single situation in which large quantities of protein, fat, carbohydrate, water, and ions are removed from the body, producing a deficit that must be constantly replenished. Lactation involves the activities of a mammotropic complex that includes estrogens, progesterone, insulin, cortisol, somatotropin, adrenocorticotropin (ACTH), and prolactin itself. Some of these hormones are directed to the cells of the mammary gland as a morphological unit; others to the production of an energy-rich secretion; the former are likely to be particularly important during pregnancy. Related to this theme, pregnant goats treated with ACTH increased lactose secretion and udder volume at or after 84 days of gestation, and between 109 and 127 days increased

their plasma glucose, mammary glucose uptake, and mammary blood flow (Stewart and Walker, 1987).

A number of hormones and metabolites are transferred into milk. Not surprising are the lipid-soluble estrogens and progesterone, the latter being available for pregnancy diagnosis in cattle, goats, and ewes. In cattle, estrogens of the systemic plasma are present in milk in similar concentrations (Gyawu and Popem, 1983). Prolactin also is secreted into bovine milk and can be recovered by milking 3–10 days before parturition (Pennington and Malven, 1985).

III. Placental and Feto-Placental Endocrine Factors

Reference may be made to a classical summary by Newton (1938) on hormones and the placenta. He believed that the placenta can be regarded as the liver of the fetus: a metabolizing as well as a transmitting organ, and a veritable storehouse of practically everything that has been sought in it. He was also convinced of its endocrine function in the control of gestation and predicted that these functions would be difficult to separate and prove in detail. To one who has contributed to this section for 30 years, all these statements still ring true. Moreover, placentas possess a remarkable autonomy, and survive removal of the fetus in rats, mice, cats, and rhesus monkeys, and such surviving placentas are viable, exert endocrine roles, and tend to be delivered at the normal expected time. However, it has not been possible to define a normal life span of the placenta in either histological or biochemical terms. Thus, the human placenta has been said to be "fully functional" at parturition, but it is unclear what criteria can be applied to it to make such a statement meaningful.

A. Steroid Hormones

1. Estrogen and Progesterone

The presence of estrogens and progesterone in the human placenta has been known from the beginning of the age of endocrinology in the late 1920s. Placental synthesis and interconversion were documented by the appearance of steroids in pregnancy urine, particularly remarkable for their number and variety in humans and equids. Although biological activity originally drew attention to these substances, most of them are biologically inert. Experience with castration suggested that the placenta and/or fetus must attain a certain maturity and mass before its functional contribution becomes sufficient to maintain pregnancy, if indeed it does so at all. The placenta lacks enzyme systems to form cholesterol from acetate. It converts maternal cholesterol to progesterone, and fetal androgens of testicular or adrenal origin to estrogen. Animals that abort after castration (the rabbit, dog, sow, goat) do not produce significant amounts of placental progesterone. Estrogens are present in the placentas of goat (doe) and sow; in the cow, mare, and ewe, both hormones are produced by the placenta, but in the cow the amounts of progesterone, at least, are insufficient to save the cow from abortion after oophorectomy.

2. Androgens and Adrenal Steroids

Testosterone and its metabolites are reported in pregnancy secretions and blood, adding to previous reports (Catchpole, 1977). In the bitch, androstenedione follows the progesterone pattern (Concannon and Castracane, 1985). Androstenedione, testosterone, and epitestosterone are present in the blood of cows in the last 6 months of pregnancy (Môstl et al., 1983), and androstenedione and testosterone are present in blood of late-pregnancy ewes. In the mare, unbound testosterone increases to 10 times the basal value at 7 months, before falling to baseline at parturition (Silberzahn et al., 1984), closely mimicking the estrogen profile. It is not clear whether these findings have any physiological significance or represent a brisk feto-placental steroid traffic tied to progesterone and fetal androgenic products.

B. Protein and Polypeptide Hormones

The placental gonadotropin of humans (hCG) has a subunit in common with pituitary LH and FSH of domestic animals and thus shares their molecular phylogeny. It appears reasonably certain that the human hormone is a luteotropin.

Equine chorionic gonadotropin (eCG) appears in the blood of pregnant mares at approximately day 40, at a fetal crown-rump length of about 2.5 cm; it achieves high concentrations between days 60 and 80, and falls to zero by day 150. It is a glycoprotein with a high carbohydrate content consisting of two polypeptide chains. The alpha chain is shared with other pituitary and placental gonadotropins, and with thyrotropic hormone; the beta chain, which confers its specificity, has also been elucidated in great part (Stewart et al., 1988).

The high molecular weight (68,000) of eCG explains to some degree its absence from the urine of the mare and its long half-life in the blood of the rabbit, as well as of the gelding in figures originally determined in the 1930s. Its source is the allantochorionic membrane, expressed as an invasive growth in the fertile horn of the mare uterus leading to formation of the "endometrial cups." This structure's life cycle parallels that of the hormone in the blood. Attempts have been made to unravel the relations of the endometrial cups and their product to the immunology both of the normal mare pregnancy allograft and the remarkable xenograft pregnancies between horses and donkeys (Allen, 1982; Antczak and Allen, 1989).

The possible local masking of trophoblast antigens by the highly sialated glycoprotein hormones, such as eCG, appears to be a viable hypothesis. In other species, nonhormonal glycoprotein products of the conceptus could substitute. A luteotropic action of eCG now appears less likely, and it also appears not to exhibit FSH activity in the mare herself. Its major role is therefore to luteinize the new ovarian follicles of early pregnancy in the mare.

1. Placental Lactogens and Luteotropins

A human placental polypeptide hormone, placental lactogen (hPL), has molecular affinities with pituitary growth and lactogenic hormones and can substitute for the latter. Its role in pregnancy is considered to be luteotropic. Placental lactogens have also been described in rats, mice, and sheep. Blood levels were measured in the ewe throughout pregnancy (Kelly et al., 1974). Differing somewhat from earlier reported results, this study showed significant amounts up to day 80, followed by a rise or plateau between days 100 and 140, with a steep fall at parturition. This curve has considerable similarity to the ewe progesterone

profile. It has already been remarked that pituitary prolactin levels in the ewe are low throughout pregnancy, but do show a small rise near term. In a contrary sense, neutralization of ovine placental lactogen for at least 12 h with an ovine placental lactogen (oPL) antibody showed no influence on plasma progesterone and no role in progesterone production in late pregnancy in the ewe (Waters et al., 1985).

2. Placental Relaxin

Placental relaxin has been identified as a product of the placenta of humans, mares, cats, pigs, rabbits, and monkeys. It is present in low quantity in endometrial and placental tissues of sheep (Renegar and Larkin, 1985) but absent in bovine placenta.

In the mare, corpora lutea regress after day 200, and the increasing amounts of serum relaxin suggest that the placenta is the sole significant source in late pregnancy and at foaling (Stewart et al., 1982a, 1982b).

Rabbits ovariectomized at day 14 of pregnancy and maintained with progesterone have the same relaxin level as controls. It is produced by the syncytiotrophoblast and is present in membrane-bound granules. In fact, relaxin appears not to be produced by the rabbit ovary.

The blood serum content of relaxin during pregnancy thus represents contributions from both ovarian and placental sources, depending on the species.

C. Pregnancy Proteins

A variety of proteins, glycoproteins, and polypeptides derived from the feto-placental unit has been isolated in recent years. They show no overt hormonal properties, but nevertheless frequently induce hormone-like responses, and have been endowed with a variety of possible functions. Some, which enter the maternal circulation, are markers for pregnancy, and have been suggested or used for this purpose. Others seem to be related to various aspects of fetal nutrition or protection not unlike the embryotrophe to which they are, indeed, a contributor, and are local in action. Others seem to be specific for early pregnancy and may be related to maternal recognition. Apart from sheep and cattle, it is not usually possible to interrelate these substances from species to species.

1. Ovine and Bovine Placental Proteins

A bovine trophoblast protein (bTP-1) is a major secretory component of cattle conceptuses and is immunologically related to, but not identical with, a principal product of sheep conceptuses in culture (oTP-1). These are proteins of molecular weight 22,000 and 24,000 (bovine) and 18,000 (sheep), and are secreted into the uterine lumen after about day 12–14 day of pregnancy. Their suggested role is that of a fetal signal for the maternal recognition of pregnancy (Helmer et al., 1987; Kazemi et al., 1988), and the suggested mode of action is local suppression of a prostaglandin. They are said not to be luteotropic (Godkin et al., 1982). What appears to be the same protein was reported under the name trophoblastin as an early pregnancy factor (EPF) that is antiluteolytic (Nancarrow et al., 1981). Other EPFs have been identified in the serum of ewes (Cerini et al., 1976; Staples et al., 1978) and sows (Morton et al., 1983).

A distinct product of 13- to 18-day cow conceptuses was a small (molecular weight less than 10,000), heat-labile, lipid-soluble substance absorbable by dextran charcoal. It stimulated progesterone synthesis by dis-

persed bovine luteal cells and its function would therefore be luteotropic (Hickey and Hansel, 1987; Hickey et al., 1989).

A protein was extracted from bovine fetal cotyledons at 50–100 days, having a molecular weight of 60,000, that stimulated progesterone synthesis in cultured cells. It was described as chorionic gonadotropic-like and is also luteotropic (Ishar and Shemesh, 1989).

Again differing from any of the above is a pregnancy-specific acidic glycoprotein (PSPB) of molecular weight 78,000 found in the uterus and sera of pregnant cows and sheep throughout pregnancy, and produced by binucleate cells of the placenta. Its biological function is not known (Laster, 1977; Sasser et al., 1986). A serological test for PSPB can be used as a pregnancy test in both ruminants (Laster, 1977).

2. Equine Placental Proteins

Two pregnancy specific proteins were detected in early pregnancy mares: a beta-1 mobile protein was found at day 30, and an alpha-2 mobile protein at day 42. The latter was identical with equine chorionic gonadotropin (Gidley-Baird et al., 1983). These pregnancy proteins, one a well-known hormone, are secreted at a time when the mare blastocyst is beginning to become fixed in the fertile cornus of the uterus, thus losing its previous mobility. This situation can be contrasted with the early implantation and early chorionic gonadotropic production in humans and other primates. An immunoprotective role has been suggested in both equids and primates.

3. Suid Placental Proteins

A major basic protein was purified and characterized in part from the culture medium of 14- to 17-day pig conceptuses. A single sugar-containing polypeptide was obtained of molecular weight about 43,000. It appears at the beginning of trophoblast expansion into filaments and is postulated to be related to maternal/fetal relationships in the preimplantation period (Baumbach *et al.*, 1988).

The author is aware of some pessimism regarding placental proteins that have been isolated in considerable profusion in humans, and the problem of assigning them functions if indeed they possess them. There is doubt also concerning which of them should be considered markers for pathological states (Chard, 1973). An antidote to these views is also available (Bischoff, 1974). Because of freedom to manipulate the conceptus, and to deal with normal pregnancies only, experiments on the domestic species have already generated valuable information and their extension appears worthwhile.

IV. Fetal Endocrine Functions

The capacity of the fetus and its membranes to secrete a variety of hormonally active and other proteins of possible significance in maternal recognition, immune reactions, and endocrine reactions was implicit in several of the preceding sections. To proteins must be added steroids that are formed in the fetus and fetal membranes and reach the maternal circulation.

Primordia of the endocrine glands are laid down early in intrauterine life, and by the time of delivery, with a few exceptions, endocrine glands possess a histological structure of recognizable adult pattern. A systematic study of the fetal endocrines would seek to ascertain (1) when the gland becomes hormonally competent, (2) if these

glands produce reactions within the fetus, (3) if fetal endocrines respond to maternal or exogenously administered hormones, and (4) if the fetus contributes to maternal hormone relations. Only some general comments will be made regarding this very extensive program. The fetal endocrine system, including the gonads, pituitary, thyroid, and adrenal glands, become functional surprisingly early. Of interest is the positive response of the fetal bovine hypophysis to injected gonadotropin-releasing hormone at 120 days. LH was secreted into the fetal serum. The gonads of the sheep, pig, and cattle are functionally active by the third week of pregnancy, and the adrenals of the sheep certainly by the prepartum. The thyrotropic–thyroid system was long ago shown to become active by the first quarter in bovine and by mid-pregnancy in the pig.

A succinct review of some surgical interventions in the macaque showed that hypophysectomy after 110 days of gestation (term is at 180 days) led to striking retardation of fetal growth and gonadal development, while hypophysectomy of the mother had no discernable effects (Gulyas, 1978), a study bearing on questions (2) and (3) above. Recent studies also touch on several of the above themes.

Secretions of LH and FSH were measured in chronically catheterized fetal sheep. Between days 79 and 140 of gestation a pulsatile secretion was detected in 39 of 51 fetuses, indicating that pituitary receptors were present and functional (Clark *et al.*, 1984). Similarly, prolactin was measured as a somewhat variable constituent of fetal sheep plasma in the final 30 days of pregnancy (McMillen *et al.*, 1983).

Estradiol was detected in developing pig trophectoderm and yolk-sac ectoderm, increasing by day 10 and most intense by day 16 of pregnancy. Thus estrogen is locally available for its antiluteolytic role frequently advocated for the pig (King and Ackerley, 1985). Ovine fetal membranes (chorioallantois and amnion) in culture at 50 days and at term convert pregnenolone to progesterone, emphasizing the potential of the fetus to contribute to the local milieu (Power and Challis, 1983).

Estrogens, including estradiol and estrone, and progesterone were present in blastocysts of pony mares between days 8 and 18, and also in the uterine lumen and in plasma. A conceptus origin and local effects were assigned (Zavy *et al.,* 1984b). It will be remembered that these blastocysts are still entirely free in the uterus.

Removal of the gonads of fetuses in the mare led to a lowering of estrogen levels in the mare blood, while progesterone levels were unchanged. Foals were delivered in very poor condition and three out of four did not survive (Pashen and Allen, 1979). Estrogens were long ago shown to be present in these fetal gonads, but the outcome of castration is still difficult to interpret.

The same is true of the fetal gonads themselves. What maternal hormone or hormones is responsible for the proliferation of interstitial cells in the gonads of both sexes in the horse fetus to a maximum at 6.5–8 months, with subsequent degeneration prior to term? None of the candidates seem to fit. Does the donkey also show the phenomenon? Aristotle failed to report on this, and the author remains curious.

V. Hormonal Aspects of Parturition

Observations on the domestic species have shown that pregnancy length may be af-fected by the age and parity of the mother and by sex, size, and genotype of the fetus (Holm, 1966). The last is strikingly illustrated by gestation spans in the equid intercrosses: horse, 340 days; hinney, 350 days; mule, 355 days; and donkey, 365 days (Short, 1960).

Parturition is preceded by a period of increased contractility of the myometrium of somewhat indefinite duration, but the final episodes are usually timely and prompt, suggesting a pharmacologically timed event that is imitated by the exogenous use of oxytocin, a stimulator of uterine contraction.

A strictly mechanical problem of parturition concerns the dimensions of the birth canal, which is solved of necessity in some species by relaxation of the symphysis pubis by secretion of the hormone relaxin. This hormone acts also to soften the musculature of the cervix.

Experiments already quoted on the persistence of the placenta after fetal removal and its delivery at or near the expected time do not necessarily negate a role for the fetus: the experiment is interesting, but it is not controlled, and indicates only that important elements of the parturition process are located at the uterine–placental locus. These influences, as they have been throughout pregnancy, are strongly hormonal in nature and revolve around intricate relationships between estrogens, progesterone, adrenal steroids, oxytocin, relaxin, and prostaglandins. These six (and possibly more) degrees of freedom have been manipulated in a variety of ways, and have not produced a convincing general explanation of parturition. Measurement of hormone levels in blood and if possible locally at successive stages of normal labor appears to offer the most convincing description of the process in individual species, and some recent examples will be presented.

A. Estrogen and Progesterone

The uterine quiescence at the beginning of pregnancy is ushered in by falling blood estrogen and rising blood progesterone titers, covering a period in which the progesterone effect becomes dominant. Reactivation of uterine contraction at the end of pregnancy takes place when the reverse process is occurring. In some species there may be active luteolysis before parturition, attributed to an unblocking of prostaglandin. However, hormone profiles show that progesterone levels may begin to fall well before parturition in many species, although remaining high in humans and mare, while estrogen titers are rising sometimes to peaks, in rat/mouse, cow, sheep, goat, and pig. In any case the presence of these two hormones provides the milieu of action of all other active factors. A reversal of the estrogen/progesterone ratio and its possible effect on uterine reactivity is therefore a constant factor at parturition. The uterus of the primate, at least, is known to be highly sensitive both to different absolute amounts as well as to different proportions of these hormones.

B. Oxytocin

It has always appeared that oxytocin should have some role in the normal course of parturition, but the nature of this role has proven to be elusive. Originally named oxytocin (= quick birth) for its specific action on the uterus, oxytocin or one of its synthetic analogs is the agent of choice for augmenting or initiating uterine contraction in

humans and it has been shown to be effective in the mare. A positive report, also in the mare, showed important interrelations between hormonal factors most often invoked to explain the nature of parturition. Levels of plasma progesterone, estradiol-17β, oxytocin, and prostaglandin PGF-2α were measured in pony mares. In the last few days prepartum there was a nonsignificant increase in the estrogen/progesterone ratio, which was nevertheless highest at delivery. Oxytocin levels were basal throughout pregnancy and increased 10-fold just before the expulsive stage of labor, and earlier than a 15-fold increase in prostaglandin. It was concluded that a sudden oxytocin surge precipitates the expulsive phase of parturition, which is mediated by prostaglandin (Haluska and Currie, 1988).

In the cow, in equally persuasive experiments, the maximal oxytocin levels occurred at actual delivery of the fetus, supporting the view that endogenous maternal oxytocin does not primarily induce parturition (Landgraf et al., 1983). Both these reported studies tend to equate maximal titers of hormone with maximal hormonal effect on the myometrium; this may not be so in the phasic circumstances of uterine contractions.

C. Prostaglandins

In dogs, prostaglandin PGF-2α (PGFM) levels beginning 48 h prepartum and until 3 h prepartum increased approximately 5-fold, associated with the onset of luteolysis, while progesterone levels fell from 2.8 to 0.7 ng/ml. Peak levels of PGFM were seen between the birth of pups. It was suggested that luteolysis was initiated by prostaglandin entering the maternal circulation and that the falling amounts of progesterone gave a further rise, which initiated uterotonic activity and expulsion of the pups (Concannon *et al.*, 1988).

Ovarian or uterine lymph was collected in pregnant cows 96–278 days postcoitus. At all stages of pregnancy, concentrations of progestins (200×) and androgens (60×) were higher in ovarian than in the uterine lymph or blood plasma. At parturition, estrogens increased somewhat in ovarian lymph and there was an abrupt increase in PGF-2 in the uterine lymph, and its release was strongly phasic (Hein *et al.*, 1988).

Injection of a 3β-hydroxysteroid dehydrogenase inhibitor in late pregnancy goats produced a fall in progesterone and cortisol to 20% of the pretreatment levels and led to premature delivery in 44 ± 2 h. Progesterone withdrawal is an important component of the mechanisms which initiate parturition in this form. Suppression of prostaglandin (PG) synthesis delayed parturition, and PG is essential for delivery although perhaps not for luteolysis (Taylor, 1987).

D. Overview

As compared with a previous report (Catchpole, 1977) prostaglandins have entered more definitely the hierarchy of hormones and factors controlling parturition, with oxytocin staging a comeback threat as an endogenous hypothalamic candidate. The position of the adrenal corticoids and the role of the fetal adrenal glands, studied most intensively in rabbits and sheep, remain undefined (Yu *et al.*, 1983). Estrogen and progesterone blood levels need to be determined for many species at close successive stages of labor, representing tedious but necessary work. Modification of either of these hormones, at any stage of pregnancy, is liable to cause abortion. The status of the uterine receptors for these hormones demands close attention.

By the end of gestation there has been a considerable maternal investment in the progeny. Three major systems we have considered—maternal, feto-placental, and fetal—have contributed to the hormonal milieu of pregnancy. This hormonal redundancy is designed to guarantee the successful outcome of pregnancy and parturition. The dynamics may vary somewhat from species to species, and in particular adaptations for parturition remain the descriptive and analytical tasks of specialists in each of these varied and fascinating areas of animal production.

References

Adair, V., Anderson, L. L., Stromer, M. H., and McDonald, W. G. (1982). *J. Anim. Sci.* **55**(Suppl. 1), 333.

Addiego, L. A. (1987). *Biol. Reprod.* **37**, 1165.

Allen, W. R. (1982). *J. Reprod. Fertil. Suppl.* **31**, 57.

Anderson, L. L., Perezgrovas, R., O'Byrne, E. M. and Steinmetz, B. C. (1982). *Ann. N.Y. Acad. Sci.* **380**, 131.

Antczak, D. F., and Allen, W. R. (1989). *J. Reprod. Fertil. Suppl.* **37**, 69.

Asdell, S. A. (1964). "Patterns of Mammalian Reproduction." Cornell University Press, Ithaca, New York.

Auletta, F. J., and Flint, A. P. F. (1988). *Endocr. Rev.* **9**, 88.

Barkley, M. S., Michael, S. D., Geschwind, I. I., and Bradford, G. E. (1977). *Endocrinology* **100**, 1472.

Basu, S. (1985). *Nord. Vet. Med.* **37**, 57.

Baumbach, G. A., Climer, A. H., Bartley, N. G., Kattesh, H. G., and Godkin, J. D. (1988). *Biol. Reprod.* **39**, 1171.

Bazer, F. W., and First, N. L. (1983). *J. Anim. Sci.* **57**(Suppl. 2), 425.

Bazer, F. W., Vallet, J. L., Harney, J. P., Gross, T. S., and Thatcher, W. W. (1986). *J. Reprod. Fertil.* **76**, 841.

Bischoff, P. (1974). "Placental Proteins." Contributions to Gynecology and Obstetrics. Karger, Basel.

Boime, I., Landefeld, T., McQueen, S., and McWilliams, D. (1978). In "Structure and Function of the Gonadotropins" (K. W. McKerns, ed.), p. 235. Plenum, New York.

Breslow, E. (1979). Annu. Rev. Biochem. 48, 251.

Bryant-Greenwood, G. D. (1985). Res. Reprod. 17, (3), 1.

Buttle, H. L. (1983). J. Physiol. Lond. 342, 399.

Catchpole, H. R. (1973). In "The Inflammatory Process" (B. W. Zweifach, L. Grant, and R. T. McCluskey, eds.), Vol. 2, p. 121. Academic Press, New York.

Catchpole, H. R. (1977). In "Reproduction in Domestic Animals" (H. H. Cole and P. T. Cupps, eds.), 3rd ed., p. 341. Academic Press, New York.

Cerini, M., Cerini, J. C., Findlay, J. K., and Lawson, R. A. S. (1976). J. Reprod. Fertil. 46, 534.

Chang, C. F., and Estergreen, V. L. (1983). Steroids 41, 173.

Chard, T. (1973). In "Pregnancy Proteins" (J. G. Grudzinskas, B. Teissner, and M. Seppälä, eds.), p. 3. Academic Press, New York.

Cheesman, K. L., Chatterton, R. T. Jr., Mehta, R. R., and Venton, D. L. (1982). Fertil. Steril. 38, 475.

Clark, J. H., Peck, E. J., and Glasser, S. R. (1977). In "Reproduction in Domestic Animals" (H. H. Cole and P. T. Cupps, eds.), 3rd ed., p. 143. Academic Press, New York.

Clark, S. J., Ellis, N., Steyne, D. N., Gluckman, P. D., Kaplan, S. L., and Grumbach, M. M. (1984). Endocrinology 115, 1774.

Concannon, P. W., and Castracane, V. D. (1985). Biol. Reprod. 33, 1078.

Concannon, P. W., Isaman, L., Frank, D. A., Michel, F. J., and Carrie, W. B. (1988). J. Reprod. Fertil. 84, 171.

Csapo, A. (1956). Am. J. Anat. 98, 273.

Fairclough, R. J., Moore, L. G., Peterson, A. J., and Watkins, W. B. (1984). Biol. Reprod. 31, 36.

Fiddes, J. C., Seeburg, P. H., Denoto, F., Hallewell, R. A., Baxter, J. D., and Goodman, H. M. (1979). Recent Prog. Horm. Res. 40, 43.

Findlay, J. K. (1978). Biol. Reprod. 19, 1076.

Findlay, J. K. (1981). J. Reprod. Fertil. Suppl. 30, 171.

Flint, A. P. F., and Sheldrick, E. L. (1982). Nature (Lond.) 297, 587.

Flint, A. P. F., Sheldrick, E. L., Jones, D. S. C., and Auletta, F. J. (1989). J. Reprod. Fertil. Suppl. 37, 195.

Ford, S. P. (1982). J. Anim. Sci. 55(Suppl. 2), 32.

Fylling, P. (1970). Acta. Endocrinol. 65, 273.

Gadsby, J. E., and Keyes, P. L. (1984). Biol. Reprod. 31, 16.

Gidley-Baird, A. A., Teisner, B., Hau, J., and Grudzinskas, J. G. (1983). J. Reprod. Fertil. 67, 129.

Godkin, J. D., Bazer, F. W., Moffatt, J., Sessions, F., and Roberts, R. M. (1982). J. Reprod. Fertil. 65, 141.

Gulyas, B. J. (1978). Res. Reprod. 10(5), 3.

Gyawu, P., and Popem, G. S. (1983). J. Steroid Biochem. 19, 877.

Haluska, G. T., and Currie, W. B. (1988). J. Reprod. Fertil. 84, 635.

Hansel, W., Stock, A., and Battista, P. J. (1989). J. Reprod. Fertil. Suppl. 37, 11.

Heap, R. B., and Flint A. P. F. (1979). In "Reproduction in Mammals" (G. R. Augber and R. V. Short eds.), Vol. 7, p. 185. Cambridge University Press, Cambridge.

Heap, R. B., Ackland, N. and Weir, B. J. (1981). J. Reprod. Fertil. 63, 477.

Hearn, J. P. (1986). J. Reprod. Fertil. Suppl. 22, 809.

Hein, W. R., Shelton, J. N., Simpson-Morgan, M. W., Seamark, R. F., and Morris B. (1988). J. Reprod. Fertil. 83, 309.

Helmer, S. D., Hanson, P. J., Anthony, R, V., Thatcher, W. W., Bazer, F. W., and Roberts, R. M. (1987). J. Reprod. Fertil. 79, 83.

Hickey, G. J., and Hansel, W. (1987). J. Reprod. Fertil. 80, 569.

Hickey, G. J., Walton, J. S., and Hansel, W. (1989). J. Reprod. Fertil. Suppl. 37, 29.

Holm, L. W. (1966). Symp. Zool. Soc. Lond. 15, 403.

Holtan, D. W., Squires, E. L., Lapin, D. R., and Gunther, O. J. (1979). J. Reprod. Fertil. Suppl. 27, 457.

Homeida, A. M., and Cooke, R. G. (1983). Prostaglandins 26, 103.

Hudson, P., Haley, J., John, M., Cronk, M., Crawford, R., Haralambidis, J., Tregear, G., Shine, J., and Niall, H. (1983). Nature (Lond.) 301, 628.

Ishar, M., and Shemesh, M. (1989). J. Reprod. Biol. Suppl. 37, 37.

Kazemi, M., Malathy, P. V., Keisler, D. H., and Roberts, R. M. (1988). Biol. Reprod. 39, 457.

Kelly, P. A., Robertson, H. A., and Friesen, H. G. (1974). Nature 248, 435.

Kenneth, J. H. (1947). Imp. Bur. Anim. Breeding Genet. Tech. Commun. (Edinburgh) 5.

King, G. J., and Ackerley, C. A. (1985). *J. Reprod. Fertil.* **73,** 361.

King, G. J., and Rajamahendran, R. (1988). *J. Endocrinol.* **119,** 111.

Knowler, J. T., and Beaumont, J. M. (1985). *Essays Biochem.* **20,** 1.

Landgraf, R., Schultz, J., Eulenberger, K., and Wilhelm, J. (1983). *Exp. Clin. Endocrinol.* **81,** 321.

Laster, D. B. (1977). *Biol. Reprod.* **16,** 682.

Marshall, F. H. A. (1910). "The Physiology of Reproduction." Longmans, Green, London.

Martin, R. D., and McLaren, A. M. (1987) *Nature* (Lond.) **313,** 220.

McDowell, K. J., Sharp, D. C., Grubaugh, W., Thatcher, W. W., and Wilcox, C. J. (1988). *Biol. Reprod.* **39,** 457.

McMillen, I. C., Jenkin, G., Robinson, J. S., and Thorburn, G. D. (1983). *J. Endocrinol.* **99,** 107.

Morton, H., Morton, D. J., and Ellendorf, F. (1983). *J. Reprod. Fertil.* **68,** 437.

Môstl, E., Choi, H. S., Holzweber, E., and Bamberg, E. (1983). *Zentralbl. Veterinarmed. [A]* **30,** 559.

Nancarrow, C. D., Wallace, A. L. C., and Grewal, A. S. (1981). *J. Reprod. Fertil. Suppl.* **30,** 191.

Newton, W. H. (1938). *Physiol. Rev.* **18,** 419.

O'Neill, C., Collier, M., Ryan, J. P., and Spinks, N. R. (1989). *J. Reprod. Fertil. Suppl.* **37,** 19.

Pashen, R. L., and Allen, W. R. (1979). *J. Reprod. Fertil.* **27,** 499.

Pashen, R. L. (1984). *Equine Vet. J.* **16,** 233.

Pennington, J. A., and Malven, P. V. (1985). *J. Dairy Sci.* **68,** 1116.

Pepe, G. J., and Rothchild, I. (1974). *Endocrinology* **95,** 275.

Popescu, L. M., Paniou, C., Nutu, O., and Toescu, E. C. (1984). *Fed. Proc.* **43,** 497.

Power, S. G., and Challis, J. R. (1983). *J. Endocrinol.* **97,** 347.

Procknor, M., Dachir, S., Owens, R. E., Little, D. E., and Harms, P. G. (1986). *J. Anim. Sci.* **62,** 191.

Renegar, R. H., and Larkin, L. H. (1985). *Biol. Reprod.* **32,** 840.

Renfree, M. B. (1981). *Nature (Lond.)* **293,** 100.

Resnick, R. (1983) *Clin. Perinatol.* **10,** 567.

Roberts, R. M., and Bazer, F. W. (1988). *J. Reprod. Fertil.* **82,** 875.

Robertson, H. A., and King, G. J. (1974). *J. Reprod. Fertil.* **40,** 133.

Sarma, V. (1983). *Nature (Lond.)* **302,** 366.

Sasser, R. G., Ruder, C. A., Ivani, K. A., Butler, J. E., and Hamilton, W. C. (1986). *Biol. Reprod.* **35,** 936.

Sasser, R. G., and Ruder, C. A. (1987). *J. Reprod. Fertil.* **34,** 261.

Sharp, D. C., McDowell, K. J., Weithenauer, J., and Thatcher, W. W. (1989). *J. Reprod. Fertil. Suppl.* **37,** 101.

Shideler, R. K., Squires, E. L., Voss, J. L., Eikenberry, D. J., and Pickett, B. W. (1982). *J. Reprod. Fertil. Suppl.* **32,** 459.

Short, R. V. (1960). *J. Reprod. Fertil.* **1,** 61.

Silberzahn, P., Zwain, I., and Martin, B. (1984). *Endocrinology.* **115,** 416.

Stanley, C. J., Paris, F., Webb, A. E., Heap, R. B., Ellis, S. T., Hamon, M., Worsfold, A. and Booth, J. M. (1986). *Vet. Rec.* **118,** 664.

Staples, L. D., Lawson, R. A. S., and Findlay, J. K. (1978). *Biol. Reprod.* **19,** 1076.

Steinetz, B. G., Goldsmith, L. T., and Lust, G. (1987). *Biol. Reprod.* **37,** 1719.

Steward, D. R., Stabenfeldt, G. H,. and Hughes, J. P. (1982a). *J. Reprod. Fertil. Suppl.* **32,** 603.

Stewart, D. R., Stabenfeldt, G. H., Hughes, J. P., and Meagher, D. M. (1982b). *Biol. Reprod.* **27,** 17.

Stewart, H. J., and Walker, F. M. (1987). *J. Dairy Res.* **54,** 179.

Stewart, H. J., Jones, D. S. C., Pascall, J. C., Popkin, R. M., and Flint, A. P. F. (1988). *J. Reprod. Fertil.* **83,** 1.

Summerlee, A. J. S., O'Byrne, K. T., Paisley, A. C., Breeze, M. F., and Porter, D. G. (1984). *Nature* **309,** 372.

Taylor, M. J. (1987). *J. Endocrinol.* **113,** 489.

Urwin, V. (1983). *J. Endocrinol.* **99,** 199.

Vernon, R. G., Clegg, R. A., and Flint, D. J. (1981). *Biochem. J.* **200,** 307.

Waters, M. J., Oddy, V. H., McCloghry, C. E., Gluckman, P. D., Duplock, R., Owens, P. C., and Brinsmead, M. W. (1985). *J. Endocrinol.* **106,** 377.

Wathes, D. C., and Swann, R. W. (1982). *Nature (Lond.)* **297,** 225.

Weber, D. M., Fields, P. A., Romvell, L. J., Tumwasorn, S., Ball, B. A., Drost, M., and Fields, M. J. (1987). *Biol. Reprod.* **37,** 685.

Wise, M. E., Nilson, J. H., Nejedlik, M. T., and Nett, T. M. (1985). *Biol. Reprod.* **33,** 1009.

Worthy, K., Escreet, R., Renton, J. P., Eckersall, P. D., Douglas, T. A., and Flint, D. J. (1986). *J. Reprod. Fertil.* **77,** 569.

Wright, L. J., Feinstein, A., Heap, R. B., Saunders, J. C., Bennett, R. C., and Wang, M.-Y. (1982). *Nature (Lond.)* **295,** 415.

Yamamoto, K. R. (1985). *Annu. Rev. Genet.* **19,** 209.

Yu, H. K., Cabalum, T., Jansen, C. A., Buster, J. E., and Nathanielsz, P. W. (1983). *Endocrinology.* **113,** 2216.

Zavy, M. T., Vernon, M. W., Asquith, R. L., Bazer, F. W., and Sharp, D. C. (1984a). *Prostaglandins* **27,** 311.

Zavy, M. T., Vernon, M. W., Sharp, D. C., and Bazer, F. W. (1984b). *Endocrinology* **115,** 214.

CHAPTER 11

Mammary Gland Development and Lactation

R. L. BALDWIN and P. S. MILLER

I. Introduction

The preparation of a general, concise treatment of mammary gland morphogenesis, growth, and the initiation and maintenance of lactation requires extensive reliance upon reference to previous treatments of the sub-ject. These provide a view of the development of knowledge in the area and furnish detailed treatments of specific aspects of mammary growth and lactation. A number of comprehensive reviews of lactation have been published (Turner, 1952; Larson and Smith, 1974, 1978; Mepham, 1983, 1987;

Neville and Daniel, 1987). These should be consulted for additional details regarding mammary gland anatomy and development, the initiation and maintenance of lactation, hormonal control mechanisms, and mammary gland metabolism. Emphasis in this chapter is placed upon summarizing information on the mammary gland and lactation, as well as critical evaluation of recent literature in selected areas of current research emphasis.

II. Anatomy, Morphogenesis, and Development of the Mammary Gland

A. Anatomy

Turner (1952), in his classic book on the gross and microscopic anatomy of mammary gland, considered cattle, other hoofed animals, and marine mammals. Others have considered the anatomy of glands of a number of additional species (Silver, 1953; Flux, 1954; Netter, 1954; Benson *et al.* 1957; Kon and Cowie, 1961; Helminen and Ericsson, 1968; Hollmann, 1974). The basic microscopic and gross anatomical structures of the mammary glands of mammals are remarkably similar. The only major differences among species are in gross anatomy associated with the removal of milk gathered in the major ducts. In the monotremes or egg-laying mammals (e.g., achidna and platypus), the major ducts draining the mammary gland lead directly to the skin surface forming mammary patches. In humans, dogs, and other species, 15–20 major ducts terminate in a nipple (Fig. 1). In cows and goats (Fig. 2) the major ducts terminate in a gland cistern, which is contiguous with a teat cistern. The teat cistern is drained by

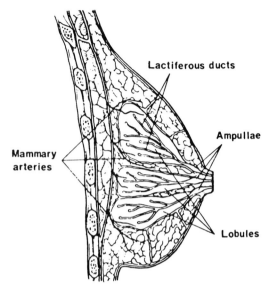

Figure 1 Schematic drawing of human breast, showing ducts leading to nipple from ampulla and the intermingling of lobuloalveolar elements with adipose tissue. From Baldwin (1974).

a streak canal. The prominence of the gland and teat cisterns varies considerably among animals, being quite large in ruminants and small in pigs and horses. In several species including pigs and horses, more than one mammary gland complex is emptied via a single teat. In these cases the teat has two streak canals, one for each gland.

The alveolus is the basic secretory element of the mammary glands of all mammals (Fig. 3). The alveoli are small vesicles or sacs made up of a single layer of secretory epithelial cells which surround the lumen. The alveoli are surrounded by a basement membrane, fine networks of capillaries, and myoepithelial cells. Alveoli are arranged in lobules (Fig. 2) and are drained by intralobular ducts. These ducts connect with interlobular ducts, which join larger ducts and provide a route for milk removal from the

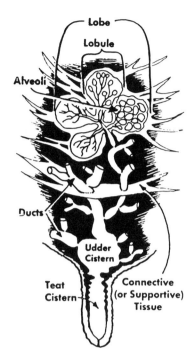

Figure 2 Schematic drawing of a cow's udder, showing the major structures of the teat, the gland cistern, the connective stroma, and lobuloalveolar structure. From Turner (1969).

gland. Lobules are surrounded by connective tissue and are arranged in lobes, also surrounded by connective tissue elements. Lymphatic capillaries are found in the interlobular connective tissue but not in the intralobular area. Fully developed mammary secretory cells have very prominent nuclei, mitochondria, and an extensive network of endoplasmic reticulum and Golgi apparatus (Fig. 4). Early speculations regarding mechanisms of milk secretion by alveolar cells were reviewed by Turner (1952). The general conclusion was that although secretory cells undergo considerable destruction as evidenced by the presence of enzymes and subcellular particles in milk, the mechanism

of secretion must be conservative or merocrine in nature. It was also considered that secretion during milk removal from the gland might be partially holocrine in type, resulting in cell decapitation.

Recent studies have clarified further, though not completely, the mechanisms of secretion of milk components (Keenan and Dylewski, 1985). Early electron microscopic and biochemical observations indicated that the membranes that surround milk fat globules are formed from the (apical) membrane (Barmann and Knoop, 1959; Keenan et al., 1974; Saacke and Heald, 1974). Recently it has been shown that the membrane structure surrounding fat globules may also be derived directly from intracellular membranes including endoplasmic reticulum, Golgi apparatus, and secretory vesicles (Dylewski et al., 1984; Powell et al., 1977; Wooding, 1977). Findings that fat globules were surrounded by a portion of the apical membrane suggested that the mechanism responsible for fat secretion into the alveolar lumen involved a form of reverse pinocytosis (exocytosis). It now appears that although exocytosis may partially account for secretion of fat globules, other mechanisms are implied by the association of nonapical membranes with secreted fat globules. The mechanisms of secretion are not entirely conservative, since cytoplasmic elements and enzymes are secreted along with the fat droplet (Mather, 1987).

Casein micelles and lactose are formed in the Golgi body and secretory vesicles of secretory cells (Ebner and Schanbacher, 1974; Larson and Jorgensen, 1974). The mechanism of intracellular transport of secretory vesicles (Fig. 4) to the apical membrane is still debatable. Currently, several theories exist explaining the translocation of secretory vesicles. One hypothesis proposes that

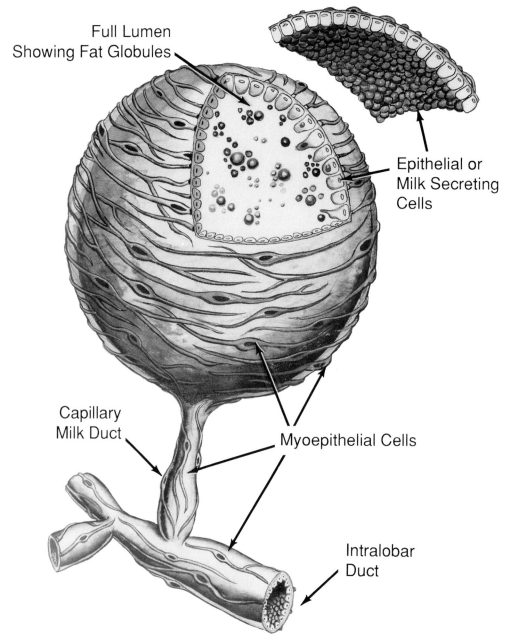

Figure 3 A schematic drawing of an alveolus, showing how the mammary secretory cells are arranged to form a saclike structure, how the lumen, into which milk is secreted by the secretory cells, is connected to an intralobar duct, the close relationship between blood capillaries and the alveolus, and the myoepithelial cells that contract to expel milk from the alveolus. From Turner (1969).

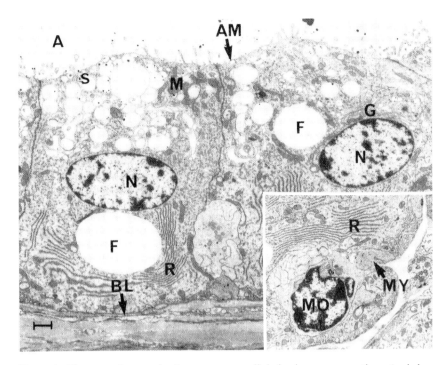

Figure 4 Electron micrograph of two secretory cells in bovine mammary tissue (scale bar, 1.0 μm). A, Alveolar lumen; AM, apical membrane; BL, basal lamina; F, fat droplets; G, Golgi apparatus; S, secretory vesicle containing protein granules; N, nucleus; R, rough endoplasmic reticulum; M, mitochondria. Inset: MO, monocyte; MY, myoepithelial cell. From Mepham (1987).

microtubules and microfilaments direct vesicles toward the alveolar membrane (Mepham, 1987). An alternative proposal links the transfer of vesicular material to a process termed *compound exocytosis,* in which secretory vesicles sequentially fuse and ultimately release their combined contents into the alveolar lumen (Keenan and Dylewski, 1985). Both of these hypotheses hold that lactose and casein are secreted either via Golgi tubules that are contiguous with the plasma membrane or via vesicles that are released by the Golgi, drift toward the plasma membrane, and, in the process of joining the plasma membrane, release lactose and

casein into the alveolar lumen. Both of these possibilities imply that secretory-cell plasma membranes are formed from Golgi membranes.

B. Morphogenesis

The basic sequences of events during morphogenesis of the mammary glands are similar in most species (Turner, 1952). Two milk lines become apparent on the abdomens of very early fetuses. Epidermal thickenings are soon formed at points at which mammary glands will be formed in the particular species examined. In the case of the bovine fetus, for example, four epidermal thickenings called *mammary buds* are formed on the milk lines in the inguinal region between the hindlimbs. The mammary buds themselves do not differentiate to form mammary structures, but rather serve as a focal point for differentiation of dermis, epidermis, and mesenchyme. Early in development, the mammary buds sink into the mesenchyme and a condensation of mesenchymal cells takes place around the bud, causing the appearance of embryonic teat hillocks. At this stage, ectodermal cells overlying the bud proliferate to form a neck of epidermis connecting the mammary buds to the epidermis at the apex of the teat hillock. In subsequent stages, the epidermal neck increases in size and forms a funnel-shaped cone at the base of which an epithelial cord or primary sprout extending the length of the fetal teat develops. The primary sprout canalizes starting at the apex of the teat during the 19-cm stage of the bovine such that when the fetus reaches 30 cm in length, structures corresponding to the streak canal and teat cistern are clearly defined. These are composed of a basement membrane with a double-layered lining of epithelial cells. Secondary and tertiary sprouts representing the beginning of the duct system of the gland penetrate the surrounding mesenchyme in various directions. During later stages the mesenchyme differentiates to form the connective tissue that will support the fully developed mammary gland, the connective glandular stroma that will ultimately surround the lobulalveolar structures of the mammary gland, and adipose tissue that surrounds the glandular elements and comprises the bulk of the udder at birth.

The duct system is still rudimentary at birth and is confined to a small area surrounding the gland cistern. The teats are well formed, while the sphincter muscle surrounding the streak canal at the apex of the teat and the smooth muscle cells surrounding the teat cistern are not clearly evident. The connective stroma and vascular and lymphatic systems are reasonably well developed.

Data summarized by Raynaud (1961) and Anderson (1974) indicate that hormones are not necessary for early mammary development. The primary hormonal effect upon fetal mammary development occurs in the male as a result of the secretion of androgens by the developing testes. The androgens inhibit growth of the teat and primary mammary sprout. Steroid hormones, especially estrogen, although not required for development, can produce alterations in the pattern of development and when injected in large doses can induce abnormal development.

C. Prepubertal Development

Development of the mammary glands from birth to puberty is characterized by generalized growth and maturation of elements not clearly defined at birth, such as the teat sphincter and smooth muscle fibers. A number of techniques have been employed to investigate changes that occur during this period. Matthews *et al.* (1949) and Swett *et al.* (1955) used gross morphological criteria such as changes in the weights, dimensions, and capacities of udders of heifers to characterize growth and development. Several investigators (Silver, 1953; Flux, 1954; Benson *et al.*, 1957) employed histological techniques for quantitative measurement of development and branching of the mammary duct system.

Udder weight in calves was highly correlated with age. Udder size as assessed by palpation of calves between 3 and 15 months of age was somewhat related to subsequent lactation performance, an observation that led to the proposal that these measurements might be employed to cull calves of low production potential at an early age. However, the usefulness of size and weight criteria for predicting future lactational performance was seriously questioned on the grounds that variance is very large and that these criteria do not distinguish between development of the mammary parenchyma and growth of associated adipose tissue. Criteria based upon measurement of increases in gland capacity or internal volume determined by injection of fluids appeared to be useful but have not been employed to assess development because of the difficult and unphysiological nature of the measurement. A steady increase in udder capacity or internal volume, presumably reflecting growth of the gland cistern and duct proliferation, has been reported for a number of species.

Measurements of mammary duct growth, degree of branching, and duct area in whole-gland mounts (Fig. 5) have been employed to characterize prepubertal mammary gland development in mice and rats (Silver, 1953; Flux, 1954; Silberstein and Daniel, 1987). Mammary gland growth is isometric with respect to rate of growth of the body as a whole and becomes allometric (faster than the overall growth rate) several weeks prior to the onset of estrous cycles. In ovariectomized females and normal and castrate males, mammary gland growth rates are isometric throughout this period.

The whole-gland mount technique is difficult to employ for the study of rats and mice after puberty because the glands start to develop in three dimensions at this time. For this reason, the technique cannot be em-ployed to study prepubertal development in species such as guinea pigs and goats whose glands have a three-dimensional structure at birth. Techniques for the study of pre- and postpubertal growth in mammary glands having a three-dimensional structure have been developed (Munford, 1961). Early application of these techniques involved tedious measurements of percentage of gland parenchyma and of average, cross-sectional areas of serial or sample sections of the gland. More recently, a computer-based image analysis approach (tomography) has been applied (Sorensen et al., 1987). Results obtained with these techniques confirm the observation that mammary growth is isometric early in life and becomes allometric prior to puberty as a result of prepubertal ovarian activity (Benson et al., 1987).

D. Postpubertal Development

Although allometric growth of glandular epithelium commences prior to puberty, puberty is generally considered the time during which rapid mammary development occurs. The smaller ducts proliferate very quickly, while estrogen levels are high and regress slowly during the luteal phase. In each sequential cycle, more ducts are formed during the proliferation period than are lost during the period of regression, with the net result that a highly branched matrix of ducts develops. This matrix represents the beginning of the development of true lobular structures. The proliferation of ducts and ductules is accompanied by, or in humans preceded by, development of the connective tissue stroma, growth of adipose tissue, and further development of vascular and lymphatic systems. At this stage species differences arise, due in part to differences in estrous cycles and

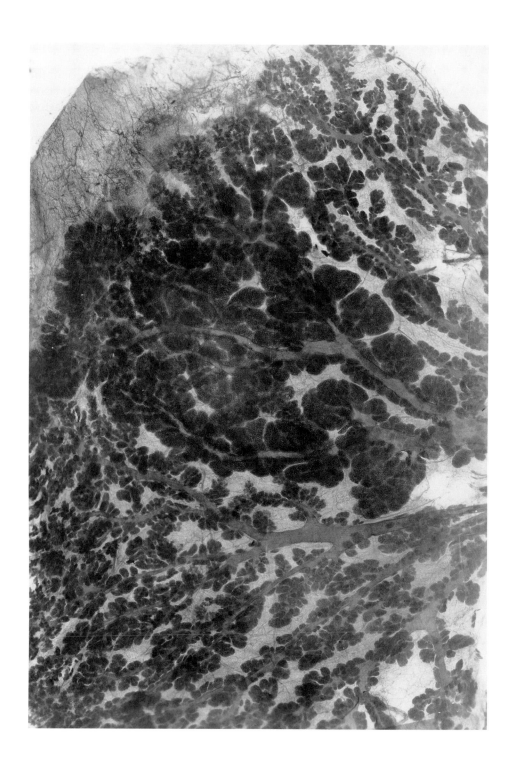

in part to inherent species differences in the responses of glandular elements to the ovarian hormones. In cows, goats, and rodents, adipose tissue growth is rapid and development of the connective tissue stroma lags behind duct growth. In humans, adipose tissue growth is slower while connective stroma develops quickly. Mayer and Klein (1961) noted that, in general, species that have short estrous cycles with a very short luteal phase exhibit primarily duct growth, while in species such as dogs where the corpus luteum is very long-lived, duct growth is accompanied by considerable lobuloalveolar development. Hence, estrogen is considered to be largely responsible, either directly or indirectly, for duct growth, while lobuloalveolar development is considered to be dependent upon progesterone.

E. Development during Pregnancy and Early Lactation

The mammary gland attains its maximum development during pregnancy or immediately thereafter. Mammary gland differentiation during this period is characterized by a relative reduction in stromal cells and a concomitant increase in ductal and lobuloalveolar tissue. Mammary gland growth as indicated by changes in total DNA accelerates exponentially throughout pregnancy in guinea pigs, goats, and cows (Table I). The development that occurs during this period is dependent upon

Table I

Mammary DNA during Pregnancy in Guinea Pigs, Goats, and Cows[a]

	DNA at day X of pregnancy/DNA at day 0 of pregnancy		
Days pregnant	Guinea pigs[b]	Goats[c]	Cows[d]
0	1	1	1
20	1.97	1.46	1.17
40	3.90	2.14	1.38
65	9.12	3.44	1.68
130	—	11.82	2.83
280	—	—	9.39

[a] Adapted from Larson (1985).
[b] $Y = e^{0.034x}$ where Y = DNA/DNA and x = day pregnant.
[c] $Y = 2^{0.019x}$.
[d] $Y = e^{0.008x}$.

estrogen, progesterone, and pituitary hormones. The role of pituitary secretions in development will be discussed in detail in Sections III,B and III,C.

Numerous early studies, summarized in a comprehensive review by Meites (1961), were undertaken with the goal of developing techniques for the artificial initiation of lactation. These studies were based on the premises that full mammary development occurs during pregnancy as a result of maintenance of an appropriate balance between progesterone and estrogen secretion and that simulation of pregnancy through administration of progesterone and estrogen in sterile cows could lead to artificially induced lactations, which would be of

Figure 5 Two sectors of mammary gland from a 5-month-old oophorectomized rabbit pretreated with 20 μg estrone plus 1 mg progesterone 5 days weekly for 4 weeks, and then after a 3-day interval with 0.1 mg (3 IU) ovine prolactin in 1 ml of 2% butanol into the right sector, and only 1 ml butanol into the left. Milk in the right sector could be seen through the skin, and could be expressed by 48 h. No other sector of any gland expressed milk. Multiplication ×6. By permission of W. R. Lyons *et al.* (1958) and Butterworths and Co., London.

benefit to the dairy industry. The most promising experimental techniques were based on administration of estrogen–progesterone combinations for prolonged periods to induce udder growth followed by administration of a "triggering" dose of estrogen and initiation of milking. In general, although estrogen–progesterone combinations have been found that induce significant lobuloalveolar development in several species, milk yields after artificial induction of lactation are variable and below normal (Meites, 1961; Tucker, 1974). Because of the time and expense required, artificial induction of lactation by simulation of pregnancy was not deemed practical (Meites, 1961) and research on induced lactation was limited for a number of years. Later, Smith and Schanbacher (1973) reported that large doses of estradiol-17β (0.1 mg/kg body weight) and progesterone (O.25 mg/kg body weight) for 7 days will initiate lactation in 60% of infertile cows. Milk yields in these artificially induced lactations were promising. These observations renewed interest in the possibility of obtaining practical benefit by inducing lactation in infertile cows. This approach has been successfully applied in the dairy industry.

III. Hormonal Requirements for Mammary Gland Development and Lactation

A. Methods of Evaluation

1. Criteria of Development

Several morphological and histological techniques that have been employed to characterize and quantitate mammary gland development were discussed in previous sections. These techniques have also been employed in studies of hormonal effects upon mammary tissue and in the study of changes occurring in mammary glands during late pregnancy and lactation. However, their usefulness during these periods is impaired by the complexity of development during this period and by accumulation of fluids that distend glandular structures. Histological techniques that utilize numbers of cell divisions as a criterion of continuing growth and development have provided considerable information. Numbers of cell divisions are usually determined by estimating frequencies of mitosis directly, by estimating frequencies of mitotic events in glands after treatment with colchicine, or by estimating numbers of radioactive nuclei present after administration of tritiated thymidine. This latter technique provides a basis of assessing the types of cells being formed. Electron microscopy has been used extensively to evaluate hormonal effects upon ultrastructure and mechanisms of milk component secretion (Hollmann, 1974; Topper and Oka, 1974). Biochemical techniques have been employed to study changes in numbers and types of cells in the mammary glands during pregnancy, lactation, and involution. A prominent index employed to study changes in cell numbers has been total gland DNA. Use of this index of cellularity implies acceptance of the assumptions that DNA content per nucleus remains constant throughout various stages and that no changes in average numbers of nuclei per cell occur. It is generally assumed that the amount of DNA per nucleus remains constant throughout pregnancy and lactation. However, Simpson and Schmidt (1970) presented data in conflict with this assumption. Mayer and Klein (1961) presented arguments indicating that numbers of multinucleate cells increase during late pregnancy and lactation. These observations indicate the mammary DNA data must be interpreted with care. However, measurement of total DNA is a very useful and convenient method for assessing changes in mammary cellularity.

Numerous types of measurements of mammary metabolic activity have been used as criteria of development. Included in these are estimates of oxygen uptake, rates of milk component synthesis, rates of oxidation of specific substrates by gland explants, slices or isolated cells incubated *in vitro,* and determinations of RNA and enzyme levels.

Recent developments in the area of molecular biology have vastly widened the range of techniques available for assessing specific actions of hormones and mammary gland development (Rosen *et al.,* 1986). For example, this new technology has provided means for discriminating among effects upon rates of transcription, mRNA translocation and translation, and posttranslational modification of proteins.

2. In Vivo *Techniques*

Studies with intact and surgically altered animals have contributed significantly to the development of knowledge of hormonal relationships in mammary growth and lactation.

In the former case, patterns of endogenous plasma hormone concentrations can be related to mammary gland development and function, or effects of exogenous hormones can be evaluated. Such studies must be interpreted with great care because of interactions among plasma hormones. Also, effects of hormones administered to intact animals upon development and function can be direct or indirect. For example, estrogens can alter prolactin secretion in addition to possibly stimulating duct growth directly and increasing the responsiveness of the mammary gland to prolactin (Meites, 1965). In order to avoid secondary endocrine effects that can arise as a result of administration of hormones to intact animals, various workers have preferred to use hypophysectomized, ovariectomized, and/or adrenalectomized animals to study hormone requirements for mammary gland development and the initiation and maintenance of lactation. Jacobsohn (1961) emphasized that judicious evaluation of the effects of these various endocrinectomies upon the general physiological status of animals must be exercised. In cases where replacement therapies are employed, one must distinguish between "permissive" hormonal effects arising from improvement of physiological status and direct hormonal effects on the tissue. Despite the fact that it is often difficult to distinguish direct from indirect hormonal effects, evaluations of hormone actions *in vivo* are essential. This is the only environment in which mammary epithelial cells fully express their capacities for proliferation, differentiation, development, and synthesis of milk components. Therefore, regulatory phenomena characterized *in vitro* can only be evaluated in terms of quantitative significance *in vivo.*

3. In Vitro *Techniques*

Tremendous progress has been made in assessing the specific hormonal requirements for mammary gland growth and differentiation through the use of *in vitro* gland explant and cell culture techniques. The clear advantage of these techniques is rigorous control of experimental conditions. The primary difficulties or limitations encountered appear to arise from the small amounts of tissue available for study from explants and rapid losses of secretory activities in explants and cells from lactating animals (Ebner *et al.*, 1961). These limitations make results difficult to interpret. Of particular concern is relating mechanisms established *in vitro* to the *in vivo* situation. Hormone actions *in vitro* leading to increased rates of milk synthesis, which are only 1 or 2% of rates observed in lactating tissue *in vivo*, may or may not be quantitatively important *in vivo*.

B. Development during Pregnancy

Several aspects of mammary gland development following puberty and during pregnancy were previously discussed. It was implied that estrogen and progesterone are the primary hormones required for development in intact animals during this period, that estrogen stimulated duct growth, and that progesterone regulated lobuloalveolar development. In general, these implications are correct, but species differences, the role(s) of hormones from the anterior pituitary, and interactions between ovarian and pituitary hormones must also be considered.

Folley (1970) considered species differences in response to ovarian hormones as representing three broad categories. Species in the first category include rats, mice, rabbits, and cats, which are described as exhibiting only duct growth when physiological doses of estrogen are administered. In these species, lobuloalveolar growth occurs only when progesterone is administered. The second category includes guinea pigs, goats, and cows, which require both estrogen and progesterone for normal duct development. Some lobuloalveolar development occurs in these species when estrogens alone are administered. Bitches have been placed in a third category because little or no mammary

development occurs when estrogen alone is administered. These species discrepancies may be due to differences in the actions of estrogens on glandular tissues, differences in the effects of estrogens on other tissues (including the ovaries and the anterior pituitary), differences in endogenous secretion of progesterone, differences in synergistic relationships between steroid hormones, prolactin, and growth hormone, or, more likely, a combination of these.

The literature concerning the role of the anterior pituitary in mammary development during pregnancy is quite extensive and will not be reviewed here in detail. Early experiments with hypophysectomized animals indicated that little or no mammary development occurred unless anterior pituitary extracts were supplied. The anterior pituitary hormones might act indirectly by restoring estrogen and progesterone secretion. A number of experiments, however, indicated that ovarian steroids administered alone were ineffective in stimulating mammary growth in hypophysectomized animals. These were interpreted as indicating that hormone(s) of the anterior pituitary act directly upon the mammary gland. Lyons (1942) and, later, Mizuno and Haito (1956) reported that intraduct injections of prolactin into rabbits caused, in addition to secretory activity, considerable localized growth of the alveolar epithelium as assessed histologically and by DNA measurement (Fig. 5).

C. Initiation of Lactation

Subsequent to the marked increases in lobuloalveolar structures and cell numbers (DNA; Table I) during pregnancy, three events occur during the immediate pre- and postpartum period that are essential to lactogenesis (the initiation of lactation). Although these events are highly interrelated, regulated by the same hormones, and often occur almost simultaneously, it is convenient to separate them for purposes of discussion and experimentation. The first of these events is the formation of functionally differentiated secretory cells. This event apparently requires cell division and is characterized by acquisition of the capacity to develop the characteristic enzyme complement of mammary secretory cells (Baldwin, 1969; Topper and Oka, 1974). The second event essential to lactogenesis is development of these newly formed, functionally differentiated cells. Part of this development appears to be inherent, requiring only an environment consistent with maintenance of the cells. Another portion of this development seems to be dependent upon specific hormonally regulated processes. The third essential event is expression of the capacity for milk synthesis that results from the occurrence of the first two events. *In vivo* experiments with hypophysectomized, ovariectomized, and/or adrenalectomized animals indicate that the primary hormones required for lactogenesis in laboratory animals are prolactin, cortisol or corticosterone, and insulin. Prolactin appears to fulfill functions essential to secretory cell proliferation and differentiation during late pregnancy (Lyons, 1942; Mizuno and Haito, 1956; Greenbaum and Slater, 1957; Lyons *et al.,* 1958; Baldwin and Martin, 1968; Baldwin, 1969; Korsrud and Baldwin, 1969; Dilley, 1971; Anderson, 1974; Topper and Oka, 1974; Tucker, 1974). In postmitotic or differentiated mammary cells, prolactin fulfills a central role in the regulation of RNA and protein synthesis, and thus is essential for the development within newly formed cells of the (enzymatic) capacity for milk component synthesis (Baldwin and Martin,

1968; Baldwin *et al.,* 1969; Korsrud and Baldwin, 1969; Topper and Oka, 1974). Glucocorticoids regulate, in part, rates of synthesis of several enzymes essential to milk biosynthesis, and, along with insulin, are essential for development of the extensive rough endoplasmic reticulum that is characteristic of fully developed mammary cells.

Mammary secretory cells are uniquely dependent upon insulin for their formation, development, survival, and function. When diabetic lactating rats are deprived of insulin therapy for 36–48 h, extensive and irreversible cell losses occur (Martin and Baldwin, 1971). Despite extensive work, the exact mechanisms of prolactin and insulin action in the regulation of secretory cell proliferation, differentiation, and development and of the mechanisms of glucocorticoid action in mammary cell development are not fully known. However, these are currently areas of intensive investigation, and continuing progress can be expected.

The third event essential to lactogenesis is expression by secretory cells of their capacity for milk synthesis. Early studies with cows indicate that, in contrast to rodents where cell proliferation, cell development, and the onset on milk synthesis occur almost simultaneously, cell proliferation and development are essentially complete several weeks prior to parturition and the onset of copious milk synthesis. These observations pose the possibility that a hormonal change associated with parturition "triggers" the onset of milk synthesis. Three possible explanations for failure of the apparently fully developed cow mammary secretory cells to synthesize milk prepartum have been considered. One is that nutrient supply to the glands is limited due to low blood flow rates. At parturition blood flow rates to the mammary glands increase markedly (Linzell, 1969). This is not likely the most limiting factor, however, since tissue concentrations of milk precursors and intermediates in the pathways of milk biosynthesis are higher pre- than postpartum (Baldwin and Cheng, 1968). Several data consistent with the possibility that milk constituents accumulated in the mammary glands prepartum inhibit milk synthesis have been published (Levy, 1964). In general, the accumulation of milk constituents is not considered a sufficient explanation, since prepartum milking does not lead to full expression of milk synthetic capacity. A final possibility is that several key biosynthetic enzymes [i.e., lactose synthetase, acetylcoenzyme A (acetyl-CoA) carboxylase] are limiting prepartum and must be induced prior to the onset of full lactation (Mellenberger *et al.*, 1973). Kuhn (1969) has argued that synthesis of these key enzymes is inhibited by progesterone prepartum. Thus, the decline in progesterone prior to parturition "triggers" the synthesis of several key enzymes, and lactation is initiated. This is an attractive hypothesis, and several data consistent with this mechanism have been published (Kuhn, 1969; Mellenberger *et al.*, 1973; Baldwin and Yang, 1974).

D. Maintenance of Lactation

1. Nonruminants

The hormonal requirements for maintenance of lactation have been reviewed in detail (Turner, 1974). The minimal hormonal requirements for the maintenance of secretory activity in hypophysectomized laboratory animals are cortisol, prolactin, and oxytocin (Table II). Some enhancement of production is obtained with addition of

Table II

Effects of Hypophysectomy and Hormone Replacement Therapies on Lactation in Rats[a]

Response	Treatment[b]				
	N	H	HC	HP	HPC
Pup gain (g)	5.6	−1.4	−0.8	−0.6	2.7
Pup deaths (no.)	0.0	3.2	3.0	1.4	0.0
Mammary gland					
Weight (g)	10.4	3.5	4.2	4.0	10.1
DNA (mg)	19.8	4.6	5.9	6.4	18.2
rRNA synthesis (cpm/g)	0.16	0.05	0.05	0.12	0.16
mRNA synthesis (cpm/g)	0.75	0.13	0.29	0.38	0.65
Casein synthesis (cpm/g)	17.5	3.8	4.5	9.6	17.3
Cystosolic protein synthesis (cpm/g)	5.4	1.6	3.1	5.0	7.0
Glucose-6-phosphate dehydrogenase (units/g)	2.0	0.28	0.26	0.28	1.25
Citrate cleavage enzyme (units/g)	1.2	0.07	0.37	0.06	1.40

[a] From Korsrud and Baldwin (1969).
[b] Treatments were N (normal 10-day lactating rats), H (hypophysectomized), HC (hypophysectomized receiving 0.5 mg 2 times/daily of cortisol), HP (hypophysectomized receiving 13 units 4 times daily of prolactin), HPC (hypophysectomized receiving cortisol plus prolactin). All animals except N received 2.5 μg thryoxin, 0.5 units oxytocin, and 5.0 μg estradiol-17β four times daily for 5 days after hypophysectomy on day 5 of lactation.

other hormomes such as long-acting insulin, estrogens, and triiodothyronine. Treatment with these hormones does not completely restore milk yields, and it is not clear whether they exert their effects directly upon secretory tissue or indirectly by affecting general metabolism. After complete hypophysectomy, oxytocin must be administered to facilitate milk ejection. Otherwise, the mammary glands degenerate due to accumulation of milk in them. Prolactin and cortisol retard but do not prevent gland degeneration due to lack of oxytocin administration and milk removal.

As mentioned previously, lactating secretory cells degenerate quickly *in vitro*. Thus, most data available on specific hormone actions during lactation result from *in vivo* studies with normal, endocrinectomized, or anti-insulin-injected animals and from short-term *in vitro* studies with tissues removed from such animals. Also mentioned was the fact that mature mammary secretory cells are extremely dependent upon insulin for normal function and survival. During acute insulin insufficiency *in vivo*, milk synthesis is drastically reduced within hours and cell degeneration becomes prominent within 1 or 2 days. The specific metabolic defects leading to cessation of milk synthesis within hours of insulin insufficiency are not known. The most limiting site of insulin action in mammary cells is not membrane transport or hexokinase function, since cell concentrations of amino acids, glucose, and glucose 6-phosphate are elevated during acute insulin insufficiency (Martin and Baldwin, 1971). Also during acute insulin insufficiency, cellular energy charge and redox state are altered (Yang and Baldwin, 1975). Adrenalectomy of mid-lactating rodents results in a 40–50% decrease in milk production as indicated by pup growth and decreases of 40–80% in activities of a number of enzymes closely associated with milk synthesis (Table III). Activities of many other enzymes are unchanged. These effects are reversed by glucocorticoid therapy. The depressions in enzyme levels are due, largely, to decreased rates of synthesis, indicating that glucocorticoids regulate, in part, rates of synthesis of key enzymes involved in milk synthesis. These decreases in rates of synthesis of key enzymes are probably mediated via glucocorticoid effects on rates on transcription, since glucocorticoids have been reported to affect rates of formation of several mRNAs *in vitro* (Mercier and Gaye, 1983). However, the complex interactions between glucocorticoids and prolactin complicate interpretation of these observations.

Table III
Effects of Adrenalectomy and Hormone Replacement Therapy on Lacation in Rats[a]

	Treatments			
	N-5	N-15	A-15	AC-15
Food intake (g/d)	—	54	37	46.0
Weight gain (g)	—	6	9	− 19.0
Pup gain (g)	—	18.6	9.8	17.0
Glucose-6-phosphate dehydrogenase (U/g)	10.0	21.0	10.0	26.0
Citrate Cleavage Enzyme (U/g)	0.5	1.0	0.3	1.1
Hexokinase (U/g)	0.2	0.2	0.2	0.2

[a]From Korsrud and Baldwin (1972a). Treatments were normal animals at days 5 and 15 of lactation (N-5, N-15), or adrenalectomized on day 5 of lactation and maintained until day 16 of lactation with (AC-15) and without (A-15) glucocorticoid therapy. Enzyme units per gram expressed as U/g.

Apparently, the depression in milk synthesis that occurs after adrenalectomy is not due to the reduced enzyme activities, since mammary metabolite data indicate that none of these become rate-limiting. Several data indicate that flux through the phosphofructokinase reaction is reduced after adrenalectomy, even though amount of this enzyme is not reduced (Korsrud and Baldwin, 1972b).

Prolactin appears to act in a general fashion upon lactating mammary tissue in laboratory species in that it is essential to maintenance of all species of RNA and synthesis of cytosolic and milk proteins (Table III). Specifically, numerous *in vitro* and a few *in vivo* observations indicate that prolactin both regulates rates of transcription of genes for the milk proteins and increases the half-lives of resulting m-RNAs. Further, glucocorticoids enhance the effect of prolactin on rates of transcription of the casein genes (Matusik and Rosen, 1978; Nagaiah *et al.,* 1981). These effects of the two hormones in regulation of casein mRNA levels are reflected by the more than additive rates of casein synthesis in hypophysectomized rats receiving prolactin plus glucocorticoid replacement therapy as compared to responses to either hormone administered alone (Table III). This contrasts with the observation that responses in rates of cytosolic protein synthesis to prolactin and a glucocorticoid administered separately sum to the response observed when both hormones are administered together (Table III). Above, citing data derived from studies with adrenalectomized, lactating rats (Table II), it was suggested that glucocorticoids may regulate rates of synthesis of several enzymes, possibly via stimulating rates of gene transcription. The data in Table III imply that this suggestion must be considered with reservation since in hypophysectomized rats, the response to the two hormones together was much more than the sum of responses to either hormone administered alone. From these observations, it must be clear that the interactions between glucocorticoids and prolactin are quite complex, making studies of their separate actions very difficult.

2. Ruminants

Several differences in hormonal requirements for the maintenance of lactation in ruminants have been reported. For example, adrenalectomy of lactating ewes does not reduce lactational performance or the activities of enzymes in mammary tissue (Ely and Baldwin, 1976) as discussed above for rats. Most important, perhaps, are observations that prolactin is not required for the maintenance of established lactation in goats and cattle, while somatotropin is (Cowie *et al.,* 1964). Further, administration of pituitary or recombinant bovine somatotropin (bST) to dairy cattle increases milk production 15–40%, while prolactin has no effect (Bauman and McCutcheon, 1986). Several biotechnology corporations are presently developing efficacy, safety, and production programs directed to marketing recombinant bST products for use in the dairy cattle industry.

The mechanisms by which bST acts to increase milk yield in dairy cattle are currently being investigated. bST action has been investigated by examining the three major factors controlling milk production:

1. Blood nutrient concentration
2. Blood flow to the udder
3. Biosynthetic capacity of the udder

Blood nutrient concentrations in cows injected with bST differ little if at all as compared to placebo-treated animals. Blood flow to the mammary gland increases with milk production, while arteriovenous differences for most nutrients in blood remain constant. Blood flow rates are regulated at the local (mammary) level by products of gland metabolism (CO_2 etc.) in venous blood leaving the gland. Therefore, it appears likely that mammary metabolic/biosynthetic capacity increases with bST treatment, producing an increase in metabolite concentrations in venous blood, which in turn produces increased blood flow to the gland.

It is not likely that the effect of bST on mammary metabolic capacity is direct, since bST receptors have not been found in mammary gland secretory cells. The most widely held hypothesis of bST action is that it stimulates insulin-like growth factor 1 (IGF-1) secretion by the liver and that IGF-1, in turn, acts upon the secretory cells. bST administration to lactating dairy cows has been shown to elicit an increase in plasma IGF-1 levels (Glimm *et al.*, 1988). Baumrucker (1986) presented preliminary work that indicated that IGF-1 stimulates mammary secretory cell proliferation and increases metabolic capacity in cow mammary tissue incubated *in vitro*. Very few additional data regarding specific mechanism(s) of bST action are currently available.

IV. Milk Synthesis

Considerable data are available regarding milk precursor–product relationships and metabolic pathways associated with energy metabolism and the synthesis of the major milk components. The data to be considered in subsequent sections were obtained through *in vivo* and *in vitro* experiments with isotope tracers, arteriovenous difference and blood flow measurements, and enzymatic and molecular biological techniques. The isotope tracer techniques enabled investigators to determine the proportions and amounts of milk components formed from specific blood metabolites and to estimate the activities of specific metabolic pathways in mammary tissue. Studies based on blood flow measurements and determinations of metabolite levels in arterial blood entering and venous blood leaving the mammary glands provide a basis for quantitative estimation of the amounts of blood metabolites removed by the mammary gland and, if coupled with isotope techniques, assessment of the amounts of milk components formed from each blood metabolite. This approach requires accurate assessment of mammary blood flow rates, and the venous blood obtained must be representative of the blood leaving the whole gland (Linzell, 1974). Almost all mammary blood flow and arteriovenous (A-V) measurements have been made on ruminants because of the size and structure of their mammary glands. Enzymatic techniques have been employed primarily to characterize the enzymes and pathways of energy metabolism and of biosynthesis of milk components. Many useful data have been obtained on ruminant and nonruminant mammary gland metabolism using these techniques. Comprehensive reviews pertaining to various aspects of milk synthesis are available (Davis and Collier, 1983; Baldwin and Smith, 1983).

Precursor–product relationships of the lactating goat udder are summarized in Table IV. The amount and composition of milk made is determined by the precursors taken up by the gland and the biochemical transformations that the precursors undergo in the secretory cells. Figures 6 and 7 represent flow diagrams of the metabolic

Table IV
Major Precursor–Product Relationships for the Lactating Goat Udder

Precursor	A-V difference (mg/100 ml blood)[a]	Uptake (mg/ml milk produced)[b]	Product	Amount (mg/ml milk produced)
Glucose	14.4	69.6	Lactose	46
			Co_2	17.4
			Glycerol	2.7
Acetate	5.7	27.5	Fatty acids	14.5
			CO_2	12.9
Triacylglycerol	6.7	32.4	Triacylglycerol	29.0
Amino acids	1.04	5.0	Casein	4.6

[a] Arteriovenous differences (A-V) were obtained by subtracting mammary venous precursor concentrations from arterial concentrations.

[b] Uptake of precursors was calculated using a ratio of mammary blood flow to milk yield of 438 (Linzell, 1974).

pathways involved in milk synthesis for rat and cow mammary glands, respectively. These two species differ greatly in precursors used for synthesizing milk components and the composition of milk produced. For example, rats make milk fat from glucose while cows use acetate as the primary precursor of fatty acids synthesized in the mammary gland. Since details regarding metabolic pathways of milk synthesis are available in many biochemistry texts and cited references (Reynolds and Folley, 1969; Falconer, 1971; Larson and Smith, 1974), the present discussion of milk synthesis will summarize data relating only to the regulation of metabolic pathways, quantitative precursor–product relationships, and some aspects of energetic efficiency.

A. Lactose Synthesis

The blood precursor of lactose is glucose (Linzell, 1974). There are two sites for regulation of lactose synthesis: glucose uptake and lactose synthetase. Evidence that glu-cose uptake is important include observations that low blood glucose concentrations result in low milk production (Linzell, 1974). Lactose synthetase catalyzes the last step in the pathway for lactose synthesis. The lactose synthetase complex consists of two proteins, galactosyltransferase and α-lactalbumin. Galactosyltransferase is present in all tissues; however, α-lactalbumin is found only in mammary tissue and is required for lactose synthesis.

A representation of lactose synthesis is shown in Figure 8. Glucose and UDP-galactose move freely through the Golgi membrane and are condensed via the lactose synthetase complex to form UDP and lactose. In as much as UDP is an inhibitor of lactose synthetase, Kuhn (1983) proposed a mechanism by which UDP can be recycled and transported back to the cytosol. According to this hypothesis, UDP is converted to UMP by nucleotide diphosphatase (NBPase), which is embedded on the Golgi membrane. UMP, unlike UDP, is able to move from the Golgi lumen into the cytosol. Once in the cy-

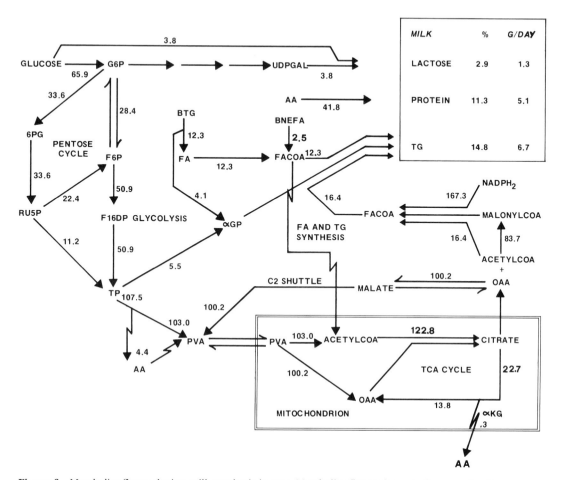

Figure 6 Metabolite fluxes during milk synthesis in rats. Metabolite fluxes through the several pathways were computed on the basis of carbon. NADH, NADPH, and ATP requirements are for synthesis of 45 g milk/day, with the composition indicated. Fluxes for each reaction and/or pathways are expressed in millimoles per day. Rounding of errors in notating fluxes on arrows leads to slight imbalances, but basically the system is in balance. Energy requirements were considered for maintenance but not for uptake, synthesis, and/or secretion of minor milk components. The primary precursors of milk were considered to be blood glucose, triglycerides (BTG), free fatty acids (BNEFA), and amino acids (AA). Abbreviations include glucose 6-phosphate (G6P), UDP galactose (UDPGAL), 6-phosphogluconate (6PG), ribulose 5-phosphate (RU5P), fructose 6-phosphate (F6P), fructose 1,6-diphosphate (F16DP), triose phosphates (TP), α-glycerol-P (αGP), pyruvate (PVA), oxaloacetic acid (OAA), fatty acid (FA), triglyceride (TG), tricarboxylic acid pathway (TCA), and fatty acyl-CoA (FACOA). From Plucinski (1976).

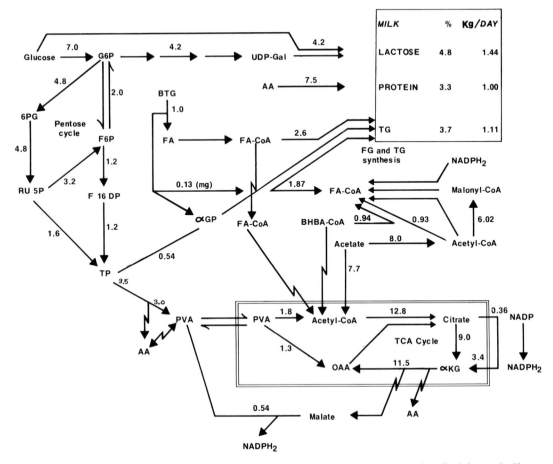

Figure 7 Metabolite fluxes during milk synthesis in cows. Fluxes were computed as described for rat in Figure 6, with the exception that the primary precursors considered were glucose, acetate, amino acids, ketone bodies (BHBA), and blood triglycerols, and that fluxes are expressed as mol/day. Milk yield was considered to be 30 kg/day. Some metabolism of amino acids was considered and the secretion of 0.36 mol/day of citrate in milk was accommodated. From Baldwin and Yang (1974).

tosol, UMP can be recharged to UTP. If substantiated, this pathway of UDP reutilization would require reevaluation of the estimates of efficiency of milk synthesis presented below (Tables VI and VII), since additional high-energy phosphates (ATP) presently not accounted for are needed.

B. Protein Synthesis

Milk crude protein (nitrogen content $\times$ 6.25 g protein/g N) is composed of the caseins (α_{s1}, α_{s2}, β, γ κ), α-lactalbumin, β-lactoglobulin, serum albumin, lactoferrin, lysozyme, immunoglobulins, and nonpro-

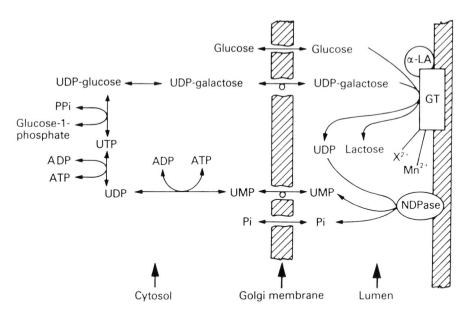

Figure 8 Proposed uridine nucleotide cycle supporting lactose synthesis. Double-headed arrows show reversible reactions or transmembrane movements via proposed ports ($\leftrightarrow$) or carriers ($\leftarrow\!\circ\!\rightarrow$). GT, galactosyltransferase; α-LA, α-lactalbumin; NDPase, nucleoside diphosphatase; X^{2+}, unknown cationic activator. From Mepham, (1987).

tein nitrogen compounds of which urea is the most prominent. The caseins, α-lactalbumin, and β-lactoglobulin comprise over 90% of milk crude protein in most species and are exclusively synthesized in the mammary gland. An exception is human milk, which is very low in protein (1.0–1.4%) as compared, for example, to cows (3.0–3.4%). Caseins plus α-lactalbumin and β-lactoglobulin constitute less than 50% of milk protein in humans. Human milk is very high in percentages of lactoferrin and nonprotein nitrogen compounds (Mepham, 1987).

Linzell (1974) summarized many data on the quantitative contributions of plasma amino acids to the synthesis of milk proteins. These experiments, based upon arteriovenous (A-V) difference studies of amino acid uptake by the mammary glands and upon radioisotope tracer studies, demonstrate that uptakes of essential amino acids by the glands are adequate to support milk protein synthesis and that these amino acids are incorporated into milk protein. Uptakes of nonessential amino acids and other N-containing compounds are quite variable but are sufficient to provide directly or indirectly (via amino acid synthesis) for milk protein synthesis.

Mechanisms involved in milk protein synthesis are similar to those in other mammalian cells that secrete proteins (Frisch, 1969; Larson and Jorgenson, 1974; Mercier and Gaye, 1983). These mechanisms are characterized by the presence on mRNAs for proteins to be secreted of "leader" or "signal" sequences of 15–30 amino acids. The leader sequences direct the nascent polypeptide chains formed during mRNA translation on the ribosomes to the endoplasmic reticulum (ER), where a channel is formed. This allows the growing peptide chain to enter the ER. Within the ER, the "signal" sequences are cleaved from the polypeptides. The resulting polypeptide chains are phosphorylated and glycosylated in the endoplasmic reticulum and Golgi apparatus to form the protein, which is finally secreted. In the case of the caseins, these posttranslational changes are extensive and quite complex, resulting in condensation with calcium and formation of the casein micelles that are essential to cheese production.

C. Milk Fat Synthesis

A great many problems have been encountered in the investigation of milk fat biosynthesis, due largely to variation in the fatty acid composition of the milk triacylglycerides, the large number of potential and real precursors of milk fat found in blood, and the multiplicity of pathways for the synthesis and alteration of fatty acids in the mammary gland. The literature pertaining to these problems is much too extensive for specific consideration within the present context, and the reader desiring details is urged to consult these reviews (Baldwin and Yang, 1974; Bauman and Davis, 1974; Davis and Bauman, 1974; Linzell, 1974).

The primary blood precursors of the fatty acids in milk fat triacylglycerides in ruminants are acetate, β-hydroxybutyrate (BHBA), and triacylglycerides of the very-low-density blood lipoproteins and chylomicra. The primary blood precursors of milk fatty acids in nonruminants are glucose and blood triacylglycerides. Table V presents a summary of numerous data on the contri-

Table V
Precursors of Milk Fat Fatty Acids[a]

Fatty acid	Mole percent in milk fat	g/100 g milk fatty acid from	
		Acetate + BHBA	Blood triacylglycerol
C_4	10	4.0	—
C_6	4	2.1	—
C_8	2	1.3	—
C_{10}	5	4.0	—
C_{12}	3	2.8	—
C^{14}	10	5.2	5.2
C_{16}	25	14.1	14.1
$C_{16:1}$	3	—	3.6
C_{18}	8	—	10.5
$C_{18:1}$	27	—	35.0
$C_{18:2}$	3	—	3.5

[a]Adapted from Smith (1970).

butions of the primary precursors of the fatty acids in cow milk triacylglycerides. Only the major fatty acid components of milk fat are represented. BHBA is a major blood precursor of butyrate and the first four carbons of C_6 to C_{10} fatty acids formed from BHBA and acetate. Acetate is a major precursor of C_4–C_{16} fatty acids, and plasma lipids are a major precursor of C_{16}–C_{18} fatty acids in milk fat.

Included in Figures 6 and 7 are the pathways of fatty acid synthesis in rat and cow mammary glands. Represented are synthesis via the malonyl-CoA pathway in rats and synthesis via the malonyl-CoA and chain elongation (BHBA + acetyl-CoA) pathways in cows. More detail on pathways of $NADPH_2$ synthesis in rats and cows (Figs. 6 and 7) can be found in the review by Bauman (1974). An intriguing feature of mammary fatty acid synthesis is formation of a mixture of fatty acids varying in chain length from C_{10} to C_{16} via the malonyl-CoA pathway. In other tissues the primary product of this pathway is palmitate (C_{16}). This difference is due to the presence in mammary tissue of thioesterases, which terminate the synthetic process at different chain length specifications (C_{10} to C_{16}). Blood triacylglycerides in chylomicra and very-low-density lipoproteins are hydrolyzed by lipoprotein lipase in the mammary capillaries prior to uptake (Linzell, 1974). Triacylglyceride glycerol in milk fat is formed from blood triacylglyceride glycerol and blood glucose.

D. Efficiency of Milk Synthesis

The data presented in Tables VI and VII were prepared to enable estimation of efficiencies of milk synthesis in the mammary gland. The calculations are based on data on uptake of nutrients by the mammary glands, milk composition, and consideration of the metabolic pathways presented in Figures 6 and 7.

Estimates obtained indicate that milk synthesis is a very efficient process. Efficiencies of milk synthesis for whole animals are less than those for the gland alone because energy is lost during substrate transformations in extramammary tissues and increased work in heart muscle, muscles of respiration, etc. Whole-animal data support the view that mammary milk synthesis occurs at

Table VI
Estimation of the Efficiency of Milk Synthesis in the Rat Mammary Gland[a,b]

	Uptake/100 g milk			Output/100 g milk	
	mmol	kcal		mmol	kcal
Glucose	155.0	104.3	Lactose	8.4	11.3
Amino acids	102.7	58.2	Fat	21.3	143.7
Triacylglycerol	11.0	83.5	Protein	102.7	58.5
Total		246.0			213.5

[a]Efficiency = (output/uptake) × 100 = (213.5/239.9) × 100 = 87%.
[b]Adapted from Plucinski (1976).

Table VII
Estimation of the Efficiency of Milk Synthesis in the Cow Mammary Gland[a,b]

	Uptake/kg milk			Output/kg milk	
	mmol	kcal		mmol	kcal
Glucose	373	251	Fat	48	304
Acetate	525	110	Lactose	140	189
BHBA	52	27	Protein	250	143
Triacylglycerol	30	213		—	—
Amino acids	250	142		—	—
Total		743			636

[a]Efficiency = (output/uptake) × 100 = (636/743) × 100 = 86%.
[b]Adapted from Baldwin and Yang (1974).

close to the theoretical (biochemical) efficiencies presented in Tables VI and VII (Canas *et al.,* 1976).

V. Milk Ejection

Milk secreted by the alveolar cells of the mammary gland cannot be removed until the myoepithelial cells surrounding the alveoli contract and force the milk from the alveoli and small ducts to the large ducts and the gland and teat cisterns. The neuroendocrine reflex activated by suckling or milking, which regulates myoepithelial cell contraction and milk ejection, has been the subject of extensive investigation. The development of knowledge concerning milk ejection was treated in a very interesting review by Folley (1970). The functional innervation of the mammary glands, the hypothalamo–neurohypophyseal system, and the role of oxytocin in milk expulsion have been reviewed in detail (Denamur, 1965; Grosvenor and Mena, 1974).

A. *Functional Innervation of the Mammary Glands*

Innervation of the mammary glands is by somatic sensory and sympathetic motor fibers arising from different segments of the spinal cord, depending upon whether a given species has thoracic, abdominal, and/or inguinal mammary glands. There has been some controversy regarding the presence or absence of parasympathetic innervation in the udders of ruminants, but at present no convincing evidence of such innervation is available (Cross, 1961). The sensory nerves are distributed, primarily, in the skin surrounding the mammary glands and are present in exceptionally large numbers in teats and nipples. The supply of sensory nerves to the mammary parenchyma appears to be very limited. The motor nerve endings found in the mammary glands supply, primarily, the blood vessels, the connective stroma, the large ducts, and the muscles of the teat and nipple (Cross, 1961). Stimulation of the peripheral extremities of the mammary nerves and administration of epinephrine cause vasoconstriction, rhythmic contractions in the teats or nipples, and relaxation of smooth muscles surrounding the teat and gland cisterns. These observations and others mentioned later support the contentions that the autonomic fibers supplying the mammary gland are sympathetic and adrenergic and that motor endings do not supply the myoepithelial cells surrounding the alveoli and small ducts. Hence, motor elements are not responsible for milk ejection.

B. Regulation of Oxytocin Release

The sensory stimuli associated with suckling or milking lead to the liberation of the neurohormone oxytocin into blood. Considerable attention has been focused on examination of the efficiency of various visual, conditioned, and physical stimuli in evoking oxytocin release and subsequent milk ejection. These observations are important to the formulation of recommendations concerning proper, practical premilking practices for dairy cattle, goats, and nursing human mothers. These studies have resulted in the realization that practices that cause adrenalin secretion prevent milk ejection. Adrenalin has two actions in this regard: it blocks oxytocin release at the posterior pituitary and causes vasoconstriction in the mammary glands, thus preventing oxytocin entry. The most effective stimulus to milk ejection is manipulation of teats or nipples during suckling or milking. It is surprising that the types of stimuli (e.g., tactile, thermal) that excite the sensory nerve endings in teats or nipples have not been characterized in detail. However, there is general agreement that stimuli associated with milking and suckling are transmitted to the neurohypophysis and cause oxytocin release. Transmission of these signals involves a complex of ascending spinal paths and possibly medullary paths. Supraoptic and paraventricular nuclei perform two functions related to the release of oxytocin. Oxytocin is synthesized in these nuclei and is transported to and stored in the neurohypophysis. Second, the supraoptic and paraventricular nuclei regulate oxytocin secretion. The role of the hypothalamus in regulation of the secretion of adenohypophseal hormones involved in lactation is described in Chapter 3.

C. Milk Ejection

Oxytocin released by the neurohypophysis is transported by the blood to the mammary glands, where it acts upon the myoepithelial cells surrounding the alveoli and ducts of the glands and causes them to contract and expel the milk. Oxytocin also causes relaxation of the smooth muscles surrounding the large ducts and gland and teat cisterns, thus providing for enlargement of these structures to accommodate the milk ejected from the alveoli. Intramammary pressure rises as a result of the forcible ejection of milk from the alveoli and small ducts, and the suckling young or the milker only have to overcome the resistance of the teat sphincter to accomplish the final stage of milk removal (Folley, 1970). In many animals the buildup of intramammary pressure is sufficiently great to overcome the resistance of the teat sphincter, causing milk to drip or spurt from the teats (Folley, 1970).

References

Anderson, R. R. (1974). *In* "Lactation: A Comprehensive Treatise" (B. L. Larson and V. R. Smith, eds.), Vol. I, pp. 97–140. Academic Press, New York.

Baldwin, R. L. (1969). *J. Dairy Sci.* **52,** 729.

Baldwin, R. L. (1974). *In* "Animal Agriculture" (H. H. Cole and M. Ronning, eds.), pp. 409–420. W. H. Freeman, San Francisco.

Baldwin, R. L., and Cheng, W. (1968). *J. Dairy Sci.* **52,** 523–528.

Baldwin, R. L., Korsrud, G. O., Martin, R. J., Cheng, W., and Schober, N. A. (1969). *Biol. Reprod.* **1,** 31.

Baldwin, R. L. and Smith, N. E.. (1983). *In* "Dynamic Biochemistry of Animal Production," (P. M. Riis, ed.), pp. 359–388. Elsevier, New York.

Baldwin, R. L., and Martin, R. J. (1968). *Endocrinology* **82,** 1209.

Baldwin, R. L., and Yang, Y. T. (1974). *In* "Lactation: A Comprehensive Treatise" (B. L. Larson and V. R. Smith, eds.), Vol. I, pp. 349–411. Academic Press, New York.

Barmann, W., and Knoop, A. (1959). *Z. Zellforsch.* **49,** 344.

Bauman, D. E., and Davis, C. L. (1974). *In* "Lactation: A Comprehensive Treatise" (B. L. Larson and V. R. Smith, eds.), Vol. II, pp 31–75. Academic Press, New York.

Bauman, D. E., and McCutcheon, S. N. (1986). *In* "Control of Digestion and Metabolism ion Ruminants" (L. P. Milligan, W. L. Grovum, and A. Dobson, eds.), pp. 436–455. Prentice-Hall, Englewood Cliffs, New Jersey.

Baumrucker, C. R. (1986). *J. Dairy Sci. Suppl.* 1, **69,** 120.

Benson, G. K., Cowie, A. T., Cox, C. P., and Goldzweig, S. A. (1957). *J. Endocrinol.* **15,** 126.

Canas, R. R., Romero, J. J., Baldwin, R. L., and Koong, L. J. (1976). *J. Dairy Sci.* **59,** 57.

Cowie, A. T., Knaggs, G. S., and Tindal, J. S. (1964). *J. Endocrinol.* **28,** 267.

Cross, B. A. (1961). *In* "Milk: The Mammary Gland and Its Secretion" (S. K. Kon and A. T. Cowie, eds), Vol. I, pp. 229–277. Academic Press, New York.

Davis, C. L., and Bauman, D. E. (1974). *In* "Lactation in Comprehensive Treatise" (B. L. Larson and V. R. Smith, eds.), Vol. II, pp. 3–30. Academic Press, New York.

Davis, S. R., and Collier, R. J. (1983). *J. Dairy Sci.* **68,** 1041.

Denamur, R. (1965). *Dairy Sci. Abstr.* **27,** 193.

Dilley, W. G. (1971). *J. Endocrinol.* **50,** 501.

Dylewski, D. P., Dapper, C. H., Valivullah, H. M., Deeney, J. T., and Keenan, T. W. (1984). *Eur. J. Cell Biol.* **35,** 111.

Ebner, K. E., Hageman, E. C., and Larson, B. L. (1961). *Exp. Cell Res.* **25,** 555.

Ebner, K. E., and Schanbacher, F. L. (1974). *In* "Lactation: A Comprehensive Treatise" (B. L. Larson and V. R. Smith, eds.), Vol II, pp. 77–113. Academic Press, New York.

Ely, L. O., and Baldwin, R. L. (1976). *J. Dairy Sci.* **59,** 491–503.

Falconer, I. R. (1971). "Lactation." Pennsylvania State University Press, University Park, Pennsylvania.

Flux, D. S. (1954). *J. Endocrinol.* **11,** 223.

Folley, S. J. (1970). *Perspect. Biol. Med.* **13,** 476.

Frisch, H. (1969). *Cold Spring Harbor Symp. Quant. Biol.* **34.**

Glimm, D. R., Baracos, V. E., and Kennelly, J. J. (1988). *J. Dairy Sci.* **71,** 2923–2935.

Greenbaum, A. L., and Slater, T. F. (1957). *Biochem. J.* **66,** 155.

Grosvenor, C. E., and Mena, F. (1974). *In* "Lactation: A Comprehensive Treatise" (B. L. Larson and V. R. Smith, eds.), Vol. I, pp. 227–276. Academic Press, New York.

Helminen, H. J., and Ericsson, J. L. E. (1968). *J. Ultrastruct. Res.* **25,** 193.

Hollmann, K. H. (1974). *In* "Lactation: A Comprehensive Treatise" (B. L. Larson and V. R. Smith, eds.), Vol. I, pp. 3–95. Academic Press, New York.

Jacobsohn, D. (1961). *In* "Milk: The Mammary Gland and Its Secretion" (S. K. Kon and A. T. Cowie, eds.), Vol. I, pp. 127–160. Academic Press, New York.

Keenan, T. W., and Dylewski, D. P. (1985). *J. Dairy Sci.* **68,** 1025–40.

Keenan, T. W., Morre, D. J., and Huang, C. M. (1974). *In* "Lactation, A Comprehensive Treatise" (B. L. Larson and V. R. Smith, eds.), Vol. II, pp. 191–233. Academic Press, New York.

Kon, S. K., and Cowie, A. T. (eds.) (1961). "Milk: The Mammary Gland and Its Secretion," Vols. I and II. Academic Press, New York.

Korsrud, G. O., and Baldwin, R. L. (1969). *Biol. Reprod.* **1,** 21.

Korsrud, G. O., and Baldwin, R. L. (1972a). *Can. J. Biochem.* **50,** 366–376.

Korsrud, G. O., and Baldwin, R. L. (1972b). *Can. J. Biochem* **50:** 377–385.

Kuhn, N. J. (1969). *J. Endocrinol.* **44,** 39.

Kuhn, N. J. (1983). *In* "Biochemistry of Lactation" (T. B. Mepham, ed.), pp. 159–176. Elsevier, New York.

Larson, B. L., and Smith, V. R. (eds.) (1974). *In* "Lactation: A Comprehensive Treatise," Vols. I, II, and III. Academic Press, New York.

Larson, B. L., and Jorgensen, G. N. (1974). *In* "Lactation, A Comprehensive Treatise" (B. L. Larson and

V. R. Smith, eds.), Vol. II, pp. 115–146. Academic Press, New York.

Larson, B. L., and Smith, V. R. (eds.) (1978). "Lactation—A Comprehensive Treatise," Vols. IV and V. Academic Press, New York.

Larson, B. L. (ed.) (1985). "Lactation." Iowa State University Press, Ames, Iowa.

Levy, R. H. (1964). *Biochim. Biophys. Acta* **84,** 229.

Linzell, J. L. (1969). *In* "Lactogenesis" (M. Reynolds and S. J. Folley, eds.), pp. 153–169. University of Pennsylvania Press, Philadelphia.

Linzell, J. L. (1974). *In* "Lactation: A Comprehensive Treatise" (B. L. Larson and V. R. Smith, eds.), Vol. I, pp. 143–225. Academic Press, New York.

Lyons, W. R. (1942). *Proc. Soc. Exp. Biol. Med.* **51,** 308.

Lyons, W. R., Li, C. H., and Johnson, R. F. (1958). *Recent Progr. Horm. Res.* **14,** 219.

Martin, R. J., and Baldwin, R. L. (1971). *Endocrinology* **88,** 863.

Mather, I. H. (1987). *In* "The Mammary Gland-Development, Regulation, and Function" (M. C. Neville and C. W. Daniel, eds.), pp. 217–267. Plenum Press, New York.

Matthews, C. A., Swett, W. W., and Fohrman, M. H. (1949). *U.S. Dept. Agric. Tech. Bull.* **993.**

Matusik, R. J., and Rosen, J. M. (1978). *J. Biol. Chem.* **253,** 2343–2347.

Mayer, G., and Klein, M. (1961). *In* "Milk: The Mammary Gland and Its Secretion" (S. K. Kon and A. T. Cowie, eds.), Vol. I, pp. 47–126. Academic Press, New York.

Meites, J. (1961). *In* "Milk: The Mammary Gland and Its Secretion" (S. K. Kon and A. T. Cowie, eds.), Vol. I, pp. 321–367. Academic Press, New York.

Meites, J. (1965). *Endocrinology* **76,** 1220.

Mellenberger, R. W., Bauman, D. E., and Nelson, D. R. (1973). *Biochem J.* **136,** 741.

Mepham, T. B. (ed.) (1983). "Biochemistry of Lactation." Elsevier, New York.

Mepham, T. B. (1987). "Physiology of Lactation." Open University Press, Milton Keynes, Philadelphia.

Mercier, J. C., and Gaye, P. (1983). In "Biochemistry of Lactation" (T. B. Mepham, ed.), pp. 177–227. Elsevier, New York.

Mizuno, H., and Haito, M. (1956). *Endocrinol. Jpn.* **3,** 227.

Munford, R. E. (1961). *Dairy Sci. Abstr.* **26,** 293.

Nagaiah, K., Bolander, F. F., Jr., Nicholas, K. R., Takemoto, T., and Topper, T. J. (1981). *Biochem. Biophys. Res. Commun.* **98,** 380–387.

Netter, F. H. (1954). "The CIBA Collection of Medical Illustrations," Vol. II. CIBA Pharmaceutical Products, Inc., Summit, New Jersey.

Neville, M. C., and Daniel, C. W. (eds.) (1987). "The Mammary Gland-Development, Regulation, and Function." Plenum Press, New York.

Plucinski, T. M. (1976). A dynamic simulation model of rat mammary metabolism. Ph.D. Thesis, University of California, Davis.

Powell, J. T., Jarlfors, U., and Brew, K. (1977). *J. Cell Biol.* **72,** 617.

Raynaud, A. (1961). *In* "Milk: The Mammary Gland and Its Secretion" (S. K. Kon and A. T. Cowie, eds.), Vol. I, pp. 3–46. Academic Press, New York.

Reynolds, M., and Folley, S. J. (eds.) (1969). "Lactogenesis: The Initiation of Milk Secretion at Parturition." University of Pennsylvania Press, Philadelphia.

Rosen, J. M., Rodgers, J. R., Couch, C. H., Bisbee, C. A, David-Inouye, Y., Campbell, S. M., and Yu, L. Y. Lee, (1986). *In* "Metabolic Regulation: Applications of Recombinant DNA Techniques." New York Academy of Science, New York.

Saacke, R. G., and Heald, C. W. (1974). *In* "Lactation: A Comprehensive Treatise" (B. L. Larson and V. R. Smith, eds.), Vol. II, pp. 147–189. Academic Press, New York.

Silberstein, G. B., and Daniel, C. W. (1987). *J. Dairy Sci.* **70,** 1981–1990.

Silver, M. (1953). *J. Endocrinol.* **10,** 35.

Simpson, A. A., and Schmidt, G. H. (1970). *Proc. Soc. Exp. Biol. Med.* **133,** 897.

Smith, K. L., and Schanbacher, F. L. (1973). *J. Dairy Sci.* **56,** 738.

Smith, N. E. (1970). Quantitative simulation analyses of ruminant metabolic functions: basal; lactation; milk fat depression Ph.D. Thesis, University of California, Davis.

Sorensen, M. T., Sejrsen, K., and Foldager, J. (1987). *J. Dairy Sci.* **70,** 265–270.

Swett, W. W., Book, J. H., Matthews, C. A., and Fohrman, M. H. (1955). *U.S. Dept. Agric. Tech. Bull.* **1111.**

Topper, Y. J., and Oka, T. (1974). *In* "Lactation: A Comprehensive Treatise" (B. L. Larson and V. R. Smith, eds.), Vol. I, pp. 327–348. Academic Press, New York.

Tucker, H. A. (1974). *In* "Lactation: A Comprehensive Treatise" (B. L. Larson and V. R. Smith, eds.). Vol. I, pp. 277–236. Academic Press, New York.

Turner, C. W. (1952). "The Mammary Gland. I. The

Anatomy of the Udder of Cattle and Domestic Animals." Lucas Brothers, Columbia, Missouri.

Turner, C. W. (1969). "Harvesting Your Milk Crop." Babson Brothers, Chicago.

Wooding, F. B. P (1977). *In* "Comparative Aspects of Lactation" (M. Peaker, ed.), pp. 1–41. Academic Press, London.

Yang, Y. T., and Baldwin, R. L. (1975). *J. Dairy Sci.* **58,** 337–343.

CHAPTER 12

Reproduction in Horses

PETER F. DAELS, JOHN P. HUGHES, and GEORGE H. STABENFELDT

An increased understanding of reproductive physiology of the mare and stallion has occurred during the last 20 years due to the development of new methodology in both the area of reproductive endocrinology and the area of clinical diagnostic techniques. As concerns diagnostic techniques, the most important development in the area of equine reproduction during the last decade is the introduction of ultrasonography for the examination of the genital tract per rectum (Palmer and Driancourt, 1980). The ability to monitor ovarian dynamics in detail and to detect pregnancy as early as 10 days after conception has allowed new insights into equine reproduction for both researchers and veterinary practitioners. The development of a technique for cannulation of the intercavernal venous sinus of the equine pituitary gland has greatly enhanced our ability to study the regulation of the hypothalamo–pituitary–gonadal axis (Irvine and Alexander, 1987). Finally, the introduction of computer programs for analysis of equine semen has given us new tools for more objective evaluation of semen quality in the stallion.

The International Symposium on Equine Reproduction, held every fourth year, has become an important meeting point for specialists in equine reproduction from all over the world and the proceedings are an important source of information. The reader is also referred to Ginther (1979) and Rossdale and Ricketts (1980) as valuable sources of information in horse reproduction.

I. Introduction

Many of the breeding problems of horses arise because of the designation of January 1 by breed associations as the date that all foals become 1 year of age. This regulation sets up the desire for foals to be born as soon after January 1 as possible because competition often begins by 2 years of age; a few months either side of 2 years of age

can make a difference in the ability of the animal to compete successfully. Reproductive activity in the mare is controlled by light, with cessation of ovarian activity in the autumn and reestablishment of activity in the spring, often March, or April in the northern hemisphere. Because the gestation period of the mare is long ($11-11\frac{1}{2}$ months), breeding of mares occurs as soon after February 14 as possible. Foaling mares are able to ovulate, involute the uterus, and conceive within 2 weeks of parturition, even if they deliver during the seasonal anestrum. The problem resides mainly with nonpregnant mares in which breeding is often attempted before the onset of the physiologic breeding season.

II. Estrous Cycle

A. Photoperiod

Most mares are seasonal polyestrous, with ovarian activity affected by photoperiod and usually ceasing under conditions of decreasing light. The periods of ovarian inactivity range from 2 to 6 months in duration, depending on the latitude and possibly the climate. The occurrence of photoperiod-induced anestrus becomes more pronounced as the latitude increases, either north or south. The seasonality of ovarian activity of mares in Australia is shown in Figure 1 (Osborne, 1966). The data were collected at a latitude of about 30°S. Seasonality does exist close to the equator (Mexico, latitudes 15–22°N) with an anestrous season produced by a maximum photoperiod change of 2 h (Saltiel et al., 1982). Photoperiod control of ovarian activity is not absolute in the sense that some mares continue to have cyclic ovarian activ-

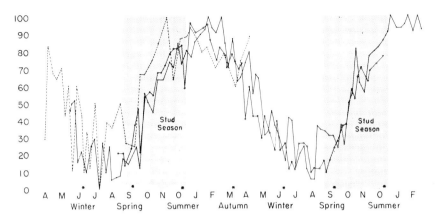

Figure 1 Frequency distribution curve for the seasonal occurrence (percentage) of ovulation in mares obtained from three surveys in Australia. o----o 1959–60 Avg. 7.5 mares/wk, ●——● 1961–63—24; o——o 1964–65—37. Note the low ovulation rate associated with the start of the breeding season. The asterisk (*) refers to the summer and winter solstice and the vernal and autumn equinox. From Osborne (1966).

ity throughout the year (Osborne, 1966; Hughes *et al.*, 1973; Saltiel *et al.*, 1982).

Light is the important factor that influences the sexual season of the mare. Although the mare responds to decreasing light by terminating cyclical ovarian activity, the response is slow and may lag 4 to 6 months after summer solstice. The pineal gland is central to the translation of photoperiod in the mare. Exposure to darkness results in an increase of melatonin secretion (Kilner *et al.* 1982). Extending the photoperiod in the evening results in a suppression of melatonin secretion and effectively stimulates ovarian activity. Extending daylight by adding more hours of light in the morning is ineffective in stimulating ovarian activity. Palmer *et al.* (1982) demonstrated that a 1-h period of light about 10 h after dark was equally effective in inducing cyclic ovarian activity. The importance of some darkness in the photoperiodic stimulation of mares was emphasized by the fact that the

stimulatory response under continuous light was less effective in stimulation of ovarian activity as compared to a light–dark photoperiod regimen. The pineal gland is involved in the long-term timing of the response to increased light in that pinealectomy delayed the onset of ovarian activity during the second year following surgery, but not the first (Grubaugh *et al.*, 1982).

Photoperiod can also influence the timing of puberty. In one study, exposure of 6 to 8 month fillies to a fixed long photoperiod (16L : 8D) or a fixed short photoperiod (9L : 15D) interfered with the onset of puberty as compared to animals exposed to the natural photoperiod (Wesson and Ginther, 1982). This contrasts with effects of a fixed long photoperiod (16L : 8D) in mature mares, which is stimulatory for ovarian activity (Kooistra and Ginther, 1975). The prepubertal perception of photoperiod by fillies is obviously different from that of the mature mare.

B. Length

During the physiological breeding season, the estrous cycle, defined as the period between two subsequent ovulations, averages 21 to 22 days in length. Estrous cycle intervals have been reported to vary from a few days to greater than 30 days (Andrews and McKenzie, 1941). These data arose from the fact that estrous cycle intervals were originally defined as the period from the start of estrus to the next estrus rather than ovulation to ovulation. In the spring, mares undergo a transitional phase of eratic sexual behavior and follicle development prior to the first ovulation at the onset of the physiological breeding season. During this transitional phase, it is not unusual for mares to exhibit estrus for one to several weeks in the absence of ovulation. During the physiological breeding season, estrous cycles as short as 19 days are considered normal. Estrous cycles that are shorter may be the result of uterine infection, which induces the secretion of prostaglandin and re-

gression of the corpus luteum (Neely et al., 1979). In addition, spontaneous prolongation of the luteal phase during the physiologic breeding season is usually the result of the presence of a persistent corpus luteum in which estrous cycle intervals can be as long as 3 months (Stabenfeldt et al., 1974a, 1984). Estrous cycle lengths of 24 to 25 days can occur in conjunction with the delayed onset of folliculogenesis observed prior to the cessation of ovarian activity at the end of the physiologic breeding season (Neely, 1979), or can be due to the occurrence of ovulation late in the luteal phase (Hughes et al., 1973; Hughes et al., 1984).

C. Estrus

The mare is usually sexually receptive for 5–6 days during the physiological breeding season. The duration of estrus decreases as the physiologic breeding season proceeds. In one study, we found the duration of estrus to be 8.4, 7.7, 5.7, and 4.5 days in March, April, May, and June, respectively

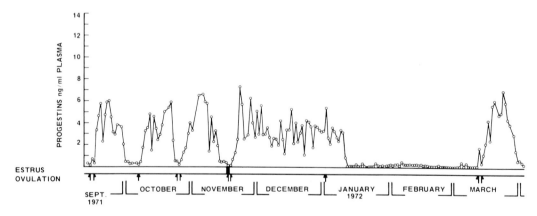

Figure 2 Peripheral plasma progestin levels in a nonpregnant mare. The black horizontal bars indicate the time of sexual receptivity. The arrows indicate the occurrence of ovulation. Note the absence of sexual receptivity (except for one day) in spite of regular ovarian activity, the occurrence of a prolonged corpus luteum (December), and the occurrence of anestrus (January to March). From Stabenfeldt et al. (1974b).

(Hughes *et al.*, 1973). This shortening of estrus represents an acceleration of folliculogenesis prior to ovulation with an increasingly favorable photoperiod. Ovulation usually occurs between 24 and 48 h prior to the end of estrus. An estrous period of normal duration is a good indication that ovulation has occurred. Estrus is preceded by a period of 1–2 days in which mares show interest in stallions, yet do not allow copulation.

Some mares fail to show estrus even though endocrine analysis of blood indicates normal cyclic ovarian activity with estrogen concentrations being normal (Stabenfeldt *et al.*, 1974b) (Fig. 2). While the suggestion has been made that mares failing to manifest estrus have lower estrogen concentrations (Nelson *et al.*, 1985), the data presented suggest the animals had unusually short intervals from luteolysis to ovulation. These animals may not have had sufficient time to manifest estrus before the onset of ovulation. Some animals fail to display estrous behavior, during the follicular phase of the estrous cycle, due to psychologic inhibition. The best example of this behavior is the nursing mare with foal at side in which maternal protective instincts override sexual receptivity. Other animals with large ovarian follicles may fail to show estrus because they are in the luteal phase of a normal cycle (Stabenfeldt *et al.*, 1972), or because they are in a persistent luteal phase with estrous manifestation blocked by the presence of progesterone (Stabenfeldt *et al.*, 1974b).

D. Folliculogenesis

An important concept concerning folliculogenesis in large domestic animal species is that significant folliculogenesis occurs during the luteal phase of the estrous cycle

prior to the onset of luteolysis. This is also true for the mare where a follicle of about 25 mm is usually present at the time of regression of the corpus luteum (Pierson and Ginther, 1987). This follicle is usually the dominant follicle and undergoes a final development to the point of ovulation within 6–7 days (Fig. 3).

Our knowledge of gonadotropin control of folliculogenesis in the mare has been greatly enhanced through the development of a technique that allows cannulation of the intercavernal venous sinus of the pituitary via the facial vein (Irvine and Alexander, 1987). This approach has allowed information to be obtained on the relationship of gonadotropin-releasing hormone (GnRH) secretion and gonadotropin release. Alexander and Irvine (1987) have found that pulsatile secretion of GnRH results in the concomitant pulsatile release of both follicle-stimulating hormone (FSH) and luteinizing hormone (LH); 91% of the GnRH pulses produced gonadotropin pulses (Fig. 4).

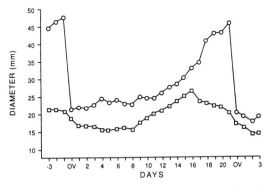

Figure 3 Mean diameter of the largest follicle and second largest follicle during the cycle. Note the divergence in diameter between the largest (—○—) and second largest (—□—) follicle after day 16 after ovulation. From Pierson and Ginther (1987).

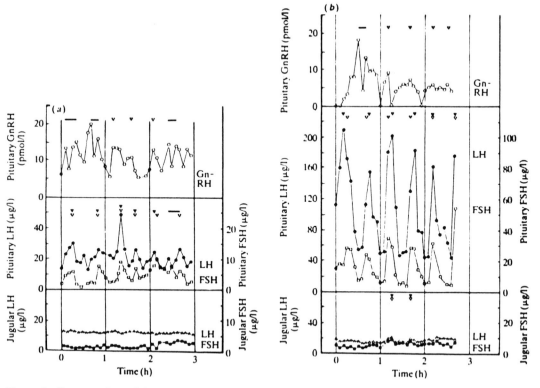

Figure 4 Concentrations of GnRH, LH, and FSH in pituitary venous effluent, and of LH and FSH in concurrent jugular blood samples collected at 5-min intervals from mare LS on (a) the day before ovulation and (b) the day of ovulation. Significant GnRH and FSH pulses are marked by open arrow heads and LH pulses by solid arrow heads. Horizontal bars indicate prolonged secretory episodes. *Sample in which insufficient plasma was collected for GnRH assay. From Irvine and Alexander (1987).

The number of follicles that develop during the ovarian cycle is surprising because the mare usually aborts when more than one fetus is *in utero*. In a study of 276 estrous periods (Hughes *et al.*, 1973), we found the following numbers of large follicles (25–50 mm) during estrus: one follicle, 106 (37%); two follicles, 115 (43%); three follicles, 33 (12%); four follicles, 22 (8%). Thus, in 63% of the ovarian cycles, more than one large follicle was present.

E. Ovulation

Most of the large domestic species ovulate at a predictable time after the onset of estrus, usually 30–36 h. This is due to the tight coupling of the onset of the gonadotropin surge and estrus, both driven by increased pulsatile release of GnRH. In the mare, the interval from the onset of estrus (and the initial increase in LH) to ovulation is often 4–5 days. This long interval makes prediction of the time of ovulation difficult in mares as compared to other domestic species. Follicles increase in size from about 25 mm to 45 mm over the last 6 days before ovulation (Hughes *et al.*, 1973). Alexander and Irvine (1987) found that both pulse rate (1.87 pulses/h) and amplitude of gonadotropin secretion accelerate on the day of ovulation (Fig. 4).

The largest follicle at the onset of luteolysis usually continues to enlarge and eventually becomes the primary, or first, ovulatory follicle (Pierson and Ginther, 1987) (Fig. 3). In one study, we found that 226/241 ovulations involved the primary follicle; only 15/241 primary ovulations involved the development of a second, and smaller, follicle that was present at the onset of estrus. The majority of ovulations occur 24 (45%) or 48 (32%) h before the end of estrus, with another 12% ovulating as early as 72 h before the end of estrus (Hughes *et al.*, 1973). While softening of the follicle occurred immediately before ovulation in only 28% of the cases, it may be most follicles soften before ovulation, but that 24-h palpation schedules by which data have been obtained do not allow this aspect of ovulation to be detected.

It is not unusual for large follicles, present at the time of ovulation of the primary follicle, to be ovulated either in conjunction with the primary follicle (within 24 h) or at some point during the luteal phase (Hughes *et al.*, 1984). Of 261 diestrous periods we studied, 64 (24%) involved luteal phase ovulations that ranged from 2 to 15 days from the initial occurrence of the primary ovulation (Hughes *et al.*, 1973). Corpora lutea from these luteal phase ovulations were normal as concerns their ability to secrete progesterone (Hughes *et al.*, 1984). These luteal-phase ovulations do not usually affect the length of the estrous cycle because $PGF_{2\alpha}$ synthesis and release occur at the normal time (Section II,F) and the newly formed corpus luteum is usually responsive to the luteolytic effects of $PGF_{2\alpha}$. A corpus luteum formed late in the luteal phase and less than 4 days old at the time of prostaglandin release may fail to regress and result in a prolonged luteal phase caused by the persistence of the the luteal phase corpus luteum. Conception is even possible during the luteal phase if an ovulating substance, such as human chorionic gonadotropin (hCG), is given and the mare is inseminated artificially (Hughes *et al.*, 1985).

We have found differences among mares as to their ovulatory rate: in one series, one animal had only 1/24 cycles with more than one ovulation, while the other extreme was one animal that had multiple ovulations in 25/34 cycles (Hughes *et al.*, 1973) (Fig. 5). The incidence of multiple ovulation in this study was 25.5% as compared to 0% (Arthur and Allen, 1972), 3.8% (Andrews and McKenzie, 1941), 10% (Wesson and Ginther, 1981), 14.5% (Osborne, 1966), 18.5% (Arthur, 1958), and 42.8% (Warszawski *et al.*, 1972) in other reports. Because of the difficulty in identifying multiple follicles by palpation per rectum, it is likely that the incidence of multiple ovulation is underestimated.

Luteotropin substances, such as hCG, have been used to advance the time of ovulation in the mare (Stabenfeldt and Hughes, 1987). Perhaps surprisingly, the administration of a $PGF_{2\alpha}$ analog during estrus also advances the time of ovulation (Savage and Liptrap, 1987). Utilizing the intercavernal sinus technique of Irvine and Alexander (1987), researchers found that

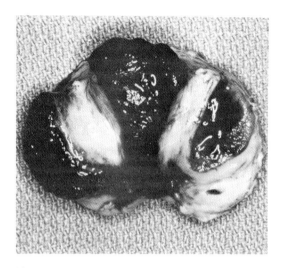

Figure 5 A cross-sectional view of the ovary of a mare 2 days after a double ovulation. Note the wedge-shaped corpora lutea at the left extremity and middle of the ovary with the constricted portion directed toward the ovulation fossa. A large follicle in cross section is on the right extremity. From Stabenfeldt and Hughes (1977).

the administration of a $PGF_{2\alpha}$ analog during the luteal phase of the estrous cycle produced an immediate increase in LH concentrations (in addition to producing luteolysis) (Jöchle *et al.*, 1987). The relationship between LH and ovulation remains somewhat uncertain, at least in part because luteal-phase ovulations occur at mid to late luteal phase at a time when LH concentrations are decreasing (Geschwind *et al.*, 1975).

The development of the fetal ovary in the horse is different from other domestic species because the cortex migrates ventrally to condense on the ventral surface of the ovary before invading the medulla (Mossman and Dukes, 1973; Walt *et al.*, 1979). While developing follicles distend the surface of the ovary during development, they only ovulate through the ventral surface of the

ovary, that is, through the ovulation fossa (Fig. 6).

F. The Corpus Luteum

The corpus hemorrhagicum, formed immediately after ovulation, often fills with blood in the mare by 10–12 h postovulation with redistension of the follicle to its preovulatory size. Progesterone secretion begins within 24 h of ovulation (Stabenfeldt *et al.*, 1975; van Rensburg and van Niekerk, 1968). Mature luteal activity is reached within 6–7 days postovulation. The normal lifespan of the corpus luteum in the nonpregnant mare is 14–15 days (Stabenfeldt *et al.*, 1972).

Regression of the corpus luteum is under control of $PGF_{2\alpha}$, which is released in pulsatile form from the uterus at about 14 days postovulation. The time for regression averages 35–40 h. The importance of the uterus in control of the corpus luteum has been shown through hysterectomy studies wherein persistence of the corpus luteum occurred in the absence of the uterus

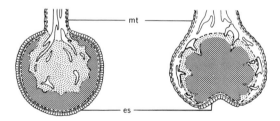

Figure 6 Diagrammatic comparison of the corticomedullary arrangement in the mature equine ovary (right) and other mammalian ovaries (left): es, epithelium superficiale; mt, mesothelium; close stipple, cortex; open stipple, medulla. Note the reversal of corticomedullary areas with only a small area of contact of the cortical area with the ovarian surface (ovulation fossa). From Mossman and Dukes (1973).

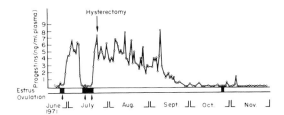

Figure 7 Progestin levels in the plasma of a mare during the estrous cycle before hysterectomy and following hysterectomy on day 4 after ovulation during the next cycle. Horizontal bars indicate the period of sexual receptivity. Vertical arrows indicate the occurrence of ovulation. From Stabenfeldt *et al.* (1974a).

(Ginther and First, 1971; Stabenfeldt *et al.*, 1974a) (Fig. 7).

It is common for mares to fail to regress the corpus luteum at the expected time of 14–15 days postovulation (Stabenfeldt *et al.*, 1974b). The problem is associated with inadequate synthesis and release of $PGF_{2\alpha}$ sufficient to cause regression of the corpus luteum (Neely *et al.*, 1979) (Fig. 8). Persistence of the luteal phase in nonpregnant mares represents a serious breeding problem in

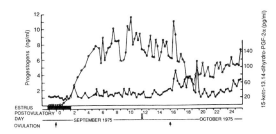

Figure 8 Peripheral plasma concentrations of the 15-keto-13,14-dihydro-$PGF_{2\alpha}$ (●) and progestogens (○) of a mare with a persistent luteal phase. Note that the increased $PGF_{2\alpha}$ secretion between days 16 and 20 postovulation did not result in luteolysis. From Neely *et al.* (1979).

that mares remain out of estrus for 2–3 months before the corpus luteum regresses (Stabenfeldt *et al.*, 1974, 1984). As many as 25% of estrous cycles can involve the development of a persistent corpus luteum (Hughes *et al.*, 1973). Acute inflammatory episodes within the endometrium due to the presence of bacteria can initiate synthesis and release of $PGF_{2\alpha}$ and premature regression of the corpus luteum (Neely *et al.*, 1979). Endogenous release of $PGF_{2\alpha}$ and regression of the corpus luteum can be induced through the infusion of acidic saline *in utero*, assuming the presence of a corpus luteum of at least 5 days duration at the time of infusion (Neely *et al.*, 1979; Pascoe *et al.*, 1989).

III. Pregnancy

After fertilization, the equine embryo is transported through the oviduct and arrives in the uterus on day 6 after ovulation. In the uterus, the blastocyst hatches and expands while maintaining its spherical shape. Until day 16 postovulation, the embryonic vesicle moves freely in the uterine lumen. This embryonic migration is believed to be an essential element of maternal recognition (Section III,B). Around day 16 postovulation, the diameter of the embryonic vesicle and the increased tone of the uterine musculature no longer allow free movement of the embryo in the uterine lumen. Fixation of the embryonic vesicle typically occurs at the base of one uterine horn, independent of the side of ovulation.

By day 36, the chorionic girdle cells invade the endometrium, where they develop into autonomous endocrine glands, the endometrial cups (Section III,C). From day 42 to 45, the chorionic membranes begin to establish a stable, microvillous contact with the adjacent endometrial epithelium. Gradually, the fetal sac starts to expand and elongate so that the entire endometrial surface is covered, and the allantochorion, the outside membrane of the placenta, is firmly attached to the endometrium by day 90–100 of pregnancy. Increasingly complex and branching chorionic villi protrude into the endometrial crypts and greatly enlarge the total area for placental exchange (Samuel *et al.*, 1975; Ginther, 1979).

During the first half of gestation, the fetus is surrounded by the amniotic membranes and remains free-floating in the allantoic fluids, attached only by a long umbilical cord (Whitwell, 1975). The twisting of the umbilical cord, which can be seen at birth, is a reflection of the frequent changes in position and orientation of the fetus that take place during the first half of pregnancy (Whitwell, 1982; Vandeplassche and Lauwers, 1986). In the last months of gestation the fetus can no longer change its orientation and remains in a dorsopubic position until parturition (Jeffcott and Rossdale, 1979). The mean length of gestation in the mare is 342 days, but individual gestation lengths can vary from 330 to 370 days without an apparent effect on the health of dam or foal. As a rule, the mare carries only one fetus, although the incidence of twin conception has been reported as high as 15% (Bowman, 1986). When a mare conceives twins, in most instances the twin pregnancy is either reduced through resorption or mummification of one dead fetus or ends prematurely in abortion of both fetuses (Jeffcott and Whitwell, 1973; Whitwell, 1980).

A. Endocrinology

1. Progesterone

In the mare, pregnancy is supported initially by the corpus luteum and later by the placenta. Following conception, maternal recognition of pregnancy prevents the release of luteolytic $PGF_{2\alpha}$ from the endometrium, and the primary corpus luteum of pregnancy continues to secrete progesterone beyond day 13–14. Progesterone secretion reaches a first peak around day 20 of gestation, then declines until day 40–50. Between day 40 and 90, secondary corporalutea develop and secrete progesterone, resulting in increased progesterone concentration (Holtan *et al.*, 1975) (Fig. 9).

After day 90, maternal progesterone concentrations decline progressively and reach low levels around day 150. The decline in progesterone concentrations coincides with the disappearance of the primary and secondary corpora lutea between day 120 and 150 of gestation (Squires and Ginther, 1975; Squires *et al.*, 1974; Holtan *et al.*, 1975). Progesterone concentrations remain low from the fifth to tenth month of gestation and increase again during the last month of gestation (Lovell *et al.*, 1975). Secondary corpora lutea, which are unique within the domestic species, develop from large follicles, by either ovulation or luteinization. It is postulated that the numerous follicles that develop during the first 3 months of pregnancy are the result of a continuation of maternal FSH secretion during that period (Evans and Irvine, 1975). Equine chorionic gonadotropin (eCG), secreted by the endometrial cups, promotes luteinization or ovulation of these large follicles. With the

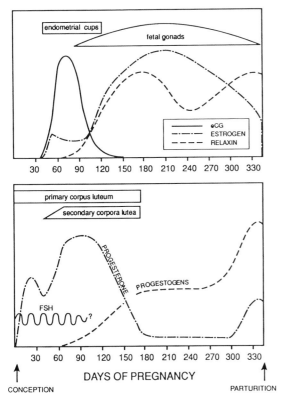

Figure 9 Summary of the temporal relationships among changes in hormonal concentrations and morphological changes throughout the gestation period of the mare.

onset of the winter anestrus season the 10-day FSH waves cease, and consequently fewer large follicles develop. This may explain why secondary corpora lutea are rare or absent in mares that conceive late in the breeding season. The large variability in both the number of secondary corpora lutea and the time at which the secondary corpora lutea develop casts doubt on their importance in the maintenance of pregnancy, especially since the equine pregnancy contin-

ues uneventfully in the absence of secondary corpora lutea (Allen *et al.*, 1987).

The placenta secretes progestogens, progesterone and progesterone metabolites, beginning around day 50 and is the sole source of progestogens after the corpora lutea have regressed (Ganjam and Kenney, 1975). The progestogens secreted by the placenta include the 5α-pregnane metabolites of progesterone, 20α-dihydroprogesterone and 17α-hydroxyprogesterone (Holtan *et al.*, 1975; Barnes *et al.*, 1975). Maternal ovariectomy can be performed as early as day 50 in some mares without loss of the fetus (Holtan *et al.*, 1979; Shideler *et al.*, 1981). Fetal gonadectomy in mid-gestation causes an immediate drop in maternal concentrations of estrogen but not progestogen, indicating that the equine placenta is capable of *de novo*-progestogen synthesis and does not need precursors from the fetal gonads (Pashen and Allen, 1979).

2. Estrogen

The equine embryo is capable of estrogen synthesis as early as day 7 after conception (Flood *et al.*, 1979; Heap *et al.*, 1982; Zavy *et al.*, 1984). Nevertheless, circulating estrogen concentrations during the first 30 days of gestation are similar to levels in nonpregnant mares during diestrus. Between days 35 and 40 of pregnancy, the concentration of conjugated estrogen increases two- to threefold in conjunction with the onset of eCG secretion by the endometrial cups (Kindahl *et al.*, 1982; Daels *et al.*, 1990a). Following this increase, estrogen secretion remains fairly constant through days 70–80, when a second rapid increase occurs (Fig. 9). Peak estrogen values are reached during the seventh and eighth month of gestation and decrease gradually toward term (Nett *et al.*, 1975; Stewart *et al.*, 1984).

Until day 70 of pregnancy, the maternal ovaries are the major source of estrogen. Following maternal ovariectomy or pharmacological elimination of the primary corpus luteum, circulating estrogen concentrations remain low until day 50–60, when placental estrogen synthesis gradually develops (Daels *et al.*, 1990a,b; P. F. Daels, personal communication). It is hypothesized that the rapid increase in estrogen secretion, observed around day 35–40, is the result of the stimulation of ovarian estrogen secretion by eCG.

After day 70, fetal estrogen production predominates. During the later half of gestation, two groups of estrogens are present: (1) estrone, estradiol, and their conjugated forms, and (2) the ring-B unsaturated estrogens, equilin and equilenin, which are unique to the horse. A true fetoplacental unit for estrogen production exists in the horse, with both the fetus and placenta being required for estrogen production (Pashen and Allen, 1979; Pashen *et al.*, 1982). The fetal gonads provide the steroid precursor, dehydroepiandrosterone, for estrogen synthesis by the placenta (Pashen, 1984). The gonads in the equine fetus, male or female, increase in size in parallel with the increase in fetal estrogen secretion (Section III,D).

The importance of the high estrogen concentrations during pregnancy is unclear. Transfer of equine embryos to ovariectomized mares maintained on progesterone only results in normal progression of pregnancy (Hinrichs *et al.*, 1986). Similarly, fetal gonadectomy in mid-gestation results in a dramatic decrease in estrogen secretion but does not terminate pregnancy (Pashen and Allen, 1979).

3. Equine Chorionic Gonadotropin

Cole and Hart (1930) first described the presence of a substance in serum of pregnant mares that stimulates ovarian activity in rats. Initially it was named "pregnant mare serum gonadotropin" (PMSG) but was later renamed "equine chorionic gonadotropin" (eCG), on the basis of its trophoblastic origin (Allen and Moor, 1972; Allen *et al.*, 1973b).

Equine chorionic gonadotropin is first detected in the maternal circulation between days 35 and 40 (Fig. 9). Concentrations increase rapidly and reach peak levels around day 70 (Ginther, 1979). Once peak concentrations have been reached, levels decline steadily and eCG is no longer detectable after day 150. The decline in eCG values coincides with the slow invasion of the endometrial cups by leukocytes.

Although eCG has both FSH- and LH-like activity in other domestic species, it appears to act primarily as a luteotropin in the pregnant mare. The first evidence that eCG had a profound LH effect was provided by Cole *et al.*, (1931), who observed that the formation of secondary corpora lutea coincided with the period of gestation when eCG was detectable in serum. In addition, Squires and Ginther (1975) observed that eCG maintained the primary corpus luteum as well as stimulated the formation of secondary corpora lutea but did not influence follicular development. These findings support the observation that in mares eCG does not have an FSH-like activity as is seen in rodents and cattle.

A number of factors have been found to govern the production of eCG: parity, size, number of fetuses present, and genotype of the fetus and its parents. Higher eCG levels were noted in serum of ponies as compared to larger breeds of horses (Cole, 1938). While age does not influence eCG production, parity is a factor. A reduction in eCG concentrations in successive pregnancies was noted in Shetland ponies but not in Thoroughbreds (Rowlands, 1963; Allen, 1969). Higher levels of eCG were found in the serum of mares carrying twin fetuses located in opposite horns, each with a set of endometrial cups (Rowlands, 1963). A mare carrying a mule fetus has much lower concentrations of the hormone than if she were carrying a horse fetus (Clegg *et al.*, 1962). Similarly, Manning *et al.* (1987) were able to attribute a portion of the variability in eCG concentrations to the dam or stallion and confirmed that fetal genotype influences eCG levels.

The endometrial cups secrete eCG seemingly independent from the conceptus and remain functional for about the same length of time, even after fetal death or surgical removal of the fetus (Allen, 1969; Mitchell and Betteridge, 1973).

4. Relaxin

In the mare, the placenta is the primary source of relaxin during pregnancy (Stewart *et al.*, 1982; Madej *et al.*, 1987). Circulating relaxin concentrations are first detected around day 70 of gestation and increase to a first peak in the fifth or sixth month (Fig. 9) (Stewart and Stabenfeldt, 1981). Closer to parturition, relaxin concentrations increase again to maximum values at birth. The time at which relaxin concentrations return to baseline is tightly correlated with expulsion of the placenta, with relaxin and prostaglandin levels remaining high in mares that retain the fetal membranes (Stewart *et al.*, 1984).

5. Oxytocin

Oxytocin secretion remains at basal levels throughout pregnancy and increases sharply only moments before fetal delivery (Allen *et al.*, 1973a; Haluska and Currie, 1988).

6. Prostaglandin

Prostaglandin $F_{2\alpha}$ is produced in increasing quantities in the last trimester of pregnancy in the mare, and constant high concentrations of 15-ketodihydro-$PGF_{2\alpha}$, a $PGF_{2\alpha}$ metabolite, are present during the 10 days preceding parturition (Haluska and Currie, 1988; Stewart *et al.*, 1984; Silver *et al.*, 1979; Barnes *et al.*, 1978).

B. Recognition of Pregnancy

Several researchers have demonstrated that only fertilized ova reach the uterus and will migrate past unfertilized ova, which are retained in the oviduct (Betteridge and Mitchell, 1974; Steffenhagen *et al.*, 1972; van Niekerk and Gerneke, 1966). These findings suggest that the equine oviduct is able to differentiate between fertilized and unfertilized ova.

Survival of the equine embryo beyond day 14 is dependent on the continued secretion of progesterone by the primary corpus luteum (Section III,A,1). Around day 14 after ovulation, the binding affinity of the equine corpus luteum for $PGF_{2\alpha}$ in pregnant mares is similar to or greater than that in nonpregnant mares and thus luteal maintenance during pregnancy is not associated with a reduction of $PGF_{2\alpha}$-binding capabilities. This emphasizes the importance of the inhibition of pulsatile secretion of $PGF_{2\alpha}$ by the endometrium in the pregnant mare (Vernon *et al.*, 1979). In the pregnant mare, the pulsatile secretion of $PGF_{2\alpha}$ that causes luteolysis in the nonpregnant mare around day 14 postovulation is absent (Kindahl *et al.*, 1982). In the pregnant pig, it has been proposed that $PGF_{2\alpha}$ release is redirected to the uterine lumen (exocrine secretion) beginning on day 14 postovulation (Baser and Thatcher, 1977). However, this mechanism has not been confirmed for the mare, instead, intraluminal $PGF_{2\alpha}$ concentrations in the pregnant uterus are depressed around day 14 after ovulation (Zavy *et al.*, 1984; Berglund *et al.*, 1982). Two mechanisms have been implied in the inhibition of pulsatile $PGF_{2\alpha}$ secretion by the endometrium of the pregnant mare: (1) secretion of an antiluteolytic substance by the embryo and (2) migration of the embryo in the uterine lumen. Co-incubation of equine embryonic membranes with endometrial explants *in vitro* results in a significant reduction of endometrial $PGF_{2\alpha}$ secretion (Berglund *et al.*, 1982). An embryonic secretory product of a molecular weight between 1000 and 12,000 has been implied in the inhibition of $PGF_{2\alpha}$ secretion (Weithenauer *et al.*, 1987). In contrast to other species, the equine conceptus does not elongate, but migrates extensively through the uterine lumen until day 17 postovulation. At the time of expected luteolysis, the migratory activity of the embryo is maximal, and it is believed that it is essential for the transmission of the antiluteolytic stimulus to the endometrium (Ginther, 1983, 1985; Leith and Ginther, 1984). When movement of the embryo is restricted to one uterine horn, luteolysis occurs at the time normally expected in cyclic mares (McDowell *et al.*, 1988).

C. Endometrial Cups

The formation of the endometrial cups is unique to the horse, and is unusual in that embryonic cells detach themselves, invade the endometrium, and temporarily establish isolated endocrine glands that secrete eCG (Allen and Moor, 1972; Allen *et al.*, 1973b; Hamilton *et al.*, 1973).

The reader is referred to Chapter 9 for a detailed discussion of the formation of the endometrial cups. By day 36, the chorionic girdle cells have developed into a discrete, pale band that surrounds the spherical conceptus in the region where the membranes of the enlarging allantoic and the regressing yolk sac adjoin (Allen *et al.*, 1973b). This chorionic girdle consists of fingerlike projections of rapidly multiplying trophoblast cells, which separate themselves from the fetal membranes and invade the underlying maternal endometrium between day 36 and 38. The cells remain clumped together to form the endometrial cups, which are seen as a series of ulcerlike endometrial protuberances arranged in a circle around the conceptus. The endometrial cups have a lifespan of 60 to 100 days, throughout which increasing numbers of leukocytes accumulate in the surrounding endometrial tissues. After day 80, the leukocytes begin to invade the cup tissue and destroy individual cells. Eventually, between day 100 and 140 of gestation, the whole necrotic cup is sloughed from the surface of the endometrium.

Despite the intimate epitheliochorial contact between the remainder of the fetal membranes and the endometrium, no accumulation of leukocytes occurs at the trophoblast–endometrium interface (Allen, 1979). Allen and co-workers (1987) have suggested that the endometrial cup reaction provides an antigenic stimulus that results in an immunoprotective response in the mare toward her fetus *in utero*. The transfer of extraspecific donkey embryos into mares has given some evidence of the immunosuppressive role of the endometrial cups. In this instance the donkey chorionic girdle fails completely to invade the endometrium of the surrogate mare and no endometrial cups are formed. Between day 80 and 95, leukocytes that have been accumulating in the endometrial tissue for several weeks begin to pass through the endometrial epithelium and actively attack the donkey trophoblast (Allen, 1982). Administration of exogenous progestogen or eCG is unable to prevent abortion of the donkey fetus (Allen *et al.*, 1987). However, immunization of the mares with lymphocytes recovered from the genetic sire and dam of the donkey fetus they were carrying resulted in the survival of four out of six fetuses. The success of lymphocyte immunization in raising fetal survival rate in donkey-in-horse pregnancy supports the hypothesis that the endometrial cups are required for immune recognition and development of an immunotolerance toward the developing fetus (Allen *et al.*, 1987).

In conclusion, the endometrial cups have two different functions during pregnancy: (1) secretion of a luteotrophic agent, eCG, and (2) induction of maternal immunotolerance toward the antigenic foreign fetus.

D. Fetal Gonads

The gonads of both the male and female equine fetus become greatly enlarged between the third and ninth month of gestation (Douglas and Ginther, 1975). They reach their maximal size around day 200 of gestation, when they are significantly larger than the maternal ovaries. After day 250 the fetal gonads decrease in size, and at birth they are approximately one-tenth of their maximal size during fetal life.

The enlargement of the fetal gonads involves a proliferation of the interstitial cells (Gonzales-Angul *et al.*, 1971; Hay and Allen, 1975). The interstitial cells are morphologically similar to luteal cells and contain all the organelles normally associated with steroid synthesis. After day 250, the degenerative changes in the interstitial cells account for the progressive decrease in size of the gonads, which parallels the decline in estrogen concentrations.

IV. Parturition

A. Signs

Distention of the udder begins about 3–4 weeks before parturition, and the major increase in size occurs in the last 2 weeks of pregnancy (Arthur, 1975; Rossdale and Ricketts, 1980). The teats usually distend with colostrum 24–48 h before foaling, and small amounts of colostrum may escape through the teat orifice, resulting in the formation of waxy material on the ends of the teat (waxing). Waxing may be seen as early as 4 days before foaling, and some mares may be dripping milk for a week or more prior to foaling (Rossdale and Ricketts, 1980). Normally the mammary secretion changes from a thick, tenacious, straw-colored material to a cloudy grey, slightly thinner secretion and then to a yellowish-white viscous consistency at foaling

(Rossdale and Ricketts, 1980). Analysis of preparturient mammary secretion has demonstrated a rise in total calcium concentration to above 10 mmol/l when foaling was imminent (Peaker *et al.*, 1979). Changes in calcium and magnesium concentrations have been used to predict the time of parturition (Ousey *et al.*, 1989). Other signs of impending parturition include relaxation of the sacrosciatic ligaments, vulvar edema, sweating, restlessness, pawing, switching the tail, and intermittent lying down and standing (Barty, 1974).

The mare seems able to exert some control over the time of day that parturition takes place, since the majority of foals are born during the night when barn activities are at a minimum (Tram, 1947; Rossdale and Short, 1967; Jeffcott, 1972; Bain and Howey, 1975).

B. Delivery

In the mare no visible signs of straining are observed during first-stage labor. Signs indicative of first-stage labor include sweating, increasing restlessness, lying down and getting up, stretching as if to urinate, and looking at the flank. These signs may be observed as early as 4 h before birth of the foal. Sometimes signs of first-stage labor are interrupted and foaling may be delayed for several hours or days (Rossdale and Ricketts, 1980). During first-stage labor the fetus begins to rotate itself from its normal dorsopubic position to the dorsosacral position (Jeffcott and Rossdale, 1979), and the lying down and rolling movements of the mare during first stage labor are believed to play a role in the positioning of the foal.

Second-stage labor begins with the rupture of the chorioallantoic membrane. As uterine contractions increase the cervix dilates and the foal passes into the birth canal, initiating the forceful abdominal contractions necessary for expulsion (Ferguson's reflex). The foal is presented at the vulva with both front legs and head extended and is often delivered covered by the amniotic membranes. The actual time of delivery is very short, often accomplished within 10–15 min once the foal has entered the birth canal. The umbilical cord of the fetus is long and usually still intact after birth. The mare normally remains recumbent after delivery, with mare and foal connected by the umbilical cord for 5 to 10 min after birth.

During third-stage labor visible straining ceases but peristaltic waves of myometrial contraction continue, resulting in separation and expulsion of the fetal membranes, generally within 30 min after birth. In the mare, retention of the fetal membranes for more than a few hours can cause systemic illness, endotoxic shock, and laminitis, and is considered a high risk that needs prompt veterinary attention.

C. Hormonal Events

The mare does not have a withdrawal of progesterone prior to parturition but delivers in the presence of high circulating progesterone concentrations (Lovell *et al.*, 1975). Nevertheless, progesterone concentrations decrease significantly and the estradiol-17β to progesterone ratio increases in the last 24 h prepartum (Haluska *et al.*, 1987; Haluska, 1989). Total estrogens gradually decrease from early mid-pregnancy values until parturition (Nett *et al.*, 1973), and there is no consistent change in the level of estradiol-17β during the last week prepartum (Haluska *et al.*, 1987). Relaxin secretion is elevated during the last half of pregnancy, increases in the hours before parturition, reaching a peak immediately after delivery, and ends with the expulsion of the placenta (Stewart *et al.*, 1982). Oxytocin and prostaglandin $F_{2\alpha}$ increase abruptly at the end of second-stage labor (Stewart *et al.*, 1984; Haluska and Currie, 1988). Approximately 2 h before delivery, 15-ketodihydro-$PGF_{2\alpha}$ concentrations increase and maximum concentrations are reached between the time of rupture of the chorioallantois and the completion of delivery. Following delivery, concentrations decline rapidly and occasionally $PGF_{2\alpha}$ peaks occur during the first days postpartum.

During delivery, oxytocin and $PGF_{2\alpha}$ stimulate the powerful uterine contractions that result in expulsion of the foal and placenta. In sheep, goat, and cow, the increase in $PGF_{2\alpha}$ secretion and the subsequent increase in uterine activity appear to be preceded by a surge of estrogen production (Hunter, 1980). No such estrogen surge occurs in the mare, and in fact estrogen concentrations are high in the mare from day 80 through term. The experiments that highlighted the role of the fetal gonads in estrogen production give the best indication of the role of estrogen at term. Fetal gonadectomy between 200 and 250 days of gestation does not necessarily terminate pregnancy but results in very low estrogen concentrations in the dam (Pashen and Allen, 1979). Although parturition commenced spontaneously in the mares carrying gonadectomized fetuses, the pattern of labor was abnormal and uterine contractions were weak and inadequate for complete expulsion of the foal. Little or no $PGF_{2\alpha}$ secretion occurred during these deliveries, suggesting that estrogen priming of uterine and placental tissues is essential for normal $PGF_{2\alpha}$ synthesis during parturition.

Closely timed blood samples taken during parturition show that the increase in oxytocin secretion precedes the $PGF_{2\alpha}$ surge and suggest that oxytocin may be the trigger for the rapid increase in $PGF_{2\alpha}$ secretion (Haluska and Currie, 1988). This is consistent with the rapid increase of $PGF_{2\alpha}$ synthesis that occurs after oxytocin administration during induced parturition (Pashen, 1980; Stewart *et al.*, 1984). It would seem that the increased secretion of oxytocin by the maternal pituitary is the trigger for second-stage labor. The maternal oxytocin secretion is under central control, and factors such as stress can decrease oxytocin secretion. This central control could be the reason the mare seems to have an unusually high degree of maternal control over the time of delivery: more than 70% foal at night, with the delivery-triggering release of oxytocin only occurring in stress-free conditions (Haluska and Currie, 1988).

D. Induction of Parturition

Elective induction of parturition has become a practical tool in the management of the brood mare. Because most mares foal at night, induction of parturition during the day has helped to assure that professional help is present or available. Although induction of parturition has been used successfully, it is critical to carefully select those mares eligible to be induced. It is imperative that the fetus is sufficiently mature to survive in the extrauterine environment. Because of the great variability in gestation length in the mare, it is impossible to arbitrarily set a gestation length at which induction can be done. To assure a mature foal a minimum gestation period of 330 days is used in combination with other criteria, including an enlarged udder, distended teat filled with yellow-white colostrum, relaxed sacrosciatic ligaments, and cervical relaxation.

1. Oxytocin

Oxytocin was used in the first reports of induction of parturition in the mare and is still the most widely used drug for this purpose (Britton, 1963). Oxytocin can be administered by intramuscular or intravenous injection of a small dose or by slow intravenous infusion. The time and intensity of the appearance of foaling signs are dose dependent. Within minutes after the administration of a low dose of oxytocin (10–40 IU), $PGF_{2\alpha}$ secretion increases dramatically, signs of second-stage labor occur within 15 min, and delivery is competed within 45–90 min (Pashen, 1980; Stewart et al., 1984).

2. Prostaglandin

The mare does not possess an active corpus luteum in her ovaries at term, and attempts to induce foaling by injection of the natural compound $PGF_{2\alpha}$ have not been successful (Alm et al., 1975). Intramuscular injection of a synthetic prostaglandin analog, such as fluprostenol, which has a longer half-life time, can induce an apparent normal birth within 1–3 h (Rossdale et al., 1979). When the $PGF_{2\alpha}$ analog was administered to mares not ready to foal (less than 344 days pregnant), none of the mares delivered as a result of the treatment.

3. Glucocorticoid

It has been reported that repeated massive doses of dexamethasone will induce parturition in the mare (Alm et al., 1974; Alm et al., 1975; First and Alm, 1977). However, postpartum complications (i.e. retained placenta and death of the foal), together with the inconvenience of the repeated dosing (100 mg/day for 4 days) and the delay between treatment and delivery has rendered this technique unpractical.

V. Postpartum Period

Among the domestic mammals, the mare has a remarkable ability to quickly restore normal cyclicity and fertility following parturition (Matthews et al., 1967). Following months of ovarian quiescence during pregnancy, the maternal ovaries start to develop follicles immediately after parturition. By day 6 postpartum a preovulatory follicle is present and the mare shows signs of estrus ("foal heat"). Most commonly, mares ovulate between days 7 and 12 after parturition, with 93% of mares having ovulated by day 15. After this foal-heat ovulation most mares resume normal cyclic activity (Loy, 1980). Lactational anestrus has not been recognized in the mare, and postpartum anestrus is rare.

Conception rate following breeding during foal heat is only slightly lower than during subsequent cycles. In the mare, involution, or restoration, of the postpartum uterus is achieved within 10–15 days. Within 30–90 min after delivery of the foal, the placenta is shed; retention of the fetal membranes for more than 6 h is considered pathological and can affect foal-heat fertility. Vulvar discharge is prominent during the first 48 h and may persist for several days. Until the first ovulation the cervix remains soft and open, facilitating the elimination of the uterine content. In normal mares, uterine diameter and endometrial morphology are similar to the pregravid uterus by 15 days postpartum. Taking into account the time the embryo remains in the oviduct (6 days), an embryo conceived on foal heat arrives in a relatively normal uterine environment, approximately 13–18 days postpartum.

Because of the relatively long gestation period, it is critical to have the mare conceive as soon as possible after foaling if she is to produce one foal per year. It has been reported that conception rate and live foal rate are slightly lower for foal-heat conceptions than for pregnancies conceived during subsequent cycles (Loy, 1980; Merkt and Gunzel, 1979). However, it is well accepted that breeding on foal heat will result in a decrease in average foaling interval and has a positive effect on the number of mares pregnant each year and on the birth date of the foals, that is, as close as possible to January 1 (Loy, 1980).

VI. Reproductive Physiology of the Stallion

The stallion represents 50% of the genetic makeup for his offspring and thus is critical to the final product of any particular mating. Since the stallion's potential to be mated to many mares exists, his overall contribution to the genetic pool is greater than an individual mare. Like the mare, however, his reproductive activity often is based on criteria other than fertility (i.e., performance, bloodlines, conformation, and color).

A. Male Reproductive System

The male reproductive organs include two testes located in paired pouches between the thighs and suspended by the spermatic cord and cremaster muscle; two epididymides consisting of a head, body, and tail; two deferent ducts (vas deferens) ending in enlarged tubular structures, the ampullae; paired seminal vesicles; a single prostate gland; paired bulbo-urethral glands, and the penis (Pickett *et al.*, 1989; Roberts, 1986).

The spermatozoa produced in the seminiferous tubules of the testes are not capable of fertilization until they undergo maturation and become motile in the head and body of the epididymis. The mature sperm are mainly stored in the tail of the epididymis with additional storage in the deferent ducts until they are ejaculated through the penis (Pickett *et al.*, 1989).

1. Testes

The testes are present within the scrotum of most newborn colts within 14 days after birth (Bergin *et al.*, 1970). When one or both testicles are retained within the inguinal canal or abdomen, the animal is termed a cryptorchid. Many consider true cryptorchids to be those in which the testicle (testes) are retained within the abdomen. Horses are sometimes presented exhibiting male sexual activity when supposedly castrated. These animals have neither visable nor palpable testes. It is important to differentiate the castrated from the cryptorchid animal if surgical removal of the retained testicle is contemplated. About 40–50% of the suspected cryptorchids are castrated animals (Cox *et al.*, 1973; Stabenfeldt and Hughes, 1987). A single testosterone analysis of plasma is usually sufficient to differentiate these conditions. Castrated animals have less than 40 pg/ml of testosterone and cryptorchids have greater than 100 pg/ml. Preparations with LH activity, hCG, induce a rise in testosterone secretion in the cryptorchid animal—that is, about a 5% increase in testosterone concentration—and this can be used as a criterion for diagnosis (Cox *et al.*, 1973; Stabenfeldt and Hughes, 1987).

The testes are oval in shape (7.5–12.5 cm long × 4–7 cm dorsoventral × 5 cm wide), with a slight compression from side to side. The long axis of the unretracted testicle is almost horizontal, with the head of the epididymis anterior, the body dorsal, and the tail in a posterior position. The deferent duct continues from the tail of the epididymis through the spermatic canal, widening into the ampulla before emptying into the pelvic urethra (Roberts, 1986; Rossdale and Ricketts, 1980).

Puberty is attained around 90 weeks of age, but only about 50 million spermatozoa with ≥10% motility are present in an ejaculate. Ejaculates of a 2-year-old may contain 3.3 billion spermatozoa, but the percentages of morphologically normal and motile spermatozoa are low (Pickett *et al.*, 1989).

The testes continue to grow beyond 5 years of age, with the testes of 7-year-olds being larger than those of younger stallions (Nishikawa, 1959). The total scrotal width is an important consideration in evaluating a stallion's potential fertility. Colorado workers found the average scrotal width over all breeds, ages, and seasons to be 108 mm and a value of 80 mm or less to be unacceptable for a sound breeder (Pickett *et al.*, 1989).

2. Accessory Sex Organs

As indicated, the epididymis is concerned with maturation, nutrition, transportation, and storage of the spermatozoa. Within the tail of the epididymides of the mature stallion are stored approximately 54 billion spermatozoa with potentially normal fertilizing capability (Amann *et al.*, 1979). Secretions from the epididymis are high in glycerylphosphorylcholine (Mann, 1969).

The ampullary secretions are characterized by the presence of ergothioneine and the absence of fructose (Mann, 1969). The ampullae are palpable rectally above the neck of the bladder and between the seminal vesicles. They are quite large, being 15–20 cm in length and approximately 2–$2\frac{1}{2}$ cm in diameter (Roberts, 1986).

The vesicular glands (seminal vesicles) are smooth, hollow structures 15–20 cm long and 2.5–5 cm in diameter that diverge from the base of the bladder along the floor of the pelvis on either side of the ampulla. The secretion from the seminal vesicles is gelatinous in texture and has a high concentration of citric and lactic acids (Mann, 1969).

The prostate is a lobulated structure with two lobes and a connecting isthmus located on the floor of the pelvis. They can be palpated with some difficulty per rectum. The prostate secretes a thin watery secretion, which is thought to cleanse the urethra before ejaculation while also contributing to the seminal plasma. The paired bulbourethral glands (Cowper's glands) can be found near the ischial arch. Not accessible to palpation per rectum, their secretion contributes to the seminal plasma.

It is tempting to speculate that the biochemical components of the seminal plasma from the epididymis, prostate, bulbourethral glands, and seminal vesicles could be analyzed and values related to fertility or freezability of the semen. Presently, while these glands contribute most of the fluid of the seminal plasma, they are not considered necessary for normal fertility of spermatozoa (Pickett *et al.*, 1989).

B. Reproductive Activity

Mares are recognized as seasonal breeders, but this fact is often overlooked in stallions since they will mate with mares at any time of the year. Total seminal volume, sperm concentration, and total number of spermatozoa are, however, lowest in January and February, and increase gradually in the spring to peak in June. Volume increases approximately 40% from low months to high months and sperm output approximately 50% during the same period (Pickett *et al.*, 1989; Johnson and Neaves, 1981).

Good libido or desire to mate with a mare is an essential ingredient to reproductive efficiency in the stallion. The stallion should be observed in the presence of a mare in estrus for premounting libido, erection of the penis, mounting, exploratory movements with the penis for the vulva, copulatory movements, "flagging" (rhythmic contractions of the tail accompanying ejaculation), ejaculation, relaxation of the penis, dismounting, and shrinkage of the glans (Bielanski, 1972). The main abnormalities one looks for are lack of libido with long periods before an erection is obtained and mounting of the mare occurs, dismounting several times before ejaculation occurs, failure of ejaculation, failure to maintain an erection, failure to complete intromission, and dismounting just as ejaculation begins.

Conditioned stallions may become aroused with the first preparatory movements to bring them out of the stall; however, sexual excitement usually begins when the stallion first sees the mare. Erection time ranged from 119 to 163 s and mounting reflex from 101 to 206 s in one study (Wierzgowski and Hafez, 1961). The time from sighting the mare until beginning copulation ranged from 3.4 min in August to a high of 13.6 min in January in another study (Pickett *et al.*, 1989). Following intromission, ejaculation usually occurs in about

13 s, while the duration of ejaculation averages 7.6 s (Tischner *et al.*, 1974). There is little difference in the mounts required per ejaculate between natural mating (1.4) and when an artificial vagina is used (1.3) (Asbury and Hughes, 1964).

Ejaculation, the expulsion of semen throughout the urethra, occurs after an average of seven (5–11) intravaginal thrusts in the form of eight (5–10) seminal jets. The first three jets contain 80% of the ejaculated spermatozoa and are emitted under high pressure (Tischner *et al.*, 1974). The remaining semen is expelled under lower pressure and declining erection as the penis is withdrawn.

C. Spermatogenesis

Spermatogenesis is a specialized process of cell division and cellular change that result in spermatozoa being produced from spermatogonia. The germ cells of the seminiferous epithelium (spermatogonia, primary spermatocytes, secondary spermatocytes, and spermatids) have well-defined associations that succeed one another in a cyclic pattern (Monesi, 1972; Swierstra *et al.*, 1974). Each cycle of the seminiferous epithelium requires 12.2 days, and the duration of spermatogenesis is about 57 days (Pickett *et al.*, 1989). Transit time of spermatozoa through the head and body of the epididymis is about 4–8 days, and this time cannot be hastened by ejaculation (Swierstra *et al.*, 1974). Thus, approximately 61–65 days is required to develop spermatozoa to the point of ejaculation. Spermatozoa are produced continuously, and those not ejaculated in inactive stallions are probably voided with urination. One gram of testicular parenchyma during the breeding season produces approximately 19 million spermatozoa per day, and 15 million in the nonbreeding season (Pickett *et al.*, 1989).

Sertoli cells along with germ cells constitute the seminiferous epithelium, while Leydig cells are interspersed within the connective tissue between the seminiferous tubules. Sertoli cells provide nourishment and support for the development of germ cells to spermatozoa (Johnson and Nguyen, 1986).

Leydig cells produce androgens, mainly testosterone, under the control of luteinizing hormone (LH) from the anterior pituitary gland. Follicle-stimulating hormone from the anterior pituitary is thought to act on Sertoli cells to produce inhibin and activin and perhaps other protein hormones (Pickett *et al.*, 1989). Little is known about these hormones in the stallion, however, and their role is still speculative. Sertoli cells and Leydig cells increase in number in the testes with age and exhibit seasonal variation–that is, their numbers are significantly greater in the breeding season (Johnson and Nguyen, 1986; Johnson and Thompson, 1983).

The stallion testes are unique in the high concentrations of estrogens produced. The source of estrogen production is unknown (Amann and Ganjam, 1981; Gaillard and Silberzahn, 1987).

Under the influence of gonadotropin-releasing hormone from the hypothalamus, the anterior pituitary produces LH and FSH, which stimulate the testes to produce spermatozoa as well as testosterone and estrogen. In a study by M. Seamans and J. Roser (personal communication) cannulation of the testicular and jugular veins showed LH and FSH levels to be similar in the two vascular systems but the concentration of testosterone was 100-fold and estrogen conjugates 23-fold higher in the testicular circulation.

Subfertile breeding stallions are a dilemma that has concerned owners and managers of these animals for many years. A great deal of research is being conducted to determine if GnRH pulses delivered over a long term are effective in treating subfertile stallions with endocrine dysfunction. While abnormal hormone profiles have been observed in stallions, the role of LH, FSH, estrogen, and inhibin in causing poor seminal quality and/or poor conception rates is unknown.

R. Whitecomb and J. Roser (personal communication, 1990) have recently identified a circulating LH antagonist from infertile stallions that may shed light on one cause for infertility.

D. Fertility Evaluation

There is no single test that can serve as an absolute indicator of fertility for the stallion. Live foals produced under good management conditions from the mares mated are the final measure of fertility. This is influenced by the breeding soundness of mares mated, conception rates per cycle, cycles per pregnancy, and pregnancy rate by month and by cycles (Kenney *et al.*, 1983; Pickett *et al.*, 1989). A marginally fertile stallion may by careful management and selection of mares obtain a pregnancy rate by the end of the breeding season equal to a more fertile stallion even though the latter has a much higher pregnancy rate per cycle.

Evaluation of fertility must take into consideration season of the year, age, conformation, condition, libido, health, mating behavior, and past reproductive history. Testicular size is a general guide to numbers of spermatozoa produced, and therefore total scrotal width and consistency of the testes is an important characteristic to be evaluated (Kenney *et al.*, 1983; Pickett *et al.*, 1989).

No judgement of potential fertility is complete without evaluating seminal characteristics. Sperm production is influenced by the season of the year, size of the testicle, frequency of ejaculation, age of the animal, and status of the testicle (normal/abnormal) (Bielanski, 1972; Kenney *et al.*, 1983; Nishikawa, 1959; Pickett *et al.*, 1989). Sperm concentration, volume, total spermatozoa, and total scrotal width are all affected by season, being greater in the physiologic breeding season versus the nonbreeding season. Progressive motility and morphology of the spermatozoa is not influenced by season. Semen is usually collected with a suitable artificial vagina. Dismount samples are not suitable for evaluating semen quality (Hurtgen, 1987). Semen characteristics that are usually measured include (1) volume (gel, gel-free, and total); (2) concentration of spermatozoa per milliliter of gel-free semen; (3) total spermatozoa (concentration × volume); (4) progressive motility (%); (5) morphology (normal and abnormal forms); (6) pH; (7) presence of white blood cells, red blood cells, and immature spermatozoa; (8) longevity of the motility of raw semen at 22°C; and (9) bacteria (Hurtgen, 1987; Kenney *et al.*, 1983; Pickett *et al.*, 1989).

Examination of the semen is conducted after 4–7 days of sexual rest on (1) two ejaculates, 1 h apart, or (2) two ejaculates collected 1 h apart and then a single ejaculate daily for 6 days. The latter technique will deplete extragonadal sperm reserves and will determine daily sperm output (Pickett *et al.*, 1989). Most fertile stallions have semen with (1) sperm in average to high numbers (6×10^9 or greater); (2) a high percentage of sperm with progessive motility (50% or more); (3) sperm with a low proportion of abnormal forms (less than 10% affecting the head or midpiece); (4) pH of 7.35–7.8; and (5) a volume of approximately 45 ml but great variability.

While the volume of the two ejaculates collected 1 h apart should be approximately equal, the second ejaculate contains about half the number of spermatozoa found in the first (Pickett *et al.*, 1989).

Many stallions with normal fertility harbor bacteria within the urethra, urethral diverticulum, or prepuce without producing recognizable disease (Hughes *et al.*, 1967; Pickett *et al.*, 1989). These stallions are a potential source of bacterial contamination to susceptible mares; however, most mares are unaffected by the bacteria introduced at copulation. Bacteria of special significance, when repeatedly isolated in significant numbers are *Pseudomonas aeruginosa*, *Klebsiella pneumoniae* (capsule types 1 and 5), and β-hemolytic streptococcus. If mares become infected after mating to a stallion and the same organism is isolated from the genitalia of both, it must be considered significant.

Blood in the semen (hemospermia) affects fertility according to the quantity of blood (Hurtgen, 1987). Hemospermia is most often associated with bacterial urethritis (Pickett *et al.*, 1989).

Stallions sometimes urinate along with ejaculation, usually detected by the yellow color, odor, and presence of crystals. Fertility is affected according to the amount of urine passed in the semen sample. When artificial insemination is practiced, the semen can be fractionated by jets and the urine-contaminated portion discarded.

E. Artificial Insemination

Artificial insemination (AI) for domestic animals was first used successfully in the mare (Bowen, 1969). Artificial insemination in the mare is a simple procedure, which will result in conception rates equal to those obtained with natural mating and superior to natural mating when dealing with mares susceptible to bacterial infection (Hughes and Loy, 1970; Kenney *et al.*, 1983; Pickett *et al.*, 1989). Except for Thoroughbreds, most breeds in the United States accept AI. Semen is frequently extended with skim milk or cream gelatin (Hughes and Loy, 1970). Freezing of stallion semen is becoming increasingly popular in many countries. It is utilized in China more extensively than in any other country of the world (110,000 mares inseminated with frozen semen from 1980 through 1985) (Pickett *et al.*, 1989).

VII. Genetic Aspects of Reproduction

A. *Normal Chromosome Complement*

Chromosome preparations can be made from various tissues. In the horse, peripheral blood for lymphocyte culture is easily obtained and excellent karyograms can be produced. The lymphocytes are purified by density-gradient procedures before culturing (Taylor and Smith, 1976). Cells undergoing division are then arrested in metaphase, swollen in hypotonic solution, fixed, and stained (Hungerford, 1965). Photographs of the chromosomes are used to arrange the chromosomal pairs according to their morphological features. Chromosomes are now more accurately paired using banding techniques (Fig. 10) (Taylor and Smith, 1976; Bowling *et al.*, 1987). The normal chromosome complement of the horse

is 64, the donkey 62, and the mule or hinny 63 (Benirschke *et al.*, 1962).

B. *Equine Hybrids*

The most common equine hybrid is the mule (donkey male × horse female). The reciprocal cross is called a hinny (horse male × donkey female). Zebras can also be crossed with horses or donkeys to produce hybrids. Hybrid offspring have chromosome numbers intermediate between those of their two parents (Benirschke *et al.*, 1962; Short, 1972). The sterility of male hybrids is due to a block in spermatogenesis during meiosis because maternal and paternal chromosomes are dissimilar in size and number, making true pairing impossible. Thus while the testes of mule hybrids secrete testosterone, they do not form spermatozoa (Short, 1972). Female mules come into estrus and ovulate on an irregular basis (Bielanski, 1965; Nishikawa, 1959). Female mules have been reported to produce offspring, but most of these observations are not proven and not confirmed by chromosome analysis (Benirschke *et al.*, 1962; Bielanski, 1955; Short, 1972). Chromosomes of an alleged mule who foaled were like those of a donkey, in spite of some mule-like characteristics of the animal (Benirschke *et al.*, 1964). According to recent findings, a female mule, substantiated by chromosomal analysis as having 63 chromosomes, produced a male foal possessing 63 chromosomes with banding patterns characteristic of a male mule, when mated to a male donkey (Ryder *et al.*, 1985). Blood typing procedures verified that the male mule qualified as the offspring of the male donkey and the female mule.

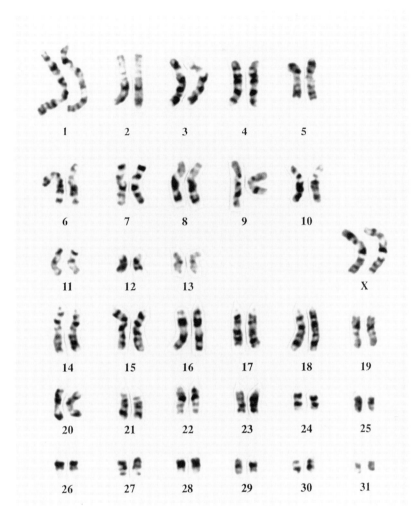

Figure 10 G-Banded karyotype from peripheral blood lymphocytes of a mare demonstrating a normal sex chromosome constitution, 62,XX.

C. Chromosomal Errors

1. Gonadal Dysgenesis

There continues to be more and more documentation of chromosomal abnormalities in infertile mares. The most commonly reported abnormality is 63,X where one of the sex chromosomes is lacking. This condition is similar to female gonadal dysgenesis in humans (Turner's syndrome). In addition to 63,X, various reports of 63,X/ 64,XX;63,X/64,XY;65,XXX and 63,X/

64,XY/65,XXY plus structural abnormality of one X chromosome (64,X,de) (XP) (Fig. 11) and autosomal trisomy along with XY testicular feminization and XY gonadal dysgenesis appear in the literature (Blue *et al.*, 1978; Trommershausen-Smith *et al.*, 1979; Power, 1986; Bowling *et al.*, 1987; Power, 1987).

Mares with gonadal dysgenesis are phenotypic females with small inactive ovaries.

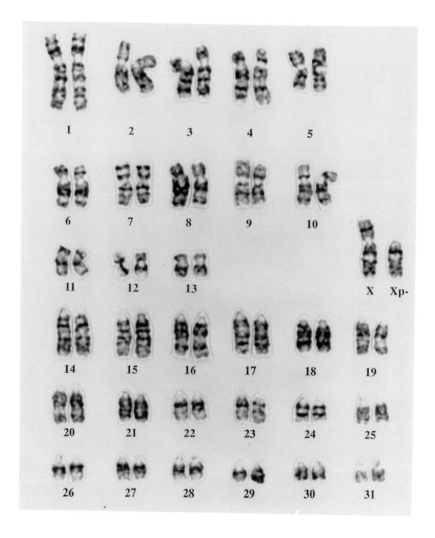

Figure 11 G-Banded karyotype from peripheral blood lymphocytes of an Arabian mare demonstrating a normal X chromosome and a second X with the short arms deleted, 64,X,del(XP).

The ovaries usually lack germ cells and consist primarily of undifferentiated ovarian stroma. These mares present problems due to chronic infertility and failure to cycle on a regular basis. The ovaries are very small (0.5 × 0.5 × 1 cm to 1 cm × 3 cm) and lack follicular activity. The cervix and uterus are flaccid on palpation per rectum. Most animals are small in size (Hughes *et al.*, 1975).

Bowling *et al.* (1987) reported on 180 mares with gonadal dysgenesis (Table I). Thirty-one percent of these mares were 63,X. The second most frequent abnormality was 64,XY (12.2%). The sex chromosomes of these phenotypic females were indistinguishable from that of a male horse (Fig. 12). The gonads on rectal palpation were small and inactive and thus indistinguishable from those of the mare with 63,X gonadal dysgenesis. Testicular feminization, another form of XY sex reversal in the mare, is very uncommon compared to XY sex reversal with gonadal dysgenesis. These mares are phenotypic females with abdominal testes (Power, 1986; Bowling *et al.*, 1987). They lack a cervix and uterus and may display masculine behavior. The fundamental defect is a lack of or decrease in receptor sites for androgens, and therefore male differentiation is not completed (Power, 1986). Its genetic basis is considered to be an X-linked *Tfm* mutant gene.

A stallion, according to his stud book record, sired 65 males and 150 females. Sixty-seven of his daughters were karyotyped and 22 (32.8%) were 64,XY. Karyotyping was requested due to infertility in some of the phenotypically female offspring. This condition does not fit a pattern of X-linked inheritance, but rather a familial hypothesis must be considered (Bowling *et al.*, 1987).

It is important to keep in mind that not all mares with small ovaries are abnormal and have chromosomal abnormalities. It requires a chromosome analysis to make the final judgment. It is imperative, however, to perform a chromosome analysis on mares with small inactive ovaries and a flaccid

Table I
Primary Infertility in 180 Mares Aged 3 Years and Older

Karyotype	Breed[a]									TOTAL
	AP	AR	MH	PP	QH	ST	TB	TWH	XX	
63,X	0	20	1	2	12	3	16	1	1	56
63,X/64,XX	1	3	0	0	0	1	1	0	0	6
63,X/64,XY	0	2	0	0	0	0	2	0	0	4
65,XXX	0	1	0	0	0	1	3	0	0	5
64,XX/65,XXY	0	0	0	0	1	0	1	0	1	3
64,X,del(Xp)	0	1	0	0	1	0	0	0	0	2
64,XY	1	5	2	0	6	0	7	0	1	22
64,XX	2	41	2	1	13	4	14	1	4	82

[a]AP, Appaloosa; AR, Arabian; MH, Morgan; PP, Peruvian Passo; ST, Standardbred; TB, Thoroughbred; TWH, Tennessee Walking Horse; XX, Crossbred or other breed. From Bowling *et al.* (1987).

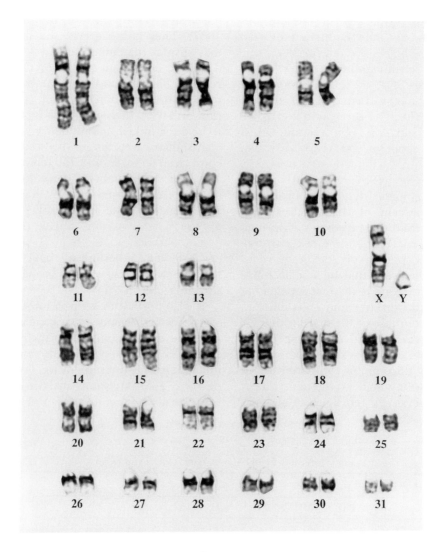

Figure 12 G-Banded karyotype from peripheral blood lymphocytes of a horse with a male genotype (normal XY sex chromosome constitution; 64,XY) with a female fenotype.

cervix and uterus to provide a judgment on future fertility.

2. Intersexuality

Intersex animals are those that possess anatomical characteristics of both sexes. These are divided into three main classes: "true" hermaphrodites, male pseudohermaphrodites, and female pseudohermaphrodites. True hermaphrodites, that is, animals having both ovarian and testicular tissue, are very rare in horses.

Pseudohermaphrodites have the gonads of one sex, but the external genitalia and other characteristics resemble those of the opposite sex. The male pseudohermaphrodite is the most common intersex in horses and characteristically has hypoplastic testicles within the abdominal cavity or inguinal canal, a penislike clitoris emerging from a rudimentary vulva, and exhibits male libido (Bouters *et al.*, 1972; Gerneke and Coubrough, 1970).

The karyotype of most male pseudohermaphrodites is that of the normal female (2n–64,XX) (Bouters *et al.*, 1972; Dunn *et al.*, 1974). Several other karyotypes have been described having the following chromosome patterns: 64,XX/64,XY; 66,XXXY; 64,XX/65,XXY; 64,XX/64,XY/65,XXY. The 66,XXXY and the 64,XX/65XXY chromosome patterns resemble some of the karyotypes noted for Klinefelter's syndrome in humans (Bouters *et al.*, 1972; Dunn *et al.*, 1974).

References

Alexander, S. L., and Irvine, C. H. G. (1987). *J. Endocrinol.* **114**, 351–362.

Allen, W. R. (1969). *J. Endocrinol.* **43**, 593.

Allen, W. R. (1979). *In* "Maternal Recognition of Pregnancy" (Ciba Fdn Series No. 64), pp. 323–352. Excerpta Medica, Amsterdam.

Allen, W. R. (1982). *J. Reprod. Fertil. Suppl.* **31**, 57–94.

Allen, W. R., and Moor, R. M. (1972). *J. Reprod. Fertil.* **29**, 313–316.

Allen, W. E., Chard, T., and Forsling, M. L. (1973a). *J. Endocrinol.* **57**, 175–176.

Allen, W. R., Hamilton, D. W., and Moor, R. M. (1973b). *Anat. Rec.* **177**, 485–505.

Allen, W. R., Kydd, J. H., Boyle, M. S., and Antczak, D. F. (1987). *J. Reprod. Fertil. Suppl.* **35**, 197–209.

Alm, C. C., Sullivan, J., and First, N. L. (1974). *J. Am. Vet. Med. Assoc.* **165**, 721.

Alm, C. C., Sullivan, J. J., and First, N. L. (1975). *J. Reprod. Fertil. Suppl.* **23**, 637–640.

Amann, R. P., and Ganjam, V. K. (1981). *J. Androl.* **2**, 132–139.

Amann, R. P., Thompson, D. L., Jr., Squires, E. L., and Pickett, B. W. (1979). *J. Reprod. Fertil. Suppl.* **27**, 1–6.

Andrews, F. N., and McKenzie, F. F. (1941). *Mo. Agric. Exp. Station Res. Bull.* **329**.

Arthur, G. H. (1958). *Vet. Rec.* **70**, 682–686.

Arthur G. H. (1975) *In* "Veterinary Reproduction and Obstetrics," 4th ed., pp. 130–144. Bailliere Tindale, London.

Arthur, G. H., and Allen, W. E. (1972). *Equine Vet. J.* **4**, 109–117.

Asbury, A. C., and Hughes, J. P. (1964). *J. Am. Vet. Med. Assoc.* **144**, 879.

Bain, A. M., and Howey, W. P. (1975). *J. Reprod. Fertil. Suppl.* **23**, 545–546.

Barnes, R. J., Nathanielsz, P. W., Rossdale, P. D., Comline, R. S., and Silver, M. (1975). *J. Reprod. Fertil. Suppl.* **23**, 617–623.

Barnes, R. J., Comline, R. S., Jeffcott, L. B., Mitchell, M. D., Rossdale, P. D., and Silver, M. (1978). *J. Endocrinol.* **78**, 201–215.

Barty, K. J. (1974). *Austr. Vet. J.* **50:** 553.

Baser, F. W., and Thatcher, W. W. (1977). *Prostaglandins* **14**, 347–401.

Benirschke, K., Brownhill, I. F., and Beath, M. M. (1962). *J. Reprod. Fertil.* **4**, 319.

Benirschke, K., Low, R. J., Sullivan, M. M., and Carter, R. M. (1964). *J. Hered.* **55**, 31.

Bergin, W. C., Gier, H. T., Marion, R. B., and Coffman, J. R. (1970). *Biol. Reprod.* **3**, 82–92.

Berglund, L. A., Sharp, D. C., Vernon, M. W., and Thatcher, W. W. (1982). *J. Reprod. Fertil. Suppl.* **32**, 335–341.

Betteridge, K. J., and Mitchell, D. J. (1974). *J. Reprod. Fertil.* **39**, 145.

Bielanski, W. (1955). *Bull. Acad. Polon. Sci.* **3**, 243.

Bielanski, W. (1972). *In* "Animal Reproduction," 2nd ed. Painstwowe Wydawnictwo Rohnicze, Lesne, Warsaw, Poland.

Blue, M. G., Bruere, A. N., and Dewes, H. F. (1978). *N.Z. Vet. J.* **26**, 137–141.

Bouters, R., Vandeplassche, M., and DeMoor, A. (1972). *Equine Vet. J.* **4**, 150.

Bowen, J. M. (1969). *Equine Vet. J.* **1**, 98.

Bowling, A. T., Millon, L., and Hughes, J. P. (1987). *J. Reprod. Fertil. Suppl.* **35**, 149–155.

Bowman, T. (1986). Proc. 32d Annu. Conv. American Association of Equine Practitioners, 35–43.

Britton, J. W. (1963) *In* "Equine Medicine and Surgery" (E. J. Catcott, ed.), p. 649. American Veterinary Publications, Inc., Santa Barbara.

Clegg, M. T., Cole, H. H., Howard, C. B., and Pigon, H. (1962). *J. Endocrinol.* **25,** 245–248.

Cole, H. H., and Hart, G. H. (1930). *Am. J. Physiol.* **93,** 57–58.

Cox, J. E., Williams, J. H. Rowe, P. H., and Smith, J. A. (1973). *Equine Vet. J.* **5,** 85–90.

Daels, P. F., Shideler, S., Lasley, B. L., Hughes, J. P., and Stabenfeldt, G. H. (1990a). *J. Reprod. Fertil.* (in press).

Daels, P. F., DeMoraes, M. J., Stabenfeldt, G. H., Hughes, J. P., and Lasley, B. L. (1990b). The corpus luteum: a major source of oestrogen during early pregnancy in the mare. *J. Reprod. Fert. (Suppl.)* Proceedings of the 5th International Symposium on Equine Reproduction, July 1–7, Deauville, France. (In press.)

Douglas, R. H., and Ginther, O. J. (1975). *J. Reprod. Fertil. Suppl.* **23,** 503–505.

Dunn, H. O., Vaughan, J. T., and McEntee, K. (1974). *Cornell Vet.* **64,** 265.

Evans, M. J., and Irvine, C. H. G. (1975). *J. Reprod. Fertil. Suppl.* **23,** 193–200.

First, N. L., and Alm, C. C. (1977). *J. Anim. Sci.* **44,** 1072.

Flood, P. F., Betteridge, K. J., and Irvine, D. S. (1979). *J. Reprod. Fertil. Suppl.* 27, 414–420.

Gaillard, J.-L., and Silberzahn, P. (1987). *J. Biol. Chem.* **262,** 5717–5722.

Ganjam, V. K., and Kenney, R. M. (1975). *Proc. 21st Annu. Conv. A.A.E.P.* 263–276.

Gerneke, W. H., and Coubrough, R. I. (1970). *Onderstepoort J. Vet. Res.* **37,** 211.

Geschwind, I. I., Dewey, R., Hughes, J. P., Evans, J. W., and Stabenfeldt, G. H. (1975). *J. Reprod. Fertil. Suppl.* **23,** 207.

Ginther, O. J. (1983). *Theriogenology* **19,** 603–611.

Ginther, O. J. (1979). *In* "Reproductive Biology of the Mare: Basic and Applied Aspects," pp. 255–320. McNaughton and Gunn, Ann Arbor, Michigan.

Ginther, O. J. (1985). *Equine Vet. J. Suppl.* **3,** 41–47.

Ginther, O. J., and First, N. L. (1971). *Am. J. Vet. Res.* **32,** 1687.

Gonzales-Angul, A., Hernandez-Jauregui, P., and Marquez-Monter, H. (1971). *J. Vet. Res.* **32,** 1661.

Grubaugh, W., Sharp, D. C., Berglund, L. A., McDowell, K. J., Kilmer, D. M., Peck, L. S., and Seamans, K. W. (1982). *J. Reprod. Fertil. Suppl.* **32,** 293–295.

Haluska, G. J. (1989). *Tierarztliche Praxis, Suppl.* **4,** 56–62.

Haluska, G. J., and Currie, W. B. (1988). *J. Reprod. Fertil. Suppl.* **84,** 635–646.

Haluska, G. J., Lowe, J. E., and Curie, W. B. (1987). *J. Reprod. Fertil. Suppl.* **35:** 553–564.

Hamilton, D. W., Allen, W. R., and Moor, R. M. (1973). *Anat. Rec.* **177,** 503–518.

Hay, F. M., and Allen, W. R. (1975). An ultrastructural and histochemical study of the interstitial cells in the gonads of the fetal horse. *J. Reprod. Fert. (Suppl.)* **23,** 557–561.

Heap, R. B., Hamon, M., and Allen, W. R. (1982). *J. Reprod. Fertil. Suppl.* **32,** 343–352.

Hinrichs, K., Sertich, P. L., and Kenney, R. M. (1986). *Theriogenology* **26,** 455–460.

Holtan, D. W. Nett, T. M., and Estergreen, V. L. (1975). *J. Reprod. Fertil. Suppl.* **23,** 419–424.

Holtan, D. W., Squires, E. L., Lapin, D. R., and Ginther, O. J. (1979). *J. Reprod. Fertil. Suppl.* **27,** 457–463.

Hughes, J. P., and Loy, R. G. (1970). *Cornell Vet.* **60,** 463.

Hughes, J. P., Asbury, A. C., Loy, R. G., and Burt, H. E. (1967). *Cornell Vet.* **57,** 53–69.

Hughes, J. P., Stabenfeldt, G. H., and Evans, J. W. (1973). *Proc. 18th Annu. Conv. A.A.E.P.,* (San Francisco, 1972) 119–151.

Hughes, J. P., Benirschke, K., Kennedy, P. C., and Trommershausen-Smith, A. (1975). *J. Reprod. Fertil. Suppl.* **23,** 385.

Hughes, J. P., Stabenfeldt, G., Cupps, P., Evans, J. W., and Kennedy, P. (1984). *Vlaams Diergeneeskundig Tijdschrift* **53,** 191–196.

Hughes, J. P., Couto, M. A., and Stabenfeldt, G. H. (1985). *Proc. Soc. Theriogenol.* 123–125.

Hungerford, D. A. (1965) *Stain Technol.* **40,** 333–338.

Hunter, R. H. F. (1980). *In* "Physiology and Technology of Reproduction in Female Domestic Animals," pp. 316–336. Academic Press, London.

Hurtgen, J. P. (1987). *In* "Current Therapy in Equine Medicine" (N. E. Robinson, ed.), Vol, 2, pp. 555–558. W. B. Saunders, Philadelphia.

Irvine, C. H. G., and Alexander, S. L. (1987). *J. Endocrinol.* **113,** 183–192.

Jeffcott, L. B. (1972). *Equine Vet. J.* **4,** 202–213.

Jeffcott, L. B., and Whitwell, K. T. (1973). *J. Comp. Pathol.* **83,** 91–106.

Jeffcott, L. B., and Rossdale, P. D. (1979). *J. Reprod. Fertil. Suppl.* **27,** 563–569.

Jöchle, W., Irvine, C. H. G., Alexander, S. L., and Newby, T. J. (1987). *J. Reprod. Fertil. Suppl.* **35,** 261–267.

Johnson, L., and Neaves, W. B. (1981). *Biol. Reprod.* **24,** 703–712.

Johnson, L., and Thompson, D. L., Jr. (1983). *Biol. Reprod.* **29,** 777–789.

Johnson, L., and Nguyen, H. B. (1986). *J. Reprod. Fertil.* **76,** 311–316.

Kenney, R. M., Hurtgen, J., Pierson, R., Witherspoon, D., and Simons, J. (1983). *J. Soc. Theriogenol.* **IX.**

Kilmer, D. M., Sharp, D. C., Berglund, L. A., Grubaugh, W., McDowell, K. J., and Peck, L. S. (1982). *J. Reprod. Fertil. Suppl.* **32,** 303–307.

Kindahl, H., Knudsen, O., Madej, A., and Edqvist, L. E. (1982). *J. Reprod. Fertil. Suppl.* **32,** 353–359.

Kooistra, L. H., and Ginther, O. J. (1975). *Am. J. Vet. Res.* **36,** 1413–1419.

Leith, G. S., and Ginther, O. J. (1984). *Theriogenology* **24,** 701–711.

Lovell, J. D., Stabenfeldt, G. H., Hughes, J. P., and Evans, J. W. (1975). *J. Reprod. Fertil. Suppl.* **23,** 449–456.

Loy, R. G. (1980). *Vet. Clin. North Am. Large Anim. Pract.* **2**(2), 345–359.

Madej, A., Kindahl, H., Nydahl, C., Edqvist, L.-E., and Stewart, D. R. (1987). *J. Reprod. Fertil. Suppl.* **35,** 479–484.

Mann, T. (1969). *In* "Reproduction in Domestic Animals" (H. H. Cole and P. T. Cupps, eds.), 2nd ed., p. 277. Academic Press, New York.

Manning, A. W., Rajkumar, K., Bristol, F., Flood, P. F., and Murphy, B. D. (1987). *J. Reprod. Fertil. Suppl.* **35,** 389–397.

McDowell, K. J., Sharp, D. C., Grubaugh, W., and Wilcox, C. J. (1988). *Biol. Reprod.* **39,** 340–348.

Merkt, H., and Gunzel, A.-R. (1979). *Equine Vet. J.* **11,** 256.

Mitchell, D., and Betteridge, K. J. (1973). *Proc. 6th Int. Congr. Anim. Reprod. and A. I. (Munich 1972)* **II,** 567.

Monesi, V. (1972). *In* "Reproduction in Mammals" (C. R. Austin and R. V. Short, eds.), Vol. 1. Cambridge University Press, Cambridge.

Moss, G. E., Estergreen, V. L., Becker, S. R., and Grant, B. D. (1979). *J. Reprod. Fertil. Suppl.* **27,** 522–519

Mossman, H. W., and Dukes, K. L. (1973). *In* "Comparative Morphology of the Mammalian Ovary." University of Wisconsin Press, Madison, Wisconsin.

Neely, D. P. (1979). Studies on the control of luteal function and prostaglandin release in the mare. Ph.D. dissertation, University of California, Davis.

Neely, D. P., Kindahl, H., Stabenfeldt, G. H., Edqvist, L.-E., and Hughes, J. P. (1979). *J. Reprod. Fertil. Suppl.* **27,** 181–189.

Nelson, E. M., Kiefer, B. L., Roser, J. F., and Evans, J. W. (1985). *Theriogenology* **23,** 241–262.

Nett, T. M. Holtan, D. W., and Estergreen, V. L. (1973). *J. Anim. Sci.* **37,** 962–970.

Nett, T. M., Holtan, D. W., and Estergreen, V. L. (1975). *J. Reprod. Fertil. Suppl.* **23,** 457–462.

Nishikawa, Y. (1959). *In* "Studies on Reproduction in Horses." Japan Racing Association, Shi62, Tamuracho Minztoku, Tokyo, Japan.

Osborne, V. E. (1966). *Aust. Vet. J.* **42,** 149.

Ousey, J. C., Delcaux, M., and Rossdale, P. D. (1989). *Equine Vet. J.* **21,** 196–200.

Palmer, E., and Driancourt, M. S. (1980). *Theriogenology* **13,** 203–216.

Palmer, E., Driancourt, M. A., and Ortavant, R. (1982). *J. Reprod. Fertil. Suppl.* **32,** 275–282.

Pascoe, D. R., Stabenfeldt, G. H., Hughes, J. P., and Kindahl, H. (1989). *Am. J. Vet. Res.* **50,** 1080–1083.

Pashen, R. L. (1980). *Equine Vet. J.* **12,** 85–87.

Pashen, R. L. (1984). *Equine Vet J.* **16,** 233–238.

Pashen, R. L., and Allen, W. R. (1979). *J. Reprod. Fertil. Suppl.* **27,** 499–509.

Pashen, R. L., Shelderick, E. L., Allen, W. R., and Flint, A. P. F. (1982). *J. Reprod. Fertil. Suppl.* **32,** 389–397.

Peaker, M., Rossdale, P. D., Forsyth, I. A., and Falk, M. (1979). *J. Reprod. Fertil. Suppl.* **27,** 555–559.

Pickett, B. W., Amann, R. P., McKinnon, A. O., Squires, E. L., and Voss, J. L. (1989). *Colorado State Univ. Anim. Reprod. Lab. Bull.* **5.**

Pierson, R. A., and Ginther, O. J. (1987). *Anim. Reprod. Sci.* **14,** 219–231.

Power, M. (1986). *Equine Vet. J.* **18,** 233–236.

Power, M. (1987). *Cytogenet. Cell. Genet.* **45,** 163–168.

Roberts, S. G. (1986). *In* "Veterinary Obstetrics and Genital Diseases (Theriogenology)." Published by the author, Woodstock, Vermont.

Rossdale, P. D., and Short, R. V. (1967). *J. Reprod. Fertil.* **13:** 341.

Rossdale, P. D., and Ricketts, S. W. (1980). *In* "Equine Stud Farm Medicine," 2d ed., pp. 213–357. Lea & Febiger, Philadelphia.

Rossdale, P. D., Pashen, R. L., and Jeffcott, L. B. (1979). *J. Reprod. Fertil. Suppl.* **27,** 521–529.

Rowlands, I. W. (1963). *In* "Gonadotropins, Their Chemical and Biological Properties and Secretory Control" (H. H. Cole, ed.) pp. 74–107. Freeman, San Francisco.

Ryder, O. A., Chemnick, L. G., Bowling, A. T., and Benirschke, K. (1985). *J. Hered.* **76,** 379–381.

Saltiel, A., Calderon, A., Garcia, N., and Hurley, D. P. (1982). *J. Reprod. Fertil. Suppl.* **32,** 261–267.

Samuel, C. A., Allen, W. R., and Steven, D. H. (1975). *J. Reprod. Fertil. Suppl.* **23,** 575–578.

Savage, N. C., and Liptrap, R. M. (1987). *J. Reprod. Fertil. Suppl.* **35,** 239–243.

Shideler, R. K., Squires, E. L., Voss, J. L., and Eikenberry, D. J. (1981). *Proc. 27th Ann. Conv. American Association of Equine Practitioners,* pp. 211–220.

Short, R. V. (1972). *In* "Reproduction in Mammals" (C. R. Austin and R. V. Short eds.), Vol. IV. Cambridge University Press, Cambridge.

Silver, M., Barnes, R. J., Comline, R. S., Fowden, A. L., and Mitchell, M. D. (1979). *J. Reprod. Fertil. Suppl.* **27,** 531–539.

Squires, E. L., and Ginther, O. J. (1975). *J. Reprod. Fertil. Suppl.* **23,** 429–433.

Squires, E. L., Douglas, R. H, Steffenhagen, W. P., and Ginther, O. J. (1974). *J. Anim. Sci.* **38,** 330–338.

Stabenfeldt, G. H., and Hughes, J. P. (1977). *In* "Reproduction in Domestic Animals," 3rd ed. (H. H. Cole and P. T. Cupps, eds.), p. 410. Academic Press, New York.

Stabenfeldt, G. H., and Hughes, J. P. (1987). *Comp. Cont. Educ. Pract. Vet.* **9**(6), 678–684.

Stabenfeldt, G. H., Hughes, J. P., and Evans, J. W. (1972). *Endocrinology* **90,** 1379–1384.

Stabenfeldt, G. H., Hughes, J. P., Wheat, J. D., Evans, J. W., Kennedy, P. C., and Cupps, P. T. (1974a). *J. Reprod. Fertil.* **37,** 343.

Stabenfeldt, G. H., Hughes, J. P., Evans, J. W., and Neely, D. P. (1974b). *Equine Vet. J.* **6,** 158–163.

Stabenfeldt, G. H., Hughes, J. P., Evans, J. W., and Geschwind, I. I. (1975). *J. Reprod. Fertil. Suppl.* **23,** 155–160.

Stabenfeldt, G. H., Hughes, J. P., Kindahl, H., Liu, I., and Pascoe, D. (1984). *Vlaams Diergeneesk. Tijdschr.* **53,** 120–130.

Steffenhagen, W. P., Pineda, M. H., and Ginther, O. J. (1972). *Am. J. Vet. Res.* **33,** 2391.

Stewart, D. R., and Stabenfeldt, G. H. (1981). *Biol. Reprod.* **27,** 17–24.

Stewart, D. R., Stabenfeldt, G. H., Hughes, J. P., and Meagher, D. M. (1982). *Biol. Reprod.* **27,** 17–24.

Stewart, D. R., Kindahl, H., Stabenfeldt, G. H., and Hughes, J. P. (1984). *Equine Vet. J.* **16,** 270–274.

Swierstra, E. E., Gebauer, M. R., and Pickett, B. W. (1974). *J. Reprod. Fertil.* **40,** 113.

Taylor, N. J., and Smith, A. T. (1976). *Proc. Equine Hematol. Symp. Am.,* 124–131. American Association of Equine Practitioners, Golden, Colorado.

Tischner, M., Kosiniak, K., and Bielanski, W. (1974). *J. Reprod. Fertil.* **41,** 329.

Tram, B. F. (1947). *Blood Horse* **27,** 106.

Trommershausen-Smith, A., Hughes, J. P., and Neely, D. P. (1979). *J. Reprod, Fertil. Suppl.* **27,** 271–276.

van Rensburg, S. J., and van Niekerk, C. H. (1968). *Onderstepoort J. Vet. Res.* **35,** 301.

van Niekerk, C. H., and Gerneke, W. H. (1966). *Onderstepoort J. Vet. Res.* **31,** 195.

Vandeplassche, M., and Lauwers, H. (1986). *Anim. Reprod. Sci.* **10,** 163–175.

Vernon, M. W., Strauss, S., Simonelli, M., Zavy, M. T., and Sharp, D. C. (1979). *J. Reprod. Fertil. Suppl.* **27,** 421–429.

Walt, M. L., Stabenfeldt, G. H., Hughes, J. P., Neely, D. P., and Bradbury, R. (1979). *J. Reprod. Fertil. Suppl.* **27,** 471–477.

Warszawski, L. F., Parker, W. G., First, N. L., and Ginther, O. J. (1972). *Am. J. Vet. Res.* **33,** 19.

Weithenauer, J., Sharp, D. C., McDowell, K. J., Davis, S. D., Seroussi, M., and Sheerin, P. (1987). *Biol. Reprod. Suppl.* **1,** 329 (Abstr.).

Wesson J. A., and Ginther, O. J. (1982). *J. Reprod. Fertil. Suppl.* **32,** 269–274.

Whitwell, K. E. (1975). *J. Reprod. Fertil. Suppl.* **23,** 599–603.

Whitwell, K. E. (1980). *Vet. Clin. North Am. Large Anim. Pract.* **2,** 313–331.

Whitwell, K. E. (1982). *J. Reprod. Fertil. Suppl.* **32,** 630–632.

Wierzgowski, S., and Hafez, E. S. F. (1961). *Proc. 6th. Int. Congr. Anim. Reprod. and A.I.* **2,** 176.

Zavy, M. T., Vernon, M. W., Sharp, D. C., and Bazer, F. W. (1984). *Endocrinology* **115,** 214–218.

CHAPTER 13

Reproduction in Cattle

T. J. ROBINSON and J. N. SHELTON

I. Introduction

Asdell (1964) has summarized the basic data concerning reproductive phenomena in cattle. As with other mammals, three cycles are involved in the processes of reproduction of each sex, namely, the life cycle, the annual breeding cycle, and the estrous cycle in females and the spermatogenic cycle in males.

II. Cycles in Reproduction

A. The Life Cycle

There are five major periods in the life cycle; fetal, prepubertal, pubertal, reproductive, and senescent. Reproductive efficiency reaches a peak early in the reproductive period, remains high for several years, and declines thereafter.

1. The Fetal Period

The chronological development of the genital organs of the bovine fetus is shown in Table I. Details of the origin, migration, and differentation of the primordial germ cells are described by Mauleon (1961).

2. The Prepubertal Period and Puberty

a. Morphology and Endocrinology At birth, the primary oocytes, some $75-160 \times 10^3$, are arranged around the periphery of the ovary. Tertiary follicles with antra appear late in fetal life, and their number increases after birth, reaching a maximum at the end of the second month (Erickson, 1966a). Prior to puberty the ovaries, despite the presence of follicles with antra, are small.

Puberty is defined broadly as a process whereby a heifer becomes capable of reproduction (Robinson, 1977) or, more narrowly, as the first occurrence of behavioral estrus followed by the development of a functional corpus luteum (see Moran et al., 1989). There is a great increase in ovarian size due to the secretion of liquor folliculi, associated with an ability of follicles to ovulate as a result of a luteinizing hormone (LH) peak. The onset of puberty is a gradual process involving maturation within the hypothalamic–hypophyseal–ovarian axis of

Table I
Sexual Differentiation of the Calf Fetus[a]

Age (days)	Crown–rump length (mm)	Males	Females
39–40		Early testicular organogenesis	Undifferentiated gonadal primordium
42–43	24–26	Definite albuginea	
45	29	Testicular interstitial cells	
47	32	First flexure of penile urethra	
48–49	34–36	Anogenital distance definitely increased	Beginning of prolonged and slow thickening of the superficial ovarian layers
		Growing Mullerian ducts in both sexes	
50–52	38–42	Upper Mullerian ducts: reduction in diameter	
56	53	Prostatic buds and seminal vesicles appear	
		Male urogenital connections	
58–60	59–66	Mullerian ducts: anterior part disappears Penis opens under the umbilicus Balanopreputial fold in organization Scrotum develops	Uterus increases in diameter
70	115	Mullerian ducts absent or regressing	Retrogression of Wolffian ducts begins
75	—		First premeiotic figures

[a]From Jost et al. (1972).

the capacity to produce specific hormones and most importantly, to respond to hormonal signals from each other. Both the juvenile pituitary and ovary can respond to exogenous hormones early in life, indicating a considerable degree of maturity prior to puberty. However, the hypothalamus of the prepubertal animal is hypersensitive to estrogen so even the small amount of ovarian estrogen is sufficient to suppress the output of gonadotropin-releasing hormone (GnRH). The small pulses of GnRH maintain pituitary secretion of LH at low levels so that ovarian follicles do not grow to ovulatory size (Kinder et al., 1987). Gradual reduction in sensitivity of the hypothalamus to estrogen, possibly due to a decline in the population of estrogen receptors, results in an increase in frequency of LH pulses (Dodson et al., 1988). This causes development of follicles that produce sufficient estrogen to induce estrus. The estrogen may also cause a preovulatory surge of LH, resulting in ovulation. Several types of incomplete cycle preceding the initiation of regular cyclic activity have been reported. In some cases, the secretion of estrogen may be insufficient to induce ovulation, resulting in anovulatory

estrous behavior. Short luteal phases also are common, which may or may not be associated with the formation of a transient corpus luteum (Berardinelli et al., 1979; Dodson et al., 1988). The role of these transient prepubertal elevated levels of progesterone in subsequent endocrine and ovarian function is unknown. They may simply reflect a transition from discrete endocrine events to an orderly interacting endocrine mechanism. In the sheep, progesterone pre-estrus does have an important functional role; it is essential for the manifestation of estrus (Robinson, 1955), and it may have a similar function in the heifer. For example, progesterone implants followed by removal have been used to advance the onset of puberty (e.g., Short et al., 1976).

In the bull calf, the approach of puberty and the production of viable spermatozoa are associated with the size of the testicles, for which there are three distinct phases of growth (Attal and Courot, 1963) characterized by a progressive development of the spermatogenic cycle (Table II). These changes are associated with changes in the pattern of pituitary LH output. Pubertyfollows a change in sensitivity of the

Table II
Development of Spermatogenic Activity in the Bull Calf[a]

Phase of testicular growth	Live weight (kg)	Types of cell present in sex cords	
1	30–100	Supporting cells ↓	Gonocytes ↓
2	100–300	Supporting cells	Spermatogonia ↓
			Spermatocytes ↓
			Spermatids ↓
3	300–800	Sertoli cells	Spermatozoa

[a]From Ortavant et al. (1969).

hypothalamus–anterior pituitary to negative feedback of a gonadal factor, probably estrogen (Amann *et al.*, 1986). After 12 weeks of age basal plasma LH concentration is elevated and pulsatile LH discharges occur at intervals of less than 2 h, with maximum activity by 4 months. The pulsatile discharge of LH induces differentiation of Leydig cells that secrete first androstenedione and later testosterone. The androgen induces differentiation of Sertoli cells and concomitant differentiation of gonocytes to pre- and A-spermatogonia.

b. Environmental Influences The environment in which a calf is reared (photoperiod, temperature, nutrition) affects the age of puberty; calves reared in spring to summer conditions of high temperatures and long photoperiod reach puberty at a younger age than do those reared in fall to winter conditions of low temperatures and short photoperiod (Schillo *et al.*, 1983). An interaction between photoperiod, plasma LH concentration, and pulse amplitude may be involved (Critser *et al.*, 1987a).

The time of onset of puberty appears to be a function more of size than of age, so low nutritional intake will delay puberty, possibly due to suppression of mean plasma concentrations of LH by low levels of dietary energy and reduced concentration of propionate in the rumen (Day *et al.*, 1986; Randel *et al.*, 1982). It is difficult to define "size" so that dimensional measurements—height at withers and body length as indicators of skeletal size—are more useful parameters than is live weight (Sorenson *et al.*, 1959) and are less susceptible to environmental influence (Baker *et al.*, 1988). Under good nutritional conditions a calf attains puberty at approximately two-thirds adult "size," which varies greatly between breeds.

The presence of mature cows influences puberty in beef heifers (Nelsen *et al.*, 1985), but the presence of bulls seems to have little effect (Roberson *et al.*, 1987), despite the suggestion of a stimulatory pheromone in bull's urine (Izard and Vandenbergh, 1982a).

In the developing male, undernutrition results in a delay in the onset of puberty due to a marked impairment of testicular development and sperm production, associated with inhibition of androgenic function and consequent fructose and citric acid production by the seminal vesicles (Mann *et al.*, 1967). If sufficiently severe it can impair sperm production permanently (VanDemark *et al.*, 1964).

There is a positive correlation between scrotal circumference and age at puberty (Lunstra *et al.*, 1978) and numbers and quality of sperm produced (Almquist and Amann, 1961; Cates, 1975). Nevertheless, within the ranges of nutrition typically used in the rearing of bulls, nutritional treatments that increase scrotal circumference do not necessarily hasten sexual development (Pruit *et al.*, 1986).

3. The Reproductive Period

Once sexual activity commences, it can continue for many years. The number of primordial follicles remains static until about the fourth year (mean 133,000) and declines thereafter until near zero between 15 and 20 years (Erickson, 1966b). Fertility increases from the first (pubertal) estrus to the third (Byerley *et al.*, 1987). However, lifetime productivity is greater in beef heifers that calve first at 2 years of age than in those that calve first at 3 years or older (Donaldson, 1968).

Testicular size is a good index of sperm production and reproductive activity. *Effi-*

Table III
Development of Sperm Production in Holstein Bulls[a,b]

Age	Number of bulls	Gross weight paired testes (g)	Daily sperm production	
			10^6/bull	10^6/g testis
0–4 mo	25	20	0	0
5–7 mo	15	97	104[c]	1[c]
8–10 mo	20	284	1750	7
11–12 mo	15	370	3300	10
17 mo	13	480	4480	10
3 yr	10	586	6040	11
4–5 yr	11	647	6530	11
>7 yr	11	806	8000	11

[a]From Amann and Almquist (1976)
[b]Daily sperm production was calculated from testicular homogenate counts using a divisor of 5.32 days.
[c]Mean for six bulls producing spermatids or spermatozoa.

ciency of sperm production per gram of testicular tissue reaches a maximum at 11 months of age and thereafter remains constant. Total *production,* however, continues to increase as a result of the doubling of the weight of the testes over the next 5 or 6 years (Table III).

4. Senescence

This is of little practical interest as few cattle (some 0.25%) are kept to 15 years of age.

B. The Annual Breeding Cycle

There is a tendency for *Bos taurus* and *Bos indicus* to calve in the spring (Hauser, 1984; Jochle, 1972a), and quiet ovulations ("silent heats") have been observed in *Bos indicus* during winter (Plasse *et al.,* 1968a, 1970). Photoperiod has been implicated in this seasonality (Thibault *et al.,* 1966), and there is good evidence of photoperiodic regulation of bovine gonadotropins (Critser *et al.,* 1987b). Nevertheless, other components of

the environment—temperature, rainfall, feed supply—can be of overriding importance (Jochle, 1972b).

The complexity of the interacting factors is shown by a study on Zebu cattle in the Mexican Gulf coast region. Pubertal heifers had a high rate of conception during the unfavorable dry season, indicating a photoperiodic constraint to provide an early spring calving. However, with successive calvings, the tendency to conceptions in the rainy season increased (Table IV), so that by the fifth calving almost the entire herd was relieved of any constraint of photoperiod. This principle applies also to *Bos taurus,* reared in tropical environments (Fallon, 1961). This overriding influence of rainfall, with consequent pasture growth and plane of nutrition, appears due to a necessity for a buildup in body condition prior to conception (Lamond, 1970), and is as important in improved dairy herds as in primitive cattle. It is discussed in the next section under postpartum anestrus.

Bulls also show a seasonal variation in

Table IV
Order of Consecutive Conceptions in Zebu Cows and Their Distribution in Relation to the Dry and Rainy Season[a]

Conception number	Dry season January–May		Rainy season June–October		Total number
	Number	Percentage	Number	Percentage	
1	190	48.6	201	51.4	391
2–4	245	36.2	431	63.8	676
5–7	139	27.5	366	72.5	505
8–10	78	27.9	202	72.1	280
Total	652	35.2	1200	64.8	1852

[a]From Jochle (1972b).

fertility in which both photoperiod and temperature are involved; high temperatures depress semen quality (Clegg *et al.*, 1965).

C. The Estrous Cycle

The characteristics of the estrous cycle of domestic and other cattle have been discussed in detail in the earlier editions of this volume. The length of the cycle is 20 days in heifers (SD 2.3 days) and 21 days in cows (SD 3.7 days). Estrus is of short duration, 12–24 h in *Bos taurus*, but the range is considerable. *Bos indicus* cattle are less sensitive to estrogen than are *Bos taurus*, and estrus is of shorter duration and is less intense (Randel, 1984). Further, it occurs later in relation to the estrogen stimulus that is responsible for the onset of estrous behavior and the subsequent release of LH. Ovulation occurs some 24 h after the onset of estrus, commonly after the end of estrous behavior in *Bos taurus* but somewhat earlier in *Bos indicus*. The resultant corpus luteum is smaller in *Bos indicus* and contains less progesterone than that of *Bos taurus*, resulting in lower plasma progesterone concentrations. Dobson and Kamonpatana (1986) have published a comparative review of female reproduction in buffalo, cows, and zebus.

Events preceding the first estrus at puberty, are discussed above. Once initiated, the events follow an orderly, well-timed sequence. For convenience, the day of estrus is commonly designated day 0, on which day the plasma concentrations of ovarian steriods are very low (Glencross *et al.*, 1973). The animal is in a transitional phase from being under the primary influence of follicular estrogen to that of luteal progesterone, the importance of which was recognized some 70 years ago by Marshall (1922).

During the estrous cycle, there is a continuous pattern of growth and atresia of follicles. Some conflict exists in the literature concerning the time scale from the preantral to the preovulatory stage, due essentially to limitations of technology. Use of modern techniques such as ultrasound and mitotic index has led to an apparent resolution of this conflict. It now appears that the definitive follicle—that destined to ovulate—takes two cycles to grow through the antral phase to preovulatory size (Lussier *et al.*, 1987), that is, from 0.13 to 8.56 mm in diameter. Earlier estimates (e.g., Scaramuzzi *et al.*, 1980) of a shorter time (22 days) started with a larger early antral follicle of 0.4 mm. The time taken for a follicle to grow to this size from 0.13 mm is about 20 days, making a total of 42 days. Waves of follicular growth are detectable by ultrasonography. In some cycles three waves are detectable, the first from day 4 to day 13, the second from day 13 to day 18, and the third from day 18 to ovulation. Associated with these waves of growth are elevated concentrations of estradiol in ovarian venous blood (Shemesh *et al.*, 1972; Ireland and Roche, 1987). Whereas Savio *et al.* (1988) and Sirois and Fortune (1988) found that three-wave cyles predominated, Ginther *et al.* (1989) found more cycles with two waves. The reasons for these differences or their significance are not apparent.

The ovulatory follicle produces a critical level of estrogen that stimulates the hypothalamus to increase the frequency and amplitude of pulses of gonadotropin-releasing hormone (GnRH). This results in an increased frequency and amplitude of LH and follicle-stimulating hormone (FSH) pulses (Walters and Schallenberger, 1984), which complete follicular maturation and result in a surge of estrogen—mainly estradiol-17β. Estrus occurs some 24h later, accompanied by an ovulatory surge of LH at or about the same time. Ovulation occurs some 30 h later. The mechanism controlling the number of follicles completing development is not fully understood, but it is clear that the circulating level of FSH is a major factor. Inhibin produced by the granulosa cells inhibits the secretion of FSH by the anterior pituitary gland (deJong and Robertson, 1985) and also may have a minor role in the control of LH secretion (Martin *et al.*, 1988). Thus the concentration of FSH in the

circulation is governed by a feedback loop consisting of steriods, inhibin, and possibly other ovarian products.

Production of progesterone within the definitive follicle commences shortly before ovulation but after the ovulatory LH peak (Dieleman and Blankenstein, 1985). Following ovulation, a corpus luteum forms and produces increasing amounts of progesterone under the primary influence of LH, although other hormones, notably prostacyclin (PGI-2) and oxytocin, of luteal origin (Hansel and Dowd, 1986), and adrenaline (Battista *et al.*, 1987) have been ascribed a possible role. The corpus luteum, which consists of two cell types that may have different functions and mechanisms of control, secretes progesterone for a finite time.

As in the ewe, $PGF_{2\alpha}$ produced by the endometrium in response to an interaction between ovarian oxytocin and estradiol (Lafrance and Goff, 1985) is the luteolytic factor (Hansel and Beal, 1979). This $PGF_{2\alpha}$ is transported from the uterus via the venous and lymphatic drainage (Heap *et al.*, 1985; Hein *et al.*, 1988) and is transferred to the ovarian artery by a countercurrent mechanism (Fig. 1). Exogenous prostaglandin, intrauterine or systemic, is highly effective as a luteolytic agent (Rowson *et al.*, 1972b).

Released from the constraints of a high level of circulating progesterone, the ensuing high level of estrogen initiates the next cycle.

These events, initially reviewed by Denamur (1972) and illustrated in Figure 2, pose a basic question concerning the fundamental nature of the estrous cycle. Is there an intrinsic pattern of growth and regression of follicles, with associated production of estrogen, which constitutes a biological "clock," and is this responsible for the accurate timing of the cycle, as in the laboratory rodents? The estrous cycles of ruminants bear a remarkable resemblance to the pseudopregnancy of rats and mice and, in a biological sense, the cycle may be regarded as a pseudopregnancy that overrides, or is

geared to, a fundamental cycle of growth and regression of follicles. Figure 2 poses another related question. Is there a missing potential developing follicle in mid-cycle, the development of which is prevented by the high plasma concentration of progesterone? If so, the cow has a basic 5-day ovarian cycle.

The physiological changes in the reproductive tract consequent upon these endocrinological events have been described by Asdell (1964) and Hansel (1959). Estrus is characterized by anatomical and behavioral changes that include increased motility of the oviducts, particularly that ipsilateral to the active ovary, for 3–5 days before estrus to 3–5 days after (Ruckebusch and Bayard, 1975; Bennett *et al.*, 1988). This increased activity is associated with sperm transport and passage of the ovum and is further evidence of localized endocrine action. Problems posed by "silent heats" and heats of short duration are not uncommon (Cupps *et al.*, 1969) and may be exacerbated, especially on the range, by malnutrition (Lamond, 1970) and stress (Wagnon *et al.*, 1972). Further, the time of estrus during the day is not random in *Bos taurus* (Trimberger, 1948) or *Bos indicus* (Howes *et al.*,

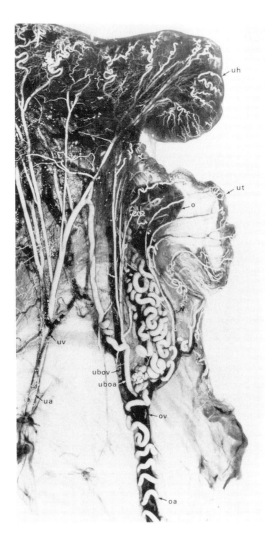

Figure 1 Ventral view of arteries (light) and veins (dark) of a uterine horn and adjacent ovary of a cow. Veins were injected with gelatin containing black ink, arteries were injected with red latex, and tissues were cleared. The venous drainage from the ovary and the uterine tube, and from much of the uterine horn, forms a common utero-ovarian vein (ovarian vein). The ovarian artery is very tortuous and is closely applied to the utero-ovarian vein. o, Ovary; oa, ovarian artery; ov, ovarian vein; ua, uterine artery; uboa, uterine branch of ovarian artery; ubov, uterine branch of ovarian vein; uh, uterine horn; ut, uterine tube; uv, uterine vein. From Ginther and Del Campo (1974), courtesy of *American Journal of Veterinary Research.*

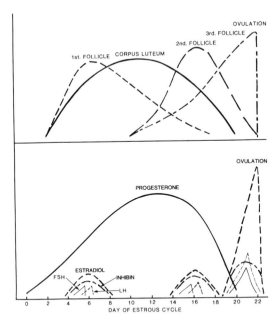

Figure 2 Diagrammatic representation of morphological and endocrinological events in the estrous cycle of cows.

1960). The impact of social and environmental factors on the efficiency of detection of estrus and of mating in cattle has not been studied as intensively as in the sheep, but pheromones produced by bulls and cows appear to intensify estrous activity (Izard and Vandenbergh, 1982a, 1982b). Bleeding occurs from the uterus in many animals during metestrus, usually about 24 h after ovulation.

D. The Spermatogenic Cycle

The liberation of spermatozoa by the testes of the bull is a continuous process, some $10-13 \times 10^9$ per day being produced by an adult (Amann and Almquist, 1976), of which about half can be harvested in successive daily collections. This process is the re-

sult of an accurately timed spermatogenic cycle of some 61 days duration (Courot *et al*, 1970). Transit of sperm through the epididymus of the bull requires 5–6 days.

III. Mating and Fertilization

A. Behavior and Libido

It is not easy to induce estrus consistently in ovariectomized cows and heifers with estrogen alone or with testosterone (see Nessan and King, 1981), with results being extremely variable. The most consistent pattern was obtained by Carrick and Shelton (1969), who found the heifer to be highly sensitive to estrogen, the median effective dose (MED) of estradiol benzoate (EB) required to induce estrus being some 120–130 μg. Although mating behavior resulted as a direct effect of estrogen, progesterone was involved but the interaction was not as clear-cut as in the sheep. Repeated injections with estrogen resulted in a refractory state that could only be removed by a course of progesterone injections. Full sensitivity to estrogen was recovered, but thereafter the heifers remained dependent upon an alternate progesterone–estrogen regime in order to retain their sensitivity (Table V). Further evidence for a role of progesterone in the manifestation of estrus comes from the observation that a short luteal phase precedes the first observed estrus at puberty (Dodson *et al.*, 1988).

Bulls show little, if any, seasonal variation in mating ability, but problems of inhibition may arise in artificial insemination centers, generally as a result of mismanagement, and reaction time may be affected by season in tropical areas (Rollinson, 1971).

Table V
Response of Spayed Heifers[a] to Estradiol Benzoate (EB) with and without Pretreatment[b] with Progesterone[c]

| Dose of EB (μg) | EB after progesterone | | | | EB alone | | | |
| | Number of heifers | | Mean time to onset (h) | Mean duration (h) | Number of heifers | | Meantime to onset (h) | Mean duration (h) |
	Treated	Estrus			Treated	Estrus		
400	14	13	19.4	5.8	14	1	21.0	3.0
200	14	14	19.7	8.6	14	2	27.0	3.0
141	14	9	25.0	4.3	14	4	26.2	7.5
100	14	4	25.5	5.2	14	1	27.0	3.0

[a]Previously made refractory to estrogen by repeated injections.
[b]At 10 mg/day for 5 days followed 3 days later by EB.
[c]From Carrick and Shelton (1969).

B. Ovulation and Fertilization

Although ovulation occurs spontaneously, it may be hastened by mating (Marion et al., 1950) and there is some evidence for a circadian rhythm (Jochle, 1966). Normally, one egg is shed, with the right ovary ovulating more frequently than the left (Asdell, 1964). The incidence of twin ovulations ranges from about 2% for dairy cattle to 0.5% for beef breeds.

The mature ovum is shed from the ovary some 10 h after the termination of estrus or 28–31 h after standing estrus begins (Hunter and Wilmut, 1984). The ovum is thought to remain viable for only 8–10 h after ovulation, whereas spermatozoa survive in the female reproductive tract for 18–24 h (Trimberger, 1984). The ovum requires about 6 h to travel to the site of fertilization at the isthmo–ampullar junction. In order to provide maximum chance of fertilization while the ovum is still viable, an adequate number of spermatozoa must be immediately available. There is some evidence for an early rapid phase of sperm transport (Hafez, 1972), but more impor-

tant is a less rapid colonization of the cervix and subsequent migration through the uterus (Robinson, 1973) and the establishment of a sperm reservoir in the caudal portion of the oviduct. This reservoir is the immediate source of viable spermatozoa at the time of ovulation, and its establishment requires about 12 h from mating or insemination (Hunter and Wilmut, 1984). Hence cows should be inseminated or mated about mid-estrus, some 13–18 h before ovulation.

Failure of fertilization in cattle exhibiting normal estrus is not a major problem (Diskin and Sreenan, 1980) except early in lactation or, possibly, after the ingestion of phytoestrogens (Thain, 1968). It is a problem, however, in cattle in which estrus has been synchronized by treatment with exogenous progestagen. Although direct evidence is lacking for cows, it seems certain, by analogy with ewes, that this is due to impairment of the normal pattern of sperm transport.

Undernutrition commonly is associated with infertility, and there is good evidence that aberrant pituitary and ovarian function are related to the changes in insulin and glucose metabolism that are caused by underfeeding (McClure *et al.*, 1978: McCann and Hansel, 1986). Less is known of the nature of subfertility associated with specific nutritional deficiencies, but it is likely that they interfere with metabolic processes essential to the provision of adequate amounts of glucose and amino acids for synthesis of releasing factors and gonadotropins by the hypothalamus and anterior pituitary.

Finally, genetic defects can prevent fertilization and constitute an important but not primary cause of subfertility (Tanabe and Almquist, 1967).

IV. Pregnancy

Despite the high percentage of healthy breeders in which eggs are fertilized following insemination (90–100%), only approximately 50–55% calve (Table VI). The progressive loss is due to embryonic and fetal death (Baier, 1972), the former being much the greater (Hawk, 1978).

Pregnancy may be divided into three stages: (1) early cleavage of the zygote, (2) differentiation and attachment, and (3) development of the placenta and fetal growth.

Table VI
Percentage of Cows Pregnant at Various Stages as Estimated by Pregnancy Diagnosis or Nonreturn to Service[a]

Interval from service				
1 month	2 months	3 months	9 months	Calving
67.8	58.4	55.7	—	52.9
74.9	68.3	65.3	58.6	53.2
78.4	—	65.9	—	53.5

[a]From Robinson (1957).

A. Period of Early Cleavage

The fertilized egg remains in the oviduct for 72–96 h and then enters the uterus at the 16-cell stage. The oviduct provides a unique environment that appears not to be dependent upon the endocrine status of the animal, nor is it species specific. Thus, a fertilized one-cell cow egg will develop to a normal blastocyst at a normal rate in the oviduct of the rabbit (Lawson *et al.*, 1972) or the ewe (Eyestone *et al.*, 1985), regardless of the stage of estrous cycle of the host animal. Oviductal epithelial cells, even when cultured *in vitro*, support development of precompaction embryos (Gandolfi and Moor, 1987). By contrast, the uterine environment is potentially hostile and survival of eggs is jeopardized if they enter the uterus prematurely (Rowson *et al.*, 1972a).

B. Period of Differentiation and Attachment

In the uterus, the egg develops through the morula to the blastocyst stage with clearly defined inner cell mass and trophoblast and hatches from the zona pellucida at day 8–9. The trophoblast enlarges rapidly and the discrete embryonic membranes become obvious. Also, subtle but important biochemical events are occurring. Maternal recognition of pregnancy is essential before day 17 to prevent luteolysis and the reinitiation of ovarian cycles, and substances produced by both the conceptus and the uterus are involved in this recognition.

As early as day 10 the embryo produces a small nonprotein molecule (MW < 10,000) that stimulates increased production of progesterone by the corpus luteum (Hansel and Hickey, 1988). These authors suggest that this molecule may be synonymous with the embryo-derived platelet activating factor described in mice by O'Neill (1985). Bovine conceptus secretory protein, referred to as bovine trophoblast protein 1, inhibits the increased production of $PGF_{2\alpha}$ in response to estradiol and oxytocin (reviewed by Thatcher et al., 1985). This protein antiluteolytic factor of MW 22,000–26,000 (Bartol et al.,1985) is produced by both the chorion and the amnion, apparently from day 15 (Godkin et al., 1988). Bovine embryos also produce steroids and prostenoids between days 13 and 16 (Shemesh et al., 1979).

Proteins produced by the endometrium, of which there are at least four specific to pregnancy (Bartol et al., 1981), may have a role in protecting the conceptus from uterine proteases and in immunosuppression (Roberts and Bazer, 1988).

By day 18 the chorionic sac fills the ipsilateral uterine horn and may have commenced growth into the other horn. There follows gradual erosion of the epithelium of the caruncles by the trophoblast cells and then active invasion leading to the development of the fetal–maternal cotyledonary placenta. Binucleate cells of the trophoblast invade the uterine epithelium as early as day 22, and close apposition of the trophoblast to the caruncular epithelium leads to mutual interdigitation of microvilli by day 24, progressing to intimate attachment by day 27 (King et al., 1980). Development of the mature placenta is a gradual process, characterized by a progressive invasion of the caruncles, and is not completed until the end of the first third of pregnancy (Amoroso, 1952).

The first 6 weeks of pregnancy is a period highly susceptible to loss (see Table VI)

with, on the average, about 31–34% of fertilized eggs failing to survive (Robinson, 1957). Embryonic mortality thus is the major cause of failure of pregnancy in cattle (Hawk, 1978). Estimates of the time of death of embryos vary between studies, ranging from 5 days or less (Wiebold, 1988) to after day 16, with differences between repeat-breeder and normal cows (Ayalon, 1978). No single factor can be implicated, nor is it possible to apportion the loss between genetic causes and failure of the maternal environment (Wilmut et al., 1986). Genetic causes include chromosomal aberrations that can result in a range of conditions from early embryonic mortality to dwarf calves (Nicholas, 1987). Maternal failure, due perhaps to an inadequate antiluteolytic signal, is important, as shown by egg transfer studies (Rowson et al., 1972a). A 2-day asynchrony between the endocrine state of the uterus and the stage of development of the fertilized egg results in embryonic death. Proximity to another embryo also reduces the chances of survival, as shown by the doubling of the chance of twins when one egg is shed from each ovary as compared to survival when both are shed from the same ovary (Gordon et al., 1962). Site of attachment to the endometrium is highly important; survival of transplanted twin embryos is enhanced if one is placed in the tip of the uterine horn ipsilateral to the functional corpus luteum (Newcomb et al., 1980).

In multiple pregnancies, adjacent trophoblasts (chorions) fuse and when the allantois, with its associated blood vessels, grows out to invest the chorion to provide the allantochorion there is early anastomosis of the blood vessels (Hafez and Rajakoski, 1964).

C. Period of the Fetus and Placenta

Differentiation into the component parts of the new organism and its membranes is effected by the end of the sixth week. Thereafter, pregnancy consists of consolidation and growth of the placenta, which does not attain its mature form until the third or fourth month (Hammond, 1927), and growth of the fetus. The development of the syndesmochorial placenta is elegantly described and illustrated by Amoroso (1952) and more recently by Perry (1981).

In addition to its role as a vehicle for the transmission of nutrients and metabolites to the fetus and the removal of waste products, the placenta—or parts of it—is credited with a role in immunoprotection of the fetus and it is a highly productive endocrine organ (see Chapter 10). However, like the goat, and unlike the sheep, the placenta of the cow does not synthesize progesterone, so that this species is dependent on ovarian progesterone for the maintenance of pregnancy (Stabenfeldt et al., 1970).

During the first two-thirds of pregnancy, the rate of growth of the placenta exceeds that of the fetus, the growth of which is so remarkably uniform that fetal age can be determined by reference to standard measurements and photographs (Winters et al., 1942). During the last one-third of pregnancy the situation is reversed, and it is during this period that nutritional stresses affecting birth weight and survivability are important.

D. Duration of Pregnancy

The mean duration of gestation in cattle is commonly cited as 270–290 days. This duration is determined by the fetus and is characteristic for each species and breed; Bos taurus cattle have a gestation period shorter by about 14 days than do Bos indicus (Plasse et al., 1968b). Twins result in pregnancies 3–6 days shorter than for singles, and male calves are carried 1–2 days longer than females (Jainudeen and Hafez, 1980).

E. Diagnosis of Pregnancy

1. Rectal Palpation

In the cow, pregnancy is routinely detected by palpation through the rectal and uterine walls for fetal membranes, amniotic vesicle, cotyledons, or fetus (Wisnicky and Casida, 1948). An experienced technician can detect pregnancy by 35 days after insemination, but accuracy is better at 40–45 days.

2. Ultrasound

Real-time ultrasonic scanning by transrectal probe can detect the bovine embryonic vesicle by 12–14 days (Pierson and Ginther, 1984), and pregnancy can be detected with confidence by an experienced operator at 30 days (White et al., 1985).

3. Biochemical Methods

a. Steroids The assay of progesterone level in plasma or milk on day 20 to 25 after insemination has proved to be a reliable test for early pregnancy, and is a potentially useful tool for studying embryonic mortality (Pope and Hodgson-Jones, 1975). Radioimmunoassay was the first technique to be used on a large scale, but the enforced restriction to licenced specialized laboratories limited commercial application. Recently, enzyme immunoassay techniques that do not require the use of radioactive tracers have been developed successfully for milk samples (Arnstadt and Cleere, 1981; Sauer *et al.*, 1986) and should make the service much more accessible. False negatives are virtually nil, while false positives can be attributed to embryonic mortality or to long estrous cycles. Practical problems associated with the collection and assay of blood plasma make the milk test more attractive.

b. Proteins Sasser *et al.* (1986) have developed a radioimmunoassay for the bovine pregnancy-specific proteins detected by Butler *et al.* (1982) and claim that, at least under laboratory conditions, it can be used as a pregnancy test after 24 days. Insemination dates are not needed. The assay has yet to be developed as a simple diagnostic test for routine use.

F. Hemopoietic Chimerism and Freemartinism

In 1917, Lillie postulated that the changes wrought in heifers born twin to a bull were due to internal secretions borne to the heifer by an anastomosed placental circulation. In 1969, the freemartin syndrome formed the central theme of a symposium on intersexuality (Perry, 1969) and has been studied intensively since.

The "freemartin" condition (free = sterile; martin = bovine) is a direct result of the form of placentation in the cow. In multiparous ruminants, adjacent trophoblasts fuse to form a common chorion. In some, such as the sheep, the blood vessels that grow out with the allantois of each embryo in the formation of the allantochorion appear not to come into close juxtaposition with those of the adjacent embryo. In the cow, by contrast, they do. They occupy the same cotyledons, and there is fusion of placental blood

vessels (Fig. 3), well established by day 39 (Jost *et al.*, 1972). This permits a degree of exchange of blood between fetuses, with consequent profound effects.

In naturally occurring twins the incidence of anastomosis has been reported as more than 90% (Marcum, 1974), but when twins of disparate genotype were produced by embryo transfer, the incidence was 100% for the Friesian–Brahman combination and only 50% for the Jersey–Brahman combination (Summers *et al.*, 1984).

In the fetal calf the testes of the normal male become histologically recognizable about day 40, whereas the female gonads remain largely undifferentiated for several weeks longer (see Table I). Disturbance in development is first apparent at about day 50 when the Mullerian ducts of males and of future freemartins commence to decrease in diameter relative to those of normal females (Fig. 4); by day 62 they are vestigial (Jost *et al.*, 1972). Over the same period there is inhibition of growth of the presumptive ovaries.

This time difference between male and female organogenesis led Jost *et al.* (1972) to suggest that the destruction of the Mullerian ducts is caused by an "antifeminine" substance or hormone (not testosterone) produced by the male twin, followed by a positive masculinization process caused by a second testicular factor that stimulates the vestigial ovaries, seminal vesicles, and other structures. More recently, a glycoprotein dimer secreted by the Sertoli cells (Tran and Josso, 1982) and known as anti-Mullerian hormone (AMH) has been identified (Picard *et al.*, 1978; 1986). In the case of mixed-sex bovine twins with conjoined circulations the effects of AMH are evident in the female as well as in the male fetus.

There has been much speculation that

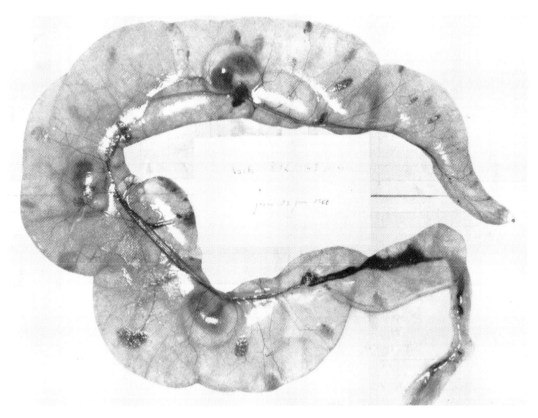

Figure 3 Fused allantochorion in a triplet set, 49 days after insemination, showing fetal components of the cotyledons and the three fetuses contained in their individual amnions. Anastomoses of both large and small blood vessels are visible. The fetus in the middle was a female, and the others were males. Scale: 5-cm large squares seen on the graduated paper. From Jost *et al.* (1972), courtesy of *Journal of Reproduction and Fertility Ltd.*

another testicular product as postulated by Jost *et al.* (1972), namely H-Y antigen, is involved in the virilization of the ovaries, but there is now convincing evidence that most, if not all, of the changes in the ovaries of the freemartin can be attributed to AMH (Vigier *et al.*, 1987).

The ovaries cease to grow, become depleted of germ cells, and are unable to produce estrogen but produce substantial amounts of testosterone and androstenedione (Pierrepoint *et al.*, 1969). This aberrant steroidogenesis, which has commenced by day 47 of fetal life (Shore and Shemesh, 1981), may be accompanied by the development of seminiferous tubules resulting in an ovo–testis. The result is a range of conditions from no apparent abnormality, other than sterility, to a considerable and progressive degree of masculinity. The external genitalia are not greatly modified, so that the freemartin remains phenotypically female.

Chorionic vascular anastomosis permits

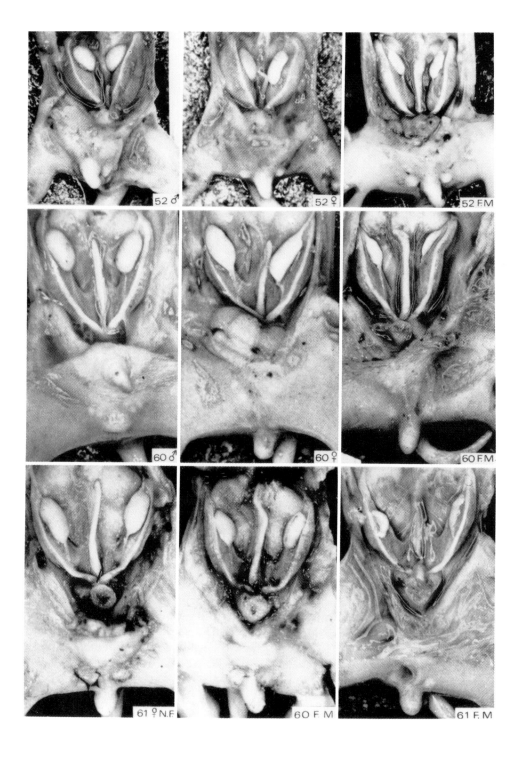

exchange between the fetuses of cellular elements as well as hormones. The exchange of hemopoietic stem cells results in twin calves having chimeric lymphocyte populations and red cell types that usually are identical (Marcum, 1974). In naturally occurring twin calves that are full sibs, chimerism of the lymphocyte population seems to be random so that the proportional representation derived from each calf may be between 0 and 100% but usually is stable and similar in both co-twins. Curiously, in Friesian–Brahman twins, produced by embryo transfer, the lymphocyte population was found always to be dominated by Friesian cells, and this dominance increased with age (Summers et al., 1984). Immunological tolerance has been demonstrated between chimeric twin calves, but this does not apply completely to skin grafted between naturally occurring Bos taurus twins (Emery and McCullagh, 1980). This is attributed to the existence of antigens that are specific to the skin and thus are not subject to the immunological tolerance invoked by hemopoietic chimerism. The situation appears to differ in Bos indicus cattle, as these will accept long-term skin grafts from Bos taurus twins produced by embryo transfer (Summers and Shelton, 1985). The question of whether chimerism in these animals ever extends to the germ cells has not been resolved, and

observations on progeny of dizygotic twins have produced only negative results (Ross and Thomas, 1978). It could occur only if primordial germ cells in the bovine fetus were transported in the bloodstream, as has been suggested (Ohno and Gropp, 1965), and for which there is one unconfirmed report (Wartenberg, 1983).

The most simple method for diagnosis of freemartinism is measurement of the length of the vagina, which is always short in affected animals (Kastli and Hall, 1978). Red-cell typing will reveal chimerism, which almost invariably is associated with freemartinism if the cotwin is male, while karyotyping of lymphocytes will demonstrate the presence of XY cells.

There is some evidence that fertility of the male twin to a freemartin is subnormal but the causative mechanism is unclear (Dunn et al., 1979).

V. Parturition

The endocrinological factors involved in parturition are discussed elsewhere (Chapter 10).

When twins consist of two genotypes of widely disparate durations of gestation, such as may be produced by embryo transfer, the calf whose genotype has the shorter

Figure 4 Dissections showing the gonads, the genital ducts, and the external genitalia of calf fetuses ($\times 3.3$ for all fetuses). Upper row: 52-day-old fetuses. On the left, normal male; notice that the testes are less elongated than the ovaries in the female and that the penis has already moved beyond the genital swellings. In the middle, normal female. On the right, freemartin (FM) with stunted ovaries. Middle row: 60-day-old fetuses. On the left, normal male; notice the scrotum and the long anogenital distance; in the middle, normal female. On the right, freemartin (FM) whose ovaries show the average degree of inhibition. Lower row: Variations in 60- or 61-day-old fetuses. On the left, normal 61-day-old female from a male–female twin pregnancy with no fusion (NF) of the chorions and no vascular connections. In the middle, 60-day-old freemartin (FM) showing slight ovarian inhibition; it belonged to a quadruplet (one male + three females) pregnancy. On the right, 61-day-old freemartin (FM) showing a severe degree of ovarian inhibition; it belonged to a triplet (one male + two females) pregnancy. From Jost et al. (1972), courtesy of Journals of Reproduction and Fertility Ltd.

duration induces parturition even when it occupies the uterine horn contralateral to the ovary with the active corpus luteum (Summers *et al.*, 1983).

Dystocia and retained placenta are the most frequent problems related to parturition in cattle, and Bellows *et al.* (1979) estimated that some 6.4% of calves die around the time of parturition, 72% of these from dystocia (Anderson and Bellows, 1967). Dystocia occurs most commonly when the sire is of a large breed or the dam is a heifer (Price and Wiltbank, 1978).

Parturition can occur within a few hours or as long as a day after the cow becomes restless and secludes herself from the herd. The first stage of labor lasts 2–6 h. Delivery lasts 30–40 min, and the placenta normally is shed 2–6 h later, but retention for up to 12 h is considered normal (Bazer and First, 1983).

VI. Postpartum Anestrus

Following parturition there is a variable period during which cows do not exhibit estrus. The duration of this postpartum anestrus is affected by a number of factors, including season of the year, level of nutrition before and after parturition, and suckling, as distinct from milking (Terqui *et al.*, 1982).

During pregnancy, the high levels of circulating progesterone and estradiol are associated with a marked reduction in the LH content of the anterior pituitary, and this is one of the initial limitations to the resumption of normal estrous cycles in postpartum cows (Nett, 1987). Removal of the negative feedback at parturition affects the hypothalamic–hypophyseal axis, increasing the releasable pool of LH and the frequency of LH pulses (Peters *et al.*, 1981). Pulse frequency increases gradually, leading to the final stages of follicular development and ovulation by 50 days postpartum in the majority of milked cows (Bulman and Lamming, 1978).

In suckling cows ovulation and estrus are further delayed. Although pituitary content of LH is normal, suckling suppresses the pulsatile release of LH (Short *et al.*, 1972; Walters *et al.*, 1982a), and there is some evidence that endogenous opioid peptides may be involved in the mediation of this effect (Gregg *et al.*, 1985). Milking, even four times a day, does not have as great an effect as does suckling (Carruthers and Hafs, 1980).

Strategies to reduce the duration of postpartum anestrus have included progestagen treatment by subcutaneous implant (Smith *et al.*, 1979, 1983) or intravaginal device (Munro and Moore, 1985), alone or in combination with equine chorionic gonadotropin (eCG) and short-term calf removal. Pulsatile injections of GnRH also have been used (Walters *et al.*, 1982b; Edwards *et al.*, 1983). Low intakes of energy (Dunn *et al.*, 1969) or crude protein (Wiltbank *et al.*, 1965) extend the duration of postpartum anestrus and modify the response to calf removal and GnRH (Whisnant *et al.*, 1985). These effects may be mediated through changes in insulin and glucose metabolism (McCann and Hansel, 1986). Further, there is evidence that exposure to bulls will reduce the duration of postpartum anestrus (Macmillan *et al.*, 1979).

VII. Control of Fertility

Several reproductive phenomena are amenable to a degree of artificial control. These include the time of ovulation and estrus, the number of eggs shed, and the time of parturition. In addition, a number of manipulative procedures such as embryo transfer, artificial insemination, *in vitro* fertilization, and the storage of frozen embryos may be used in combination with them.

A. Synchronization of Estrus

Control of the time of estrus allows more efficient use of artificial insemination and has become an integral part of embryo transfer programs. Similar techniques may be used to induce early postpartum estrus. Essentially, two approaches have been used in the cyclic cow, namely, artificial prolonging of diestrus using exogenous progestagen or shortening it with a luteolysin (see Mauleon and Ortavant, 1975).

Developments using progestagens may be divided into a number of phases (Robinson, 1977). Initially, progesterone and a number of analogs were administered by injection or feeding for 3 weeks; daily injections were impractical and the time of estrus after cessation of feeding was imprecise. Later, administration by subcutaneous implant or intravaginal pessary was investigated. The intravaginal progesterone releasing device (PRID; Ceva Chemicals, Sydney) provided a simple method of administration and removal and resulted in physiological levels of plasma progesterone over a long period of time. Following insertion, the plasma progesterone level rose to 5–10 ng/ml within an hour and remained high for 3–5 days. By 21 days it had declined to 2 ng/ml (Mauer *et al.*, 1975). Subcutaneous and intravaginal treatments resulted in an acceptable incidence and timing of the controlled estrus, but conception rates generally were lower than in untreated animals. There followed attempts to combine progestational treatments with estrogen or gonadotropin (Hansel and Beal, 1979). Estrogen played a negative role but, as in sheep, eCG has a useful role (Chupin *et al.*, 1975; Ortavant and Thibault, 1970). PRIDs for 14 days, accompanied by eCG injected 24 h before withdrawal, have resulted in good pregnancy rates (Munro and Moore, 1985) even when used with fixed-time insemination (Munro, 1987).

Following the demonstration that $PGF_{2\alpha}$ was luteolytic in the cow (Rowson *et al.*, 1972b), a new approach was taken. Destruction of an existing corpus luteum was followed rapidly by a fresh ovulation and estrus. To overcome the problem that luteolysis could only be induced reliably in corpora lutea more than 5 days old, two treatments, 11 days apart, were needed (Graves *et al.*, 1974). An alternative approach was to inseminate all cows in estrus over a period of 5 days. Cows that failed to show estrus then received $PGF_{2\alpha}$ and were inseminated at the ensuing estrus. Methods that combine progestagen and luteolytic treatments also are used. One involves administration of progestagen, either by PRID or a controlled intravaginal drug-releasing device (CIDR, AHI Plastic Moulding Company New Zealand,), or by subcutaneous silastic implant for 7–10 days followed by injection of $PGF_{2\alpha}$ at withdrawal or 24 h earlier. Such treatment has resulted in pregnancy rates, to fixed-time insemination, equal to those in control animals (Smith *et al.*, 1984). Another technique is to lyse the corpus luteum at the beginning of treatment with an injection of estrogen and follow with a progestagen implant for 10 days (Wiltbank *et al.*, 1971). Injection of eCG at the time of removal of the implant is advantageous (Mauleon *et al.*, 1978).

Table VII
Data from Experimental Synchronization of Estrus

Animals			Duration of progestagen (days)	Insemination time after end of progestagen treatment (h)	Calved to synchronized estrus (%)	Source
Type	Number	Treatment				
Friesian cows	252	PRID with estrogen capsule	12	52	27	Mawhinney and
calved >35 days	254	PRID with estrogen capsule	12	56	37	Roche (1978)
	241	PRID with estrogen capsule	12	60	37	
Saler cows	219	Norgestomet implant, estradiol by	7	48 and 72	59	Mauleon *et al.*
suckling calves		injection day 0, eCG at implant				(1978)
		removal				
	1092	Same as above	9	48 and 72	54	
	114	Same as above	11	48 and 72	46	
	166	Same as above	13–16	48 and 72	29	
Holstein heifers	242	PRID plus PG day 6	6 or 7	At observed estrus	72–82[a]	Smith *et al.* (1984)
Holstein heifers	274	PRID plus PG day 6	7	60	66[a]	Smith *et al.* (1984)
Friesian cows	288	PRID plus 500 or 750 IU eCG	14 or 16	54–58	33	Munro and
4–7 weeks						Moore (1985)
postpartum						
Friesian heifers	68	PRID plus 500 IU eCG at removal	14	48 or 56	75	Munro and
						Moore (1985)
Cross-bred heifers	88	Norgestomet implant, norgestomet	9	At observed estrus	59	Whittier *et al.*
		and estradiol injected day 0				(1986)
	89	Norgestomet implant, PG at	7	At observed estrus	58	
		removal				

[a]Pregnant 45–60 days after insemination.

Table VII presents a summary of some recent results using these several techniques.

B. Twinning

Since cattle are predominantly monotocous, there has been sustained interest in the induction of twinning. The use of exogenous gonadotropin has the disadvantage that the rate of transuterine migration of ova in cows is low (Scanlon, 1972), and the presence of one embryo in each uterine horn is necessary for a high success rate of twinning (Rowson *et al.*, 1971). Further, there is an enormous variation in response to eCG (reviewed by Robinson, 1977). Nevertheless, acceptable results have been achieved using eCG following a short-term treatment with progestagen (Mulvihill and Sreenan, 1978).

Recently, attention has been turned to other methods of inducing multiple ovulations using immunization against steroids (Wise and Schanbacher, 1983; Price *et al.*, 1987) or inhibin (Bindon *et al.*, 1988) but it is not yet clear whether these methods will achieve better control than use of eCG.

A number of experiments have been conducted to examine the feasibility of inducing twin pregnancies by embryo transfer, initially by surgical and later by nonsurgical methods. The latter offers the most viable commercial procedure. There are two approaches. The first is to transfer an embryo to a previously inseminated cow and the second is to transfer two embryos to an unmated cow at the appropriate time after estrus. In either case, best results generally have been obtained when one embryo is in each uterine horn (Rowson *et al.*, 1971; Scanlon 1972), although Sreenan and Diskin (1989) observed no difference in embryo survival and twin births in inseminated cows that received a transferred embryo in the uterine horn either ipsilateral or contralateral to the corpus luteum. However, the outcome of these strategies is affected by the dependence of embryo survival in the contralateral uterine horn, on the presence of a viable embryo in the ipsilateral horn (Christie *et al.*, 1979), and by the site of placement of the embryo in this horn (Newcomb *et al.*, 1980; Table VIII).

A major limitation to the use of embryo transfer in the production of twin calves is the availability of an adequate number of embryos. Recent experiments on *in vitro* fertilization of cattle oocytes matured *in vitro* offer the exciting possibility of establishing large supplies of embryos for use in commercial twinning (Lu *et al.*, 1987).

Although the incidence of twin ovulations in cattle is low (see above), results of ongoing experiments suggest that useful progress in enhancing fecundity can be achieved by selection (reviewed by Morris, 1984).

C. Embryo Manipulation

Embryo transfer and related reproductive technologies are of growing importance in animal production. The capacity of embryo transfer to increase fecundity and re-

Table VIII

Fetal survival at Day 42 in Heifers after Transfer on Day 7 of a Single embryo to the tip or base of the uterine horn ipsilateral or contralateral to the corpus luteum[a]

Group[b]	A	B	C	D
Site of transfer				
Ipsilateral horn	Tip	Tip	Base	Base
Contralateral horn	Tip	Base	Tip	Base
Recipients	20	20	20	20
Pregnant	18	17	15	13
Viable fetuses				
Ipsilateral horn	16	15	12	12
Contralateral horn	13	14	11	11
Total	29	29	23	23
Percent survival	72.5	72.5	57.5	57.5
Recipients with				
Viable twin fetuses	11	12	8	10
Percent of recipients	55.0	60.0	40.0	50.0

[a]From Newcomb *et al.*, (1980).

[b]*Significance* Highest pregnancy rates and overall fetal survival occurred when the ipsilateral transfer was to the tip of the uterine horn. Pregnancy: Groups A & B; 35/40; C & D; 28/40 ($P = 0.05$). Survival: Groups A & B; 58/80; C & D; 46/80 ($P < 0.05$).

duce generation interval constitutes a powerful tool in cattle breeding programs (Smith, 1988). The rate of genetic progress can be further enhanced by the splitting of embryos to produce monozygous twins (Williams *et al.*, 1983; Nicholas and Smith, 1983). Yet another dimension will be introduced with the improvement of the technique for production of large clones from embryonic cells (Robl *et al.*, 1986). The next step, that of gene transfer, may prove extremely useful in increasing the frequency of single desirable genes. Research on the production of transgenetic cattle will be greatly stimulated by the above-mentioned developments in the technology of *in vitro* maturation of oocytes (Lu, *et al.*, 1987), which may result in availability of the large numbers required for research and development in this area.

D. Induction of Parturition

The use of exogenous estrogen, prostaglandin, or corticosteroid for the induction of parturition is considered in Chapter 10.

References

Almquist, J. O., and Amann, R. P. (1961). *J. Dairy Sci.* **44,** 1668–1678.

Amann, R. P., and Almquist, J. O. (1976). *Proc. 6th Tech. Conf. A.I. and Reproduction, Milwaukee, 1976,* N.A.A.B., Columbia, Missouri, pp. 1–10.

Amann, R. P., Wise, M. E., Glass, J. D., and Nett, T. M. (1986). *Biol. Reprod.* **34,** 71–80.

Amoroso, E. C. (1952). *In* "Marshall's Physiology of Reproduction" (A. S. Parkes, ed.), 3rd ed., Vol. 2, pp. 127–311. Longmans, Green, London.

Anderson, D. C., and Bellows, R. A. (1967). *J. Anim. Sci.* **26,** 941 (abstr).

Arnstadt, K. I., and Cleere, W. F. (1981). *J. Reprod. Fertil.* **62,** 173–180.

Asdell, S. A. (1964). "Patterns of Mammalian Reproduction," 2nd ed. Cornell University Press (Comstock), Ithaca, New York.

Attal, J., and Courot, M. (1963). *Ann. Biol. Anim. Biochim. Biophys.* **3,** 219–241.

Ayalon, N. (1978). *J. Reprod. Fertil.* **54,** 483–493.

Baier, W. (1972). *In* "Riproduzione Animale e Fecondazione Artificiale," pp. 17–22. Edagricola, Bologna.

Baker, J. F., Stewart, T. S., Long, C. R., and Cartwright, T. C. (1988). *J. Anim. Sci.* **66,** 2147–2158.

Bartol, F. F., Thatcher, W. W., Bazer, F. W., Kimball, F. A., Chenault, J. R., Wilcox, C. J., and Roberts, R. M. (1981). *Biol. Reprod.* **25,** 759–776.

Bartol, F. F., Roberts,. R. M., Bazer, F. W., Lewis, G. F., Godkin, J. D., and Thatcher W. W. (1985). *Biol. Reprod.* **32,** 681–693.

Battista, P. J., Poff, J. P., Deaver, D. R., and Condon, W. A. (1987). *J. Reprod. Fertil.* **80,** 517–522.

Bazer, F. W., and First, N. L. (1983). *J. Animal Sci.* **57**(Suppl. 2), 425–460.

Bellows, R. A., Short, R. E., and Staigmiller, R. B. (1979). *In* "Animal Reproduction" (H. W. Hawk, C.

A. Kiddy and H. C. Cecil, eds.), p. 3. Allenheld, Osmun, Montclair, New Jersey.

Bennett, W. A., Watts, T. L., Blair, W. D., Waldhalm, S. J., and Fuquay, J. W. (1988). *J. Reprod. Fertil.* **83,** 537–543.

Berardinelli, J., Dailey, R. A., Butcher, R. L., and Inskeep, E. K. (1979). *J. Anim. Sci.* **49,** 1276–1280.

Bindon, B. M., O'Shea, T., Miyamoto, K., Hillard, M. A., Piper, L. R., Nethery, R. D., and Uphill, G. (1988). *Proc. Aust. Soc. Reprod. Biol.* **20,** 28 (abstr.).

Bulman, D. C., and Lamming, G. E. (1978). *J. Reprod. Fertil.* **54,** 447–458.

Butler, J. E., Hamilton, W. C., Sasser, R. G., Ruder, C. A., Hass, G. M., and Williams, R. J. (1982). *Biol. Reprod.* **26,** 925–933.

Byerley, D. J., Staigmiller, R. B., Berardinelli, J. G., and Short, R. E. (1987). *J. Anim. Sci.* **65,** 645–650.

Carrick, M. J., and Shelton, J. N. (1969). *J. Endocrinol.* **45,** 99–109.

Carruthers, T. D., and Hafs, H. D. (1980). *J. Anim. Sci.* **50,** 919–925.

Cates, W. F. (1975). *Proc. Annu. Meet. Soc. Theriogenol.,* 1–11.

Christie, W. B., Newcombe, R. and Rowson, L. E. A. (1979). *J. Reprod. Fert.* **56,** 701–706.

Chupin, D., Pelot, J., and Thimonier, J. (1975). *Ann. Biol. Anim. Biochim. Biophys.* **15,** 263–271.

Clegg, M. T., Weir, W. C., and Cole, H. H., (1965). *U.S. Dept. Agric. Misc. Publ.* **1005**.

Courot, M., Hochereau-de Reviers, M. T., and Ortavant, R. (1970). *In* "The Testis" (A. D. Johnson, W. R. Gomes, and L. L. VanDemark, eds.), Vol. 1, pp. 339–432. Academic Press, New York.

Critser, J. K., Block, T. M., Folkman, S., and Hauser, E. R. (1987a). *J. Reprod. Fertil.* **81,** 29–39.

Critser, J. K., Lindstrom, M. J., Hinshelwood, M. M., and Hauser, E. R. (1987b). *J. Reprod. Fertil.* **79,** 599–608.

Cupps, P. T., Anderson, L. L., and Cole, H. H. (1969). *In* "Reproduction in Domestic Animals" (H. H. Cole and P. T. Cupps, eds.), 2nd ed., pp. 217–250. Academic Press, New York.

Day, M. L., Imakawa, K., Pennel, P. L., Zalesky, D. D., Clutter, A. C., Kittok, R. J., and Kinder, J. E. (1986). *Biol. Reprod.* **62,** 1641–1648.

deJong, F. H., and Robertson, D. M. (1985). *Molec. cell. Endocrinol.* **42,** 95–103.

Denamur, R. (1972). *Proc. Int. Congr. Anim. Reprod., 7th, Munich* **1,** 19–44.

Dieleman, S. J., and Blankenstein, D. M. (1985). *J. Reprod. Fertil.* **75,** 609–615.

Diskin, M. G., and Sreenan, J. M. (1980). *J. Reprod. Fertil.* **59,** 463–468.

Dobson, H., and Kamonpatana, M. (1986). *J. Reprod. Fertil.* **77,** 1–36.

Dodson, S. E., McLeod, B. J., Haresign, W., Peters, A. R., and Lamming, G. E. (1988). *J. Reprod. Fertil.* **82,** 527–538.

Donaldson, L. E. (1968). *Aust. Vet. J.* **44,** 493–495

Dunn, H. O., McEntee, K., Hall, C. E., Johnson, R. H., Jr., and Stone, W. H. (1979). *J. Reprod. Fertil.* **57,** 21–30.

Dunn, T. G., Ingalls, J. E., Zimmerman, D. R., and Wiltbank, J. N. (1969). *J. Anim. Sci.* **29,** 719–726.

Edwards, S., Roche, J. F., and Niswender, G. D. (1983). *J. Reprod. Fertil.* **69,** 65–72.

Emery, D. and McCullagh, P. (1980). *Transplantation* **29,** 4–9.

Erickson, B. H., (1966a). *J. Anim. Sci.* **24,** 568–583).

Erickson, B. H., (1966b). *J. Anim. Sci.* **25,** 800–805.

Eyestone, W. H., Northey, D. L., and Leibfried-Rutlidge M. L. (1985). *Biol. Reprod.* **32,** (Suppl.), 100 (abstr.).

Fallon, G. R. (1961). *Proc. Int. Congr. Anim. Reprod., 4th, The Hague* **2,** 180–185.

Gandolfi, F and Moor, R. M. (1987). *J. Reprod. Fertil.* **81,** 23–28.

Ginther, O. J., and Del Campo, C. H. (1974). *Am. J. Vet. Res.* **35,** 193–203.

Ginther, O. J., Knopf, L., and Kastelic, J. P. (1989) *J. Reprod. Fert.* **87,** 223–230.

Glencross, R. G., Munro, I. B., Senior, B. E., and Pope, G. S. (1973). *Acta Endocrinol.* **73,** 374–384.

Godkin, J. D., Lifsey, B. J., Jr., and Baumbach, G. A. (1988). *Biol. Reprod.* **39,** 195–204.

Gordon, I., Williams, G. L., and Edwards, J. (1962). *J. Agric. Sci.* **59,** 143–198.

Graves, N. W., Short, R. E., Randel, R. D., Bellows, R. A., Kaltenbach, C. C., and Dunn, T. G. (1974). *J. Anim. Sci.* **39,** 208–209.

Gregg, D. W., Moss, G. E., Hudgens, R. E., and Malven, P. V. (1985). *J. Anim. Sci.* **61** (Suppl. 1), 418 (Abstr.).

Hafez, E. S. E. (1972). *In* "Riproduzione Animale e Fecondazione Artificiale," pp 107–123. Edagricola, Bologna.

Hafez, E. S. E., and Rajakoski, E. (1964). *Anat. Rec.* **150,** 303–316.

Hammond, J. (1927). "The Physiology of Reproduction in the Cow." Cambridge University Press, London.

Hansel, W. (1959). *In* "Reproduction in Domestic Animals" (H. H. Cole and P. T. Cupps, eds.), Vol. 1, pp. 223–265. Academic Press, New York.

Hansel, W., and Beal, W. E. (1979). *In* "Animal Reproduction," Beltsville Symposium in Agricultural Research 3 (H. W. Hawke, C. A. Kiddy, and H. C. Cecil, eds.), pp. 91–110. Allenheld, Osmun, Montclair, New Jersey.

Hansel, W., and Dowd, J. P. (1986). Hammond Memorial Lecture. *J. Reprod. Fertil.* **78,** 755–768.

Hansel, W. and Hickey, G. J. (1988). *Ann. N.Y. Acad. Sci.* **541,** 472–484.

Hauser, E. R. (1984). *Theriogenology* **21,** 150–169.

Hawk, H. W. (1978). *In* "Animal Reproduction," pp. 19–29. Beltsville Symposia in Agricultural Research. John Wiley and Sons, New York.

Heap, R. B., Fleet, I. R., and Hamon, M. (1985). *J. Reprod. Fertil.* **74,** 645–656.

Hein, W. R., Shelton, J. N., Simpson-Morgan, M. W., Seamark, R. F., and Morris, B. (1988). *J. Reprod. Fertil.* **83,** 309–323.

Howes J. R., Warnick, A. C., and Hentges, J. F. (1960). *Fertil. Steril.* **11,** 508–517.

Hunter, R. H. F., and Wilmut, I. (1984). *Reprod. Nutr. Develop.* **24,** 597–608.

Ireland, J. J., and Roche, J. F. (1987). In "Follicular Growth and Ovulation Rate in Farm Animals" (J. F. Roche and D. O'Callaghan, eds.), pp. 1–18. Martinus Nijhoff, The Hague.

Izard, M. K., and Vandenbergh, J. G. (1982a). *J. Anim. Sci.* **55,** 1160–1168.

Izard, M. K., and Vandenbergh, J. G. (1982b). *J. Reprod. Fertil.* **66,** 189–196.

Jainudeen, M., and Hafez, E. S. E. (1980). *In* "Reproduction in Farm Animals" (E. S. E. Hafez, ed.), 4th ed., pp. 247–283. Lea and Febiger, Philadelphia.

Jochle, W. (1966). *In* "Reproduction in the Female Mammal" (G. E. Lamming and E. C. Amoroso, eds.), pp. 267–281. Butterworths, London.

Jochle, W. (1972a). *Proc. Int. Congr. Anim. Reprod., 7th, Munich* **1,** 97–124.

Jochle, W. (1972b). *Int. J. Biometerol.* **16,** 131–144.

Jost, A., Vigier, B., and Prepin, J. (1972). *J. Reprod. Fertil.* **29,** 349–379.

Kastli, F., and Hall, J. G. (1978). *Vet. Rec.* **102,** 80–83.

Kinder, J. E., Day, M. L., and Kittok, R. J. (1987). *J. Reprod. Fertil. Suppl.* **34.** 167–186.

King, G. J., Atkinson, B. A., and Robertson, H. A. (1980). *J. Reprod. Fertil.* **59,** 95–100.

Lafrance, M., and Goff, A. K. (1985). *Biol. Reprod.* **32** (Suppl.), 41 (abstr.).

Lamond, D. R. (1970). *Anim. Breed. Abstr.* **38,** 359–372.

Lawson, R. A. S., Rowson, L. E. A., and Adams, C. E. (1972). *J. Reprod. Fertil.* **28**, 313–315.

Lillie, F. R. (1917). *J. Exp. Zool.* **23**, 371–452.

Lu, K. H., Gordon, I., Gallagher, M., and McGovern, H. (1987). *Vet. Rec.* **121**, 259–260.

Lunstra, D. D., Ford, J. J., and Echternkamp, S. E. (1978). *J. Anim. Sci.* **46**, 1054–1062.

Lussier, J. G., Matton, P., and Dufour, J. J. (1987). *J. Reprod. Fertil.* **81**, 301–307.

Macmillan, K. L., Allison, A. J., and Struthers, G. S. (1979). *N. Z. J. Exp. Agric.* **7**, 121–124.

McCann, J. P., and Hansel, W. (1986). *Biol. Reprod.* **34**, 630–641.

McClure, T. J., Nancarrow, C. D., and Radford, H. M. (1978). *Aust. J. Biol. Sci.* **31**, 183–186.

Mann, T., Rowson, L. E. A., Short, R. V., and Skinner, J. D. (1967). *J. Endocrinol.* **38**, 455–468.

Marcum, J. B. (1974). *Anim. Breed. Abstr.* **42**, 227–242.

Marion, G. B., Smith, V. R., Wiley, T. E., and Barrett, G. R. (1950). *J. Dairy Sci.* **33**, 885–889.

Marshall, F. H. A. (1922). "The Physiology of Reproduction," 2nd ed. Longmans, Green, London.

Martin, J. B., Price, C. A., Thiery, J.-C., and Webb, R. (1988). *J. Reprod. Fertil.* **82**, 319–328.

Mauer, R. E., Webel, S. K., and Brown, M. D. (1975). *Ann. Biol. Anim. Biochim. Biophys.* **15**, 291–296.

Mauleon, P. (1961). *Ann. Biol. Anim. Biochim. Biophys.* **1**, 1–9.

Mauleon, P., and Ortavant, R. (eds.) (1975). Colloquium, "Control of Sexual Cycles in Domestic Animals," *Ann. Biol. Anim. Biochim. Biophys.* **15**.

Mauleon, P., Chupin, D., Pelot, J., and Aguer, D. (1978). *In* "Control of Reproduction in the Cow" (J. M. Sreenan, ed.), pp. 531–545. Martinus Nijhoff, The Hague.

Mahwhinney, S., and Roche, J. F. (1978). *In* "Control of Reproduction in the Cow" (J. M. Sreenan, ed.), pp. 511–530. Martinus Nijhoff, The Hague.

Moran, C., Quirke, J. F., and Roche, J. F. (1989). *Anim. Reprod. Sci.* **18**, 167–182.

Morris, C. A. (1984). *Anim. Breed. Abstr.* **52**, 803–819.

Mulvihill, P. and Sreenan, J. M. (1978). In "Control of Reproduction in the Cow" (J. M. Sreenan, ed.) pp. 486–510. Martinus Nijhoff, The Hague.

Munro, R. K. (1987). *Aust. Vet. J.* **64**, 62–63.

Munro, R. K., and Moore, N. W. (1985). *Aust. Vet. J.* **62**, 228–234.

Nelsen, T. C., Short, R. E., Phelps, D. A., and Staigmiller, R. B. (1985). *J. Anim. Sci.* **61**, 470–473.

Nessan, G. K., and King, G. J. (1981). *J. Reprod. Fertil.* **61**, 171–178.

Nett, T. M. (1987). *J. Reprod. Fertil. Suppl.* **34**, 201–203.

Newcomb, R., Christie, W. B., and Rowson, L. E. A. (1980). *J. Reprod. Fertil.* **59**, 31–36.

Nicholas, F. W. (1987). "Veterinary Genetics." Clarendon Press, Oxford.

Nicholas, F. W., and Smith, C. (1983). *Anim. Prod.* **36**, 341–353.

Ohno, S., and Gropp, A. (1965). *Cytogenetics* **4**, 251–261.

O'Neill, C. (1985). *J. Reprod. Fertil.* **75**, 375–380.

Ortavant, R., and Thibault, C. (1970). *Ann. Biol. Anim. Biochim. Biophys.* **10**, Suppl.

Ortavant, R., Courot, M., and Hochereau, M. T. (1969). *In* "Reproduction in Domestic Animals" (H. H. Cole and P. T. Cupps, eds.), 2nd ed., pp. 251–276. Academic Press, New York.

Perry, J. S. (1969). *J. Reprod. Fertil. Suppl.* **7**.

Perry, J. S. (1981). *J. Reprod. Fertil.* **62**, 321–335.

Peters, A. R., Lamming, G. E., and Fisher, M. W. (1981). *J. Reprod. Fertil.* **62**, 567–573.

Picard, J. Y., Tran, D., and Josso, N. (1978). *Molec. Cell. Endocrinol.* **12**, 17–30.

Picard, J. Y., Goulut, C., Bourrillon, R., and Josso, N. (1986). *FEBS Lett.* **195**, 73–76.

Pierrepoint, C. G., Stewart, J. S. S., and Rack, J. (1969). *J. Reprod. Fertil. Suppl.* **7**, 63–72.

Pierson, R. A., and Ginther, O. J. (1984). *Theriogenology* **22**, 225–233.

Plasse, D., Warnick, A. C., and Koger, M. (1968a). *J. Anim. Sci.* **27**, 94–100.

Plasse, D., Warnick, A. C., Reese, R. E., and Koger, M. (1968b). *J. Anim. Sci.* **27**, 101–104.

Plasse, D., Warnick, A. C., and Koger, M. (1970). *J. Anim. Sci.* **30**, 63–72.

Pope, G. S., and Hodgson-Jones, L. S. (1975). *Vet. Rec.* **96**, 154.

Price, C. A., Morris, B. A., and Webb, R. (1987). *J. Reprod. Fertil.* **81**, 149–160.

Price, T. D., and Wiltbank, J. N. (1978). *Theriogenology* **9**, 195–219.

Pruit, R. J., Corah, L. R., Stevenson, J. S., and Kiracofe, G. H. (1986). *J. Anim. Sci.* **63**, 579–585.

Randel, R. D. (1984). *Theriogenology* **21**, 170–185.

Randel, R. D., Rutter, L. M., and Rhodes, R. C. (1982). *J. Anim. Sci.* **54**, 806–810.

Roberson, M. S., Ansotegui, R. P., Berardinelli, J. G., Whitman, R. W., and McInerney, M. J. (1987). *J. Anim. Sci.* **64**, 1601–1605.

Roberts, R. M., and Bazer, F. W. (1988). *J. Reprod. Fertil.* **82**, 875–892.

Robinson, T. J. (1955). *J. Endocrinol.* **12**, 163–173.

Robinson, T. J. (1957). *In* "Progress in the Physiology of Domestic Animals" (J. Hammond, ed.), Vol. 3, pp. 793–904. Butterworths, London.

Robinson, T. J. (1973). *J. Reprod. Fertil. Suppl.* **18,** 103–109.

Robinson, T. J. (1977). *In* "Reproduction in Domestic Animals" (H. H. Cole and P. T. Cupps, eds.), 3rd ed., pp. 433–454. Academic Press, New York.

Rollinson, D. H. L. (1971). *Anim. Breed. Abstr.* **39,** 407–427.

Robl, J. M., Prather, R., Eyestone, W., Barnes, F., Northey, D., Gilligan, B., and First, N. L. (1986). *Theriogenology* **25,** 189 (abstr.).

Ross, D. S., and Thomas, W. J. (1978). *Anim. Blood Groups Biochem. Genet.* **9,** 3–8.

Rowson, L. E. A., Lawson, R. A. S., and Moor, R. M. (1971). *J. Reprod. Fertil.* **25,** 261–268.

Rowson, L. E. A., Lawson, R. A. S., Moore, R. M., and Baker, A. A. (1972a). *J. Reprod. Fertil.* **28,** 427–431.

Rowson, L. E. A., Tervit, R., and Brand, A. (1972b). *J. Reprod. Fertil.* **29,** 145 (abstr.).

Ruckebusch, Y., and Bayard, F. (1975). *J. Reprod. Fertil.* **43,** 193–196.

Sasser, R. G., Ruder, C. A., Ivani, K. A., Butler, J. E., and Hamilton, W. C. (1986). *Biol. Reprod.* **35,** 936–942.

Sauer, M. J., Foulkes, J. A., Worsfold, A., and Morris, B. A. (1986). *J. Reprod. Fertil.* **76,** 375–391.

Savio, J. D., Keenan, L., Boland, M. P., and Roche, J. F, (1988). *J. Reprod. Fertil.* **83,** 663–671.

Scanlon, P. F. (1972). *J. Anim. Sci.* **34,** 791–794.

Scaramuzzi, R. J., Turnbull, K. E., and Nancarrow, C. D. (1980). *Aust. J. Biol. Sci.* **33,** 63–69.

Schillo, K. K., Hansen, P. J., Kamwanya, L. A., Dierschke, D. J., and Hansen, E. R. (1983). *Biol. Reprod.* **28,** 329–341.

Shemesh, M., Ayalon, N., and Lindner, H. R. (1972). *J. Endocrinol.* **55,** 73–78.

Shemesh, M., Milaguir, F., Ayalon, N., and Hansel, W. (1979). *J. Reprod. Fertil.* **56,** 181–185.

Shore, L., and Shemesh, M. (1981). *J. Reprod. Fertil.* **63,** 309–314.

Short, R. E., Bellows, R. A., Moody, E. L., and Howland, B. E. (1972). *J. Anim. Sci.* **34,** 70–74.

Short, R. E., Bellows, R. A., Carr, J. B., and Staigmiller, R. B. (1976). *J. Anim. Sci.* **43,** 1254–1258.

Sirois, J., and Fortune, J. E. (1988). *Biol. Reprod.* **39,** 308–317.

Smith, C. (1988). *Theriogenology* **29,** 203–212.

Smith, M. F., Burrell, W. C., Shipp, L. D., Sprott, L. R., Songster, W. N., and Wiltbank, J. N. (1979). *J. Anim. Sci.* **48,** 1285–1294.

Smith, M. F., Lishman, A. W., Lewis, G. S., Harms, P. G., Ellersieck, M. R., Inskeep, J. N., Wiltbank, J. N., and Amoss, M. S. (1983). *J. Anim. Sci.* **57,** 418–424.

Smith, R. D., Pomerantz, A. J., Beal, W. E., McCann, J. P., Pilbeam, P. E., and Hansel, W. (1984). *J. Anim. Sci.* **58,** 792–800.

Sorenson, A. M., Hansel, W., Hough, W. H., Armstrong, D. T., McEntee, K., and Bratton, R. W. (1959). *Cornell Agric. Exp. Sta. Bull.* **936.**

Sreenan, J. M. and Diskin, M. G. (1989). *J. Reprod. Fertil.* **87,** 657–664.

Stabenfeldt, G. H., Osburn, B. I., and Ewing, L. L. (1970). *Am. J. Physiol.* **218,** 571–575.

Summers, P. M., and Shelton, J. N. (1985). *Aust. J. Exp. Biol. Med. Sci.* **63,** 329–332.

Summers, P. M., Shelton, J. N., and Edwards, J. (1983). *Anim. Reprod. Sci.* **6,** 79–89.

Summers, P. M., Shelton, J. N., Morris, B., and Bell, K. (1984). *Aust. J. Exp. Biol. Med. Sci.* **62,** 27–45.

Tanabe, T. Y., and Almquist, J. O. (1967). *Pa. Agric. Expt. Sta. Bull.* **736.**

Terqui, M., Chupin, D., Gautier, D., Perez, N., Pelot, J., and Mauleon, P. (1982). *In* "Factors Influencing Fertility in the Postpartum Cow," pp. 384–408. Martinus Nijhoff, The Hague.

Thain, R. I. (1968). *Aust. Vet. J.* **44,** 218–222.

Thatcher, W. W., Knickerbocker, J. J., Bartol, F. F., Bazer, F. W., Roberts, R. M., and Drost, M. (1985). *Theriogenology* **23,** 129–143.

Thibault, C., Courot, M., Martinet, L., Mauleon, P., du Mesnil du Buisson, F., Ortavant, R., Pelletier, J., and Signoret, J. P. (1966). *J. Anim. Sci. Suppl.* **25,** 119–142.

Tran, D., and Joso, N. (1982). *Endocrinology* **111,** 1562–1567.

Trimberger, G. W. (1948). *Nebr. Univ. Agr. Exp. Sta. Res. Bull.* **153.**

VanDemark, N. L., Fritz, G. R., and Mauger, A. E. (1964). *J. Dairy Sci.* **47,** 898–904.

Vigier, B., Watrin, F., Magree, S., Tran, D., and Josso, N. (1987). *Development* **100,** 43–55.

Wagnon, K. A., Rollins, W. C., Cupps, P. T., and Carroll, F. D. (1972). *J. Anim. Sci.* **34,** 1003–1010.

Walters, D. L., and Schallenberger, E. (1984). *J. Reprod. Fertil.* **71,** 503–512.

Walters, D. L., Kaltenbach, C. C., Dunn, T. G., and Short, R. E. (1982a). *Biol. Reprod.* **26,** 640–646.

Walters, D. L., Short, R. E., Convey, E. M., Staigmiller, R. B., Dunn, T. G., and Kaltenbach, C. C. (1982b). *Biol. Reprod.* **26,** 655–662.

Wartenberg, H. (1983). *Bibl. Anat.* **24,** 93–110.

Whisnant, C. S., Kiser, T. E., Thompson, F. N., and Hall, J. B, (1985). *Theriogenology* **24,** 565–573.

White, I. R., Russel. A. J. F., Wright, I. A., and Whyte, T. K. (1985). *Vet. Rec.* **117,** 5–8.

Whittier, J. C., Deutscher, G. H., and Clanton, D. C. (1986). *J. Anim. Sci.* **63,** 700–704.

Wiebold, J. L. (1988). *J. Reprod. Fertil.* **84,** 393–399.

Williams, T. J., Elsden, R., and Seidel, G. E., Jr. (1983). *Proc. Ann. Conf. A.I. & Embryo Transfer in Beef Cattle,* Denver, Colorado, January 1983, pp. 45-51.

Wilmut, I., Sales, D. I., and Ashworth, C. J. (1986). *J. Reprod. Fertil.* **76,** 851–864.

Wiltbank, J, N., Bond, J., Warwick, E. J., Davis, R. E., Cook, A. C., Reynolds, W. L., and Hazen, M. W. (1965). *USDA Tech. Bull.* **1314.**

Wiltbank, J, N., Sturges, J. C., Wideman, D., LeFever, D. G., and Faulkner, L. G. (1971). *J. Anim. Sci.* **33,** 600–606.

Winters, L. M., Green, W. W., and Comstock, R. E. (1942). *Minn. Univ. Agric. Exp. Sta. Tech. Bull.* **151.**

Wise, T. H., and Schanbacher, B. D. (1983). *J. Reprod. Fertil.* **69,** 605–612.

Wisnicky, W., and Casida, L. E. (1948). *J. Am. Vet. Med. Assoc.* **113, 451–452.**

CHAPTER 14

Reproduction in the Pig

PHILIP DZIUK

I. Introduction

The pig has some unique reproductive characteristics that set it somewhat apart from other species. The uterine horns are very long and convoluted to accommodate the relatively large litter contained within a relatively short abdominal cavity. Insemination is intrauterine with a relatively large volume of semen. Pregnancy is maintained on the basis of the proportion of the uterus occupied. Time of initiation of parturition seems to be determined by an endocrine contribution by each of the fetuses, with each mature fetus contributing more than each of the less developed litter mates. With the development of several experimental techniques in recent years the pig has

achieved considerable attention as a respectable research animal. This chapter will discuss some of the ideas we now consider current knowledge. Emphasis will be on those areas of study with which I am most familiar. Any exclusions do not connote lesser importance but the discussion cannot be all inclusive. This chapter is intended to complement other chapters that deal with specific aspects of reproduction in greater depth.

II. Puberty

Onset of puberty in the pig is not a discrete, distinct event but a gradual attainment of the ability to reproduce. In spite of considerable research effort to unravel the complexities of those factors associated with puberty, there is still much to be understood. Genetic diversity contributes much to variation in age at puberty. Some prolific Chinese breeds are fertile as early as 90 days of age, whereas typical occidental breeds are pubertal at about 210 days of age (Wang *et al.*, 1988). The mechanisms controlling this diversity are unknown. Although these Chinese breeds also frequently have a high number of ovulations, there is no evidence that high ovulation rate is necessarily linked to early puberty (Legault, 1985). Because not all prolific breeds are also pubertal at a young age, there is reason to think that age at puberty and ovulation rate can be separate.

The endocrine and neuroendocrine status of pigs at puberty may be influenced by events far in advance of puberty, perhaps extending back into fetal stages (Meijer *et*

al., 1985). Feedback mechanisms that function at puberty may be established in the fetus at day 35–40 of gestation when concentrations of steroids in fetal fluids are relatively high and may differ between sexes (Ponzilius *et al.*, 1986).

Gradual changes in the production, metabolism, and sensitivity to steroids and gonadotropins occur from birth to puberty (Dial *et al.*, 1984). The gilt can apparently metabolize estrogen less readily as she matures, even though she may not be producing much estrogen during early growth. The growth of the uterus is a sensitive biological measure of circulating estrogen. The uterus of an intact gilt at 100 days of age will be the same length and weight as one from a gilt that has been ovariectomized much earlier. Only after 100 days of age does the intact gilt have a heavier and longer uterus than her ovariectomized contemporary, indicating that biologically effective concen-

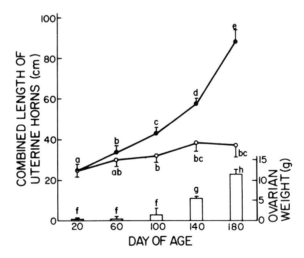

Figure 1 Growth patterns (mean ± SEM) of uterine length and ovarian weight in intact gilts (●) and uterine length in ovariectomized gilts (○) from 20 to 180 days of age. Means with different letters differ, $p < 0.05$. From Wu and Dziuk (1988).

trations of estrogen are not reached at an earlier age (Wu and Dziuk, 1988). Fig 1

Gilts do not develop follicles following injections of exogenous gonadotropins such as equine chorionic gonadotropin (eCG) before 100 days of age. From about 120 to 180 days of age the sensitivity of the ovaries to gonadotropins increases (Christenson et al., 1985). The sensitivity of the hypothalamic–pituitary system also increases as evidenced by increased response to exogenous estrogen. Gilts induced to ovulate at 145 days of age will show estrus, eggs will be fertilized, and embryos will develop, but pregnancies will rarely carry to term (Dziuk and Gehlbach, 1966). Attempts to maintain pregnancy in these young gilts with exogenous estrogen, which is normally very luteotropic, have not succeeded. Administration of exogenous progestin in the form of implants, injections, or orally active materials has maintained pregnancy in some cases (Ellicott et al., 1973). This would indicate that the function of the corpora lutea is in some way inadequate but that the uterus can function. Adolescent infertility, for whatever cause, effectively prevents any significant proportion of prepubertal gilts from completing gestation. As gilts approach puberty at the normal age, or in older gilts with delayed puberty, a single injection of eCG will induce follicular development, estrus, and ovulation with fertilization and a normal pregnancy (Dziuk and Dhindsa, 1969). The ovary, uterus, and the mechanisms for maintenance of pregnancy develop gradually, and release of pituitary gonadotropins is perhaps the final limitation just at puberty.

The pituitary can be stimulated to secrete the follicle-inducing gonadotropins not only by estrogen but by a number of environmental factors (Dial et al., 1983). The smell, sight, sound, and touch of a mature boar make up a powerful signal to the gilt on the verge of puberty. Puberty may be delayed for several weeks in gilts kept isolated from boars or in close confinement (Caton et al., 1986). Daily exposure of gilts near puberty to a boar will result in estrus and ovulation in a significant proportion in about 5 days, with a normal pregnancy following mating. Several studies have been directed toward understanding the relative importance of transport of gilts, new environment, number of females per group, and the smell, sight, and presence of a boar (Cronin, 1983; Eastman et al., 1986; Killian et al., 1987). Each of these factors can have some influence, but it appears that odor associated with the boar's breath and urine is a primary factor in inducing puberty (Claus and Weiler, 1985; Diekman and Grieger, 1988; Lee et al., 1987).

In addition to the social environment, the physical environment can influence the age at the onset of puberty. Season of birth seems to influence age at puberty, with gilts born in fall coming into estrus in the spring at a younger age than their spring-born sisters. Research to understand the influence of day length has given somewhat equivocal results. The sensitivity of the gilt to different light intensities and light–dark periods is minimal (Awotwi and Anderson, 1985; Simoneau et al., 1988). The influence of temperature on onset of puberty is also minimal, although anecdotal reports indicate that severe weather may temporarily inhibit estrus and pleasant weather may stimulate it.

Plane of nutrition may influence age at puberty; however, the pig seems to be influenced less by body weight and plane of nutrition than some other species (Hughes, 1982). Very severe restriction of diet can delay puberty, but an excess of nutrients does not seem to hasten the onset. Under some circumstances, the number of ovulations can be increased by a rising plane of nutrition, but effect on age at puberty is not great.

In summary, the age at puberty is governed primarily by the genotype of the gilt and the presence of a mature boar, with some lesser influence by other factors.

III. Estrous Cycle

The estrous cycle of the gilt is normally about 21 days. The variation in length is not great; thus the date of a recurring estrus can be predicted with reasonable confidence. Mated gilts that return to estrus have one peak of frequency of estrus at 21 days after mating and a second peak at 25–30 days. This second peak seems to be the result of the influence of a small number of embryos that can delay the regression of the corpora lutea but are incapable of maintaining pregnancy. Pregnancy normally interrupts estrous cycles but not always. Observations of recurring estrus and matings together with dates of farrowing reveal that some farrowings are at intervals from mating that indicate pregnancy existed at the time of second mating.

The behavior of the gilt throughout the estrous cycle has certain characteristics modified by individual behavior patterns. The successful herdsman needs to be keenly alert to not only the more obvious signs that nearly all gilts exhibit but also the subtle differences peculiar to each gilt (Hemsworth, 1982). During diestrus the gilt rejects all advances by the boar. In the early stages of proestrus, she may be attracted to the boar and show more than usual interest in him

and the herdsman who may be checking for estrus. This interest may extend to fighting rather than to retreat, and she may have a male-like attitude toward other gilts. The vulva may be swollen and pink with a slight discharge of cloudy mucus. As estrus approaches the gilt actively seeks the boar, nuzzling and smelling and emitting a peculiar growling. During estrus the gilt will stand near the boar and when pushed will resist moving, ears will be perked, and she will act as if in a trance. Attempts to move an estrous gilt at this time only make her more determined to stand solidly. As estrus progresses interest in the boar wanes until the boar is again rejected. Gilts that are receptive to one boar often reject a second boar or may stand for back pressure by the herdsman but not accept a particular boar. The characteristic that renders one boar more attractive than another is unknown. It may be odor, behavior of the boar, or the result of early experience by the gilt.

An accounting of the behavior of a sow that had been kept for many years and had borne many litters is both interesting and revealing. The sow was termed "sagacious" in the ways of opening gates and finding her way to and from places of interest. When her fancy turned to thoughts of a boar she was reported to open gates and circumvent fences until she reached her object. When her needs had been satisfied she returned in the same manner without any human intervention (White, 1902).

In managing a breeding herd using a hand-mating system, it is easier and more effective to allow the gilt to express her interest rather than trying to force the boar to elicit a response. Gilts and sows know when they are in estrus. Boars only discover it when the gilt tells them by body language—they stand rather than run. Gilts allowed access to contact with boars on a periodic routine daily basis will readily run to the boar and stand in his presence. The sagacious sow was a good example. The attitude of the herdsman as reflected by his interaction with the animals has a profound effect on the behavior of the animals toward the herdsman and consequently the reproductive efficiency of the herd. Happy pigs are more fertile pigs. Fearful pigs are apprehensive of all manipulations by the herdsman, including those associated with reproduction. Herds with a high reproductive efficiency have frequent favorable contact with the herdsman. When hand mating is practiced, the proportion of animals detected in heat, mated, and conceiving is highest when checks for estrous behavior are frequent and conducted by a conscientious and considerate herdsman.

The concentrations of hormones change dramatically and quite predictably during the estrous cycle (Fig. 2). Estrogen rises during the follicular phase, triggering a peak of luteinizing hormone (LH) at the onset of estrus. (Hansel *et al.*, 1973). Ovulation occurs about 40 h after the LH surge. This interval from LH surge to ovulation was determined by recovery of recently ovulated eggs after injection of human chorionic gonadotropin (Dziuk and Baker, 1962). Concentrations of progesterone rise gradually for the first 10 days after ovulation. By day 16 progesterone declines and the cycle repeats.

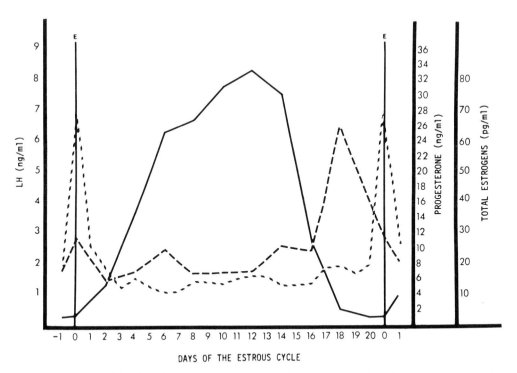

Figure 2 Peripheral levels of luteinizing hormone (-----), progesterone, (——) and estrogens (–––) during the porcine estrous cycle. From Hansel *et al.* (1973).

Corpora lutea regress and follicles begin development near day 15 after ovulation in a nonpregnant gilt. During the luteal phase estrogens are very luteotropic. An injection or oral administration of an estrogen will prolong the life of the corpora lutea for several weeks (Long *et al.*, 1988). Hysterectomy or ingestion of estrogenic mycotoxins such as zealeralone will also extend the life of the corpora lutea with consequent extended diestrus (Flowers *et al.*, 1987). These conditions often led the herdsmen to consider gilts pregnant when they are pseudopregnant. Progesterone concentration is maintained under these conditions and further supports such an erroneous conclusion.

IV. Fertilization

Successful fertilization of the eggs is a result of the optimal combination of numbers of normal sperm and time of insemination relative to ovulation. Eggs and sperm both have finite lives only in an environment that meets many rigid requirements. A converse truism is that deposition or storage of sperm in a hostile environment or at a time or place that does not permit sperm to meet an egg within a few hours after ovulation will not result in normal fertilization. Pig breeders nearly always want to optimize fertility. What principles and practices will lead to maximum fertility?

The ejaculate of the boar may contain 10 to 100×10^9 sperm in 50–400 ml of fluid. Insemination is intrauterine with the counterclockwise, corkscrew tip of the boar's penis locking into the posterior one-third of the cervix. Copulation may extend over a period of several minutes while periodic waves of sperm and accessory secretion are ejaculated. Within a few hours after mating the uterus is relatively free of semen and a reservoir of sperm is established at the uterotubal junction. Sperm can be recovered from the ampulla of the oviduct within a few minutes after insemination, but mating normally precedes ovulation by several hours; thus the vanguard of sperm may not serve any useful purpose. Sperm in the reservoir move up to the site of fertilization in the ampulla at the time of ovulation (Hunter *et al.*, 1987). Relatively few sperm are in the vicinity of the recently ovulated eggs when the ratio of recovered eggs to recovered sperm is nearly 1 : 1. As the interval from ovulation increases, so does the number of sperm in the vicinity of the egg and embedded in the zona pellucida increase. This indicates that there is a decreasing gradient in the number of sperm present from the infundibulum to the uterus. There is an increasing gradient of sperm in the ampulla from the time of ovulation forward. These gradients due to time and anatomy serve to increase the chance that some sperm will arrive at the propitious moment but prevent the chance of polyspermy from too many sperm penetrating the egg.

The time of insemination relative to ovulation has an important effect on the efficiency of fertilization and consequently on conception rate and litter size. A number of studies have noted that both conception rate and litter size are low when inseminations are done at the very onset of estrus. They rise to a maximum when insemination is done near 12 h before ovulation and decline to near zero when insemination is a few hours after ovulation. Gilts were inseminated twice with boars whose offspring were distinguishable, first with one boar and then six hours later to a second boar. About 85%

of piglets were sired by the second boar when the interval from the second boar to ovulation was 12 h or more. The majority of piglets were sired by the first boar when his mating was 12 h or less from ovulation. This is corroborating evidence that the optimum time for insemination is 12 h before ovulation (Dziuk, 1970).

The number of sperm at the site of fertilization capable of fertilizing eggs is also critical. This number can be influenced by one or more of several factors: fertility of the male, storage conditions, concentration of sperm, and volume of semen. The large uterus of the pig requires that the volume of semen placed intracervically be from 50 to 100 ml for adequate transport of sperm. Insemination by surgery of as little as 0.2 ml of semen at the uterotubal junction is adequate to establish a reservoir and produce fertilized eggs. The concentration of sperm is most critical. The number of sperm at the site of fertilization is directly related to the concentration of sperm inseminated without regard for the volume, provided the volume was adequate for transport (Baker *et al.*, 1968).

Although it is important to have an adequate number of sperm at the site of fertilization at the time of ovulation, it is equally important that an excess number is not present. Normal fertilization requires one sperm per egg, no more. If an egg is penetrated simultaneously by several sperm, the resulting polyspermy is lethal. Polyspermy occurs more frequently when insemination is done after ovulation or when the restriction of the uterotubal junction and isthmus is overridden by surgical or endocrine manipulation (Hunter and Nichol, 1988). Apparently the block to polyspermy does not have a chance to work when too many sperm are available and penetrate the zona

pellucida within a short period. As important as it is that some sperm reach the site of fertilization, it is perhaps a greater challenge to reduce the 50,000,000,000 inseminated down to the 10 to 20 needed for fertilization.

The differences between boars in the ability to fertilize eggs are reflected in both conception rate and litter size. Because failure to conceive is equivalent to a litter size of zero, the two measures can really be combined. Fertility from natural and artificial insemination has been found to vary considerably between boars. When semen from boars is mixed or when gilts are mated within a few minutes to two boars whose offspring are distinguishable, one boar usually sires a preponderance of piglets. The boar that sires most of the piglets from heterospermic insemination also has the higher conception rate and greater litter size when used alone (Martin and Dziuk, 1977). Evidence from work with mice indicates that sperm from the more fertile male attach to and penetrate eggs more quickly than sperm from the less fertile male (Robl and Dziuk, 1987). Because both eggs and sperm have a finite fertilizable life span, the slower sperm may not fertilize the egg before either or both the egg and sperm are aged beyond their ability to produce normal fertilization. This delay may produce an abnormal embryo, or the embryo may be retarded and not implant or compete successfully with normal littermates for uterine resources (Dziuk, 1987).

A useful general admonition to swine breeders is to mate gilts often during estrus to different boars. This ensures that at least one mating is likely to be near the optimum interval to ovulation and to at least one fertile boar.

V. Pregnancy

A. Maintenance

The mated gilt must recognize if she is pregnant and should maintain the corpora lutea and the pregnancy, or if she is not pregnant and should return to estrus, ovulate and start over. Not only does she have to decide between pregnant and not pregnant by day 15, but she has to also consider if she is carrying about 10 embryos or two or three embryos. A set of twins in pigs is not efficient reproduction. The mechanisms that determine the numbers of embryos needed to continue the pregnancy are not completely understood, but results of several studies have given some insight into possible mechanisms. Embryos at the four-cell stage are moved through the oviduct into the uterus at about 48 h after ovulation. They remain in the upper uterus until day 6, when they are moved slowly down each uterine horn. At about day 9 embryos reach the uterine body, where they continue to move into the horn opposite from the one of origin (Dhindsa *et al.*, 1967). Embryos from each horn pass each other and mix so that about three-eighths of the embryos in a horn originate from the horn opposite to the one in which they finally implant (Dziuk *et al.*, 1964). When one ovary is removed before mating, all embryos must necessarily originate from one horn. In gilts with only one ovary, embryos are distributed throughout both horns and pregnancy is maintained normally. When embryos are restricted from entering and implanting in one-third of the length of the uterus by ligating the uterus at day 4, only 50% of pregnancies continue regardless of the number

of embryos (Fig. 3) (Dziuk, 1968b). When the proportion of uterus with no embryos is 50% or greater, essentially no pregnancies continue (Dhindsa and Dziuk, 1968). The particular segment of the uterus that is unoccupied seems to make little difference. If one complete uterine horn is unoccupied, the effect is the same as if half of each horn is unoccupied. After unilateral ovariectomy or oviduct ligation, embryos enter from one oviduct. Because embryos from one horn do not enter the other horn until day 9, the decision to maintain pregnancy or return to estrus is not made before day 9. The decision to maintain the pregnancy should be made before day 15, when the corpora lutea normally regress and new follicles develop for the next ovulation at day 21. A study restricting embryos to the upper half of each uterine horn by ligation at day 4 and then removing the ligatures at day 8, 9, 10, 11, 12, or 13 has given evidence that narrow the day of decision to near day 12. When ligatures were removed at day 8, 9, 10, or 11, pregnancy was maintained and embryos moved and implanted beyond the point where ligatures had been. No pregnancies continued when ligatures were removed at day 12 or 13 unless supplemental progesterone was given. Even though ligatures were removed at day 13 and pregnancy was maintained, embryos did not move beyond the point of previous ligation (Polge and Dziuk, 1970). The embryos are not defective but the corpora lutea regress when a part of the uterus is unoccupied.

Once pregnancy has been established it is possible to remove or kill all embryos in one horn as early as day 14 and pregnancy will be maintained in the one remaining horn with embryos. Killing all embryos at day 30 does not disrupt the pregnancy, which in re-

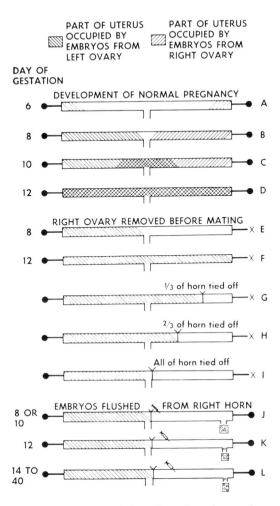

Figure 3 Migration of pig embryos in early gestation. From Dziuk (1968b).

B. Migration of Embryos

By transferring embryos recovered from a black breed to the tip of one horn and embryos from a white breed to the other horn at day 4 and examining the uterus and fetuses at day 85 when skin color was distinguishable, embryos from each horn were found to migrate past each other and mix in each horn (Dziuk *et al.*, 1964). Distribution of embryos uniformly throughout both uterine horns is important not only for maintenance of pregnancy but also utilize the available uterine space. Embryos are moved throughout the uterus and implant more or less equidistant from each other (Dziuk, 1985). When several embryos are restricted to less than normal total space each will occupy an equal space, but it will of necessity be less than normal for each embryo. Each fetus requires a certain minimum space to survive. When space is marginal the fetus may survive but will be smaller than normal. Unequal distribution of embryos in the uterus can lead to adequate space for some embryos but inadequate space for others. Embryos themselves apparently have some influence on spacing. This ability to acquire or produce space may develop as the embryos develop and would appear to be most important near day 12 when the embryo is growing at a very rapid rate. An embryo that is somewhat slower in developing than its littermate at day 12 may find the available space is already occupied when it tries to implant a few hours later (Dziuk, 1987). The embryo and the uterus must be in close synchrony throughout pregnancy. An embryo out of synchrony with the uterus not only finds space is already occupied by littermates but also does not receive nor send the timely signals that

ality is now a pseudopregnancy (Webel *et al.*, 1975). This helps explain how pregnancy can be established, then lost, with no outward signs of estrus. The producer considers the animal to be pregnant on the basis of failure to return to estrus but no litter is produced.

go between mother and embryo for maintenance of pregnancy.

C. Space for Embryos

When spacing of embryos in the uterus is perfectly equidistant, all fetuses would have equal opportunity to survive and grow. Spacing is, however, rarely perfect, and consequently some embryos have more or less space than litter mates. When space is below a certain minimum, embryos may temporarily survive but then succumb and become mummified (Wu et al., 1988a). When space is just above minimum for survival, fetuses live but do not grow to normal size, resulting in runts. Runts are classified as pigs whose weights are at least one standard deviation below the average for the litter. Weight of fetuses is positively related to the space available but only up to a certain point. Excess space will not allow the fetus to grow beyond a certain limit.

The absolute space available to each fetus is that fraction of the uterus resulting from dividing the length of the uterus by the number of fetuses. If the number of fetuses increases and the uterine length remains constant, a point will be reached when the uterus can no longer accommodate all of the potential inhabitants (Fig. 6) (Wu et al., 1987). This observation prompted several studies to determine when the uterus grows and to what extent the embryo can cause growth of the uterus. The ovary apparently does not influence the length of the uterus until the prepubertal gilt is about 100 days of age (Wu and Dziuk, 1988). The uterus grows gradually until puberty, then essentially doubles in length and weight at the first estrus. Embryos cause growth of the uterus during pregnancy beginning at day 18, continuing until about day 30 (Wu et al.,

1988b). From day 18 to 30 the uterus doubles in length but grows relatively little during the remainder of gestation. Fluid in the amnion and allantois accumulates, permitting detection of pregnancy by ultrasound by day 23 at the earliest (Lindahl et al., 1975). Growth of the uterus is stimulated by each embryo and is limited to that section of the uterus occupied by an embryo. The length of each pregnant uterine horn is dependent on the number of fetuses within that horn, independent of the number in the opposite horn. Each fetus causes the uterus to grow about 11 cm, and each fetus occupies at least 21 cm of uterus. When fetuses are so crowded that the uterus cannot grow sufficiently to accommodate all fetuses, some will mummify or be runts. The mean number of fetuses in domestic occidental breeds reaches a plateau at about 14 regardless of the potential based on number of ovulations. This limitation has also been demonstrated by unilateral ovariectomy and

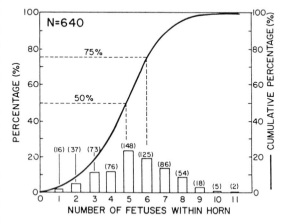

Figure 4 Frequency and cumulative distributions of the number of fetuses in a single uterine horn. Numbers in parentheses represent the number of uterine horns observed. In total 640 uterine horns from 320 pregnant pigs were observed. From Wu et al. (1987).

hysterectomy (Dziuk, 1968a). This model provides only half the uterine space but the normal number of ovulations due to ovarian compensation. Under these conditions one uterine horn can support a mean of seven fetuses. Thus litter size is limited first by number of embryos, second by the proportion of embryos that survive, and then by the resources available to the fetus in the form of uterine space.

The stage of gestation when uterine resources (space) begin to affect potential litter size has been the subject of several studies (Fig. 5) (Webel and Dziuk, 1974). The degree of limitation may also influence the stage at which space limits have an effect. Very crowded fetuses may die at an earlier stage than those with a bit more space. When recipient gilts received either 12 or 24 embryos the proportion of embryos surviving to day 30 was the same regardless of the number transferred (Pope *et al.*, 1972). By ligating the uterus at day 4 in a manner that

restricted embryos in one horn to half the space of that available to embryos from the opposite side, survival was found to be unaffected before day 30 by that restriction (Dziuk, 1968a). Embryonic losses before day 30 are apparently influenced little by uterine space under the usual range of conditions, but fetal losses do occur from limited uterine space after day 30.

Embryonic loss has been arbitrarily classified as the proportion of ovulation points (corpora lutea) not represented by live embryos. The cutoff point for stage of development of embryonic loss is about day 25–30 of gestation when subsequent losses are classified as fetal. The pig is a useful model for study of embryonic loss because the relatively large litter allows for some loss within a litter while still maintaining pregnancy. Much research has been devoted to understanding possible causes and applying potential cures. Thus far there is little consistency in either cause or cure. Many factors

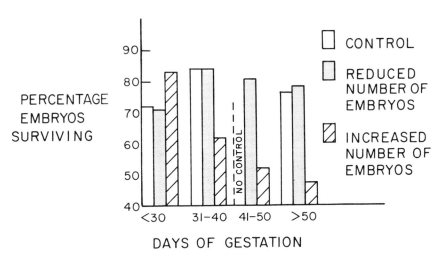

Figure 5 Effect of altering the proportion of uterine space available to embryos on their survival during early gestation. From Webel *et al.* (1975).

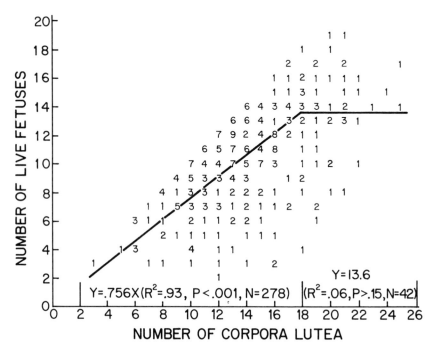

Figure 6 Relationship between the number of corpora lutea and the number of live fetuses per female. Numbers represent the number of females observed at that point. From Wu *et al.* (1987).

have been shown to or suspected of influencing embryonic loss, including toxins, plane of nutrition, specific nutrients, chromosomal aberrations, genetic defects, fertility of the boar, semen handling, temperature of the environment, endocrine imbalances, prolonged ovulation, and infectious disease. Each factor may be responsible for some part of loss in some cases, but there is little conclusive evidence that any one cause is major, and thus there cannot be a certain prescribed cure applicable in all cases (Dziuk, 1987).

D. Endocrinology of Gestation

The present state of knowledge on endocrinology of gestation is probably quite elementary and possibly so simplistic that it could be in error. Gilts can readily conceive and bear a litter at the first estrus. There does not seem to be a requisite of a practice or preliminary estrous cycle, as there is little evidence that conception rate of a gilt fed and managed properly increases with each subsequent estrus (Connor and Van Lunen, 1988). The composite profile of the concentrations of hormones indicates estrogen rising before estrus as a result of follicular growth from the stimulus of follicle-stimulating hormone (FSH) and LH. As estrogen rises, a surge of LH occurs at the onset of estrus triggering ovulation, oocyte maturation, and formation of corpora lutea. Progesterone rises gradually for the first 10 days after ovulation, and its concentration is related positively to the mass of luteal tissue, which in turn is related to the number of corpora lutea. Measurement of the concentration of progesterone at a precisely determined stage of the luteal phase can give a rough estimate of the number of ovulations (Dial and Dziuk, 1983). This relationship exists only during the first 9 days after ovulation. As the cycle progresses the mated gilt with embryos will maintain corpora lutea beyond day 15 based on presence and distribution of embryos at day 12 as discussed previously. The concentration of progesterone will continue to rise from day 15 to about day 30 and then decline slightly. Studies to determine the influence of the number of fetuses on concentration of progesterone have found that once pregnancy has been established the number of fetuses has little or no effect on the concentration of progesterone in the maternal plasma (Webel *et al.*, 1975). After day 25 the number of corpora lutea has little effect on concentration of progesterone in maternal plasma. Reduction of the number of corpora lutea by removing one ovary or removing certain corpora lutea by electrocautery has not produced a commensurate reduction in progesterone (Martin *et al.*, 1977). Apparently a very wide margin of safety exists in the number of corpora lutea and the concentration of progesterone for maintenance of pregnancy. Pregnancy is maintained with as few as one or two corpora lutea, provided the reduction to that number is done in a stepwise fashion (Thomford *et al.*, 1984). Reduction to less than five corpora lutea in one step results in abortion in 48 h. On the basis of maintaining pregnancy by exogenously administered progesterone or by reducing number of corpora lutea, it appears that a minimum concentration of about 6 ng/ml is needed for maintenance of pregnancy (Ellicott and Dziuk, 1973).

Pigs are almost unique in having a very great rise in the concentration of estrone sulfate in maternal plasma beginning at about day 17. Estrone sulfate rises to a peak near day 30 and declines at day 35. The fetus can be assumed to be the source of this estrogen metabolite because of the relationship between the number of fetuses and concentration of estrone sulfate in maternal plasma. When a blood sample is taken at a known and specified day of gestation between day 22 and 28 and the concentration of estrone sulfate is determined it is possible not only to distinguish between pregnant and nonpregnant sows with certainty but also to get a rough estimate of the number of fetuses present (Horne *et al.*, 1983).

E. General Strategy of Pregnancy

Pigs have evolved several strategies to maintain pregnancy and reproduce successfully. The embryos of the pig enter the tip of the uterine horns at 48 h after ovulation at the four-cell stage. Each uterine horn can be 200 cm in length. The embryos must be distributed more or less uniformly throughout the uterus to be able to grow and survive. The migration, distribution, and spacing of embryos and the effect on maintenance of pregnancy and survival and growth of fetuses have been the subject of several studies over a period of several years. Embryos have no means of propelling themselves through the uterus; thus they must rely on the peristaltic and antiperistaltic contractions of the uterus for transport. A remarkable aspect of embryo transport is that such a small, fragile, growing unit of life can be moved so effectively such distances. It is all the more remarkable when one considers the size of the uterus and the relationship of the size of the embryo to the distance it travels. Each uterine horn can be as long as 2000 mm. An embryo less than 1 mm in diameter can originate in the tip of one horn and implant in the tip of the opposite horn at a distance 4000 times its diameter. Not only do embryos move to occupy the available space, but they also in some manner signal the mother, telling her what proportion of her uterus is unoccupied so that the mother will maintain the pregnancy or abandon it. The embryos also must decide how many are present and measure the uterus to divide the available space more or less equally. If one or more of them is slow in developing, their voice may not be heard and the uterine space will be divided among the more advanced embryos, leaving the slow ones with too little space to survive.

The mechanisms that control these events are partly understood. Several separate observations lead to the conclusion that the pig decides to continue pregnancy on the basis of the extent that the uterus is unoccupied at day 12 or 13, but once the decision is made then later reduction of embryos or fetuses does not terminate pregnancy or pseudopregnancy. The position of embryos in the uterus not only affects maintenance of pregnancy but also influences the destiny of the fetus. If spacing of embryos is uneven or if an excess of embryos causes uterine crowding, some fetuses will have insufficient space to grow and survive. The position of the fetus in the uterus seems to influence the space available to it. Fetuses at the tip of the uterine horn are generally heavier, longer, and occupy more space than fetuses in other locations. They are heavier and longer because more space

is available. Crowding leads to death and resorption or mummification. The space available to each embryo or fetus near the body of the uterus is different as gestation progresses. An apparent increase in space available to fetuses near the body of the uterus from day 25 to day 40 is due to early resorption or mummification of some embryos in that region. Fetuses in the midsection of the uterus between the tip and the body sometimes have space available that is marginal, that is, sufficient to survive, but less than enough to grow normally. Runts arise from those conditions. Unless runts are given heroic special treatment at birth, they rarely survive. As the fetuses at the tip of the horn are rarely runts or mummies it would seem to be the safe place to implant. However, most stillbirths arise from the last few piglets born coming from the tip of the uterine horn. This causes a dilemma of where to implant in the uterus. Growing in the tip may lead to being stillborn, the fetus in the midsection of the uterus has an increased chance of being a runt, and those fetuses near the body may be resorbed very early in gestation.

How many fetuses can the uterus support? The number differs from animal to animal depending on uterine length. The number of live fetuses at birth rises with an increased number of ovulations until the mean number of fetuses reaches about 14. At that point the number of fetuses does not increase with increased number of ovulations. The length of the uterus determines the number of survivors. In gilts with one ovary and one uterine horn, the number of fetuses is about seven. Very prolific pigs apparently have longer uterine horns that can accommodate more fetuses.

VI. Parturition

Litter-bearing animals have some peculiar hormonal and physical requirements for parturition. In the pig, as in many species, the fetus rather than the mother determines length of gestation. Initiation of parturition is controlled by messages from each fetus acting in concert. Not only must the mother respond to an individual fetus but to the litter as a whole. The birth process is complicated by the need to maintain the intrauterine integrity of part of the litter while the other part is in the birth process or has been born. The concentrations of hormones near parturition follow a fairly reliable pattern. Estrogen rises, progesterone falls, relaxin and prolactin peak, and oxytocin and cortisol rise at the time of birth (Silver *et al.*, 1979). The concentration of estrogen near parturition is higher in sows with large litters than in sows with small litters (Martin *et al.*, 1977a). This is strong evidence that the fetal placental unit is the source of estrogen. The observation that shorter gestation periods are associated with higher concentrations of estrogen and larger litters would also indicate that estrogen can influence initiation of parturition. Concentration of estrogen declines sharply after parturition, providing further evidence that the fetal placental unit produces the estrogen. Progesterone declines just before parturition, presumably in response to luteolysis of corpora lutea by prostaglandins. Parturition can also be induced by inhibiting synthesis of progesterone. Prostaglandin can induce parturition by its luteolytic action, and parturition can be delayed by inhibiting synthesis and secretion of prostaglandins. It is apparently not critical that prostaglandins

reach the ovaries via a local countercurrent mechanism, because pregnancy is maintained and terminated normally in gilts with ovaries transplanted to the body wall. The life span of corpora lutea of pregnancy is not intrinsic beyond a 14-day period. New corpora lutea induced late in gestation at day 89, in the absence of other corpora lutea, can maintain pregnancy and will regress at the end of the normal gestation period. However, corpora lutea induced less than 14 days from the last day of a normal gestation period at day 109 do not regress and gestation is prolonged (Martin *et al.*, 1977b).

Relaxin rises gradually until about 16 h before parturition. Then a sharp peak occurs associated with exogenous or endogenous prostaglandins (Sherwood *et al.*, 1975). A rise in relaxin occurs in pseudopregnant, hysterectomized gilts at the time when gestation would have ended if fetuses had been present. This indicates the possibility of dual control over the release of relaxin from the corpora lutea if the usual source of prostaglandin is only uterine. Relaxin helps in cervical relaxation and may be very important in normal delivery of live fetuses. Gilts ovariectomized after mating and given exogenous progesterone to maintain pregnancy do not farrow normally. Farrowing is very prolonged and many piglets are stillborn. Administration of relaxin to these sows corrects the problem.

Endocrine signals involving the fetal pituitary initiate parturition. A litter of decapitated or hypophysectomized fetuses has a prolonged gestation (Stryker and Dziuk, 1975). The fetal pituitary stimulates the adrenal, thyroid, gonads, and other glands that produce secretions that can influence the onset of parturition. An understanding of the complex interrelationships between maternal ovaries, uterus, fetus, placenta, and parturition is not yet complete (Taverne *et al.*, 1979). When an intact fetus is the sole occupant of the uterus, parturition is initiated at the usual stage of development. When a uterus has approximately equal numbers of intact and decapitated fetuses, the intact fetuses apparently exert an overwhelming effect with parturition occurring at the normal time. In a litter of one intact fetus and four decapitated fetuses gestation is prolonged. The strength of the signal to maintain the gestation by the decapitated fetuses must be about one-fourth of the intact fetus who is trying to initiate parturition (Dziuk, 1979; Bazer and First, 1983). In a litter with fetuses of varying stages of maturity, a minority of the more mature fetuses could cause the birth of the entire litter. From an evolutionary view this arrangement would favor the more rapidly develop-

ing fetuses and stabilize the length of gestation. If length of gestation were controlled by the slowest developing member of a litter, the more advanced fetuses could develop beyond the size and stage of maturity for normal delivery, jeopardizing not only their life but also that of the mother. Natural selection for the slowest maturing fetus would also tend to extend the gestation period. Nature usually favors the survival of the most rapidly developing individuals.

Only in the last few decades have some of the secrets been discovered about the mechanics of parturition. By labeling fetuses *in utero* with distinguishing dyes and then observing their birth it has been possible to learn the order of birth in relation to the position in the uterus. Fetuses more or less alternate from each horn, but exceptions occur. One horn can empty completely before any fetuses come from the other horn

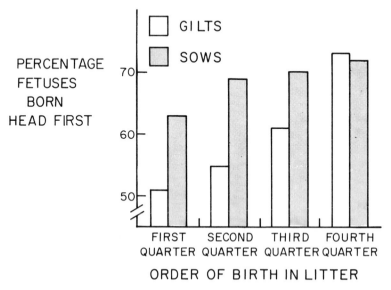

Figure 7 The relationship between anteroposterior presentation of fetuses and their order of birth. From Reimers *et al.* (1973).

(Dziuk and Harmon, 1969). Most of the fetuses maintain the same head–tail orientation in the uterus from day 40 of gestation to parturition. In the relatively few cases of reversed orientation at birth it appears that one explanation may be that during birth a fetus goes from one uterine horn to the other, bypassing the cervix and vagina. Later the fetus would be born but with the orientation opposite to the original uterine one—a little like backing a car out of a driveway before heading down the street. This explanation would also help account for the rare case of a fetus being born out of order within a uterine horn. One fetus goes into the opposite horn, and its near former neighbor moves down the horn and is born ahead (Taverne *et al.,* 1977).

About 7% of fully formed fetuses are stillborn. Stillborn piglets are not the small runts but often the larger fetuses. This loss of a piglet or two per litter translates into more than 10,000,000 deaths per year when multiplied by the number of farrowings in the United States alone. When this death loss is again multiplied by the dollar value

of a piglet at birth, the resultant cost is staggering (Sprecher *et al.,* 1974).

A greater percentage of fetuses are stillborn in litters larger than 14 and smaller than 5. In litters of 8, 9, or 10 there are very few stillborns. Fetuses in small litters often must traverse sections of previously unoccupied uterus, thus slowing delivery. Fetuses can survive in the uterus and vagina for only a few minutes after the umbilicus is broken or compressed or when the placenta is detached from the uterus. It is imperative that delivery be prompt. A uterine horn with 7 or 8 fetuses would of necessity be at least 200 cm in length. The umbilicus is much shorter than 200 cm. The uterus then must shorten with the delivery of each fetus so the next fetus is close to the exit or the fetus must be moved very rapidly through the uterus, cervix, and vagina. A fetus at the tip of the uterus is at great risk. When the interval between delivery of fetuses increases, so does the risk of a stillbirth. Most stillbirths in large litters occur when the interval between births is extended, and when the fetuses are located in the tip of the uterus.

Numerous attempts have been made to reduce the incidence of stillbirths by supervision of farrowing or by administering compounds known to stimulate uterine contractions (Guthrie, 1985). These attempts have not had a consistently beneficial effect. This may be due to the possible different causes of stillbirths, so that a treatment succeeds under one circumstance but fails under another.

Reproduction in the pig has many interesting and unique characteristics. Not only does greater knowledge satisfy an academic motive, but it can lead to more efficient food production. Both goals are worthy and provide challenges that will be met in the future.

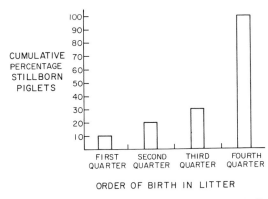

Figure 8   The relationship between incidence of stillbirths and order of birth. From Dziuk *et al.* (1972).

References

Awotwi, E. K., and Anderson, L. L. (1985). *Anim. Reprod. Sci.* **9**, 51–61.

Baker, R. D., Dziuk, P. J., and Norton, H. W. (1968). *J. Anim. Sci.* **27**, 88–93.

Bazer, F. W., and First, N. L. (1983). *J. Anim. Sci.* **57**, 425–460.

Caton, J. S., Jesse, G. W., Day, B. N., and Ellersieck, M. R. (1986). *J. Anim. Sci.* **62**, 1203–1209.

Christenson, R. K., Ford, J. J., and Redmer, D. A. (1985). *J. Reprod. Fertil. Suppl.* **33**, 21–36.

Claus, R., and Weiler, U. (1985). *J. Reprod. Fertil. Suppl.* **33**, 185–197.

Connor, M. L., and Van Lunen, T. A. (1988). *Int. Cong. Anim. Reprod. and A. I.* **2**, 86–88.

Cronin, G. M. (1983). *Anim. Reprod. Sci.* **6**, 51–57.

Dhindsa, D. S., and Dziuk, P. J. (1968). *J. Anim. Sci.* **27**, 122–126.

Dhindsa, D. S., Dziuk, P. J., and Norton, H. W. (1967). *Anat. Rec.* **159**, 325–330.

Dial, G. D., and Dziuk, P. J. (1983). *J. Anim. Sci.* **47**, 1260–1269.

Dial, G. D., Dial, O. K., BeVier, G. W., Glenn, S. D., and Dziuk, P. J. (1983). *Biol. Reprod.* **29**, 1047–1056.

Dial, G. D., Dial, O. K., Wilkinson, R. S., and Dziuk, P. J. (1984). *Biol. Reprod.* **30**, 289–299.

Diekman, M. A., and Grieger, D. M. (1988). *Anim. Reprod. Sci.* **16**, 295–301.

Dziuk, P. J., and Gehlbach, G. D. (1966). *J. Anim. Sci.* **25**, 410–413.

Dziuk, P. J. (1968a). *J. Anim. Sci.* **27**, 673–676.

Dziuk, P. J. (1979). *Anim. Reprod. Sci.* **2**, 335–342.

Dziuk, P. J. (1987). *In* "Manipulating Pig Production" (APSA Committee, eds.), pp. 28–39. Australian Pig Science Association, Werribee, Australia.

Dziuk, P. J. (1968b). *Ill. Res.* **10**, 18.

Dziuk, P. J. (1970). *J. Reprod. Fertil.* **22**, 277–282.

Dziuk, P. J. (1985). *J. Reprod. Fertil. Suppl.* **33**, 57–63.

Dziuk, P. J., and Baker, R. D. (1962). *J. Anim. Sci.* **21**, 697–699.

Dziuk, P. J. and Dhindsa, D. S. (1969). *J. Anim. Sci.* **29**, 39–40.

Dziuk, P. J., and Harmon, B. G. (1969). *Am. J. Vet. Res.* **30**, 419–421.

Dziuk, P. J., Sprecher, D. J., Webel, S. K., and Harmon, B. G. (1972). *J. Anim. Sci.* **35**, 240.

Dziuk, P. J., Polge, C., and Rowson, L. E. (1964). *J. Anim. Sci.* **23**, 37–42.

Eastman, P. R., Dyck, G. W., and Cole, D. J. A. (1986). *Anim. Reprod. Sci.* **12**, 31–38.

Ellicott, A. R., Dziuk, P. J., and Polge, C. (1973). *J. Anim. Sci.* **37**, 971–973.

Ellicott, A. R., and Dziuk, P. J. (1973). *Biol. Reprod.* **9**, 300–304.

Flowers, B., Cantley, T., and Day, B. N. (1987). *J. Anim. Sci.* **65**, 1576–1584.

Guthrie, H. D. (1985). *J. Reprod. Fertil. Suppl.* **33**, 229–244.

Hansel, W., Concannon, P. W., and Lukaszewska, J. H. (1973). *Biol. Reprod.* **8**, 222.

Hemsworth, P. H. (1982). "Control of Pig Reproduction" (D. Cole and G. Foxcroft, eds.), pp. 585–601. Butterworths, London.

Horne, C., Chew, B. P., Wiseman, B. S., and Dziuk, P. J. (1983). *Biol. Reprod.* **29**, 56–62.

Hughes, P. E. (1982). "Control of Pig Reproduction" (D. Cole and G. Foxcroft, eds.), pp. 117–138. Butterworths, London.

Hunter, R. H. F., Fléchon, B., and Fléchon, J. E. (1987). *Tissue Cell* **19**, 423–436.

Hunter, R. H. F., and Nichol, R. (1988). *Gamete Res.* **21**, 255–266.

Killian, D. B., Kiesling, D. O., Wulff, F. P., and Stewart, A. N. V. (1987). *J. Anim. Sci.* **64**, 231–236.

Lee, K. H., Diekman, M. A., Brandt, K. E., Grieger, D. M., and Allrich, R. D. (1987). *J. Anim. Sci.* **64**, 1110–1116.

Legault, C. (1985). *J. Reprod. Fertil. Suppl.* **33**, 151–166.

Lindahl, I. L., Totsch, J. P., Martin, P. A., and Dziuk, P. J. (1975). *J. Anim. Sci.* **40**, 220–222.

Long, G. G., Diekman, M. A., and Scheidt, A. B. (1988). *J. Anim. Sci.* **66**, 452–458.

Martin, P. A., and Dziuk, P. J. (1977). *J. Reprod. Fertil.* **49**, 323–329.

Martin, P. A., BeVier, G. W., and Dziuk, P. J. (1977a). *Biol Reprod.* **16**, 633–637.

Martin, P. A., Norton, H. W., and Dziuk, P. J. (1977b). *Biol. Reprod.* **17**, 712–717.

Meijer, J. C., Colenbrander, B., Poot, P., and Wensing, C. J. G. (1985). *Biol. Reprod.* **32**, 137–143.

Polge, C., and Dziuk, P. J. (1970). *J. Anim. Sci.* **31**, 565–566.

Ponzilius, K. H., Parvizi, N., Elsaesser, F., and Ellendorff, F. (1986). *Biol. Reprod.* **34**, 602–612.

Pope, C. E., Christenson, R. K., Zimmerman-Pope, V. A., and Day, B. N. (1972). *J. Anim. Sci.* **35**, 805–808.

Reimers, T. J., Dziuk, P. J., Bahr, J., Sprecher, D. J., Webel, S. K., and Harmon, B. G. (1973). *J. Anim. Sci.* **37**, 1212.

Robl, J. M., and Dziuk, P. J. (1987). *J. Exp. Zool.* **242,** 181–187.

Sherwood, O. D., Chang, C. C., BeVier, G. W., and Dziuk, P. J. (1975). *Endocrinology* **97,** 834–837.

Silver, M., Barnes, R. J., Comline, R. S., Fowden, A. L., Clover, L., and Mitchell, M. D. (1979). *Anim. Reprod. Sci.* **2,** 305–322.

Simoneau, C., Estrada, R., Matton, P., Roy, P. E., and Dufour, J. J. (1988). *J. Anim. Sci.* **66,** 2606–2613.

Sprecher, D. J., Leman, A. D., Dziuk, P. J., Cropper, M., and Dedecker, M. (1974). *J. Am. Vet. Med. Assoc.* **165,** 698–701.

Stryker, J. L., and Dziuk, P. J. (1975). *J. Anim. Sci.* **40,** 252–287.

Taverne, M. A. M., van der Weyden, G. C., Fortijne, P., Ellendorff, F., Naaktgeboren, C., and Smidt, D. (1977). *Am. J. Vet. Res.* **38,** 1761–1764.

Taverne, M. A. M., Naaktgeboren, C., Elsaesser, F., Forsling, M. L., van der Weyden, G. C., Ellendorff, F., and Smidt, D. (1979). *Biol. Reprod.* **21,** 1125–1134.

Thomford, P. J., Sander, H. K. L., Kendall, J. Z., Sherwood, O. D., and Dziuk, P. J. (1984). *Biol. Reprod.* **31,** 494–498.

Wang, R. X., Zhao, S. J., Huang, M. Y., Sun, Y., Jio, S. X., and Chen, Y. M. (1988). *Int. Cong. Anim. Reprod. and A. I.* **2,** 73–75.

Webel, S. K., Reimers, T. J., and Dziuk, P. J. (1975). *Biol. Reprod.* **13,** 177–186.

Webel, S. K., and Dziuk, P. J. (1974). *J. Anim. Sci.* **38,** 960–963.

White, G. (1902). *In* "The Natural History of Selborne" (G. Allen, ed.), p. 290. John Lane, London.

Wu, M. C., Hentzel, M. D., and Dziuk, P. J. (1988a). *J. Anim. Sci.* **66,** 3203–3207.

Wu, M. C., Shin, W. J., and Dziuk, P. J. (1988b). *J. Anim. Sci.* **66,** 1721–1726,

Wu, M. C., and Dziuk, P. J. (1988). *J. Anim. Sci.* **66,** 2893–2898.

Wu, M. C. Hentzel, M. D., and Dziuk, P. J. (1987). *J. Anim. Sci.* **65,** 762–770.

CHAPTER 15

Reproduction in the Sheep and Goat

D. R. LINDSAY

I. Introduction

Sheep and goats are distributed the most widely of all domestic ruminants. The first domestication probably began in West Asia (Ryder, 1983) and spread from the tropics to the arctic circle. This vast spread has resulted in animals that have adapted reproductively to enable them to survive and reproduce in a wide variety of environments. In some cases they have adopted seasonal breeding to enable them only to reproduce at a time likely to allow the young to survive. In others they have adopted opportunistic breeding strategies to cope with periods of drought, shortage of feed, and extreme temperatures. A few genotypes produce relatively large litters to compensate for expected losses of offspring. Others seldom produce more than one offspring, except in the most favorable conditions, to ensure the best chances of survival of both mother and young. All have the capacity to store and make later use of energy reserves as a normal part of their reproductive strategy. This is seen most obviously in some of the fat-tailed breeds of sheep in Africa and Asia.

As ruminants, sheep and goats have traditionally occupied a niche in the spectrum of domestic animals that has made use of their ability to forage for, and utilize, feedstuffs that are unavailable to other species. Thus they need not compete with humans or other domestic animals like poultry and pigs for food. Nevertheless, in some societies in recent times changing prices and tastes have seen the development of intensive feeding and breeding programs. These have encouraged a search for ways of uncoupling inherent survival mechanisms from natural reproductive physiology so that breeders can exert absolute control over the timing and success of reproduction.

Modern sheep and goat husbandry thus has two reproductive goals: improving the efficiency of extensive production without substantially increasing the costs of input, and controlling the reproductive process for intensive production where control of the environment—hormonal, nutritional, social, and physical—is economically acceptable. A prerequisite for both goals is that we understand the reproductive physiology of these species. In this chapter we look at the state of our understanding of those aspects of reproductive physiology that have potential for development in sheep and goat husbandry, and our capacity to use them.

II. The Female—The Ewe and Doe

A. *The Estrous Cycle*

Ewes and does give very little overt evidence of reproductive activity except that they will usually accept mating by a male about the time that they ovulate—a period that is designated as estrus. There is little evidence of proestrus or metestrus and, unlike the cow in the absence of a male, ewes show no sign of estrus either. From external observation the estrous cycle can therefore be divided into the period of estrus and the interestrus period, or diestrus. At the ovarian level the cycle can be divided into the follicular phase covering the 2–3 days in which ovulatory follicles grow and produce their ova, and the luteal phase dominated by the

presence of one or more corpora lutea on the ovary. The length of the luteal phase determines the length of the estrous cycle or the period between successive estrous events. In the ewe the length of the cycle is about 17 days, and in does 20–21 days. Cyclical female goats in the late luteal or early follicular phase may have their estrous cycles shortened by the sudden introduction of a buck (Chemineau, 1983). There is recent evidence that such a phenomenon may also exist in certain sheep breeds (Hassan *et al.*, 1988).

If no corpus luteum eventuates from an ovulation, then the female experiences a short cycle of only 6–8 days corresponding to what is considered to be the normal cycle in the rodent. Short cycles are characteristic of both ewes and does at the beginning (Robinson, 1959) and the end of the breeding season (Land *et al.*, 1973), and in does they are often found during the breeding season as well (Camp *et al.*, 1983). Ovulation without estrus, or silent ovulation, is common in both species (Pretorius, 1973) and is characteristic of short cycles, especially in the ewe. Estrus without ovulation is less common (Thimonier, 1981) and is confined mainly to young animals (Edey *et al.*, 1977) or some adult ewes just after parturition.

The two events, estrus and ovulation, are usually closely linked, but not always. The hormonal stimulus for estrus is estradiol in both species. Dose-response lines to injected estradiol can be demonstrated in ovariectomized females where the response is either behavioral estrus (Robinson *et al.*, 1956) or length of estrus (Scaramuzzi *et al.*, 1971; Fletcher and Lindsay, 1971a; Land *et al.*, 1972). In the ewe a period of 6–8 days of progesterone priming is essential for the female to be fully sensitive to estrogen. Without progesterone she becomes increasingly refractory to exogenous estradiol (Dauzier and Wintenbergen, 1952; Dutt, 1952; Robinson, 1952; Robinson, 1954). When entire ewes ovulate without having had a corpus luteum from a previous ovulation such as at the first ovulation of the breeding season, or after a short cycle, there is generally no behavioral estrus (Robinson, 1954). Progesterone from a previous corpus luteum is also essential for survival and normal development of embryos (Moore and Miller, 1984). By contrast, does do not require a long period of progesterone prior to estradiol. They have normal fertility after short cycles (Chemineau, 1983), and ovariectomized females do not become refractory to repeated doses of estradiol, so that alternate treatments with progesterone have no effect on the response to estradiol (Sutherland, 1988).

1. Estrus

The duration of estrus in both species is between 24 and 36 h, but this is highly variable, being shorter in younger females than in older and longer in mid-season than at the beginning or the end of the season (McKenzie and Terrill, 1937; Parsons and Hunter, 1967).

The ewe is usually considered to be a spontaneous ovulator, but she is in fact a semi-induced ovulator. The continual presence of males reduces the duration of receptivity by about 12 h (Parsons and Hunter, 1967; Van der Westhuysen *et al.*, 1970; Fletcher and Lindsay, 1971b) and advances the time of ovulation by about the same amount (Lindsay *et al.*, 1975). This is of particular importance in AI program where females may have little contact with males. The use of vasectomized rams after AI to hasten ovulation has improved fertility in at least one study (Restall, 1961).

The relationship between the beginning and the end of estrus and the exact time of ovulation is not entirely consistent. Most ewes and does ovulate toward the end of estrus (McKenzie and Terrill, 1937), but at a variable time from its beginning, so it is difficult to predict accurately when a female will ovulate.

Experiments that have sought to define an optimum time during estrus when insemination leads to maximum fertility have produced equivocal results. Some suggest that the optimum time is during the first half of estrus (e.g., Morrant and Dun, 1960), and others have suggested that it is at later stages (e.g., Restall, 1961). Amir and Schindler (1972) found no difference in fertility following inseminations at any time from 4 to 40 h after the beginning of estrus. In ewes mating naturally throughout the receptive period the most successful inseminations were between 9 and 15 h after the onset of estrus (Jewell *et al.*, 1986).

2. Ovulation

The ewe and doe distinguish themselves from the other main domestic herbivores like the cow and the mare by being capable of multiple ovulations at estrus. In fact, increasing ovulation rate is one of the most effective ways of improving prolificacy in these species. The potential for ovulation in the form of primordial follicles is enormous, and from about 20 days of fetal life there is a continuing process of new follicles commencing growth (Turnbull *et al.*, 1977; Cahill and Mauleon, 1980). Almost all of these die or undergo atresia at some stage of their development before they reach the point of ovulation, and only a very few remain healthy until they eventually produce ova (Peters *et al.*, 1975). Ovulation is therefore not so much the result of the development of one or two follicles through their many stages, as it is the prevention of destruction, at some point, of a very few of the thousands of follicles in the process of development. Folliculogenesis in sheep and goats takes about 6 months from the recruitment of an oocyte from its primordial population to final ovulation. Most of this time is spent in the slow, preantral stage during which there is little atresia (Turnbull *et al.*, 1977; Cahill *et al.*, 1979). When follicles enter the antral phase development is more rapid but atresia is more frequent. The numbers of follicles undergoing development at this stage bear little relationship to the eventual rate of ovulation. For example, during anestrus when there are no ovulations at all, there are more follicles entering the growth phase than during the breeding season (Cahill and Mauleon, 1980). Highly fecund breeds of sheep have fewer total oocytes than breeds of low fecundity when young (Land, 1970; Cahill *et al.*, 1979) but appear to reverse this when adult (Cahill *et al.*, 1982).

The development of follicles in the antral phase appears to depend heavily on the ability to synchronize estradiol from androgen precursors. Growth *in vitro* can be accelerated by the addition of estradiol and follicle-stimulating hormone (FSH) (McNatty *et al.*, 1979). *In vivo*, large but atretic follicles have a much higher ratio of androgen to estradiol than large healthy follicles, reflecting their inability to aromatize androgens to estradiol (Baird, 1983). FSH is essential in this aromatization process (Leung and Armstrong, 1980), without which androgens will hasten atresia (Louvet *et al.*, 1975). In addition, FSH and estradiol together are needed for the production of luteinizing hormone (LH) receptors in the granulosa cells of the follicle. Without a good population of receptors the follicle cannot respond to the

preovulatory surge of LH and ovulate (Richards, 1979).

At any stage in the ovary of a mature female there are 5–24 antral follicles (McNatty, 1982) that have developed by gonadal stimulation. After luteolysis one, or only a few, of these develop further or are "selected." The selected follicles dominate the others, which become atretic. At this time in the cycle there is a modest fall in the plasma concentration of FSH, probably caused by negative feedback to the pituitary as a result of the selected follicle or follicles secreting large quantities of estradiol and inhibin. The dominant follicle then deprives the others of this limited FSH, their granulosa cells become incapable of converting androgens to estradiol, and their demise is accelerated (Baird *et al.*, 1975; Baird and McNeilly, 1981; Armstrong *et al.*, 1981). In addition there are local inhibiting factors produced by the dominant follicle or follicles (Cahill, 1984) that restrict the growth of adjoining, unselected follicles. Steroid-free ovine follicular fluid can stimulate atresia in the absence of feedback in hypophysectomized ewes whose follicles are supported by small doses of equine chorionic gonadotropin (eCG) (Cahill *et al.* 1985).

If eCG, or more recently FSH (Wright *et al.*, 1981), is given at this time the follicles are able to continue to produce estradiol and the atretic process can be prevented (Peters *et al.*, 1975; Mauleon and Mariana 1977). This technique has been used regularly for over 50 years to increase ovulation rate in ewes and does (Cole and Miller, 1935). The number of ovulations in response to eCG is partly dose dependent but principally depends on the number of antral follicles available to be saved at the time of administration. This is highly variable (McNatty, 1982), and so is the response.

3. Control of Ovulation Rate

Several new means have emerged in the last decade for improving the ovulation rate of sheep and goats. Injection of eCG or FSH preparations continues to be the main method but requires control or monitoring of the estrous cycle so that they can be administered on or just before the follicular phase of the cycle. Variability in response and cost of application remain as the main obstacles to its widespread use.

Like eCG, all of the new techniques appear to work by reducing atresia in medium to large follicles, thereby leaving more of them to ovulate. In most cases the exact mode of action by which they achieve this has not been fully explained. In fact some have been discovered accidentally. A case in point is the technique of immunizing against steroids, which was found during studies into luteal repression in ewes using passive immunization against estradiol (Scaramuzzi, 1975). Later it was found that active immunization against estrone, androstenedione, testosterone (Cox et al., 1976; Scaramuzzi et al., 1977; Land et al., 1982), and even progesterone (Thomas et al., 1987) all increase ovulation rate. After considerable refinement of the techniques of immunization and dosage to reduce undesirable side effects, a commercial technique involving immunization against androstenedione has become available (Geldard and Dow, 1984). Immunization against estradiol should reduce the amount of feedback restricting the release of LH, and this offers a plausible explanation for its action in preventing atresia, but androgens, including androstenedione, are not involved in the feedback mechanism and indeed FSH concentrations do not change (Campbell et al., 1988), so the immunization against them

must work in another way. Some increases in the rate of pulsing of LH during the luteal phase only have been recorded, but the relationship of the pulses to ovulation rate is tenuous. Removal of free androgen may directly reduce atresia, but a precise explanation of this phenomenon is still awaited. By contrast to the sheep, cattle respond very poorly when immunized against steroids, suggesting that sheep may control their ovulation rate much differently from cattle (Webb and Morris, 1988). So far, little work has been done in this area with goats.

Since inhibin specifically reduces secretion of FSH (de Jong, 1979), it is likely that immunization against inhibin or follicular fluid containing large quantities of inhibin will increase secretion of FSH and therefore ovulation rate. This is indeed the case in sheep (O'Shea et al., 1982, 1984; Henderson et al., 1984) and opens up another potential method for the routine improvement of ovulation rate in this species. Unfortunately, the response is still too highly variable to be reliable (Cummins et al., 1986). This may be because of the impurity of preparations of follicular fluid, and more purified inhibin or its alpha subunit may give better responses (Forage et al., 1987). A curious corollary to immunization against inhibin is the "rebound effect" of administration of inhibin (McNeilly and Wallace, 1987). In this case inhibin suppresses output of FSH while it is being administered but upon its withdrawal there is a temporary overcompensation of FSH production with a resulting increase in ovulation rate.

The composition and amount of feed eaten by sheep and goats before and around ovulation can influence the number of follicles that ovulate. The concept of "flushing" ewes or giving them a "rising plane of nutrition" has been known and used for most of this century for increasing twinning in sheep. The concept has been refined in recent years. There is a clear relationship between live weight and ovulation rate (Edey, 1968; Morley *et al.,* 1978), but the amount and the availability of the food eaten by a female just before ovulation can also influence the ovulation rate (Knight *et al.,* 1975; Lindsay, 1976). A change to an improved diet only a few days before ovulation can improve ovulation rate in ewes (Oldham and Lindsay 1984; Stewart and Oldham 1986). Despite these relationships, the physiological link between nutrition and reproduction has remained elusive. There is very little discernable change in the normal reproductive hormones in response to feeding. Production of luteinizing hormone does not change (Radford *et al.,* 1980; Knight *et al.,* 1981), and in some studies FSH has increased slightly but significantly in response to high quality feed (Brien *et al.,* 1976; Davis *et al.,* 1981; Nottle *et al.,* 1988). The responses due to changes in nutrition are usually no more than 0.1–0.5 ovulations per ewe, which is less than could be expected using hormonal techniques, so it is feasible that subtle changes in FSH just before ovulation may be sufficient to produce these responses.

Severe chronic undernutrition may produce long-term effects, probably as happens in the rat (Bronson 1986; Bronson and Manning 1988). In these cases ovulation of any sort is in question, rather than the number of ovulations. There is some evidence that long-term undernutrition may reduce the number of cycles the following breeding season (Fletcher, 1971; Oldham, 1980), but little definitive work on the physiological mechanisms involved in long-term undernutrition on ovulation has been done in sheep and goats.

There are wide genetic differences in ovulation rate and a number of genotypes: both temperate (Finnish Landrace, Romanov, and Booroola Merino) and semitropical or tropical (Barbados Blackbelly, D'man, and Javanese Thin-Tail) regularly produce many ova per ovulation. Inheritance of the capacity for multiple ovulation also varies from classical additive gene transfer, for example, in the Romanov and Finnish Landrace (Ricordeau, 1988), to major gene segregation in breeds such as the Booroola Merino (Piper and Bindon, 1982) and Javanese Thin-Tail (Bradford *et al.* 1986).

As with differences in ovulation rate due to nutrition, differences due to genotypes are frustratingly poorly correlated with differences in reproductive endocrinology (Bindon *et al.,* 1985). Among fecund genotypes, high ovulation rates are apparently achieved through different mechanisms (Land *et al.,* 1988). Nevertheless, in at least one genotype, the Booroola Merino, there have been small but potentially important differences in the FSH concentrations in the plasma of sheep carrying the major gene compared to noncarriers (McNatty *et al.,* 1987).

4. The Breeding Season

Among domestic farm mammals, sheep and goats are the most conspicuous seasonal breeders. The evolutionary reason for this is undoubtedly that, in regions where climate and nutrition change throughout the year, offspring are born at the most favorable time to survive and make their early growth. From studies on breeds from temperate regions sheep and goats have been called "short-day breeders" because they breed in fall and give birth in spring—a mechanism that protects the young from the harsh winters and the lack of feed at this time (Yeates,

1949). Photoperiodic changes act only as signals to allow animals to predict the availability of food in the future. Animals may respond in different ways, most probably according to the pattern of availability of food in their regions of origin. For example, the Romanov and the Solognote are both highly seasonal breeds and both respond to short days. The breeding season for the Romanov begins at the same time as the Solognote but ends about 2 months later (Land *et al.*, 1973).

But modern genotypes of sheep have originated from latitudes ranging from the frigid zones to the equator. In many environments winter is not the harshest time of the year, and near the equator there are no "short" or "long" days to signal the beginning and the end of a breeding season. Nevertheless, seasonal breeding in low latitudes may still be a phenomenon in both goats (Cognié *et al.*, 1981; Chemineau, 1986) and sheep (Ishaq, 1982). In this case it is probably feed supply, albeit a result of the climate, rainfall rather than temperature, that determines when breeding takes place (Thimonier and Chemineau 1988). In seasonal Pakistani breeds of sheep, Ishaq (1982) recognized two distinct breeding seasons, one in fall and one in spring, clear evidence that day length has little effect on these breeds.

So animals such as sheep and goats that attempt to match their breeding with the most likely period of abundant nutrients rely less on day length and more on other signals—pheromonal, nutritional, or social—as their centers of origin move from high to low latitudes (Bronson 1988). Temperature has also been suggested as a signal, but there is little evidence that it is used by animals to shift periods of breeding activity (Wodzicka-Tomaszewska *et al.*, 1967). Animals from

high latitudes do not lose their photoresponsiveness when moved to tropical latitudes or subjected to constant photoperiods (Thwaites, 1965; Carles and Kipragero, 1986; Thimonier and Chemineau, 1988), even after several generations. Conversely, aseasonal animals from low latitudes when subjected to photoperiodic light regimes resembling those of high latitudes continue to ovulate throughout the year (Fletcher and Putu, 1985; Lahlou-Kassi and Boukhliq, 1988), although estrous activity may cease during long days. Photoresponsiveness is therefore an innate adaptation of breeds from high latitudes that is poorly developed in breeds originating near the tropics. In between these extremes lies a large population of sheep and goats that show elements of photoresponsiveness that may be overridden or intensified by other cues. All animals originating in a band stretching from 35°N to 35°S are probably in this category. Fifty-seven percent of the world's sheep population is found in this zone, and an even higher proportion of the goat population (McDowell, 1972). The most numerically important of these is the Merino. This breed has a breeding season in fall that corresponds roughly to that of other breeds of mid- or high latitudes (Wheeler and Land, 1973; Oldham, 1980), but its nonbreeding season can be completely overridden by other factors, notably the ram effect, in which the introduction of males to anovular and previously sexually isolated females can induce ovulation and reestablish the breeding season. Thus signals used by sheep and goats to define their breeding and nonbreeding seasons may be photoperiodic, nutritional, social, or combinations of any of these, and the importance of these signals varies according to the origin of breed of animal in question.

a. Hormonal Control of Seasonal Breeding There is strong evidence that the hormonal controls within the animal that cause it to enter or leave a period of breeding activity are identical regardless of the signals that initiate these events. The breeding season is characterized by a capability in the female of regulating the pulsatile release of LH at frequencies and amplitudes appropriate to bring about the luteal and follicular phases of the estrous cycle. When these pulses can no longer be generated and modulated, the female lapses into anestrus. Anestrus is thus characterized by regular but very infrequent pulses of LH (Scaramuzzi and Martensz, 1975). Simply, the transition to the breeding season involves development of the ability by the pulse-generating mechanism in the brain to reinstate pulsatile activity followed by the normal ovarian–hypothalamic interplay that leads to ovulation. The key to the mechanism of this switch was provided by Legan and Karsch (1979), who showed that in the nonbreeding season the pulse-generating system becomes extremely sensitive to the action of estradiol—so much so that even the limited quantities being produced by the small and intermediate follicles of the anestrous ovary are sufficient to maintain a very low frequency of LH pulses. Long days may dampen the frequency of pulsing in their own right (Goodman *et al.*, 1982), but this effect is far less quantitatively important than the control by sensitivity to estradiol. Once the pulse-generating mechanism overcomes its extreme sensitivity to estradiol, it can produce sustained increases in LH even in the face of increased production of estradiol from the ovary and finally evoke the first ovulation of the breeding season. The effect of short days (Karsch *et al.*, 1984), the male effect, and possibly improved nutri-

tion (Scaramuzzi and Radford, 1983) is to render the mechanism less sensitive to estradiol and to allow it to increase the frequency of tonic release of LH. Such increases in frequency have been seen to precede the onset of the natural breeding season in sheep (I'Anson, 1983), and almost instantaneous increases are found following the introduction of males (Oldham *et al.*, 1978). During the breeding season the pulse-generating system remains relatively insensitive to estradiol and is free to stimulate continual cyclical activity. Then under the influence of lengthening days, poor nutrition, or other external factors, the sensitivity to estradiol returns and the female enters anestrus.

b. Photoperiod as a Controlling Mechanism In those genotypes in which day length affects the timing of episodes of reproduction, or breeding season (Mauleon and Rougeot, 1962), the perception by the animals of "short days" following "long days" is assumed to be the trigger that reduces sensitivity of the pulse generator to the inhibitory action of estradiol (Thimonier, 1981). Except in experimental situations this is invariably the late summer and fall. Conversely, long days after short days restore sensitivity and inhibit reproduction (Lincoln and Short, 1980). However, it is not simply a process in which females commence breeding after the length of day passes below a certain limit in the fall and stop when it becomes longer in spring. The absolute number of hours of daylight at which sheep begin to breed in the fall may be greater than when the same ewes stop breeding in spring (Robinson and Karsch, 1984). This may not, of course, be the case for other breeds, because all breeds do not respond identically to the same light regime

(Land *et al.*, 1973). The concept that day length or change in day length actually induces breeding activity has been replaced by one that proposes that the sheep or goat has an innate circannual rhythm that determines, at least approximately, when it will start breeding regardless of the light regime under which it is running (Ducker and Bowman, 1970; Howles *et al.*, 1982; Almeida and Lincoln, 1984; Robinson and Karsch, 1988). So animals from high latitudes brought to the equator or kept in rooms with constant light, or blinded animals, none of which can perceive changes in day length, all continue to show seasonal cyclical activity although it becomes increasingly erratic in successive years, and asynchronous among individual animals. Thus the animals' perception of daylight does not force them into breeding activity but entrains or "fine tunes" an already active endogenous rhythm to initiate the breeding season (Karsch and Wayne, 1988; Robinson and Karsch, 1988).

For the same reasons, it is believed that day length may not be the principal controlling factor in the determination of the end of the breeding season and the time of transition into anestrus. In fact, it has been proposed that females become photorefractory, or unresponsive to the photoperiod that they are experiencing, and cease breeding more or less spontaneously because constant short days cannot sustain unlimited breeding seasons (Howles *et al.*, 1982; Karsch and Wayne, 1988).

The thyroid has an enigmatic role in the process of "switching off" the breeding season. When ewes are thyroidectomized there is no interference with the onset of breeding but the animals continue to breed for much longer than control animals regardless of the light regime to which they have been subjected (Nicholls *et al.*, 1988). The thyroid is thus clearly involved in the cessation of the breeding season but seemingly in a permissive role, because intact animals do not show any significant changes in thyroid secretion that could be linked to the period of transition from the breeding season to anestrus (Nicholls *et al.*, 1988).

Not only do sheep and goats have a circannual rhythm that acts as a blueprint for the breeding season, but they probably have another endogenous rhythm to allow them to perceive daily changes in illumination. This circadian rhythm, proposed originally by Bunning (1960), consists of two phases. In the first phase the animal is insensitive to light, but during the second it is highly sensitive. Changes from dark to light in the early part have little effect, and only those made in the sensitive phase are detected by the animal. This has been used to manipulate breeding activity in both sheep and goats by subjecting them to "phantom" days (Ravault and Ortavant, 1977; Ortavant *et al.*, 1978, 1988; Thimonier *et al.*, 1978; Schanbacher and Crouse, 1981). By providing 8 h of light per day in the form of 7 h of continuous light and a 1-h "flash" at varying times during the dark phase it can be shown that there is a photoinduceable period during which the "flash" is highly effective. The animals read the length of day as the total period between the beginning of the 7 h of light and the end of the 1-h "flash." The period of dark between these two is effectively ignored. Long days can therefore be simulated in animals in the fall and winter by providing them with a short period of artificial lighting during natural darkness and creating artificially long days. These have been used experimentally to prime both sheep and goats with long days preparatory to changing them back to short days to stimulate the early onset of the breeding season (Chemineau *et al.,* 1988). The endogenous circannual rhythm can be manipulated artificially by changing the frequency and timing of alternating long and short day signals. By this means it is possible to reverse the natural cyclicity of reproductive activity (Ortavant *et al.,* 1964) or accelerate it to produce 8-month, 6-month, or even 3-month reproductive "years" (Karsch *et al.,* 1984; Ravault and Thimonier, 1988). As we have already seen, the short days serve to entrain the endogenous rhythm to produce breeding activity, and long days are necessary to prime animals that have become photorefractory before they will respond again to the photoinductive influence of short days. If periods of short days and periods of long days are alternated at a rate that is faster than the animals can become photorefractory, a continuous breeding season should ensue. This is indeed the case for rams and bucks (Chemineau *et al.,* 1988; Pelletier *et al.,* 1988) but is yet to be demonstrated in females. The link between signals associated with day length and the hormonal control of breeding seasons is the pineal gland, which transduces photoperiodic messages to chemical messages (Bittman, 1978). The pineal is innervated from the superior cervical ganglia, through which it receives regulatory photoperiodic messages. In response it produces a number of peptide hormones including seratonin, 5-methoxytryptamine, and melatonin. Melatonin appears to be the most active of these endocrinologically for the control of breeding activity. It is synthesized and released only during the hours of darkness (Rollag and Niswender, 1976), and its production can be inhibited by artificial light given at night or can be abolished completely by placing animals in controlled light. So the amount of darkness perceived

by the animal regulates the duration of high levels of melatonin in its blood (Rollag *et al.,* 1978; Kennaway *et al.,* 1983; Arendt, 1986), and it seems probable (Goldman, 1983) that the duration of this nocturnal production of melatonin is the means by which photoperiodic messages are passed to the brain in both the sheep (Karsch *et al.,* 1988) and the goat (Maeda *et al.,* 1988), although this concept does not fully explain the response to flashes of light that tend to break up the solid nocturnal bolus of melatonin (Ravault and Thimonier, 1988).

The functional response by the brain is a modification of the pulses of luteinizing-hormone releasing hormone (LHRH), but the manner in which melatonin influences the secretion of LHRH is still unknown. However, it has been demonstrated that endogenous opioid peptides that are intimately involved in the control of production of LHRH also vary in response to melatonin. It is proposed, but by no means confirmed at this stage, that melatonin may have its effect by activating the endogenous opioid peptide mechanism in the hypothalamus.

The relationship between daily length of darkness and the secretion of melatonin has encouraged research into controlling the breeding season by supplying exogenous melatonin by injection, implants (English *et al.,* 1986), sponges (Nowak and Rodway, 1985), or orally (Kennaway and Seamark, 1980). Melatonin can mimic the effects of short days by advancing the breeding season and also by decreasing the output of prolactin in both sheep and goats (Kennaway *et al.,* 1980; Nett and Niswender, 1982; Arendt *et al.,* 1983; Symons *et al.,* 1983). Originally treatment was given only in the late afternoon so that the exogenous melatonin would overlap with endogenous night-time secretion and provide the animal with a daily profile resembling short days. Later it was found that administration of melatonin continuously gave the same response as timed daily doses and the breeding season could be advanced using chronic implants (English *et al.*, 1986; Poulton *et al.*, 1987). Despite this, melatonin is unable to induce breeding at all stages of anestrus, and positive results from its use are confined to the period immediately before the expected breeding season. In addition, in breeds such as the Merino, which is less strongly seasonal, results have been very variable. A precondition for adequate response to melatonin appears to be that the animal must first have experienced a period of long days (Chemineau *et al.*, 1988). Early in the anestrous period this condition may not have been met. By subjecting goats to artificially long days in winter Chemineau *et al.* (1986) induced does to respond strongly to melatonin in spring with normal and persistent ovarian cyclicity and behavior. This response could not be induced without the "priming" treatment of long days.

A curious, potentially useful, but as yet unexplained side effect of the treatment with melatonin is that in short-term treatment it increases the ovulation rate and lambing percentage of treated ewes over those of untreated animals (Kennaway *et al.*, 1984, 1987). Even more curious is that long-term treatment appears to reduce the rate of ovulation below that of treated animals (Jordan *et al.*, 1988).

c. The Male Effect The "male effect" is a social stimulus that acts to initiate the onset of the breeding season in both sheep and goats. Its existence has been known for almost half a century (Underwood *et al.*, 1944; Schinckel, 1954; Radford and Watson,

1957), but it is only recently that its potential for controlling the breeding season for management purposes has been studied. Nonovulating females, when introduced to males after a period of separation, commence rapid LH pulses that lead to a preovulatory surge of LH and ovulation about 48 h later (Martin *et al.*, 1980, 1983). This is normally not accompanied by estrus, but the cyclical activity thus begun can lead to normal estrus and ovulation 17 days later if a corpus luteum (CL) is formed from the first ovulation. In about 50% of animals no CL is formed after the first ovulation and there is a short cycle resulting in the animal reovulating in about 6 days, again without estrus. The resultant CL this time is always normal, and normal estrus and ovulation follow 7 days later. The short cycles can be prevented in sheep by a single injection of 20 mg progesterone given around the time at which the ewe becomes exposed to the ram (Cognié *et al.*, 1982). In the doe Chemineau (1985) found that a single injection of the synthetic progestagen fluorogestone acetate prevented short cycles, and conception rate in treated animals was 85% compared with 10% in controls.

Short cycles seem to be associated with a lack of progesterone from the anestrous ovary at the beginning of the breeding season. The "male effect" is effective in inducing breeding activity in goats as well as sheep, whether anestrus is due to photoperiod (Shelton and Morrow, 1965; Ott *et al.*, 1980; Restall, 1988) or to nutrition (Chemineau, 1983; Mgongo, 1988). The efficacy of the "male effect" in inducing anestrous animals to initiate breeding cycles seems to be inversely related to the degree of latitude of the origin of the genotype. In other words, the more an animal is photoreceptive, less widespread its response to the male. Sheep and goats from high latitudes already under heavy control from day length do not ovulate in response to males except during the few weeks immediately preceding the light-induced breeding season. Nonetheless, even in these breeds the male can increase the rate of pulsatile release of LH from ewes in deep anestrus, but this does not continue through to a preovulatory surge or ovulation (Signoret and Lindsay, 1982).

d. Nutrition Nutritional control of the breeding season is obviously more noticeable in tropical latitudes than in temperate latitudes because in the absence of photoperiodic cues it is one of the few signals available to the female. Whether it interacts with photoperiod to modify either the length or timing of the breeding season in high latitudes is not known, although there is some evidence (Smith, 1965; Oldham, 1980) that poor nutrition of ewes can shorten the length of their breeding season 6 months later.

B. Pregnancy

1. Insemination

The cervix of the doe is shorter and less convoluted than that of the ewe and therefore allows deep cervical or even intrauterine insemination. In the ewe this is normally impossible except in old, multiparous animals. The doe has considerable potential advantages when using frozen semen because the only consistently successful results follow intrauterine insemination, which in the ewe usually involves laproscopy (Killeen and Caffrey, 1982). For the same reason the possibility of using nonsurgical recovery and transfer of embryos is also much greater in the doe (Riera. 1984).

2. Fertilization

After ovulation the ova are quickly gathered into the fallopian tubes, where fertilization takes place in the ampulla of the oviduct. The processes of fertilization and subsequent growth of the fetus have been covered in Chapter 8, but certain features of fertilization and pregnancy in sheep and goats differ from those of other species. Once the basic conditions of normality of spermatoza and ova and the presence of both in the ampulla about the same time have been met, fertilization is usually highly successful and is seldom a significant source of reproductive loss (Kelly, 1984). This is emphasized by the fact that following multiple ovulations the fertilization of one ovum seems to be invariably accompanied by fertilization of the others (Restall et al., 1976). Sheep grazing pastures high in estrogenic isoflavones are an exception in that they have low rates of fertilization (Lightfoot and Wroth, 1974) due to poor transport of spermatozoa (Lightfoot et al., 1967), caused by altered production and consistency of cervical mucus (Smith, 1971; Adams, 1975).

3. Survival of Embryos

By contrast with fertilization, losses of embryos, particularly those in the preimplantation and early implantation phase—between ovulation and 30 days of gestation—can be considerable. Much of this loss may be attributed to a failure to meet the complex hormonal and micronutritional demands of the conceptus during this period (see Chapters 9 and 10). However, association of losses with management factors has not been clearly established. The relationship between management and embryo loss has been reviewed by Kelly (1984), and only the age of the female stands out as a consistent factor in embryonic loss. Young ewes, especially ewe lambs, have very high losses of cleaved ova compared with older animals, even when these ova are transferred into older ewes (Quirke and Hanrahan, 1977).

The hormonal milieu, especially the concentration of progesterone, has been implicated in survival of embryos but the results are equivocal. Progesterone has been shown to be the only gonadal hormone required for the maintenance of pregnancy after the first 2 days, and a wide range of doses can support pregnancy in ovariectomized ewes (Foote *et al.*, 1957; Moore and Rowson, 1959; Bindon, 1971) or ovariectomized and adrenalectomized ewes (Cumming *et al.*, 1974). In both the sheep and the goat the source of this progesterone is the corpus luteum of pregnancy. Castration of the doe at any time during pregnancy results in abortion, but in the ewe pregnancy is also supported by progesterone produced by the conceptus and this secondary source of progesterone is of sufficient quantity that castration after about 50 days does not result in abortion (Denamur *et al.*, 1973). The maintenance of the early corpus luteum and its development into the corpus luteum of pregnancy involves the interplay of a wide range of luteotrophic and luteolytic substances. Hypophysectomy of ewes up to 50 days of pregnancy and goats at any time during pregnancy results in the demise of the corpus luteum and abortion, suggesting that the pituitary provides at least one source of luteotropin. But the embryo itself, as early as the single-cell stage after fertilization, produces an astonishing array of substances, some of which are believed to be luteotropic and act long before there is any physical association between the embryo and the uterus (Heap *et al.*, 1988; Hansel, 1988). In addition it is now well established

that the conceptus overrides the normal luteolytic activity seen in the nonpregnant uterus that produces prostaglandin $F_{2\alpha}$ (Thatcher *et al.*, 1988, for review).

At the production level the extent of losses due to breakdown of any or all of these mechanisms has not been established, and consequently neither have any reliable adjustments to management to reduce early embryonic losses.

By contrast with early pregnancy, loss of fetuses from the time of complete attachment of the placenta at about 30 days until parturition is relatively low, and undernutrition, excessive heat, and a host of other stressors do not lead to significant failure of pregnancy at this time. The fetal cotyledons and the caruncles or maternal cotyledons in both sheep and goats are opposed in a cotyledonary syndesmo–chorial placentation. As the number of fetuses increases, the total number of active caruncles increases (Alexander, 1964). In the ewe this number varies from about 60 to 120; in the doe it is almost double, from 160 to 200. In does there are more in the gravid horn but in ewes both the gravid and the nongravid horn have caruncles in equal numbers (Amoroso, 1952).

One other form of loss that often remains hidden is partial failure of multiple ovulation in which ewes or does remain gravid but the number of fetuses is less than the number of ovulations. Losses from this source increase with the number of ovulations (Kelly, 1984) and suggest that the female may exert some form of uterine control over numbers of offspring. The possibility of such control has been suggested for breeds of sheep such as the Barbados Blackbelly and the East Java Fat-Tail, which have a narrower than usual distribution of birth types based around an unusually high proportion of twins compared with triplets and single lambs (Lindsay, 1982). There is no evidence from better-known breeds that have been induced to superovulate, for example, of such precise uterine control and superovulation by whatever means is normally characterized by a wide variation in numbers of offspring per female.

4. Diagnosis of Pregnancy

The embryo becomes biochemically active from the moment of fertilization, so it is theoretically possible to detect pregnancy almost as soon as the two pronuclei fuse to form the diploid organism. Tests such as the rosette inhibition test for assaying "early pregnancy factor" (Morton et al., 1979) have been used at laboratory level to diagnose pregnancy at all stages up to implantation (Nancarrow et al., 1981). They are slow and expensive, and such early diagnosis carries little advantage for sheep and goat husbandry, particularly as it tests animals before the major period of expected loss of embryos. The earliest practical diagnostic is the concentration of progesterone in plasma or milk at days 17–19 in the ewe, or 20–22 in the doe (Robertson and Sarda, 1971; Thimonier, 1973). If animals are nonpregnant they will have no corpus luteum of pregnancy and should be in their follicular phase at this time. As a result the concentration of progesterone would be very low (less than 0.5 μg/ml compared with pregnant animals (> 1 μg/ml). Multiple ovulating animals will have more corpora lutea and therefore should secrete more progesterone, but the relationship is not entirely reliable and in any case each corpus luteum in a pregnant female may not be represented by a viable embryo. Among field tests the simplest but most unreliable is the recording of the "returns to service." Males, vasectomized or entire, are put with potentially pregnant fe-

males and those that mate are assumed to be nonpregnant. There are two sources of error. Some pregnant ewes show variable degrees of estrus and are therefore diagnosed as nonpregnant, and, depending on the time of the year and the seasonality of the breed, nonpregnant ewes may enter anestrus and be diagnosed as pregnant. There are various techniques for sensing the presence of the fetus, including simple ballotment with the aid of a rectal probe (Hulet, 1972), laparotomy (Cutten, 1970), machines that use the Doppler effect to detect fetal heart beat (Lindhal, 1971), X-ray machines (Rizzoli et al., 1976), and real-time ultrasound (Fowler and Wilkins, 1984). Of these, the most commonly used in recent years have been the ultrasound machines. They are the most effective but also the most expensive. Using external probes both pregnancy and, within pregnant ewes, multiple pregnancy can be diagnosed accurately from about day 45. Using internal rectal probes pregnancy can be diagnosed from day 20, and multiple pregnancy can be determined accurately from day 27 (Wilkins, 1988). Pregnancy diagnosis using this technique in flocks with a high proportion of multiple births can help substantially with the economic distribution of available nutrients according to the needs of the pregnant females.

C. Parturition and Maternal Care

The ewe has been the principal model for the study of mammalian parturition and this subject is covered fully in Chapter 10. In grazing animals like the sheep and goat where the offspring are relatively well advanced at birth, the early and selective recognition of the young by the mother is essential to its survival. The complex group of activities, collectively called "maternal behavior," that results in this recognition and care is an important part of the reproductive process. The hormones of late pregnancy and parturition all serve to prime the female for the care and feeding of her young. In particular, estrogen, which reaches very high concentrations at the time of parturition, appears to be important in facilitating the onset of maternal behavior (Poindron *et al.*, 1988). Experiments using supplementary doses of estradiol have improved lactation and growth of offspring in goats (Terqui, 1978) and prolonged the period during which parturient ewes will accept lambs (Poindron *et al.*, 1979). By contrast, prolactin appears to have little effect on this response (Poindron *et al.*, 1980). The other hormone that is intimately involved in maternal behavior is the neurohormone oxytocin. The specific stimulus for the release of oxytocin is genital stimulation from the lamb at the point of parturition. At this time the ewe changes her behavior very suddenly, becomes less aggressive, is attracted to amniotic fluids, and is stimulated to lick and to encourage suckling of her own lamb (Keverne, 1988; Poindron *et al.*, 1988). Most of these behaviors are suppressed by peridural anesthesia just before lambing, which obliterates the neural signal responsible for releasing oxytocin. Oxytocin from the neurohypophysis or from exogenous injections does not cross the blood–brain barrier (Kendrick *et al.*, 1986), so it is the central oxytocin produced by the brain itself that is effective. Increases in concentration of oxytocin in the cerebrospinal fluid (CSF) have been reported in sheep during both vaginocervical stimulation and natural parturition (Kendrick *et al.*, 1986). In addition, short-term maternal responses have been reported after intracerebral injections of oxytocin in the nonpregnant estrogen-primed ewe (Baldwin *et al.*, 1986; Kendrick *et al.*, 1987). The control of a mechanism as complex as maternal behavior undoubtedly involves a wide range of exocrine and endocrine factors, but estrogen and oxytocin remain as the two outstanding controlling factors.

D. Intersexuality and Freemartinism

Intersexuality is a phenomenon seen frequently in goats but rarely in sheep (Riera, 1984). Genetic females are usually masculinized to a greater or lesser degree, and males have abnormalities associated with degrees of cryptorchidism. Both are usually sterile. Polledness in some but not all breeds of goats is associated with intersexuality, for reasons that have not yet been elucidated (Taneja, 1982).

Freemartinism is a characteristic of females from multiple pregnancies that have been influenced by male twins *in utero* either hormonally during gonadal formation or through blood chimeras, or both. These are common in cattle but rare in sheep, despite the fact that multiple pregnancies and therefore opportunities for freemartinism are much higher in sheep than in cattle.

E. The Postpartum Period

Resumption of reproductive capacity after parturition depends on the reinstatement of three distinct factors. First the uterus must involute with regeneration of the uterine epithelium and reduction in size. Then ovarian activity must begin, followed by estrous behavior, which must be synchronized with ovulation. Some sheep show estrous activity immediately after lambing (Barker and Wiggins, 1964; Land,

1971), but this appears to be due to the large quantity of circulating estrogen at parturition following the progesterone of pregnancy, and is not accompanied by any ovarian activity (Robertson, 1977). The phenomenon does not appear to have been reported in goats. Involution probably has to be complete before the first complete estrus and ovulation (Novoa, 1984). This period has been variously estimated at between 60 and 70 days and is prolonged by active suckling (Honmade, 1977). After parturition there is a period of ovarian quiescence, and the return to cyclical activity depends principally on the season and, to a much lesser extent, lactation. In breeds of both sheep and goats of tropical origin that are therefore aseasonal, fertile mating during lactation is very common (Obst *et al.*, 1980; Riera, 1982). Breeds that experience anestrus and lamb or kid, as they normally do in the anestrous period, do not begin cyclical activity until the following season, but this should not be confused with lactational anestrus. When ewes or does give birth out of season—that is, at the beginning of their normal breeding season—lactation has less effect or a more indirect effect than has often been assumed (Hunter, 1968). Mauleon and Dauzier (1965) showed that ewes from which their lambs had been weaned at 48 h began cycling earlier than ewes that were hand-milked, which in turn cycled earlier than suckling ewes. The difference was one of degree, and lactation certainly does not prevent estrus (Land, 1971; Restall, 1971; Skevah *et al.*, 1974). Lactation imposes a severe nutritional stress on ewes, and suckling, because it promotes a higher level of milk production than hand-milking, may impose an even higher stress. It has been suggested that lactation acts indirectly on estrus in this way (Hunter and van Aarde,

1973). Nutrition and season have been shown to interact with lactation in determining the postpartum interval to first estrus (Corbett and Furnival, 1976; Restall and Starr, 1977). The beginning of cyclical activity after parturition is usually accompanied by one or more ovulations without estrus (Mauléon and Dauzier, 1965) and often by short cycles (Restall, 1971). This is identical with the picture seen at the beginning of the normal breeding season, and the evidence suggests that the two phenomena are similarly controlled through a gradual decrease in sensitivity to negative feedback from estrogen and a consequent increase in frequency in pulses of LH (Wright *et al.*, 1980, 1981, 1983). High levels of prolactin seen particularly in suckling ewes and does are negatively correlated with concentrations of LH (Kann and Martinet, 1975; Kann *et al.*, 1977). This offers an explanation for the inhibitory effect of suckling on return to estrous activity. However, the mechanism by which prolactin works is still unclear, as it appears to have no direct effect on the sensitivity of the hypothalamic–hypophyseal axis to feedback from estrogen.

III. The Male—The Ram and the Buck

A. Seasonality in the Male

As with females, rams and bucks may alter their capacity for reproduction according to the season of the year. It is easier to measure seasonal changes in males than in females because production of spermatozoa is a continuous process unlike the production of ova, and the volume of the testicular tissue is relatively easy to measure. Testicular volume and daily production of spermatozoa are highly correlated (Ortavant, 1958; Knight, 1973), so that testicular volume is a useful indicator of changes in reproductive capacity. The degree of influence of photoperiod on testicular volume seems to be related to the degree of latitude of the center of origin of individual breeds. For example, in the Ile de France breed, Pelletier (1971) has shown that testicles of rams under constant nutrition vary from 190 g each in the spring to 300 g in the fall. The testicles of a breed from an even higher latitude, the Soay, vary even more with season, and its sexual behavior and aggressiveness vary in parallel (Lincoln and Davidson, 1977). By contrast, Moule *et al.* (1966), using Merinos, found relatively small differences between the production of spermatozoa in spring and in the fall when animals were on a constant diet. Breeds originating from equatorial regions show no seasonality that can be attributed to photoperiod (Chemineau, 1986). In genotypes that are only mildly influenced by photoperiod, variation in nutrition is rapidly reflected in variation in testicular volume (Oldham *et al.*, 1978). When the animals are grazing, testicular volume is influenced by the availability of feed regardless of photoperiodic changes. Masters and Fels (1984) showed that the testicular volume in spring when feed was of good quality and quantity was twice that in the fall when the feed supply was poor.

Seasonal changes in testicular volume and production of spermatozoa are correlated with a number of other important male characteristics. Libido in both rams (Lindsay and Ellsmore, 1968; Lincoln and Davidson, 1977) and bucks (Corteel, 1973) also changes with season. The fertilizing capacity of semen collected in spring is lower than that of semen collected in the fall (Colas *et al.*, 1985; Corteel, 1977), even after

eliminating seasonal differences in fertility of ewes by using frozen, stored semen and inseminating them all at the one period. Thus when attempting to breed animals originating from mid- or high latitudes out of season it is equally as important to ensure that males are fertile as it is to induce females to undergo reproductive cycles. When seasonality is the result of a fluctuating supply of forage, supplements containing moderately high levels of protein have produced rapid and substantial improvement in testicular volume when given during the period of poor nutrition (Oldham *et al.*, 1978; Lindsay *et al.*, 1977). Low reproductive capacity as a result of "long day" photoperiodic episodes is not as readily overcome. By subjecting animals to a light regime of decreasing daylight the unfavorable influence of spring days can be eliminated (Colas *et al.*, 1985), and the technique has been used in artificial insemination centers for sheep in France. The problem is that, like the female, males become photorefractory and after a time cease to be stimulated by a regimen of short days, with the result that their testicles begin to regress. If the frequency of switching the light regime from long to short days and back again is increased, males receive the stimulatory effects of short days sufficiently often to be reactivated before photorefractoriness sets in. Some evidence of this can be seen in 3-month cycles (Pelletier *et al.*, 1985), but if the light regime is changed every 2 months or even every 30 days there is no regression in testicular volume even after several years of such treatment (Pelletier *et al.*, 1988). This concept has been further developed to alternate short days with "skeleton" long days (7 h light : 8 h dark : 1 h light : 8 h dark), which also successfully eliminates the period of normal seasonal regression and maintains continuous testicular activity in both rams and bucks (Pelletier *et al.*, 1988). When semen is collected for artificial insemination and females are treated to induce them to breed out of season this form of management of rams appears ideal to ensure a continual supply of high-quality semen.

The effect of short days may be simulated in males by using implants of melatonin (Lincoln and Ebling, 1985; Chemineau *et al.*, 1988). Melatonin works more reliably if periods of treatment (usually continuous implants) are interspersed with periods of long days—real or "skeletal" (Chemineau *et al.*, 1988). This becomes particularly important when the frequency at which the implants are given is relatively rapid in order to simulate a light rhythm with a frequency of only a few months. Melatonin may act by changing the sensitivity of the hypothalamus to the inhibitory effect of endogenous peptide opioids (Lincoln *et al.*, 1987; Lincoln, 1988), resulting in an increased frequency of release of pulses of LH (Schanbacher, 1982, 1985) and a consequent resurgence of testicular activity.

B. Reproductive Capacity in Males

Setchell (1978) gives the range of ratios of testicular weight to live weight in a large number of species as 0.02–0.50% and rams and bucks occupy the upper limit of this range. There is a functional significance in certain species having large testicles (Harcourt *et al.*, 1981) and in the case of rams and goats it is to mate large number of males and females in a very short time (Lindsay, 1988). In the wild, annual breeding seasons are often very short and even under domestication there are often good economic and management reasons why breeding activity should not be protracted. Mating ratios of males to females are normally low (1 : 30–40) in recognition of the high reproductive activity of males, but individual animals are capable of impregnating many more females than this in the space of one sexual cycle (Haughey, 1959; Allison, 1975). A measure of this capacity would be very useful for making the best use of good males and for planning management at mating, especially where estrus in females is synchronized and artificial insemination is not used. The first step is to consider the minimum amount of semen required by the female to ensure normal fertility. This number has been variously estimated in artificial insemination (AI) studies in sheep to be from 120×10^6 (Salamon, 1962) to 400×10^6 (Allison and Robinson, 1971), but higher figures generally relate to ewes that have been artificially induced to breed (Langford and Marcus, 1982). In naturally mating sheep with ewes in natural estrus only 60×10^6 spermatozoa seem to be required (Fulkerson *et al.*, 1982). If we contrast these figures with the estimated daily output of spermatozoa from a ram of 300,000 or $400,000 \times 10^6$ (Cameron, 1987) there seem to be ample spermatozoa to spare. However, there are two problems in this raw estimation. First, the ram does not distribute his spermatozoa evenly among available ewes (Synnott *et al.*, 1981), and second, his capacity for sustained production of spermatozoa may not persist for long. Rams prefer certain individual ewes to others, and these preferences are consistent between rams (Rouger, 1974; Tilbrook and Lindsay, 1987). Thus, given a free choice among several ewes in estrus at the same time, rams mate some ewes to the exclusion of others. Fortunately, when several rams are used together the subordinate rams are obliged to mate with the less favored ewes and all ewes have a chance of conceiving (Synnott *et al.*, 1981).

Published figures for the number of spermatozoa produced per ejaculate by male domestic animals are usually gross overestimates. In the case of the ram and the buck this is given at around $2000–3000 \times 10^6$ but the true figure in actively mating animals is very much lower than this (Cameron, 1987; Lindsay and Thimonier, 1988). The numbers of spermatozoa per ejaculate and the volume of the ejaculate are inversely proportional to the numbers of ejaculates per day and the number of days that males have been working. In AI programs where both of these factors can be controlled, large numbers of ejaculates not only reduce the numbers of spermatozoa per ejaculate but also the total number of spermatozoa produced (Cameron, 1987). In the ram the total sperm output falls as the number of collections per day increases above four. In natural mating, unrestricted individual rams and bucks may exceed 30 or 40 ejaculates per day so that individual ejaculates and total output are probably extremely small. In one study individual ejaculates were less than the minimum required for normal fertility (Synnott *et al.*, 1981); in another (Cameron *et al.*, 1987), where rams were well husbanded, individual ejaculates were around 160×10^6, which is enough for normal fertility. Whatever the case, two or more matings at estrus result in higher fertility in ewes than a single mating (Mattner and Braden, 1967; Knight and Lindsay, 1973). Sustained mating activity results in rapid reduction in testicular volume and hence capacity to produce spermatozoa (Lindsay *et al.*, 1977). This may be due to a physiological effect of mating per se leading to a lack of support of the seminiferous tubules (Knight

et al., 1987). More importantly, the fact that males spend little time eating during the breeding season can severely influence testicular volume, as this is highly sensitive to nutrition (Lindsay *et al.*, 1977; Oldham *et al.*, 1978). In wild Soay sheep the debilitating effect of the breeding season and the energy used to establish a dominance hierarchy among rams significantly reduce their chances of surviving the harshness of the following winter relative to that of ewes in the same territory (Grubb and Jewell, 1966).

Males vary greatly in their serving capacity (Fowler, 1984) and in competitive conditions those that serve the most females are likely to be the most successful. Apart from supplying additive nutrition to males in poor condition no techniques have so far been advanced for increasing the serving capacity of individuals. Testosterone and other steroids can induce male behavior in castrates (Fulkerson *et al.*, 1981) or even females (Signoret *et al.*, 1982/1983), but the role of these hormones appears to be permissive rather than quantitative because serving capacity in active rams or bucks cannot be enhanced, at least in the short term, by supplementary injections of steroids.

The most active individuals can be identified using short-term tests of serving capacity (Mattner *et al.*, 1971; Chenoweth, 1981; Blockey, 1983; Fowler, 1984). The predictive worth of such tests has been challenged (Purvis *et al.*, 1984; Cahill *et al.*, 1975; Kelly *et al.*, 1975; Allison, 1978), but some interesting correlations between serving capacity and pregnancy rates (Fowler, 1984) suggest that they may be useful under some circumstances. Fowler (1984) has proposed that in the final analysis a measure of testicular volume that reflects the capacity for production of semen and some estimate of capacity for frequent service are the two most important predictors of the ability of individual males to impregnate large numbers of females in a short time.

References

Adams, N. R. (1975). *J. Reprod. Fertil.* **43,** 391–392.
Alexander, G. E. (1964). *J. Reprod. Fertil.* **7,** 189–305.
Allison, A. J. (1975). *N.Z. J. Agric. Res.* **18,** 1–18.
Allison, A. J. (1978). *N.Z. J. Agric. Res.* **21,** 187–195.
Allison, A. J., and Robinson, T. J. (1971). *Aust. J. Biol. Sci.* **24,** 1001–1008.
Almeida, O. F. X., and Lincoln, G. A. (1984). *Biol. Reprod.* **30,** 143–158.
Amoroso, E. C. (1952). *In* "Marshall's Physiology of Reproduction," Vol. II, 3rd ed., pp. 127–311. Longmans, New York.
Amir, D., and Schindler, H. (1972). *J. Reprod. Fertil.* **28,** 261–264.
Arendt, J. (1986). *In* "Oxford Reviews of Reproductive Biology," pp. 266–320. Clarendon Press, Oxford.
Arendt, J., Symons, A. M., Land, C. A., and Pryde, S. J. (1983). *J. Endocrinol.* **97,** 395–400.
Armstrong, D. T., Weiss, T. J., Selstam, G., and Seamark, R. F. (1981). *J. Reprod. Fertil. Suppl.* **30,** 143–154.
Baird, D. T. (1983). *J. Reprod. Fertil.* **69,** 343–352.
Baird, D. T., Baker, T.G., McNatty, K. P., and Neal, P. (1975). *J. Reprod. Fertil.* **45,** 611.
Baird, D. T., and McNeilly, A. S. (1981). *J. Reprod. Fertil. Suppl.* **30,** 119–133.
Baldwin, R. A., Kendrick, K. M., and Keverne, E. B. (1986). *J. Physiol.* **381,** 80.
Barker, H. B., and Wiggins, E. L. (1964). *J. Anim. Sci.* **23,** 967–972.
Bindon, B. M. (1971). *Aust. J. Biol. Sci.* **24,** 149–158.
Bindon, B. M., Piper, L. R., Cummins, L. J., O'Shea, T., Hillard, M. A., Findlay, J. K., and Robertson, D. M. (1985). *In* "Genetics of Reproduction in Sheep" (R. B. Land and D. W. Robertson, eds.), pp. 3–18. Butterworths, London.
Bittman, E. L. (1978). *Science* **202,** 648—650.
Blockey, M. A. de B. (1983). *Univ. Sydney Postgrad. Comm. Vet. Sci. Proc.* **67,** 119–127.
Bradford, G. E., Quirke, J. F., Setorus, P., Inocuni, I., Tresnamurti, B., Bell, F. L., Fletcher, I., and Torell, D. T. (1986). *J. Anim. Sci.* **63,** 418–431.
Brien, F. D., Baxter, R. W., Findlay, J. K., and Cum-

ming, I. A. (1976). *Proc. Aust. Soc. Anim. Prod.* **11,** 237–240.
Bronson, F. H. (1986). *Endocrinology* **118,** 2483–2487.
Bronson, F. H. (1988). *Reprod. Nutr. Dévelop.* **28**(2B), 335–347.
Bronson, F. H., and Manning, J. (1988). *11th Int. Congr. Anim. Reprod. AI, Dublin* **5,** 110–116.
Bunning, E. (1960). *Cold Spring Harbour Symp. Q. Biol.* **25,** 249–256.
Cahill, L. P. (1984). *In* "Reproduction in Sheep" (D. R. Lindsay and D. T. Pearce, eds). Cambridge University Press.
Cahill, L. P., and Mauleon, P. (1980). *J. Reprod. Fertil.* **58,** 321–328.
Cahill, L. P., Blockey, M. A. de B., and Parr, R. A. (1975). *Aust. J. Exp. Agric. Anim. Husb.* **15,** 337–341.
Cahill, L. P., Driancourt, M. A., Chamley, W. A., and Findlay, J. K. (1985). *J. Reprod. Fertil.* **75,** 599–607.
Cahill, L. P., Loel, T. A., Turnbull, K. E., Piper, L. R., Bindon, B. M., and Scaramuzzi, R. J. (1982). *Proc. Aust. Soc. Reprod. Biol.* **14,** 76(Abstr.).
Cahill, L. P., Mariana, J. C., and Mauleon, P. (1979). *J. Reprod. Fertil.* **55,** 27–36.
Cameron, A. W. N. (1987). The production of spermatozoa by rams and its consequences for flock fertility Ph.D. thesis, University of Western Australia.
Cameron, A. W. N., Tilbrook, A. J., Lindsay, D. R., Fairnie, I. J., and Keogh, E. J. (1987). *Anim. Reprod. Sci.* **13,** 105–115.
Camp, M. J., Wildt, D. E., Howard, P. K., Stuart, L. D., and Chakraborty, P. K. (1983). *Biol. Reprod.* **28,** 673–681.
Campbell, B. K., Scaramuzzi, R. J., and Evans, G. (1988). *J. Reprod. Fertil. (Abstr. Ser 1),* 26.
Carles, A. B., and Kipragero, W. A. K. (1986). *Anim. Prod.* **43,** 447–457.
Chemineau, P. (1983). *J. Reprod. Fertil.* **67,** 65–72.
Chemineau, P. (1985). *Anim. Reprod. Sci.* **9,** 87–94.
Chemineau, P. (1986). *Réprod. Nutr. Dévelop.* **26,** 441–452.
Chemineau, P., Normant, E., Ravault, J. P., and Thimonier, J. (1986). *J. Reprod. Fertil.* **78,** 497–504.
Chemineau, P., Pelletier, J., Guérin, Y., Colas, G., Ravault, J. P., Touré, G., Almeida, G., Thimonier, J., and Ortavant, R. (1988). *Réprod. Nutr. Dévelop.* **28**(2B), 409–422.
Chenoweth, P. J. (1981). *Theriogenology* **16,** 155–177.
Cognié, Y., Houix, Y., and Logeay, B. (1981). *Proc. 2nd Int. Conf. Elevage Caprin,* Tours, pp. 345–350.

Cognié, Y., Gray, S. J., Lindsay, D. R., Oldham, C. M., Pearce, D. T., and Signoret, J.-P. (1982). *Proc. Aust. Soc. Anim. Prod.* **14,** 519–522.

Colas, G., Guérin, Y., Clonet, V., and Solari, A. (1985). *Reprod. Nutr. Dévelop.* **25,** 101–111.

Cole, H. H., and Miller, R. F. (1935). *Am. J. Anat.* **57,** 39–97.

Corbett, J. L., and Furniva, E. P. (1976). *Aust. J. Exp. Agric. Anim. Husb.* **16,** 256.

Corteel, J. M. (1973). *World Rev. Anim. Prod.* **9,** 73–99.

Corteel, J. M. (1977). *In* "Management of Reproduction in Sheep and Goats," pp. 41–57. University of Wisconsin, Madison.

Cox, R. I., Wilson, P. A., and Mattner, P. E. (1976). *Theriogenology* **6,** 607.

Cumming, I. A., Baxter, R. and Lawson, R. A. S. (1974). *J. Reprod. Fertil.* **40,** 443–446.

Cummins, L. J., O'Shea, T., Al-Obaidi, S. A. R., Bindon, B. M., and Findlay, J. K. (1986). *J. Reprod. Fertil.* **77,** 365–372.

Cutten, I. N. (1970). *Proc. Aust. Soc. Anim. Prod.* **8,** 388.

Dauzier, L., and Winterbergen, S. (1952). *Ann. Zootech.* **IV,** 49–52.

Davis, I. F., Brien, F. D., Findlay, J. K., and Cumming, I. A. (1981). *Anim. Reprod. Sci.* **4,** 19–28.

de Jong, F. H. (1979). *Mol. Cell. Endocrinol.* **13,** 1–10.

Denamur, R., Kann, C., and Short, R. V. (1973). *In* "The Endocrinology of Pregnancy and Parturition" (C. G. Pierrepoint, ed.), pp. 2–4. Alpha Omega Alpha, Cardiff.

Ducker, M. J., and Bowman, J. C. (1970). *Anim. Prod.* **12,** 465–471.

Dutt, R. H. (1952). *J. Anim. Sci.* **11,** 792 (abstr.).

Edey, T. N. (1968). *Proc. Aust. Soc. Anim. Prod.* **7,** 173–189.

Edey, T. N., Chu, T. T., Kilgour, R., Smith, J. F., and Tervit, H. R. (1977). *Theriogenology* **7,** 11–15.

English, J. E., Poulton, A. L., Arendt, J., and Symons, A. M. (1986). *J. Reprod. Fertil.* **77,** 321–327.

Fletcher, I. C. (1971). *Aust. J. Agric. Res.* **22,** 321–330.

Fletcher, I. C., and Lindsay, D. R. (1971a). *J. Endocrinol.* **50,** 685–696.

Fletcher, I. C., and Lindsay, D. R. (1971b). *J. Reprod. Fertil.* **25,** 253–259.

Fletcher, I. C., and Putu, I. G. (1985). *Proc. Aust. Soc. Reprod. Biol.* **17,** 34 (abstr.).

Foote, W. D., Gooch, L. D., Pope, A. L., and Casida, L. E. (1957). *J. Anim. Sci.* **58,** 565–580.

Forage, R. G., Brown, R. W., Oliver, K. J., Atrache, B. T., Devire, P. C., Hudson, G. C., Goss, N. H., Bertram, K. C., Tolstoshev, P., Robertson, D. M., de Kretser, D. M., Doughton, B., Burger, H. G., and Findlay, J. K. (1987). *J. Endocrinol.* **114,** R1-R4.

Fowler, D. G. (1984). *In* "Reproduction in Sheep" (D. R. Lindsay and D. T. Pearce eds.), pp. 39–46. Cambridge University Press, Cambridge.

Fowler, D. G., and Wilkins, J. F. (1984). *Livestock Prod. Sci.* **11,** 437–450.

Fulkerson, W. J., Adams, N. R. and Gherardi, P.B. G. (1981). *Appl. Anim. Ethol.* **7,** 57–66.

Fulkerson, W. J., Synnott, Anthea, L., and Lindsay, D. R. (1982). *J. Reprod. Fertil.* **66,** 129–132.

Geldard, H., and Dow, G. J. (1984). *10th Int. Congr. Anim. Reprod. A. I.,* Illinois. IV, VIII–28–VIII–31.

Goldman, B. D. (1983). *In* "Pineal Research Reviews" (R. J. Reiter, ed.), vol. 1, pp. 145–181. A. R. Liss, N.Y.

Goodman, R. L., Bittman, E. L., Foster, D. L., and Karsch, F. J. (1982). *Biol. Reprod.* **27,** 580.

Grubb, P., and Jewell, P. A. (1966). *Symp. Zool. Soc. Lond.* **18,** 170–210.

Hansel, W. (1988). *Proc. 11th Int. Congr. Anim. Reprod.* **AI5,** 62–70.

Harcourt, A. H., Harvey, P. H., Larson, S. G., and Short, R. V. (1981). *Nature (Lond.)* **293,** 55–57.

Hassan, F. A., El Nahkla, S. M. Aboul-Ela, M. B., and Aboul-Naga, A. M. (1988). *Proc. 11th Int. Congr. on Anim. Reprod. A. I., Dublin* 407 (abstr.)

Haughey, K. G. (1959). *N.Z. Sheep Farming Annu.* **17,** 17–26.

Heap, R. B., Davis, A. J., Fleet, I. R., Goode, J. A., Hamon, M., Nowak, R. A., Stewart, H. J., Whyte, A., and Flint, A. D. F. (1988). *Proc. 11th Int. Congr. Anim. Reprod. A. I., Dublin* **5,** 56–60.

Henderson, K. M., Franchimont, P., Lecomte-Yerna, M. J., Hudson, N., and Ball, K. (1984). *J. Endocrinol.* **102,** 305–309.

Honmade, D. (1977). *Zhivatmovodstvo* **3,** 62–63. Cited in *Anim. Breeding Abstr.* **45,** 384.

Howles, C. M., Craigon, J., and Haynes, N. B. (1982). *J. Reprod. Fertil.* **65,** 439–446.

Hulet, C. V. (1972). *J. Anim. Sci.* **35,** 814.

Hunter, G. L. (1968). *Anim. Breed. Abstr.* **36,** 347–378.

Hunter, G. L., and Van Aarde, I. M. R. (1973). *J. Reprod. Fertil.* **32,** 1–8.

I'Anson, H. (1983). *Biol. Reprod.* **28,**(1), 64.

Ishaq, S. M. (1982). *Proc. Int. Semin. Sheep and Wool.* Pakistan Agricultural Research Council, Islamabad, pp. 177–198.

Jewell, P. A., Hall, S. J. G., and Rosenberg, M. M. (1986). *J. Reprod. Fertil.* **77,** 81–89.

Jordan, B., Hanrahan, J. P., and Roche, J. F. (1988). *Proc. 11th Int. Congr. Anim. Reprod. A. I. Dublin* abstr. 409.

Kann, G., and Martinet, J. (1975). *Nature (Lond.)* **257,** 63–64.

Kann, G., Martinet, J., and Schirar, A. (1977). *In* "Prolactin and Human Reproduction" (P. G. Crosignani and C. Robyn, eds.), pp. 47–59. Academic Press, London.

Karsch, F. J., and Wayne, Nancy, L. (1988). *Proc. 11th Int. Congr. Anim. Reprod. A. I. Dublin* **5,** 221–227.

Karsch, F. J., Bittman, E. L., Foster, D. L., Goodman, R. L., Legan, S. J., and Robinson, J. E. (1984). *Recent Prog. Horm. Res.* **40,** 185–232.

Karsch, F. J., Malpaux, B., Wayne, N. L., and Robinson, T. J. (1988). *Reprod. Nutr. Dévelop.* **28**(2B), 459–472.

Kelly, R. W. (1984). *In* "Reproduction in Sheep" (D. R. Lindsay and D. T. Pearce, eds.), pp. 127–133. Cambridge University Press, Cambridge.

Kelly, R. W., Allison, A. J., and Shackell, G. H. (1975). *Proc. N.Z. Soc. Anim. Prod.* **35,** 204–211.

Kendrick, K. M., Keverne, E. B., and Baldwin, B. A. (1987). *Neuroendocrinology* **46,** 56–61.

Kendrick, K. M., Keverne, E. B., Baldwin, B. A., and Sharman, D. F. (1986). *Neuroendocrinology* **44,** 149–156.

Kennaway, D. J., and Seamark, R. F. (1980). *Aust. J. Biol. Sci.* **33,** 349–353.

Kennaway, D. J., Dunstan, E. A., and Staples, L. D. (1987). *J. Reprod. Fertil. Suppl.* **34,** 187–199.

Kennaway, D. J., Dunstan, E. A., Gilmore, T. A., and Seamark, R. F. (1984). *Proc. 7th Int. Congr. Endocrinol. Quebec* abstr. 21.

Kennaway, D. J., Hooley, R. D. and Seamark, R. J. (1980). *Proc. XI Int. Congr. Int. Soc. Psychoneuroendocrinol. Florence* abstr. 143.

Kennaway, D. J., Sandford, L. M., Godfrey, B., and Friesen, H. G. (1983). *J. Endocrinol.* **97,** 229–242.

Keverne, E. B. (1988). *Psychoneuroendocrinology* **13,** 127–141.

Killeen, I. D., and Caffery, G. J. (1982). *Aust. Vet. J.* **59,** 95.

Knight, T. W. (1973). A study of factors which affect the potential fertility of the ram. Ph.D. Thesis, University of Western Australia, Perth, Western Australia.

Knight, T. W., and Lindsay, D. R. (1973). *Aust. J. Agric. Res.* **24,** 579–585.

Knight, T. W., Gherardi, S., and Lindsay, D. R. (1987). *Anim. Reprod. Sci.* **13,** 105–115.

Knight, T. W., Oldham, C. M., and Lindsay, D. R. (1975). *Aust. J. Agric. Res.* **26,** 567–575.

Knight, T. W., Payne, E., and Peterson, A. J. (1981). *Proc. Aust. Soc. Reprod. Biol.* **13,** 19 (abstr.).

Lahlou-Kassi, A., and Boukhliq, R. (1988). *In* "Isotope-Aided Studies on Livestock Productivity in Mediterranean and North African Countries," pp. 131–139. International Atomic Energy Agency, Vienna.

Land, R. B. (1970). *J. Reprod. Fertil.* **21,** 517–521.

Land, R. B. (1971). *J. Reprod. Fertil.* **24,** 345–352.

Land, R. B., Bodin, L., Driancourt, M. A., Haley, C. S., and McNeilly, J. R. (1988). *Proc. 3rd World Congr. Sheep Beef Cattle Breeding, Paris* **2,** 611–620.

Land, R. B., Morris, B. A., Baxter, G., Fordyce, M., and Forster, J. (1982). *J. Reprod. Fertil.* **66,** 625–634.

Land, R. B., Pelletier, J., Thimonier, J., and Mauléon, P. (1973). *J. Endocrinol.* **58,** 305–317.

Land, R. B., Thompson, R., and Baird, D. T. (1972). *J. Reprod. Fertil.* **30,** 39–44.

Langford, G. A., and Marcus, G. H. (1982). *J. Reprod. Fertil.* **675,** 325–329.

Legan, S. J., and Karsch, F. J. (1979). *Biol. Reprod.* **20,** 74–85.

Leung, P. C. K., and Armstrong, D. T. (1980). *Am. Rev. Physiol.* **42,** 71–82.

Lightfoot, R. J., and Wroth, R. H. (1974). *Proc. Aust. Soc. Anim. Prod.* **10,** 130–134.

Lightfoot, R. J., Croker, K. P., and Neil, H. G. (1967). *Aust. J. Agric. Res.* **18,** 755–765.

Lincoln, G. A. (1988). *Reprod. Nutr. Dévelop.* **28,** 527–539.

Lincoln, G. A., and Davidson, W. (1977). *J. Reprod. Fertil.* **49,** 267–276.

Lincoln, G. A., and Ebling, F. J. P. (1985). *J. Reprod. Fertil.* **73,** 241–254.

Lincoln, G. A., and Short, R. V. (1980). *Recent Prog. Horm. Res.* **36,** 1–52.

Lincoln, G. A., Ebling, F. J. P., and Martin, G. B. (1987). *J. Endocrinol.* **115,** 425–438.

Lindhal, I. C. (1971). *J. Anim. Sci.* **32,** 922.

Lindsay, D. R. (1976). *Proc. Aust. Soc. Anim. Prod.* **11,** 217–224.

Lindsay, D. R. (1982). *In* "Future Developments in the Genetic Improvement of Animals," pp. 89–101. Academic Press, Sydney.

Lindsay, D. R. (1988). *Aust. J. Biol. Sci.* **41,** 97–102.

Lindsay, D. R., and Ellsmore, J. (1968). *Aust. J. Exp. Agric. Anim. Husb.* **8,** 649–652.

Lindsay, D. R., and Thimonier, J. (1988). *Proc 3rd*

World Congr. Sheep Beef Cattle Breeding Paris 547–565.

Lindsay, D. R., Cognié, Y., Pelletier, J., and Signoret, J. P. (1975). *Physiol. Behav.* **5,** 423–426.

Lindsay, D. R., Gherardi, P. B., and Oldham, C. M. (1977). *In* "Sheep Breeding" (G. J. Tomes, D. E. Robertson, and R. J. Lightfoot, eds.), pp. 294–298. WAIT Press, Perth.

Louvet, J. P., Harman, S. M., Schrieber, J., and Ross, G. T. (1975). *Endocrinology* 97, 336–372.

McKenzie, F. F., and Terrill, C. E. (1937). *Missouri Agric. Exp. Sta. Res. Bull.* **264.**

Maeda, K-I., Mori, Y., and Kano, Y. (1988). *Reprod. Nutr. Dévelop.* 28(2B), 487–497.

Martin, G. B., Oldham, C. M., and Lindsay, D. R. (1980). *Anim. Reprod. Sci.* **3,** 125–132.

Martin, G. B., Scaramuzzi, R. J., and Lindsay, D. R. (1983). *J. Reprod. Fertil.* **67,** 47–55.

Masters, D. G., and Fels, H. E. (1984). *Proc. Aust. Soc. Anim. Prod.* **15,** 444–447.

Mattner, P. E., and Braden, A. W. H. (1967). *Aust. J. Exp. Agric. Anim. Husb.* **7,** 110–116.

Mattner, P. E., Braden, A. W. H., and George, J. M. (1971). *Aust. J. Exp. Agric. Anim. Husb.* **11,** 473–477.

Mauléon, P., and Dauzier, L. (1965). *Ann. Biol. Anim. Biochim. Biophys.* **5,** 131–143.

Mauléon, P., and Mariana, J. C. (1977). *In* "Reproduction in Domestic Animals" (H. H. Cole and P. T. Cupps, eds.), 3rd ed., pp. 175–198. Academic Press, New York.

Mauléon, P., and Rougeot, J. (1962). *Ann. Biol. Anim. Biochim. Biophys.* **2,** 209–222.

McDowell, R. E. (1972). "Improvement of Livestock Production in Warm Climates." W. H. Freeman, San Francisco.

McKenzie, F. F., and Terrill, C. E. (1937). *Res. Bull. Missouri Agric. Exp. Sta.* **264.**

McNatty, K. P. (1982). *In* "Follicular Maturation and Ovulation" (R. Rolland, E. V. van Hall, S. G. Hillier, K. P. McNatty, and J. Schoemaker, eds.), pp. 1–19. Excerpta Medica, Amsterdam.

McNatty, K. P., Hudson, K. M., Henderson, K. M., Gibbs, M., Morrison, L., Ball, K., and Smith, P. (1987). *J. Reprod. Fertil.* **80,** 577–588.

McNatty, K. P., Makris, A., DeGrazia, C., Osathanondh, R., and Ryan, K. J. (1979). *J. Clin. Endocrinol. Metab.* **49,** 687–699.

McNeilly, A. S., and Wallace, J. M. (1987). *In* "Follicular Growth and Ovulation Rate in Farm Animals" (J. F. Roche and D. O'Callagan, eds.), pp. 119–127. Martinus Nijhoff, Dordrecht.

Mgongo, F. O. K. (1988). *Proc. 11th Int. Congr. Anim. Reprod. A. I. Dublin* abstr. 48.

Moore, N. W., and Miller, B. G. (1984). *In* "Reproduction in Sheep" (D. R. Lindsay and D. T. Pearce, eds.), pp. 112–114. Cambridge University Press, Cambridge.

Moore, N. W., and Rowson, E. A. (1959). *Nature* **189,** 1410.

Morley, F. H. W., White, P. H., Kenney, P. A., and Davis, I. F. (1978). *Agric. Syst.* **3,** 27–45.

Morrant, A. J., and Dun, R. B. (1960). *Aust. Vet. J.* **36,** 1–7.

Morton, H., Nancarrow, C. D., Scaramuzzi, R. J., Evison, B. M., and Clunie, G. J. A. (1979). *J. Reprod. Fertil.* **56,** 75–80.

Moule, G. R., Braden, A. W. H., and Mattner, P. E. (1966). *Aust. J. Agric. Res.* **17,** 923–931.

Nancarrow, C. D., Wallace, A. L. C., and Grewal, A. S. (1981) *J. Reprod. Fertil.* 30, 191–199

Nett, T. M., and Niswender, G. D. (1982). *Theriogenology* **17,** 645–652.

Nicholls, T. J., Follett, B. K., Goldsmith, A. R., and Pearson, H. (1988). *Reprod. Nutr. Dévelop.* **28**(2B), 375–385.

Nottle, M. B., Seamark, R. F., and Setchell, B. P. (1988). *J. Reprod. Fertil. Abstr.* **1,** 31 (abstr.).

Novoa, C. (1984). *Proc. 10th Int. Congr. Anim. Reprod. A. I., Illinois.* IV, VII-24–VII-30.

Nowak, R., and Rodway, R. G,. (1985). *J. Reprod. Fertil.* **74,** 287–293.

O'Shea, T., Al-Obadi, S. A. R., Bindon, B. M., Cummins, L. J., Findlay, J. K., and Hillard, M. A. (1984). *In* "Reproduction in Sheep" (D. R. Lindsay and D. T. Pearce, eds), pp. 335–337. Cambridge University Press, Cambridge.

O'Shea, T., Cummins, L. J., Bindon, B. M., and Findlay, J. K. (1982). *Proc. Aust. Soc. Reprod. Biol.* **14,** 85 (abstr.).

Obst, J. M., Boyes, T., and Chaniago, T. (1980). *Proc. Aust. Soc. Anim. Prod.* **13,** 321–324.

Oldham, C. M. (1980). A study of sexual and ovarian activity in Merino sheep. *Ph.D. thesis, University of Western Australia.*

Oldham, C. M., Adams, N. R., Gherardi, P. B., Lindsay, D. R., and Mackintosh, J. B. (1978). *Aust. J. Agric. Res.* **29,** 173–179.

Oldham, C. M., and Lindsay, D. R. (1984). *In* "Reproduction in Sheep" (D. R. Lindsay and D. T. Pearce, eds.), pp. 274–276. Cambridge University Press, Cambridge.

Oldham, C. M., Martin, G. B., and Knight, T. W. (1978). *Anim. Reprod. Sci.* **1,** 283–290.

Ortavant, R. (1958). Le cycle spermatogenetique chez lebélier. Thèse, University of Paris, France.

Ortavant, R., Mauléon, P., and Thibault, C. (1964). *Ann. N.Y. Acad. Sci.* **117,** 157–193.

Ortavant, R., Bocquier, F., Pelletier, J., Ravault, J. P., Thimonier, J., and Vollant-Nail, P. (1988). *In* "Controlled Breeding in Sheep—A Tribute to T. J. Robinson" (J. K. Findlay, ed.), pp. 69–85. CSIRO, Australia.

Ortavant, R., Pelletier, J., Ravault, J. P., and Thimonier, J. (1978). *In* "Environmental Endocrinology" (I. Assenmacher and D. S. Farmer, eds.), pp. 75–78. Springer-Verlag, Berlin.

Ott, R. S., Nelson, D. R., and Hixon, J. E. (1980). *Theriogenology* **13,** 183–190.

Parsons, S. D., and Hunter, C. L. (1967). *J. Reprod. Fertil.* **14,** 61–70.

Pelletier, J. (1971). Influence du photopériodisme et des androgènes sur la synthèse et la libération de L. H. chez le bélier. Thèse Doctorate ès Sci. Nat., Université Paris.

Pelletier, J., Chemineau, P., and Delgadillo, J. A. (1988). *Proc. 11th Int. Conf. Anim. Reprod. A. I. Dublin* **5,** 211–219.

Pelletier, J., Brien, V., Chesneau, D., Pisslet, C., and de Reviers, M. T. (1985). *C. R. Hebd. Acad. Sci., Paris* **301,** 665–668.

Peters, H., Byskov, A. G., Himelstein-Braw, R., and Faber, M. (1975). *J. Reprod. Fertil.* **45,** 559–566.

Piper, L. R., and Bindon, B. M. (1982). *Proc. 1st World Congr. Sheep Beef Cattle Breeding* **1,** 315–331.

Poindron, P., Lévy, F., and Krehbiel, D. (1988). *Psychoneuroendocrinology* **13,** 99–125.

Poindron, P., Martin, G. B., and Hooley, R. D. (1979). *Physiol. Behav.* **23,** 1081–1087.

Poindron, P., Orgeur, P., Le Neindre, P., Kann, G., and Raksanyi, I. (1980). *Horm. Behav.* **14,** 173–177.

Poulton, A. L., English, J., Symons, A. M., and Arendt, J. (1987). *J. Endocrinol.* **112,** 103–111.

Pretorius, P. S. (1973). *Agroanimalia* **5,** 55–58.

Purvis, I. W., Edey, T. N., Kilgour, R. J., and Piper, L. R. (1984). *In* "Reproduction in Sheep" (D. R. Lindsay and D. T. Pearce, eds), pp. 59–61. Cambridge University Press, Cambridge.

Quirke, J. F., and Hanrahan, J. P. (1977). *J. Reprod. Fertil.* **51,** 487–489.

Radford, H. M., and Watson, R. H. (1957). *Aust. J. Agric. Res.* **8,** 460–470.

Radford, H. M., Donegan, S., and Scaramuzzi, R. J. (1980). *Proc. Aust. Soc. Anim. Prod.* **13,** 457.

Ravault, J. P., and Ortavant, R. (1977). *Ann. Biol. Anim. Biochim. Biophys.* **17,** 459–473.

Ravault, J. P., and Thimonier, J. (1988). *Reprod. Nutr. Dévelop.* **28**(2B), 473–486.

Restall, B. J. (1961). *Aust. Vet. J.* **32,** 70–72.

Restall, B. J. (1971). *J. Reprod. Fertil.* **24,** 145–146.

Restall, B. J. (1988). *Proc. Aust. Soc. Anim. Prod.* **17,** 302–305.

Restall, B. J., and Starr, B. J. (1977). *J. Reprod. Fertil.* **49,** 297–303.

Restall, B. J., Brown, G. H., Blockey, M. A. de B., Cahill, L., and Kearins, R. (1976). *Aust. J. Exp. Agric. Anim. Husb.* **16,** 329–335.

Richards, J. S. (1979). *Recent Prog. Horm. Res.* **35,** 343–373.

Ricordeau, G. (1988). *Proc 3rd World Congr. Sheep Beef Cattle Breeding,* Paris, 567–588.

Riera, G. S. (1984). *Proc. 10th Int. Congr. Anim. Reprod. A. I.,* Illinois, IV, VII-1–VII-7.

Riera, G. S. (1982). *Proc. 3rd Int. Conf. Goat Prod. Dis. Tucson* 162–174.

Rizzoli, D. J., Winfield, C. G., Howard, T. J., Englund, I. K. J. and Goding, J. R. (1976). *J. Agric. Sci. Camb.* **87,** 671.

Robertson, H. A. (1977). *In* "Reproduction in Domestic Animals" (H. H. Cole and P. T. Cupps, eds.), pp. 475–498. Academic Press, New York.

Robertson, H. A., and Sarda I. R. (1971). *J. Endocrinol.* **49,** 407–419.

Robinson, J. E., and Karsch, F. J. (1984). *Biol. Reprod.* **31,** 656–663.

Robinson, J. E., and Karsch, F. J. (1988). *Reprod. Nutr. Dévelop.* **28**(2B), 365–374.

Robinson, T. J. (1952). *Nature (Lond.)* **170,** 373–374.

Robinson, T. J. (1954). *J. Endocrinol.* **10,** 117–124.

Robinson, T. J. (1959). *In* "Reproduction in Domestic Animals," Vol. I (H. H. Cole and P. T. Cupps, eds.), pp. 291–333. Academic Press, New York.

Robinson, T. J., Moore, N. W., and Binet, F. E. (1956). *J. Endocrinol.* **14,** 1–7.

Rollag, M. D., and Niswender, G. D. (1976). *Endocrinology* **98,** 482–489.

Rollag, M. D., O'Callaghan, P. L., and Niswender, G. D. (1978). *Biol. Reprod.* **18,** 279–285.

Rouger, Y. (1974). Étude des interactions de l'environnement et des hormones dans la régulation du compontement sexuel des Bovidae. Thèse Doctorat ès Sci. Nat., Université de Rennes.

Ryder, M. L. (1983). "Sheep and Man." Duckworth, London.

Salamon, S. (1962). *Aust. J. Agric. Res.* **13,** 1137–1150.

Scaramuzzi, R. J. (1975). *In* "Physiological Effects of Immunity against Reproductive Hormones" (R. G

Edwards and M. H. Johnson, eds.), pp. 67–90. Cambridge University Press, Cambridge.

Scaramuzzi, R. J., and Radford, H. M. (1983). *J. Reprod. Fertil.* **69**, 353–367.

Scaramuzzi, R. J., and Martensz, N. D. (1975). *In* "Immunization with Hormones in Reproductive Research" (E. Nieschlag, ed.), pp. 141–152. North Holland, Amsterdam.

Scaramuzzi, R. J., Davidson, W. R., and Van Look, P. F. A. (1977). *Nature (Lond.)* **269**, 817–818.

Scaramuzzi, R. J., Lindsay, D. R., and Shelton, J. N. (1971). *J. Endocrinol.* **50**, 345–346.

Schanbacher, B. D. (1982). *J. Androl.* **3**, 41–42.

Schanbacher, B. D. (1985). *Dom. Anim. Endocrinol.* **2**, 67–75.

Schanbacher, B. D., and Crouse, J. D. (1981). *Am. J. Physiol.* **241**, E1–E5.

Schinckel, P. G. (1954). *Aust. J. Agric. Res.* **5**, 465–469.

Setchell, B. P. (1978). "The Mammalian Testis." Paul Elek, London.

Shelton, M. and Morrow, T. (1965). *Rep. Texas Agric. Exp. Sta.*, pp. 20–21.

Signoret, J. P., and Lindsay, D. R. (1982). *In* "Olfaction and Endocrine Regulation" (W. Breiphol, ed.), pp. 63–70. IRC Press, London.

Signoret, J. P., Fulkerson, W. J., and Lindsay, D. R. (1982/1983). *Appl. Anim. Ethol.* **9**, 37–45.

Skevah, Y., Black, W. J. M., Carr, W. R., and Land, R. B. (1974). *J. Reprod. Fertil.* **38**, 369–378.

Smith, J. F. (1971). *Aust. J. Agric. Res.* **22**, 513–519.

Smith, I. D. (1965). *World Rev. Anim. Prod.* **4**, 95–101.

Stewart, R., and Oldham, C. M. (1986). *Proc. Aust. Soc. Anim. Prod.* **16**, 367–370.

Sutherland, S. R. D. (1988). *Proc. 4th AAAP Anim. Sci. Congr., Hamilton, New Zealand.* 259.

Symons, A. M., Arendt, J., and Land, C. A. (1983). *J. Endocrinol.* **99**, 41–46.

Synnott, Anthea L., Fulkerson, W. J., and Lindsay, D. R. (1981). *J. Reprod. Fertil.* **61**, 355–361.

Taneja, G. C. (1982). *Proc. III Int. Conf. Goat Prod. Dis. Tucson.* 49.

Terqui, M. (1978). Contribution à l'étude des oestrogenos chez la brebis et la truie. Thèse de Doctorat d'Etat ès-Sciences Naturelles, Université Paris VI.

Thatcher, W. W., Hansen, P. J., and Bazer, F. W. (1988). *Proc. 11th Int. Congr. Anim. Reprod. A. I. Dublin* **5**, 45–54.

Tilbrook, A. J., and Lindsay, D. R. (1987). *Appl. Anim. Behav. Sci.* **17**, 129–138.

Thimonier, J. (1973). *Recl. Med. Vet.* **149**, 1303–1318.

Thimonier, J. (1981). *J. Reprod. Fertil. Suppl.* **30**, 33–45.

Thimonier, J., and Chemineau, P. (1988). *Proc. 11th Int. Congr. Anim. Reprod. A. I., Dublin* **5**, 230–237.

Thimonier, J., Ravault, J. P., and Ortavant, R. (1978). *Ann. Biol. Anim. Biochim. Biophys.* **9**, 233–250.

Thomas, G. B., Oldham, C. M., Hoskinson, R. M., Scaramuzzi, R. J., and Martin, G. B. (1987). *Aust. J. Biol. Sci.* **40**, 307–313.

Thwaites, C. J. (1965). *J. Agric. Sci.* **65**, 57–64.

Turnbull, K. E., Braden, A. W. H., and Mattner, P. E. (1977). *Aust. J. Biol. Sci.* **30**, 229–241.

Underwood, E. J., Shier, F. L., and Davenport, N. (1944) *J. Agric. (West Aust.)* **11**, 135–143.

Van der Westhuysen, J. M., Van Niekerk, C., and Hunter, G. L. (1970). *Agroanimalia* **2**, 131–138.

Webb, R., and Morris, B. A. (1988). *Proc. 11th Int. Conf. Reprod. A. I. Dublin* **5**, 183–191.

Wheeler, A. G., and Land, R. B. (1973). *J. Reprod. Fertil.* **35**, 583–584.

Wilkins, J. F. (1988). *Proc. Aust. Soc. Anim. Prod.* **17**, 480.

Wodzicka-Tomaszewska, M., Hutchinson, J. C. D., and Bennett, N. W. (1967). *J. Agric. Sci. Camb.* **68**, 61–67.

Wright, P. J., Geytenbeek, P. E., Clarke, I. J., and Findlay, J. K. (1980). *J. Reprod. Fertil.* **60**, 171–176.

Wright, P. J., Geytenbeek, P. E., Clarke, I. J., and Findlay, J. K. (1981). *J. Reprod. Fertil.* **61**, 97–102.

Wright, P. J., Geytenbeek, P. E., Clarke, I. J., and Findlay, J. K. (1983). *J. Reprod. Fertil.* **67**, 257–262.

Wright, R. W., Bondioli, K., Grammer, J., Keezan, F., and Menino, A. (1981). *J. Anim. Sci.* **52**, 115–118.

Yeates, N. T. M. (1949). *J. Agric. Sci.* **39**, 1–43.

CHAPTER 16

Reproduction in the Dog and Cat

PATRICK W. CONCANNON

I. The Domestic Bitch

A. Overview

Dogs are monoestrous, usually nonseasonal, polytocous, spontaneous ovulators (Evans and Cole, 1931; Anderson and Simpson, 1973; McDonald, 1989). They have one or two ovulatory cycles per year at intervals of 5–13 months. The average interval is 7 months. Pregnancy lasts 9 weeks and lactation 6 weeks. They may have one or two litters per year. In the absence of pregnancy, luteal function persists for about 2 months. Interestrus intervals in pregnant and nonpregnant bitches are similar (Concannon et al., 1975; Stabenfeldt and Shille, 1977).

B. Stages of the Canine Ovarian Cycle

The stages of the dog cycle are *proestrus, estrus, metestrus,* and *anestrus* (Evans and Cole, 1931). Ovulations occur synchronously about 40–50 h after the LH surge that occurs in late proestrus or early estrus (Concannon et al., 1977a; Phemister et al., 1973). Corpus luteum formation is spontaneous and the luteal phase of the nonpregnant cycle is similar in duration to the 9 weeks of pregnancy (Smith and McDonald, 1974; Concannon et al., 1975). The major events and endocrine changes associated with the stages of the nonpregnant and pregnant ovarian cycles of the bitch are summarized in Figures 1, 2, and 3 and Table I.

1. Proestrus

Proestrus is a response to increasing levels of estrogen. Proestrus is marked by discharges of uterine serosanguinous fluid through the vagina, and edematous swelling of the vulva and perineum. Throughout the

rise in estrogen there are progressive increases in vulval edema, vaginal cornification, and pheromonal attraction of males—each of which becomes maximal near the end of proestrus. Attraction of males involves the presence of increased methylhydroxybenzoate in the vaginal mucus (Goodwin et al., 1979). During most of proestrus the bitch refuses to permit males to mount by turning and growling. Near the end of proestrus the bitch may permit mounting, but usually prevents intromission by sitting or lying down, and fails to show the reflex sex behavior characteristic of estrus. Proestrus can be as brief as 3 days or as long as 3 weeks (Figure 3); the average is 9 days (Concannon, 1983a; Christie and Bell, 1972).

2. Estrus

Estrus in the bitch is characterized by reflex and sometimes spontaneous stiffening and deviation of the tail, standing firmly to permit a male to mount, and reflex lordosis-like presentation of the vulva and perineum (Concannon et al., 1979b; Beach et al., 1982). Estrous behavior in dogs is a response to a decline in estrogen, and is facilitated by a rise in progesterone, both of which usually occur near the onset of estrus (Concannon et al., 1979b; Beach et al., 1982). The onset of estrus most often occurs within a day or two of the preovulatory luteinizing hormone (LH) surge (Nett et al., 1975; Jones et al., 1973), which also occurs in response to a decline in the estrogen to progesterone ratio (Concannon et al., 1979a). However, estrus onset and the preovulatory LH surges may not be as synchronous, with estrus onset occurring as early as 4 days before the LH surge or as late as 6 days after the LH surge (Concannon, 1983a, 1986a). A bitch may not display estrous behavior in response to a particular

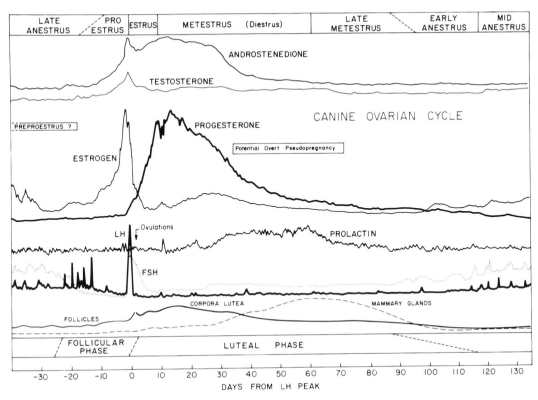

Figure 1 Schematic representation of typical changes in serum or plasma levels of estrogen, progesterone, LH, FSH, prolactin, testosterone, and androstenedione reported or presumed to occur during the canine ovarian cycle, and their temporal relation to classical stages and functional phases of the estrous cycle. Basal and peak serum levels of steroids are, respectively, 5–10 and 50–100 pg/ml for estradiol, 0.2–0.5 and 15–90 ng/ml for progesterone, 0.1 and 1 ng/ml for testosterone, and 0.2 and 2 ng/ml for androstenedione. Typical basal and preovulatory peak levels of LH based on the cLH standard LER1685-1 are 0.8 and 25 ng/ml, respectively. Prolactin levels based on the cPRL standard AFP-2451-B range from basal levels of 1.5 ng/ml to late metestrus peaks of 3–5 ng/ml. From Concannon (1986b).

male despite normal changes in hormone levels. Estrus may be as brief as 3 days or last for several weeks, but it is usually 4–12 days in length; the average is 9 days. Estrous behavior can be observed in a false estrus during the decline in estrogen following a nonovulatory follicular phase. Intromission and mating may not occur until the second or third day of estrous behavior, after some decrease in vulval turgidity. Mating is usu-

ally completed by a 5- to 45-min copulatory lock, during which the male dismounts and stands beside or faces away from the bitch, with the penis in the vagina. The lock withstands tugging and pulling, and is terminated by detumescence of the penis.

3. Metestrus

Metestrus represents the remainder of the luteal phase once the bitch stops

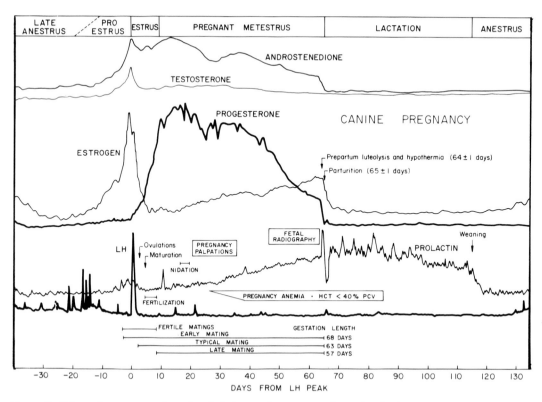

Figure 2 Schematic representation of typical changes in serum or plasma levels of estrogen, progesterone, LH, prolactin, testosterone, and androstenedione reported or presumed to occur during pregnancy and lactation in the bitch, and their relation to indicated events considered important for breeding programs and clinical management of pregnancy. Typical basal and peak serum levels of steroids are, respectively, 5–10, and 50–100 pg/ml for estradiol, 0.2–0.5 and 15–90 ng/ml for progesterone, 0.1 and 1 ng/ml for testosterone, and 0.2 and 2 ng/ml for androstenedione. Basal and preovulatory peak levels of LH based on the cLH standard LER1685-1 are 0.8 and 25 ng/ml, respectively. Prolactin levels based on the cPRL standard AFP-2451-B range from basal levels of 1.5 ng/ml to peak prepartum and lactation levels of 35 ng/ml.

accepting the male (Evans and Cole, 1931; Andersen, 1970). The term *diestrus* is used synonymously with metestrus in the bitch (Olson *et al.,* 1989); Holst and Phemister, 1974). Various parameters have been used as indicators of metestrus in nonpregnant dogs and to denote its duration. These have included 1–2 months of waning vulval enlargement, the 120–140 days of secretory endometrium observed histologically (An-

derson and Simpson, 1973), 2–4 months of modest to obvious palpable mammary enlargement, 55–110 days of elevated levels of progesterone ≥1 ng/ml in serum or plasma and, more often, the 2 months corresponding to pregnancy or the 3.5 months corresponding to pregnancy and lactation combined (Concannon, 1986a). The end of metestrus is poorly defined since the luteal phase subsides slowly. There is no acute lu-

Table I

Durations and Characteristics of the Stages of the 5- to 13-Month Canine Ovarian Cycle

	Proestrus	Estrus	Metestrus	Anestrus
Duration (days)				
Range	3–27	4–30	50–100	50–300
Most	4–12	4–12	60–80	100–200
Average	9	9	70	120
Estradiol level	Increasing	Decreasing	Moderate	Variable
Early	20–30 pg/ml	40–80 pg/ml	10–30 pg/ml	5–30 pg/ml
Late	40–110 pg/ml	10–30 pg/ml	10–30 pg/ml	5–30 pg/ml
Progesterone levels	Low	Increasing	Decreasing	Low
Early	0.2–0.6 ng/ml	0.8–2.0 ng/ml	15–80 ng/ml	0.8–1.0 ng/ml
Late	0.5–1.1 ng/ml	5–80 ng/ml	1–2 ng/ml	0.3–0.5 ng/ml
Vaginal mucosa	Enlarging	Enlarged	Receding	Flat
	Edematous	Prominent	Irregular	Thin
	Smooth	Wrinkled	Smooth	Smooth
	Pink–pale	Pale–white	Mottled	Red

teolytic mechanism, and the transition from metestrus to anestrus is a gradual one.

4. Anestrus

Anestrus is the period of apparent ovarian quiescence between metestrus and the subsequent proestrus. Anestrus can last from 2 to 10 months; the average is 4 months.

C. Puberty, Cycle Intervals, Seasonality

Age at puberty (first ovarian cycle) in the bitch can range from 5 to 18 months. Most occur from 6 to 15 months, and the average varies with breed (Christiansen, 1984). Smaller breeds tend to have earlier puberty than larger breeds, with puberty occurring 0–3 months after reaching adult body weight. In 8–15 kg beagles the range is 5–13 months, average 9 months. In 16–22 kg mongrels, the range may be about 10–16 months (Concannon et al., 1988c). Unobserved, weak or silent estrus at first cycles are not uncommon. Therefore, ages reported for first estrus may be greater than actual age at puberty. Age at first estrus is affected by breed, genetics, sire, and housing conditions (Christiansen, 1984).

The variation in estrus intervals is probably as great within most breeds as it is among breeds. Some bitches are very consistent, and others are not. Differences among breeds in average interval have been suggested, ranging from 26 weeks in Alsatians to 36 weeks in collies, and body weight has no effect (Christie and Bell, 1972; Christiansen, 1984). Average intervals may be shorter in mongrels than in purebred dogs, progressively lengthened with age, and about 3 weeks longer on average for pregnant cycles than for nonpregnant cycles, perhaps depending on the time of weaning. Wild dogs and dingos may, like wolves, be spring breeders, but evidence is lacking. The basenji cycles once a year in the autumn (Stabenfeldt and Shille, 1977). Dogs housed outdoors cycle year round (Andersen,

1970), as do dogs housed under constant lighting of 12L : 12D (Concannon, 1986a).

Subtle aspects of canine sex behavior have been studied in detail, including the proceptive, playful, teasing behavior of proestrus, and the receptive, sometimes sexually aggressive and initiating behavior of estrus, as well as the spinal reflexes to investigation by males (Beach *et al.*, 1982; Concannon *et al.*, 1979b). The proceptive behavior of proestrus is a response to an elevation in estrogen, while the positive reflexes and receptivity of estrus are a response to estrogen withdrawal (Concannon *et al.*, 1979b). Experimentally, estrous behavior in ovariectomized dogs was potentiated by a prerequisite rise in estradiol, but only elicited by a decline in estradiol or rise in progesterone. Estrus onset and duration were both facilitated by a simultaneous rise in progesterone and decline in estradiol (Figure 4).

D. Vaginal, Genital, and Uterine Changes

1. Reproductive Tract

The dog has a long vagina, short uterine fundus, and long uterine horns (Anderson and Simpson, 1973). The ovaries are attached by a suspensary ligament to the diaphragm and kidneys. Exposure at laparotomy requires tearing of the ligament. The ovary is bean-shaped and enclosed in a complete bursa. The fimbriated end of the oviduct proliferates through the bursal slit during proestrus and estrus. The ovaries are rarely clearly visible except in younger bitches with a transparent bursa. The external cervical os is hidden among surface folds, on the surface of a large cervical protuberance or papilla that extends into the vagina ventrally and often laterally. The

cervical lumen has 1-mm-deep irregular folds, and cervical cannulation is difficult. There is a prominent, anatomically permanent, dorsal-median fold of vaginal mucosa on the roof of the caudal third of the vagina that connects to the dorsal aspect of the external cervix. The enlargement of the dorsal-median fold during proestrus and estrus nearly occludes the vaginal lumen, and its posterior aspect is often mistaken for the cervix and is termed the pseudocervix (Pineda *et al.*, 1973). The distance from the vulva to the pseudocervix in a medium-sized beagle ranges from 12 to 25 cm (Lindsay and Concannon, 1986).

The entire tubular reproductive tract increases in diameter, length, and wall thickness under the influence of estrogen (Evans and Cole, 1931; Anderson and Simpson, 1973; Concannon, 1986a). During proestrus and estrus the greatly enlarged cervix is usually palpable through the abdomen. Edematous hypertrophy and turgidity of the vulva and perineum increase progressively throughout the proestrus rise in estrogen. The enlarged vulva normally softens during the decline in estrogen and rise in progesterone associated with the LH surge and ovulation. Vaginal hyperplasia and cornification can be monitored by vaginal cytology or gross vaginoscopic observation (Concannon and DiGregorio, 1986; Lindsay and Concannon, 1986). Changes during the periovulatory period are summarized in Table I and Figure 3.

2. Vaginal Cornification and Vaginal Smears

Vaginal mucosa growth and cornification follow the proestrus rise in estrogen with a delay of about 3–4 days, and continue until completed over a period of 5–10 days (Concannon and DiGregorio, 1986). Because of

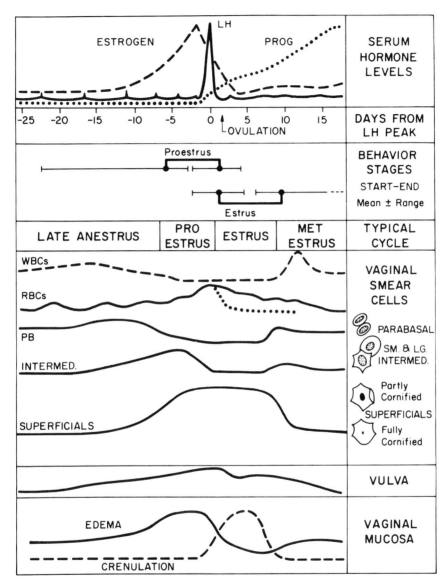

Figure 3 Schematic summary of the temporal relationships among the periovulatory endocrine events, behavioral and vulval changes, and changes in vaginal smears during proestrus, estrus and early metestrus in the bitch. From Concannon and Lein (1989).

the variation in the length of the follicular phase, vaginal cornification may be maximal and complete by 5 or more days before the LH surge, or not until 1 or 2 days after the LH surge, based on exfoliate vaginal cytology. Vaginal smears are usually collected by wiping the vaginal mucosa with a saline-dampened cotton swab, using a speculum to prevent contamination with vestibular cells or debris, transfer of cells by gentle rolling to a slide, and staining with a modified Wright's stain after air-drying or fixation. Relying on that technique and differential cell morphology, maximum cornification was observed 1–5 days before the LH peak (Concannon and DiGregorio, 1986). Using more complex stains that provide an eosinophilic index of the superficial cells may be more accurate in delineating the time of ovulation (Linde and Karlsson, 1984).

The terminology for exfoliated vaginal epithelial cells is based on the size, morphology, and position of cells within the fully cornified stratified squamous epithelium of the estrous bitch (Concannon and Di-Gregorio, 1986). *Superficial cells* are large, angulated, 45–75 μm squamous cells with evidence of pyknosis or karyolysis. Transitional or *large intermediates* are similar but have large, healthy-appearing, 8–18 μm nuclei. Rounded or *small intermediates* are non-angulated, smooth-border, variable-sized 22–60 μm cells. *Parabasals* are small, rounded, often darker-stained, 9–20 μm cells. Smears may also contain red blood cells (RBCs) from the uterus after diapedesis through the endometrium, and neutrophils that migrate through the noncornified vaginal mucosa. During the proestrus rise in estrogen the total number of superficial cells in the smear increases. RBCs are usually present. The numbers of superficials as a percent of epithelial cells increase, those of

parabasals decrease, and those of small, round intermediates initially increase and then decrease. Depending on the final extent of cornification, the number of large intermediates may follow the pattern of small intermediates or may remain throughout proestrus and estrus as a proportion of the population of angulated, squamous cells. Vaginal fluids remain serosanguinous throughout proestrus and estrus. The numbers of red blood cells may decline or remain prominent around the time of ovulation. Neutrophils are common in small numbers in very early proestrus. They are virtually absent for most of proestrus and estrus, for 1 week or longer, before ovulation, and about a week after ovulation. The first reappearance of neutrophils may be a meager or large influx, and the timing is not very precise between day 6 and 15. More precise is the delcine in cornification, in which the percentages of superficials decline and of parabasals increases between 6 and 11 days, average 8 days, after the LH surge. This metestrus or diestrus shift occurs, on average, 6 days after ovulation (Holst and Phemister, 1974, 1975; Concannon and DiGregorio, 1986).

Vaginal cornification can be monitored on the gross level by vaginoscopy (Lindsay, 1983; Lindsay and Concannon, 1986). The mucosa appears pink, smooth, soft, and moist in early proestrus. The longitudinal folds become progressively more swollen, prominent, and pale. They are thick, white, firm, and dry-looking in late proestrus and slightly wrinkled. This wrinkling reflects a decline in estrogen-induced water retention. During and following the LH surge and ovulation they become progressively more wrinkled and dehydrated in appearance. Delamination of cornified layers becomes evident toward the end of the fertile period of oocytes, about 6–9 days after the LH surge. The mucosa then rapidly becomes smooth and patchy-colored with red, pink, pale, and white areas typical of metestrus. In anestrus the mucosa appears thin and red, due to the visibility of capillaries beneath the thin epithelium. Electrical resistance of the vaginal mucosa increases in proestrus and declines at the end of estrus (Gunzel *et al.*, 1986).

3. Uterine Changes

The endometrium and myometrium enlarge in response to estrogen during proestrus, and enlarge further under the influence of progesterone during estrus and early metestrus. The horns lengthen and become tortuous. Progesterone has a proliferative as well as secretory effect on the dog endometrium. As a result, cystic endometrial hyperplasia is common in dogs and may result in mucometra or pyrometra. The synergistic effects of progesterone on the growth of the estrogen-primed uterus have been used to bioassay progestagenic and antiprogestagenic compounds (Concannon *et al.*, 1988a). Return of the endometrium to the anestrous state occurs around days 120–140 of the cycle (Anderson and Simpson, 1973).

E. Hormone Levels and Endocrine Mechanisms

1. Estradiol

During anestrus serum estradiol levels are variable but generally low, near 5–20 pg/ml, but transient elevations of 30–50 pg/ml may be detected (Olson *et al.*, 1982). Estradiol is increased to 20–40 pg/ml about 3–5 days prior to proestrus and reaches peaks of 50–110 pg/ml in late proestrus or early estrus 0–3 days before the preovulatory surge in LH (Edqist *et al.*, 1975; Concannon *et al.*, 1975; Hadley, 1975; Wildt *et al.*, 1979a). Estradiol declines during estrus to reach 5–20 pg/ml in early metestrus and may increase slightly during metestrus before returning to low variable values during anestrus. Estrone levels follow the same pattern at concentrations similar to those of estradiol (Chakraborty, 1987).

2. Progesterone

Serum progesterone levels are about 0.5 ng/ml during anestrus and may increase slightly to 0.8 ng/ml during proestrus. Progesterone increases sharply above 1 ng/ml during the preovulatory LH surge on day 0 (Concannon *et al.*, 1977b), rises to reach peak levels of 15–80 ng/ml between day 12 and 30, and then slowly declines to 2–10 ng/ml by day 45, and less than 1 ng/ml by days 60–110 (Christie *et al.*, 1971; Smith and McDonald, 1974; Concannon *et al.*, 1975; Graf, 1978; Olson *et al.*, 1984c; Chakraborty, 1987). Normal luteal progesterone secretion in both nonpregnant and pregnant dogs is dependent on both LH and prolactin for luteotropic support throughout the luteal phase. Progesterone secretion is terminated by hypophysectomy (Concannon, 1980), suppression of prolactin by the dopamine agonist bromocriptine (Concannon *et al.*, 1987), or administration of a gonadotropin-releasing hormone (GnRH) antagonist (Vickery *et al.*, 1989), and is depressed during passive immunization against LH (Concannon *et al.*, 1987). There is no acute luteolytic mechanism, and hysterectomy does not affect the duration of luteal function or cycle intervals (Olson *et al.*, 1984c; Okkens *et al.*, 1985a). However, prostaglandin (PG) $F_{2\alpha}$ is luteolytic if administered in sufficient doses (20–200 μg/kg) two or more times a day for several days (Concannon and Hansel, 1977; Paradis *et al.*, 1983; Lein *et al.*, 1989; Oettle *et al.*, 1988). During the first 30 days of the cycle, compared to late in the cycle, corpora lutea are much more resistant to the luteolytic effects of bromocriptine (Concannon *et al.*, 1987), GnRH antagonists (Vickery *et al.*, 1989), or PGF$_{2\alpha}$ (Lein *et al.*, 1989; Concannon and Hansel, 1977). Luteal autonomy during the first 2–3 weeks of the cycle has been suggested (Okkens *et al.*, 1986). Luteal regression in dogs is characterized by a decrease in mean cell size from 35 μm at day 30 to 17 μm at day 120, a decrease in cell numbers due to scattered destruction of cells, and an accumulation of cell lipid between days 30 and 60 (Dore, 1989).

3. Androgens

Androgen levels during the follicular phase parallel those of estradiol, reaching peaks of 0.3–1.0 and 0.6–2.3 ng/ml, respectively (Concannon and Castracane, 1985; Olson *et al.*, 1984b). During the luteal phase androstenedione is the predominant androgen (0.8 ± 0.1 ng/ml) and follows the pattern of progesterone, while testosterone remains low (<0.1 ng/ml), in nonpregnant and pregnant bitches.

4. LH

Levels of LH and other pituitary hormones reported in dog serum or plasma vary with the reference standards used in different laboratories, but relative changes observed within studies seem clear (Smith and McDonald, 1974; Concannon *et al.*, 1975; 1989b; Nett *et al.*, 1975; Chakraborty *et al.*, 1980). During anestrus LH pulses of 2–25 ng/ml (average 8 ng/ml) occur at intervals of 2–8 hours. Levels are 0.2–1.2 ng/ml between pulses and mean levels are 1.5 ng/

ml (Concannon *et al.*, 1989b). Pulse intervals of 60–90 min occur during the 1–2 weeks just prior to the onset of proestrus, and mean LH levels are elevated near 3 ng/ml for several days. During proestrus, LH levels are very low and pulses are nondetectable due to estradiol negative feedback. LH increases 20- to 40-fold to 8–50 ng/ml (average 20 ng/ml) during the 1- to 2-day preovulatory surge. LH than declines and remains very low during 1–2 weeks of pituitary LH depletion (Fernandes *et al.*, 1987). Levels increase slightly during late metestrus and are again variable during anestrus (Olson *et al.*, 1984c). Major regulation of LH release in dogs is by estradiol negative feedback. Preovulatory surge release is a response to a rapid fall in the estrogen : progesterone ratio at the end of proestrus. In estrogen-treated ovariectomized dogs, preovulatory-like surge release of LH was delayed by continual increases in estrogen, initiated by estrogen withdrawal, and facilitated by administration of progesterone (Concannon *et al.*, 1979a). Naloxone-reversible opioid suppression of LH release exists throughout most of the cycle, and LH release in response to naloxone increases prior to proestrus (Concannon and Temple, 1988).

5. FSH

Serum follicle-stimulating hormone (FSH) levels are elevated during anestrus (Olson *et al.*, 1982). FSH levels were around 300 ng/ml in anestrus and reduced to 100 ng/ml during proestrus due to estrogen and inhibin negative feedback. FSH increases during the preovulatory LH surge (Reimers *et al.*, 1978), reaches peaks similar to levels in anestrus by 0–2 days after the LH peak (Olson *et al.*, 1982, 1989), declines to intermediate levels, and increases again toward the end of metestrus.

6. Prolactin

Serum prolactin levels are usually low (1–2 ng/ml) during anestrus, and show no consistent changes in proestrus, estrus, or early metestrus. Levels are usually slightly elevated (3–4 ng/ml) during mid to late metestrus, during the decline in progesterone (Concannon et al., 1989b; Graf, 1978; DeCoster et al., 1983). Indirect evidence suggests that prolactin secretion in dogs may be stimulated by withdrawal of progesterone. Overt lactational pseudopregnancy has been observed following abrupt declines in progesterone at the end of metestrus (Smith and McDonald, 1974) and overt pseudopregnancy can be caused by ovariohysterectomy during the luteal phase of a nonpregnant cycle (Christiansen, 1984; Concannon, 1986a).

F. Timing of Ovulation and Fertility

Because dogs are monestrous, the detection of the time of ovulation is critical in breeding management (Fig. 3). Oocytes are immature at ovulation, about 40–50 h after the preovulatory LH surge (Phemister et al., 1973). Oocyte maturation occurs in the oviduct about 2 days after ovulation. Tsutsui estimated that oocyte maturation occurs at 48–60 h after ovulation. Superfecundation was not obtained if the mating with the second male was later than 60 h after ovulation (Tsutsui, 1989). Dog sperm can remain fertile in the uterus or oviducts for up to 6 or 7 days (Concannon et al., 1983), but the average life span is probably only 2 or 3 days since fertility declines with matings more than 1 day before the LH surge (Holst and Phemister, 1974, 1975). It is not known if dog sperm routinely penetrate oviductal immature oocytes, as they do with unovulated oocytes in vitro and await the formation of the female pronucleus. Van der Stricht concluded that immature, dictyate, oviductal dog oocytes could be penetrated by sperm at any time after ovulation (Van der Stricht, 1923). However, sperm from matings 48–60 h after ovulation have been reported to contribute to fertilization along with sperm from matings 2–3 days earlier (Tsutsui, 1989). Some mature oocytes can also remain fertile for several days. Fertile matings have been obtained in the author's colony up to 9 days after the LH surge (Concannon et al., 1989b). However, the average life span of the mature oocyte is probably only 2–3 days, because fertility and litter size decline rapidly with matings later than 6 days after the LH surge (Concannon et al., 1989b; Holst and Phemister, 1975) or 4 days after ovulation (Tsutsui, 1989). The average onset of estrus is 1 day after the LH surge and the average time of ovulation is therefore day 2 of estrus. However, because of variations in the onset of estrous behavior relative to the LH surge, ovulations in dogs can occur before behavioral estrus, early in estrus or in late estrus (Concannon et al., 1983, 1989b).

The period of peak fertility for natural matings ranges from 1 day before the preovulatory LH surge (day 0) until day 5 or 6 after the LH surge (Holst and Phemister, 1974; Concannon et al., 1989b). Breeding twice, between days 0 and 4, is probably ideal. Determination of the time to breed or arrange artificial insemination is based on methods for estimating the time of the LH surge, including transition from proestrus to estrous behavior, distinct softening of the vulva, cytological evidence of complete vaginal cornification for 3 or more days, and, in some cycles, a decline in the amount, RBC content, or coloration of the serosanguinous discharge (Fig. 3). Accuracy is improved by monitoring the rise in serum progesterone or observing the progressive wrinkling of the vaginal mucosa, which begins prior to ovulation (Jeffcoate and Lindsay, 1989). A practical approach is to breed on alternate days until smears show the decreased cornification 6–11 days after the LH surge.

Follicle diameters are 2–3 mm in mid-proestrus, increase slightly in late proestrus, and rapidly enlarge to 8–10 mm between the LH surge and ovulation (Concannon et al., 1977b). Follicle enlargements can be observed by abdominal ultrasound of the anechoic antrum (England and Allen, 1989), but the abrupt increase in size or ovulation may not be distinct due to the thickening of the wall associated with preovulatory luteinization. At ovulation antral cavities may become ecogenic and indistinct, perhaps due to hemorrhage or further luteinization (Wallace et al., 1989).

G. Pseudopregnancy and Mammary Development

The long luteal phase of the nonpregnant ovarian cycle has been termed a physiological pseudopregnancy (Stabenfeldt and Shille, 1977) due to its similarity to pseudopregnancy in rats, rabbits, or cats following sterile mating. However, the long luteal phase in dogs occurs spontaneously and is not dependent on mating-induced ovulation or activation of luteal function as in other species. The term physiological pseudopregnancy may be appropriate in dogs, since the luteal phase of every nonpregnant cycle results in some degree of mammary gland enlargement, which can be detected by palpation by day 35 or 40, is maximal around day 80, and regresses during anestrus. In one study, mean weights of paired mammary glands in normally cycling beagle bitches were 44 ± 5, 130 ± 25, 132 ± 30, 49 ± 8, and 30 ± 12 g at days 1–25, 26–55, 56–100, 101–230, and 231–0, respectively (Concannon, 1986a). In older, particularly parous bitches, the mammary gland may not regress to the juvenile state during anestrous. Overt or clinical pseudopregnancy in dogs refers to extensive mammary development and behavioral changes typical of pregnancy and lactation, which occurs with variable incidences in bitches of many breeds.

H. Pregnancy and Parturition

Gestation length in dogs in most instances is a consistent 64-, 65-, or 66-day interval from the preovulatory LH surge to parturition (Concannon et al., 1983). However, the interval from a mating to parturition varies from 56 to 69 days because of the variable onset and duration of estrus, the potential 7 or more days of postinsemination life of some sperm, and the survival of some oocytes for up to 4 or 5 days after maturation (Concannon et al., 1983; Holst and Phemister, 1974; Phemister et al., 1973; Doak et al., 1967). Apparent gestation lengths of up to 72 days can occur based on the first of several matings (Concannon et al., 1983). Obviously the opportunity for superfecundation (multiple sires) is considerable in dogs.

The timing of events of pregnancy is probably very consistent relative to the preovulatory LH surge, but many aspects have only been studied in relation to the time of mating (Concannon, 1986b; Christiansen, 1984). Estimates based on observations by the author and others are summarized in Table II.

The exact timing and mechanism for entry of morulae or blastocysts into the uterus

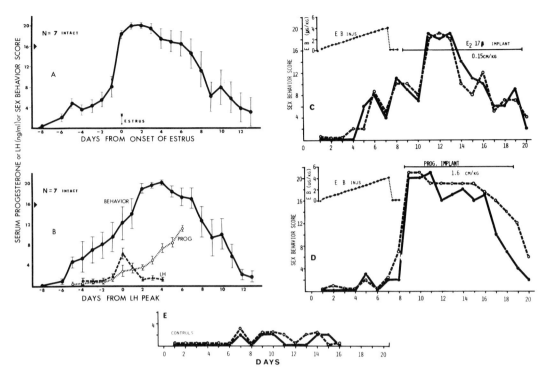

Figure 4 (A,B) Mean (±SEM) sex behavior scores of intact beagle bitches during proestrus and estrus aligned to a common time of estrus onset (A) or a common time of LH peak occurrence (B) with concomitant mean (±SEM) serum levels of LH and progesterone. (C,D) Sex behavior scores of individual ovariectomized beagle bitches injected im every 8 h with increasing doses of estradiol benzoate for 6 days prior to the sc administration of silastic implants containing estradiol-17β (C) or progesterone (D). Scores of 0–1, 2–4, 5–12, and 16–21 represent behavior characteristic of anestrus, early proestrus, late proestrus, and estrus, respectively. From Concannon *et al.* (1979).

between day 9 and 12 have not been studied. Implantations occur about day 19 or 20. No biochemical test for pregnancy exists, but assay of the pregnancy-specific appearance of relaxin in serum could be used (Steinetz *et al.*, 1987). Ultrasound can detect implantation vesicles by day 19–20, embryonic masses by day 23–25, and fetal heart beats by day 24 or 25 (Concannon *et al.*, 1989b). Fetal skeletons do not become radiopaque until day 45 or 46 (Concannon and Rendano, 1983). Uterine enlargements at sites of individual fetuses can be palpated

through the abdomen in many bitches, first as 1-cm beads at day 20, more readily at day 25, and then as 3-cm swellings at day 30–35 (Arthur *et al.*, 1982; Concannon, 1983a; 1986a). Thereafter, palpation becomes less useful as sites become more confluent, softer, and less distinct.

Body weight may increase 20–55%, mostly between day 35 and term (Concannon *et al.*, 1977a; Arthur *et al.*, 1982). A corresponding increase in blood volume contributes to the normochromic, normocytic anemia that develops over the same period.

Table II

Timing of Selected Events of the Fertile Ovarian Cycle and Pregnancy of the Domestic Dog in Relation to the Day of the Preovulatory LH Peak and to Potential Times of Fertile Matings

Selected reproductive events	Days after LH peak[b]	Days after fertile mating[c]
Onset of proestrus	−25 to −3	
Full vaginal cornification reached	−8 to +3	
Onset of estrus behavior	−4 to +6	
Estradiol peak	−3 to −1	
Decreased vaginal edema	−2 to 0	
LH surge and sharp rise in progesterone	0	−9 to +3
First fertile mating	−3 to +9	0
Initial crenulation of vaginal mucosa begins	−1 to +1	
Peak vaginal crenulation	2	
Ovulation of primary oocytes	2	−7 to +5
Oviductal oocytes		
Resumption of meiosis	3	
Extrusion of first polar body	4	−4 to +7
Sperm penetration	2 to 9	0 to 7
Fertilization/pronucleus formation	4 to 9	0 to 7
Loss of unfertilized ova	6 to 9	
Two cell embryo	6 to 10	1 to 12
Loss of vaginal crenulation	6 to 10	0 to 9
Reduced vaginal cornification	6 to 11	1 to 9
Return of leukocytes to vaginal smear	5 to 13	
Morulae (8–16 cells) seen in oviduct	8 to 10	
Blastocysts (32–64 cells) in uterus	9 to 11	3 to 14
Intracornual migration (1-mm blastocysts)	10 to 13	
Transcornual migration (2-mm blastocysts)	12 to 15	
Attachment sites established	16 to 18	9 to 21
Swelling of implantation sites	17 to 19	9 to 22
Amniotic cavities ultrasound detectable	19 to 22	10 to 25
Uterine swellings of 1 cm diameter palpable	20 to 25	12 to 28
Fetal heartbeat ultrasound detectable	24 to 25	15 to 28
Onset of pregnancy anemia	25 to 30	
Relaxin detectable	25 to 30	
Uterine swellings x-ray detectable	30 to 32	
Reduced palpability of 3-cm swelling	32 to 34	26 to 38
Hematocrit below 40% PCV	38 to 40	30 to 43
Hematocrit below 35% PCV	48 to 50	40 to 53
Fetal skull and spine radiopaque	44 to 46	36 to 49
Radiographic diagnosis of pregnancy	45 to 48	38 to 50
Fetal pelvis becomes radiopaque	53 to 57	45 to 60
Fetal teeth radiopaque	58 to 61	50 to 64
Prepartum luteolysis and hypothermia	63 to 65	55 to 68
Parturition	64 to 66	57 to 69

[a]Adapted from Concannon (1986b).

[b]Conservative estimates based on published and unpublished observations.

[c]Based on fertile single matings from 3 days before to 9 days after the LH peak.

Percent packed cell volume (PCV) is reduced from the normal range of 45–55% to below 40% by day 35, and to below 35% at day 60 (Concannon, et al., 1977a). PCV values below 30% are not uncommon at parturition. Hematocrits slowly return to near normal values by 30–90 days postpartum. Pregnancy-specific increases in blood levels of fibrinogen and coagulation factors have also been reported (Gentry and Liptrap, 1981). Mammary gland enlargement is usually obvious by day 40, and serous fluid can often be expressed from the glands by day 55, and milk may appear 1–4 days before parturition (Christiansen, 1984; Arthur et al., 1982). During the last half of pregnancy food consumption increases by 50% but drops rapidly for 1–2 days before parturition (Bebiak, 1988). The dry-matter energy requirement increases about 40%, and the absolute requirement for carbohydrates and protein also increases (Christiansen, 1984; Romsos et al., 1981). Pregnancy may result in moderate or severe insulin resistance (Siegel, 1977; Concannon, 1986b; McCann et al., 1988). The uterine production of PGE_2 and PGI_2 is considerable during most of late pregnancy, and that of $PGF_{2\alpha}$ negligible (Gerber et al., 1979, 1981; Olson et al., 1984a). The high PGE to PGF ratio may protect luteal function and maintain uterine tone.

Parturition of a litter involves a series of stage-II labors after stage-I dilation of the cervix. The most common signs of impending parturition in dogs are seclusion, decreased food intake, nesting with scratching or digging, and restlessness (Long et al., 1978). A transient drop in body temperature $\geq 1°C$ occurs during the 24 h prior to parturition, probably in response to the fall in progesterone during prepartum luteolysis (Concannon et al., 1977a; Concannon

and Hansel, 1977; van der Weyden et al., 1989). The intervals between pups can range from several minutes to a few hours. An average litter can be born in 2–4 h or take 12–18 h without incident (Concannon, 1986a, b; 1989). A copious green discharge is normal shortly before, during, and after parturition (Arthur et al., 1982). It results from breakdown of the marginal hematomas of the zonary, circumferential, endothelio–chorial placenta. There is a 78% frequency of alternation of right and left horns in the sequencing of delivery in dogs (van der Weyden et al., 1981). Pups may be born within or free of fetal membranes. Bitches commonly eat the membranes and placentas and chew the umbilical cords (Arthur et al., 1982).

Calcium-dependent uterine contractions show patterns during midpregnancy that are not present in nonpregnant bitches (van der Weyden et al., 1989; Wheaton et al., 1988), but they are not locally influenced by placental sites. The protracted uterine electromyogram (EMG) episodes lasting 3–10 min each occur two to three times per hour in mid-pregnancy and decrease in frequency during the last week of pregnancy. At the same time, the briefer EMG episodes of 1–3 min or less increase in frequency during the decline in progesterone. The EMG burst rates were about 2, 3, 7, and 7 per hour around 72, 36, 18, and 9 h prepartum, respectively, and were continuous during expulsions of pups (van der Weyden et al., 1981).

Normal litter size is 2–13 pups and can be as high as 23 pups. It varies within and between breeds, averaging 4 pups in many terriers, 6 in beagles, 8 in Alsatians and retrievers, and 10 in bloodhounds and pekinese (Christiansen, 1984). Immediately postpartum, pups experience a combined respiratory and metabolic acidosis more severe than in other species. Subnormal body temperature for several hours may have a protective effect against asphyxial damage (Van der Weyden *et al.*, 1989). Fetal losses have been estimated to average 25% percent, stillbirths 5%, and neonatal mortality about 20–30% (Tsutsui, 1975; van der Weyden, 1989; Christiansen, 1984). Eyes open at 10 days. Lactation normally lasts 6 weeks, during which energy requirements of the bitch may increase an additional 100% (Bebiak, 1988).

I. Endocrinology of Pregnancy

Mean levels of progesterone and estrogen are slightly, but not significantly, higher during the last half of pregnancy than in the nonpregnant bitch. Secondary increases in progesterone after implantation may be more dramatic in some bitches than in others (Smith and McDonald, 1974; Jones *et al.*, 1973; Concannon *et al.*, 1975, 1977a; Graf, 1978). In one study mean progesterone was higher in nonpregnant dogs than in pregnant dogs (Chakraborty, 1987). Pregnancy-specific increases in progesterone and estrogen may not be observed because concentrations in serum or plasma are diluted by the increase in blood volume associated with the anemia of pregnancy (Concannon *et al.*, 1977a). Increased metabolism in pregnancy has not been studied. Serum progesterone falls from peaks of 15–80 ng/ml (average 30 ng/ml) around day 25 to levels of about 5–15 ng/ml around day 55 and remains at 3–16 ng/ml until day 63–65, that is, 1 or 2 days before parturition (Concannon *et al.*, 1977b, 1978; Van der Weyden *et al.*, 1989).

Corpora lutea remain the source of progesterone and are required for the maintenance of pregnancy throughout gestation.

Pregnancy is terminated by ovariectomy at any stage (Sokolowski, 1974; Anderson and Simpson, 1973). LH and prolactin remain required luteotropins throughout gestation, and prostaglandin remains luteolytic. Administration of leuteolytic doses of erogocryptine or prostaglandin will cause resorption or abortion (Concannon *et al.*, 1987; Conley and Evans, 1984). The minimum level of progesterone required to support pregnancy is around 2 ng/ml. Parturition is dependent on a prepartum luteolysis beginning 18–36 h prepartum, causing progesterone to fall to about 1 ng/ml before parturition. Maintenance of progesterone above 5 ng/ml prevented normal parturition (Concannon *et al.*, 1977a), and administration of an antiprogestagen starting on day 57 induced early parturition in 1–3 days (Van der Weyden *et al.*, 1989). The prepartum luteolysis is probably the result of uterine or placental release of prostaglandin $F_{2\alpha}$. Maternal plasma levels of 13,14-dihydro-15-keto-$PGF_{2\alpha}$, the major metabolite of $PGF_{2\alpha}$, increased 6-fold, from 0.3 ± 0.1 to 1.9 ± 0.6 ng/ml, during the 36 h prior to parturition, peaked at 2.1 ± 0.6 ng/ml during delivery of the litter, and then declined (Concannon *et al.*, 1988b).

The signal for parturition is presumed to be fetal in origin, involving maturation of the fetal pituitary–adrenal axis as described for other species (Concannon *et al.*, 1977a, 1988b). Elevations in maternal cortisol levels are usually seen 1 day prepartum, but not during parturition (Concannon *et al.*, 1978). Since there is no acute rise in estrogen at parturition, it is unclear whether fetal cortisol would cause release of prostaglandin in dogs by altering placental steroid hormone levels, or by some other mechanism (Concannon *et al.*, 1988b).

Prolactin levels increase above basal levels

of about 4–5 ng/ml starting around day 30–35, and reach four- to five fold higher levels (average 25 ng/ml) by day 60 (DeCoster *et al.*, 1983). Prolactin levels surge during the prepartum fall in progesterone and reach peak levels of about 100 ng/ml at or shortly after parturition (Concannon *et al.*, 1978). Prolactin levels in some animals are decreased for 1 or 2 days postpartum. Levels are elevated but variable in response to suckling-dependent release throughout lactation, in general decline during the last half of lactation, and fall to basal levels within 1–4 days after natural or premature weaning (Concannon *et al.*, 1978). Prolactin is required for lactation, and lactation is inhibited by suppression of prolactin with dopamine agonists (Jochle *et al.*, 1989). Increases in plasma oxytocin levels have been measured in response to suckling (Uvnas-Moberg *et al.*, 1985).

Relaxin levels are detectable by day 25–30 and are nondetectable in nonpregnant dogs. Peak relaxin levels of about 5 ng/ml at 40–50 days are followed by slight declines before parturition, severe declines after parturition, and low but detectable levels (0.5–2 ng/ml) during early lactation in some dogs (Steinetz *et al.*, 1987). The placenta is the major source of relaxin in dogs, but the ovary may also be a source (Steinetz *et al.*, 1989).

J. Fertility Regulation

Because of the role of dogs as companion animals and the resulting excess population that can result, interest in methods to prevent ovarian cycles or to intercept pregnancy has been prominent in dog reproduction research (Christiansen, 1984; Concannon 1983b; Olson *et al.*, 1989). Ovarian cyclicity is suppressed by maintaining progesterone levels over 2 ng/ml via subcutaneous implants or by repeated long-acting depot injection of synthetic progestagen such as medroxyprogesterone acetate or proligestone (McCann *et al.*, 1987; Evans and Sutton, 1989). However, chronic administration of a progestagen in dogs promotes the development of uterine disease, mammary tumors, acromegaly, and insulin resistance (Concannon *et al.*, 1980a; McCann *et al.*, 1987).

Cycles are also suppressed by repeated injection of testosterone or by daily oral administration of the synthetic androgen mibolerone (Concannon, 1983; Burke, 1986). Cycles can be postponed by single injections of a depot progestagen or by several days of oral administration of progestagen, such as megesterol acetate in late anestrus or early proestrus (Evans and Sutton, 1989). Ovarian cycles are suppressed by constant administration of a GnRH agonist in prepubertal or adult dogs (Concannon *et al.*, 1988c; Vickery *et al.*, 1989), but initial administration can induce estrus in anestrous adults (Concannon, 1989). Attempts to induce infertility by immunization against pituitary gonadotropins (Faulkner *et al.*, 1975), GnRH conjugates (Gonzalez *et al.*, 1989), or zona pellucida protein (Mahi-Brown *et al.*, 1985) have had variable results.

Unwanted pregnancies can be prevented by the single administration of the long-acting estrogen estradiol cypionate once during estrus, or by repeated doses of diethylstilbestrol (Concannon, 1983b). However, estrogen doses that are routinely efficacious also increase the incidence of uterine disease (Bowen *et al.*, 1985). Pregnancy can be terminated by a series of twice daily injections of $PGF_{2\alpha}$, using high doses of 200 μg/kg starting around day 15 (Oettle *et al.*,

1988), or moderate doses of 30–100 μg/kg after day 30 (Concannon and Hansel, 1977; Paradis *et al.*, 1983; Lein *et al.*, 1989). Such doses often produced clinical side effects including vomiting, defecation, and ataxia. Pregnancy is terminated after day 30 by luteolytic suppression of prolactin with dopamine agonists including bromocriptine (Concannon *et al.*, 1987) and cabergoline (Jochle *et al.*, 1989). Luteolytic suppression of LH by injection of GnRH antagonist terminated pregnancy when given after day 25, but before day 15 was effective only if given in combination with a prostaglandin analog (Vickery *et al.*, 1989). Pregnancy can also be terminated by the progesterone receptor antagonist mifepristone (Concannon *et al.*, 1990), and by suppression of progesterone synthesis with an inhibitor of hydroxysteroid dehydrogenase, epostane (Keister *et al.*, 1989).

K. Anestrus and Induction of Estrus and Ovulation

Anestrus is characterized by elevated FSH, variable estradiol (Olson *et al.*, 1982), and infrequent episodes of LH release. Anestrus is naturally terminated by increased episodic LH release (Concannon *et al.*, 1986; Concannon, 1989), which results in elevated mean levels of LH shortly before the onset of proestrus. The mechanisms involved in altering the LH secretion pattern are not known. Reduced opioid inhibition of LH release could be involved (Concannon and Temple, 1988). Prolactin may also play a role in interestrous intervals. Suppression of prolactin with dopamine agonists can cause an extreme shortening of anestrus (Okkens *et al.*, 1985b; van Haaften *et al.*, 1989) or induction of estrus in bitches with a prolonged anestrus (Arbeiter *et al.*, 1988; Jochle *et al.*,

1989). Acute induction of estrus during anestrus has been studied using gonadotropins, GnRH, and GnRH agonist.

Induction of estrus with exogenous gonadotropins in anestrous bitches has usually resulted in ovulation failure, short luteal phases, underdeveloped embryos, and low fertility (Table III). Results with equine chorionic gonadotropin (eCG) have usually been better than with FSH, but fertility rates are usually 20% or less. Prolonged daily administration often results in premature luteinization or hyperestrogenism (Shille *et al.*, 1984, 1989; Arnold *et al.*, 1989). Administration of eCG (20 IU/kg/day) for 5 rather than 10 days reduced the incidence of hyperestrogenism and resulting pathologies and increased fertility at induced estrus to 50% (Arnold *et al.*, 1989). Controlled studies of more restricted eCG administration have not been reported. Use of estrogen priming followed by serial injections of LH and then FSH was very successful in one study, but not in another, which substituted human chorionic gonadotropin (hCG) for LH (Moses and Shille, 1988; Shille *et al.*, 1989). Induction of fertile estrus has also been obtained by pulsatile administration of GnRH at doses of 75–500 ng/kg every 90 min for 6–12 days, with success rates of 38–85% (Vanderlip *et al.*, 1987; Cain *et al.*, 1989). Stimulation of endogenous gonadotropin release by constant infusion of a GnRH agonist has also resulted in fertile spontaneous ovulation during the induced estrus (Concannon, 1989). Infusions of a GnRH agonist over 25 times more potent than GnRH, at a dose of 24 μg/day for 14 days, resulted in consistently timed onsets of proestrus (3–6 days) and estrus (7–10) days), spontaneous preovulatory LH surges (Day 8–13), and ovulation in 18 of 24 dogs, and parturition (71–79 days) in 50% of the ovulating bitches.

II. The Domestic Queen

Table III
Incidence of Induced Estrus, Ovulation, and Resulting Litters in Selected Reports Providing Details of Estrus-Induction Protocols Tested in Anestrous Bitches

Ref.[a]	Dogs (n)	Folliculotrophic regimen	Days	Ovulating hormone	Responses (%)			
					PE[b]	Estrus	Ovulation	Litters
1	(25)	eCG (20–500 IU/day)	×10	hCG, 500 IU	50	50	50	NA[c]
2	(8)	eCG (44 IU/kg/day)	×9	hCG, 500 IU	NA	70	50	NA
3	(9)	FSH (4 mg/day) post DES	×10	hCG, 100 IU	90	90	30	NA
4	(11)	eCG (250 IU/day)	×14	hCG, 500 IU	75	60	75	NA
5	(15)	eCG (500 IU/day)	×10	hCG, 500 IU	100	87	NA	20
6	(5)	FSH (10 mg/day)	×1	—	80	40	40	20
6	(4)	FSH (1–10 mg/day)	×10	—	75	50	50	0
7	(15)	eCG (44 U/kg/day)	×9	hCG, 500 IU	87	74	67	20
8	(7)	DES (5–10 mg/day) LH (5 mg), day 0 FSH (10 mg), days 4 and 6	×7–14	—	100	100	100	100
9	(17)	eCG (20 IU/kg/day)	×10	hCG, 500 IU	100	100	100	0
9	(6)	eCG (20 IU/kg/day) and hCG (300 IU) once	×5 day 5	—	100	100	100	50
10	(8)	GnRH (2–8 µg/90 min)	×6–12	—	75	62	50	38
11	(7)	GnRH (1–25 µg/90 min)	×11–13	—	100	100	85	85
12	(24)	GnRH agonist (24 µg/day)	×14	—	88	79	75	38

[a](1) Thun *et al.*, 1977; (2) Archbald *et al.*, 1980; (3) Olson *et al.*, 1981; (4) Wright, 1982; (5) Chaffaux *et al.*, 1984; (6) Shille *et al.*, 1984; (7) Nakao *et al.*, 1985; (8) Moses and Shille, 1988; (9) Arnold *et al.*, 1989; (10) Vanderlip *et al.*, 1987; (11) Cain *et al.*, 1988; (12) Concannon, 1989.

[b]PE = proestrus

[c]Not addressed.

[d]DES = diethylstibestrol

II. The Domestic Queen

A. Overview

Cats are seasonally polyestrous, reflex ovulators. Cycles in northern temperate zones typically occur from February to September with 4- to 8-day periods of estrus interrupted by nonestrous periods of 5–12 days. Pregnancy lasts 9 weeks. Lactation lasts about 6 weeks. Cats may produce and raise one to three litters per year (Dawson, 1952; Stabenfeldt and Shille, 1977; Christiansen, 1984; Goodrowe *et al.*, 1989; Herron, 1977). A sterile mating can result in a 4- to 5-week luteal-phase pseudopregnancy and prolonged interestrous interval. Terminology for reproductive stages in the queen has not been standardized. Patterns of polyestrous cycles are affected by photoperiod, breed, and housing.

B. Polyestrous Cycles

In unmated queens, nonovulatory waves of follicular development and atresia result in excursions of blood estradiol levels and periods of behavioral estrus interrupted by sexual nonreceptivity (Fig. 5). The stages of the cycle are *proestrus*, *estrus*, and *interestrus*.

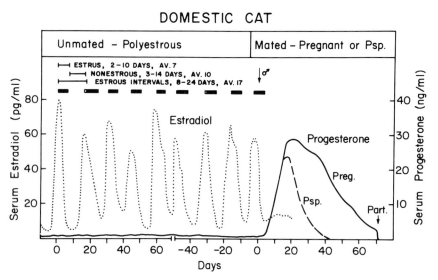

Figure 5 Schematic representation of endocrine changes typically observed during poly-estrus cyclic activity in unmated queens and during pregnancy or pseudopregnancy (Psp) following fertile matings or vaginal stimulation resulting in ovulation. Solid bars indicate periods of estrus; open portions indicate potential proestrous periods. From Concannon and Lein (1983).

The interestrous periods have also been termed metestrus or anestrus in nonovulatory cycles, metestrus or diestrus in nonovulatory or ovulatory cycles, or nonestrus (Concannon and Lein, 1983; Concannon et al., 1989a; Stabenfeldt and Shille, 1977; Goodrowe et al., 1989; Leyva et al., 1989a). Anestrus usually refers to prolonged periods of nonestrus, either seasonal or lactational. Estrus is characterized by overt display of sexual behavior and acceptance of males. Estrus lasts 2–10 days. The average is about 7 days. Mating and ovulation early in estrus may shorten estrus by 1 or 2 days but not necessarily (Shille et al., 1979, 1983; Goodrowe et al., 1989; Scott, 1970). Some queens may have extended periods of persistent estrus, perhaps due to overlapping waves of follicular growth (Christiansen, 1984). Proestrus precedes estrus by 1–3 days and is poorly characterized in the ab-sence of a male and vaginal smears (Stabenfeldt and Shille, 1977; Shille et al., 1979). Much of the spontaneous behavior characteristic of estrus is observed during proestrus, including attraction of males. However, the queen will not permit mounting and/or intromission. Being brief, proestrus is often not recognized, studied, or monitored. Interestrous or nonestrous periods tend to range from 3 to 14 days and average 10 days. The intervals between estrus onset in unmated cats can range from 7 to 40 days, most are 8–24 days, and over 50% are 14–21 days (Jemmett and Evans, 1977; Wildt et al., 1978a; 1981; Goodrowe et al., 1989). The average in the author's colony is about 17 days.

Follicular development associated with increased estradiol secretion and proestrus occurs rapidly over 1–3 days (Wildt and Seager, 1980). About 3–7 primary follicles develop to 2 mm in diameter (Foster and Hisaw, 1935). Estradiol levels increase from basal levels of 5–20 pg/ml to peaks of 40–100 pg/ml, and then decline over the next several days (Verhage *et al.*, 1976; Shille *et al.*, 1979; Wildt *et al.*, 1981; Schmidt *et al.*, 1983). The onset of estrus has varied among studies, occurring before, at, or after the peak in estradiol. However, studies varied in attention to proestrus versus estrus behavior. Whether full estrous behavior is elicited by the decline in estrogen, as proposed in dogs, has not been studied in cats. Estrous cycles in breeding colonies of group-housed queens are usually monitored based on observations of overt estrouslike behavior in the absence of males.

C. Reproductive Tract

The queen has a reproductive tract similar to the bitch, except that the vagina is relatively short. There are tubular cervical glands and vestibular Bartholin glands (Christiansen, 1984). The ovarian bursa is open, permitting clear observation of the ovary (Wildt and Seager, 1980).

The cat oviductal and uterine epithelia have been used to study the synergistic and antagonistic effect of estrogen and progesterone (Bareither and Verhage, 1980). Estrogen causes rapid secretory-cell development and ciliation of the oviduct. The effect is rapidly reversed by progesterone, to the point of atrophy. In contrast, the effect of progesterone on the estrogen-primed cat endometrium is further hypertrophy and secretory activity. During estrus vulval swelling is not observed, vaginal secretions are negligible, and erythrocytes are rare in the smear. Vaginal cornification is progressive during proestrus and is lost at the end of estrus (Colby, 1980; Shille *et al.*, 1979).

Smears are usually obtained either by lavage or with swabs inserted carefully to avoid induction of ovulation. Smears in interestrus or anestrus consist of a mixture of parabasal, intermediate, and nucleated superficial cells. The rise in estradiol in proestrus produces clearer smears with decreased debris, increasing numbers of superficial cells, decreases in intermediates, and a loss of parabasals (Shille *et al.*, 1979). By early estrus the smear is dominated by anuclear superficials (40%) and nucleated superficials (50%), with only small numbers of intermediates (10%).

D. Sexual Behavior

The distinctive behavior of proestrus and estrus includes increased purring and preening, persistent episodes of "calling" vocalizations, rolling, rubbing against inanimate objects, and treading of the hind legs (Dawson, 1952; Michael, 1961; Wildt *et al.*, 1978a; Concannon *et al.*, 1989a). The vaginal pheromone causing male attraction has been reported to be valeric acid (Christiansen, 1984). Responses to an acceptable male include treading and standing in place, tail deviation, and lordosis in response to his approach or mounting (Michael, 1961; Voith, 1980). In the absence of a male, a fully estrous queen will usually display a full complement of sex behavior if grasped firmly by the skin on the back of the neck and stroked on the back, tail-head, or genitalia (Concannon and Lein, 1983). Ingestion of catnip can induce estruslike behavior but not physiological estrus (Colby, 1980).

A typical mating sequence involves the male approaching from the side or back, establishing a neck bite, mounting, adjusting position to align genitalia, pelvic thrusting, and intromission and ejaculation

(Concannon *et al.,* 1989a). The queen lowers her chest, elevates her pelvis, treads with the hind feet, raises the vulva, and deviates the tail to the side. Upon intromission the queen emits an often alarming cry and attempts to terminate the mating by turning, rolling, and striking at the male. An experienced male immediately withdraws and observes the queen's coital afterreaction. The afterreaction includes bouts of erratic and often violent rolling and thrashing, obsessive licking of the vulva, and rebuttal of approaches by the male. Positioning by the male ranges from 20 s to 9 min, intromissions last 1–27 s and average 8 s, and female afterreaction lasts 1–17 min and averages 2.5 min. Intervals between matings range from 3 min to 5 h, average about 20 min for the first 2 h, and 1 h or longer thereafter. Copulations may number 3–8 at 4 h, 7–20 at 12 h, and 20–36 at 36 h of unrestricted mating (Concannon *et al.,* 1989a). Mating frequency is regulated by female receptivity. Frequency of mounting attempts by males was two to six times greater than copulation frequency. For matings delayed until mid-estrus (day 4–5) there may be increased submission to mounting and duration of mounting, and increased incidence of ovulation compared to matings in early (day 1–2) estrus (Glover *et al.,* 1985).

E. Reflex LH Release and Ovulation

Ovulation in cats occurs in response to adequate reflex LH release. In more than half the cases, single copulations were not sufficient to release ovulatory amounts of LH. Multiple copulations result in higher elevations in LH for longer periods of time, and LH patterns differ greatly among animals subjected to regimented mating schemes. In one study (Concannon *et al.,*

1980b), when LH release after single copulations peaked under 10 ng/ml in less than 1 h and declined to basal levels of 1 ng/ml in less than 4 h, it failed to cause ovulation (Fig. 6). If levels continued to increase over 20 ng/ml at 1 h, and remained above base-

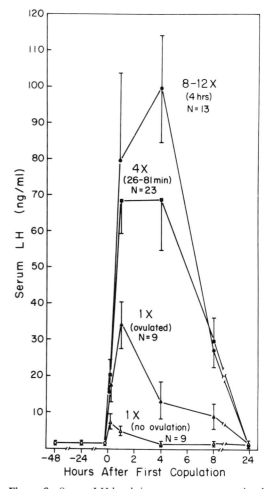

Figure 6 Serum LH levels in estrus queens permitted single (1 X) or multiple (4 and 8 to 12 X) copulations. Single copulations induced LH release sufficient to cause ovulations in only one-half the queens. Multiple copulations caused a greater release of LH, and all queens ovulated. From Concannon *et al.* (1980b).

line at 4 h, ovulation occurred. When 4–12 unrestricted matings were allowed in a 4-h period, mean levels of LH were three to six times higher than for single matings and all animals ovulated. Mounting without intromission released small, nonovulatory amounts of LH. Mating at limited, predetermined intervals on individual or sequential days of estrus can result in reflex LH release of variable incidence, magnitude, duration, and potency (Banks and Stabenfeldt, 1982; Wildt et al., 1978a, 1980). This may contribute to the variation in reported gestation lengths. Uninterrupted matings not only result in maximal amounts of LH release within 1–2 h, they also fail to result in significant additional LH release after 2–4 h (Concannon et al., 1989a). LH levels continue to decline and remain low throughout the next 2 or 3 days of mating (Fig. 7). Re-

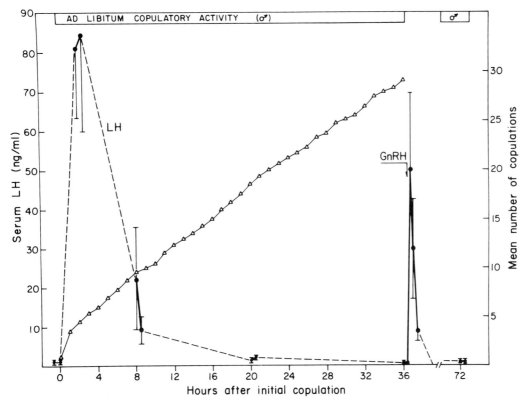

Figure 7 Mean (±SEM) serum luteinizing hormone (LH) levels (●) shortly before and 30 min after (solid lines) individual copulations occurring approximately 2, 8, 20, and 36 h after the initial copulation in 5 estrous cats during 36 h of *ad libitum* copulatory activity; at 15, 30, and 60 min after a gonadotropin-releasing hormone (GnRH) injection (3–15 μg/kg) given 30 min after the last unrestricted copulation; and before and 30 min after a final copulation permitted at 72 h of study. The mean total number of copulations at hourly intervals during the 36 h of continuous observations are also indicated (△). Reprinted with permission from Concannon *et al.* (1989).

flex LH release in the cat therefore appears to be subject to self-inhibition at the level of the hypothalamus and/or pituitary. Whether the mechanism involves depletion of GnRH, ultra-short-loop feedback, or pituitary down-regulation has not been determined. Some queens may not release adequate amounts of LH or ovulate following repeated mating very early in estrus, but will do so late in estrus (Glover et al., 1985). Data on the incidence and cause of this phenomenon are not available. A practical approach to mating is to keep the queen away from the tomcat until mid or peak estrus, 3–4 days after initial signs of proestrus, and then to permit 3–10 copulations over 2–12 h.

Germinal vesicle breakdown occurs at 12 h and first polar body formation at 22 h. Ovulations occur 24–36 h after the mating-induced LH surge and are detectable laparoscopically at 48–60 h (Dawson and Friedgood, 1940; Wildt et al., 1981). The 1–2 mm antral follicles enlarge to 3–5 mm, become hemorrhagic at ovulation, and develop into 4–5 mm corpora lutea (Wildt and Seager, 1980). Serum progesterone levels remain under 1 ng/ml during the LH surge and ovulation, and increase over 1 ng/ml on day 2–4, shortly after ovulation (Verhage et al., 1976; Wildt et al., 1981; Schmidt et al., 1983). Estradiol levels decline to basal values by the end of estrus. Oocytes remain viable for only 25 h after ovulation. Fertilization does not result from matings that occur more than 50 h after injection of hCG to induce ovulation (Hammer et al., 1970).

F. Pseudopregnancy

Nonfertile mating or mechanical stimulation of the vagina results in formation of corpora lutea that secrete progesterone for 30–45 days. Such pseudopregnancies result in suppression of estrous cycles and interestrous intervals of 35–70 days, average 45 days (Verhage et al., 1976; Wildt et al., 1981; Christiansen, 1984). Late in the breeding season a pseudopregnancy may be followed by anestrus and have an apparent duration of several months. Progesterone levels reach peaks of 15–90 ng/ml by day 15–25, and decline below 1 ng/ml by day 30–50 (Paape et al., 1975; Verhage et al., 1976; Wildt et al., 1981). This is followed in 5–10 days by the next follicular phase. Repeated injection of high doses of $PGF_{2\alpha}$ caused only transient declines in progesterone and failed to cause complete luteolysis in pseudopregnant cats (Shille and Stabenfeldt, 1979; Wildt et al., 1979b). Levels of estradiol, LH, and prolactin remain basal during pseudopregnancy (Paape et al., 1975; Verhage et al., 1976; Wildt et al., 1981). Induction of pseudopregnancy for purposes of suppressing ovarian cycles can be accomplished by repeated mechanical stimulation of the vagina with a glass rod sufficient to evoke afterreactions (Concannon and Lein, 1983), or by administration of hCG (Goodrowe et al., 1989). The incidence of spontaneous ovulation in cats may vary with housing and social conditions, being rare or nonexistent in some research facilities and common in others. There are no data on whether apparently spontaneous ovulations result from spontaneous LH release or from reflex LH release caused by handling, interaction with other queens, or autostimulation of genitalia. Spontaneous pseudopregnancy could be involved in reports of prolonged interestrous intervals or periods of nonseasonal anestrus (Christiansen, 1984; Herron, 1977).

G. Pregnancy

Most pregnancies last 64–68 days (Christiansen, 1984; Schmidt *et al.*, 1983; Herron, 1986). The average in the author's colony is 65 days. Reports of intervals from observed matings to parturition ranging from 52 to 71 days (Jemmett and Evans, 1977; Christiansen, 1984) may include intervals based on matings before or after the mating that induced the ovulation. The timing of various events of feline pregnancy are summarized in Table IV, based on estimates from several sources (Stabenfeldt and Shille, 1977; Christiansen, 1984; Herron and Sis, 1974; Dawson, 1952; Concannon and Lein, 1983). Implantation occurs around day 12. Obvious mammary development around day 40 is usually indicative of pregnancy, since pseudopregnancy in cats is uncommon and rarely overt.

As in pseudopregnancy, progesterone levels in pregnancy rise above 1 ng/ml shortly after ovulation, reach peaks of 15–90 ng/ml by day 15–25, and then decline. However, progesterone remains elevated and variable throughout the 65 days of pregnancy, may increase to secondary peaks of 10–60 ng/ml between day 35 and 45, and decline slowly through day 60. Progesterone is maintained at 3–12 ng/ml for the last week of pregnancy, and declines below 1 ng/ml at parturition (Verhage *et al.*, 1976; Banks *et al.*, 1983; Schmidt *et al.*, 1983). Corpora lutea appear to remain functional throughout gestation and regress at parturition with some evidence to rejuvenation histologically at 1 week postpartum in response to suckling and prolactin (Dawson, 1952). The capacity of the placenta to produce progesterone *in vitro* increases in late gestation (Malassine and Ferre, 1979) and may be able to support pregnancy in the absence of the ovaries after day 50 (Scott, 1970).

Table IV

Events of Pregnancy in Cats Relative to the Day of Fertile Mating and Reflex Preovulatory LH Surge

Events	Days from fertile mating
Mating and preovulatory LH surge	0
Ovulation of mature oocytes	1–2
Fertilization	1–2
Serum progesterone over 1 ng/ml	2–3
Blastocyst entry to uterus	6–7
Transcornual migration	8–11
Implantation	12–13
Palpable 1-cm implantation sites	16–18
Radiography of uterine swellings	17–18
Palpable 3-cm swellings	28–30
Increase in serum relaxin	30–35
Increase in serum prolactin	35–40
Obvious mammary development	38–45
Fetus radiopaque	40–42
Prepartum increase in estradiol	62–64
Parturition and peripartum decline in progesterone	65–66

The relative contributions of the placenta and corpora lutea to progesterone levels during late pregnancy are not known. Luteotropin requirements during pregnancy or pseudopregnancy have not been characterized. Prolactin may not be a required luteotropin during pseudopregnancy (Concannon and Lein, 1983). However, prolactin is required to maintain progesterone secretion during the last half of pregnancy. Administration of prolactin-suppressing doses of dopamine agonist for 3–5 days between days 35 and 45 terminates pregnancy (Jochle et al., 1989).

During the last week of pregnancy, estradiol levels become elevated to proestruslike levels of about 40–80 pg/ml. Estradiol declines at parturition and increases again after weaning (Verhage et al., 1976; Schmidt et al., 1983). Prolactin levels gradually increase about fivefold between day 40 and 60, from 4–12 ng/ml to 20–45 ng/ml, and increase to 35–55 ng/ml 2–3 days before parturition (Banks et al., 1983). Prolactin is elevated the first 4 weeks of lactation, declines somewhat due to decreased sucking in weeks 5 and 6, and declines to basal levels by 1–2 weeks after weaning. Relaxin becomes detectable around day 25 of pregnancy, increases to peak levels of 4–11 ng/ml around day 40–50, decreases 50% between day 50 and parturition, falls rapidly at parturition, and is nondetectable 1 day after parturition (Stewart and Stabenfeldt, 1985).

Parturition in cats is similar to that in dogs except that litter size is usually more consistent (two to six kittens) and coloration of the placental breakdown discharge is brown. Intervals between deliveries are normally 5–60 min, but interruption of parturition for up to 11–24 h between kittens is not unexpected (Christiansen, 1984). The physiology of cat parturition has not been investigated. Weaning normally occurs at 6–7 weeks.

H. Lactation Anestrus

Estrous behavior and mating may occur during pregnancy. Resulting superfecundation and births of litters many days apart have not been well documented (Scott and Lloyd Jacob, 1955; Goodrowe et al., 1989). Postpartum estrus is uncommon, but has been reported in colony cats (Colby, 1980). Lactation normally produces a postpartum anestrus with return to estrus 2–8 weeks after weaning (Schmidt et al., 1983; Colby, 1980). In one study of cats weaned at 7 weeks postpartum, first estrus occurred 15–60 days (average 30 days) after weaning, and about 10–14 weeks postpartum (Concannon and Lein, 1983). Following weaning at 1–2 days postpartum, estrus occurred in 7 days. After spontaneous abortions around day 40 of pregnancy, estrus occurred in 4–5 days (Fig. 8). Breeding at the first postlactation estrus may result in corpora lutea with subnormal progesterone secretion (Schmidt et al., 1983).

I. Seasonality and Puberty

Most queens are seasonally polyestrous, long-day breeders, cycling from January or February until September or October in northern temperate zones. The season is longer at equatorial latitudes and shorter at more northern latitudes (Christiansen, 1984). In the author's colony, most cats cycle year round while exposed to a continuous 14L : 10D photoperiod supplemented with natural light, but a few may show winter anestrus. Constant 12L : 12D or 14L : 10D results in an equal incidence of estrus year round in laboratory colonies (Hurni, 1981).

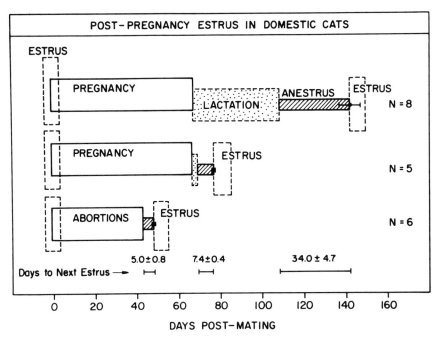

Figure 8 Diagrammatic representation of durations of estrus, pregnancy, lactation, and nonestrus intervals prior to the postpregnancy estrus following lactation (or pregnancy) in queens allowed a normal 7-week lactation, queens weaned 3 days postpartum, and queens that aborted their kittens. The respective intervals from the end of pregnancy to the next estrus averaged 12 weeks, 1 week, and less than 1 week. From Concannon and Lein (1983).

Most cats show continued polyestrus in the natural breeding season. Some may have periods of contiguous cycles interrupted two or three times a year by periods of nonseasonal anestrus (Herron, 1986). The occurrence of silent or weak-estrous cycles (Herron, 1986; Wildt *et al.*, 1978a; Shille *et al.*, 1979) or spontaneous pseudopregnancy could be involved. Breed can also affect seasonality, even for cats housed indoors. Long-haired breeds tend to become anestrous periodically throughout the year, whereas shorthaired breeds, in particular Siamese, often cycle year round. In laboratory cats acclimated to 14L : 10D photoperiods, constant light (24L : 0D) increased numbers of large follicles and estradiol levels, and lengthened interestrus intervals (Leyva *et al.*, 1989a). Exposure to short days (8L : 16D) immediately inhibited ovarian cycles, and estradiol levels were lower than in normal interestrus. Photoperiod effects on pineal secretion of melatonin, and possibly prolactin secretion, appear to be involved. Plasma levels of melatonin and prolactin are inversely proportional to artificial daylength, and melatonin administration blocks follicular development and ovarian cyclicity for 2 months (Leyva *et al.*, 1989b).

The age at first estrus in female cats is very variable, depending on breed and season of birth relative to the natural breeding season. Most cycle around 7–9 months of age after reaching adult body weight of 2.3–2.5 kg, but the normal range is 4–12 months (Christiansen, 1984; Stabenfeldt and Shille, 1977). Thus, barn cats born in August might cycle in February at 6 months of age, while those born in December–March might cycle either at 4–5 months or after the seasonal anestrous at 12–14 months of age. Some long-haired breeds do not cycle until 12–18 months (Christiansen, 1984). Fertility can last for 14–20 years but declines after 5 years (Christiansen, 1984).

J. Induction of Estrus and Ovulation

Induction of fertile estrus in cats is readily achieved by injection(s) of eCG or FSH followed by mating or hCG to induce ovulation (Colby, 1970; Cline *et al.*, 1970; Wildt *et al.*, 1978b; Goodrowe *et al.*, 1989). eCG has a long half-life and daily eCG may cause ovarian hyperstimulation (Wildt *et al.*, 1978b). In anestrus cats, single injections of 100 IU eCG 5–7 days before hCG (50 IU) results in normal ovulation and pregnancy rates. However, daily eCG for 4–5 days yields a lower pregnancy rate and reduced kitten survival (Cline *et al.*, 1980). In contrast, single injections of FSH are not productive. Daily injections of 2 mg FSH for 5–7 days induced estrus and 72% of the queens allowed to mate ovulated and had normal pregnancies (Wildt *et al.*, 1978b).

K. Cat Contraception

Ovarian cycles of cats can be suppressed by administration of many progestagens or the androgen mibolerone (Lein and Concannon, 1983; Christiansen, 1984), but concerns about safety and side effects have limited their application. Megestrol acetate is used in some countries to prevent estrous cycles in cats. Early pregnancy can be terminated prior to implantation by administration of estrogen to block tubal transport of ova within 1 week of mating (Herron and Sis, 1974). Induction of pseudopregnancy is also used to cause a temporary suppression of estrous behavior.

III. Male Dogs

A. Genitalia, Prostate, and Testes

Male dogs are sexually active year round. Possible influences of seasonality are probably minimal. Most breeds are presumed to approach sexual maturity at an average age 0–2 months after females, often at 9–12 months of age (Christiansen, 1984; Stabenfeldt and Shille, 1977). Beagles were observed to become pubertal around 6 months of age with ejaculates of 0.2×10^9 sperm per ejaculate. However, they did not reach sexual maturity until 15–16 months with 0.7×10^9 sperm per ejaculate (Aman, 1986). The prostate is the only accessory gland in dogs. It is globular, divided by a dorsal-median furrow, and encircles the cranial uterthra caudal to the neck of the bladder (Christiansen, 1984). Benign prostatic hypertrophy is common in older dogs. Dogs have frequently been used to study the etiology of prostate disease (Cochran *et al.*, 1981). For much of its length, the glans penis contains an os penis with a fibrocartilage process that protrudes into the tip. The proximal portion of the external penis has more erectile tissue than the distal portion (pars longa), producing a spherical enlargement, the bulbus glandus. Testicular descent usually occurs with a few weeks of birth, and descent later than 8 weeks is considered abnormal, but it can be delayed to 6 or 8 months (Christiansen, 1984; Meyers-Wallen, 1989). Testis size, which determines daily sperm production, semen volume, and sperm counts, is related to body size. Efficiency of sperm production in normal dogs has been estimated to be $15–19 \times 10^6$ sperm/day/g of testis (Aman, 1986). In large (25–40 kg) versus small (5–15 kg) dogs there were two- to threefold differences in average testis weight (44 versus 16 g), daily sperm production (750 versus 280×10^9/g/day), nonprostatic ejaculate volume (5.4 versus 2.4 ml), and total sperm per ejaculate (1.4 versus 0.5×10^9).

B. Ejaculation

The stud dog normally mounts from the side, clasps the bitch in the lumbar region with his front legs, and rests his chest or neck on her back. One or more mounts and bouts of pelvic thrusting with a semierect penis continue in an apparently random pattern until intromission is achieved. Erection is completed intravaginally. The stud then "steadies" or treads with the hind legs for 15–60 s and then dismounts by stepping one hind leg over the back of the bitch. The penile shaft is held in the vagina and the bulbus in the vestibule during the 5–40 min copulatory lock. The dismount causes a 180° flexion of the penis but not torsion (Grandage, 1972). The lock is terminated by detumescence of the penis.

Ejaculation may last 1–20 min and have a volume of 1–22 ml in dogs <20 kg and 2–45 ml in dogs >20 kg (Dubiel, 1976). There are three distinct fractions (Andersen, 1980). The *presperm fraction* is a low-volume (0.1–2 ml), brief or protracted (5–90 s) emission of clear fluid containing few or no sperm which is released during pelvic

thrusting and attempted intromissions. The second *sperm rich fraction* is milky in color, variable in volume (0.2–4 ml), and ejaculated in 5–200 s, usually 30–60 s (Morton and Bruce, 1989), while the dog steadies himself after achieving complete intromission and full erection. The mean duration of the sperm-rich fraction may be shorter for collections with an artificial vagina (5 s) than for manual collection (54 s) Christiansen, 1984). The final *prostatic fluid fraction* is usually the largest (1–30 ml) and longest duration (1–45 min). It is comprised of clear prostatic fluid with few sperm and released in pulses during the copulatory lock while the stud is dismounted and turned from the bitch or collector (Christiansen, 1984; Morton and Bruce, 1989); Andersen, 1980).

Normally semen is collected directly into a clear container or via a latex sheath by constricting the unsheathed penis with the thumb and forefinger central to the bulbus and exerting considerable pressure. Firm squeezing and pulling, with the bulbus in the palm of the hand, result in erection and ejaculation and permit the dog to step over the arm of the collector and face away as if in a coital lock. The presence of a teaser female often facilitates collection and may improve ejaculate quality. Daily ejaculation for 15 days reduces sperm counts to 25–30% of those in rested animals (Aman, 1986). Studs should therefore not be used for extended periods more frequently than 2- to 4-day intervals.

C. Semen and Sperm

Sperm counts in fertile dogs can range from 22×10^6 to 6000×10^6 (Christiansen, 1984; Chatterjee *et al.*, 1976; Morton and Bruce, 1989). Averages reported include 500×10^6 (Christiansen, 1984) and $900 \times$

10^6 (Morton and Bruce, 1989). Sperm counts under 0.2×10^9 in medium-size dogs are considered subnormal. Expected sperm motility is 75–85%; progressive motility is 70–85%, with reported values varying greatly with methodology and definition of normal. The number of sperm per insemination required for normal fertilization has been estimated to be between 50 and 200×10^6 sperm (Gill *et al.*, 1970). Biochemical analyses of the composition of the separate portions of the ejaculate have been reviewed (Christiansen, 1984). Dog semen has a pH of 6–8. The sperm tolerate a pH of 5–10. Concentrations of sodium and chloride are high and those of calcium, magnesium, and potassium are low. Fructose levels are below 2 mg%, possibly due to the absence of seminal vesicles. Activities of aminotransferases and of phosphatase vary greatly between the sperm-rich fractions and prostatic fluid. Acid phosphatase activity increases with sexual maturity, and along with alkaline phosphatase is an index of prostatic function. These enzymes may affect sperm quality and fertility.

Dog sperm are 55–70 μm long, with 5×7 μm head, 11 μm midpieces, and 50 μm tails (Christiansen, 1984). Motility under 60% is considered abnormal. *In vitro* dog sperm can penetrate immature and degenerate oocytes as well as mature oocytes, and require about 7 h for capacitation (Mahi and Yanagamachi, 1976).

D. Male Hormone Patterns

LH secretion in stud dogs is episodic, with peaks occurring at intervals of 1.5–5 h, each usually followed in 60 min by a peak in testosterone. During the day LH ranges from 0.2 to 11.0 ng/ml. Testosterone ranges from 0.4 to 6.0 ng/ml and averages 2.5 ng/ml (De Palatis *et al.*, 1978; DeCoster *et al.*, 1983). LH secretion is inhibited more by estradiol than by testosterone (Winter *et al.*, 1982). Regulation of LH secretion involves aromatization of androgen to estrogen in the brain and effects of systemic estrogen on the pituitary (Winter *et al.*, 1982; Aman, 1986).

IV. Male Cats

A. Reproductive Organs and Puberty

The testes are descended at birth but may not remain permanently in the scrotum until 4–6 months (Christiansen, 1984). The scrotum is not pendulous as it is in dogs, but protuberant just ventral to the anus. Cats have a prostate and bulbourethral glands, but no seminal vesicles. Tom cats are sexually active year round, at least in the laboratory with 12–14 h of light per day. Puberty occurs at 7–18 months of age but mostly by 8–12 months and about 1–2 months later than queens. Sex behavior may occur at 4 months, initial spermatogenesis at 5 months, mature Leydig cells at 6 months, and fertile ejaculates at 7 months. The penis is conical, located ventral to the scrotum, and directed to the posterior. There is a short, ungrooved os penis. The surface of the penis has 100–200 cornified papillae or spines that appear at around 6 months, are androgen dependent, and regrass after castration (Stabenfeldt and Shille, 1977). The penile spines presumably facilitate vaginal stimulation and reflex ovulation but are not needed for fertile mating. Tom cats are territorial within catteries and laboratory colonies. Breeding success is greater when females are placed in the male's territory or in a designated breeding area familiar to the male.

B. Ejaculates, Semen, and Sperm

Natural intromission and ejaculation are completed in 1–27 s (average 8 s) (Concannon *et al.*, 1989a). Semen can be collected into small tubes via a latex artificial vagina made from a soft, 2-cm^3 pipette bulb. Toms are trained and collected using a teaser queen (Sojka *et al.*, 1970). Electroejaculation under general anesthesia has also been used (Platz and Seager, 1978). Both methods yielded ejaculates containing 15–130 × 10^6 sperm and 60–95% motility (Platz *et al.*, 1978; Goodrowe *et al.*, 1989). Collections by artificial vagina generally yield smaller volumes (34–40 versus 100–223 μl) and greater numbers of sperm (57–61 × 10^6 versus 11–30 × 10^6) compared to electroejaculation (Goodrowe *et al.*, 1989). Cat semen has a pH of 7–8. The osmolality of the seminal plasma is 340 ± 2 mmol/kg. Cat sperm are sensitive to abrupt changes in pH. Motility is irreversibly depressed by urine contamination, and motility after 90 min at 37°C is 0% for undiluted semen and 20% for extended semen (Goodrowe *et al.*, 1989). Volume and sperm number decrease with daily, but not three times a week, manual collection (Sojka *et al.*, 1970). For *in vitro* fertilization cat sperm required about 2 h capacitation in the uterus (Hamner *et al.*, 1970) or *in vitro* (Goodrowe *et al.*, 1988b).

C. Hormone Levels

Testosterone secretion in tom cats is episodic and levels can range from 0.1 to 3.3 ng/ml within a 6-h period. LH secretion is presumably episodic, although variation between males (3–29 ng/ml) was greater than within males (Goodrowe *et al.*, 1985). GnRH doses of 10 μg/cat can be used to test the pituitary–gonadal axis, with sixfold increases in LH at 30 min and peaks in testosterone (6 ± 1 ng/ml) at 60 min.

V. Artificial Insemination, IVF, and Embryo Transfer

A. Fresh Semen AI in Dogs and Cats

In dogs, vaginal insemination with freshly collected semen can yield fertility nearly equal to natural service. In a large study of several breeds, correctly timed fresh semen insemination resulted in 84% fertility (Linde-Forsberg and Forsberg, 1989). All or a portion of the sperm-rich fraction of ejaculate (≥200 × 10^6 sperm) is deposited into the anterior vagina along the dorsal-median fold using a plastic insemination pipet and syringe. Elevation of the hindquarters of the bitch to prevent loss of inseminate and digital stimulation of the genitalia are recommended. Storage of fresh semen requires 1 : 1 or greater dilution in an extender at room temperature and cooling to 5°C. This can provide over 50% motility for 2–8 days using 20% egg-yolk buffered extenders (with glycerol having some suppressive effect), but only for a few hours using skim-milk extenders (Foote, 1964; Concannon and Battista, 1989). Use of an extender is important. Sperm survival at 4°C for 3 days was 87% for extended semen and 18% for raw semen (Morton and Bruce, 1989).

In cats, fresh semen AI has been accomplished with 100 μl semen diluted with saline (Sojka *et al.*, 1970). Insemination of 5–50 × 10^6 sperm at the time hCG (50 IU) was injected to induce ovulation yielded at 50% pregnancy rate. A second insemination at the time of ovulation 24 h later increased success rate to 75%. Fertility rates were reduced to 0–33% when less than 5×10^6 sperm were used.

B. Frozen Semen AI

In dogs, frozen semen artificial insemination has resulted in variable success rates and results within, as well as among, methods (Concannon and Battista, 1988). With vaginal deposition, fertility rates of 40–92% have been reported for semen extended in 11% lactose, 4% glycerol, 20% egg yolk, and frozen in pellets (Platz and Seager, 1977; Lees and Castleberry, 1977). The fertility rate in these studies depends on numbers of sperm ($150–800 \times 10^6$) inseminated, and numbers (three to nine) of inseminations, if not the site of insemination (Concannon and Battista, 1988). Others have used liquid nitrogen vapors to freeze dog semen in straws after extension in a Tris (2–3%)–citric acid (13–17%)–glucose or fructose (8–13%)–egg yolk (15–20%)–glycerol (3–10%) extender. Fertility rates for intravaginal deposition ranged from 0% (Andersen, 1980; Farstad, 1984; Farstad and Andersen-Berg, 1989; Gill et al., 1970) to 25% (Olar, 1984). However, the same semen has yielded 72% fertility when deposited through the cervix into the uterus (Farstad and Andersen-Berg, 1989).

For freezing dog semen in straws, studies on postthaw motility suggest that perhaps Tris-citrate is preferred to lactose extender, that dimethyl sulfoxide (DMSO) cannot improve upon glycerol, that rapid thawing is preferred to slow, and that straws are worse than pellets when lactose extender is used (Battista et al., 1988; Olar, 1984; Yubi et al., 1987). Initial dilution, cold equilibration, and freezing each contributed to the 35%

incidence of postthaw acrosome deformation of dog sperm in Tris-citrate–yolk and frozen straws (Oettle, 1986). Dog sperm appear to exhibit greater postthaw thermolability than other species, and motility approaches 0–11% by 1–4 h (Concannon and Battista, 1989). Therefore the timing and site of insemination are major considerations with frozen dog semen.

Intrauterine insemination of frozen semen has yielded fertility rates of 67–78% using Tris-citrate extenders and manually guided transcervical placement of a metal insemination catheter with an offset tip one to three times during estrus (Andersen, 1975, 1980; Farstad and Andersen-Berg, 1989; Fergusen et al., 1989). In studies using a different extender, rates were 46% with surgical intrauterine deposition via laparotomy twice in estrus (Smith, 1984), and 30% with vaginoscope-guided transcervical placement of plastic catheters (Battista et al., 1988). Timing of inseminations of short-lived frozen-thawed semen is critical. Success appears to require deposition between day 4 and 7 after the LH surge (Smith, 1984; Concannon and Battista, 1989), which can be monitored using daily assay of progesterone (Jeffcoate and Lindsay, 1989) or assay of LH (Madej et al., 1989). Intrauterine insemination on day 6 and again 1–3 days later yielded an 83% pregnancy rate with frozen–thawed dog semen (Fergusen et al., 1989).

In cats, frozen semen AI has also been successful. Semen was extended and cold-equilibrated in glycerolated–lactose–yolk diluent, frozen in pellets, thawed in 37°C saline, and inseminated vaginally in doses containing $50–100 \times 10^6$ motile sperm. Fertility was 11% in cats in natural or induced estrus that ovulated in response to hCG or a vasectomized male (Platz et al., 1978).

C. Embryo Transfer and IVF in Dogs

Healthy, immature cumulus intact dog oocytes removed from ovaries at random stages of the cycle will mature *in vitro* in medium-199, 20% fetal calf serum, 5% CO_2. At 48 and 72 h of culture, 33 and 48%, respectively, had undergone germinal vesicle breakdown while 22 and 25% reached metaphase I or II (Mahi and Yanagamachi, 1976; 1978). Dog oocytes have a high concentration of lipidlike material, and must be cleared and stained to study maturation, pronucleus formation, or syngamy in detail. Dog sperm incubated with mature or immature dog oocytes, in contrast to reports for other species, can penetrate and undergo decondensation in immature dictyate oocytes as well as mature oocytes. Zona pellucida penetration occurred after a 7-h delay for sperm capacitation and vitellis penetration occurred by 24 h. *In vitro* fertilization and culture to syngamy or cleavage has not been reported for dogs. Surgical transfer of uterine embryos from 26 naturally cycling bitches was successful in 3 of 7 naturally cycling recipients, with 4 of 32 transferred embryos resulting in pups (Kraemer *et al.*, 1979). Collections and transfers were done at 10–17 days after onset of estrus with bitches within 4 days of one another.

D. Embryo Transfer and IVF in Cats

In cats, surgical collection of uterine embryos at 6–9 days after hCG injection or mating provided 47 embryos in 17 collections from naturally mated queens. Pregnancy occurred in four of nine sterile mated or hCG-injected recipients (Kraemer *et al.*, 1979). Pregnancies have resulted from embryo transfers following natural estrus or estrus induced with five daily injections of 2 mg FSH and ovulation induced by natural mating and hCG injections in donors and recipients, respectively (Goodrowe *et al.*, 1988a; Schmidt, 1986).

Pregnancies from transfer of frozen-thawed cat embryos have also been reported in 5 of 11 recipient queens (Dresser *et al.*, 1987, 1989). Embryos were collected from cats superovulated with six daily injections of FSH totaling 3–8 mg followed by two daily injections of hCG. Commercial cattle embryo freezing techniques were used. Embryos were thawed at various temperatures and cultured overnight in a 20% fetal calf serum medium prior to transfer into queens in which estrus was induced by 5 days of FSH (0.2 mg sc), 1 day of FSH (0.1 mg) and hCG (750 IU im), and 1 day of hCG (750 IU).

Cat oocytes are opaque, like those of carnivores, but the lipidlike material can be displaced to form a cap by centrifugation (Goodrowe *et al.*, 1989). *In vitro* fertilization rates of 48–80% have been obtained with recently ovulated oviductal ova incubated with sperm capacitated *in utero* (Hamner *et al.*, 1970). Two- or four-cell embryos produced by *in vitro* fertilization resulted in term pregnancies in five of six recipients following transfer into the oviducts (Goodrowe *et al.*, 1988b, 1989). Unovulated oocytes were obtained by laparoscopic aspiration of follicles 25–27 h following an hCG (100–200 IU) injection given 72–84 h after eCG (150 IU), and were fertilized during incubation with swim-up processed sperm for 18–20 h in 5% CO_2 at 37°C. Embryos were transferred into oviducts of recipients after hyaluronidase treatment, another 6–10 h of culture, and examination.

References

Aman, R. P. (1986). *In* "Current Therapy in Theriogenology" (D. Morrow, ed.), pp. 532–538. W. B. Saunders, Philadelphia.

Andersen, A. C. (1970). *In* "The Beagle As An Experimental Animal" (A. C. Anderson, ed.), pp. 31–39. Iowa State University Press, Ames, Iowa.

Andersen, A. C., and Simpson, M. E. (1973). "The Ovary and Reproductive Cycle of the Dog (Beagle)." Geron-X Press, Los Altos, California.

Andersen, K. (1980). *In* "Current Therapy in Theriogenology" (D. Morrow, ed.), pp. 661–665. W. B. Saunders, Philadelphia.

Arbeiter, K., Brass, W., Ballabio, R., and Jochle, W. (1988). *J. Small Anim. Pract.* **29,** 781–788.

Archbald, L. F., Baker, B. A., Clooney, L. L., and Godke, R. A. (1980). *Vet. Med. Small Anim. Clin.* **75,** 228–238.

Arnold, S., Arnold, P., Concannon, P. W., Weilenmann, R., Hubler, M., Casal, M., Dobeli, M., Fairburn, A., Eggenberger, E., and Rusch, P. (1989). *J. Reprod. Fertil. Suppl.* **39,** 115–122.

Arthur, G., Noakes, D., and Pearson, H. (1982). "Veterinary Reproduction and Obstetrics." Bailliere Tindall, London.

Banks, D. H., and Stabenfeldt, G. H. (1982). *Biol. Reprod.* **16,** 603–611.

Banks, D. R., Paape, S. R., and Stabenfeldt, G. H. (1983). *Biol. Reprod.* **28,** 923–932.

Bareither, M. L., and Verhage, H. G. (1981). *Am. J. Anat.* **162,** 107–111.

Battista, M., Parks, J., and Concannon, P. W. (1988). *Proc. 11th Int. Congr. Anim. Reprod. and Artificial Insemination* **3,** 229–231.

Beach, F. A., Dunbar, I. F., and Buehlar, M. G. (1982). *Horm. Behav.* **16,** 414–442.

Bebiak, D. M. (1988). *In* "Proceedings of the Annual Meeting of the Society for Theriogenology, (Orlando)," pp. 167—173. Society for Theriogenology, Hastings, Nebraska.

Bowen, R. A., Olson, P. N., Behrendt, M. D., Wheeler, S. L., Husted, P. W., and Nett, T. M. (1985). *J. Am. Vet. Med. Assoc.* **186,** 783–788.

Burke, T. J. (1986). *In* "Current Therapy in Theriogenology" (D. Morrow, ed.), pp. 528–531. W.B. Saunders, Philadelphia.

Cain, J. L., Cain, G. R., Feldman, E. C., Lasley, B. L., and Stabenfeldt, G. (1988) *Am. J. Vet. Res.* **49,** 1993–1996.

Cain, J. L., Lasley, B. L., Cain, G. R., Feldman, E. C., and Stabenfeldt, G. H. (1989). *J. Reprod. Fertil. Suppl. 39,* 143–147.

Chaffaux, S., Locci, D., Pontois, M., Deletang, F., and Thibier, M. (1984). *Br. Vet. J.* **140,** 191–195.

Chakraborty, P. (1987). *Theriogenology* **27,** 827–840.

Chakraborty, P. K., Panko, W. B., and Fletcher, W. S. (1980). *Biol. Reprod.* **22,** 227–232.

Chatterjee, S. N., Meenakshi Sharma, R. N., and Kar, A. B. (1976). Semen characteristics of normal and vasectomized dogs. *Indian J. Exp. Biol.* **14,** 411–414.

Christiansen, I. J. (1984). "Reproduction in the Dog And Cat." Bailliere Tindall, London.

Christie, D. W., and Bell, E. T. (1972) *Anim. Behav.* **20,** 621–631.

Christie, D. W., Bell, E. T., Horth, C. E., and Palmer, R. F. (1971). *Acta Endocrinol.* **68,** 543–550.

Cline, E. M., Jennings, L. L, and Sojka, N. J. (1980). *Lab. Anim. Sci.* **30,** 1003–1005.

Cochran, R. C., Ewing, L. L., and Niswender, G. D. (1981) *Invest. Urol.* **19,** 142–147.

Colby, E. D. (1970). *Lab. Anim. Care* **20,** 1075.

Colby, E. D. (1980). *In* "Current Therapy in Theriogenology" (D. Morrow, ed.), pp 861–864. W. B. Saunders, Philadelphia.

Concannon, P. W. (1980). *J. Reprod. Fertil.* **58,** 407–410.

Concannon, P. W. (1983). *In* "Current Veterinary Therapy, Small Animal Practice," Vol. VIII (R. W. Kirk, ed.), pp. 886–900. W. B. Saunders Company, Philadelphia.

Concannon, P. W., and Lein, D. H. (1983). *In* "Current Veterinary Therapy, Small Animal Practice" (R. W. Kirk, ed.), Vol. VIII, pp. 901–909. W. B. Saunders, Philadelphia.

Concannon, P. W. (1986a). *In* "Small Animal Reproduction and Infertility" (T. Burke, ed.), pp. 23–77. Lea and Febiger, Philadelphia.

Concannon, P. W. (1986b). *Vet. Clin. North Am. (Small Anim. Pract.).* **16,** 453–475.

Concannon, P. W. (1989). *J. Reprod. Fertil. Suppl.* **39,** 149–160.

Concannon, P. W., and Battista, M. (1989). *In* "Current Veterinary Therapy, Small Animal Practice" (R. Kirk, ed.), pp. 1247–1259. W. B. Saunders Company, Philadelphia.

Concannon, P. W., and Castracane, V. D. (1985). *Biol. Reprod.* **33,** 1078–1083.

Concannon, P. W., and DiGregorio, G. B. (1986). *In* "Small Animal Reproduction and Infertility" (T. Burke, ed.), pp. 96–111. Lea and Febiger, Philadelphia.

Concannon, P. W., and Hansel, W. (1977). *Prostaglandins* **13(3),** 533–542.

Concannon, P. W., and Lein, D. H. (1983). *In* "Current Veterinary Therapy, Small Animal Practice" (R. W. Kirk, ed.), Vol. VIII, pp. 932–936. W. B. Saunders Company, Philadelphia.

Concannon, P. W., and Lein, D. H. (1989). *In* "Current Veterinary Therapy, Small Animal Practice" (R. W. Kirk, ed.), Vol. X, pp. 1269–1282. W. B. Saunders Company, Philadelphia.

Concannon, P. W., and Rendano, V. (1983). *Am. J. Vet. Res.* **44(8),** 1506–1511.

Concannon, P. W., and Temple, M. (1988). *Biol. Reprod.* **36** *(Suppl. 1),* 101.

Concannon, P. W., Hansel, W., and Visek, W. (1975). *Biol. Reprod.* **13,** 112–121.

Concannon, P. W., Powers, M. E., Holder, W., and Hansel, W. (1977a). *Biol. Reprod.* **16,** 517–526.

Concannon, P. W., Hansel, W., and McEntee, K. (1977b). *Biol. Reprod.* **17,** 604–613.

Concannon, P. W., Butler, W. R., Hansel, W., Knight, P. J., and Hamilton, J. M. (1978). *Biol. Reprod.* **19,** 1113–1118.

Concannon, P. W., Cowan, R. G., and Hansel, W. (1979a). *Biol. Reprod.* **20,** 523–531.

Concannon, P. W., Weigand, N., Wilson, S., and Hansel, W. (1979b). *Biol. Reprod.* **20,** 799–809.

Concannon, P. W., Altszuler, N., Hampshire, J., Butler, W. R., and Hansel, W. (1980a). *Endocrinology* **106(4),** 1173–1177.

Concannon, P. W., Hodgson, B., and Lein, D. (1980b). *Biol. Reprod.* **23,** 111–117.

Concannon, P. W., Whaley, S., Lein, D., and Wissler, R. (1983). *Am. J. Vet. Res.* **44(10),** 1819–1821.

Concannon, P. W., Whaley, S., and Anderson, S. P. (1986). *Biol. Reprod.* **34,** 119 (Abstr.).

Concannon, P. W., Weinstein, R., Whaley, S., and Frank, D. (1987). *J. Reprod. Fertil.* **81,** 175–180.

Concannon, P. W., Dillingham, L., and Spitz, I. M. (1988a). *Acta Endocrinol. (Copenh.).* **118,** 389–398.

Concannon, P. W., Isaman, L., Frank, D. A., Michel, F. J., and Currie, W. B. (1988b). *J. Reprod. Fertil.* **84,** 71–77.

Concannon, P. W., Montanez, A., and Frank, D. (1988c). *Proc. 11th Int. Congr. Anim. Reprod. Artificial Insemination* **4,** 427–429.

Concannon, P. W., Lein, D. H., and Hodgson, B. G. (1989a). *Biol. Reprod.* **40,** 1179–1187.

Concannon, P. W., McCann, J. P., and Temple, M. (1989b). *J. Reprod. Fertil. Suppl.* **39,** 3–25.

Concannon, P. W., Yeager, A., Frank, D., and Iyampillai, A. (1990). *J. Reprod. Fertil.* **88,** 99–104.

Conley, A., and Evans, L. (1984). *Proc. 10th Int. Congr. Anim. Reprod. Artificial Insemination* 504. (Abstract).

Dawson, A. B. (1952). *In* "Care and Breeding of Laboratory Animals" (E. J. Farris, ed.), pp. 202–233. John Wiley and Sons, New York.

Dawson, A. B., and Friedgood, H. B. (1940). *Anat. Rec.* **76,** 411–429.

DeCoster, R., Beckers, J. F., Beerens, D., and DeMey, J. (1983). *Acta Endocrinol.* **103,** 473–478.

DePalatis, L., Moore, J., and Falvo, R. E. (1978). *J. Reprod. Fertil.* **52,** 201–207.

Doak, R. L., Hall, A., and Dale, H. E. (1967). *J. Reprod. Fertil.* **13,** 51–58.

Dore, M. A. P. (1989). *J. Reprod. Fertil. Suppl.* **39,** 41–53.

Dresser, B. L., Sehlhorst, C. S., Wachs, K. B., Keller, G. L., Gelwicks, E. J., and Turner, J. L. (1987). *Theriogenology* **28,** 915–927.

Dresser, B. L., Gelwicks, E. J., Wachs, K. B., and Keller, G. L. (1989). *J. Reprod. Fertil. Suppl.* **39,** 332.

Dubiel, A. (1976). In *Proc. VIII Int. Congr. Anim. Reprod., Krakow* **1,** 75.

Edquist, L. E., Johansson, E. D. B., Kasstrom, H., Olsson, S. E., and Richkind, M. (1975). *Acta Endocrinol.* **78,** 554–564.

England, G. C. W., and Allen, W. E. (1989). *J. Reprod. Fertil. Suppl.* **39,** 91–100.

Evans, H. M., and Cole, H. H. (1931). *Memoirs Univ. Calif.* **9(2),** 65–103.

Evans, J. M., and Sutton, D. J. (1989). *J. Reprod. Fertil. Suppl.* **39,** 163–173.

Farstad, W. (1984) *J. Small Anim. Pract.* **25,** 561–565.

Farstad, W., and Andersen-Berg, K. (1989) *J. Reprod. Fertil. Suppl.* **39,** 289–292.

Faulkner, L. C., Pineda, M. H., and Reimers, T. J. (1975). *In* "Immunization With Hormones in Reproduction Research" (E. Neischlag, ed.), pp. 199–214. North Holland, Amsterdam.

Ferguson, J. M., Renton, J. P., Farstad, W., and Douglas, T. A. (1989) *J. Reprod. Fertil. Suppl.* **39,** 293–298.

Fernandes, P. A., Bowen, R. A., Kostas, A. C., Sawyer, H. R., Nett, T. M., and Olson, P. N. (1987). *Biol. Reprod.* **37,** 804–811.

Foote, R. H. (1964). *Cornell Vet.* **54,** 89–97.

Foster, M. A., and Hisaw, F. L. (1935). *Anat. Rec.* **62,** 75–93.

Gentry, P. A., and Liptrap, R. M. (1981). *J. Small Anim. Pract.* **22,** 185–194.

Gerber, J. G., Hubbard, W. C., and Nies, A. S. (1979). *Prostaglandins* **17,** 623–627.

Gerber, J. G., Payne, N. A., Murphy, R. C., and Nies, A. S. (1981). *J. Clin. Invest.* **67,** 632–636.

Gill, H. P., Kaufman, C. F., Foote, R. H., and Kirk, R. W. (1970) *Am. J. Vet. Res.* **31,** 1807–1813.

Glover, T. E., Watson, P. F., and Bonney, R. C. (1985). *J. Reprod. Fertil.* **75,** 145–152.

Gonzalez, A., Allen, A. F., Post, K, Mapletoft, R. J., and Murphy, B. D. (1989). *J. Reprod. Fertil. Suppl.* **39,** 189–198.

Goodrowe, K. L., Chakraborty, P. K., and Wildt, D. E. (1985). *J. Endocrinol.* **105,** 175–181.

Goodrowe, K. L., Howard, J. G., and Wildt, D. E. (1988a). *J. Reprod. Fertil.* **82,** 553–561.

Goodrowe, K. L., Miller, A. M., and Wildt, D. E. (1988b). *Proc. 11th Internat. Cong. Anim. Reprod. Artif. Insemin. 3,* 245.

Goodrowe, K. L., Howard, J. G., Schmidt, P. M., and Wildt, D. E. (1989). *J. Reprod. Fertil. Suppl.* **39,** 73–90.

Goodwin, M., Gooding, K. M., and Regnier, F. (1979) *Science* **203,** 559–561.

Graf, K. J. (1978). *J. Reprod. Fertil.* **52,** 9–14.

Grandage, J. (1972). *Vet. Rec.* **91,** 141–147.

Gunzel, A., Koivisto, P., and Fougner, J. (1986). *Theriogenology* **25,** 559–570.

Hadley, J. C. (1975). *J. Reprod. Fertil.* **44,** 453–460.

Hamner, C. E., Jennings, L. L., and Sojka, N. J. (1970). *J. Reprod. Fertil.* **23,** 477–480.

Herron, M. A. (1977). *Vet. Clin. N. Am.* **7,** 715–722.

Herron, M. A. (1986). *In* "Small Animal Reproduction and Infertility" (T. J. Burke, ed), pp. 13–23. Lea and Febiger, Philadelphia.

Herron, M. A., and Sis, R. F. (1974). *Am. J. Vet. Res.* **35,** 1277–1279.

Holst, P. A., and Phemister, R. D. (1974). *Am. J. Vet. Res.* **35,** 401–406.

Holst, P. A., and Phemister, R. D. (1975). *Am. J. Vet. Res.* **36,** 705–706.

Hurni, H. (1981). *Lab. Anim.* **15,** 229–233.

Jeffcoate, I. A., and Lindsay, F. E. F. (1989). *J. Reprod. Fertil. Suppl.* **39,** 277–287.

Jemmett, J. E., and Evans, J. M. (1977). *J. Small Anim. Pract.* **18,** 31–37.

Jochle, W., Arbeiter, K., Post, K., Ballabio, R., and D'Ver, A. S. (1989). *J. Reprod. Fertil. Suppl.* **39,** 199–207.

Jones, G. E., Boynes, A. R., Cameron, E. H. D., Bell, E. T., Christie, D. W., and Parkes, M. F. (1973). *J. Reprod. Fertil.* **35,** 187–189.

Keister, D. M., Gutheil, R. F., Kaiser, L. D., and D'Ver, A. S. (1989). *J. Reprod. Fertil. Suppl.* **39,** 241–249.

Kraemer, D. C., Flow, B. L., Schriver, M. D., Kinney, G. M., and Pennycook, J. W. (1979). *Theriogenology* **11,** 51–62.

Lees, G. E., and Castleberry, M. W. (1977) *J. Am. Anim. Hosp. Assoc.* **13,** 382–386.

Lein, D. H., and Concannon, P. W. (1983). *In* "Current Veterinary Therapy, Small Animal Practice" (R. W. Kirk, ed.), pp. 936–942. W. B. Saunders, Philadelphia.

Lein, D. H., Concannon, P. W., Hornbuckle, W. E., Gilbert, R. O., Glendening, J. R., and Dunlap, H. L. (1989). *J. Reprod. Fertil. Suppl.* **39,** 231–240.

Leyva, H., Madley, T., and Stabenfeldt, G. H. (1989a). *J. Reprod. Fertil. Suppl.* **39,** 125–133.

Leyva, H., Madley, T., and Stabenfeldt, G. H. (1989b). *J. Reprod. Fertil. Suppl.* **39,** 135–142.

Linde, C., and Karlsson, I. (1984). *J. Small Anim. Pract.* **25,** 77–82.

Linde-Forsberg, C., and Forsberg, M. (1989). *J. Reprod. Fertil. Suppl.* **39,** 299–310.

Lindsay, F. E. F. (1983). *In* "Current Veterinary Therapy, Small Animal Practice" (R. W. Kirk, ed.), Vol. VIII, pp. 912–921. W. B. Saunders, Philadelphia.

Lindsay, F. E. F., and Concannon, P. W. (1986). *In* "Small Animal Reproduction and Infertility" (T. Burke, ed.), pp. 112–120. Lea and Febiger, Philadelphia.

Long, D., Mezza, R., and Krakowa, S. (1978). *Lab. Anim. Sci.* **28,** 178–181.

Madej, A., Linde-Forsberg, C., and Garnum, F. (1989). *J. Reprod. Fertil. Suppl.* **39,** 329 (abstract)

Mahi, C. A., and Yanagamachi, R. (1976). *J. Exp. Zool.* **196,** 189–196.

Mahi, C. A., and Yanagimachi, R. (1978). *Gamete Res.* **1,** 101–109.

Mahi-Brown, C. A., Yanagimachi, R., Hoffman, J. C., and Huang, T. T. F., Jr. (1985). *Biol. Reprod.* **32,** 761–772.

Malassine, A., and Ferre, F. (1979). *Biol. Reprod.* **21,** 965–971.

McCann, J. P., Altszuler, N., Hampshire, J., and Concannon, P. W. (1987). *Acta Endocrinol.* **116,** 73–80.

McCann, J. P., Temple, M., and Concannon, P. W. (1988). *Proc. 11th Int. Congr. Anim. Reprod. Artificial Insemination* **2,** 103–105.

McDonald, L. E. (1989). *In* "Veterinary Endocrinology and Reproduction" (L. E. McDonald, ed.), pp. 460–486. Lea and Febiger, Philadelphia.

Meyers-Wallen, V. N., and Patterson, D. F. (1989). *J. Reprod. Fertil. Suppl.* **39**, 57–64.

Michael, R. P. (1961). *Behaviour* **18**, 1–24.

Morton, D. B., and Bruce, S. G. (1989). *J. Reprod. Fertil. Suppl.* **39**, 311–316.

Moses, D. L., and Shille, V. M. (1988). *J. Am. Vet. Med. Assoc.* **192**, 1541–1545.

Nakao, T., Aoto, Y., Fukushima, S., Moriyoshi, M., and Kawata, K. (1985). *Jpn. J. Vet. Sci.* **47**, 17–24.

Nett, T. M., Akbar, A. M., Phemister, R. D., Holst, P. A., Reichert, L. E., Jr., and Niswender, G. D. (1975). *Proc. Soc. Exp. Biol. Med.* **148**, 134–139.

Oettle, E. E. (1986). *Anim. Reprod. Sci.* **12**, 145–150.

Oettle, E. E., Bertschinger, H. J., Botha, A. E., and Marais, A. (1988). *Theriogenology* **29**, 757–763.

Okkens, A. C., Dieleman, S., Bevers, M., and Willemse, A. (1985a). *Vet. Q.* **7**, 169–172.

Okkens, A. C., Bevers, M., Dieleman, S., and Willemse, A. (1985b). *Vet. Q.* **7**, 173–176.

Okkens, A. C., Dieleman, S. J., Bevers, M. M., Lubberink, A. A. M. E., and Willemse, A. H. (1986). *J. Reprod. Fertil.* **77**, 187–192.

Olar, T. T. (1984). Cryopreservation of dog spermatozoa. Ph.D. Thesis, Colorado State University.

Olson, P. N., Bowen, R. A., and Nett, T. M. (1981). *Proc. 118th Annu. Mtg. Am. Vet. Med. Assoc.* **96** (Abstr.).

Olson, P. N., Bowen, R. A., and Behrendt, M., Olson, J. D., and Nett, T. M. (1982). *Biol. Reprod.* **27**, 1196–1206.

Olson, P. N., Bowen, R. A., Behrendt, M. D., Olson, J. D., and Nett, T. M. (1984a). *Am. J. Vet. Res.* **45**, 119–124.

Olson, P. N., Bowen, R. A., Behrendt, M. D., Olson, J. D., and Nett, T. M. (1984b). *Am. J. Vet. Res.* **45**, 145–148.

Olson, P. N., Bowen, R. A., Behrendt, M. D., Olson, J. D., and Nett, T. M. (1984c). *Am. J. Vet. Res.* **45**, 149–153.

Olson, P. N., Nett, T. N., Bowen, R. A., Sawyer, H. R., and Niswender, G. D. (1989). *J. Reprod. Fertil. Suppl.* **39**, 27–40.

Paape, S. R., Shille, V. M., Seto, H., and Stabenfeldt, G. H. (1975). *Biol. Reprod.* **13**, 470–474.

Paradis, M., Post, K., and Mapletoft, R. (1983). *Can. Vet. J.* **24**, 239–242.

Phemister, R. D., Holst, P. A., Spano, J. S., and Hopwood, M. L. (1973). *Biol. Reprod.* **8**, 74–82.

Pineda, M. H., Kainer, R. A., and Faulkner, L. C. (1973). *Am. J. Vet. Res.* **34**, 1487–1491.

Platz, C. C., and Seager, S. W. J. (1977). *Lab. Anim. Sci.* **27**, 1013–1016.

Platz, C. C., and Seager, S. W. J. (1978). *J. Am. Vet. Med. Assoc.* **173**, 1353–1355.

Platz, C. C., Wildt, D. E., and Seager, S. W. J. (1978). *J. Reprod. Fertil.* **52**, 279–282.

Reimers, T., Phemister, R., and Niswender, G. (1978). *Biol. Reprod.* **19**, 673–679.

Romsos, D., Palmer, H., Muiruri, M., and Bennink, M. (1981). *J. Nutr.* **3**, 678–689.

Schmidt, P. M. (1986). *Vet. Clin. North Am.* **16**, 435–452.

Schmidt, P. M., Chakraborty, P. K., and Wildt, D. E. (1983). *Biol. Reprod.* **28**, 657–671.

Scott, P. P. (1970). *In* "Reproduction and Breeding Techniques for Laboratory Animals" (E. S. E. Hafex, ed.), pp. 192–208. Lea and Febiger, Philadelphia.

Scott, P. P., and Lyoyd-Jacob, M. A. (1955). Some interesting features in the reproductive cycle of the cat. *In* "Studies in Fertility," (R. G. Harrison, ed.), Chap. 12. Lea and Febiger, Philadelphia.

Shille, V. M., and Stabenfeldt, G. H. (1979). *Biol. Reprod.* **21**, 1217–1223.

Shille, V. M., Lundstrom, K. E., and Stabenfeldt, G. H. (1979). *Biol. Reprod.* **21**, 953–963.

Shille, V. M., Munro, C., Farmer, S. W., Papkoff, H., and Stabenfeldt, G. H. (1983). *J. Reprod. Fertil.* **68**, 29–39.

Shille, V. M., Thatcher, M. J., and Simmons, K. J. (1984). *Am. Vet. Med. Assoc.* **184**, 1469–1473.

Shille, V. M., Thatcher, M. J., Lloyd, M. L., Miller, D. D., Seyfert, D. F., and Sherrod, J. D. (1989). *J. Reprod. Fertil. Suppl.* **39**, 103–113.

Siegel, E. T. (1977). "Endocrine Diseases of the Dog." Lea and Febiger, Philadelphia.

Smith, F. O. (1984). Cryopreservation of canine semen: Technique and performance. Ph.D. Thesis, University of Minnesota.

Smith, M. S., and McDonald, L. E. (1974). *Endocrinology* **94**, 404–412.

Sojka, N. J., Jennings, L. L., and Hamner, C. E. (1970). *Lab. Anim. Care* **20**, 198–204.

Sokolowski, J. (1974). *Lab. Anim. Sci.* **21**, 696–699.

Stabenfeldt, G. H., and Shille, V. M. (1977). *In* "Reproduction in Domestic Animals" (H. H. Cole, and P. T. Cupps, eds.), pp. 499–527. Academic Press, New York.

Steinetz, B., Goldsmith, L., and Lust, G. (1987). *Biol. Reprod.* **37**, 719–725.

Steinetz, B. G., Goldsmith, L. T., Harvey, H. J., and Lust, G. (1989). *Am. J. Vet. Res.* **50**, 68–71.

Stewart, D. R., and Stabenfeldt, G. H. (1985). *Biol. Reprod.* **32,** 848–854.

Thun, R., Watson, P., and Jackson, G. L. (1977). *Am. J. Vet. Res.* **38,** 483–486.

Tsutsui, T. (1975). *Jpn. J. Anim. Reprod.* **21,** 98–101.

Tsutsui, T. (1989). *J. Reprod. Fertil. Suppl.* **39,** 269–275.

Uvnas-Moberg, K., Stock, S., Eriksson, M., Linden, A., Einarsson, S., and Kunavongkrit, A. (1985). *Acta Physiol. Scand.* **124,** 391–398.

Van der Stricht, O. (1923). *Arch. Biol.* **33,** 231–300.

van der Weyden, G., Taverne, M., Okkens, A., and Fontijne, P. (1981). *J. Small Anim. Pract.* **22,** 503–510.

van der Weyden, G. C., Taverne, M. A. M., Dieleman, S. J., Wurth, Y., Bevers, M. M., and van Oord, H. A. (1989). *J. Reprod. Fertil. Suppl.* **39,** 211–224.

Vanderlip, S., Wing, A., Linke, D., Rivier, J., Concannon, P. W., and Lasley, B. (1987). *Lab. Anim. Sci.* **37(4),** 459–464.

van Haaften, B., Dieleman, S. J., Okkens, A. C., Bevers, M. M., and Willemse, A. H. (1989). *J. Reprod. Fertil. Suppl.* **39,** 330–331 (abstr.).

Verhage, H. G., Beamer, N. B., and Brenner, R. M. (1976). *Biol. Reprod.* **14,** 579–585.

Vickery, B. H., McRae, G. I., Goodpasture, J. C., Sanders, L. M. (1989). *J. Reprod. Fertil. Suppl.* **39,** 175–187.

Voith, V. E. (1980). *In* "Current Therapy in Theriogenology" (D. Morrow, ed.), pp. 839–843. W. B. Saunders, Philadelphia.

Wallace, S. S., Mahaffey, M. B., Miller, D. M., and Thompson, F. N. (1989). *J. Reprod. Fertil. Suppl.* **39,** 331 (abstr.).

Wheaton, L. G., Pijanowski, G. J., Weston, P. G., and Burke, T. J. (1988). *Am. J. Vet. Res.* **49,** 82–86.

Wildt, D. E., and Seager, S. W. J. (1980). *In* "Current Therapy in Theriogenology" (D. Morrow, ed.), pp. 828–832. W. B. Saunders, Philadelphia.

Wildt, D. E., Guthrie, S. C., and Seager, S. W. J. (1978a). *Horm. Behav.* **10,** 251–257.

Wildt, D. E., Kinney, G. M., and Seager, S. W. J. (1978b). *Lab. Anim. Sci.* **28,** 301–307.

Wildt, D. E., Panko, W. B., Chakraborty, P., and Seager, S. W. (1979a). *Biol. Reprod.* **20,** 648–658.

Wildt, D. E., Panko, W. B., and Seager, S. W. J. (1979b). *Prostaglandins* **18,** 883–892.

Wildt, D. E., Seager, S. W. J., and Chakraborty, P. K. (1980). *Endocrinology* **107,** 1212–1217.

Wildt, D. E., Chan, S., Seager, S. W. J., and Chakraborty, P. (1981). *Biol. Reprod.* **25,** 15–28.

Winter, M., Pirmann, J., Falvo, R. E., Schanbacher, B. D., and Miller, J. (1982). *J. Reprod. Fertil.* **64,** 449–455.

Wright, P. J. (1982). *Aust. Vet. J.* **59,** 123–124.

Yubi, A. C., Ferguson, J. M., Renton, J. P., Harter, S., Harney, M. J. A., Bagyenji, B., and Douglas, T. A. (1987). *J. Small Anim. Pract.* **28,** 753–761.

CHAPTER 17

Reproduction in Poultry

JANICE M. BAHR and PATRICIA A. JOHNSON

I. Introduction

The reproductive physiology of the chicken is fascinating because unlike the extended estrous cycle of the mammal, the events of the ovulatory cycle of the chicken are compressed into a number of hours. The alternations between the follicular and luteal phases that occur during the mammalian reproductive cycle are absent in the chicken. Whereas the ovarian steroid and pituitary hormones are similar to those of the mammal, the signaling mechanism by which the ovarian steroid hormones control the hypothalamic–hypophyseal system in the chicken may be different from that of the mammalian system. This relationship, which is well understood in mammals, requires additional study in the chicken. In this chapter we will present the anatomy and physiology of the reproductive system of the female and male chicken. We will also discuss briefly molting and changes in reproduction with age.

II. The Female Reproductive System

A. Anatomy

In the chicken, unlike the mammal, only the left ovary and oviduct are functional. In the embryo, both right and left ovary and oviduct develop initially. However, the production of the Müllerian inhibiting substance by the ovary results in the regression of the right duct but not the left duct. The left duct is apparently spared because it has a higher number of estrogen receptors and is thus more responsive to estrogen than the right duct. It appears that estrogen suppresses the negative action of the Müllerian inhibiting substance (Hutson et al., 1983).

The left ovary is attached to the mesovarian ligament, ventral to the aorta and cranial to the kidney. The avian ovary differs morphologically from the mammalian ovary in that four to six preovulatory follicles are arranged in a distinct hierarchy and attached to the ovary by follicular stalks.

The ovary also contains numerous small yellow follicles (SYF; 4–10 mm in diameter), large white follicles (LWF; 2–4 mm in diameter), and small white follicles (SWF; less than 2 mm in diameter) and several postovulatory follicles (POF) (Fig. 1). A rich extensive vascular system supplies the ovary and its numerous follicles. The neural components of the ovary have been examined most thoroughly by Gilbert (1965, 1968, 1969), Freedman (1968), and Dahl (1970a,b). The ovary is well innervated with a neural supply derived from an extensive network of ganglia, nerve cells, and nerves lying adjacent to and within the ovarian stalk. Both cholinergic and adrenergic nerves are associated with the blood vessels

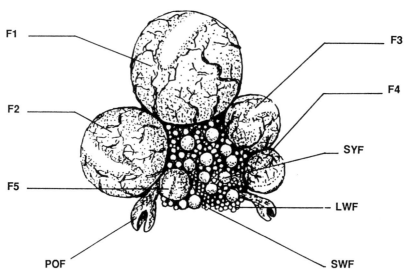

Figure 1   Ovary of the chicken. The preovulatory follicles of the hierarchy are identified according to size with the F_1 follicle being the largest follicle and the next follicle to ovulate, followed by the F_2 follicle, the second largest follicle, etc. The postovulatory follicle (POF) is a saclike structure containing all the cell layers present in the preovulatory follicle. Small follicles are classified according to size: small yellow follicle (SYF; 4–10 mm in diameter), large white follicles (LWF; 2–4 mm in diameter), and small white follicles (SWF; less than 2 mm in diameter).

and smooth muscles of the medulla and cortex of the ovary. Nerves extend to the theca layer of developing and mature follicles.

In the immature ovary, there are thousands of oocytes. At the time of sexual maturity (18–20 weeks of age), four to six of these oocytes increase in diameter and the follicular hierarchy is established. During the subsequent ovulatory cycles, only the largest follicle will ovulate, followed on successive days by the second, third, and fourth largest follicle, each of which enlarges to assume the size of its ovulated predecessor. It is not known what controls this follicular hierarchy or how and why certain follicles become part of the hierarchy while other follicles will mature in subsequent cycles. It has been postulated that certain follicles are selected to grow because of their proximity to the vascular system, which provides hormones, nutrients, and lipovitellin.

The follicle consists of an oocyte and the surrounding layers, namely, the vitelline membrane and zona radiata (innermost layer), perivitelline membrane, granulosa cells, basal lamina, theca interna and external layers, loose connective tissue, and epithelium (Fig. 2). Smooth muscle bundles are present only in the part of the follicle where it attaches to the ovary. Nerves and blood vessels penetrate the theca layer, whereas the granulosa layer lacks both neural and

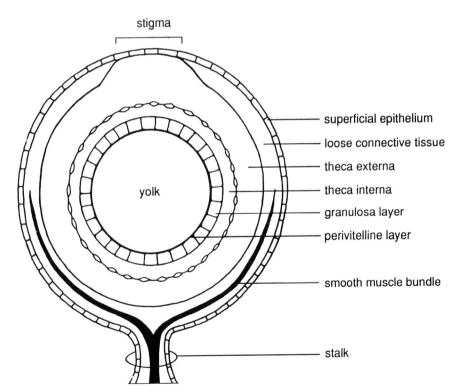

Figure 2 Anatomical structure of the preovulatory follicle.

vascular components (Dahl, 1970a,b; Gilbert, 1965). Through electron microscopy studies, Dahl (1970a,b) has identified afferent terminals of both sympathetic (adrenergic) and parasympathetic (cholinergic) nerves in membranous contact with "steroid-producing" cells. As the follicle increases in size, the stigma region becomes visible on the follicle. The stigma region, a specialized area of the follicle where the split occurs at ovulation, is a pale band approximately 2–3 mm wide and is supplied with very few veins and arteries (Nalbandov and James, 1949). Also during the development of the follicle, there is a very rapid accumulation of yellow yolk (Gilbert, 1971a, 1971b) during the 7–11 days before ovulation.

The oviduct, a long tortuous tube, extends from the ovary to the cloaca. The oviduct consists of five segments each having a separate function (Fig. 3). The segments are the infundibulum, magnum, isthmus, shell gland, and vagina. The ovum spends varying amounts of time in each segment of the oviduct, that is, 0.25–0.5 h in the infundibulum, 2.0–3.0 h in the magnum, 1.25 h in the isthmus, and 18–20 h in the uterus (Warren, and Scott, 1935). The oviduct has a rich nerve supply from both sympathetic and parasympathetic divisions, with the shell gland and shell gland–vaginal junction being most densely innervated.

The infundibulum, located at the anterior end of the oviduct, has a funnel-shaped opening. At the time of ovulation, the infundibulum becomes very active as it tries to engulf the ovum. These wavelike movements are caused by vascular engorgement and muscular contractions of the infundibulum that are probably under neural and hormonal control. Whenever these mechanisms fail, the freshly ovulated ovum is de-

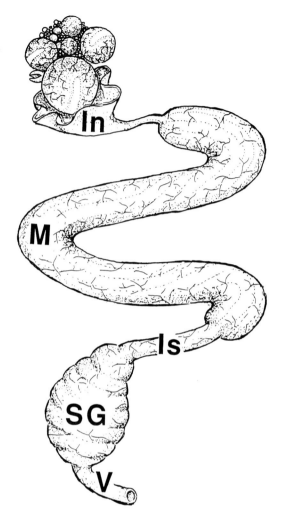

Figure 3 Oviduct of the chicken. In, infundibulum; M, magnum; Is, isthmus; SG, shell gland; V, vagina.

posited in the abdominal cavity (internal laying) and is gradually reabsorbed.

The magnum, the longest part of the oviduct, is very conspicuous because of its dull white color. The thick wall of the magnum contains glandular tissue, which secretes copious amounts of albumen as the ovum moves through the magnum in a rotating motion.

The isthmus, the next section of the chicken's oviduct, secretes the shell membranes around the ovum, which has been coated with albumen. One of the principal constituents of the shell membranes is ovokeratin (Romanoff and Romanoff, 1949).

The shell gland is a large expanded part of the oviduct. The egg is detained approximately 20 h or more in the shell gland as the shell is laid down. Porphyrins, which are secreted by the shell gland epithelium, cause a distinct egg coloring in some avian species.

The vagina is a short tube that probably has no role in the formation of an egg. Sperm deposited by either natural mating or artificial insemination are stored in the vaginal glands, which are short, simple tubules near the shell gland–vaginal junction.

B. Secretion of Albumen and Shell Formation

For extensive reviews of the chemical nature and biological synthesis of egg albumen, see Baker (1968) and Feeney and Allison (1969).

Egg albumen contains numerous proteins. Lush (1961) identified 19 major components of egg whites. Some of these proteins possess bacteriocidal activity. The biological function of other proteins is still obscure. Feeney and Allison (1969) have suggested that some proteins are found in the albumen to provide bulk material and to ensure correct amino acid milieu for embryonic development. Ovalbumin, the most abundant protein in egg white (54%), contains all essential amino acids. Other principal proteins are ovotransferrin (13%), which binds polyvalent metals, ovomucoid

(11%), an inhibitor of proteases, and lysozyme (4%), an enzyme.

Whereas yolk proteins are formed in the liver and subsequently transported to the yolk, albumen is synthesized in the oviductal tissue (O'Malley et al., 1968). Though the oviduct may store albumen for 2 days, approximately 45% of the albumen can be readily produced as the egg passes through the magnum (Fertuck and Newstead, 1970; Smith et al., 1959).

Albumen formation is under hormonal control. Gilbert (1971a,b) suggests three different possibilities for the regulation of albumen secretion: (1) direct mechanical stimulation by the ovum as it passes through the oviduct, (2) a humoral agent, and (3) a neural coordinating mechanism. There is considerable evidence for a direct mechanical stimulation of the oviduct by the ovum because albumen can be secreted around foreign objects placed in the oviduct.

In the formation of a shell, two fibrous membranes are first laid down by the isthmus. Then the outer shell, which consists almost totally of calcium carbonate, is secreted by the shell gland. At the same time there is watery secretion by the shell gland, which "plumps" the eggs. During shell formation, Ca^{2+} and other inorganic components of the shell are actively removed from the blood (Schraer et al., 1965). The formation of the shell and also the egg, which is almost a daily occurrence during the ovulatory cycle, places tremendous metabolic demands on the chicken.

At this time, it would be informative to point out that many of the activities involved in egg formation are under hormonal control, specifically the ovarian steroid hormones. Estrogen has several important functions in the metabolic economy of the laying chicken. Estrogen has a regulator role in Ca^{2+} metabolism by altering directly or indirectly the activities of the enzymes in the biosynthetic pathway of vitamin D. When laying commences, about 2–3 g $CaCO_3$ is required daily for the building of the shell. This Ca^{2+} is mobilized from the gut and from the bone. Equally important is the action of estrogen on the liver (and to some extent perhaps on the fat deposits), where the lipoproteins essential for yolk formation are synthesized and transported through the peripheral circulation to the growing follicles. Because the follicle is one of the fastest growing biological structures (it grows from a weight of micrograms to a weight of about 24 g in 8 days), the amount of lipoprotein deposited in the growing follicle may reach 2 g or more in 24 h. The amount of lipoprotein mobilized by estrogen and present in the blood makes the laying chicken have the highest level of hyperlipemia known in any animal. Estrogen is also essential for the development of oviducts in the growing pullet. These estrogen-built and maintained secretory glands when acted upon by progesterone cause the magnum to secrete ovalbumen.

C. Ovulatory Cycle

1. Sequence

The ovulatory cycle of the domestic chicken is characterized by the sequence or clutch. A sequence consists of a number of days on which an egg is laid, followed by a pause day. Sequences may range in length from as few as 1 or 2 eggs to over 200 eggs. When a chicken first comes into production and begins to lay eggs at about 20 weeks of age, very long sequences occur. After several months of production, the sequence length often decreases. On a conventional lighting schedule of 14 h light and 10 h dark with lights off during the night, the first egg of the sequence is generally laid early in the morning (Fig. 4). The first egg of the sequence is termed C_1 and the succeeding eggs are named accordingly (C_2, C_3, etc.). Eggs on succeeding days are laid at progressively later times of the day throughout the sequence until the last egg is laid quite late in the day and then a pause day intervenes. In very long sequences, the interval between ovipositions averages 24 h but in shorter sequences, the interval is approximately 26–27 h. The deviation from a pattern of 24-h laying has been termed "lag" by Fraps (1954). This unusual laying pattern is the case despite the fact that the total light/dark cycle is 24 h. We still do not know the cause for the restriction of egg laying to the daylight hours, but information about the control of hormones involved in ovulation will be presented later in the chapter.

Each oviposition (laying of the egg) in the chicken is preceded approximately 24–26 h by an ovulation. The oviposition is the outward sign of the previous ovulation and is the basis for determination of rate of egg production. Ovulation generally occurs 15–45 min after an oviposition, except for the last oviposition of the sequence, when ovulation is delayed until the next day (Warren and Scott, 1935).

When chickens are placed under constant lighting conditions, with no other cues such as regular times of feeding and watering, they will lay eggs at random throughout the day (Warren and Scott, 1936). That is, chickens will continue to lay at approximately 24- to 26-h intervals but they will not be cued by the light/dark cycle. Therefore, egg laying will not be restricted to the daylight hours as in conventionally raised chickens. As such, eggs will not be restricted to

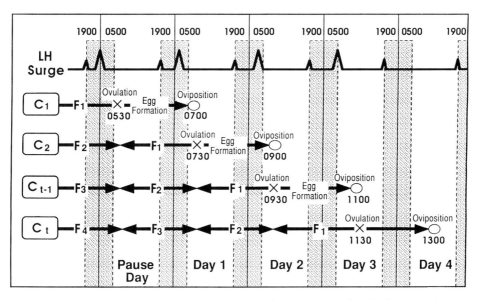

Figure 4 Diagrammatic description of events during the ovulatory cycle of a chicken laying a sequence of four eggs. The shaded area indicates the time of lights off. C_1 (first egg laid), C_2 (second egg laid), C_{t-1} (third egg laid) and C_t (terminal egg laid) refer to the position of the egg in the sequence. LH concentrations in the blood are elevated slightly at the time of lights off (crepuscular) followed by the LH surge. The F_1 follicle ovulates 6 h later and is oviposited approximately 26.5 h later. The LH surge occurs at a later time each night until eventually no LH surge occurs. Consequently, ovulation and oviposition do not occur. The day no egg is laid is called pause day.

sequences. Warren and Scott (1936) also used differences in light intensity to show that darkness did not of itself restrict the laying pattern. They demonstrated that chickens can lay eggs at any time during the day and that the dark phase merely served as a daily cue to phase the ovulatory rhythm. Other environmental cues such as temperature fluctuations (Bhatti and Morris, 1977) and feeding rhythms (McNally, 1947) can also phase the laying pattern. Under constant light, it was found that laying occurred during the warmer part of the day when temperature rhythms were used to cue the chicken or during the time of feeding when regular feeding cycles were used.

2. Control of the Reproductive Cycle by Ovarian Steroids and Pituitary Hormones

Many of the endocrine events associated with follicular maturation and ovulation have been defined. With the technique of radioimmunoassay, it has been determined that preovulatory peaks of estrogen (E), progesterone (P_4), and luteinizing hormone (LH) occur simultaneously approximately 4–6 h prior to ovulation (Johnson and van Tienhoven, 1980a,b; Wilson and Sharp, 1973) (Fig. 5). A preovulatory peak of testosterone occurs about 2–4 h prior to the peaks of the other hormones (Etches and Cunningham, 1977). In addition, a small

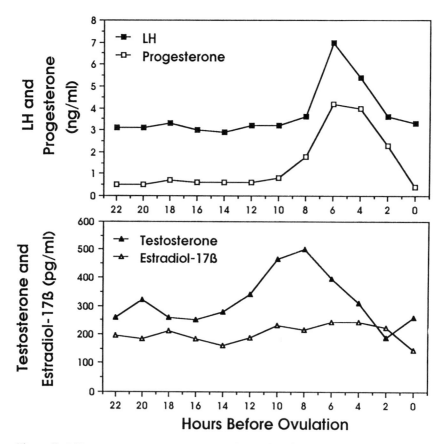

Figure 5 LH, progesterone, testosterone, and estradiol-17β concentrations in blood of chicken during the ovulatory cycle. Values were obtained from Johnson and van Tienhoven (1980a).

peak of corticosterone occurs at the time of ovulation (Beuving and Vonder, 1981) and a large peak at oviposition (Beuving and Vonder, 1977; Johnson and van Tienhoven, 1981).

The steroid hormones have been investigated with respect to their role in the initiation of ovulation. Progesterone is the primary ovarian steroid responsible for inducing an LH surge and ovulation in the chicken (Etches and Cunningham, 1976;

Fraps, 1955; Johnson and van Tienhoven, 1984; Lang et al., 1984). Convincing proof that progesterone is the most important steroid came from experiments using a drug (aminoglutethimide) that blocks steroid hormone synthesis. In the presence of this drug, it was shown that progesterone is capable of inducing a preovulatory surge of LH and ovulation in the absence of any increase of testosterone or estrogen (Johnson and van Tienhoven, 1984). Estrogen is important in priming of the hypothalamus–pituitary system (Wilson and Sharp, 1976) and in the formation of the egg.

Earlier bioassays had detected three peaks of LH in the blood of the chicken during an ovulatory cycle (Bullock and Nalbandov, 1967; Nelson *et al.*, 1965). These peaks occurred at approximately 8, 14, and 20 h prior to ovulation. With radioimmunoassay and serial blood sampling, one main preovulatory peak was detected at approximately 4–7 h before ovulation (Furr *et al.*, 1973; Johnson and van Tienhoven, 1980a,b; Wilson and Sharp, 1973). In addition, a small rise in LH (a crepuscular peak) was detected at the time of lights-off on nights when an ovulation occurred as well as on nights when no ovulation occurred (Johnson and van Tienhoven, 1984; Wilson and Sharp, 1973). It was proposed that this crepuscular rise in LH may initiate the series of events that results in the preovulatory surge of LH and ovulation (Johnson and van Tienhoven, 1984; Scanes *et al.*, 1978). It is interesting that the secretion of LH in the female chicken does not occur in pulses, whereas distinct LH pulses have been measured in the male chicken (Wilson and Sharp, 1973). Currently there is no explanation for this unique difference in the pattern of LH secretion between the female and male chicken and the mammal.

There have been few reports concerning the role of circulating avian follicle-stimulating hormone (FSH), primarily because of the poor availability of pure avian FSH. One study that reported blood levels of FSH in chickens showed an increase of FSH at approximately 14–15 h prior to ovulation (Scanes *et al.*, 1977a). It is assumed that FSH is involved in follicular selection and growth in the chicken because of the presence of FSH receptors and a responsive FSH ade-

nylyl cyclase system in the ovary (Calvo and Bahr, 1983; Ritzhaupt and Bahr, 1987). However, the precise role of FSH in preovulatory events and ovulation needs to be elucidated.

Similar to mammals, the avian gonadotropins are controlled by gonadotropin-releasing hormones (GnRH). In fact, two forms of GnRH have been isolated from the chicken hypothalamus (King and Millar, 1982; Miyamoto *et al.*, 1984). Both GnRH I and II stimulate the release of LH and FSH in chickens (Hattori *et al.*, 1986).

The role of prolactin in chickens has not received much attention. This is in contrast to the situation in turkeys that develop broodiness, which is dependent on prolactin. Commercial laying chickens are quite refractory to the development of broodiness. However, prolactin has been shown to inhibit gonadotropin-stimulated ovulation in the chicken (Scanes *et al.*, 1977b; Tanaka *et al.*, 1971), and a decrease in plasma prolactin has been found before and during the preovulatory LH surge (Scanes *et al.*, 1977b). Some inhibitory effects of prolactin on estrogen production at the ovarian level have been demonstrated (Zadworny *et al.*, 1989). Further work needs to be done to determine the role of prolactin at the ovary and centrally in the chicken.

3. Theories of the Ovulatory Cycle

As mentioned above, the approximate 26-h ovulatory cycle of the chicken is rather unusual. Most biological rhythms with lengths between 22 and 26 h become entrained to a 24-h rhythm if given appropriate cues (Saunders, 1977). In contrast, the laying hen defies these cues and maintains an ovulatory rhythm greater than 24 h. Several hypotheses have been proposed to explain the unusual timing of the ovulatory cycle (Bastian and Zarrow, 1955; Fraps, 1954; Nalbandov, 1959). These theories relate to daily or ovulatory-related changes at the level of the hypothalamus or ovary that may interact to cause ovulation. More recent hypotheses have focused on the possible role of a daily rhythm in corticosterone (Wilson and Cunningham, 1980) to set the phase of the ovulatory rhythm or the crepuscular peak of LH to initiate progesterone secretion from the ovary (Fraps, 1954; Johnson and van Tienhoven, 1984; Scanes *et al.*, 1978). The theory of Fraps (1954) that there is a daily rhythm in the sensitivity of the hypothalamus to progesterone feedback has received the most attention. Although this theory has been neither proven nor disproven and we have learned much about the endocrinology of ovulation in the chicken through the testing of the various hypotheses, we still have no conclusive explanation for the unusual ovulatory pattern.

D. Ovulation

Approximately 6 h after the LH surge and 15–45 min after oviposition, ovulation occurs in the chicken (Fig. 6). The biochemical events induced by LH that alter the structural components of the follicular wall to result in rupture and subsequent expulsion of the egg are not fully understood. The stigma region, the area of the follicular wall that ruptures, is anatomically different from the nonstigma part of the follicular wall. The stigma region is composed of epithelium, theca externa and interna, and granulosa layers (see Fig. 1; Fujii and Yoshimura 1979). The theca externa, which comprises most of the stigma region, consists of fibroblasts and an extracellular matrix of collagen fibers and ground substance. It has been postulated that the stigma region has lower tensile strength because collagen fibers run parallel with the axis of the stigma region in ordered bundles, whereas collagen fibers are extensively intertwined in the nonstigma region (Fujii and Yoshimura, 1979). Prior to ovulation, the stigma region increases in width and becomes transparent. The collagen in the theca externa changes from dense, ordered bundles to loose, widely dispersed fibrils with an increase in the intercellular space.

Figure 6 Ovulation of the F_1 follicle in the chicken. The stigma region, an anatomically unique region of the follicular wall, ruptures. The yolk surrounded by the perivitelline layer and containing the oocyte is engulfed by the infundibulum.

Structural alterations are restricted to the stigma region (Fujii *et al.*, 1980).

Recent evidence suggests a role for proteolytic enzymes and collagenase in the dissociation of collagen fibers prior to ovulation in the domestic chicken (Fuji *et al.*, 1981; Ogawa and Goto, 1984). Follicular collagenase activity increases with follicular maturation (Fuji *et al.*, 1981), and the theca layer from the stigma region produces increased amounts of LH-stimulable collagenase activity *in vitro* (Ogawa and Goto, 1984). The dispersion of follicular-wall collagen fibers into individual fibrils may be due in part to a change in the matrix or ground substance that holds the fibrils together (Yoshimura and Koga, 1982). The biochemical and structural changes that occur in the stigma region of the follicle of the chicken before ovulation are similar to those changes observed in the mammalian follicle (Bjersing and Cajander, 1974; Espey, 1967, 1974; Okamura *et al.*, 1980). The chicken is an excellent animal model to investigate ovulation because of the size of the preovulatory follicle and the temporal link between oviposition and ovulation.

E. Oviposition

Approximately 24–26 h after ovulation and after the egg has been formed in the oviduct, oviposition occurs. The chicken displays distinctive behavioral patterns around the time that she will lay the egg. These include pacing, nesting behavior, and calling. The hormonal control of oviposition has received a good deal of research attention and certain aspects are quite well understood. Prostaglandins and hormones of the posterior pituitary gland are the most important hormonal substances involved in oviposition.

The two main types of prostaglandins involved in oviposition are $PGF_{2\alpha}$ and prostaglandins of the E series (PGE_1 and PGE_2). Several investigators (Hertelendy *et al.*, 1974; Shimada and Asai, 1979) have demonstrated that injections of these prostaglandins into the chicken cause premature oviposition and an increase in muscular contractions of the shell gland. The drug indomethacin is an inhibitor of prostaglandin synthesis. Therefore, in the presence of this drug, the amount of prostaglandins present in the circulation should be reduced. It has been found that injection of indomethacin before a mid-sequence oviposition causes a delay in oviposition by decreasing prostaglandins and thereby the contractions of the shell gland necessary for oviposition (Day and Nalbandov, 1977; Shimada *et al.*, 1986; Shimada and Asai, 1978). In addition to this evidence for the role of prostaglandin in oviposition, it has been found that the concentration of a metabolite of $PGF_{2\alpha}$ increases in the blood around the time of oviposition (Olson and Hertelendy, 1981).

The original observation that removal of the postovulatory follicle resulted in a delay of oviposition suggested that the postovulatory follicle may have a role in oviposition (Rothchild and Fraps, 1944). Subsequent experiments have confirmed these findings and have demonstrated that the largest preovulatory follicle also has a role in oviposition (Tanaka, 1976; Tanaka *et al.*, 1987).

The main posterior pituitary (neurohypophyseal) hormone involved in oviposition in the chicken is arginine vasotocin. This hormone is released from the posterior pituitary, and similar to its mammalian counterpart, oxytocin, causes contraction of the shell gland. It has been found that the increase in shell-gland contractions associated with the increase in prostaglandins causes an increase in the secretion of arginine vasotocin and thereby further increases contractions of the shell gland and expulsion of the egg (Shimada *et al.*, 1987). All of the mechanisms involved in the release of arginine vasotocin in the chicken have not been identified. A summary of physiological events associated with follicular growth and ovulation, albumen secretion, and shell formation and oviposition is presented in Figure 7.

F. Ovarian Function

1. Steroidogenesis and Its Regulation

One of the major functions of the ovary is the production of steroids that are essential for the growth and function of the reproductive system. The major steroids in the chicken are progesterone, androgens, and estrogens. Several specific actions of these steroids in the chicken are the following: progesterone—albumen secretion, induction of the LH surge; androgens—secondary sex characteristics (comb, wattles); estrogens—yolk synthesis by the liver, calcium mobilization from medullary bones to the shell gland. These steroids are produced by the small follicles (less than 10 mm in diameter) that have not entered the hierarchy, the preovulatory follicles of the hierarchy, and the postovulatory follicle. The small follicles, which have been investigated less extensively, are the primary producers of estrogens. According to Armstrong (1982, 1984) and Robinson and Etches (1986), these small follicles produce over 50% of the estrogens secreted by the ovary. Apparently, these small follicles produce a con-

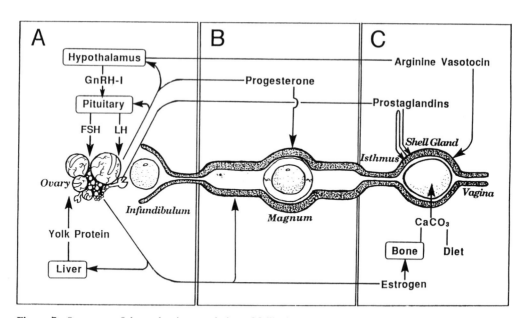

Figure 7 Summary of the endocrine regulation of follicular growth and ovulation (A), albumen secretion (B), and shell formation and oviposition in the chicken (C). Note: Panels A, B, and C represent three separate events in that only one yolk is present in the reproductive tract at one time.

stant supply of estrogen, which is essential for egg formation. The preovulatory follicles secrete progesterone, androgens (androstenedione, testosterone), and estrogens (estrone and estradiol-17β). In contrast to the mammal, the granulosa cells are the primary source of progesterone and small amounts of androgens, whereas the theca cells produce androgens and estradiol-17β.

As the follicle progresses from a small preovulatory follicle (fifth largest, F_5) to the largest (F_1), the granulosa layer produces ever-increasing amounts of progesterone, whereas the theca layer synthesizes decreasing amounts of androgens and estradiol-17β (Fig. 8) (Bahr et al., 1983; Etches and Duke, 1984). Following ovulation, a postovulatory follicle is formed that contains functional granulosa cells, which secrete progesterone. Unlike the mammal, granulosa cells do not luteinize because there is no need for a corpus luteum, a structure associated with pregnancy.

The regulation of steroidogenesis in the chicken is regulated by a number of hormones. The pituitary hormones, follicle-stimulating hormone (FSH) and luteinizing hormone (LH), are necessary and provide a macrocontrol system, whereas at the level of the ovary locally produced hormones, that is, growth factors and vasoactive intestinal peptide, may provide a microcontrol system to modulate the actions of the gonadotropins. The granulosa cells of smaller preovulatory follicles are the primary target of FSH as indicated by in vitro and in vivo studies. Granulosa cells of small preovulatory follicles (F_5–F_3 follicles) have a larger number of FSH receptors, have a more responsive FSH adenylyl cyclase system, and produce more progesterone in response to FSH than do granulosa cells of the larger preovulatory follicles (F_2–F_1) (Calvo and Bahr, 1983; Hammond et al., 1981; Ritzhaupt and Bahr, 1987). In contrast, the granulosa cells of the larger preovulatory follicles are primarily regulated by LH (Calvo et al., 1981; Hammond et al., 1981). Therefore, as the preovulatory follicles of the hierarchy mature, they progress from an FSH-dominated phase to that of an LH-dominated phase. This alternation in responsiveness from FSH to LH is similar to that observed in mammals. However, in the chicken, these changes in gonadotropin support are easier to discern because of the hierarchial arrangement of the follicles.

Recent evidence suggests that ovarian steroids may modulate the actions of FSH and LH on the preovulatory follicle. The perifusion of granulosa layers from F_1 follicles of different maturities with a low dose of LH resulted in the production of similar amounts of progesterone. However, the co-incubation of theca and granulosa layers

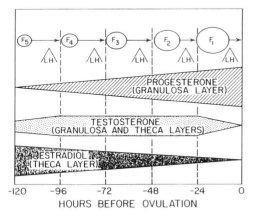

Figure 8 Diagrammatic presentation of changes in steroid concentrations in theca and granulosa layers as the preovulatory follicles mature and approach ovulation. From Bahr et al. (1983).

from an immature F_1 follicle (F_1 follicle for approximately 8 h) resulted in a suppression of progesterone secretion, whereas coincubation of theca and granulosa layers from a mature F_1 follicle (F_1 follicle for approximately 32 h) did not (Johnson et al., 1987). These results suggested that some factor(s) from the theca layer regulates the production of progesterone by the granulosa layers. This finding may explain why the largest follicle must have been the largest follicle for approximately 10 h before it can ovulate. In subsequent studies, we demonstrated that androgens and estradiol-17β, major steroidal products of the theca layer, suppress the production of progesterone by the granulosa layer by specifically decreasing the activities of the enzymes that convert cholesterol to pregnenolone and pregnenolone to progesterone (Lee and Bahr, 1989, 1990). Therefore, the gradual decrease in androgen and estradiol-17β production by the theca layer during maturation of the preovulatory follicle may be necessary to enable the granulosa layer to secrete increasing amounts of progesterone, which are necessary to trigger the LH surge and ovulation.

Follicle-stimulating hormone and LH acting via the generation of the second messenger, cAMP, is one means by which follicular maturation and steroidogenesis are regulated. However, there is increasing evidence that products of the phosphoinositol cycle have a role in regulating follicular function. Activation of protein kinase C alters progesterone and androstenedione production by granulosa and theca cells, respectively, of the F_1 follicle (Johnson and Tilly, 1988; Tilly and Johnson, 1989). Therefore, the second messenger of protein kinase C is another means by which steroidogenesis is regulated.

2. Ovarian Nonsteroidal Hormones—Cellular Source and Action

Various nonsteroidal hormones are produced by the ovary and regulate follicular development, ovulation, and oviposition. The major hormones are catecholamines, prostaglandins, plasminogen, activator, and inhibin. The role of prostaglandins in oviposition was discussed earlier in this chapter.

The catecholamines, dopamine, epinephrine, and norepinephrine, are localized in the theca layer with the highest amounts measured in the F_1 follicle (Bahr et al., 1986; Moudgal and Razdan, 1983). Norepinephrine concentrations were approximately 6- and 30-fold greater than those of epinephrine and dopamine, respectively. The concentrations of epinephrine and norepinephrine are relatively constant during the cycle, with the exception of a significant increase 6 h before ovulation, which is the time of the LH surge. Previous work by Kao and Nalbandov (1972) demonstrated that injection of the F_1 follicle with α-adrenergic blocking agents delayed ovulation. Induction of ovulation by perfusing the chicken ovary with norephinephrine suggests that norepinephrine may trigger one of the critical steps in the cascade of biochemical events resulting in ovulation (Tanaka and Nakada, 1974).

Inhibin has also been identified in the chicken ovary (Akashiba et al., 1988; Johnson, 1989; Tsonis et al., 1988). Inhibin is produced by the granulosa cells with the highest amounts found in granulosa cells of the F_1 follicle. Based on mammalian studies, it is assumed that inhibin regulates FSH secretion in the chicken.

Another ovarian peptide that may have

an important role in follicular growth and ovulation is plasminogen activator (PA). This neutral serine protease converts plasminogen into plasmin, which in turn converts latent collagenase into an active form (Espey, 1980). Plasminogen activator is contained in and secreted by granulosa cells, with the highest amounts found in granulosa cells of the F_1 follicle. The regulation of PA production is apparently under the control of two messenger systems. Activation of the adenylyl cyclase system results in a decrease in the production of PA, whereas activators of protein kinase C, such as, addition of the tumor-promoting phorbol ester, phorbol 12-myristate 13-acetate to granulosa cells, increase the amount of PA (Tilly *et al.*, 1989).

III. Molting

Generally once a year, for approximately a month, chickens stop laying eggs and molt. Molting in a chicken is preceded by decreased egg production, which is an external indicator of an increased interval between ovulations and a slower rate of follicular maturation. As will be discussed in the next section, the cause for the reduction in ovarian activity may be a loss of a photosensitivity of the hypothalamus to a stimulatory daylength. Molting is characterized by a cessation of egg production and regression of the reproductive tract and ovary and decreased liver weight (Brake and Thaxton, 1979). Hormone changes include a decrease in progesterone and possibly LH and an increase in thyroxine (Advis *et al.*, 1985; Dickerman and Bahr, 1989). Whereas molting occurs naturally in chickens, it can be induced by removal of feed, dietary changes (diets with low calcium or zinc), or administration of high levels of progesterone or gonadotropin-releasing hormone agonist (GnRH-A) (Dickerman and Bahr, 1989; Wolford, 1984). With the exception of the use of progesterone or GnRH-A, the precise mechanism by which these other agents cause a chicken to molt has not been defined. Infusion of a potent GnRH-A for several days apparently suppresses release of LH, which results in ovarian regression. There is a rapid increase in thyroxine and decrease in progesterone concentrations in the blood. The elevated thyroxine may be the cause for feather loss because thyroxine stimulates the growth of new feathers. However, the elevated levels of thyroxine may also make the molting chicken photorefractory because she is unresponsive to photostimulation. Data obtained from quail and starling studies indicate that the elevation of thyroxine during molt produces a photorefractoriness (Follett and Nicholls, 1985; Goldsmith and Nicholls, 1984). It is possible that this photorefractoriness allows the development of postmolt photosensitivity, which results in increased egg production as compared to the premolt period. In the commercial production of eggs, chickens are molted when egg production drops to 50–60%. Following a molt, egg production will increase to approximately 80–85% for a period of time, after which egg production will decrease gradually. Even though molting is extensively used as a method to increase egg production, the endocrine control of molting is not understood. Many studies have been done using the removal of feed as an approach to induce molt. However, this approach causes severe physiological disruptions, which prevent the elucidation of the basic endocrine mechanisms of molting. The use of a GnRH-A that has a specific site of action without the concomitant physiological interruptions is a more appropriate model to study the endocrine mechanisms of molting (Dickerman and Bahr, 1989).

IV. Changes in Reproduction with Age

The chicken lays her first egg at approximately 20 weeks of age. Within 6–8 weeks, she obtains a peak egg production of 85–90%, which is maintained for several months. Gradually egg production decreases so that by the end of the first year of lay, egg production rates are 50–60%. As the chicken lays fewer eggs, the size of the egg usually increases and egg shell thickness and egg quality decrease. There is also a higher incidence of soft-shelled eggs as the chicken ages. In some cases, the chicken may stop laying eggs, which is usually the result of a failure of the ovulated egg to enter the oviduct.

The decrease in egg production is an external indicator that a decrease in ovulation rate is occurring. An intriguing question is why the frequency of ovulation decreases with age. Williams and Sharp (1978b) observed that preovulatory follicles of old chickens mature at a slower rate and ovulate at a larger size compared to the preovulatory follicles of young hens (Fig. 9).

The average interval between ovulations in young chickens of high egg production is 24–25 h, whereas the interval between ovulations in old chickens of low egg production is 25–28 h or more. This increase in the interval between ovulations as the chicken ages is due to a slower rate of maturational changes, which allow the follicle to acquire

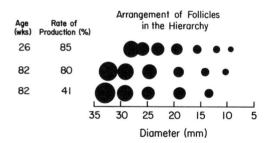

Figure 9 A diagrammatic comparison of the arrangement, size, and numbers of yellow-yolky follicles in young (26 weeks) and old (82 weeks) hens of a commercial egg-producing strain. The 82-week-old hens were divided into good and poor layers; the poor layers laid at about half the rate of the good layers. Each circle represents the mean size of a follicle in a given position in the follicular hierarchies of between 18 and 26 hens. The follicular diameters are one-tenth of the scale size. The rate of egg production was calculated for each bird from the number of eggs laid during the 28th day preceding death. From Williams and Sharp (1978b).

a competency to ovulate (Etches *et al.*, 1983; Johnson *et al.*, 1986).

Another change with age that results in a greater interval between ovulations and a decline in egg production is the decrease in the recruitment of follicles into the hierarchy and increased atresia of small follicles (Palmer and Bahr, 1988; Waddington *et al.*, 1985; Williams and Sharp, 1978b). The precise causes for these changes are not known; however, several hypotheses can be proposed.

First, increased atresia with age may be the result of a lack of adequate gonadotropic support. This hypothesis is supported by the observation that treatment of old chickens with FSH decreases the number of atretic follicles and increases the number of small yellow yolky follicles and the number of follicles in the hierarchy (Palmer and Bahr, 1988).

A second hypothesis is that the rate of

maturation of smaller follicles and rate of entrance of these follicles into the hierarchy decrease with age. The cause of these changes could be changes in either the amounts or ratios of FSH and LH in the chicken as she ages. This change in gonadotropic stimulation of the ovary may be responsible for a slower maturation of ovarian follicles. Experiments done in our laboratory suggest that follicles progress less quickly from an FSH-dominated phase to an LH-dominated phase as the chickens age. Follicles must be responsive to LH to ovulate. To test this hypothesis, frequent measurements of serum FSH and LH will need to be made in young and old chickens. The presence of a higher number of FSH receptors in granulosa cells of small preovulatory follicles and a greater response of these cells to FSH as indicated by elevated progesterone secretion supports this hypothesis (Palmer, 1989).

A third hypothesis is that a decrease in estrogen production causes slower follicular maturation. The higher incidence of atresia of small follicles, which are the primary source of estrogen, could result in lower intraovarian levels of estrogen. If one can assume that basic physiological processes are similar in mammals and birds, it is possible that lower estrogen levels in old chickens could result in a decrease in follicular recruitment and growth.

A fourth hypothesis is that the sensitivity of the hypothalamus–pituitary axis to the positive feedback of steroids decreases with age so that higher levels of progesterone may be required to trigger the LH surge. Williams and Sharp (1978a) reported that a larger dose of progesterone was required to induce an LH surge in old chickens compared to young chickens.

The increased incidence of soft-shelled and shell-less eggs with age is a major problem in the poultry industry. Whereas the occurrence of this problem is well documented, few answers are available to explain the causes. Proper eggshell formation is dependent upon the interaction of two major endocrine systems, that of $1\alpha,25$-dihydroxyvitamin D_3 [$1\alpha,25(OH)_2D_3$] and estrogen production. One cause may be a decrease in the synthesis of $1\alpha,25(OH)_2D_3$, which is essential for normal calcium metabolism and egg shell deposition. Chickens that produce eggs with thick shells have higher serum levels of $1\alpha,25(OH)_2D_3$ than do chickens that produce eggs with thin shells (Soares and Ottinger, 1988). The synthesis of $1\alpha,25(OH)_2D_3$ and its actions are regulated primarily by estrogen (Baksi and Kenny, 1977; Pike et al., 1978). Therefore, decreased estrogen production by aging chickens may result in lower levels of $1\alpha,25(OH)_2D_3$ and formation of thin-shelled eggs. However, rigorous studies are required to document this hypothesis.

Finally, internal laying or the inability of the fimbria to pick up the ovulated egg increases with age. As ovulation approaches, the infundibulum becomes engorged with blood and moves in a waving motion in the vicinity of the ovulating follicle. The follicle is engulfed by the infundibulum immediately before ovulation. This phenomenal process is under hormonal and neural control. Apparently, as the chicken ages, there is some failure in the coordination of this fine-tuned activity of the oviduct. As a result, many ovulated eggs enter the body cavity instead of the oviduct. Therefore, the decrease in or even cessation of egg production in the old chicken can be the result of internal laying.

In summary, the decrease in egg production with age is a major reproductive problem. The physiological changes associated with aging are an increase in the interval between ovulations, a decrease in recruitment of follicles into the hierarchy and an increase in atresia, a deterioration of egg shell thickness and egg quality, and an increased incidence of internal laying. An identification of the basic mechanism causing these marked physiological changes is essential and would be of significant value to the commercial producers.

V. The Male Reproductive System

A. Anatomy of the Reproductive Tract

The structure and location of the avian male reproductive tract is significantly different from the tract of most mammals. Unlike the scrotal mammals, the paired testes are attached to the dorsal wall at the anterior end of the kidney. The accessory reproductive glands (i.e., prostate, seminal vesicles) that are found in most mammals are absent in the cockerel. Furthermore, the pampiniform plexus of blood vessels, which is necessary for temperature regulation of the mammalian testis, is also lacking in the avians (Waites, 1970). In contrast, the avian testes function at a body temperature of 43°C.

Because the testes lack connective-tissue septa, the tubules are not divided into distinct lobules (Lake, 1957). The seminiferous tubules, having a large diameter and high fluid content, are surrounded by blood vessels and Leydig cells. Only a small epididymis, attached to the dorsal surface of the testis, is present. The vas deferens, a long, extensive, convoluted tube, runs posteriorly

along the medioventral surface of the kidney, parallel with the ureters, and terminates in the cloacal wall.

Chickens lack an intromittent organ, which is present in mammals. There is a small erectile phallus in the ventral part of the cloaca, which becomes erect when it is engorged by lymphlike fluid. Since the bird lacks accessory glands, the seminal plasma is produced primarily by the seminiferous tubules and the epididymis. There is also a transparent fluid, an exudate from the lymph folds in the cloaca, which is expelled when semen is collected from the cockerel. It is possible that some of this fluid is also released during the normal mating process. For a more detailed discussion of the reproductive tract in cockerels and the physical and chemical composition of the seminal plasmas, the reader is directed to a review by Lake (1971).

B. Testicular Steroidogenesis

The functions of the testes are the production of spermatozoa and the secretion of hormones. These sex steroids are necessary for the growth and maintenance of the accessory reproductive organs and the regulation of courtship behavior. Estrogens, androgens, and progestins have been identified in the cockerel. Testosterone is the primary androgen present in the adult cockerel. There is evidence that both the Δ^4 pathway and Δ^5 pathway exist in the avian testis (Galli et al., 1973; Nakamura and Tanabe, 1972). At the present time, it is not clear which pathway is preferred.

The avian testis consists of two steroidogenic cell types, the Leydig cells and Sertoli cells. It is assumed that the Leydig cells are the site of testosterone production. However, unlike the mammal, the specific sites of steroidogenesis in the avian testis have not been identified.

The testis is stimulated directly by gonadotropins. There is a correlation between LH concentrations in the blood and testicular size. The presence of LH receptors in the avian testis has not been reported, however, as in mammals it is assumed that the Leydig cells are the target tissue for LH. FSH receptors have been measured in the testis (Ishii and Adachi, 1977; J. M. Bahr, unpublished results). The precise cellular site of the FSH receptors in the testis has not been determined, nor is the function of FSH in the testis known. Again extrapolating from the mammalian literature, a potential function for FSH is the regulation of spermatogenesis.

In conclusion, it is readily apparent that our knowledge of avian male reproduction, especially the endocrine aspects, is very sparse. There is an extensive literature on semen, collection of sperm, etc. The availability of LH and FSH receptor assays and radioimmunoassays, specific antibodies against the P-450 steroidogenic enzymes, and new methods to obtain specific cell types should advance our understanding of the regulation of avian male reproductive endocrinology.

Acknowledgments

This work was supported in part by NSF grants BMS-75-18994, DCB-84-21061, DCB-88-02032, NIH grant HD-16328, and USDA grants 85-02577, 86-01150, and 89-01666.

We thank Hiroaki Nitta for the art work and Cathie Godwin for typing the manuscript.

References

Advis, J. P., Conlyock, A. M., and Johnson, A. L. (1985). *Biol. Reprod.* **32,** 820.

Akashiba, H., Taya, K., and Sasamoto, S. (1988). *Poult. Sci.* **67,** 1625.

Armstrong, D. G. (1982). *J. Endocrinol.* **93,** 415.

Armstrong, D. G. (1984). *J. Endocrinol.* **100,** 81.

Bahr, J. M., Wang, S.-C., Huang, M. Y., and Calvo, F. O. (1983). *Biol. Reprod.* **29,** 326.

Bahr, J. M., Ritzhaupt, L. K., McCullough, S., Arbogast, L. A., and Ben-Jonathan, N. (1986). Biol. Reprod. **34,** 502.

Baker, C. M. A. (1968). *In* "Egg Quality: A Study of the Hen's Egg" (T.C. Carter, ed.), p. 67. Oliver and Boyd, Edinburg.

Baksi, S. N., and Kenny, A. D. (1977). *Endocrinology* **101,** 1216.

Bastian, J. W., and Zarrow, M. X. (1955). *Poult. Sci.* **34,** 776.

Beuving, G., and Vonder, G. M. A. (1977). *J. Reprod. Fertil.* **51,** 169.

Beuving, G., and Vonder, G. M. A. (1981). *Gen. Comp. Endocrinol.* **44,** 382.

Bhatti, B. M., and Morris, T. R. (1977). *Br. Poult. Sci.* **18,** 391.

Bjersing, L., and Cajander, S. (1974). *Cell Tissue Res.* **153,** 15.

Brake, J., and Thaxton, P. (1979). Poult. Sci., **58,** 707.

Bullock, D. W., and Nalbandov, A. V. (1967). *J. Endocrinol.* **38,** 407.

Calvo, F. O. and Bahr, J. M. (1983). *Biol. Reprod.* **29,** 542.

Calvo, F. O., Wang, S.-C., and Bahr, J. M. (1981). *Biol. Reprod.* **25,** 805.

Dahl, E. (1970a). *Zellforsch. Z. Mikrosk. Anat.* **109,** 195.

Dahl, E. (1970b). *Zellforsch. Z. Mikrosk. Anat.* **109,** 212.

Day, S. L., and Nalbandov, A. V. (1977). *Biol. Reprod.* **16,** 486.

Dickerman, R. W., and Bahr, J. M. (1989). *Poult. Sci.* **68,** 1402.

Espey, L. L. (1967). *Endocrinology* **81,** 267.

Espey, L. L. (1974). *Biol. Reprod.* **10,** 216.

Espey, L. L. (1980). *Biol. Reprod.* **22,** 73.

Etches, R. J., and Cunningham, F. J. (1976). *J. Endocrinol.* **71,** 51.

Etches, R. J., and Cunningham, F. J. (1977). *Acta Endocrinol.* **84,** 357.

Etches, R. J., and Duke, C. E. (1984). *J. Endocrinol.* **103,** 71.

Etches, R. J., MacGregor, H. E., Morris, T. F., and Williams, J. B. (1983). *J. Reprod. Fertil.* **67,** 351.

Feeney, R. E., and Allison, R. G. (1969). *In* "Evolutionary Biochemistry of Proteins." Wiley, New York.

Fertuck, H. C., and Newstead, J. D. (1970). *Zellforsch. Z. Mikrosk. Anat.* **103,** 447.

Follett, B. K., and Nicholls, T. J. (1985). *J. Endocrinol.* **107,** 211.

Fraps, R. M. (1954). *Proc. Natl. Acad. Sci. USA* **40,** 348.

Fraps, R. M. (1955). *In* "Progress in the Physiology of Farm Animals," Vol. 2 (J. Hammond, ed.), Chapter 15. London, Butterworths.

Freedman, S. L. (1968). *Acta Anat.* **69,** 18.

Fujii, S., and Yoshimura, Y. (1979). *J. Fac. Appl. Biol. Sci.* **18,** 185.

Fujii, S., Yoshimura, Y., and Tamura, T. (1980). *J. Fac. Appl. Biol. Sci.* **19,** 161.

Fujii, M., Tojo, H., and Koga, K. (1981). *Int. J. Biochem.* **13,** 1043.

Furr, B. J. A., Bonney, R. C., England, R. J., and Cunningham, F. J. (1973). *J. Endocrinol.* **57,** 159.

Galli, F. E., Irusta, O., and Wassermann, G. F. (1973). *Gen. Comp. Endocrinol.* **21,** 262.

Gilbert, A. B. (1965). *Q. J. Exp. Physiol.* **50,** 437.

Gilbert, A. B., Reynolds, M. E., and Lorenz, F. W. (1968). *J. Reprod. Fertil.* **17,** 305.

Gilbert, A. B. (1969). *Q. J. Exp. Physiol.* **54,** 404.

Gilbert, A. B. (1971a). *In* "Physiology and Biochemistry of the Domestic Fowl," Vol. 3 (D. J. Bell and B. M. Freeman, eds.), Chapter 54. Academic Press, New York.

Gilbert, A. B. (1971b). *In* "Physiology and Biochemistry of the Domestic Fowl," Vol. 3 (D. J. Bell and B. M. Freeman, eds.), Chapter 58. Academic Press, New York.

Goldsmith, A. R., and Nicholls, T. J. (1984). *Gen. Comp. Endocrinol.* **54,** 256.

Hammond, R. W., Koelkebeck, K. W., Scanes, C. G., Biellier, H. V., and Hertelendy, F. (1981). *Gen. Comp. Endocrinol.* **44,** 400.

Hattori, A., Ishii, S., and Wada, M. (1986). *Gen. Comp. Endocrinol.* **64,** 446.

Hertelendy, F., Yeh, M., and Biellier, H. V. (1974). *Gen. Comp. Endocrinol.* **22,** 529.

Hutson, J. M., MacLaughlin, D. T., Ikawa, H., Budzik, G. P. and Donahoe, P. K. (1983). *In* "Reproductive Physiology IV" (R. O. Greep, ed.), p. 177. University Park Press.

Ishii, S., and Adachi, T. (1977). *Gen. Comp. Endocrinol.* **31,** 287.

Johnson, P. A., Dickerman, R. W., and Bahr, J. M. (1986). *Biol. Reprod.* **35,** 641.

Johnson, P. A., Stoklosowa, S., and Bahr, J. M. (1987). *Biol. Reprod.* **37,** 1149.

Johnson, P. A. (1989). *Biol. Reprod. (Suppl.* 1) **40,** 119.

Johnson, A. L., and van Tienhoven, A. (1980a). *Biol. Reprod.* **23,** 386.

Johnson, A. L., and van Tienhoven, A. (1980b). *Biol. Reprod.* **23,** 910.

Johnson, A. L., and van Tienhoven, A. (1981). *J. Endocrinol.* **89,** 1.

Johnson, A. L., and van Tienhoven, A. (1984). *Endocrinology* **114,** 2276.

Johnson, A. L., and Tilly, J. (1988). *Biol. Reprod.* **38,** 296.

Kao, L. W. L., and Nalbandov, A. V. (1972). *Endocrinology* **90,** 1343.

King, J. A., and Millar, R. P. (1982). *J. Biol. Chem.* **257,** 10729.

Lake, P. E. (1957). *J. Anat.* **91,** 116.

Lake, P. E. (1971). *In* "Physiology and Biochemistry of the Domestic Fowl" (D. J. Bell and B. M. Freeman, eds.), Vol. 3, p. 1411. Academic Press, New York.

Lang, G. F., Etches, R. J., and Walton, J. S. (1984). *Biol. Reprod.* **30,** 278.

Lee, H. T., and Bahr, J. M. (1989). *Endocrinology* **125,** 760.

Lee, H. T., and Bahr, J. M. (1990). *Endocrinology* (1990). **126,** 779.

Lush, I. E. (1961). *Nature (Lond.)* **189,** 981.

McNally, E. H. (1947). *Poult. Sci.* **26,** 396.

Moudgal, R. P., and Razdan, M. N. (1983). *Br. Poult. Sci.* **24,** 173.

Miyamoto, K., Hasegawa, Y., Nomura, M., Igarashi, M., Kangawa, K., and Matsuo, H. (1984). *Proc. Natl. Acad. Sci. USA* **81,** 3874.

Nakamura, T., and Tanabe, Y. (1972). *Gen. Comp. Endocrinol.* **19,** 432.

Nalbandov, A. V. (1959). *In* "Comparative Endocrinology" (A. Gorbman, ed.), p. 161. John Wiley, New York.

Nalbandov, A. V., and James, M. F. (1949). *Am. J. Anat.* **85,** 347.

Nelson, D. M., Norton, H. W., and Nalbandov, A. V. (1965). *Endocrinology* **77,** 889.

Ogawa, K., and Goto , K. (1984). *J. Reprod. Fertil.* **71,** 545.

Okamura, H., Tokenaka, A., Yajima, Y., and Nishimura, T. (1980). *J. Reprod. Fertil.* **58,** 153.

Olson, D. M., and Hertelendy, F. (1981). *Biol. Reprod.* **24,** 496.

O'Malley, B. W., Kirscher, M. A., and Barden, C. W. (1968). *Proc. Soc. Exp. Biol. Med.* **127,** 521.

Palmer, S. S. (1989). Follicle Stimulating Hormone and Steroidogenesis in Ovarian Granulosa Cells during Aging in the Domestic Hen. Ph.D. disseration, University of Illinois, Urbana.

Palmer, S. S., and Bahr, J. M. (1988). *Biol. Reprod.* **38**(Suppl. 1), 184.

Pike, J. W., Spanos, E., Colston, K. W., MacIntyre, I., and Hausler, M. R. (1978). *Am. J. Physiol.* **235,** 338.

Ritzhaupt, L. K., and Bahr, J. M. (1987). *J. Endocrinol.* **115,** 303.

Robinson, F. E., and Etches, R. J. (1986). *Biol. Reprod.* **35,** 1096.

Romanoff, A. L., and Romanoff, A. J. (1949). *In* "The Avian Egg." Wiley, New York.

Rothchild, I., and Fraps, R. M. (1944). *Proc. Soc. Exp. Biol. Med.* **56,** 79.

Saunders, D. S. (1977). *In* "An Introduction to Biological Rhythms." Blackie, Glasgow.

Scanes, C. G., Godden, P. M. M., and Sharp, P. J. (1977a). *J. Endocrinol.* **73,** 473.

Scanes, C. G., Sharp, P. J., and Chadwick, A. (1977b). *J. Endocrinol.* **72,** 401.

Scanes, C. G., Chadwick, A., Sharp, P. J., and Bolton, N. J. (1978). *Gen. Comp. Endocrinol.* **34,** 45.

Schraer, H., Hohman, W., Ehrenspeck, G., and Schraer, R. (1965). *J. Cell Biol.* **27,** 96A.

Shimada, K., Neldon, H. L., and Koike, T. I. (1986). *Gen. Comp. Endocrinol.* **64,** 362.

Shimada, K., Saito, N., Itogawa, K., and Koike, T. I. (1987). *J. Reprod. Fertil.* **80,** 143.

Shimada, K., and Asai, I. (1978). *Biol. Reprod.* **19,** 1057.

Shimada, K. and Asai, I. (1979). *Biol. Reprod.* **21,** 523.

Smith, A. H., Court, S. A., and Martin, E. W. (1959). *Am. J. Physiol.* **197,** 1041.

Soares, J. H., Jr., and Ottinger, M. A. (1988). *Poult. Sci.* **67,** 1322.

Tanaka, K. (1976). *Poult. Sci.* **53,** 2120.

Tanaka, K., Kamiyoshi, M., and Tanabe, Y. (1971). *Poult. Sci.* **50,** 63.

Tanaka, K., and Nakada, T. (1974). *Poult. Sci.* **53,** 2120.

Tanaka, K., Li, Z. D., and Ataka, Y. (1987). *J. Reprod. Fertil.* **80,** 411.

Tilly, J. L., and Johnson, A. L. (1989). *Dom. Anim. Endocrinol.* **6,** 155.

Tsonis, C. G., Sharp, P. J., and McNeilly, A. S. (1988). *J. Endocrinol.* **116,** 293.

Waddington, D., Perry, M. M., Gilbert, A. B., and Hardie, M. A. (1985). *J. Reprod. Fertil.* **74,** 399.

Waites, G. M. H. (1970). *In* "The Testis" (A. D. Johnson, W. R. Gomes, and N. L. Van Demark, eds.), p. 241. Vol. 1, Academic Press, New York.

Warren, D. C., and Scott, H. M. (1935). *Poult. Sci.* **14,** 195.

Warren, D. C., and Scott, H. M. (1936). *J. Exp. Zool.* **74,** 137.

Williams, J. B., and Sharp, P. J. (1978a). *J. Reprod. Fertil.* **53,** 141.

Williams, J. B., and Sharp, P. J. (1978b). *Br. Poult. Sci.* **19,** 387.

Wilson, S. C., and Cunningham, F. J. (1980). *Br. Poult. Sci.* **21,** 351.

Wilson, S. C., and Sharp, P. J. (1973). *J. Reprod. Fertil.* **35,** 561.

Wilson, S. C., and Sharp, P. J. (1976). *J. Endocrinol.* **71,** 87.

Wolford, J. H. (1984). *World's Poult. Sci. J.* **40,** 66.

Yoshimura, Y. and Koga, O. (1982). *Cell Tissue Res.* **224,** 349.

Zadworny, D., Shimada, K., Ishida, H., and Sato, K. (1989). *Gen. Comp. Endocrinol.* **74,** 468.

Nutritional Influences on Reproduction

C. L. FERRELL

Influences of nutrition on reproductive functions have been recognized for many years. These influences include effects on attainment of puberty, duration of postpartum anestrus, gameteogenesis, conception rate, embryonic mortality, prenatal development, and sexual behavior. The ultimate result of improper nutrition is to reduce the number of viable offspring produced. Although many of the nutritional influences on reproductive functions have all been well characterized, underlying mechanisms are not well understood. In this chapter, some of the consequences of improper nutrition as related to reproductive functions of farm animal species will be described. Some of the mechanisms of these effects, as they are now understood, will be indicated.

I. Nutritional Influences on Sexual Development and Puberty

Age at puberty is an important production trait in most species of farm animals. In cattle, many currently used management systems require that heifers be bred at 14–16 months of age so they calve at approximately 24 months of age. In addition, the breeding season is usually restricted. It is important, especially under these types of management systems, that heifers attain puberty at a young age and conceive early in the breeding season. The number of estrous cycles prior to breeding is positively related to conception rate (Hare and Bryant, 1985; Byerley *et al.*, 1987). In addition, heifers that conceive early in the first breeding season have a greater probability of weaning more and heavier calves during their lifetimes (Burris and Priode, 1958; Lesmeister *et al.*, 1973; Spitzer *et al.*, 1975). Attainment of puberty at an early age is perhaps even more important in seasonal breeding species, such as sheep, where failure to attain puberty may result in a year delay in production of the first progeny, thus reducing lifetime production (Allison, 1974; Donaldson, 1968).

A. Development of the Reproductive Organs and Puberty

1. Under- and Overfeeding

In cattle, age at first behavioral estrus (puberty) is markedly influenced by level of nutrition or postweaning rate of gain, and varies among breeds (Reid *et al.*, 1951; Joubert, 1954, 1963; Wiltbank *et al.*, 1966, 1969; Short and Bellows, 1971; Laster *et al.*, 1972, 1976, 1979; Dufour, 1975; Ferrell, 1982). The extent to which age at puberty may be influenced by level of feed was clearly illustrated by Reid *et al.* (1951). In that study, Holstein heifers fed at 65, 100, or 140% of "Morrison Standards" attained puberty at 526, 344, and 289 days of age, respectively. In the tropics, under conditions of very poor nutrition and high de-

grees of environmental stress, some heifers may, in fact, not attain puberty until 3 years of age (Lammond, 1970). In a less extreme example, Angus–Hereford crossbred heifers fed to gain 0.27, 0.45, or 0.68 kg/day reached puberty at an average age of 433, 411, and 388 days, respectively (Short and Bellows, 1971). Although these differences are relatively small, pregnancy rates after a 60-day breeding season were 50, 86, and 87%, respectively.

Observations that weight is less variable than age at puberty have led several authors to conclude that size or weight is more important than age in determining time of onset of puberty (Rattray, 1977). However, data reported by several researchers (Joubert, 1954; Sorenson *et al.*, 1959; Wiltbank *et al.*, 1969; Dufour, 1975) indicate that weight, as well as age, at puberty is influenced by level of nutrition or postweaning rate of gain. Ferrell (1982) reported that average age at puberty was 387, 365, and 372 days and average weight at puberty was 301, 311, and 322 kg when heifers were fed to gain 0.4, 0.6, or 0.8 kg/day, respectively, postweaning. These data, as well as those from other studies, indicate that heifers fed to gain weight less rapidly after weaning attain puberty at a lighter weight but greater age than those fed to gain weight more rapidly (Arije and Wiltbank, 1974; Greer *et al.*, 1983).

Both age and weight at puberty differ substantially among breeds of cattle (Laster *et al.*, 1972, 1976, 1979; Stewart *et al.*, 1980; Sacco *et al.*, 1987) and both are decreased by heterosis. Within beef breeds, those having larger mature size tend to attain puberty at an older age and heavier weight. However, *Bos indicus* heifers generally reach puberty at an older age than *Bos taurus* heifers, and heifers from higher milk-producing breeds are generally younger at puberty than those from breeds having lower milk production. Some of those differences may be due to direct maternal effects expressed through higher rates of preweaning gain by calves from high milk-producing breeds (Plasse *et al.*, 1968; Arije and Wiltbank, 1974). Laster *et al.* (1979) reported a correlation of −0.88 between age at puberty and milk production among beef breeds. Optimum pre- and postweaning nutrition level or rate of gain may differ among breeds (Ferrell, 1982; Greer *et al.*, 1983).

These and other data support the concept of a genetically determined threshold age and weight at which heifers attain puberty. The "target weight" concept has been incorporated into many management systems (Spitzer *et al.*, 1975; Dziuk and Bellows, 1983; Wiltbank *et al.*, 1985). Simply stated, the concept is to feed heifers to attain a preselected or target weight at a given age. Excessive feeding should be avoided. In addition to increasing feed costs, overfeeding that results in excessive fatness may have detrimental effects on expression of behavioral estrus, conception rate, embryonic and neonatal survival, calving ease, milk production, and productive life (Swanson, 1960; Sorenson *et al.*, 1959; Moustgaard, 1969; Arnett *et al.*, 1971; Ferrell, 1982).

Recent data (M. Roberson and J. E. Kinder, unpublished data, 1989) show that contact with a mature bull, as well as energy intake, affects age at puberty. When heifers had contact with a bull, mean age at initiation of luteal function occurred 52 days earlier in heifers that gained 0.77 kg/day and 28 days earlier in heifers that gained 0.67 kg/day compared with heifers that had no contact with bulls.

Consequences of under- or overfeeding prior to puberty in sheep are similar to those in cattle (Moustgaard, 1969; Rattray, 1977); however, results are complicated by their seasonal breeding pattern. Ewe lambs growing at faster rates will exhibit their first estrus and are more likely to conceive at an earlier age and heavier body weight than ewe lambs growing at slower rates (Allen and Lamming, 1961; Burfening *et al.*, 1971; Southam *et al.*, 1971; Dyrmundsson, 1973, 1981; Quirke, 1979). If born late in the season or if grown too slowly, ewe lambs may not attain puberty during the first breeding season after birth. Southam *et al.* (1971), for example, citing unpublished data of Hulet, indicated the incidence of puberty during the first year was increased from 12 to 88% by increasing feed allowances following weaning to achieve a body weight of 45 kg by the middle of the breeding season. In other experiments (Casida, 1964), mild nutrient restriction of ewe lambs did not delay onset of puberty but decreased ovulation rate. Conversely, increased nutrient supply for a short interval prior to breeding (flushing) has been associated with increased ovulation rate (Bellows *et al.*, 1963; Casida, 1964; Dziuk and Bellows, 1983), increased conception rate (Keane, 1975), or increased numbers of offspring born (Moustgaard, 1969).

Average age and weight at puberty differ considerably among breeds of sheep (Dickerson and Laster, 1975; Quirke *et al.*, 1985) and are decreased by heterosis. Ewe and ram lambs born or raised as singles attain puberty at a younger age and heavier weight than lambs born and/or raised as twins (Dyrmundsson, 1973; Dickerson and Laster, 1975). This suggests nutrient intake prior to weaning in sheep influences attainment of puberty.

In domestic swine, a practical objective is generally to stimulate the attainment of puberty such that gilts may be mated at second estrus and as young as possible. Several reviews have considered relationships between attainment of puberty of gilts and age, live weight, growth rate, and nutrition (Anderson and Melampy, 1972; Hughes, 1982; Aherne and Kirkwood, 1985). Influences of these factors on time of onset of puberty in swine are not as clear as in sheep and cattle. Data have been reported, for example, indicating gilts may be as young as 85 days (Aherne *et al.*, 1976) or as old as 350 days of age (Brooks and Smith, 1980) at puberty. Weight at puberty in gilts may vary from as little as 53 kg (Burger, 1952) to as much as 120 kg or more (Hughes and Cole, 1975). Similarly, data are available to indicate that increased growth rate, level of nutrition, or energy intake influence time of puberty attainment positively, negatively, or not at all (Hughes, 1982; Aherne and Kirkwood, 1985).

Several factors contribute to differences in conclusions regarding relationships between age, weight, nutrition level, or energy intake and time of attainment of puberty in swine. Genotype in swine, as in cattle and sheep, is a major source of variation in age and weight at puberty (Christenson and Ford, 1979). Environmental factors such as confinement (Jensen *et al.*, 1970; Christenson, 1981), season of the year (Christenson, 1981), photoperiod (Hacker *et al.*, 1979), and boar contact (Brooks and Cole, 1970; Zimmerman *et al.*, 1974; Thompson and Savage, 1978) also contribute substantially to differences in observed time of expression of first behavioral estrus in gilts. These factors often have not been adequately controlled when nutritional influences on puberty have been evaluated. In many studies, nutritional levels evaluated in swine have been greater than those evaluated in cattle

or sheep. Generally, control gilts fed *ad libitum* have been compared with gilts fed a limited amount of feed. However, the degree of limitation has often been small relative to treatments typically imposed in other species. In fourteen studies reviewed by Anderson and Melampy (1972), average metabolizable energy (ME) intake of restricted gilts was 5.7 Mcal/day (over two times the ME required for maintenance). In nine of the fourteen studies, puberty was delayed by 16 days, whereas in five of the studies age at puberty was decreased by 11 days by limited feeding. Levels of intake in those studies were at least equivalent to levels generally considered adequate for cattle or sheep. In some studies, onset of puberty may have been delayed by the *ad libitum* feeding used for controls. More recent reviews have reported similar results (Hughes, 1982; Aherne and Kirkwood, 1985). In studies in which feed or energy restrictions have been more severe, that is, to 55% or less of *ad libitum* intake, results have agreed more closely with those reported in cattle or sheep. These studies suggest that age at puberty is increased and weight at puberty is decreased in gilts when nutrient restriction is sufficient to substantially limit somatic growth (Burger, 1952; Friend, 1976; Friend *et al.*, 1981).

Numerous data in several species indicate that neither age nor weight is a reliable index or reproductive development, but that threshold values for these traits must be attained before puberty can occur. In addition, the concept that a threshold level of body condition or fatness must also be attained before puberty can occur has received considerable support (Kirkwood and Aherne, 1985). Body fat, like body weight, is a static measure that reflects dynamic processes of tissue metabolism—that is, body fatness is a reflection of previous net flux of energy. That net flux of energy rather than fatness per se modulates the attainment of puberty is suggested by several studies. Kennedy and Mitra (1963) indicated that a positive energy balance was necessary for onset of puberty in rats. These data are supported by those of Frisch *et al.* (1977). It is well established that athletic training will delay onset of puberty (menarche) in girls. This may occur even though measurable body fat is within the normal range (Warren, 1980). Upon cessation of training, puberty may occur prior to a measurable change in body fat. In gilts, injection of porcine growth hormone, resulting in an increased rate of gain and decreased fatness, may delay or impair attainment of puberty (Bryan *et al.*, 1989). However, effects of growth hormone appear to be reversed shortly after administration of exogenous hormone is discontinued.

2. Protein

Some studies in which nutritional effects on puberty have been evaluated have assumed that variations in age at puberty were due to variations in energy intake. However, in many of these studies, plane of nutrition was varied by altering the amount of a complete diet fed, and thus intake of other nutrients as well as energy was varied. The influence of specific nutrients on attainment of puberty has received less attention than total diet effects. However, in addition to general relationships among age, weight, growth rate, and plane of nutrition, several specific nutrients have been shown to influence the onset of puberty. Protein deficiency results in delayed puberty in most species. However, within ranges of protein in the diet compatible with an adequate growth rate, dietary protein does not appear to substantially affect the onset of pu-

berty. Other observations suggest that effects of protein on puberty are primarily related to its relationship to somatic growth. In ruminant species, intake of specific amino acids is not generally of concern, providing protein intake is adequate, because of the symbiotic relationship of rumen microbes to the host. Amino acids supplied by the rumen microbes are adequate in terms of quantity and balance to meet the animal's needs for normal reproductive function. In nonruminant species, amino acid balance is of greater concern; amino acid imbalance may result in delay of attainment of puberty (Larsson *et al.*, 1966; Friend, 1973). Of particular concern in typical corn–soybean diets for swine are lysine and methionine. More recent evidence in dairy cattle (Ferguson and Chalupa, 1989; Swanson, 1989) indicates excessive protein intake is detrimental to reproductive function.

3. Vitamins and Minerals

Roles of vitamins and minerals in reproduction have been discussed recently in an excellent review by Hurley and Doane (1989). As noted by those authors, all vitamins and essential minerals are required for reproduction because of their cellular roles in metabolism, maintenance, and growth. In addition, many of these nutrients also have specific roles and requirements in reproductive tissues, and these requirements may change during puberty, estrous cycles, pregnancy, parturition, and lactation. Vitamins of particular concern include vitamin A and its metabolites or precursors, vitamin D and its metabolites, and vitamin E. Vitamin C or ascorbic acid, although not an essential dietary nutrient, except for guinea pigs and primates, has important roles especially in ovarian follicular development and luteal function and may be of concern in certain dietary situations. The B vitamins serve as cofactors in many major metabolic pathways and thus are required for maintenance, growth, and reproduction as well as for fetal development. Folic acid, riboflavin, pantothenic acid, choline, cobalamin, niacin, thiamine, and pyridoxine have roles in various aspects of reproduction; however, their functions and requirements have not been well defined. Ruminants generally have no dietary requirement for B vitamins because of rumen bacterial synthesis, but deficiencies may occur when high levels of antimicrobial agents are fed or when high-grain diets are fed if lactic acidosis results, that is, when dietary conditions result in reduced viability of the rumen microbial populations. Several minerals, including calcium, phosphorus, zinc, copper, cobalt, molybdenum, iodine, sodium, potassium, manganese, selenium, and magnesium have essential cellular and subcellular roles in animal function and metabolism. Minerals are involved in numerous aspects of reproductive function; however, many of the details of those involvements are only beginning to be understood. More details of the current status of knowledge regarding involvement of vitamins and minerals in reproductive functions have been presented by Hurley and Doane (1989).

B. Endocrine Functions

In order to set the framework for subsequent discussion, the endocrine regulation of puberty will be briefly described. Moran et al. (1989) and Foster (1988) have provided thorough reviews of the area. Pituitary release of the gonadotropins, luteinizing hormone (LH), and follicle-stimulating hormone (FSH) is regulated by neurohormones from the hypothalamus. Gonadotro-

pins released from the pituitary stimulate the growth of ovarian follicles. Growing follicles synthesize a wide range of steroid hormones, including estradiol, which in the prepubertal and anestrous female exerts a negative feedback on the hypothalamus and pituitary, which in turn inhibits release of pituitary gonadotropins. Progesterone, produced by the corpus luteum during the luteal phase of the estrous cycle, exerts a negative feedback action on pituitary release of LH. During late proestrus or early estrus increasing level of estradiol trigger a massive release of pituitary gonadotropins. This preovulatory surge of LH completes the final growth, maturation, and rupture of the follicle.

Sexual maturation that culminates in attainment of puberty occurs in a gradual manner. It is initiated prenatally and continues throughout the prepubertal and peripubertal periods (Quirke, 1981; Moran et al., 1989). Several components of the system are functional long before puberty. For example, the pituitary releases gonadotropins in response to hypothalamic secretagogues, and the ovaries respond to exogenous gonadotropins well before puberty. It is believed that sexual maturation is modulated through changes in hypothalamic inhibition, resulting in little or no stimulation of the release of gonadotropins from the anterior pituitary until just before puberty (Imakawa et al., 1986; Kinder et al., 1987; Stumph et al., 1989). The ovary has a primary role in inhibiting gonadotropin secretion during the peripubertal period. Responsiveness to negative feedback effects of estrogen decreases during the peripubertal period and secretion of gonadotropins increases such that ovarian follicular growth is stimulated. Estrogen secretion by follicles is then enhanced and in turn the preovulatory surge of gonadotropins is induced by increasing estrogens. There is also an increased secretion of ovarian progesterone during the peripubertal period; its role in the maturation process appears to relate primarily to its "priming" effect on the ovary (Moran *et al.*, 1989).

The age at the onset of puberty in heifers and lambs is affected by level of nutrition through modulation of gonadotropin secretion. The frequency of LH pulses was reduced and estradiol continued to exert the negative feedback action on secretion of LH when growth was retarded in ewe lambs (Foster *et al.*, 1986). When ewe lambs that had been fed restricted amounts of feed were fed *ad libitum*, sensitivity to estradiol negative feedback was decreased and LH secretion increased (Foster and Olster, 1985). Similar responses were observed in ovariectomized ewe lambs, indicating effects of undernutrition can occur independently of feedback effects of ovarian steroids (Foster *et al.*, 1989). These effects seem to be specific for pituitary gonadotropin secretion, since undernutrition of ovariectomized ewe lambs caused decreased frequency of FSH and LH release, had no effect on serum prolactin (PRL) concentrations, and increased serum growth hormone (GH) concentrations (Foster *et al.*, 1989). In addition, *ad libitum* feeding of previously restricted ovariectomized ewe lambs was associated with increased pituitary concentrations of mRNA for LH and FSH subunits, decreased mRNA for GH, and no change in mRNA for PRL, reflecting changes in LH, FSH, GH, and PRL concentrations in serum (Landefeld *et al.*, 1989). Restricted food intake also resulted in a suppression of the preovulatorylike surge of LH secretion that occurred after estradiol administration. This suggests that pituitary reserves of LH

were less, or the release of the hypothalamic neuropeptide luetinizing-hormone releasing hormone (LHRH) was less, or pituitary sensitivity to LHRH was reduced in lambs fed restricted levels of energy. Both the frequency and amplitude of LH pulses were depressed in heifers fed restricted amounts of feed during the prepubertal period. Response to LHRH administration was also depressed (Day *et al.*, 1986). These and other reports (see, for example, the review of Foster, 1988) suggest that, in ewe lambs and heifers, influences of nutritional level on puberty attainment appear to be mediated, at least partially, by sensitivity of the hypothalamus and pituitary of estradiol feedback, by pituitary responsiveness to LHRH, and by pituitary expression of gonadotropin genes.

The mechanisms by which these responses are altered by nutrition are not known. However, several mechanisms that potentially contribute to altered reproductive function have been suggested. Naturally occurring steroids are oxidatively metabolized by NADPH-dependent mixed-function oxidase enzymes in liver microsomes (Kuntzman *et al.*, 1965). Increased intake of dietary energy or protein results in increased hepatic concentration of these enzymes (Campbell and Hays, 1974; Thomas *et al.*, 1987; Thomford and Dziuk, 1988). Increased enzyme concentration as well as increased liver size resulting from increased plane of nutrition (Ferrell, 1988) could potentially result in increased liver capacity for oxidative metabolism of steroids, which may, in turn, result in decreased negative feedback on the hypothalamic and pituitary systems discussed above. Absolute and relative amounts of fat have been proposed to govern the rate of sexual maturation through regulation of sex steroid hormones and their metabolism (Frisch, 1984). Changes in circulating concentrations of certain amino acid precursors to brain neurotransmitters may reflect changes in nutritional status and may result in altered gonadotropin release. Insulin, thyroxine, and insulinlike growth factor-I (IGF-I) respond to energy and protein intake, and roles for these hormones in hypothalamic, pituitary, and ovarian function have been proposed but not well defined. Roles for epidermal growth factor (EGF), transforming growth factor-β (TGF-β), fibroblast growth factor (FGF), and inhibin have also been suggested (Griffen *et al.*, 1987; Radford *et al.*, 1987; Hammond *et al.*, 1988; Meunier *et al.*, 1988; Ronge *et al.*, 1988; Maiter *et al.*, 1989). These are potentially fruitful areas for future research.

II. Nutritional Influences on Postpartum Estrus, Ovulation, and Conception

Adequate nutrition before and during the postpartum period is essential if acceptable estrus and rebreeding performance are to be achieved in cattle, sheep, and swine (Dziuk and Bellows, 1983). Adequate nutrition during these intervals is even more critical in primiparous animals because of the nutritional requirements for growth, in addition to those for lactation during the postpartum period. Although deficiencies, imbalances, or ingestion of toxic substances can influence aspects of reproductive performance, energy and protein intake are generally of primary concern. Poor nutritional status before the beginning of the postpartum period accentuates the need for adequate nutrition during this interval.

Consequences of inadequate or improper nutrition during this period include delayed postpartum estrus, "silent" estrus, delayed ovulation, decreased ovulation rate, low conception rate, and increased embryonic mortality.

A. *Postpartum Estrus*

Time from parturition until the first postpartum estrus in which ovulation occurs (postpartum interval) varies among animals. In ewes, the postpartum interval may extend from parturition in the spring until resumption of sexual activity in the fall, for most breeds. Swine generally are anestrous during lactation, and first postpartum estrus does not occur until shortly after suckling ceases. Cattle show the greatest variation in duration of the postpartum anestrous period. In this species, the postpartum interval may vary from 15 days upward.

The economic importance of postpartum interval in sheep is dependent upon the management system that is used. In intensively managed farm flocks, in particular those managed to produce more than one lamb crop per year, duration of the postpartum anestrous period is quite important, whereas in extensively managed operations, with late fall or early winter breeding, the postpartum interval is generally of little importance. In swine, the interval from weaning until estrus is of less importance than the failure of sows to show estrus after weaning. The economic importance of the postpartum interval of anestrous in dairy cattle is debatable. Failure to rebreed is of economic concern, but efforts to maintain a 12- to 13-month calving interval may not be economically justified. In the suckled beef cow, postpartum interval is of major concern. If the postpartum interval of anestrus extends beyond 80 days, which often occurs, a 365-day calving interval cannot be maintained. Calves born to cows with extended postpartum intervals of anestrus are younger and weigh less at weaning. In addition, cows with extended postpartum interval of anestrus become cyclic late in the breeding season or fail to exhibit estrus at all, and thus have less opportunity to rebreed.

Length of the postpartum interval in ewes is influenced by season, breed, ram contact, and possibly by lactation (Hunter, 1968). Several studies have been reported that indicate little or no effect of various nutritional restrictions on onset of the breeding season (Hafez, 1952; Hunter and Lishman, 1967; Ducker and Boyd, 1974; Allison, 1977; Dufour and Wolynetz, 1977). Other studies, however, are available that show that poor body condition or submaintenance feeding may result in delayed estrus, suppressed ovulation, or ovulation without behavioral estrus (Smith, 1964; Gunn and Doney, 1975; Shevah et al., 1975). Negative effects of poor nutrition or poor body condition are accentuated by lactation (Shevah et al. , 1974; Dunn and Kaltenbach, 1980) and are further accentuated in primiparous relative to multiparous ewes (Rattray, 1977; Dunn and Kaltenbach, 1980).

Most sows exhibit complete lactational (suckling) anestrus (Edgerton, 1980). Generally, sows express estrus within 4–7 days after piglets are weaned, but with lactational periods of less than 42 days, the postweaning anestrous interval increases and becomes more variable (Dunn and Kaltenbach, 1980; Varley, 1982). Varley and Cole (1976), for example, observed the interval from weaning to estrus to be 8.2 (SD = 2.8) and 4.5 (SD = 0.5) days in sows weaned at

7–10 and 42 days postpartum, respectively. Season and breed also have marked influences on the interval from weaning to estrus and the proportion of anestrus sows (Dyck, 1972; Christenson, 1980). These and other confounding factors have complicated the study of nutritional influences on the postpartum interval in swine. It now seems evident, however, that sows that lose excessive amounts of body weight or condition during lactation have extended remating intervals and an increased incidence of anestrus (Brooks and Cole, 1972; Reese et al., 1982; Aherne and Kirkwood, 1985). Digestible energy intakes during lactation substantially below 12 Mcal/day frequently result in delayed estrus or failure to express estrus, whereas intakes above that amount do not result in additional improvement. Inadequate protein or amino acid, particularly lysine, intake during gestation and lactation results in delayed resumption of normal estrous activity (Svajgr et al., 1972; O'Grady and Hanrahan, 1975) Data reported by Reese et al., (1984) suggest that loss of body fat during lactation was more important than weight loss in causing delayed return or nonreturn to estrus. This observation probably indicates that, compared with change in body weight, change in body fat is a better indicator of energy balance of sows during lactation (Whittemore et al., 1980), but may also indicate that a minimum level of body energy reserves is necessary for resumption of normal estrus (Kirkwood and Aherne, 1985) after weaning.

Increased nutritional level following weaning of piglets has decreased or not affected the interval from weaning until resumption of normal estrous cycles (Brooks and Cole, 1972; Dyck, 1974; Brooks et al., 1975; Fahmy and Dufour, 1976). The differences in observed responses are probably due to differences in nutritional treatments among studies during the preceding gestation and lactation, resulting in differing body condition or energy reserve status at initiation of treatments. Increased dietary intake may result in decreased time to return to estrus and decreased proportions of anestrous sows if the sows are in poor condition, or if piglets are weaned at less than 42 days. However, increased intake does not affect these variables if body condition at the end of the suckling period is adequate. Although results have not been consistent, primiparous sows tend to respond to postweaning nutrition to a greater extent than do multiparous sows.

Duration of the postpartum interval in cattle is affected by nutritional status, breed, suckling, temperature, and season (Dunn and Kaltenbach, 1980; Butler and Smith, 1989; Swanson, 1989). Because of the very different management practices used with dairy cattle as compared to beef cattle, primary concerns regarding nutritional influences differ somewhat. In most dairy cows, the interval from parturition to onset of ovulatory estrous cycles ranges from 17 to 42 days (Butler and Smith, 1989). However, first-service conception rates have declined in recent years in association with increased milk production, whereas second-service conception rates have increased. These findings have been interpreted to indicate that conception rates increase with increasing numbers of estrous cycles preceding insemination (Butler and Smith, 1989) and indicate an advantage to early resumption of normal estrous cycles.

Underfeeding energy or protein during gestation results in increased postpartum interval of anestrus (Dunn and Kaltenbach, 1980). Conversely, cows that are obese at

parturition have greater incidence of metabolic, infectious, digestive, and reproductive disorders (Morrow, 1976; Fronk *et al.*, 1980; Erb *et al.*, 1985). Excessive loss of body weight or fat during the early postpartum period results in increased duration of the postpartum interval of anestrus in dairy cows. Likewise, high levels of milk production have been associated with extended postpartum intervals of anestrus. Available information shows that the duration of the postpartum interval of anestrus is negatively associated with energy balance during the early postpartum period (Butler *et al.*, 1981; Ducker and Morant, 1984; Ducker *et al.*, 1985; Butler and Smith, 1989). Likewise, protein deficiency delays resumption of estrous cycles following parturition (Moustgaard, 1969). More recently, the influence of excessive protein intake on reproductive performance has received greater attention (Treacher *et al.*, 1976; Jordan and Swanson, 1979; Swanson, 1989). In these studies the postpartum interval of anestrus was decreased by feeding high levels of protein, but interval from parturition to conception was increased. Ferguson and Chalupa (1989) concluded that excessive protein can negatively affect reproduction by toxic effects of ammonia and its metabolites on gametes and early embryos, by deficiencies of amino acids, and by exacerbations of negative energy balances.

In beef cattle, due to the less intensive management generally employed, overfeeding energy or protein occurs less frequently than in dairy cattle. Consequences of overfeeding in beef cattle are similar to those in dairy cattle. The duration of the postpartum interval of anestrus is increased in beef cows fed low levels of energy during late gestation and/or early lactation (Wiltbank *et al.*, 1962, 1964, 1965; Morris, 1970; Corah *et al.*, 1974; Dunn *et al.*, 1969; Bellows and Short, 1978; Dunn and Kaltenbach, 1980; Echternkamp *et al.*, 1982; Wetteman *et al.*, 1982; Rutter and Randel, 1984; Henricks *et al.*, 1986). Response to low levels of energy intake prepartum or weight change prepartum is dependent upon body condition at calving. Cows that are in good body condition at calving are affected little by either pre- or postpartum weight changes. Postpartum interval is increased by weight loss in cows that are in thin to moderate body condition prior to calving. This problem is exacerbated by deficient levels of energy intake and weight loss postpartum. Effects of poor body condition at calving can be partly overcome by high levels of feeding postpartum. However, the postpartum period is a period of high metabolic demand due to the requirements for lactation. Thus, it is difficult to feed enough energy to cows to compensate for poor body condition at calving. This problem is intensified in heifers because of the additional nutrient needs for growth during the lactational period. Thus, it is evident that energy status at calving and energy balance postpartum interact to influence the postpartum interval of anestrus in cows.

The duration of the postpartum interval is longer in suckled cows than in milked or nonlactating cows (Dunn and Kaltenbach, 1980; Edgerton, 1980). The delay in initiation of estrous cycles following parturition appears to result primarily from calf contact rather than suckling or lactation per se (Short *et al.*, 1972; Viker *et al.*, 1989). In addition, evidence is available that indicates the calf stimulus interacts with nutritional status of the cow such that the postpartum interval of anestrus is increased by suckling to a greater extent in cows in poor condition than in those in good condition (Dunn and Kaltenbach, 1980). Early weaning of calves (Laster *et al.*, 1973), short-term weaning (Wiltbank and Spitzer, 1978), or partial weaning, such as once per day suckling (Randel, 1981), have reduced the postpartum interval of anestrus in beef cows; however, successful use of any of these approaches requires high levels of management and other inputs (Dziuk and Bellows, 1983).

It is obvious that nutritional status can have an impact on the reproduction via the endocrine system. When nutritional status is adequate, nutrition is not the limiting factor; other factors are then the primary modulators of the reproductive system. During pregnancy, high circulating levels of progesterone and estradiol result in prolonged negative feedback on the hypothalamic–hypophyseal axis, and synthesis of LH by the pituitary is inhibited. Because synthesis is inhibited for an extended period of time, stores of LH are depleted and release is diminished (Nett, 1987). Circulating concentrations of progesterone and estradiol decrease precipitously at parturition, and recovery of the hypothalamic– hypophyseal–gonadal axis begins to occur. When nutrition is inadequate, recovery of the endocrine system is suppressed. In this situation (anestrus), the endocrine status of the postpartum female closely resembles that of the prepubertal female. In situations of poor nutrition, the hypothalamus remains hyperresponsive to negative feedback of estradiol, and secretion of LHRH, and thus LH and gonadal steroids, is low. The mechanisms by which the reproductive endocrinological functions are modulated by nutritional status are not known. The involvement

of numerous factors, as mentioned previously with regard to puberty, has been
suggested.

B. Ovulation and Ovulation Rate

In species that usually give birth to a single progeny, such as the cow, nutrition has
little if any effect on natural ovulation rate
but, as discussed earlier, can certainly influence whether an animal ovulates and exhibits estrus. Short-term increases in energy
intake result in increased incidence of ovulation in thin cows, but generally do not appear to overcome totally the adverse effects
of chronic deficiencies (Dunn and Kaltenbach, 1980). Response to superovulation
treatments is less in heifers fed low levels of
energy than in heifers fed adequately (Harrison and Randel, 1986).

In sheep, ovulation and ovulation rate
vary due to photoperiod, temperature, age,
and breed, as well as weight and/or nutritional status (Cumming, 1977; Christenson,
1980; Dyrmundsson, 1981; Bradford and
Quirke, 1986; Vesely and Swierstra, 1986;
Thomas et al., 1987). These and other confounding factors as well as interactions
among factors often complicate comparison
among studies and make quantitative assessment of the importance of the various ef-
fectors difficult. It is evident from numerous studies, however, that increased body
weight or body condition (reflecting chronic
nutritional status) of ewes at mating or increased nutrient intake at or shortly before
mating (acute nutritional status; flushing),
frequently result in increased ovulation
rates (Coop, 1966a, 1966b; Cumming,
1977; Rattray, 1977; Gunn et al., 1979;
Haresign, 1981; Thomas et al., 1987). Coop
(1966a) has termed these influences static

and dynamic nutritional effects, respectively.

A review of literature (Rattray, 1977) indicated an increase in twinning rate of 5–
20% for each 5 kg increase in body weight
of ewes at mating. A major part of this increase is likely attributable to increased body
condition rather than to body weight per se
(Gunn et al., 1972). Flushing may increase
ovulation rate by 20% or more, but body
condition influences the response to flushing. Several reports have indicated no response to flushing of ewes in good body condition but a positive response of thin ewes
(Rattray, 1977). Results of this nature suggest that in ewes in which ovulation rate is
depressed, because of their nutritional status, flushing may result in an increase to
"normal" ovulation rate. Responses to
flushing have been elicited by both protein
and energy supplementation (Cruickshank
et al., 1988); however, the results of Torell
et al. (1972) indicated the response to each
was inversely proportional to the amount
supplied by the basal ration. Those results
indicate the need of supplying a balanced
diet to achieve the greatest flushing response. Greatest responses to flushing occur
when increased feeding is begun 3–4 weeks
prior to mating. Ovulation rate is increased
less in ewe lambs or yearlings than in mature ewes following flushing.

The influence of nutritional status on
ovulation rate in swine is well documented
(Anderson and Melampy, 1972; Zimmerman, 1972; Aherne and Kirkwood, 1985).
Anderson and Melampy (1972) concluded
from their extensive review of the subject
that increased energy intake prior to mating
increased ovulation rate, and that the optimum time interval for increased feed intake
was 11–14 days. Variable results have been

obtained when gilts were given increased feed or energy on the day before or on the day of mating, and no effects were evident if energy was increased on the day after mating. Anderson and Melampy (1972) also noted that the flushing response was greater in studies in which the difference in energy intake between control and flushing gilts was greater. In a study reported by Self *et al.* (1955), flushing of restricted-fed gilts resulted in ovulation rates comparable to rates in gilts fed *ad libitum*. These and other results (Aherne and Kirkwood, 1985) support the conclusion that flushing relieves a depression in ovulation rate induced by limited feeding.

At protein levels of 12.5–16%, source or amount of protein has little effect on ovulation rate. Over a short period, protein deprivation has little effect on ovulation rate but prolonged feeding of a protein-free diet reduces ovulation rate.

It is clear that feeding very low levels of energy will compromise the female's ability to ovulate. In females of species that are monotocous, this appears to be essentially an all-or-none phenomenon. Data from females of species that are polytocous suggest that an intermediate area exists between maximum ovulation (as defined by genetic potential) and anovulation. At this time, the mechanisms by which ovulation and ovulation rate are modulated by nutrition is a matter of speculation. Increased nutritional levels may result in increased hepatic mixed-function oxidase activity, thereby increasing the rate of degradation of steroids and lessening the inhibition of the hypothalamic–pituitary axis and resulting in increased gonadotropin secretion (Kirkwood and Aherne, 1985; Thomas *et al.*, 1987). Several other mechanisms by which nutri-

tional effects on ovulation may be mediated through the hypothalamus and pituitary have been proposed (Kirkpatrick *et al.*, 1967; Cooper *et al.*, 1973; Wise *et al.*, 1986; Nett, 1987). Alternatively, or in addition, nutritional levels may influence ovarian response to circulating gonadotropins via various stimulatory (insulin, IGF-I, IGF-II, TGF) or inhibitory (EGF, follicular regulatory protein, inhibin, Mullerian inhibiting substance) regulators (Harrison and Randel, 1986; Cox *et al.*, 1987; Draincourt and Fry, 1988; Hammond et al., 1988; Mondschein *et al.*, 1988; McArdle and Holtorf, 1989). Several of these and possibly other factors may interact to produce observed responses.

C. Conception and Embryonic Mortality

In mated ewes, sows, and cows, 20–40% or more of ovulations are not represented by live births (Wilmut *et al.*, 1986; Ayalon, 1978; Scofield, 1972). These losses represent substantial economic losses as they lead to a reduction in litter size in swine and prolific sheep and in cattle and sheep with only one ovulation, an increased interval between births. These losses consist primarily of fertilization failure and embryonic mortality, that is, losses during the period from fertilization until completion of the stage of differentiation (30–35 days). Frequently this embryonic mortality occurs before maternal recognition of pregnancy (day 12–16, depending on species), and in monotocous species does not increase the interval from mating to return to estrus. Fertilization failure and embryonic mortality are often not readily distinguishable, although they are recognized as lower conception rates or lower birth rates. Embryonic losses after maternal recognition of pregnancy cause a delayed return to estrus and can usually be identified by careful observation. In species that have multiple ovulations, the combination of fertilization failure and embryonic mortality is observed as fewer embryos or offspring born than number of corpora lutea present on the ovary or simply as a reduced birth rate.

Fertilization rate can be extremely variable (Ayalon, 1978; Lunstra and Christenson, 1981), but in primiparous cattle, sheep, and swine it is generally greater than 95% (Ayalon, 1978; Pope and First, 1985). Rates of successful fertilization are somewhat lower in multiparous animals, that is, about 87% (King, 1985; Thatcher *et al.*, 1985). Nutritional factors have not been evaluated.

Embryonic mortality is a major source of reproductive losses in most species of animals. It may range from 0 to 40% or more of ova shed (Pope and First, 1985). In addition to abnormalities produced by errors at meiosis or fertilization (King, 1985), adverse nutritional effects (excess nutrients, toxic compounds, or specific deficiencies) can produce abnormal development that may cause death of the embryo. Teratogenic effects have been reported following consumption of toxic compounds or deficient diets (Ferguson and Chalupa, 1989; Hurley and Doane, 1989). For example, excess dietary protein results in decreased reproductive performance, in part through toxic effects of ammonia on gametes and embryos.

The maternal environment may be unable to support normal development because the reproductive tract is abnormal or hormonal patterns are inappropriate. Specific nutritional deficiencies may cause abnormalities in the reproductive tract and/or extend the duration of uterine involution (Hurley and Doane, 1989). The effects of

protein or energy levels appear to be minimal (Kiracofe, 1980). Although "inappropriate hormone patterns" have not been well defined, evidence suggests that progesterone and estrogen are of major importance. Pregnancy depends on a specific sequence of concentrations of progesterone and estrogen, and embryonic loss may be caused by excesses or inadequate amounts (Wilmut *et al.*, 1986; Archibong *et al.*, 1987). Level of feed intake affects progesterone and estrogen concentrations, and an intermediate level appears to be optimum for embryonic survival (Bennett *et al.*, 1964; Edey, 1966; Blockey *et al.*, 1974; Rattray, 1977; Wilmut *et al.*, 1986). However, other data are not consistent with those conclusions (Edey, 1970; Cummings *et al.*, 1975; Edey, 1976; Parr *et al.* 1982; Wiebold, 1988). While involvement of ovarian steroid hormones secreted before, during, and after estrus in survival and development of embryos has been clearly shown, the manner in which those hormones exert their effects and the role of nutrition in modulating those effects has not been well defined.

Survival may be prejudiced by a failure of several aspects of the relationship between the embryo and mother, despite the fact that both are normal. Embryos may die because they are not at the correct stage of development for the particular uterine environment, or they are inappropriately distributed within the uterus, or the immunological response provoked by the embryo may be inadequate for full development (Wilmut *et al.*, 1985; Pope, 1988). The role of nutrition on embryonic–maternal relationships has not been defined.

Although nutritional influences on fertilization or embryonic mortality have been inconsistent and the specific influences have not been well defined, evidence supports that nutrition has important influences on conception rate, as indicated by services per conception or number of progeny produced (Arnett *et al.*, 1971; Torell *et al.*, 1972; Shevah *et al.*, 1975; Leaver, 1977; Rattray, 1977; Topps, 1977; Smith *et al.*, 1980; Tomes and Nielsen, 1982; Aherne and Kirkwood, 1985; Butler and Smith, 1989). For farm animal species, an intermediate, optimumal level of energy and protein intake promotes maximum conception rate, and excessive under- or overfeeding tends to depress conception rate. In addition, in multiple ovulating species increased ovulation rate resulting from overfeeding is often associated with increased embryonic mortality. Thus, for example, although flushing increases ovulation rate, embryonic mortality may also increase, and may result in reduced or no net gain in numbers of progeny born.

III. Nutritional Influences during Pregnancy on Growth, Development, and Survival of the Fetus and Neonate

The influence of fetal growth on animal production is manifest in several ways. Birth weight typically is related to neonatal mortality by a U-shaped curve (Alexander, 1984). Less than optimum birth weights are associated with reduced energy reserves, lowered thermoregulatory capability, and increased perinatal mortality. Low birth weight is correlated with low rates of postnatal growth and decreased mature size. Conversely, birth weights greater than optimum result in increased dystocia and perinatal mortality and also decreased rebreeding performance of the dam.

A. *Prenatal Growth and Development*

Fetal growth is influenced to varying degrees by numerous factors including fecundity, sex, parity, breed or breed cross, heat or cold stress, and maternal nutrition. Their relative importance to fetal growth may differ among species. In general, however, birth weight of each fetus decreases with increased numbers of fetuses, is greater for males than for females, and increases with increased parity of the dam. Birth weight is decreased by heat stress or poor maternal nutrition and increased by cold stress. Influences of maternal nutrition on fetal growth and some of the physiological factors involved will be emphasized in this section. Several review articles are available that relate to fetal nutritional physiology and growth (Meschia, 1983; Munro *et al.*, 1983; Battaglia and Hay, 1984; Jones *et al.*, 1985; Spencer, 1985; Bell *et al.*, 1987a; Mellor, 1987; Ferrell, 1989a).

Fetal growth, in general, follows an exponential pattern (Koong *et al.*, 1975; Ferrell *et al.*, 1976) in cattle and sheep (Fig. 1). Similar patterns are seen in other species. Absolute fetal weight increases slowly during early gestation but quite rapidly during the later stages; in cows and ewes about 90% of birth weight is achieved during the later 40% of gestation. In addition to rapid change in weight, percentage dry matter, protein, and fat increases by twofold or more during this interval. The rapid changes in both fetal weight and composition suggest the need for substantial changes in maternal circulation and metabolism as well as major changes in uteroplacental function to meet fetal needs for growth and development.

In view of the small size of the fetus during early gestation, most of the research efforts regarding nutritional influences on fe-

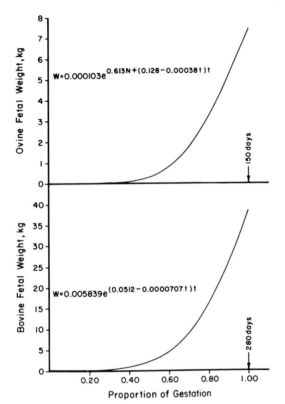

Figure 1 Relationship of fetal weight to stage of gestation in the sheep and cow.

tal growth have centered on the later stages of pregnancy. Some evidence, however, indicates that severe postmating deficiencies may retard growth of the conceptus during the embryonic stage in sheep. Parr *et al.* (1982) reported reduced weight of embryos at 11 and 21 days postmating in ewes fed at 25% of maintenance from mating, and El-Sheikh *et al.* (1955) reported chorionic but not fetal weight was reduced in undernourished ewes at 40 days postmating. Wallace (1948) found no decrease in fetal weight at 90 days in ewes that lost 7% of their body weight during the first 90 days of gestation,

and Everitt (1964) observed a 10% reduction in fetal weight and a 30% reduction in cotyledonary weight at 90 days in ewes that lost 12% of their body weight. Curet (1973) likewise observed that fetal weight at 90 days was adversely affected by protein undernutrition in early pregnancy. Thus, although deficiencies affecting fetal growth are most commonly observed during late pregnancy, severe deficiencies during early pregnancy may result in depressed rates of fetal growth in sheep. These effects have not been observed in swine (Anderson, 1975) and have not been reported in cattle. Compensation during late gestation for the effects of malnutrition during early or mid gestation is dependent upon its degree and duration (Mellor, 1987). Total compensation for the effects of severe deficiencies during early gestation in sheep does not seem to occur (Everitt, 1966, 1967; Alexander and Williams, 1971).

Severe energy or protein malnutrition during late gestation has resulted in marked reductions in fetal growth, as indicated by birth weight in the ewe, cow and sow (Hight, 1968; Pond *et al.*, 1969; Tudor, 1972; Kroker and Cummings, 1979; Levillet *et al.*, 1979; Charlton and Johengen, 1985; Holst *et al.*, 1986). Of these species, the ewe is most sensitive to the deficiencies and the sow is least sensitive. In a series of experiments, Anderson and co-workers (Anderson, 1975; Anderson and Dunseth, 1978; Anderson *et al.*, 1979; Hard and Anderson, 1983) observed minimal influence of total inanition of sows for 30–40 days during early, mid, or late gestation on fetal weight, birth weight, or pre- or postnatal survival. Protein deprivation of sows, however, has resulted in substantial reductions in birth weight and survival (Pond *et al.*, 1969; Levillet *et al.*, 1979). Numerous reports are available that demonstrate reduction in birth weight of lambs from ewes that received restricted energy, protein, or total ration during mid and/or late gestation (Wallace, 1948; Robinson, 1977; Robinson and McDonald, 1979; Holst *et al.*, 1986; Mellor, 1987). Available evidence suggests that the detrimental effects of low maternal protein intake is accentuated by low levels of energy intake (McClelland and Forbes, 1968; Sykes and Field, 1972). The effects of deficiencies are more prevalent and severe in ewes with twin or multiple fetuses than in ewes with single fetuses and are more likely to occur in immature than in mature females.

In addition to influences on birth weight and neonatal survival, inadequate food intake during late pregnancy is associated with weak labor, increased dystocia, reduced milk production and growth of progeny, extended postpartum anestrus, and lower rebreeding performance in cows (Kroker and Cummins, 1979). In ewes, the additional problem of pregnancy toxemia may result in death of both the mother and fetus.

Conversely, gross overfeeding during early and mid-gestation can cause reductions in lamb birth weight of up to 40% (Robinson, 1977). Although birth weight was not affected by obesity in cattle (Arnett *et al.*, 1971), dystocia and neonatal death loss were increased substantially. A serious consequence of overfeeding dairy cattle during gestation has become known as the "fat cow syndrome" (Morrow, 1976; Butler and Smith, 1989). It is characterized by increased occurrence of metabolic, infectious, digestive, and reproductive disorders. Reproductive disorders include retained fetal membranes, metritis, slow uterine involution, and delayed rebreeding.

Considerable progress has been made in recent years toward understanding how maternal nutrition influences fetal growth, particularly in sheep and cattle. Growth of the fetus at any time results from the balance between its genetic potential for growth, reflected in its demand for nutrients, and the limits imposed on the supply of substrates by the placental and maternal systems. It has been suggested that the fetus rarely expresses its full potential for growth because of placental and maternal constraints. The fetus can be undernourished in well-fed mothers when placental size or function is inadequate to meet fetal demands. Conversely, even though the mother is undernourished, the maternal and placental systems may compensate such that fetal malnutrition is minimal. Functional capacity of the placenta clearly has a central role in fetal nutrition, and hence fetal growth, and therefore will be discussed briefly.

In contrast to fetal growth, the pattern of placental growth differs considerably among species (Ferrell, 1989a). In sheep, placental growth ceases or slows considerably soon after mid-gestation (Bell, 1987b), whereas in cattle, placental weight continues to increase throughout gestation (Prior and Laster, 1979), but at a slower rate than fetal weight. Placental transport of oxygen, glucose, and urea and placental clearance of highly diffusible solutes continue to increase in proportion to fetal weight in both sheep and cattle (Bell *et al.*, 1986; Reynolds *et al.*, 1986). Uteroplacental oxidative metabolism is extensive throughout gestation, and even in late gestation when the fetus is several times larger than the placenta, the uteroplacenta consumes 30–50% of uterine oxygen uptake (Sparks *et al.*, 1983; Reynolds *et al.*, 1986). Uteroplacental glucose consumption accounts for at least 70% of uterine glucose uptake, even in late gestation. Thus, even though part of the placental glucose need is met by extraction from the fetal circulation (Bassett, 1986), the high rate of glucose use by uteroplacental tissues has a major impact on the maternal glucose economy.

Although placental weight is a rather crude index of functional capacity of the placenta, it is highly and positively correlated to fetal weight in normal pregnancies (Mellor, 1987), in placental ablation studies (Mellor, 1983; Owens et al., 1986), and under conditions of heat stress (Bell et al., 1987c). Likewise, uterine glucose and oxygen uptakes are positively related to placental weight during late gestation (Bell et al., 1987a). Umbilical and uterine blood flows as well as placental clearance of highly diffusible solutes are also correlated to placental size in carunclectomized or heat-stressed ewes (Owens et al., 1986; Bell et al., 1987c). These associations have been considered to indicate that placental size is an important determinant of rate of fetal growth.

Placental size and function are regulated by complex interrelationships between the fetal and maternal systems that are not well understood. A detailed discussion of the potential mechanisms is beyond the scope of this chapter, but many have been discussed in review articles cited previously. A large amount of natural variation in placental size within breed and between breeds of domestic livestock is evident (e.g., Ferrell, 1989b). Environmental factors such as heat or cold stress (Bell, 1984; Reynolds et al., 1985), parity of the dam, and number of fetuses, as well as maternal nutritional status, influence placental size and function. In general, placental size is reduced by heat stress, increased numbers of fetuses, or poor maternal nutrition.

The placenta functions as the site of exchange of metabolites, heat, respiratory gases, and water between maternal and fetal circulations and as a site of hormone synthesis and secretion. The latter has been discussed elsewhere in this volume and will not be discussed here. Mechanisms by which nutrients enter and cross the placenta include passive diffusion (e.g., oxygen, carbon dioxide, ammonia), facilitated diffusion (e.g., glucose), active transport (e.g., L-amino acids), and solvent drag (Munro et al., 1983). Quantification of rates of transport of many nutrients is complicated by their extensive metabolism by uteroplacental tissues. In addition to the functional size of the placenta, uterine and umbilical blood flows (i.e., placental perfusion) are important determinants of nutrient availability to the fetus (Meschia, 1983; Ferrell, 1989a) and metabolite exchange. Uterine and umbilical blood flows increase exponentially as gestation advances (Bell et al., 1986; Reynolds et al., 1986). Uterine blood flow increases less rapidly and umbilical blood flow increases somewhat more rapidly than fetal weight. The mechanisms of homeorrhetic regulation of placental perfusion are not known (Meschia, 1983).

It is clear that fetal growth may be limited, especially during late gestation, by an inadequate nutrient supply. Although the mechanisms are not fully understood, the following possibility may be hypothesized. Inadequate maternal nutrition reduces rates of uterine blood flow (Munro et al., 1983; Bell, 1984) and decreases nutrient concentrations in maternal arterial blood. These changes decrease nutrient delivery (especially oxygen, glucose, and branched-chain and basic amino acids) and nutrient uptake by the uterus, thereby reducing nutrients available to the uteroplacenta and fetus. Evidence now available suggests the fetus has a higher priority for nutrients than does the placenta; thus in times of insufficiency, the fetus is maintained, to a degree, at the expense of uteroplacental tissues. In a recent study, for example, although oxygen consumption of Charolais fetuses in Charolais and Brahman cows was similar at 227 days of gestation, the net oxygen consumption of the uteroplacenta in Brahman cows was about 40% of that of the uteroplacenta in Charolais cows (C. L. Ferrell, unpublished data). This phenomenon seems to decrease subsequent functional capabilities of the placenta (Munro et al., 1983), which in turn limits the fetal nutrient supply, hence fetal growth. In the aforementioned study, fetal weights of Charolais fetuses did not differ between Charolais and Brahman cows at 232 days (12.9 versus 13.2 kg, respectively), but differed by 13.1 kg (47.0 versus 33.9 kg) at 270 days of gestation.

B. Nutrient Requirements

Quantitative aspects of nutrient requirements for pregnancy in cattle, sheep and swine have been reviewed by Moustgaard (1969), Lodge (1972), and ARC (1980), and will not be discussed in detail in this chapter. Current estimates of the nutrient requirements for pregnancy are based on accretion patterns of energy, protein, and several minerals (Jakobsen, 1956; Jakobsen et al., 1957; Field and Suttle, 1967; Langlands and Sutherland, 1968; Moustgaard, 1969; Lodge, 1972; Rattray et al., 1973, 1974; Ferrell et al., 1976, 1982; McDonald et al., 1979; Noblet et al., 1985) in the fetus, amnionic and allantoic fluids, placenta, and uterus during gestation. Estimates of efficiencies of accretion of most minerals have been assumed to depend only upon efficiency of digestion and absorption. Similarly, efficiencies of protein use for conceptus devel-

opment have primarily been based on efficiencies of protein digestion and absorption as well as estimated biological value of the protein source.

Several reports describe the efficiencies of utilization of metabolizable energy for pregnancy in cattle, sheep, swine (Brody, 1938; Moe and Tyrrell, 1972; Sykes and Field, 1972; Lodge and Heaney, 1973; Rattray et al., 1973, 1974; Ferrell et al., 1976; Robinson et al., 1980; Noblet and Etienne, 1987). Efficiencies of utilization of metabolizable energy for pregnancy in cattle and sheep range from 10 to 20% and average about 13%. Some evidence indicates that efficiency is linearly related to the metabolizable energy concentration in the diet (Robinson et al., 1980). Efficiency of energy utilization in swine has been reported to be 30–40% (Close et al., 1985; Noblet and Etienne, 1987). The lower estimated efficiencies in ruminants as compared to swine are perhaps due, in part, to the greater need for gluconeogenesis in ruminants. The above estimates of efficiency of energy use for pregnancy include increases in maternal energy expenditures associated with pregnancy and maintenance costs of the tissues of the gravid uterus as well as energy costs of growth of those tissues. Oxygen consumption and heat production of gravid uterine and fetal tissues of cattle (Ferrell and Reynolds, 1987), sheep (Bell et al., 1987), and swine (Reynolds et al., 1985) have been reported. Gross efficiency of gravid uterine energy accretion (calculated as energy retention/energy retention + heat production) were about 30, 40, and 50% in cattle, sheep, and swine, respectively. Further, gross efficiency of the accretion of fetal energy in the cow was about 40%. These data indicate that fetal growth per se is a relatively efficient process energetically, but it is not readily observable because of the relatively low efficiency of energy accretion in the uteroplacental tissues that are required to support fetal growth directly, and because of the apparent increase in maternal metabolism, which is required to support fetal growth less directly.

IV. Nutritional Influences on Reproduction in the Male

Sexual behavior, consisting of libido or sex drive and mating ability, and semen quantity and quality are the primary factors of concern in male reproduction. All of these components can be adversely affected by malnutrition. Negative effects are more evident in younger than in older animals, and are more likely to be permanent. The mature male of most species is remarkably resistant to nutritional stress and infertility problems of nutritional origin are not often encountered (Rattray, 1977). Severe under- or overfeeding and deficiencies of specific nutrients, particularly vitamin A, are the most common causes of impaired reproductive capability in the male.

A. Sexual Development

Malnutrition, particularly low energy intake, in the male reduces rates of growth, delays puberty and if severe enough, can permanently impair sperm output (Bratton *et al.*, 1959; VanDemark *et al.*, 1964; Moustgaard, 1969; Hochereau-de Reviers *et al.*, 1987). Deficient nutrient intake is associated with reduction in testicular weight, secretory output of the accessory sex glands, sperm concentration, and sperm motility. Conversely, the reproductive potential of young males may be impaired by overfeeding. Coulter and Kozub (1984) reported scrotal circumference, epididymal sperm reserves, and seminal quality were reduced in bulls fed a diet containing excessive energy as compared with those fed adequately. These effects may be breed related; Pruitt and Corah (1985) concluded that it may be more likely to underfeed breeds of large mature size than to overfeed breeds of smaller mature size.

The mechanisms responsible for the influence of under- or overfeeding on reproductive function in the male are not well documented. As in the female, the influence of energy intake appears to be exerted via the hypothalamic–pituitary–gonadal axis (Howland, 1975; Amann and Schanbacher, 1983; Echternkamp and Lunstra, 1984). Patterns of LH secretion may be altered and serum concentrations decreased by reduced energy intake. Circulating testosterone concentrations are often decreased by reduced feed intake. These low concentrations may be caused by reduced gonadotropin concentrations and/or reduced response of the testes to gonadotropins. These effects are coupled with reduced sensitivity of the accessory sex glands to testosterone (Mann and Lutwak-Mann, 1981).

In the male, vitamin A deficiency is associated with degeneration of the germinal epithelium, resulting in reduction or cessation of spermatogenesis (Hurley and Doane, 1989). Bulls fed vitamin A-deficient diets have delayed puberty, reduced libido, and reduced spermatogenesis. Vitamin E deficiency may result in testicular degeneration, thereby reducing spermatogenesis. Of the minerals, zinc appears to be of primary importance to adequate reproductive function in the male. Deficiency may result in retarded testicular growth, atrophy of tubular epithelium, reduced pituitary gonadotropin output, and reduced androgen production. Zinc deficiency may also influence gonadotropin activity by influencing the gonadotropin receptor complex (Hurley and Doane, 1989). High-molybdenum, low-copper diets have resulted in low libido and degeneration of the seminiferous tubules and testicular interstitial tissue. Iodine deficiency is associated with depressed libido and deterioration of semen quality. Cobalt deficiency, hence cobalamin deficiency, is associated with anemia, general lack of thriftiness, and reduced libido.

B. Mating Behavior

Mating behavior is an important aspect of male reproductive function as it has a direct bearing on the number of females mated. In general, a common sign of a severe energy or protein deficiency in the male is the suppression of endocrine, rather than exocrine, testicular function, coupled with diminished libido, and arrest of growth and secretory activity of accessory sex glands (Moustgaard, 1969; Rattray, 1977; Mann and Lutwak-Mann, 1981; Wodzicka-Tomaszewska et al., 1981). In other studies, however, an energy or protein deficiency had little or no effect on libido (James, 1950; Warnick et al., 1961). Conversely, overfeeding and obesity may result in diminished sexual activity (Okolski, 1975; Wierzbowski, 1978). Overall, it is evident that unless males are severely deprived, there is little effect on the sexual responses and efficiency of the mating responses. Overly fat males may become less willing and able to inseminate a female. Specific nutrient deficiencies may result in lowered physical ability to mate in addition to the specific effects noted in the preceding section.

C. Spermatogenesis and Semen Quality

Nutritional influences on spermatogenesis and semen quality have been reviewed in detail in previous editions of this book (Moustgaard, 1969; Rattray, 1977) and by Leathem (1970) and Mann and Lutwak-Mann (1981). In addition, effects of specific nutrient deficiencies have been recently reviewed by Hurley and Doane (1989). As a result, nutritional effects on these components of male reproduction will be discussed only briefly. Prolonged, severe malnutrition particularly energy or protein deficiencies or water insufficiency, leads to a depression or cessation of spermatogenesis and a decrease in semen quality in most species. These effects are accompanied by decreased size of the testes and accessory sex glands. Atrophy of the interstitial and Sertoli cell populations may accompany these changes (Hochereau-de Reviers et al., 1987). These changes are also accompanied by reduced plasma concentrations of testosterone and LH. It has been suggested (Mann and Lutwak-Mann, 1981) that underfeeding, protein malnutrition, and certain hypovitaminoses somehow interfere with the regulation of the hypothalamic–pituitary axis, thereby altering gonadal steroid production. In addition, responses of accessory sex glands to testosterone are reduced.

References

Aherne, F. X., Christopherson, R. J., Thompson, J. R., and Hardin, R. T. (1976). *Can. J. Anim. Sci.* **56,** 681–692.

Aherne, F. X., and Kirkwood, R. N. (1985). *J. Reprod. Fertil. Suppl.* **33,** 169–183.

Alexander, G. (1984). *In* "Reproduction in Sheep" (D. R. Lindsay and D. T. Pearce, eds.), pp. 199–209. Cambridge University Press, Cambridge.

Alexander, G., and Williams, D. (1971). *J. Agric. Sci. (Camb.)* **76,** 53–72.

Allen, D. M., and Lamming, G. E. (1961). *J. Agric. Sci. (Camb.)* **57,** 87–95.

Allison, A. J. (1974). *Proc. N. Z. Soc. Anim. Prod.* **34,** 45–50.

Allison, A. J. (1977). *Theriogenology* **8,** 19–31.

Amann, R. P., and Schanbacher, B. D. (1983). *J. Anim. Sci.* **59**(Suppl. 2), 380–403.

Anderson, L. L. (1975). *Am. J. Physiol.* **229,** 1687–1694.

Anderson, L. L., and Dunseth, D. W. (1978). *Am. J. Physiol.* **234,** E190–E196.

Anderson, L. L., Hard, D. L., and Kertiles, L. P. (1979). *Am. J. Physiol.* **236,** E335–E341.

Anderson, L. L., and Melampy, R. M. (1972). *In* "Pig Production" (D. J. A. Cole, ed.), pp. 329–366. Butterworths, London.

ARC. (1980). "The Nutrient Requirements of Ruminant Livestock." Agricultural Research Council. Commonwealth Agricultural Bureaux, Slough, England.

Archibong, A. E., England, D. C., and Stormshak, F. (1987). *J. Anim. Sci.* **64,** 474–478.

Arije, G. F., and Wiltbank, J. N. (1974). *J. Anim. Sci.* **38,** 803–810.

Arnett, D. W., Holland, G. L., and Totusek, R. (1971). *J. Anim. Sci.* **33,** 1129–1136.

Ayalon, N. (1978). *J. Reprod. Fertil.* **54,** 483–493.

Bassett, J. M. (1986). *Proc. Nutr. Soc.* **45,** 1–10.

Battaglia, F. C., and Hay, W. W. (1984). *In* "Fetal Physiology and Medicine," 2d ed. (R. W. Beard and P. W. Nathanielsz, eds.), pp. 601–628. Butterworths, London.

Bell, A. W. (1984). In "Reproduction in Sheep" (D. R. Lindsay and D. T. Pearce, eds.), pp. 144–152. Cambridge University Press, Cambridge.

Bell, A. W., Kennaugh, J. M., Battaglia, F. C., Makowski, E. L., and Meschia, G. (1986). *Am. J. Physiol.* **250,** E538–E544.

Bell, A. W., Battaglia, F. W., and Meschia, G. (1987a). *In* "Energy Metabolism of Farm Animals" (P. W. Moe, H. F. Tyrrell, and P. J. Reynolds, eds.), pp. 294–297. Rowman and Littlefield, Totowa, New Jersey.

Bell, A. W., Bauman, D. E., and Currie, W. B. (1987b). *J. Anim. Sci.* **65**(Suppl. 2), 186–212.

Bell, A. W., Wilkening, R. B., and Meschia, G. (1987c). *J. Dev. Physiol.* **9:** 17–29.

Bellows, R. A., Pope, A. L., Meyer, R. K., Chapman, A. B., and Casida, L. E. (1963). *J. Anim. Sci.* **22,** 93–100.

Bellows, R. A., and Short, R. E. (1978). *J. Anim. Sci.* **46,** 1522–1528.

Bennett, D., Axelsen, A., and Chapman, H. W. (1964). *Proc. Aust. Soc. Anim. Prod.* **5,** 70–72.

Blockey, M. A. de B., Cumming, I. A., and Baxter, R. W. (1974). *Proc. Aust. Soc. Anim. Prod.* **10,** 265–269.

Bradford, G. E., and Quirke, J. F. (1986). *J. Anim. Sci.* **62,** 905–909.

Bratton, R. W., Musgrave, S. D., Dunn, H. O., and Foote, R. H. (1959). *Cornell Univ. Agric. Sta. Bull.* **940.**

Brody, S. (1938). *Univ. Mo. Agric. Sta. Res. Bull.* **283.**

Brooks, P. D., and Cole, D. J. A. (1970). *J. Reprod. Fertil.* **23,** 435–440.

Brooks, P. H., and Cole, D. J. A. (1972). *Anim. Prod.* **15,** 259–264.

Brooks, P. H., Cole, D. J. A., Rowlinson, P., Croxson, V. J., and Luscombe, J. R. (1975). *Anim. Prod.* **20,** 407–412.

Brooks, P. H., and Smith, D. A. (1980). *Livestock Prod. Sci.* **7,** 67–78.

Bryan, K. A., Hammond, J. M., Canning, S., Mondschein, J., Carbaugh, D. E., Clark, A. M., and Hagen, D. R. (1989). *J. Anim. Sci.* **67,** 196–205.

Burfening, P. J., Hoversland, A. S., Drummond, J., and van Horn, J. L. (1971). *J. Anim. Sci.* **33,** 711–714.

Burger, J. F. (1952). *Onderspoort J. Vet. Res. Suppl.* **2,** 1–218.

Burris, M. J., and Priode, B. M. (1958). *J. Anim. Sci.* **17,** 527–533.

Butler, W. R., Everett, R. W., and Coppock, C. E. (1981). *J. Anim. Sci.* **53,** 742–748.

Butler, W. R., and Smith, R. D. (1989). *J. Dairy Sci.* **72,** 767–783.

Byerley, D. J., Staigmiller, R. B., Berardinelli, J. G., and Short, R. E. (1987). *J. Anim. Sci.* **65,** 645–650.

Campbell, T. C., and Hayes, J. R. (1974). *Pharmacol. Rev.* **26,** 171–180.

Casida, L. E. (1964). *Proc. 6th Int. Congr. Nutr.,* Edinburgh 1963, pp. 366–378. Livingstone, Edinburgh.

Charlton, V., and Johengen, M. (1985). *Biol. Neonate* **48,** 125–142.

Christenson, R K., and Ford, J. J. (1979). *J. Anim. Sci.* **49,** 743–751.

Christenson, R. K. (1980). *J. Anim. Sci.* **51**(Suppl. 2), 53–67.

Christenson, R. K. (1981). *J. Anim. Sci.* **52,** 821–829.

Close, W. H., Noblet, J., and Heavens, R. P. (1985). *Br. J. Nutr.* **53,** 267–279.

Coop, I. E. (1966a). *J. Agric. Sci. (Camb.)* **67,** 305–323.

Coop, I. E. (1966b). *World Rev. Anim. Prod.* **4,** 69–78.

Cooper, K. J., Brooks, P. H., Cole, D. J. A., and Hayes, N. B. (1973). *J. Reprod. Fertil.* **32,** 71–78.

Corah, L. R., Quealey, A. P., Dunn, T. G., and Kaltenbach, C. C. (1974). *J. Anim. Sci.* **39,** 380–385.

Coulter, G. H., and Kozub, G. C. (1984). *J. Anim. Sci.* **59,** 432–440.

Cox, N. M., Stuart, M. J., Althen, T. G., Bennett, W. A., and Miller, H. W. (1987). *J. Anim. Sci.* **64,** 507–516.

Cruickshank, G. C., Smith, J. F., and Fraser, D. G. (1988). *Proc. N. Z. Soc. Anim. Prod.* **48,** 77–79.

Cumming, I. A. (1977). *Aust. J. Exp. Agric. Anim. Husb.* **17,** 234–241.

Cumming, I. A., Blockey, M. A. deB., Winfield, C. G., Parr, R. A., and Williams, A. H. (1975). *J. Agric. Sci. (Camb.)* **84,** 559–565.

Curet, L. B. (1973). *In* "Foetal and Neonatal Physiology" (K. S. Comline, K. W. Cross, G. S. Dawes, and P. W. Nathanielsz, eds.), pp. 342–345. Cambridge University Press, Cambridge.

Day, M. L., Imakawa, K., Zalesky, D. D., Kittock, R. J., and Kinder, J. E. (1986). *J. Anim. Sci.* **62,** 1641–1648.

Dickerson, G. E., and Laster, D. B. (1975). *J. Anim. Sci.* **41,** 1–9.

Donaldson, L. E., and Takken, A. (1968). *Proc. Aust. Soc. Anim. Prod.* 180–182.

Draincourt, M. A., and Fry, R. C. (1988). *J. Anim. Sci.* **66**(Suppl.2), 9–20.

Ducker, M. J., and Boyd, J. S. (1974). *Anim. Prod.* **18,** 159–167.

Ducker, M. J., Haggett, R. A., Fisher, W. J., Morant, S. V., and Bloomfield, G. A. (1985). *Anim. Prod.* **41,** 1–12.

Ducker, M. J., and Morant, S. V. (1984). *Anim. Prod.* **38,** 9–14.

Dufour, J. J. (1975). *Can. J. Anim. Sci.* **55,** 93–100.

Dufour, J. J., and Wolymetz, M. (1977). *Can. J. Anim. Sci.* **57,** 169–176.

Dunn, T. G., and Kaltenbach, C. C. (1980). *J. Anim. Sci.* **51**(Suppl. 2), 29–39.

Dunn, T. G., Ingalls, J. E., Zimmerman, D. R., and Wiltbank, J. N. (1969). *J. Anim. Sci.* **29,** 719–726.

Dyck, G. W. (1972). *Can. J. Anim. Sci.* **52,** 570–572.

Dyck, G. W. (1974). *Can. J. Anim. Sci.* **54,** 277–285.

Dymundsson, O. R. (1973). *Anim. Breed. Abstr.* **41,** 273–290.

Dymundsson, O. R. (1981). *Livestock Prod. Sci.* **8,** 55–65.

Dziuk, P. J., and Bellows, R. A. (1983). *J. Anim. Sci.* **57,**(Suppl. 2), 355–379.

Echternkamp, S. E., Ferrell, C. L., and Rone, J. D. (1982). *Theriogenology* **18,** 283–295.

Echternkamp, S. E., and Lunstra, D. D. (1984). *J. Anim. Sci.* **59,** 441–453.

Edey, T. N. (1966). *J. Agric. Sci. (Camb.)* **67,** 287–302.

Edey, T. N. (1970). *J. Agric. Sci. (Camb.)* **74,** 181–186, 187–192, 193–198.

Edey, T. N. (1976). *Proc. N. Z. Soc. Anim. Prod.* **36,** 231–239.

Edgerton, L. A. (1980). *J. Anim. Sci.* **51,**(Suppl. 2), 40–52.

El-Sheikh, A. S., Hulet, C. V., Pope, A. L., and Casida, L. E. (1955). *J. Anim. Sci.* **14,** 919–929.

Erb, H. N., Smith, R. D. Oltenacu, P. A., Guard, C. L., Hillman, R. B., Powers, P. A., Smith, M. C., and White, M. E. (1985). *J. Dairy Sci.* **68,** 3337–3349.

Everitt, G. C. (1964). *Nature (Lond.)* **201,** 1341–1342.

Everitt, G. C. (1966). *Proc. Aust. Soc. Anim. Prod.* **6,** 91–101.

Everitt, G. C. (1967). *Proc. N. Z. Soc. Anim. Prod.* **27,** 52–68.

Fahmy, M. H., and Dufour, J. J. (1976). *Anim. Prod.* **23,** 103–110.

Ferrell, C. L. (1982). *J. Anim. Sci.* **55,** 1272–1283.

Ferrell, C. L. (1988). *J. Anim. Sci.* **66,**(Suppl. 3), 23–34.

Ferrell, C. L. (1989a). *In* "Animal Growth Regulation" (D. R. Campion, G. J. Hausman, and R. J. Martin, eds.), pp. 1–17. Plenum Press, New York.

Ferrell, C. L. (1989b). *Proc. Midwest. Sect. Am. Soc. Anim. Sci.,* p. 143.

Ferrell, C. L., Garrett, W. N., Hinman, N., and Grichting, G. (1976). *J. Anim. Sci.* **42,** 937–950.

Ferrell, C. L., Laster, D. B., and Prior, R. L. (1982). *J. Anim. Sci.* **54,** 618–624.

Ferrell, C. L., and Reynolds, L. P. (1987). *In* "Energy Metabolism of Farm Animals" (P. W. Moe, H. F. Tyrrell, and P. J. Reynolds, eds.), pp. 298–301. Rowman and Littlefield, Totowa, New Jersey.

Ferguson, J. D., and Chaulpa, W. (1989). *J. Dairy Sci.* **72,** 746–766.

Field, A. C., and Suttle, N. F. (1967). *J. Agric. Sci. (Camb.)* **64,** 417–423.

Foster, D. L. (1988). *In* "The Physiology of Reproduction" (E. Knobil and J. Neill, eds.), pp. 1739–1762. Raven Press, New York.

Foster, D. L., Ebling, F. J. P., Micka, A. F., Vannerson, L. A., Buckholtz, D. C., Wood, R. I. (1989). *Endocrinology* **125,** 342–350.

Foster, D. L., Karsch, F. J., Olster, D. H., Ryan, D. R., and Yellon, S. M. (1986). *Recent Prog. Horm. Res.* **42,** 331–384.

Foster, D. L., and Olster, D. H. (1985). *Endocrinology* **116,** 375–381.

Friend, D. W. (1973). *J. Anim. Sci.* **37,** 701–707.

Friend, D. W. (1976). *J. Anim. Sci.* **43,** 404–412.

Friend, D. W., Lodge, G. A., and Elliot, J. I. (1981). *J. Anim. Sci.* **53,** 118–124.

Frisch, R. E. (1984). *Biol. Rev.* **59,** 161–188.

Frisch, R. E., Hegsted, D. M., and Yoshinaga, K. (1977). *Proc. Natl. Acad. Sci. USA* **74,** 379–383.

Fronk, T. J., Schultz, L. H., and Hardie, A. R. (1980). *J. Dairy Sci.* **63,** 1080–1090.

Greer, R. C., Whitman, R. W., Staigmiller, R. B., and Anderson, D. C. (1983). *J. Anim. Sci.* **56,** 30–39.

Griffen, S. C., Russel, S. M., Katz, L. S., and Nicoll, C. S. (1987). *Proc. Natl. Acad. Sci.* **84,** 7300–7304.

Gunn, R. G., and Doney, J. M. (1975). *J. Agric. Sci. (Camb.)* **85,** 465–470.

Gunn, R. G, Doney, J. M., and Russel, A. J. F. (1972). *J. Agric. Sci. (Camb.)* **79,** 19–25.

Gunn, R. G., Doney, J. M., and Smith, W. F. (1979). *Anim. Prod.* **29,** 17–23.

Hacker, R. R., King, G. J., Ntunde, B. N., and Narendran, R. (1979). *J. Reprod. Fertil.* **57,** 447–451.

Hafez, E. S. E. (1952). *J. Agric. Sci. (Camb.)* **42,** 199–265.

Hammond, J. M., Hsu, C.-J., Mondschein, J. S., and Canning, S. F. (1988). *J. Anim. Sci.* **66**(Suppl. 2), 21–31.

Hard, D. L., and Anderson, L. L. (1983). *Biol. Reprod.* **29,** 799–804.

Hare, L., and Bryant, M. J. (1985). *Anim. Reprod. Sci.* **8,** 41–52.

Haresign, W. (1981). *Anim. Prod.* **32,** 197–202.

Harrison, L. M., and Randel, R. D. (1986). *J. Anim. Sci.* **63,** 1228–1235.

Henricks, D. M., Rone, J. D., Ferrell, C. L., and Echternkamp, S. E. (1986). *Anim. Prod.* **43,** 557–560.

Hight, G. K. (1968). *N. Z. J. Agric. Res.* **11,** 71–84, 477–486.

Hochereau-de Reviers, M. T., Monet-Kuntz, C., and Courot, M. (1987). *J. Reprod. Fertil. Suppl.* **34,** 101–114.

Holst, P. J., Killeen, I. D., and Cullis, B. R. (1986). *Aust. J. Agric. Res.* **37,** 647–655.

Howland, B. E. (1975). *J. Reprod. Fertil.* **4a,** 429–436.

Hughes, P. E. (1982). *In* "Control of Pig Reproduction" (D. J. A. Cole and G. R. Foxcroft, eds.), pp. 117–138. Butterworths, London.

Hughes, P. E., and Cole, D. J. A. (1975). *Anim. Prod.* **21,** 183–189.

Hunter, G. L. (1968). *Anim. Breed. Abstr.* **3,** 347–355.

Hunter, G. L., and Lishman, A. W. (1967). *J. Reprod. Fertil.* **14,** 473–480.

Hurley, W. L., and Doane, R. M. (1989). *J. Dairy Sci.* **72,** 784–804.

Imakawa, K., Day, M. L., Garcia-Winder, M. Zalesky, D. D., Kittok, R. J., Schanbacher, B. D., and Kinder, J. E. (1986). *J. Anim. Sci.* **63,** 565–571.

Jakobsen, P. E. (1956). *Proc. 7th Int. Congr. Anim. Husb.* **6,** 115–126.

Jakobsen, P. E., Sorensen, P. H., and Larsen, H. (1957). *Acta Agric. Scand.* **7,** 103–112.

James, J. P. (1950). *Proc. N. Z. Soc. Anim. Prod.* **10,** 84–88.

Jensen, A. H., Yen, J. T., Gehring, M. M., Baker, D. A., Becker, D. E., and Harmon, B. G. (1970). *J. Anim. Sci.* **31,** 745–750.

Jones, C. T., Rolph, T. P., Lafeber, H. N., Gu, W., Harding, J. E., and Parer, J. T. (1985). *In* "Physiological Development of the Fetus and Newborn" (C. T. Jones and P. W. Nathanielsz, eds.), pp. 11–20. Academic Press, London.

Jordon, E. R., and Swanson, L. V. (1979). *J. Dairy Sci.* **62,** 58–63.

Joubert, D. M. (1954). *J. Agric. Sci. (Camb.).* **44,** 164–172.

Joubert, D. M. (1963). *Anim. Breed. Abstr.* **31,** 295–306.

Keane, M. G. (1975). *Ir. J. Agric Res.* **14,** 91–98.

Kennedy, G. C., and Mitra, J. (1963). *J. Physiol.* **166,** 408–411.

Kinder, J. E., Day, M. L., and Kittock, R. J. (1987). *J. Reprod. Fertil. Suppl.* **34,** 167–186.

King, W. A. (1985). *Theriogenology* **23,** 161–174.

Kiracofe, G. W. (1980). *J. Anim. Sci.* **51**(Suppl. 2), 16–28.

Kirkwood, R. N., and Aherne, F. X. (1985). *J. Anim. Sci.* **60,** 1518–1529.

Kirpatrick, H. L., Howland, B. E., First, N. L., and Casida, L. E. (1967). *J. Anim. Sci.* **26,** 358–364.

Koong, L. J., Garrett, W. N., and Rattray, P. V. (1975). *J. Anim. Sci.* **41,** 1069–1076.

Kroker, G. A., and Cummins, L. J. (1979). *Aust. Vet. J.* **55,** 467–474.

Kuntzman, R., Lawrence, D., and Convey, A. H. (1965). *Mol. Pharmacol.* **1,** 163–167.

Lammond, D. R. (1970). *Anim. Breed. Abstr.* **38,** 359–372.

Landefeld, T. D., Ebling, F. J. P., Suttie, J. M., Vannerson, L. A., Padmanabhan, V., Beiters, I. Z., and Foster, D. L. (1989). *Endocrinology* **125,** 351–356.

Langlands, J. P., and Sutherland, H. A. M. (1968). *Br. J. Nutr.* **22,** 217–227.

Larsson, S., Nilsson, T., and Olsson, B. (1966). *Acta Vet. Scand.* **7,** 47–53.

Laster, D. B., Glimp, H. A., and Gregory, K. E. (1972). *J. Anim. Sci.* **34,** 1031–1036.

Laster, D. B., Glimp, H. A., and Gregory, K. E. (1973). *J. Anim. Sci.* **36,** 734–740.

Laster, D. B., Smith, G. M., and Gregory, K. E. (1976). *J. Anim. Sci.* **43,** 63–70.

Laster, D. B., Smith, G. M., Cundiff, L. V., and Gregory, K. E. (1979). *J. Anim. Sci.* **48,** 500–508.

Leathem, J. H. (1970). *In* "The Testes" (A. D. Johnson, S. R. Gomes, and N. L. Vandemark, eds.), Vol. III, "Influencing Factors," pp. 169–205. Academic Press, New York.

Leaver, J. D. (1977). *Anim. Prod.* **25,** 219–224.

Lesmeister, J. L., Burfening, P. J., and Blackwell, R. L. (1973). *J. Anim. Sci.* **36,** 1–6.

Levillet, M., Entienne, M., and Salmon-Legagneur, E. (1979). *Ann. Biol. Anim. Biochim. Biophys.* **19,**(1B), 217–223.

Lodge, G. A. (1972). *In* "Pig Production" (D. J. A. Cole, ed.), pp. 399–416. Pennsylvania State University Press, University Park.

Lodge, G. A., and Heaney, D. P. (1973). *Can. J. Anim. Sci.* **53,** 479–489.

Lunstra, D. D., and Christenson, R. K. (1981). *J. Anim. Sci.* **53,** 458–466.

Maiter, D., Fliesen, T., Underwood, L. E., Maes, M., Gerard, G., Davenport, M. L., and Ketelslegers, J. M. (1989). *Endocrinology* **124,** 2604–2611.

Mann, T., and Lutwak-Mann, C. (1981). "Male Reproductive Function and Semen," p. 19. Springer-Verlag, New York.

McArdle, C. A., and Holtorf, A. P. (1989). *Endocrinology* **124,** 1278–1286.

McClelland, T. H., and Forbes, J. T. (1968). *Recent Agric. Res. Northern Ir.* **17,** 131–138.

McDonald, I., Robinson, J. J., Fraser, C., and Smart, R. I. (1979). *J. Agric. Sci. (Camb.)* **92,** 591–603.

Mellor, D. J. (1983). *Br. Vet. J.* **139,** 307–324.

Mellor, D. J. (1987). *Proc. Nutr. Soc.* **46,** 249–257.

Meschia, G. (1983). *In* "Handbook of Physiology," Section 2: Circulation, Vol. 3. (S. R. Geiger, exec. ed.), pp. 241–269. American Physiology Society, Bethesda, Maryland.

Meunier, H., Rivier, C., Evans, R. M., and Vale, W (1988). *Proc. Natl. Acad. Sci. USA* **85,** 247–251.

Moe, P. W., and Tyrrell, H. F. (1972). *J. Dairy Sci.* **55,** 480–483.

Mondschein, J. S., Canning, S. F., and Hammond, J. M. (1988). *Endocrinology* **123,** 1970–1976.

Moran, C., Quirke, J. F., and Roche, J. F. (1989). *Anim. Reprod. Sci.* **18,** 167–182.

Morris, J. G. (1970). *J. Agric. Sci. (Camb.)* **75,** 479–484.

Morrow, D. A. (1976). *J. Dairy Sci.* **59,** 1625–1629.

Moustgaard, J. (1969). *In* "Reproduction in Domestic Animals" (H. H. Cole and P. T. Cupps eds.), 2nd ed., pp. 489–516. Academic Press, New York.

Munro, H. N., Philistine, S. J., and Fant, M. E. (1983). *Annu. Rev. Nutr.* **3:** 97–124.

Nett, T. M. (1987). *J. Reprod. Fertil. Suppl.* **34,** 201–213.

Noblet, J., Close, W. H., Heavens, R. P., and Brown, D. (1985). *Br. J. Nutr.* **53,** 251–266.

Noblet, J., and Etienne, M. (1987). *Livestock Prod. Sci.* **16,** 243–257.

O'Grady, J. F., and Hanrahan, T. J. (1975). *Ir. J. Agric. Res.* **14,** 127–135.

Okolski, A. (1975). *Acta Agrar. Silvestria Zootech.* **15,** 101–121.

Owens, J. A., Falconer, J., and Robinson, J. S. (1986). *Am. J. Physiol.* **250,** R427–R434.

Parr, R. A., Cumming, I. A., and Clark, I. J. (1982). *J. Agric. Sci. (Camb.)* **38,** 39–46.

Plasse, D., Warnick, A. D., and Koger, M. (1968). *J. Anim. Sci.* **27,** 94–112.

Pope, W. F. (1988). *Biol. Reprod.* **39,** 999–1003.

Pope, W. F., and First, N. L. (1985). *Theriogenology* **23,** 91–105.

Pond, W. G., Strachan, D. N., Sinha, Y. N., Walker, E. F., Jr., Dunn, J. A., and Barnes, R. H. (1969). *J. Nutr.* **99,** 61–67.

Prior, R. L., and Laster, D. B. (1979). *J. Anim. Sci.* **48,** 1546–1553.

Pruitt, R. J., and Corah, L. R. (1985). *J. Anim. Sci.* **61,** 1186–1193.

Quirke, J. F. (1979). *Anim. Prod.* **28,** 297–307.

Quirke, J. F. (1981). *Livestock Prod. Sci.* **8,** 37–53.

Quirke, J. F., Stabenfeldt, G. H., and Bradford, G. E. (1985). *J. Anim. Sci.* **60,** 1463–1471.

Radford, H. M., Panaretto, B. A., Avenell, J. A., and Turnbull, K. E. (1987). *J. Reprod. Fertil.* **80,** 383–393.

Randel, R. D. (1981). *J. Anim. Sci.* **53,** 755–757.

Rattray, P. V. (1977). *In* "Reproduction in Domestic Animals" (P. T. Cupps and H. H. Cole, eds.), 3rd ed., pp. 553–575. Academic Press, New York.

Rattray, P. V., Garrett, W. N., East, N. E., and Hinman, N. (1973). *J. Anim. Sci.* **37,** 853–857.

Rattray, P. V., Garrett, W. N., East, N. E., and Hinman, N. (1974). *J. Anim. Sci.* **38,** 383–393.

Reid, J. T., Trimberger, G. W., Asdell, S. A., Turk, L. K., and Smith, S. E. (1951). *J. Dairy Sci.* **34:**510.

Reese, D. E., Moser, B. D., Peo, E. R., Jr., Lewis, A. J., Zimmerman, D. R. Kinder, J. E., and Stroup, W. W. (1982). *J. Anim. Sci.* **55**, 590–597.

Reese, D. E., Peo, E. R., Jr., and Lewis, A. J. (1984). *J. Anim. Sci.* **58**, 1236–1244.

Reynolds, L. P., Ferrell, C. L., Robertson, D. A., and Ford, S. P. (1986). *J. Agric. Sci. (Camb.)* **106**, 437–444.

Reynolds, L. P., Ferrell, C. L., Nienaber, J. A., and Ford, S. P. (1985). *J. Agric. Sci. (Camb.)* **104**, 289–297.

Reynolds, L. P., Ford, S. P., and Ferrell, C. L. (1985). *J. Anim. Sci.* **61**, 968–974.

Robinson, J. J. (1977). *Proc. Nutr. Soc.* **36**, 9–16.

Robinson, J. J., and McDonald, I. (1979). *Ann. Biol. Anim. Biochim. Biophys.* **19**(1B), 225–234.

Robinson, J. J., McDonald, I., Fraser, C., and Gordon, J. G. (1980). *J. Agric. Sci. (Camb.)* **9a**, 331–338.

Ronge, H., Blum, J., Clement, C., Jans, F., Leuenberger, H., and Binder, H. (1988). *Anim. Prod.* **47**, 165–183.

Rutter, L. M., and Randel, R. D. (1984). *J. Anim. Sci.* **58**, 265–224.

Sacco, R. E., Baker, J. F., and Cartwright, T. C. (1987). *J. Anim. Sci.* **64**, 1612–1618.

Self, H. L., Grummer, R. H., and Casida, L. E. (1955). *J. Anim. Sci.* **14**, 573–592.

Scofield, A. M. (1972). *In* "Pig Production" (D. J. A. Cole, ed.), pp. 367–384. Butterworths, London.

Shevah, Y., Black, W. J. M., Carr, W. R., and Land, R. B. (1974). *J. Reprod. Fertil.* **38**, 369–378.

Shevah, Y., Black, W. J. M., and Land, R. B. (1975). *J. Reprod. Fertil.* **45**, 289–299.

Short, R. E., and Bellows, R. A. (1971). *J. Anim. Sci.* **32**, 127–131.

Short, R. E., Bellows, R. A., Moody, E. L., and Howland, B. E. (1972). *J. Anim. Sci.* **34**, 70–74.

Smith, I. D. (1964). *Aust. Vet. J.* **40**, 199–205.

Smith, M. F., Shipp, L. D., Songster, W. N., Wiltbank, J. N., and Carroll, L. H. (1980). *Theriogenology* **14**, 91–104.

Sorenson, A. M., Hansel, W., Hough, W. A., Armstrong, D. T., McEntee, K., and Bratton, R. W. (1959). *Cornell Univ. Agr. Exp. Sta. Bull.* **936**.

Southam, E. R., Hulet, C. V., and Botkin, M. P. (1971). *J. Anim. Sci.* **33**, 1282–1287.

Sparks, J. W., Hay, W. W., Jr., Meschia, G., and Battaglia, F. C. (1983). *Eur. J. Obstet. Gyn. Reprod. Biol.* **14**, 331–340.

Spencer, G. S. G. (1985). *In* "Swine in Biomedical Research" (M. E. Tumbleson, ed.), pp. 1205–1213. Plenum Press, New York.

Spitzer, J. C., Wiltbank, J. N., and LeFever, D. C. (1975). Colorado State University Exp. Sta., Ft. Collins, Gen. Series 949.

Stewart, T. S., Long, C. R., and Cartwright, T. C. (1980). *J. Anim. Sci.* **50**, 808–820.

Stumph, T. T., Day, M. L., Wolfe, M. W., Clutter, A. C., Stotts, J. A., Wolfe, P. L. Kittok, R. J., and Kinder, J. E. (1989). *Biol. Reprod.* **41**, 91–97.

Svajgr, A. J., Hammel, D. L., Degeeter, M. J., Hays, V. W., Cromwell, G. L., and Dutt, R. H. (1972). *J. Reprod. Fertil.* **30**, 455–458.

Swanson, E. W. (1960). *J. Dairy Sci.* **43**, 377–387.

Swanson, L. V. (1989). *J. Dairy Sci.* **72**, 805–814.

Sykes, A. R., and Field, A. C. (1972). *J. Agric. Sci. (Camb.)* **78**, 127–133.

Thatcher, W. W., Knickerbocker, J. J., Bartol, F. F., Bazer, F. W., Roberts, R. M., and Drost, M. (1985). *Theriogenology* **23**, 129–144.

Thomas, D. L., Thomford, P. J., Crickman, J. G., Cobb, A. R., and Dziuk, P. J. (1987). *J. Anim. Sci.* **64**, 1144–1152.

Thomford, P. J., and Dziuk, P. J. (1988). *J. Anim. Sci.* **66**, 1446–1452.

Thompson, L. H., and Savage, J. S. (1978). *J. Anim. Sci.* **47**, 1141–1144.

Tomes, G. J., and Nielsen, H. E. (1982). *In* "Control of Pig Reproduction" (D. J. A. Cole and G. R. Foxcroft, eds.), pp. 527–539. Butterworths, London.

Topps, J. H. (1977). *World Rev. Anim. Prod.* **18**, 43–49.

Torell, D. T., Hume, I. D., and Weir, W. C. (1972). *J. Anim. Sci.* **34**, 479–482.

Treacher, R. J., Little, W., Collis, K. A., and Stark, A. J. (1976). *J. Dairy Res.* **43**, 357–369.

Tudor, G. D. (1972). *Aust. J. Agric. Res.* **23**, 389–395.

VanDemark, N. L., Fritz, G. R., and Marger, R. E. (1974). *J. Dairy Sci.* **47**, 798–802.

Varley, M. A. (1982). *In* "Control of Pig Reproduction" (D. J. A. Cole and G. R. Foxcroft, eds.), pp. 459–478. Butterworths, London.

Varley, M. A., and Cole, D. J. A. (1976). *Anim. Prod.* **22**, 71–77.

Vesely, J. A., and Swierstra, E. E. (1986). *J. Anim. Sci.* **62**, 1555–1562.

Viker, S. D., McGuire, W. J., and Kiracofe, G. H. (1989). *Midwest. Sect. Am. Soc. Anim. Sci. Abstr.*, 138.

Wallace, L. R. (1948). *J. Agric. Sci. (Camb.)* **38**, 93–401.

Warnick, A. C., Meacham, T. N., Cunha, T. J., Loggins, P. E., Hentges, J. F., and Shirley, R. L. (1961).

Proc. 4th Int. Congr. Anim. Reprod. (The Hague) **II,** 202–211.

Warren, M. P. (1980). *J. Clin. Endocrinol. Metab.* **5,** 1150–1156.

Wetteman, R. P., Lusby, K. S., and Turman, E. J. (1982). *Okla. Agric. Exp. Sta. MP* **112,** 12–14.

Whittemore, C. T., Franklin, M. F., and Pearce, B. S. (1980). *Anim. Prod.* **31,** 183–190.

Wiebold, J. L. (1988). *J. Reprod. Fertil.* **84,** 393–399.

Wierzbowski, S. (1978). *Appl. Anim. Ethol.* **4,** 55–60.

Wilmut, I., Sales, D. I., and Ashworth, C. J. (1985). *Theriogenology* **23,** 107–119.

Wilmut, I., Sales, D. I., and Ashworth, C. J. (1986). *J. Reprod. Fertil.* **76,** 851–864.

Wiltbank, J. N., Bond, J., Warwick, E. J., Daves, R. E., Cook, A. C., Reynolds, W. L., and Hazen, M. W. (1965). *USDA Tech. Bull.* **1314.**

Wiltbank, J. N., Gregory, K. E., Swiger, L. A., Ingalls, J. E., Rothlisberger, J. A., and Koch, R. M. (1966). *J. Anim. Sci.* **25,** 744–751.

Wiltbank, J. N., Kassen, C. W., and Ingalls, J. E. (1969). *J. Anim. Sci.* **9,** 602–605.

Wiltbank, J. N., Roberts, S., Nix, J., and Rowden, L. (1985). *J. Anim. Sci.* **60,** 25–34.

Wiltbank, J. N., Rowden, W. W., Ingalls, J. E., Gregory, K. E., and Koch, R. M. (1962). *J. Anim. Sci.* **21,** 219–225.

Wiltbank, J. N., Rowden, W. W., Ingalls, J. E., and Zimmerman, D. R. (1964). *J. Anim. Sci.* **23,** 1049–1053.

Wiltbank, J. N., and Spitzer, J. C. (1978). *World Anim. Rev.* **27,** 30–39.

Wise, M. E., Glass, J. D., and Nett, T. M. (1986). *J. Anim. Sci.* **62,** 1021–1026.

Wodzicka-Tomasewska, M., Kilgour, R., and Ryan, M. (1981). *Appl. Anim. Ethol.* **7,** 203–238.

Zimmerman, D. R. (1972). *J. Anim. Sci.* **34**(Suppl. 2), 57–66.

Zimmerman, D. R., Carlson, R., and Lantz, B. (1974). *J. Anim. Sci.* **39,** 230.

CHAPTER 19

Genetic Variation and Improvement in Reproduction

G. E. BRADFORD, J. L. SPEAROW, and J. P. HANRAHAN

I. Introduction

Reproductive rates and patterns in domestic animals vary greatly among species, among breeds or strains within species, and among individuals within breeds or strains. Individual sows range in litter size from about 5 to more than 25, and breed mean litter sizes for swine vary from 7 or 8 to 14 or 15. Sheep breeds may have a sexual season ranging from a few weeks per year to year round, and cattle breeds have mean ages at puberty from 8 to 18 months or more.

Species and breed differences clearly represent genetic variation, while the variation among individuals within populations may range from all nongenetic, as between identical twins or among individuals of a homozygous inbred line, to largely genetic, as between Merino ewes homozygous for the Booroola gene *(FF)* or its normal allele (+ +).

The objective of livestock producers is usually to maximize the number of viable offspring per breeding animal per unit time, which suggests that maximum prolificacy and minimum parturition intervals will be optimum. Actually, however, reproductive traits are most often characterized by intermediate optima. Effects of number of young per parturition on viability, and to a lesser extent on growth of individual

young, usually result in disadvantages for excessively large litters. Short parturition intervals may lead to reproduction being out of phase with feed resources, for example in grazing sheep. The existence of intermediate optima has significant implications for utilization of genetic variation for improvement of animal productivity.

The importance of reproductive rate is well illustrated in Figure 1 (Dickerson, 1982). The figure depicts the feed energy cost of producing a kilogram of edible meat protein from different species of meat ani-

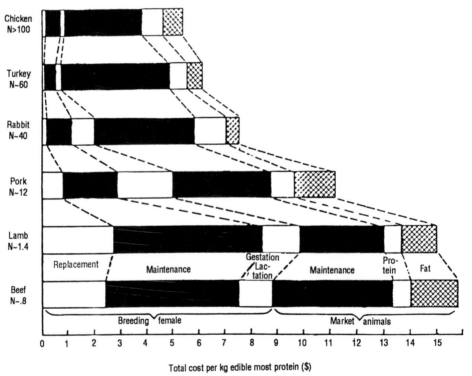

Figure 1 Total life-cycle costs per unit of edible meat protein output, partitioned by components of energy use, for meat animal species ($). Nonfeed costs are assumed to be 150% of breeding female energy costs plus 80% of market animal maintenance energy for nonruminants but only 70 and 50%, respectively, for ruminants. Prices per MJ of feed ME for breeding females and for growing market animals, respectively, were 0.956 and 1.076c for nonruminants, and 0.478 and 0.717c for ruminants. From Dickerson (1982).

mals. As this shows, differences between species in cost of the market animal itself are relatively small, while differences among breeding animal costs are very large, due to differences among the species in numbers of young produced per breeding animal. Thus, for species with relatively low numbers of offspring per unit time, increases in net reproductive rate have very high potential for increasing efficiency of production.

From an evolutionary perspective, optimum reproductive rate is that which maximizes the probability of survival of the group or species. While increases in fecundity may contribute to increasing the probability of survival of a particular group, it is instructive to reflect that larger, longer-lived species have fewer offspring than smaller, shorter-lived species. This is particularly striking as one goes from invertebrates to vertebrates. For example, an oyster female may produce 20–50 million eggs, two or three times a year, while the cow, with rare exceptions, is limited to one offspring per year. It seems clear that as one goes "up" the evolutionary scale, from less to more advanced species in terms of evolutionary time, fecundity tends strongly to decrease. Thus, efforts to increase genetic potential for prolificacy may be countered by natural selection. While this does not preclude desired genetic change, it generally means that the environment must be modified to maintain adequate survival rates as the level of prolificacy is increased above that normal for the population.

Notwithstanding the factors of intermediate optima and strong natural selection for reproductive traits, the amounts and kinds of genetic variation in such traits in domestic animals and the potential for effective utilization of this variation to improve animal productivity are very large.

This chapter is written primarily from the perspective of domestic livestock such as cattle, sheep, and swine, with reference to information from other mammalian species where this adds insight into the topic in question. Many examples are drawn from sheep because of the authors' experience, and because the sheep has a range in litter size from one to five (or more), thus in some respects providing more information on sources of variation in prolificacy than species that consistently give birth either to singles or to large litters.

II. Basis of Genetic Variation

Since genetic transmission occurs via genes (DNA sequences) caried by chromosomes, genetic variation can result from variation in either genes or chromosomes. Of the two, variation in DNA sequences is more important for improvement, but we will deal first with two aspects of chromosomal variation.

A. Effects of Chromosomal Abnormalities on Reproduction

Abnormalities in chromosome number or in the number of copies of individual chromosomal segments due to deletions or unbalanced translocations alter the gene dosage, which can result in high gametic, zygotic, embryonic, or fetal mortality (Gustavsson, 1984; Eldridge, 1985). Abnormalities in chromosome number can be due to polyploidy (multiples of N such as $3N$ rather than the normal $2N$), aneuploidy ($2N \pm 1$), or unbalanced translocations, as well as to mosaics or chimeras.

Polyploidy results from abnormal fertilization by polygyny or polyandry, fusion of cells, or suppression of the first cleavage division (Gustavsson, 1984). Aneuploidy results from nondisjunction of chromosomes during meiosis and to a lesser extent from nondisjunction during mitosis. Monosomics ($2N - 1$) are aneuploids with one copy of one chromosome and two copies of the remainder of the chromosomes. Trisomics ($2N + 1$) are aneuploids with three copies of one chromosome and two copies of the remaining chromosomes.

Polyploidy and aneuploidy of gametes or the conceptus have been shown to be major causes of fertilization failure and embryonic death loss. Either polyploidy or monosomy for any autosome is usually lethal, with many losses occurring during the early embryonic stages of development. Recent evidence in humans also shows that polyploidy/aneuploidy can be a major cause of fertilization failure (Djalali *et al.*, 1988). While most trisomics are also prenatal lethals, a portion of the trisomies of certain chromosomes survive to term. Where aneuploids do survive to term, they are associated with malformations, increased mortality, and complete or near complete infertility if they survive to adulthood. For example, in humans the most frequent autosomal aneuploid surviving to term is trisomy 21, which results in Down's syndrome. Trisomy 21 causes mental retardation, elevated age-specific mortality, and complete infertility in all but a very few females. Most autosomal aneuploidies are also lethal in livestock. For example, in cattle most trisomies die, with only a few grossly malformed trisomies such as calves with trisomy 17 or 18 surviving to birth. Since most polyploid and aneuploid gametes or conceptuses die and are lost during development, the later chromosomal abnormalities are studied, the lower their observed incidence. Thus, while about 10% of the day 10 blastocysts of normal sows have

abnormal numbers of chromosomes (McFeely, 1967), the incidence of autosomal aneuploidies or polyploidies in this and other farm animal species is very low at birth. Recent studies also support the concept that some chromosomally abnormal gametes have a lower chance of completing gametogenesis (Stewart-Scott and Bruere, 1987) and have a lower fertilization rate (Djalali *et al.*, 1988).

Several sex chromosome aneuploids do survive to term, but most are subfertile or infertile as adults. For example, in humans the sex chromosome aneuploids that survive include XXY (Klinefelter's syndrome), XO (Turner's syndrome), and XXX. These sex chromosome aneuploids are also generally subfertile or sterile in livestock species. Klinefelter's or XXY males have been described in cattle, sheep, and pigs (Hutt and Rasmusen, 1982; Eldridge, 1985). Karyotypic analysis of 71 infertile heifers showed that eight were sex chromosome aneuploids or mosaics, and five were 1/29 Robertsonian translocation heterozygotes (Swartz and Vogt, 1983). While few infertile mares have been karyotyped, several of those that have are XO, or XO mosaics (Eldridge, 1985). The two exceptions to this generalization of infertilty of sex chromosome abnormalities are (1) the XYY male, which at least in humans is fertile due to the loss of the extra Y by some spermatogenic stem cells, and (2) the chimeric or mosaic XY/XX males, which are normally fertile. One should also note the reports of elevated percentages of female offspring sired by a few specific XY/XX males (Eldridge, 1985).

Several abnormalities in chromosome structure such as translocations and inversions allow normal development when balanced but impair reproductive performance through the formation of gametes with abnormal amounts of chromosomal material. Reciprocal translocations result from the crossing over and exchange of chromosomal segments between nonhomologous chromosomes during meiosis. For example, a reciprocal translocation between chromosomes 7 and 11 in swine, rcp (7q−;11q+), results in the connection of chromosome 7 to a portion of the long arm of chromosome 11, and the reciprocal connection of chromosome 11 to the long arm of chromosome 7 (Gustavsson, 1984). In this case, during meiosis the homologous regions of chromosome 7 pair with each other as do the homologous regions of chromosome 11, on each fused rcp (7q−;11q+) chromosome. Thus, reciprocal translocations form quadrivalents during meiosis between two pairs of nonhomologous chromosomes, rather than the normal bivalents, which only form between homologous chromosomes. Since these quadrivalents segregate abnormally at anaphase, about half of the gametes formed will be unbalanced or partial aneuploids. Many of these gametes complete gametogenesis and compete with chromosomally normal sperm for fertilizing eggs. Boars with several different reciprocal translocations show a 25–50% reduction in litter size when mated to normal sows (Gustavsson, 1984; Makinen and Remes, 1986). Robertsonian translocations, or centric fusions, result from the fusion of two one-armed chromosomes to form a single fused two-armed chromosome with either one or two centromeres. Many different Robertsonian translocation have been found in cattle, sheep, goats, and pigs, with at least 25 already described in cattle (Gustavsson, 1984; Eldridge, 1985). While many different chromosomes have been found to be involved in centric fusions, certain Robertsonian translocations are more frequent than others.

For example, the 1/29 centric fusion in cattle has been found in at least 30 breeds of cattle, and showed an incidence of 12–14% in Swedish red and white cattle before selection against it was initiated (Gustavsson, 1984).

While monocentric Robertsonian translocation chromosomes segregate normally when carried in the homozygous state, they show a slightly elevated incidence of nondisjunction when carried in the heterozygous state (Eldridge, 1985; Stewart-Scott and Bruere, 1987). Fertility is adversely affected. For example, bulls carrying a 1/29 translocation chromosome have a 4.5–6% lower 56-day nonreturn rate (Gustavsson, 1984). In contrast, rams heterozygous for one to three translocation chromosomes had 3.9–5.3% aneuploid secondary spermatocytes, but no reduction in fertility. Thus, the degree of reduction in litter size or fertility caused by a particular translocation may depend on the species and the ability to select against a particular unbalanced translocation during gametogenesis.

Unbalanced chromosomal aneuploids arise *de novo* due to nondisjunction during gametogenesis. While such aneuploids show major physiological abnormalities and usually die during embryogenesis, those that do survive are generally infertile. Thus, culling such aneuploid animals will affect reproductive fitness in the generation in which they occur but not in the following generation.

In contrast, individuals with balanced reciprocal or Robertsonian translocations generally appear normal phenotypically. However, due to their frequent production of chromosomally unbalanced gametes, carriers of certain translocations have a lower embryonic survival, lower conception rate, and/or lower litter size. Since these reciprocal or Robertsonian translocations are transmitted to the next generation as balanced translocation heterozygotes, calling individuals carrying such translocations will affect the fertility of the present and of subsequent generations. Such translocation heterzygotes should be suspected in cases of low conception rate in monotocous species or low litter size in polytocous species from phenotypically normal-appearing animals, but can only be definitively detected by cytogenetic evaluation.

It is important to realize that not all Robertsonian translocations are detrimental and that such centric fusions may be an important evolutionary mechanism for reducing the number of chromosomes. While Robertsonian translocation chromosomes in the heterozygous state can cause lowered fertility, they show a normal low incidence of nondisjunction when carried in the homozygous state. Sheep homozygous for Robertsonian translocations have normal low aneuploid frequencies (Chapman and Bruere, 1975). Several stocks of mice with homozygous Robertsonian translocations have normal reproductive performance and are commercially available. Many of the normal metacentric chromosomes found in several species may be the products of centric fusions. For example, several lines of evidence suggest that *Bos taurus* cattle ($N = 30$), goats ($N = 30$), and modern sheep ($N = 27$) evolved from a common ancestor with the fusion of six acrocentric chromosomes to form three Robertsonian translocation chromosomes in sheep but not in goats or cattle.

B. Sex Determination and Variation in Gonadal Function

1. Sex Determination and Differentiation

In mammals, an individual's sex is genetically determined by the presence of a Y chromosome. Thus, individuals with a normal Y chromosome, including normal XY individuals, abnormal XYY's, and abnormal, infertile XXY's develop testes and are considered males. Individuals lacking a Y chromosome, including the normal XX as well as the abnormal infertile XO, and XXX individuals, develop as females. In contrast, in birds and in most fish, amphibians, and reptiles, the heterogametic sex is female (the female is ZW, and the male is ZZ).

Mammalian sexual development can be broken down into two major phases: (1) an early component of sexual determination dependent on the expression of a normal complement of sex determining genes which program the formation of a testis or an ovary, and (2) a later phase of sexual differentiation during which the gonad secretes hormones to direct differentiation and development into a sexually differentiated phenotype.

During the development of both male and female embryos the germ cells migrate from the yolk sac to the genital ridge to form the indifferent gonad (Byskov, 1982). Each fetus also develops an indifferent external genitalia, as well as both a Mullerian and a Wolffian duct system.

In the case of normal males, a gene on the Y sex chromosome causes the indifferent gonad to develop into a testis with Sertoli and Leydig cells. This gene is referred to as the testis-determining factor (TDF) in man or the testis-determining gene-Y chromosome (Tdy) in the mouse. While the gene coding for H-Y antigen was once thought to be the testis-determining factor, genetic recombinations between the testis-determining gene and H-Y clearly show that H-Y is not the testis-determining gene (Goodfellow and Darling, 1988). Sertoli cells of the developing testis produce anti-Mullerian hormone (AMH, which is also known as Müllerian-inhibiting hormone, factor, or substance; MIH, MIF, MIS), which causes the regression of the developing Mullerian ducts. Leydig cells produce large amounts of testosterone during mid-pregnancy, which induces the development of the Wolffian duct to form the epididymis, vas deferens, and seminal vesicles, as well as masculinizing the external genitalia to form the scrotum, penis shaft, and glans penis (Short, 1982; Goodfellow and Darling, 1988).

In the case of normal females, which lack a Y chromosome and therefore lack TDF or Tdy, the indifferent gonad develops into an ovary. The Mullerian ducts remain and develop into the oviduct, uterus, and cervix. In the absence of the masculinizing testosterone, the external genitalia develops into the vagina, vulva, and labia characteristic of normal females. Gonadectomy of an early fetus, regardless of its genetic sex, results in the development of an internal female duct system and an external female genitalia (Jost et al., 1973). However, if aberrations occur in one or more of the genetic or developmental switches involved in sexual determination or differentiation, an individual's sex can be altered or reversed, leading to the development of an intersexual form, that is, male pseudohermaphrodite (an individual with a male gonad but female external genitalia), female pseudohermaphrodite (an individual with a female gonad but male external genitalia), or true hermaphrodite (an individual with both male and female gonads).

Many pseudohermaphrodites and a few true hermaphrodites have been described in livestock species including cattle, goats, pigs, and horses, as well as in humans and laboratory animal species. While some of these in-

tersex individuals are sex chromosome aneuploid mosaics or chimeras, most have a normal karyotype. In several cases endocrine defects have been shown to be the cause. In humans, genetic deficiencies in the activity of adrenal steroidogenic enzymes such as 21-hydroxylase result in very high levels of adrenal androgen production, which cause genetic females to develop a masculinized external genitalia (White, 1987). A genetic deficiency of androgen receptors caused by the testicular feminization gene (Tfm) in humans, mice, rats, cattle, and horses (Eldridge, 1985) results in male pseudohermaphrodites since the testis produces normal amounts of testosterone but the tissues cannot respond to this androgen.

Studies on individuals with altered sexual phenotypes have provided much of our understanding of the genetic and hormonal basis of sexual determination and differentiation (Goodfellow and Darling, 1988). Normally, a small portion of the Y, known as the pseudoautosomal region (PAR), pairs and crosses over with the X chromosome during meiosis in males. The pairing of the PAR of the Y with the X is obligate for normal meiotic segregation of an X or a Y chromosome to the resulting sperm. However, since other regions of the Y chromosome contain TDF, and other genes control male differentiation, it is important for normal sex determination that the remainder of the Y not cross over with the X.

Recent gene mapping studies have shown that most of the human XX males (which are all sterile) result from crossovers that occur beyond the PAR and into the regions containing the TDF, that is, TDF crossing over from the Y to the X (Ferguson-Smith et al., 1987; Goodfellow and Darling, 1988). These XX males do not express H-Y antigen. The sex reversal (Sxr) gene discovered by Cattanach (1971) in mice involves a translocation of the mouse testis-determining region, Tdy, to the pseudoautosomal region of the Y, so that the Tdy regularly crosses over with the X (Goodfellow and Darling, 1988). The mating of Sxr carrier males to normal females results in normal XY males, normal XX females, fertile Sxr carrier XY males, and sterile Sxr XX individuals with a male phenotype. A subsequent mutation in the Sxr region, called Sxr', results in infertile XX Sxr' males that lack H-Y antigen. Since both male XX Sxr' mice and male XX humans lack H-Y antigen, these recombinant individuals show that H-Y antigen is not the testis-determining factor (Goodfellow and Darling, 1988; McLaren, 1988). The Sxr region clearly contains the genes necessary for development of a testis, yet lacks some of the Y-linked genes needed for spermatogenesis and male fertility. Further studies are needed to establish more precisely the role of H-Y and other Y sequences on testicular function. There is also evidence for the involvement of autosomal loci in sex determination. The polled gene in goats is an autosomal recessive that is linked to a gene that results in infertile XX intersexes (Eldridge, 1985; Van Vleck et al., 1987). This locus in goats, as well as loci on mouse chromosome 17 (Erickson et. al., 1987) and on human chromosome 6, may be responder or modifier genes that operate in conjunction with normal Y sequences for proper sex determination and development.

2. Freemartinism

The freemartin condition is a serious problem resulting in female sterility in cattle, and has also been found on rare occasions in sheep, goats, and pigs (Hutt and Rasmusen, 1982; Bearden and Fuquay,

1984). The freemartin condition results from the in utero fusion of the placentas of co-twin heifer and bull fetuses and the subsequent transfer of blood-borne cells and hormones through placental anastomoses. While the sexual development of the bull fetus is not greatly impaired, the male cells and elevated Mullerian inhibiting factor and androgen result in sterility of the twin heifer. Freemartin heifers show a wide range of gonadal and reproductive-tract abnormalities, with gonads ranging from almost normal ovaries, to streak gonads, to ovi-testes, to testes. The reproductive tract of the freemartin is frequently masculinized with varying degrees of development of Wolffian duct structures caused by elevated androgens secreted from the male gonad or from the female ovo-testis. The Mullerian ducts are vestigial, and fail to form tubular oviducts, uterus, cervix, or anterior vagina, due to exposure to Mullerian inhibiting factor. The external genitalia is somewhat normal, except for a frequently enlarged clitoris. However, the vagina/vestibule is very short and ends in a blind pouch. While the vagina of a normal newborn heifer is 12–15 cm long, that of the freemartin is only 5–6 cm long (Bearden and Fuquay, 1984). Thus, the freemartin condition can be detected by carefully measuring the length of the vagina of newborn heifers with a sterile 16×125 mm test tube, which may be inserted fully into the vagina of a normal heifer, but only half way into that of a freemartin. The mammary gland of freemartins also shows poor development.

It should be noted that about 8% of the heifers born co-twin to bull calves are not freemartins, presumably due to the lack of formation of anastomoses between the placental circulations. Furthermore, if a male co-twin dies in utero after the formation of placental anastomoses, a freemartin can be born as a single.

3. Gonadal and Uterine Hypoplasia

Genetic disorders resulting in gonadal hypoplasia have been described in several mammalian species. These include genetic deficiencies of the gonadal steroidogenic enzymes 3-β-Hydroxysteroid dehydrogenase, and 17α-hydroxylase/17-20-desmolase in humans. In mice, the hypogonadal (hyg) mutation causes gonadal hypoplasia and infertility due to a deletion of the gene coding for gonadotropin hormone-releasing hormone (GnRH) (Mason *et al.*, 1986). Other less well understood genetic disorders are known to cause decreases in livestock reproductive performance. Hereditary gonadal hypoplasia is a major problem in both sexes of Swedish Highland cattle. Seven years of selection against this condition reduced its frequency from 25 to 8% (Hutt and Rasmusen, 1982). White Heifer disease in Shorthorn cattle is a condition characterized by females with normally functioning ovaries and external genitalia, but with an incompletely developed uterus and vagina suggestive of a defect in Mullerian duct development. This condition has a much higher incidence in white than in roan or red Shorthorn cattle (Hutt and Rasmusen, 1982).

C. Variation in DNA Sequence

Ultimately, the basis of genetic variation is variation in the DNA contained in the chromosomes and in the extranuclear mitochondria. Very rapid advances are being made in knowledge of the structure and the regulation of function of the DNA, and in ability to isolate and characterize individual genes. Variation occurs not only in structural genes but also in regulatory genes (promoters, enhancers) that control the timing and nature of structural gene function (Lewin, 1987; Brent *et al.*, 1989; Day and Mauer, 1989). Variation may occur as a result of single base pair substitutions, frameshifts in coding regions, deletions, and other deviations from exact duplication of the parental DNA. The effect of these events varies from little or none—for example, from base substitutions resulting in conservative amino acid changes or deletions/insertions in noncritical intron regions—to severe effects such as lethality from complete loss of activity of a key enzyme.

Genes controlling differences in reproduction may affect a very large number of processes, including rate of synthesis and release of gonadotropic hormones, number and function of gonadotropin or steroid receptors, gonadal–hypothalamic–hypophyseal feedback regulation, uterine size and function, and many others. Genetic variation in important end points such as ovulation rate or litter size is undoubtedly ultimately due to changes in DNA, each resulting initially in change in amount or function of a specific gene product. However, because of the many genes involved and their interactions, effective analysis and utilization of much of the genetic variation depend on use of methods for handling quantitative variation.

D. Measurement of Genetic Variation

1. Continuously Distributed Traits

The observed variation in any trait is a reflection of both genetic and environmental influences. Thus individuals may differ with respect to any particular characteristic (trait) because they have a different array of alleles at loci affecting the trait or because the complex of environmental influences experienced by the individuals was different. Usually we cannot know or determine the precise pattern of effects for any individual. However, by analyzing the variation within a population of individuals for which genetic relationships are known, the contribution of genetic and nongenetic effects to the total variation can be estimated. The simplest model for the phenotypic value of an individual (P) is

$$P = G + E$$

where G is the effect of the genotype and E is the effect of environment.

With the assumption that G and E are independent, the variation in P is $(\sigma_P^2) = \sigma_G^2 + \sigma_E^2$. The ratio of σ_G^2 to σ_P^2 represents the importance of genetic effects in causing variation in the trait.

The magnitude of the genetic component of variation depends not only on the size of the effects of the genes that influence the trait but also on gene frequency. The simplest model of genetic variation is to consider a trait affected by the genes at one locus and with only two possible forms of the gene, A_1 and A_2 (alleles). There are three possible genotypes in the population: A_1A_1, A_1A_2, and A_2A_2. Let the average value of individuals of these genotypes, expressed as deviations from the mean of the two homozygotes, be a, d, and $-a$ and let the frequencies of A_1 and A_2 be p and q, respectively, which, with random mating, yields p^2, $2pq$, and q^2 for the frequencies of the genotypes A_1A_1, A_1A_2 and A_2A_2. The population mean is

$$M = p^2a + 2pqd - q^2a = a(p-q) + 2pqd$$

The genetic variation is

$$p^2(a-M)^2 + 2pq(d-M)^2 + q^2(-a-M)^2$$
$$= 2pg [a + d (q-p)]^2 + (2pqd)^2$$

which shows how gene frequency and the effects of each genotype contribute to the variation.

Since it is genes and not genotypes that are passed from one generation to the next, it is important to examine the variation in performance in this context. This can be usefully visualized in terms of the breeding value of an individual, which is defined as twice the difference between the genotypic value of the progeny of the individual and the population mean. It can be shown that the variance of breeding values in a random mating population is $2pq[a + d (q-p)]^2$, which is usually referred to as the additive genetic variance. When this expression is compared with that given previously for the total genetic variance, the latter is greater by an amount $(2pqd)^2$, which is attributable to the difference (d) between heterozygotes and mean of the two homozygotes. This is referred to as the dominance variance. Since only the additive effects (breeding values) are transmitted to the next generation, the most important genetic variation in terms of selection is this additive portion. If the breeding value of an individual is denoted by A then the expected phenotypic value of this individual's offspring may be expressed as $\frac{1}{2}A$ + the population mean. The regression of this phenotypic value on the phenotype of the parent is $\frac{1}{2}\sigma_A^2/\sigma_P^2 = \frac{1}{2}h^2$ where h^2 is the heritability. In a similar way it can be shown that the regression of offspring phenotype on the mean of the parents is h^2. This expression is directly relevant to quantifying what happens when selection is applied—which is choosing individuals to be parents of the next generation. The regression of offspring on parent allows the prediction of response to selection. For an individual pair of parents the expected deviation of the progeny from the mean is h^2 (mean superiority of the parents). This can be averaged over all selected parents and will be h^2S, where S is the mean superiority of the group of parents. Thus h^2S is an estimate of the response to selection. When a trait is normally distributed this response to selection can be written as $h^2i\sigma_P$, where i is the selection intensity (the superiority of the selected parents, in σ_P units).

In summary, the response from one round of selection is $h^2i\sigma_P$ and is thus dependent on heritability, the selection intensity, and the phenotypic variance. It is often useful to consider the response relative to the mean of the population as this allows comparisons among traits and is $h^2i(CV)$, where CV is the coefficient of variation. This form of the expression highlights the fact that the proportional increase due to selection can be the same for traits with very different heritabilities due to the corresponding magnitude of the CV. Many reproductive traits have a relatively high CV compared with growth rate or milk yield and hence, despite a low heritability, the response relative to the mean can be just as high as for growth or milk production.

2. All-or-None Traits

Many reproductive traits have a discrete rather than a continuous distribution, which affects the magnitude of heritability and other genetic parameters and needs to be taken into account when calculating expected response to selection. It is assumed that while such traits are discrete with often only two classes, these reflect an underlying continuous distribution of genetic plus environmental effects with thresholds that determine the observable phenotype. If the underlying unobservable phenotype is assumed to be normally distributed, then the relationships between heritability on the observed scale and on the underlying scale can be readily derived.

Consider a trait with only two classes such as conception rate or twinning. In terms of the underlying distribution with one threshold the frequency of the two classes allows one to determine where the threshold is located relative to the mean of the underlying distribution (Fig. 2). The relationship between heritability on the two scales depends on the correlation between the phenotypes on each scale and may be derived as follows:

Let h^2_u be heritability on the underlying scale, h^2_o be heritability on the observable scale, and i be the intensity of selection, and assign values of 0 and 1 to the observed phenotypes. The correlation between phenotypic values on the observed scale and phenotypic values on the underlying scale is $d = i\sqrt{p/(1-p)}$. The correlation between the phenotype on the observed scale and breeding value is h_o, and this is equal to dh_u since the underlying phenotype completely determines the observed phenotype. Thus,

$$h^2_o = \frac{i^2p}{1-p}h^2_u$$

This expression shows that heritability on the observed scale is dependent upon the

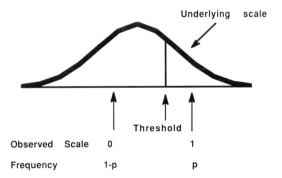

Figure 2 Relationship between observed and underlying distribution for threshold traits.

frequency of the discrete classes. For example, with the same underlying genetic variation (h^2_u) the heritability on the observed scale would be $0.2h^2_u$ for $p = 0.05$ and $0.6h^2_u$ for $p = 0.5$. Thus information on heritability of such discrete traits must take into account the frequency of occurrence before meaningful comparisons are possible.

Similar arguments apply to traits with more than two classes, and appropriate calculations can be made to obtain h given the frequency distribution of the observable phenotype. As the number of classes on the observable scale increases, the correlation between the observed and underlying scales approaches unity, so that in such cases the heritability of the observed scale is essentially equal to heritability on the underlying scale. For example, ovulation rate in prolific sheep breeds like the Finnish Landrace with a fairly wide range in ovulation rate would be essentially unaffected by scale effects, whereas breeds with low ovulation rate like the Merino would be subject to large-scale effects. Calculations for the Finn breed show that the correlation between ovulation rate and a normally distributed underlying scale is approximately 0.96, whereas for a breed with a mean ovulation rate of 1.7 (singles or twins) the correlation is 0.76. If these factors are applied to estimates of the heritability of ovulation rate in Finn (0.50) and Galway (0.32) sheep (Hanrahan and Quirke, 1985), the underlying scale heritabilities become 0.54 and 0.55, respectively. From this point of view it is evident that there is considerable genetic variation in ovulation rate in sheep populations. This is consistent with the situation with this trait in pigs and mice, where the large number of discrete values effectively removes scale effects. No reliable estimates for heritability of

ovulation rate exist in cattle, but it seems likely that a similar level of genetic variation would be found. An examination of the heritability of litter size in cattle and the relationship between litter size and ovulation rate in this species (Hanrahan, 1983) suggests a heritability of 0.3 for ovulation rate in cattle on the underlying scale.

Most of the variation in reproductive traits is quantitative in nature as described above, and the tools developed for working with such variation have been and are being used effectively to make genetic change in performance. A good introduction to these methodologies is Falconer (1989). There is also some variation amenable to manipulation by conventional Mendelian genetics techniques. This is variation due to "major" genes, that is, genes with a sufficiently large effect (one to several phenotypic standard deviations) that differences between alleles at a single locus can be identified and manipulated. A striking example of a major gene is the Booroola (F) gene in Merino sheep (Piper and Bindon, 1982, 1987), which increases ovulation rate by approximately 3 standard deviations per copy, with a very large increase in litter size also. There are apparently several such genes affecting ovulation rate in other sheep populations, and increasing evidence of genes with large effects on other traits.

A second development affecting the tools used for making genetic change is the very rapid advance in ability to identify and characterize specific DNA sequences, using restriction fragment length polymorphism (RFLP), polymerase chain reaction (PCR), and other molecular techniques. Such techniques can also be used to identify particular chromosome segments that have genes or groups of genes with measurable effects on quantitative traits [marker assisted selection for quantitative trait loci (QTLs); Lander and Botstein, 1989]. This enhanced ability to identify individual genes will move many genes from the quantitative category (polygenes) to major genes, in the sense that they can be manipulated individually. This will greatly increase the specificity of analysis of genetic variation. In principle, it should be possible ultimately to describe the genetic variation in any trait completely in terms of DNA sequences, though that time is probably some distance in the future for complex reproduction traits. Even if that point is reached, understanding all of the interactions among all of the gene products involved in producing variation among animals in fertility and fecundity will be a major scientific challenge.

III. Genetic Variation in Components of Reproduction

A. Fertility

Fertility of a herd or other population of animals is usually measured as the proportion of females conceiving in a defined period of time such as a year, or a specified breeding season, which in livestock would often be a period of 1–3 months. The term is also used with respect to individual animals, both male and female.

Genetic variation in fertility may be the result of differences among males, females, or interactions between the two. Genetic causes of infertility or subfertility include chromosomal abnormalities such as translocations, major genes, and quantitative variation, as described in Section II. Heritability of fertility in livestock species is usually low, reflecting the fact that many environmental factors—for example, age, nutri-

tional and health status, and ambient temperature—affect fertility. There is quantitative genetic variation, as evidenced by breed differences and by nonzero heritabilities. Estimates of heritability of fertility traits in dairy cattle are typically very low, 5% or less (Hansen et al., 1983; Standberg and Danell, 1988), but values of 5–10% have been reported for conception rate in sheep (Mohd-Yussuf et al., 1988) and of 17 and 7% for age at first calving in beef cattle (Meacham and Notter, 1987). The value for first calving may be affected by variation in age at puberty as well as in fertility. Heritability estimates of 5–10% have been reported for age at first farrowing and farrowing interval in pigs (Johansson and Kennedy, 1982).

The low heritability of fertility is not unexpected, as natural selection will have tended to fix alleles favoring high fertility, and the all-or-none nature of the trait reduces heritability on the observed scale as described in the previous section.

B. Ovulation Rate

Number of eggs ovulated at a fertile estrus sets the upper limit to number of young born per parturition or litter size, except for the negligibly rare occurrence of identical twins. Thus ovulation rate is a critical parameter in the determination of reproductive rate.

There is substantial genetic variation in ovulation rate in multiple ovulating animal species. This variation is evident as breed and strain differences in mean numbers of corpora lutea, ranging from about 10 to over 20 in pigs, and from 1.2 to 4.5 in sheep. Also, intrapopulation heritability estimates are moderate to high, for example 0.07 to 0.50 with a mean of 0.29 in sheep

(Hanrahan, 1987) and 0.42 in swine (Cunningham *et al.*, 1979). In addition, there are several reports of large differences in ovulation rate within breeds or strains, due apparently to segregation of major genes (Piper and Bindon, 1982; Hanrahan and Owen, 1985; Bradford *et al.*, 1986; Mahieu *et al.*, 1989). Thus in these two species (and in other litter-bearing species including goats, rabbits, rats, and mice) it is possible to manipulate genetic variation to set mean ovulation rate at any specified value over a two- to threefold or greater range.

Since some species normally ovulate only one ovum at each estrus, there is less apparent variation, phenotypic or genetic, in ovulation rate within these species than in litter-bearing species. However, evidence is accumulating for appreciable genetic variation in ovulation rate in cattle, as reviewed by Morris and Day (1986).

C. Prenatal Survival and Litter Size

Number of young born per parturition (litter size) is the product of ovulation rate and prenatal survival. Since it is virtually impossible to measure fertilization rate in pregnancies allowed to go to term, any loss due to fertilization failure is usually included in prenatal loss.

Genetic variation in prenatal survival rate may be due to genotype of the dam or of the embryo/fetus, with the sire and the dam both contributing to the latter. As for other traits, genetic variation may be evidenced as differences among breeds or inbred lines, as segregating genes with large effects, or as quantitative variation within populations.

Genes known to have large effects on prenatal survival are mostly in the detrimental category. An example is the A^y gene in mice, lethal when homozygous due to de-

fects of the trophoblast cells and failure of the embryo to implant. Chromosomal abnormalities may also cause early death of the embryo. One gene with beneficial effects, or at least a locus with alleles contributing to high and lower survival rates, is the *ped* gene in mice, which maps within the MHC complex (Warner, 1986).

Differences among breeds and strains and heritabilities within breeds both appear to be much lower for prenatal survival than for ovulation rate, across a range of laboratory and livestock species. This suggests that genetic change in litter size can be effected much more readily by changing ovulation rate than by changing prenatal survival (Hanrahan, 1987). However, there is genetic variation in prenatal survival in mice (Bradford *et al.*, 1980), in pigs (Bolet, 1986), and in sheep (Hanrahan, 1986). The low heritability means that response to selection in livestock species is likely to be slow, but it has been shown in mice (Bradford, 1969) that selection can be used to increase prenatal survival substantially. In that experiment a strain of mice with 88–90% of all corpora lutea represented by normal young at term, compared to the more typical 75–80%, was produced by 14 generations of selection, suggesting that Bishop's (1964) conclusion, that a large proportion of prenatal loss is due to deleterious genetic variation arising each generation and is thus inevitable, is perhaps unduly pessimistic.

It is important to note that high prenatal survival strains or breeds have more uniform litter size than those with lower survival. This uniformity contributes to higher postnatal survival than in populations with comparable mean but greater variation in litter size. Thus an increase in litter size achieved via an increase in prenatal survival, though more difficult to achieve, may be more valuable than the same increase via an increase in ovulation rate. It may also be noted here that there is genetic variation in the amount of variability in ovulation rate; for example, the Romanov and Finnsheep breeds of sheep have very similar mean ovulation rates but the Romanov has a much lower coefficient of variation (Hanrahan, 1986). Populations in which a major gene such as the *F* gene is segregating have very high coefficients of variation. Such information may also be used to modify variation in litter size.

The effect of these two variables on litter size is complicated by a strong negative but nonlinear relationship between the two. Using data from different sheep breeds and crosses, Hanrahan (1982) developed an equation relating number of young born *(Y)* to ovulation rate *(X)*: $Y = 0.15 + 0.926X - 0.0763X^2$. From this, survival values in ewes that conceive with two, three, four, and five ova are estimated to be 0.85, 0.75, 0.60, and 0.57, respectively. This suggests that increasing ovulation rate will increase litter size, but at a diminishing rate for each additional ovum. The same pattern appears, from breed comparisons, to hold for other species such as pigs, with much higher ovulation rates and litter sizes. However, selection within populations for ovulation rate alone in mice (Bradford, 1969) and pigs (Cunningham *et al.*, 1979) has so far failed to produce an increase in litter size. This suggests that selection, at least in species with usually high prenatal mortality, such as pigs, should be practiced for either prenatal survival or for litter size, or some index including these, as well as ovulation rate, if litter size is to be improved (Johnson *et al.*, 1984).

With regard to the relative contributions of genotype of the dam and of the fetus, the data available suggest that both contribute, with the genotype of the dam contributing much more of the total genetic variation in most populations. These conclusions are derived from results of strain and inbred line crossing, and selection and embryo transfer experiments (Bradford, 1972, 1979; Eisen, 1986).

Some work on the mode of action of genes affecting litter size and its components has been done. In most studies of most species, ovulation rate has been shown to be an additive trait, which is consistent with the higher heritability of this component. Genes affecting prenatal survival, on the other hand, show a strong degree of dominance (Bradford and Nott, 1969), resulting in substantial heterosis for this trait and for litter size (Eisen, 1986). Knowledge of the mode of gene action and of the mean component values for different breeds or strains can be used to identify stocks that will produce crosses with much above average heterosis in litter size (Bradford, 1979). That model was developed from data on mice, and has apparently not been tested extensively in other species. However, Legault and Caritez (1983) report performance of different crosses among several breeds of pigs that yield estimates of heterosis in litter size from near zero to over 40%. From the limited information available on ovulation rate and embryo survival in some of the breeds involved, it appears quite possible that the same model applies in pigs.

Heritability of litter size has been estimated for several species. As expected, based on the fact that one of the component traits has a low heritability, these estimates are fairly low, averaging about 10% in pigs (Hill and Webb, 1982) and in sheep (Hanrahan, 1987). Realized heritability has also been estimated in a number of selection experiments with mice, pigs, and sheep, with some effort in other species including evaluations of effects of selection for multiple births in cattle currently in progress in several countries. Selection has been quite effective in mice, with increases over multigeneration experiments from 25% to near 100% of the base population mean (Eklund and Bradford, 1977; Eisen, 1986). In swine, traditional selection, within relatively small population sizes, has generally been ineffective (Ollivier and Bolet, 1981), while screening a very large population for "hyperprolific" animals has produced a significant increase (Legault *et al.*, 1981). In sheep, substantial genetic change has been effected in several populations (Bradford, 1985). Results in cattle are very preliminary, but encouraging.

In those cases where the components of litter size have been assayed in lines selected for litter size, most if not all of the response can be attributed to an increase in ovulation rate (Falconer, 1960; Packham and Trifitt, 1966). However, as pointed out earlier, response to selection for increased ovulation rate in several studies has not been accompanied by an increase in litter size. This represents a rather striking example of asymmetrical correlated response (Bohren *et al.*, 1966).

D. Body Size, Litter Size, and Gestation Period

The genetic relationships between litter size and body size between and within species are also relevant to genetic variation in reproduction. In general, larger species have fewer young per parturition, while there is a tendency for larger breeds, strains, and individuals within species to have more young. A plausible hypothesis

for this has been developed by St. C. S. Taylor and was elaborated by Land (1977). Young weighing more than about 10% of the mother's adult weight suffer high mortality due to difficult parturition, while those weighing less than 3% tend to be lost because they are too weak to nurse or too small to thermoregulate in the perinatal period in a normal environment. Thus individual birth weight in the range of 5–7% of mother's weight appears to be optimum for a wide range of species.

Total birth weight in mammals varies with mature size of the species, with the average relationship described by the equation log litter weight (g) = 0.8 log maternal weight (g) − 0.33. This means that total birth weight as a percent of maternal weight falls as maternal weight increases, a relationship illustrated in Figure 3, adapted from Leitch *et al.* (1959) by Land (1977). In this figure the straight line indicates a birth weight equal to 5% of maternal weight, and it may be noted that the birth weight of many single-young-bearing mammalian species is in the region where this line crosses the interspecies curve. The higher relative total birth weight of smaller species therefore permits such species to have more than one young, consistent with the fact that most litter-bearing mammals have mean adult female weights of less than 100 kg. The domestic pig is a notable exception, but may have evolved relatively recently from a smaller species, and in any case differs markedly from other common mammals in many biological parameters (Taylor, 1985).

A species by definition is an interbreeding population, one consequence of which is uniformity in gestation period within the species; intraspecies coefficients of variation are typically 1.5–2.0%, compared to 10–15% for growth parameters and 30–40%

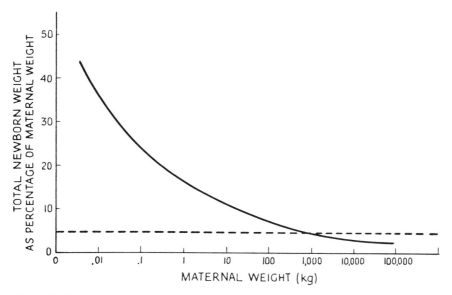

Figure 3 The relationship between maternal weight and total newborn litter weight (adapted from Leitch *et al.*) together with a 5% individual birth weight (broken line) From Land, 1977.

for ovulation rate and litter size. As stated by Land (1977):

With limits to fetal growth in a given period of time, large strains within a species would give birth to relatively small young and small strains to relatively large ones. This is borne out by the literature (Donald and Russell, 1970). The range of body weight between strains within a species will, therefore, be restricted at the upper end of the distribution by the ability of the fetus to grow fast enough for it to be more than around 3% of maternal weight, and hence able to survive; and, at the lower end of the distribution by the ability of the fetus to grow sufficiently slowly for it to be less than around 10% of maternal weight, and hence not to suffer from difficulties at parturition. With large strains giving birth to relatively small young, and small strains to relatively large ones, there would be spare "conceptus capacity" in the large strains, but inadequate capacity in the small ones, so that natural selection would fa-

vor the birth of numerically larger litters by the heavier strains and vice versa.

Thus the relationships among body weights at birth and maturity, litter size, and gestation period can account for the tendency for larger species to have fewer young but larger breeds or strains within species to be more prolific.

The positive relationship between body weight and number of young born within species is apparently mediated largely through the positive genetic correlation between body weight and ovulation rate (Brien, 1986). The correlation of body weights with other reproduction traits is much less consistent. Results from some selection experiments for growth rate or body weight suggest strong negative associations between body weight and fertility, prenatal survival, or postnatal survival (Barria and Bradford, 1981; Lasslo *et al.*, 1985), while others suggest little or no correlation (Pattie, 1965; Horstgen-Schwark *et al.*, 1984; Brien and Hill, 1986). The differences may be attributable in part to amount of prior selection for growth and the intensity and duration of selection during the experiment. The general impression from surveying the results in a number of species is that, on average, selection for growth rate results in a significant decline in net reproductive rate, and that in some cases the decline is quite severe.

E. Postnatal Survival

Postnatal survival is very much affected by a number of environmental factors, including litter size, nutrition, and protection from exposure to inclement weather or disease. Heritability therefore is low, with most estimates less than 10% (Cundiff et al., 1982; Baker and Steine, 1986). However, there is evidence of quite large differences in survival among breeds in both direct and maternal genetic effects in cattle (Cundiff *et al.*, 1982), sheep (Dickerson, 1977), and swine (Gaugler *et al.*, 1984), and substantial favorable heterosis, especially in swine (Johnson, 1981).

Number of young weaned is determined by litter size and their survival rate; birth weight is markedly affected by litter size and in turn has a large effect on survival. Thus the relationships among litter size, birth weight, and viability are important to an understanding of the possibilities of manipulating net reproductive rate by genetic means.

These relationships will be illustrated from data on sheep. Birth weight of lambs from litters of two, three, and four are ap-

proximately 78, 65, and 57%, respectively, those of single lambs from adequately fed ewes of the same breed, age, and management system. This general pattern was described initially by Dickinson *et al.* (1962), and has been shown to hold across a quite wide range of breeds and environments. [The prolific Romanov breed and its crosses appear to produce multiple birth lambs with somewhat higher birth weights, relative to singles, than the above values (Razungles *et al.*, 1985).] Survival is curvilinearly related to birth weight, with maximum survival above the mean but below the maximum, for most breeds (Shelton, 1964; Hinch *et al.*, 1985). Because twins are lighter than singles on average, they suffer higher mortality. However, singles and twins of the same birth weight have similar survival rates (Purser and Young, 1964). Triplets, on the other hand, suffer substantially greater mortality than can be explained on the basis of their lower birth weight alone. Bradford (1985) estimated that in an environment in which survival of single lambs is 88%, twins would be expected to have a survival rate of 82% and triplets 70%, based solely on expected differences in birth weight. Estimates of survival from an experimental flock (Iniguez *et al.*, 1986) were 88, 82, and 54% for litter sizes of one, two, and three, respectively, while Hinch *et al.* (1985) reported data from three flocks that gave average values of 90, 81, 55, and 46 for litters of one, two, three, and four, respectively. These results indicate that milk supply, maternal behavior, etc., as well as lower birth weights and attempts to simultaneously deliver multiples through the birth canal, contribute to mortality of triplets and higher multiples.

The average of the values above yield estimates of 0.89, 1.63, and 1.63 lambs weaned for ewes giving birth to one, two, and three lambs; that is, the number weaned does not increase as litter size increases from two to three, suggesting an optimum litter size of two (Bradford, 1985). These values apply to moderately good management conditions, and the higher mortality of multiple births can be greatly accentuated under harsher conditions, even to the point that fewer young will be weaned by twin-bearing than single-bearing dams. On the other hand, under intensive management, with fostering and/or artificial rearing, litter sizes weaned might be 0.95, 1.80, and 2.40. Thus optimum litter size born is very sensitive to variation in environment and management.

F. Seasonality and Age at Puberty

In several species, including sheep, goats, horses, and camels, females have a well-defined breeding season, and thus will produce young only at certain times of the year. This subject has received considerable research attention in sheep and to a lesser extent in goats and other seasonal breeders. The primary environmental cue controlling season is photoperiod (Karsch *et al.*, 1984; Karsch and Wayne, 1988). Thus, strains of seasonal breeding species that have evolved in the tropics breed year round (Iniguez *et al.*, 1990). Among the seasonal breeds, there is considerable variation in both date of onset and duration of the breeding season (Quirke and Hanrahan, 1985; Quirke *et al.*, 1988). There are rather few estimates of heritability of date of onset or cessation, or duration of the breeding season, but the evidence available suggests a moderate heritability (Hanrahan and Quirke, 1986).

Reduction of seasonality in seasonal breeding species may be desirable to permit marketing a more regular year-round supply of product or to utilize seasonally available feed or labor. Desired change may be effected in some cases by changing date of onset or cessation of the breeding season, but ultimately the most useful change would be elimination of seasonality. Knowledge of the nature of genetic variation in this parameter lags greatly behind that for litter size and its components, and seasonality represents an area where much more research is needed.

Age at puberty also shows genetic variation in the form of breed differences; less is known about intrabreed genetic variation, although there is evidence of low to moderate heritabilities. In cattle, breeds specialized for dairy purposes tend to have early puberty, while *Bos indicus* breeds are characteristically older at puberty than *Bos taurus* breeds, either beef or dairy. In sheep and swine, highly prolific breeds, such as Finnsheep and Romanov, and prolific Chinese breeds of pigs, tend to have very early puberty. Both prolific breeds and dairy breeds are kept in environments where they have received much closer supervision and more intensive feeding from humans than pastoral or nomadic breeds, and it is not surprising that the former reach puberty earlier while the latter are typically late.

IV. Exploitation of Genetic Variation for Improvement of Animal Performance

A. Basis of Genetic Change

Genetic change depends on replacing genes with others that differ in their effect on the phenotype of interest. This may involve major genes, for example, the replacement of a standard ("+") gene with a superior (e.g., *F*) allele, or a defective gene with a normal allele. Alternatively, it may involve selection, by conventional means or by RFLP or other markers, to increase the frequency of favorable alleles at quantitative trait loci. With major genes, the genetic change may be entirely or largely effected in one generation, while for polygenic traits, an essentially linear response to selection may continue for as many as 20 or more generations.

Reproductive performance may also change as a result of changes in frequency of different genotypes, without a change in gene frequency. An increase in proportion of homozygous loci is typically associated with a decrease in performance (inbreeding depression), while increased reproductive fitness often results from an increase in heterozygosity (heterosis).

B. Objectives

The first step in any genetic improvement program is to define the breeding objective. As indicated in the introduction, the goal of genetic change in reproduction of domestic animal populations is not necessarily to maximize litter size or other components of performance. It is rather to maximize the number of viable young, of desired production potential for other performance traits, produced per unit of input. Often, this will mean maximizing number of young weaned per breeding female per year. Achieving this objective means matching genetic potential to the production environment of interest.

Nitter (1987) estimated the economic value from an increase in mean litter size in

sheep, for different production environments. His results, depicted in Figure 4, illustrate two important points: (1) in all systems, the value of increasing number born is strongly dependent on current mean litter size, being high at mean litter sizes under 1.5 and decreasing sharply at values above that, and (2) the value of an extra unit (0.1 lamb) differs depending on the production environment, principally on mortality rates and production costs. The method outlined in Nitter's paper can be readily modified to accommodate different assumptions and different coefficients of variation in litter size, and represents a very useful method of quantifying the effects of a change in prolificacy, i.e. of defining the breeding goal for this trait.

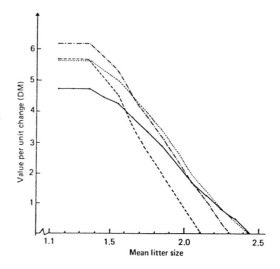

Figure 4 Economic value per unit change of litter size (0.1 lambs) depending on the intensity of production environment and on average litter size of the population. —— System 1; ⋯ system 2; - - - system 3; ⋯⋯ system 4. From Nitter (1987).

C. Use of Breed Resources

1. Breed Replacement, Partial or Complete

Large, permanent genetic change may be effected by replacing one breed with another, or by replacing a breed with a crossbred based in part on the original breed. An example of successful complete breed replacement is provided by pig production operations in the United States replacing traditional U.S. breeds, weaning 12–14 pigs per sow per year, with crossbred sows based on European Landrace and Large White breeds, capable of raising about 20 pigs per year (Bichard and David, 1986). While some changes in management may be required to achieve this increase, a change in genetic potential is an essential first step.

In sheep there are a number of highly prolific breeds available, including the Finnsheep, Romanov, D'Man, and several others (excluding breeds carrying major genes for prolificacy, discussed below). The three breeds mentioned all appear to transmit their prolificacy additively to crosses—that is, crosses with less prolific breeds have a mean litter size intermediate between those of the two parents (Ricordeau et al., 1976; Dickerson, 1977; Lahlou-Kassi et al., 1989). This makes it possible to set mean litter size at any desired level, by varying the proportion of prolific and nonprolific breed inheritance in the cross. Interestingly, there are differences in the coefficient of variation in litter size in these three breeds, which are transmitted to their crossbred daughters; the Romanov breed is least variable, and the D'Man most variable.

Inheritance from prolific breeds is being used in several countries to increase the net reproductive rate of commercial sheep flocks, most commonly in the form of "synthetic" or composite breeds carrying 25–50% inheritance of the prolific breed, with the balance from a locally adapted breed. Investigations of the possibility of using highly prolific Chinese breeds of pigs in the same manner are being carried out, but to date the deficiencies of such breeds in growth rate and carcass leanness appear to outweigh their advantages in prolificacy (Gueblez et al., 1987).

The primary difference between prolific and nonprolific breeds is in ovulation rate. Since prolific breeds are often deficient in other traits such as growth rate and carcass quality, Hanrahan (1987) has recommended selection of a prolific breed to increase its ovulation rate even higher, so that a lesser proportion of inheritance from that breed is necessary to achieve a specified increase in prolificacy in the crossbred population.

In cattle there are substantial differences between breeds in fertility and calf crop percent raised. In favorable environments, breeds of European origin are usually substantially superior to *Bos indicus* breeds, while in tropical and subtropical environments this difference may be reduced or the ranking even reversed. In such environments the *Bos taurus* × *Bos indicus* crossbred is usually superior to both parental types—that is, there is substantial heterosis in net reproductive rate (Koger, 1980). This has led to widespread use of first crosses and synthetics from such crosses in these environments.

In all cases where breed replacement is being considered, not only reproduction but such other traits as adaptation to the environment, growth rate, meat or milk composition, and behavior must be considered.

Adaptation is a particularly critical issue in transferring breeds from temperate to tropical areas. Impaired reproduction is one of the first and surest indicators of lack of adaptation. A breed may have high fertility and good viability in its native temperate climate, and long parturition intervals and high mortality in a tropical environment. This topic is discussed in the review by Trail (1986), which includes a useful list of references.

Provided adaptability of introduced breeds or their crosses is adequately taken into account, breed replacement, partial or complete, is a very effective means of making genetic change in animal populations. The approach is more generally applicable in swine and poultry, for which the environment is more closely controlled, than for species of grazing animals. For all species, an important general guideline, essential in tropical countries, is to evaluate thoroughly the performance of breeds or types that have evolved in that environment, and to use the life-cycle performance of such groups as a reference point in assessing any introduced breed or cross.

2. Heterosis

Another approach to utilizing genetic variation between breeds is to develop systematic cross-breeding programs, involving two, three, or sometimes more breeds or distinctive strains, to exploit heterosis. Heterosis is defined as superiority of the cross between two breeds or other genetically distinct stocks such as inbred lines, to the mean of the two parents. The genetic explanation for heterosis is dominance; that is, performance of individuals heterozygous at certain loci is equal or near to that of the superior homozygote for each of several loci. If two stocks differ in the frequency of favorable dominant alleles at different loci, the F_1 mean will be superior to the mean of the parent stocks.

Heterosis is usually greater in traits of low heritability, and thus is important for reproductive traits such as fertility, litter size, and viability. Levels of heterosis for such components of total reproduction in domestic animals are often found to be in the range of 3–6%, leading to increases in composite traits such as number of young weaned per breeding female per year of 10–20%; increases of more than 30% have been found in Bos taurus × Bos indicus crosses (Koger, 1980).

Maintaining a systematic two-, three-, or four-breed crossing program is operationally difficult with some species and in some environments. It is a greater problem in species with lower reproductive rates and overlapping generations, such as cattle and sheep. This leads to interest in use of a single interbreeding population based on the cross that produced the superior performance. Quantitative genetic theory indicates that one-half the heterosis resulting from dominant loci and expressed in the F_1 will be lost in the F_2, and that subsequent generations produced by inter se matings will stabilize at this level. Experimental evidence on this point from domestic animal species is rather scarce and not very consistent. However, it is clear that a substantial part of the F_1 superiority is retained in at least some cases. Furthermore, the theory indicates that a higher percentage will be retained in crosses of more than two breeds; the loss in heterozygosity (and thus heterosis) is predicted to be $1/n$, where n is the number of breeds. This suggests use of as many adapted, high-performance breeds as are available in creating a composite base population for a new breed.

3. Complementarity

In some cases crossing breeds with complementary characteristics may be advantageous whether or not heterosis as defined above is achieved. For example, crossing males that transmit good viability and rapid growth rate in an additive manner to their progeny, with females with high litter size, can result in greater total productivity than from matings within either group, even in the absence of heterosis in the component traits. Exploitation of such complementarity in general requires systematic (repeat) crossing programs, rather than utilization of a new, composite breed.

D. Within-Breed Improvement

1. Response to Selection

Use of breed resources for replacement, partial or complete, or for exploitation of heterosis and complementarity provides a rapid, one-step increase in productivity. Further improvement of genetic potential depends on selection within breeds.

The annual rate of genetic improvement is the ultimate measure of selection response. This is given by the response to a single round of selection divided by the generation interval (L). It is necessary to divide by generation interval, as this is the time interval between when parents are selected and when their progeny are available for selection. Thus genetic gain per year equals

$$ih^2\sigma_{\mathrm{p}}/L$$

The magnitude of i and L are connected by the reproductive capacity of the species. Generally speaking increasing the selection intensity (i) will increase the value of L and a balance must be sought to maximize genetic gain. Practices that increase reproductive capacity or lower the effective age at first parturition will reduce L or allow an increase in i. Selection schemes aided by the use of artificial insemination (AI) or by superovulation of selected females combined with transfer of embryos to recipients (multiple ovulation and embryo transfer, MOET) can substantially increase the annual rate of genetic improvement. AI permits a very large increase in i for males, while MOET allows a substantial increase in i for female parents while decreasing L (Land and Hill, 1975; Nicholas and Smith, 1983; Smith, 1986; Woolliams and Wilmut, 1989). Such MOET-based genetic improvement programs can increase the annual rate of genetic progress by a factor of 1.5–2 compared with more conventional selection schemes. In general the benefits will be larger in cattle than in sheep, with relatively minor application in pigs. However, an additional benefit of MOET schemes is that by their nature they need to be based on a centralized breeding nucleus, and this will in itself facilitate the application of more expensive or technically difficult measurement techniques that can improve the accuracy of selection further. Such schemes will also tend to reduce the amount of nongenetic variation with consequent increase in heritability.

2. Repeated Measurements

If a trait is expressed and measured a number of times in an individual's lifetime, then a more accurate estimate of the animal's phenotypic merit is obtained. This will mean that the effective heritability of the trait is increased. For example, ovulation rate can be measured at successive estrous cycles. There is an additional benefit for discretely distributed traits since the sum or mean of a set of observations will have a distribution that is closer to a continuous distribution and hence the complications of scale effects will be reduced. The variance of the average of n observations is

$$\frac{1 + (n - 1)r}{n} \sigma_P^2$$

where r is the repeatability of the trait, that is, the correlation between repeated expressions.

Since r is less than 1, this variance is less than σ_P^2, and since the genetic variance among individuals is not influenced by taking the mean of n observations the heritability will be increased and becomes

$$\frac{nh^2}{1 + (n - 1)r}$$

for the mean (or the sum) of n observations.

When r is moderate to low there will be substantial increase in heritability from increasing n. For $r = 0.3$ heritability increases by a factor if 1.5, 1.9, and 2.1 as n increases to 2, 3, and 4, respectively. The gains show diminishing returns to increasing n, and furthermore, if there is a substantial interval between successive observations (e.g., the interval between parturitions), extra measurements can only be obtained at the expense of increasing the generation interval, and this may be sufficient to outweigh the increased heritability. The optimum balance must be determined in each case.

3. Indirect Selection

Much effort has been devoted to finding traits that are genetically correlated with reproductive performance and may be used for selection to change reproductive rate (Land, 1977). This search has been motivated by the fact that reproductive traits are (usually) by definition sex-limited, they are often all-or-none traits, and candidates for selection must have reached puberty before

reproductive performance can be assessed. Because of the sex-limited expression of most reproductive traits, individuals of the sex that does not exhibit a trait of interest can only be evaluated by using information on relatives of the appropriate sex (e.g., parent, sibs, progeny). In such cases a trait measured directly on the individual and that is genetically correlated with the trait of interest can make an important contribution to a selection program.

The merits of a secondary trait as an indirect assessment of an individual's breeding value for the primary trait is obtained by considering the response to selection in the primary trait when selection decisions are based on the phenotype for the secondary trait. This is given by the expression

$$i \, h_1 \, h_2 \, r_a \, \sigma_{p1}$$

where h_1^2 is the heritability of the primary trait, h_2^2 is the heritability of the secondary trait, and r_a is the correlation between breeding values for the two traits (the additive genetic correlation). We have already seen the expression for the response when selection is based on the primary trait ($i \, h_1^2 \, \sigma_{p1}$). The ratio of these responses is

$$\frac{h_2}{h_1} r_a$$

and this is >1 when $h_2 r_a > h_1$.

In such a case selection on the secondary trait will be better than selection on the primary trait. If the primary trait can only be measured in females, then the response to selection based solely on the primary trait will be

$$\tfrac{1}{2} i \, h_1^2 \, \sigma_{p1}$$

and the ratio of this response to selection based on observation of the secondary trait in both sexes becomes

$$\frac{h_2}{\frac{1}{2}h_1} r_a$$

which is >1 provided $h_2 r_a > \frac{1}{2}h_1$. It is obvious that h_2^2 must usually be greater than h_1^2 for a secondary trait to be more useful than a primary trait, and that the closer r_a is to unity, the smaller h_2^2 can be relative to h_1^2. However, other considerations also have a bearing on assessment of the utility of secondary traits. If such traits can be measured at a young age, the generation interval can be reduced and the selection intensity may be increased because the available facilities can be used to assess more animals than would be possible if reproductive performance had to be measured directly. The impact of all of these considerations must be evaluated in a final assessment of the utility of indirect selection in any particular program.

Some estimates of the potential benefits of using correlated traits have been made by Haley *et al.* (1987), indicating the possibility of more than doubling rates of genetic change in litter size. A recent report (Purvis *et al.*, 1988) shows a significant positive genetic correlation between testis size and ovulation rate in sheep, suggesting that selecting young rams for testis diameter could be used as an aid to selection for increased ovulation rate in sheep. The situation in pigs appears to be less clear (Young *et al.*, 1986).

An example of the potential advantages of using a correlated trait is ovulation rate. As pointed out earlier, ovulation rate has a higher heritability than litter size and is highly correlated with litter size, at least over the lower portion of its range. Ovulation rate can be measured at a younger age than litter size, and can be measured (by laparoscopy) two or more times per reproduction cycle, while litter size can be measured only once per cycle. From these factors

Hanrahan (1987) has estimated that selection for ovulation rate could increase litter size in sheep about twice as rapidly as direct selection for litter size. As indicated earlier, selection for ovulation rate in mice and pigs, while effective, did not produce much response in litter size. There is limited information as yet on the effects on litter size of selecting for ovulation rate in sheep, but preliminary results are encouraging (Hanrahan, 1990).

4. Measurements on Relatives

Information on relatives of the subjects being considered for selection can also be used to increase the accuracy of selection and to overcome the sex-limited nature of many reproductive traits. The increased response to selection that can be achieved by using information on parents, collateral relatives, or progeny can be most efficiently examined by the methods of selection index theory and will not be considered here. The principles already examined in relation to rate of response to selection apply—increased information will improve the accuracy of selection and thus the genetic gain per generation and also per unit time provided there is not a compensating shift in selection intensity or generation interval. The principles of the use of information on relatives are considered in detail in various textbooks (e.g., Falconer, 1989; Nicholas, 1987). Applications of this approach to improving litter size in pigs have been presented in some detail by Avalos and Smith (1987). As a general guide it may be stated that progeny test selection for reproductive traits would not usually be a method of choice due to excessive increase in the generation interval relative to extra accuracy achieved. Information on collateral relatives (e.g., female sibs of males) will be more useful but may increase the generation interval. In many cases selection of males, in programs involving selecting for female reproductive performance, is based on the reproductive performance of the mother and possibly her relatives.

In the future, selection programs will involve the application of best linear unbiased prediction (BLUP) methodology, which allows the estimation of the breeding values of all individuals in the population while simultaneously adjusting for fixed effects such as year of record, age, etc. These estimates of breeding values are what selection index procedures produce when the complications of fixed effects are removed. In principle BLUP will optimally combine all of the information available to provide an estimate of breeding value for the candidate of selection at the time when selection decisions are to be made. The increasing availability of appropriate software will make such procedures the method of choice in the future.

5. Major Genes

The existence of genes with large effects on reproductive traits offers the possibility of making rapid genetic change, and the Booroola (F) gene in Merino sheep (Piper and Bindon, 1982) has received widespread attention. Piper and Bindon (1987) reported litter sizes of 1.48, 2.38, and 2.74 for $++$, $F+$, and FF ewes. Clearly, one copy of this gene, which can be obtained in a flock by mating normal ($++$) ewes to FF rams, that is, in one generation, produces a genetic change that would require many decades of selection within a breed lacking such a gene. Where a large, immediate increase in litter size is desired, this may be the method of choice. However, there are some potential drawbacks.

One problem is that the mean of nearly 2.4 may be too high (see Fig. 4). As pointed out earlier, prolific breeds that transmit their high prolificacy additively can be used to contribute a proportion of their inheritance to achieve a specified mean litter size, such as 1.6, 1.8, or 2.0. The advantage of the major gene is that it can be transferred into any breed, and the background genotype (currently Merino) can be removed by back-crossing, although this is not a trivial task (Smith, 1985). With quantitatively inherited prolificacy, one has to accept other genes from the prolific breed in proportion to the amount needed to produce the desired increase in prolificacy.

A second problem with a gene such as the Booroola is that the range in litter size may accentuate the problems of the higher-than-desired mean; $F+$ ewes produce many triplets, and some quadruplets (Hinch *et al.*, 1986). A third problem is that maintaining a flock of 100% $F+$ ewes requires maintenance of a flock of $++$ ewes to breed replacements. This may be feasible, where a stratified production system exists, but often will be an added cost. The alternatives are to fix the gene in the flock (FF), with attendant higher mean and proportion of litters >3, or to allow the gene to segregate in the flock, which will increase variance in litter size further (Nitter, 1987).

Another interesting example of a single gene related to reproduction is the effect of the color locus in sheep on litter size. The allele for white at the A locus results in a significant reduction in litter size (Adalsteinsson, 1970; Ricordeau *et al.*, 1982). Thus retention of colored animals in a segregating population will result in an increase in litter size.

Major genes are of great interest from a research standpoint because of the new light they may shed on the regulation of important physiological processes. Genes such as the Booroola gene may be directly useful for improving genetic potential in some environments. Systematic search for additional major genes is justified for both research purposes and for their potential utility. However, to be useful for improving animal productivity, their advantages must outweigh any adverse effects on fitness.

6. Potential Use of Transgenics for Improving Reproduction Performance

The recent development and integration of gene cloning, gene transfer, and embryo manipulation techniques allows the transfer of cloned genes from one or more species into the genetic material of other species (Anderson, 1986; Wagner, 1986; Capecchi, 1989; Pursel *et al.,* 1989; Sambrook *et al.,* 1989). Genes have been transferred into cultured cells and embryos using gene injection, electroporation, and viral vectors. While the methods for producing such transgenic animals have evolved and will continue to evolve, it is now possible to transfer into the DNA of laboratory and farm animal species any one of thousands of structural genes and regulatory elements that have been cloned or synthesized. These gene transfer methods allow the rapid introduction of genes that have the potential for producing major differences in reproductive performance.

While the genes transferred to livestock species to date have impaired productive and reproductive performance rather than improving it (Pursel *et al.,* 1989), there is certainly potential for improving reproduction in the future. Structural genes from one species can be fused with regulatory sequences from other genes or species. Recent developments in molecular biology tech-

niques, especially oligonucleotide-directed mutagenesis and polymerase chain reaction, make possible the production of desired alterations in the sequence of structural and regulatory genes (Higuchi, 1989; Sambrook *et al.,* 1989; Brent *et al.,* 1989). Transfer of these altered genes back into cultured cells by electroporation has allowed rapid assay of effects of each alteration in structural and regulatory gene function. Since the inactivation of genes caused by the insertion of transgenes (insertional mutagenesis) has been a major problem, techniques are also being developed for site-specific integration of cloned genes into homologous cellular genes (Capecchi, 1989). While our understanding of the effects of specific base and amino acid changes on gene function and regulation is far from complete, it seems clear that molecular genetic methods will permit transfer of genes across species, and eventually production of a variety of genetic alterations desired in the genes coding for reproductive traits.

One of the largest remaining problems for using transgenic methods to improve reproductive performance is the question of which gene(s) to transfer. As indicated throughout this text, reproduction is a very complex trait. For successful reproduction to occur, many genes must function at very specific times and levels to produce a normal endpoint. A defect or an imbalance in any component can cause subfertility or infertility. For example, transfer of an additional copy of a cloned P-450 side chain cleavage enzyme gene (which converts cholesterol to pregnenolone—the most limiting step in progesterone biosynthesis) might help prevent early pregnancy losses by elevating serum progesterone. However, unless this transgene were under the control of pregnancy-specific regulatory elements, infertility could result due to constantly elevated progesterone inhibiting LH release, normal estrous cycles, or parturition.

Optimal engineering of reproductive performance using transgenes, or using naturally occurring molecular genetic markers, will require an integration of knowledge of reproductive physiology and molecular genetics. Reproduction of animals involves coadapted complexes. Continued selection for marker loci or for conventional phenotypic performance will be needed to maximize reproductive performance in transgenic populations.

V. Overview

Efficiency of reproduction, or output of offspring per unit of input into the breeding population, is the most important factor affecting overall efficiency of production in most domestic animal species. High reproductive rates facilitate genetic progress in other traits. However, components of reproduction are subject continuously to natural selection, and as a result much of the additive genetic variation has been fixed. The resulting low heritabilities led to the conclusion on the part of geneticists, in the early years of the development of the science of quantitative genetics, that effecting genetic change in reproduction traits would be difficult or impossible. This in turn led to relative neglect of such traits by geneticists for some time. However, two important facts are overlooked in this view. First, there is a very large amount of variation among genetically separate groups of animals such as livestock breeds in most reproductive traits; this variation can be used to effect rapid, permanent genetic change in a wide variety

of situations. Second, genetic change within populations is a function of the product of heritability and phenotypic variation. While the first of these is low for reproductive traits, compared to other economically important traits, the second is generally much higher. Thus, substantial genetic change from intrapopulation selection is quite feasible for many reproductive traits.

With the realization of these two facts, rapid progress has been made in the last two to three decades in understanding the genetics of reproduction and in using genetic variation to effect useful change in reproduction. Use of breed resources has contributed to much of this change. At the same time, there has been increased appreciation of the fact that adaptation resulting from many generations of natural selection, especially in difficult environments, is an extremely important attribute of a breed, and that conservation of breed resources is important. Selection within populations is being used to increase number of young per female in sheep and pigs. Much progress has been made in understanding the physiological basis of genetic variation, and genetic variation is being used widely in studies of physiological mechanisms. Developments in reproductive biology technology such as artificial insemination, embryo transfer, and gene cloning and transfer are making or have the potential to make major increases in rates of genetic change.

Principally within the last decade, individual genes with large effects on reproduction have been identified and studied. These not only contribute to increased understanding of the control of reproduction, but in some cases have good potential for direct utilization to increase reproductive rates. The development of transgenic technology opens up the possibility of using such genes in species other than the one in which they are found.

The rapidly increasing availability of techniques for precise identification and characterization of small segments of DNA, that is, of individual genes, will almost certainly permit far greater precision in the analysis and utilization of genetic variation in reproduction in the future.

Choice of areas for research emphasis for future genetic improvement of reproduction will vary to some extent with species. In cattle, the development of genetic potential for twinning would be valuable for some of the more favorable environments and intensive management systems. For all cattle production systems, development and application of more effective selection methods to increase herd fertility and decrease calving intervals would be beneficial. In part this requires only the measurement of fertility and application of known methodologies for the improvement of low-heritability traits, but research to identify good indicator traits in both sexes at an early age could provide the means for much more effective selection.

In sheep, a better understanding of genetic and physiological control of the breeding season is needed. In both sheep and swine, there is apparently useful genetic variation between breeds in prenatal survival and in degree of variability in ovulation rate, both of which contribute to variation in litter size. Developing effective means for utilizing this variation and discovery of other means of reducing variation in litter size would make a very valuable contribution. In pigs, prenatal loss is a major source of reproductive wastage and is an area deserving special attention.

In all species, perinatal and postnatal mortality cause large losses. Because of the very low additive genetic variance in these traits, effecting genetic change is a special challenge. However, there are methods for improving such traits by selection, as long as heritability is not zero. Furthermore, postnatal mortality is strongly influenced by birth weight, which has a fairly high heritability, and birth weight in turn is strongly affected by gestation period, which is also highly heritable. There is a need for imaginative approaches to utilizing all sources of genetic variation affecting reproduction to effect improvement in these important traits.

References

Adalsteinsson, S. (1970). *J. Agric. Res. Iceland* **2**:3.

Anderson, W. F. (1986). "Genetic Engineering of Animals: An Agricultural Perspective" (J. W. Evans and A. Hollaender, eds.), p. 7 Plenum Press, New York. Basic Life Sciences Volume 37.

Avalos, E., and Smith, C. (1987). *Anim. Prod.* **44**:153.

Baker, R. L., and Steine, T. (1986). *Proc. 3rd World Congr. Genetics Appl. to Livestock Production* **XI,** 84.

Bania, N., and Bradford, G. E. (1981). *J. Anim. Sci.* **52,** 739.

Bearden, H. J., and Fuquay, J. (1984). "Applied Animal Reproduction," 2nd Ed. Reston Publishing, Reston, Virgina.

Bichard, M., and David, P. J. (1986). *J. Anim. Sci.* **63,** 1275.

Bishop, M. H. W. (1964). *J. Reprod. Fertil.* **7,** 383.

Bohren, B. B., Hill, W. G., and Robertson, A. (1966). *Genet. Res.* **7,** 44.

Bolet, G. (1986). *Curr. Top. Vet. Med. Anim. Sci.* **24,** 12.

Bradford, G. E. (1969). *Genetics* **61,** 905.

Bradford, G. E. (1972). *J. Reprod. Fertil. Suppl.* **15,** 23.

Bradford, G. E. (1979). *J. Anim. Sci.* **49**(Suppl. II), 66.

Bradford, G. E. (1985). "Genetics of Reproduction in Sheep" (R. B. Land and D. W. Robinson, eds.), p. 3 Butterworths, London.

Bradford, G. E., and Nott, C. F. G. (1969). *Genetics* **63,** 907.

Bradford, G. E., Barkley, M. S., and Spearow, J. L. (1980). "Selection Experiments in Laboratory and Domestic Animals" (A. Robertson, ed.), p. 161. Commonwealth Agricultural Bureaux, Farnham Royal, U.K.

Bradford, G. E., Quirke, J. F., Sitorus, P., Inounu, I., Tiesnamurti, B., Bell, F. L., Fletcher, I. C., and Torell, D. T. (1986). *J. Anim. Sci.* **63,** 418.

Brent, G. A., Harney, J. W., Chen, Y., Warne, R. L., Moore, D. D., and Larsen, R. (1989). *Molec. Endocrinol.* **3**(12), 1996.

Brien, F. D. (1986). *Annu. Br. Abstr.* **54,** 975.

Brien, F. D. and Hill, W. G. (1986). *Anim. Prod.* **42,** 397.

Byskov, A. G. (1982). "Reproduction in Mammals: Book 1—Germ Cells and Fertilization," 2nd ed. (C. R. Austin and R. V. Short, eds.), pp. 1–16. Cambridge University Press, Cambridge.

Capecchi, M. R. (1989). *Science* **244,** 1288.

Cattanach, B. M., Pollard, C. E., and Hawkes, S. G. (1971). *Cytogenetics* **10**:318.

Chapman, H. M., and Bruere, A. N. (1975). *J. Reprod. Fertil.* **45,** 333.

Cundiff, L. V., Gregory, K. E., and Koch, R. M. (1982). *Proc. 2nd World Congr. Genetics Applied to Livestock Production* **V,** 310.

Cunningham, P. J., England, M. E., Young, L. D., and Zimmerman, D. R. (1979). *J. Anim. Sci.* **48,** 509.

Day, R. N., and Mauer, R. A. (1989). *Mole. Endocrinol.* **3,** 3.

Dickerson, G. E. (1977). North Central Regional Publication 246. University of Nebraska, Lincoln.

Dickerson, G. E. (1982). *Proc. 2nd World Congr. Genetics Applied to Livestock Production* **V,** 252.

Dickinson, A. G., Hancock, J. L., Hovell, G. S. R., Taylor, St. C. S., and Weiner, G. (1962). *Anim. Prod.* **4,** 64.

Djalali, M., Rosenbusch, B., Wolf, M., and Sterzik, K. (1988). *J. Reprod. Fertil.* **84,** 647.

Donald, H. P., and Russell, W. S. (1970). *Anim. Prod.* **12,** 273.

Eisen, E. J. (1986). *Proc. 3rd World Congr. Genetics Applied to Livestock Production* **XI,** 153.

Eklund, J., and Bradford, G. E. (1977). *Genetics* **85,** 529.

Eldridge, F. E. (1985). "Cytogenetics of Livestock." AVI Publishing, Westport, Connecticut.

Erickson, R. P., Durbin, E. J., and Tres, L. L. (1987). *Development* **101**(Suppl.), 25.

Falconer, D. S. (1960). *J. Cell. Comp. Physiol.* **56**(Suppl. 1), 153.

Falconer, D. S. (1989). "Introduction to Quantitative Genetics," 3rd ed. Longman, New York.

Ferguson-Smith, M. A., Affara, N. A., and Magenis, R. E. (1987). *Development* **101**(Suppl.), 41.

Gaugler, H. R., Buchanan, D. S., Hintz, R. L., and Johnson, R. R. (1984). *J. Anim. Sci.* **59,** 941.

Goodfellow, P. N., and Darling, S. M. (1988). *Development* **102,** 251.

Gueblez, R., Bruel, L., and Legault, C. (1987). 19ièmes Journées de 1a Recherche Porcine en France, p. 25. Institut Technique du Porc, 1e Rheu, France.

Gustavsson, I. (1984). *Proc. 10th Int. Congr. Anim. Reprod. and Artificial Insemination* **VI,** 1.

Haley, C. S., Cameron, N. D., Slee, J., and Land, R. B. (1987). "New Technologies in Sheep Production" (I. Fayez, M. Marai, and J. B. Owen, eds.), p. 113. Butterworths, London.

Hanrahan, J. P. (1982). *Proc. 2nd World Congr. Genetics Appl. to Livestock Production* **V,** 294.

Hanrahan, J. P. (1983). *Theriogenology* **20,** 3.

Hanrahan, J. P. (1986). "Exploiting New Technologies in Animal Breeding. Genetic Developments" (C. Smith, J. W. B. King, and J. C. McKay, eds.), p 59. Oxford University Press, New York.

Hanrahan, J. P. (1987). "New Techniques in Sheep Production" (I. Fayez, I. F. Marai, and J. B. Owen, eds.), p. 37. Butterworths, London.

Hanrahan, J. P. (1990). *Proc. Br. Soc. Anim. Prod.* (in press).

Hanrahan, J. P., and Owen, J. B. (1985). *Proc. Br. Soc. Anim. Prod.,* paper 38.

Hanrahan, J. P., and Quirke, J. F. (1986). *Proc. 3rd World Congr. Genetics Appl. to Livestock Production* **XI,** 30.

Hansen, L. B., Freeman, A. E., and Berger, P. J. (1983). *J. Dairy Sci.* **66,** 281.

Higuchi, R. (1989). "PCR Technology: Principles and Applications for DNA Amplification" (H. A. Erlich, ed.), p. 61. Stockton Press, New York.

Hill, W. G., and Webb, A. J. (1982). "Control of Pig Reproduction" (D. J. A. Cole and G. R. Foxcroft, eds.), p. 541. Butterworths, London.

Hinch, G. N., Crosbie, S. F., Kelly, R. W., Owens, J. L., and Davis, G. H. (1985). *N. Z. J. Agric. Res.* **28,** 31.

Horstgen-Schwark, G., Eisen, E. J., Saxton, A. M., and Bandy, T. R. (1984). *J. Anim. Sci.* **58,** 846.

Hutt, F. B., and Rasmusen, B. A. (1982). "Animal Genetics," 2nd ed., pp. 1–582. John Wiley & Sons, New York.

Iniguez, L. C., Sanchez, M., and Ginting, S. (1990). *Small Ruminant Res.* (in press).

Iniguez, L. C., Bradford, G. E., and Okeyo-Mwai, A. (1986). *J. Anim. Sci.* **63,** 715.

Johansson, K., and Kennedy, B. W. (1982). *Proc. 2nd World Congr. Genetics Appl. to Livestock Production* **VII,** 503.

Johnson, R. K. (1981). *J. Anim. Sci.* **52,** 906.

Johnson, R. K., Zimmerman, D. R., and Kittok, R. J. (1984). *Livestock Prod. Sci.* **11,** 541.

Jost, A., Vigier, B., Prepin, J., and Perchellet, J. P. (1973). *Recent Progr. Horm. Res.* **29,** 1.

Karsch, F. J., Bittman, E. L., Foster, D. L., Goodman, R. L., Legan, S. J., and Robinson, J. E. (1984). *Recent Progr. Horm. Res.* **40,** 185.

Karsch, F. J., and Nancy L. Wayne (1988). *Proc. 11th Int. Congr. Anim. Reprod. AI,* Dublin, **5,** 221.

Koger, M. (1980). *J. Anim. Sci.* **50,** 1215.

Lahlou-Kassi, A., Berger, Y. M., Bradford, G. E., Boukhliq, R., Tibary, M., Derqaoui, L., and Boujenane, I. (1989). *Small Ruminant Res.* **2,** 225.

Land, R. B. (1977). "Reproduction in Domestic Animals" (H. H. Cole and P. T. Cupps, eds.), p. 577. Academic Press, San Francisco.

Land, R. B., and Hill, W. G. (1975). *Anim. Prod.* **21,** 1.

Lander, E. S., and Botstein, D. (1989). *Genetics* **121,** 185.

Lasslo, L. L., Bradford, G. E., Torell, D. T., and Kennedy, B. W. (1985). *J. Anim. Sci.* **61,** 387.

Legault, C., and Caritez, J. C. (1983). *Génét. Sél. Evol.* **15,** 225.

Legault, C., Gruand, J., and Bolet, G. (1981). *Journées de la Recherche Porcine en France,* p. 255. L'Institut Technique du Porc, Paris.

Leitch, J., Hytten, F. E., and Billewicz, W. Z. (1959). *Proc. Zool. Soc. Lond.* **133:**11.

Lewin, B. (1987). "Genes III." John Wiley. Inc. New York.

Mahieu, M., Jego, Y., Driancourt, M. A., and Chemineau, P. (1989). *Anim. Reprod. Sci.* **19,** 235.

Makinen, A., and Remes, E. (1986). *Hereditas* **104,** 223.

Mason, A. J., Hayflick, J. S., Zoeller, R. T., Young, W. S., III, Phillips, H. S., Nikolics, K., and Seeburg, P. H. (1986). *Science* **234,** 1366.

McFeely, R. A. (1967). *J. Reprod. Fertil.* **13,**(3), 579.

McLaren, A. (1988). *Trends Genet.* **4**(6), 153.

Meacham, N. S., and Notter, D. R. (1987). *J. Anim. Sci.* **64,** 700.

Mohd-Yussuf, M. K., Dickerson, G. E., and Young, L. D. (1988). *Proc. 3rd World Congr. on Sheep & Beef Cattle Breeding* **2,** 509.

Morris, C. A., and Day, A. M. (1986). *Proc. 3rd World Congr. Genetics Applied to Livestock Production* **XI,** 14. Lincoln, Nebraska.

Nicholas, F. W. (1987). "Veterinary Genetics." Oxford University Press, New York.

Nicholas, F. W., and Smith, C. (1983). *Anim. Prod.* **36,** 341.

Nitter, G. (1987). "New Technologies in Sheep Production" (I. Fayez, M. Marai, and J. B. Owen, eds.), p. 271. Butterworths, London.

Ollivier, L., and Bolet, G. (1981). *Journées de la Recherche Porcine en France,* p. 261. L'Institut Technique du Porc, Paris.

Packham, A., and Triffitt, L. K. (1966). *Aust. J. Agric. Res.* **17,** 515.

Pattie, W. A. (1965). *Austr. J. Exp. Agric. Anim. Husb.* **5,** 361.

Piper, L. R., and Bindon, B. M. (1982). "Proceedings of the World Congress on Sheep and Beef Cattle Breeding," Vol. 1 (R. A. Barton and W. C. Smith, eds.), p. 395. Dunmore Press, Palmerston North.

Piper, L. R., and Bindon, B. M. (1987). "Proceedings of the 2nd International Conference on Quantitative Genetics" (B. S. Weir, E. J. Eisen, M. M. Goodman, and G. Namkoong, eds.), p. 270. Sinauer Associates, Sunderland, Massachusetts.

Pursel, V. G., Pinkert, C. A., Miller, K. F., Bolt, D. J., Campbell, R. G., Palmiter, R. D., Brinster, R. L., and Hammer, R. E. (1989). *Science* **244,** 1281.

Purser, A. F., and Young, G. B. (1964). *Anim. Prod.* **6,** 32.

Purvis, I. W., Piper, L. R., Edey, T. N., and Kilgour, R. J. (1988). *Livestock Prod. Sci.* **18,** 35.

Quirke, J. F., and Hanrahan, J. P. (1985). "Endocrine Causes of Seasonal and Lactational Anestrus in Farm Animals" (E. Ellendorf and F. Elaesser, eds.), p. 29. Martinus Nijhoff, The Hague.

Quirke, J. F., Stabenfeldt, G. H., and Bradford, G. E. (1988). *Anim. Reprod. Sci.* **16,** 39.

Razungles, J., Tchamitchian, L., Bibe, B., Lefevre, C., Brunel, J. C., and Ricordeau, G. (1985). "Genetics of Reproduction in Sheep" (R. B. Land and D. W. Robinson, eds.), p. 39. Butterworths, London.

Ricordeau, G., Razungles, J., Eychenne, F., and Tchamitchian, L. (1976). *Ann. Génét. Sél. Anim.* **8,** 25.

Ricordeau, G., Tchamitchian, L., Razungles, J., Lefevre, C., Brunel, J. C., and Lajous, D. (1982). *Proc. 2nd World Congr. Genetics Appl. to Livestock Production* **VII,** 596–598.

Sambrook, J., Fritsch, E. F., and Maniatis, T. (1989). "Molecular Cloning: A Laboratory Manual," 2nd ed., Vols. 1, 2, and 3. Cold Spring Harbor Laboratory Press, Cold Spring Harbor, New York.

Shelton, J. M. (1964). *J. Anim. Sci.* **23,** 360.

Short, R. V. (1982). "Reproduction in Mammals: Book

2, Embryonic and Fetal Development," 2nd ed. (C. R. Austin and R. V. Short, eds.), p. 70. Cambridge University Press, Cambridge.

Smith, C. (1985). "Genetics of Reproduction in Sheep" (R. B. Land and D. W. Robinson, eds.), p. 151. Butterworths, London.

Smith, C. (1986). *Anim. Prod.* **42,** 81.

Stewart-Scott, I. A., and Bruere, A. N. (1987). *J. Hered.* **78,** 37.

Strandberg, E., and Dannell, B. (1988). *Proc. Word Conf. on Anim. Prod. Helsinki,* 505.

Swartz, H. A., and Vogt, D. W. (1983). *J. Hered.* **74,** 320.

Taylor, St. C. S. (1985). *J. Anim. Sci.* **61**(Suppl. 2), 118.

Trail, J. C. M. (1986). *Proc. 3rd. World Congr. Genetics Appl. to Livestock Prod.* **XI,** 474.

Van Vleck, L. D., Pollack, E. J., and Oltenacu, E. A. B. (1987). "Genetics for the Animal Sciences." W. H. Freeman, New York.

Wagner, T. E. (1986). "Genetic Engineering of Animals: An Agricultural Perspective" (J. W. Evans and A. Hollaender, eds.), Basic Life Sciences Vol. 37, p. 151. Plenum Press, New York.

Warner, C. M. (1986). *J. Anim. Sci.* **63,** 279.

White, P. C. (1987). *Recent Progr. Horm. Res.* **43,** 305.

Woolliams, J. A., and Wilmut, I. (1989). *Anim. Prod.* **48,** 3.

Young, L. D., Leymaster, K. A., and Lunstra, D. D. (1986). *J. Anim. Sci.* **63,** 17.

Influence of Infectious Diseases on Reproduction

R. H. BONDURANT

I. Introduction

In livestock species, a relationship between infectious disease agents and reproductive failure has been established for many decades. Since Bang's first description of the abortifacient capabilities of *Brucella abortus* bacteria in 1897 (cited by Wilfert, 1984), a wide variety of agents has been implicated in pregnancy wastage of domestic animals. However, the relationship between specific microbial agents and reproductive failure is often not nearly as strong as that described by Bang for *B. abortus*. That is, the mere presence of a microbe in the environment of a pregnant female does not always lead to loss of a conceptus. The picture is a bit more complex.

II. Disease Concepts

In nearly all cases, diseases in populations of animals have more than a single cause; that is, while an animal may be exposed to a certain organism that is known to be associated with specific disease, the exposed animal may or may not ever show signs of that disease. Classically, epidemiologists refer to a "triad" of determinants for animal disease (Schwabe *et al.*, 1977), including (1) the presence of a pathogenic (disease-causing) organism, (2) the environment in which the host lives, and (3) the condition (susceptibility) of the host. Using this concept, it is usually necessary to involve all three determinants in order to cause disease.

A. The Organism

Known reproductive pathogens occur in nearly every major category of microbe, including bacteria (aerobic, micro-aerophilic, and anaerobic), fungi, rickettsiae, chlamydiae, protozoa, and viruses. Table I lists some of the well-known reproductive pathogens

Table I
Examples of Reproductive Pathogens and Their Consequences

Host	Class of organism	Genus/species	Consequences of infection
Cattle	Bacteria	*Brucella abortus*	Mid- to late-term abortion; udder infection/milk contamination
		Campylobacter fetus venerealis	Early embryonic death/infertility; occasional abortion
		Leptospira interrogans, hardjo	Mid–late abortion; "flaccid mastitis"
		Listeria monocytogenes	Cause of sporadic abortion; associated with improperly cured silage; placentitis, microabscesses in fetus
		Hemophilus somnus	Occasionally cited as a cause of infertility or abortion; more often associated with respiratory or central nervous disease
		Salmonella dublin *Salmonella typhimurium*	Sporadic abortion, probably due to effect of endotoxin-induced $PGF_{2\alpha}$ on CL; occasional fetal infection as well
	(Mycoplasma)	*Ureaplasma diversum*	Lower and upper reproductive tract inflammation; early embryonic death? occasional abortion; strain differences in pathogenicity
	Fungi	*Aspergillus fumigatus*	Sporadic abortion, especially in humid environments; marked placentitis; occasional skin lesions in late-term fetus
	Protozoa	*Tritrichomonas foetus*	Infertility, early embryonic death following venereal transmission; occasional abortion

(continued)

Table I
Continued

Host	Class of organism	Genus/species	Consequences of infection
	Viruses	Bovine herpes virus I (infectious bovine rhinotracheitis)	Abortion at any stage; rapidly fatal for fetus; (virus can be identified in some tissues in spite of autolysis)
		Bovine virus diarrhea (BVD)	Abortion if infected early in gestation; teratology and/or immune tolerance if infected 90–140 days; little or no effect if infected after 150 days; significant strain differences exist
Sheep, goats	Bacteria	*Campylobacter fetus* var. *fetus*	Abortion storms are not uncommon; usually late gestation; focal liver necrosis in fetus is common; not a venereal pathogen; rare in goats
		Listeria monocytogenes	(See cattle disease)
		Brucella ovis (*B. melitensis* in goats)	Unusual cause of abortion (late term); chronic placentitis lesions
	Chlamydiae	*Chlamydia psittaci*: enzootic abortion of ewes (EAE)	Can cause abortion storms; long incubation period (>50 days); placentitis; common in goats
	Rickettsia	*Coxiella burnetii* (Q fever)	Infection is common; abortion is less common, but can occur in storms; intercotyledonary exudate
	Protozoa	*Toxoplasma gondii*	Can cause widespread losses; effects are stage-dependent (resorption, mummification, abortion, stillbirth, depending on gestational age at infection); cats are definitive hosts
	Viruses	Bluetongue virus	Resorption, mummification, abortion; or CNS teratology (hydranencephaly) if infected in first trimester; spread by biting gnat vector; significant serotype and strain differences; rarely reported for goats
		Border disease	Embryonic and fetal death if infected first trimester; mummification or abortion if infected second trimester; "hairy shaker" lamb if infected later, but before day 130; virus related to BVD
Swine	Bacteria	*Brucella suis*	Abortion (any stage), infertility; can be venereally transmitted
		Leptospira interrogans serovar *pomona* (and others)	Late-term abortions; some of the litter may be born alive
		(Enteric bacteria)	(See *Salmonella* for cattle)
	Viruses	Enteroviruses, porcine parvoviruses	Effects are stage dependent (embryonic death→ mummification→ abortion→ stillbirths→ live, weak pigs); older sows nearly always immune
		Pseudorabies (Aujesky's) virus	Resorption, mummification, abortion (stage-dependent); sick adults commonly seen

and their hosts. It is important to note that significant differences in pathogenicity exist within a particular genus, and even within species of some organisms. So-called "strain differences" for many bacteria and viruses have been documented, which sometimes make it difficult to interpret laboratory diagnostic results. Some agents are highly adapted to their hosts, and therefore cause little pathology, while others, which may be indistinguishable by many laboratory procedures, are less host-adapted and more pathogenic. A good example of this "strain difference" occurs with the bacteria *Leptospira interrogans*, serovar *hardjo* (see below, Section VI,A,3).

Moreover, the *dose* (the number of organisms) to which the host is exposed can greatly influence the result of infection. Massive infection of even minor pathogens can overwhelm host defenses, leading to serious clinical consequences; likewise, exposure to low doses of some relatively "serious" pathogens may lead only to subclinical (inapparent) infection, which is dealt with effectively by the host.

B. Environment

The housing, feeding, and maintenance of domestic species can have a profound effect on their susceptibility to disease. As an example, animals that feed from a common bunker or drink from a common trough can share respiratory and gastrointestinal pathogens, many of which may also act on the conceptus. Crowding of maternity areas for dairy cattle or brood mares can increase contamination levels such that opportunistic organisms (ones that do not cause disease under normal circumstances) build to huge numbers and infect the reproductive tract.

Another important environmental in-

fluence is the presence of *vectors*, that is, organisms or objects that can pass pathogens from an infected to an uninfected host. The mere presence of the pathogenic microorganism in the environment may not spread disease without the involvement of the appropriate vector. In livestock species, some of the most important vectors are arthropods, including biting flies, mosquitoes, gnats, and ticks. In many cases, there is considerable host and pathogen specificity among vectors. Examples of vector-borne diseases that influence livestock reproduction include bluetongue disease of sheep (Kirkbride, 1984) and epizootic bovine abortion (foothill abortion; Kimsey, 1986).

C. The Host

Given the presence of potential pathogens and an environment conducive to contamination, the host must have effective defense mechanisms in order to avoid reproductive consequences of infection. The immunological capabilities available to the host includes nonspecific and specific mechanisms.

1. Nonspecific Mechanisms

In some cases, the host has anatomical barriers to infection. In the mare, the tightly closed vulva restricts exposure of the reproductive tract to ascending infection from the environment; in ruminants, the best example of anatomical protection may be the cervical barrier, in which the tortuous cervical canal is tightly closed and coated with a microbe-trapping viscous mucus for much of the reproductive cycle and for all of pregnancy. Such an arrangement makes it difficult for opportunistic organisms to reach the upper reproductive tract once the female is in her luteal phase. There is evidence that uterine immune defense mechanisms are enhanced in the follicular phase, a phenomenon that may provide protection against ascending infection of the reproductive tract acquired at the time of coitus or artificial insemination. Such local immune systems include (1) an increase in the influx and phagocytic function of neutrophilic leukocytes in the estrous uterus (Hawk, 1971) and (2) increased migration of lymphoblasts from mesenteric lymph nodes into the cervical and vaginal mucous membranes during estrus (McDermott *et al.*, 1980).

2. Specific Immune Mechanisms

It is known that antibodies to a variety of agents can be elaborated locally by lymphocytes in the reproductive tissues (or in the lymph nodes at sites remote from those tissues). In addition, circulating serum antibodies can be sequestered in the reproductive tract in response to infection (Winter, 1982). Circulating antibodies can also neutralize potential reproductive pathogens before they arrive at the reproductive tract. Since most pathogens arrive by the hematogenous route, the presence of circulating antibodies may represent a "last line of defense" against infection of the uterus, oviducts, placenta, and ultimately the fetus. Other specific mechanisms include cell-mediated immune (CMI) responses, by which infecting agents are encountered either directly by cytotoxic lymphocytes, or indirectly through the action of mediator lymphocytes, such as T helper cells, T suppressor cells, etc. In many types of infectious diseases, combinations of the above immune mechanisms are brought into play in order to protect the host.

Table II
Development of Specific Immunity in the Bovine Fetus[a]

	IgM	IgG
Immunoglobulin production		
Unstimulated fetus, term	0.11 gm/dl	0.16 mg/dl
Stimulated with BVD virus antigens	0.11–0.31	0.50–4.20
Stimulated with agent(s) of EBA (foothill abortion)	0.17–2.22	1.48–8.11
Normal levels after colostrum intake	1.01–3.01	2.39–24.0

Antigen recognized by immunoglobulins	Fetal age at which recognition occurs
Anaplasma marginale	141
IBR virus (killed)	165
BVD virus	190
Campylobacter fetus	235
Brucella abortus	newborn

[a]Adapted from Osburn (1986).

The concept of the "host immune status" also includes the conceptus. In most cases, the fetus acquires—as a function of gestational age—the ability to defend itself against a variety of agents (see Table II).

In any case, the above review is intended to dispel the commonly held concept of "one organism–one disease." Broadening the perception of host–agent relationships will make it easier to understand the current status of our knowledge of infectious diseases of reproductive significance.

III. Definitions of Pregnancy Wastage

A. Early Embryonic Death

This is the death of the conceptus before organogenesis is complete. In most ungulates, this occurs after maternal recognition of pregnancy, but before attachment of the fetal membranes to the endometrium ("implantation") is complete. Generally, early embryonic death (EED) is clinically inapparent, except that the dam will not show estrus at the expected interval; rather, she will show a prolonged interestrus interval.

Agents that arrive in the upper reproductive tract by the venereal route (i.e., subsequent to breeding to an infected male or infected semen) are well known for their ability to cause EED. Since they are introduced into the reproductive tract concurrently with sperm, it is not surprising that they are usually able to work their pathogenic effect early in gestation. Classic examples of such pathogens include the bacterium *Campylobacter fetus* var. *venerealis* (cattle), the protozoan *Tritrichomonas foetus* (cattle), and the bacterium *Taylorella equigenitalia* (horses). In addition, some viruses

have been shown experimentally to induce EED (Miller *et al.*, 1987), although the presence of such viruses in semen/secretions of males is not commonly reported.

The defense of the dam against venereal agents includes both circulating antibody and local antibody responses. With *Campylobacter*, for example, the dam is able to mount a local antibody response in the vagina and uterus against the causative bacterium (Corbeil *et al.*, 1981). The vaginal response is greater in the immunoglobulin A (IgA) isotype of antibody, while the uterine response is more pronounced in the IgG class. The fact that clearance of the organism occurs first in the uterus suggests that the IgG response, which is more effective than IgA in "opsonizing" (processing bacteria for phagocytosis by macrophages and neutrophils), is an important means of defense. The practical implications of this are apparent when one realizes that systemic vaccination with killed *C. fetus* organisms provokes an IgG response from the host that is apparently effective in preventing the consequences of infection (versus preventing infection itself). Apparently this circulating IgG gains access to the upper reproductive tract, and neutralizes the organisms there (Winter, 1982).

A few agents that cause EED can arrive in the upper reproductive tract by the hematogenous route. In ruminants, all of whom are dependent on a functional corpus luteum (CL) for maintenance of early pregnancy, these agents probably act on the CL rather than on the conceptus itself; direct access to the early conceptus is hampered by the fact that implantation is not complete at this time, so there is no direct route for "crossing the placenta" to the conceptus. Infectious bovine rhinotracheitis (IBR) virus has been shown to be capable of invading

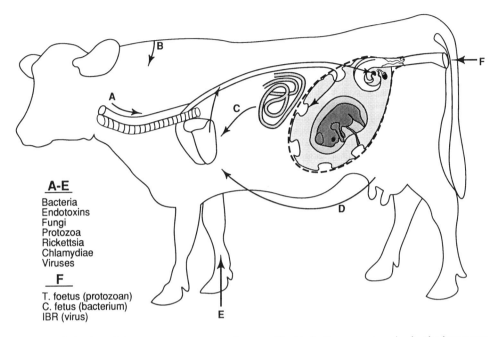

A-E

Bacteria
Endotoxins
Fungi
Protozoa
Rickettsia
Chlamydiae
Viruses

F

T. foetus (protozoan)
C. fetus (bacterium)
IBR (virus)

Figure 1 Infection of the conceptus or corpus luteum. Most infectious agents arrive by the hematogenous route, having gained access to the host's bloodstream via (A) the respiratory or upper gastrointestinal (GI) tract, (B) arthropod vectors, (C) the lower GI tract, (D) the mammary gland, (E) wounds in the skin, or any disruption of mucous membranes. A few agents are venereally transmitted (F), that is, passed from the male to the female at the time of coitus. Hematogenously delivered organisms can infect the maternal and fetal side of the placenta (and subsequently the fetus itself); some can infect the corpus luteum (arrows). Venereal organisms usually do not interrupt fertilization, but cause early embryonic or fetal death.

the CL via a hematogenous route, and of causing significant luteal destruction (Miller and van der Maaten, 1986) (see Fig. 1).

B. Abortion

By definition, abortion is the loss of a fetus, that is, pregnancy loss after organogenesis, and in most cases after implantation. In a practical sense, it is the loss of pregnancy after a positive diagnosis of pregnancy has been established. It is the most recognizable but perhaps not the most important contributor to pregnancy wastage. As Table III illustrates, in a healthy population of animals, EED accounts for nearly twice as much pregnancy wastage as either frank abortion or fertilization failure (Roberts, 1986). In

Table III
Relative Contribution to Pregnancy
Wastage in the Cow

Fertilization failure	5–15%
Early embtyonic death	20–35%
Abortion	5–10%
Survival to term	40–70%

any case, the high visibility of an aborted conceptus and the potential for contagious spread to other animals warrant the attention that has been given to this category of loss.

IV. Relationship of Time of Infection to Outcome

With few exceptions, most abortifacients arrive at the uterus by the hematogenous route. They must overwhelm the dam's immune system such that they can colonize the maternal side of the placenta, and eventually invade the fetal side. Once there, the consequences of fetal infection depend on the interplay between the virulence of the agent and the developmental stage of the fetus. As an example, bovine viral diarrhea (BVD) virus can have a variety of effects on a given pregnancy, depending on the timing of fetal infection (see Fig. 2). If fetal infection occurs early in gestation (less than 70–100 days), the conceptus is frequently killed and aborted or mummified (Roberts, 1986). Infection at 90–140 days of gestation commonly results in teratologies (developmental anomalies) of the central nervous system, including cerebellar hypoplasia, retinal dysplasia, or congenital cataracts (Baker, 1987; Bolin, 1985; Done *et al.*, 1980; Kirkbride, 1984). Infection at a slightly later stage, after CNS development is essentially complete, but before the fetal immune system is BVD-competent, leads to the birth of a physically normal but chronically infected, immune-tolerant fetus that may complete gestation and enter postnatal life as a "carrier cow" (Radostits and Littlejohns, 1988; Roeder and Drew, 1984). So-called "noncytopathic" strains of BVD virus are most often associated with such fetal infections,

whereas "cytopathic strains" apparently cannot establish a prolonged fetal viremia (Brownlie *et al.*, 1989). One of the significant consequences of congenital infection is that postnatal exposure to cytopathic strains of BVD can lead to superinfection and acute disease or death. It can also supply a herd with a "carrier" cow that becomes a source of infection for all other cows in the herd (Roeder and Drew, 1984).

Finally, fetal infection with BVD virus after 150 days of gestation usually has few clinical consequences, except for the fact that a normal calf is born with circulating antibodies to BVD virus (Roberts, 1986; Kirkbride, 1984).

The ability of the fetus to mount an immunological response to infectious agents is gradually acquired over gestation, and is not an "all-or-none" event. Some antigens are recognized by fetal calves as early as 150 days of gestation, while others (e.g., BVD) may not be recognized until later, and still others may not be effectively dealt with at all by the fetus (e.g., *Brucella abortus*). Table III lists some known reproductive pathogens, and the gestational age at which fetal immunological recognition occurs.

V. Pathological Outcome: Relationship to Physiology of Pregnancy Maintenance

In cases where infection results in fetal death, the conceptus is usually expelled. Occasionally, when the fetus is killed in early pregnancy (i.e., first half), and when there is little or no bacterial contamination or degradation of the fetal tissues, the fetus may mummify within the uterus. In this case, the fetal fluids are reabsorbed by the dam, and

FETAL BVD INFECTION
(Noncytopathic strains)

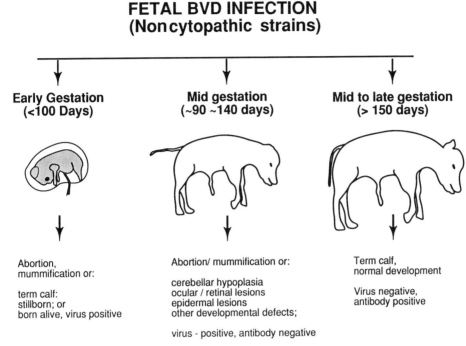

**Early Gestation
(<100 Days)**

**Mid gestation
(~90 ~140 days)**

**Mid to late gestation
(> 150 days)**

Abortion,
mummification or:

term calf:
stillborn; or
born alive, virus positive

Abortion/ mummification or:

cerebellar hypoplasia
ocular / retinal lesions
epidermal lesions
other developmental defects;

virus - positive, antibody negative

Term calf,
normal development

Virus negative,
antibody positive

Figure 2   Effect of gestational age at the time of infection. BVD infection of the bovine conceptus (by the hematogenous route) at less than 100 days of gestation usually results in fetal death, with either mummification or abortion. A few cases may survive to term and be stillborn, or born alive, with chronic BVD virus infection. Infection after 100 days, but before 150 days, can result in teratology (congenital defects), due to the virus's effects on developing organ systems, especially the central nervous system (CNS). These calves may complete gestation and be born alive. Infection after the CNS is fully developed, but before immune competence to BVD is obtained, results in the birth of calves that are chronically infected and immune tolerant to BVD virus. Infection after the fifth to sixth month usually results in a normal term calf that has circulating antibodies to BVD virus at the time of birth.

the dehydrated fetus/membranes remain in the uterine lumen indefinitely (Arthur *et al.*, 1989). The animal usually remains under the influence of the CL of pregnancy, even though she obviously does not have a viable conceptus.

In the more common case of fetal expulsion, the fetus may be in either an advanced state of autolysis, or in some cases it may be quite fresh. To a great extent, the differ-

ence is a function of the speed with which the agent dispatches the fetus and the means by which pregnancy is maintained and terminated in the species in question. In animals in which pregnancy is CL-dependent for most or all of gestation (pigs, goats, cows), rapid, infection-induced death of the conceptus usually results in the abortion of a rather autolyzed fetus; this is due to the delay between the time of fetal death, and the lysis of the CL of pregnancy. The delay can be more than 48 h, which is adequate time for significant autolysis to occur. A well-known example of such a disease is bovine herpesvirus I abortion of cattle (Kirkbride, 1984).

In species that are CL-independent for significant portions of gestation (e.g., ewes, mares), fetal death results in immediate cessation of the fetal contribution to pregnancy maintenance (i.e., placental progesterone). When circulating levels fall below threshold levels, the fetus is expelled, often in a reasonably "fresh" state. Examples include chlamydial (enzootic) abortion of ewes, and equine herpes I viral abortion (Arthur, 1989).

A third mechanism involves the stimulation of those physiological events in the fetus that lead to termination of pregnancy. In this case, the fetus must be mature enough, and survive infection long enough, to mount a "stress" response, including the production and release of adrenal cortisol. This in turn would trigger the hormonal cascade responsible for pregnancy termination. Under these circumstances, the fetus might even be delivered alive, albeit weak and heavily compromised. An example of the latter phenomenon is epizootic bovine abortion (EBA, or foothill abortion; Kimsey, 1986).

VI. Specific Diseases Causing Embryonic or Fetal Loss

A. Bacterial Agents

1. Brucellosis

The best known bacterial cause of pregnancy loss in domestic animals is brucellosis, caused in cattle by the gram-negative bacterium, *Brucella abortus*. The organism is of historical interest because of its zoonotic potential (ability to cause serious disease in animals *and* humans). This intracellular parasite is usually acquired by susceptible cows through mucous membranes (i.e., the oral, nasal, or ocular mucosae). Sources of the organism include infected milk and placental fluids from aborted pregnancies. (These are often licked or sniffed by other cows.) Lymph nodes are quickly colonized after infection, and become the source of organisms for the rest of the body. The nonpregnant uterus is quite resistant to infection, but the organism readily colonizes the pregnant uterus. The presence in the fetal chorioallantoic membranes of high levels of erythritol, a growth stimulant for Brucellae, probably explains the predilection of this pathogen for the pregnant uterus (Roberts, 1986).

Placental infection leads to a marked inflammation of the fetal membranes (placentitis), and eventually to generalized fetal infection, including the development of bronchopneumonia (Roberts, 1986; Nicoletti, 1984). Abortion is most common in mid to late gestation, and in susceptible herds can occur in "storms." Placental and fetal fluids are highly infected with *B. abortus,* and represent a significant public health hazard, as well as a threat to the rest of the herd. Following parturition or abortion, the organism is expelled from the uterus in about 2 weeks.

Although infection of the reproductive tract of bulls with *B. abortus*—leading to epididymitis—is well documented, venereal transmission of brucellosis in cattle is probably not a common means of transmission. This follows from the above observation that the pregnant—but not the nonpregnant—uterus is susceptible to infection.

Control of brucellosis in most developed countries is based on serological testing of cattle and on vaccination of females. The presence of antibodies in unvaccinated cows strongly suggests prior exposure to the organism, and because of the chronic infective nature of this bacterium, "prior exposure" (i.e., the presence of antibodies) is interpreted as "present infection," and the animal is eliminated from the herd. Vaccination of young females with modified live strains of *B. abortus* is still commonly practiced. While the vaccines do not show a complete ability to protect a host from infection, they are quite efficacious in reducing the reproductive consequences of infection (Manthei, 1968; National Research Council, 1977; Roberts, 1986). Therefore, in countries in which prevalence of *B. abortus* is significant, vaccination is likely to remain an important part of the control strategy for brucellosis. In countries where prevalence is very low, control is based solely on slaughter of animals with detectable antibodies to *B. abortus*.

Sheep, goats, and swine each have their own species-specific Brucellae, with slight epidemiological and clinical differences from *B. abortus*. It is possible, however, to infect some of these other species with *B. abortus*.

2. Genital Campylobacteriosis ("Vibriosis")

This is a venereal disease, caused by the appropriately named bacterium, *Campylobacter fetus* var. *venerealis*. The organism resides in the folds of the preputial and penile epithelium of the bull; the bull, which can carry the organism for life, suffers no ill effects of infection, but transmits the organism to a heifer or cow at coitus. Older bulls seem to be more likely to become chronic carriers than young bulls, probably because of the increased surface area and deeper epithelial crypts of the preputial/penile skin. The organism can survive in ejaculates collected for artificial insemination, although commercial bull studs take great pains to screen bulls for vibriosis prior to collection, and extended ejaculates are incubated with appropriate antibiotics before semen is commercially frozen.

The upper reproductive tract of infected females is quickly colonized by this highly motile bacterium, and the early embryo may be killed by the dam's inflammatory response before implantation. In such cases, there is a delayed return to estrus. In a smaller percentage of cases, persistently infected dams may abort a fetus from 2 to 7 months gestational age. Regardless of the outcome of the upper reproductive tract infection, the vagina remains infected for several weeks, or even months. In addition, there is a gradual increase in the secretion of local specific antibody (mostly of the IgA isotype) in the vaginal mucosa, over a period of 40–80 days following infection (Corbeil *et al.*, 1975, 1981).

Diagnosis is based on demonstration of the organism by bacterial culture of vaginal/uterine secretions, or by demonstration of specific antibodies in those secretions, indicating prior exposure to *C. fetus*. In the less common case of vibrionic abortion of an older fetus, diagnosis can be achieved by culture of fetal fluids, especially fetal stomach contents. Diagnosis of infection of the bull is based on bacterial culture or immunodiagnostic assay of preputial secretions (smegma). Control of vibriosis involves detection and treatment or removal of infected bulls, or initiating the practice of artificial insemination, as well as the immunization of males and females with an effective bacterin (killed bacterial vaccine).

Current recommendations call for immunizing the female herd twice at 1-month intervals, with the second injection a week before exposure to bulls. Bulls can be successfully vaccinated as well (Clark *et al.*, 1974); surprisingly, infected bulls can be cleared of *C. fetus* by administration of vaccine (Bouters *et al.*, 1973). The latter observation is counter to most other infectious diseases, in that vaccination in the presence of infection is usually futile.

3. Leptospirosis

In the cow, this disease is caused by infection with the spirochete *Leptospira interrogans* serovar *hardjo*, and occasionally by other serovars of this species. The *hardjo* serovar is particularly well adapted to cattle, and—after entering through almost any mucous membrane in the body—arrives via the blood stream at the kidney and the reproductive tract, where it establishes a long-term residence (Ellis *et al.*, 1986). Immune recognition by the cow is apparently somewhat diminished, as an infected cow's antibody titers are usually quite low relative to her titers for other *Leptospira* serovars (Kirkbride, 1984). Some transudation of antibodies into the reproductive tract from the circulation of the infected cow may occur, allowing for neutralization of leptospires at the site of their pathological effect (Ellis, 1986).

Leptospira interrogans serovar *hardjo* is further subdivided into distinguishable biotypes, which—although they may have identical serological profiles—may have entirely different genomic "fingerprints" (Thiermann *et al.*, 1986). Such is apparently the case with *L. (interrogans) hardjo prajitno* and *L. (interrogans) hardjo bovis;* the former is often used in vaccine production, while the latter is apparently able to cause reproductive loss in pregnant females vaccinated with *L. hardjo prajitno* (Bolin *et al.*, 1989).

In any case, infected cows often abort in mid to late gestation, without the slightest signs of sickness prior to abortion. In some cases, there may have been a recent and temporary sudden decline in milk production of dairy cows, with blood-tinged milk and flaccid udders (as opposed to the hard edematous udders of common bacterial mastitis). There are no specific lesions in the fetus or placenta.

Diagnosis is based on demonstrating the organism in fetal tissues, a feat more easily described than accomplished, given the fastidious nature of the organism. In cases where the fetus was infected late in gestation, it may be possible to demonstrate fetal antibodies specific for *L. hardjo*, a finding that would strongly implicate leptospirosis as the cause of abortion. Serological examination of the dam's blood is not particularly rewarding, since—as stated above—the organism has been residing in the dam for some time before abortion, and her immune response to it has been minimal. Moreover, the use of the classical "paired sera" serological tests (where two samples are taken 2 weeks apart to demonstrate a rise in antibody titer) is unrewarding in many cases, because no change in titer is likely to occur in chronic infections.

Leptospiral abortion in pigs is more often due to infection with *L. interrogans* serovar *pomona*, although other serovars, including *grippotyphosa, icterohemorrhagiae,* and *canicola,* are sometimes implicated. The spirochetes enter through breaks in the skin or mucous membranes, and quickly establish a leptospiremia, from which they cross the placentas to the fetuses. As with cattle, there are usually no specific lesions in aborted litters of pigs, and the sows very often show no signs of illness prior to abortion. Occasionally, an infected litter may be born alive but weak.

Control of leptospirosis is based on immunization of dams before they are bred, and in some cases again after pregnancy has been established. Commercial bacterins are available for at least five serovars of the organism. The immunity to these bacterins is relatively short-lived, such that animals may have to be inoculated as often as three times per year in areas that have experienced significant leptospiral abortion problems. A common scheme in dairy cattle is to vaccinate at the time of postpartum examination, again after pregnancy has been diagnosed, and again shortly before the end of the cow's lactation. Others simply immunize three times per year, based on evenly spaced calendar dates. The ability of commercial hardjo vaccines to protect against hardjo-induced abortion is still debated.

4. Nonspecific Enteric Infection

Infection of almost any tissue by members of the gram-negative enteric group of bacteria can lead to loss of pregnancy, especially in CL-dependent pregnancies. The precise mechanisms by which such infections destroy the conceptus is still being studied, but it is known that the release of endotoxins from such bateria can induce a systemic prostaglandin ($PGF_{2\alpha}$) response, which results in luteolysis and subsequent pregnancy failure. Cows, goats, and early gestation mares have all been shown to be susceptible to endotoxin-induced abortion (Giri et al., 1989; Fredricksson et al., 1987; Daels et al., 1989).

B. Rickettsial Agents

Coxiella burnetii, the agent of so-called "Q fever," is known to cause abortions in cattle and goats, and may cause abortions in sheep as well. Sources of infection include milk and placental fluids. There is often a marked placentitis, usually in the intercotyledonary areas (Arthur et al., 1989). Diagnosis is based on identifying the organisms in smears of the placenta or fetal organs, and upon demonstration of a recent immune response of the dam to C. burnetii (see Section VII). This agent also has significant zoonotic potential (Houwers and Richardus, 1987), so suspect animals and tissues should be handled with great care.

C. Chlamydial Agents

Enzootic abortion of ewes (EAE), caused by Chlamydia psittaci (ovis), is the best-described abortifacient among the chlamydiae. In spite of the name, the organism can cause significant pregnancy losses in goats as well as sheep. This organism has both intracellular and extracellular phases, and is capable of causing polyarthritis and pneumonia, in addition to causing fetal infection and abortion. It can enter the body by almost any route, including ingestion, inhalation, and perhaps by venereal transmission (Arthur, 1989).

Like Brucella, the chlamydial organism is present in the body for long periods before the pregnant uterus is infected. Placental colonization results in a marked and somewhat characteristic placentitis, with thickened intercotyledonary areas, necrotic cotyledons, and a yellow pasty discharge on the chorionic membrane. Most of the aborted lambs are fresh (see Section V) and show few characteristic lesions. Diagnosis is based on demonstration (by culture, staining, or immunodiagnostic methods) of the causative organism in the placenta (Storz, 1984). Control is based on reducing exposure, by avoiding overcrowded, contaminated conditions, and by serological screening of all new entrants into a flock. Current serological tests may not detect all carriers. Vaccines, using killed organisms, are available in some areas where the disease is prevalent (Arthur *et al.*, 1989). They should be administered prior to the breeding season; it probably does not prevent infection, but, like *Brucella* vaccines, protects against the reproductive consequences of infection, by providing for a maternal antibody barrier to uterine/placental infection.

D. Protozoal Agents

1. Trichomoniasis

The best known protozoal reproductive pathogen is *Tritrichomonas foetus*, a venereal agent of cattle that causes significant EED and to a lesser extent abortion. The organism is carried asymptomatically (without signs) by the bull, and is deposited in the vagina of susceptible females at the time of coitus. Fertilization is not prevented, but in most cases the conceptus is killed sometime in the first 30–70 days (Skirrow and BonDurant, 1988). Occasionally, the fetus is not killed until well into the second trimester (Rhyan *et al.*, 1988), after which it may be bacterially decomposed, and lead to a pyometra and/or macerated fetus.

The organism may have a predilection for tissues of the conceptus, as it typically causes very little pathological response in the dam's reproductive tract (Parsonson *et al.*, 1976). The dam exhibits some immune response following infection, in that she is somewhat resistant to the consequences of reexposure, albeit for a very short time (perhaps only a few months). Cows that are infected in one breeding season can be reinfected the next season as well (Clark *et al.*, 1977).

Control of trichomoniasis is based on detection and elimination of infected males. Infected females eventually clear the organism after several estrous cycles, requiring as much as 30 weeks in some cases (Skirrow and BonDurant, 1990). There is no approved treatment in either bulls or cows. A provisionally licensed vaccine has recently been made available (Fort Dodge Co., Fort Dodge, Iowa), but its efficacy is unknown.

2. Toxoplasmosis

This protozoan parasite has a world-wide distribution, and a relatively complex life cycle, in which members of the cat family contaminate feedstuffs or pasture with feces containing oocysts of *Toxoplasma gondii*. Ingestion of oocysts by sheep or goats (or many other species) allows the organism to multiply asexually in the new "host's" tissues. If the sheep or goat is not pregnant, the disease will probably be inapparent, but in pregnant animals the conceptus is especially vulnerable. As with other pathogens, the effect of fetal infection is stage-dependent. Infection in the first trimester usually causes fetal resorption, while mid-gestation infection results in abortion or mummification and late-gestation infection may result in stillbirths or the birth of weak lambs or kids. When animals are grouped for seasonal breeding or feeding purposes, abortions due to toxoplasmosis can occur in very large numbers.

As with many infectious abortions in small ruminants, toxoplasmosis is associated with a significant degree of placental in-flammation. In fact, the severe placentitis seen with toxoplasmosis—evident especially in the fetal cotyledon—is rather characteristic of the disease. Diagnosis of toxoplasma abortion is dependent on observation of characteristic lesions of the placenta, demonstration of the organism in placental tissues (using Giemsa staining, immunoperoxidase methods, or mouse inoculation), or demonstration of a rising host immune response to the agent (Dubey, 1984). In the latter case, several serological tests are available. Where infection has occurred in the last several weeks of gestation, serological testing of fetal fluids (blood, cerebrospinal fluid, thoracic and peritoneal fluid) is helpful, because the fetus may have been able to immunologically respond to the organism by this stage of gestation.

Control of toxoplasmosis is based on the presumption that there is little or no horizontal (ewe-to-ewe) transmission. Feral and domestic cat populations are controlled, and new additions to the flock/herd are exposed to the contaminated areas before the breeding season, in order to establish premunition immunity. Vaccines are being investigated, but none are commercially available as of this writing.

E. Viral Agents

1. Herpes Viruses

Nearly every species has at least one, and sometimes many more, herpes viruses that are host adapted. A characteristic common to all such viruses is their chronic infective nature, in that the organism remains in an "occult" state in the host for long periods, often for life. The route of entry of these agents can be ingestion, inhalation, other mucous membrane exposure (e.g., the eye), and venereal contact. In most cases, the organism is host adapted enough that the adult is not killed by infection, but the organism is rapidly fatal to the conceptus.

a. Equine Herpes I (Rhinopneumonitis) This virus is exceptionally contagious among grouped horses, and can be passed rapidly within a stable or brood farm. An initial mild respiratory disturbance in adults may go unnoticed, but abortion usually occurs 20–30 days after infection. This usually coincides with the time of year in which mares are in the latter stages of pregnancy, such that fetuses are aborted during the last 4 months of gestation. The rate of abortion in a susceptible stable of mares can be devastatingly high. Aborted fetuses are usually fresh (see Section V) and show pinpoint focal areas of liver necrosis, in addition to increased amounts of thoracic fluid and pericardial fluid. Diagnosis is based on observation of these lesions, and the isolation of the virus from fetal fluids. Assays for maternal antibodies are not particularly rewarding, since titers may be stable, or even falling, at the time of abortion (Arthur, 1989). Control is based on exposure prior to breeding, preferably to effective vaccines, although exposure of young stock to "wild virus" from an outbreak can confer immunity as well.

b. Bovine Herpes Virus I (Infectious Bovine Rhinotracheitis Virus, IBR) This virus is one of the more commonly diagnosed causes of bovine abortion, in part because it is more readily detected by diagnostic laboratories than many other pathogens (Kirkbride, 1984). IBR virus exerts a variety of effects, including encephalitis, rhinotracheitis, conjunctivitis, vulvo-vaginitis, and abortion. The possibility exists that the different effects attributed to this virus may be a function of different strains (biotypes) of the organism (Miller *et al.*, 1988). As with all herpes viruses, IBR can chronically infect a cow, and can recrudesce from a "latent" phase when animals are subsequently stressed or immunocompromised (Miller *et al.*, 1987). Invasion of the pregnant uterus, via the bloodstream of the dam, quickly results in the death of the fetus. Following luteolysis, the now-autolyzed fetus is expelled, showing few specific lesions. Diagnosis is based on identification of the virus by virus culture methods or by immunofluorescence of fetal tissues, especially lungs, liver, and adrenal glands. Due to the chronic nature of infection in the dam, serological results are difficult to interpret. Control of IBR abortion is based on immunization of all female cattle before their first breeding. Modified live virus vaccines (injectable or intranasally administered) are available that essentially confer life-long immunity, if given to a heifer *after* her maternal (colostral) immunity has waned (Kahrs, 1981; Todd, 1976); this is typically at about 6 months of age (Roberts, 1986). Killed vaccines are also available, and must be given twice initially (at monthly intervals), followed by annual revaccination. In theory, the dam's circulating immunoglobulins provide a protective barrier against uterine or placental infection.

c. Porcine Pseudorabies Virus (PRV; Aujesky's Disease) This organism can cause severe clinical signs in the infected adult, including serious central nervous system disturbances; but the greatest mortality is seen in suckling pigs (Roberts, 1986). As with other viral abortifacients, the effect on pregnancy is stage dependent: When infec-

tion occurs at 30 days of gestation, all fetuses are killed and are reabsorbed. If infection occurs between 40 and 60–80 days, some or all of the fetuses may mummify. After 60–80 days, infection of fetuses may result in abortion, mummification, stillbirth, or premature delivery of live, weak piglets carrying antibodies to PRV (Iglesias and Harkness, 1988).

Control of PRV is based on serological identification and slaughter of infected swine. In some areas, vaccination with attenuated or subunit vaccines is advocated; the latter is particularly beneficial if laboratories are able to distinguish between serum antibodies to vaccine antigens and antibodies to "wild" virus. Currently, vaccine manufacturers are designing and marketing subunit vaccines with "markers" such that serological responses to these immunogens can be distinguished (Thawley and Morrison, 1988).

2. Bovine Viral Diarrhea (BVD Virus)

This organism is a member of the pestivirus family, and is responsible for gastrointestinal disease in postnatal cattle, and abortion, mummification, and teratologies in the conceptus. The most notorious form of postnatal disease is mucosal disease, in which there is severe gastrointestinal necrosis and inflammation, with acute diarrhea, shock, and death. Adult cattle are infected by ingestion of virus. As noted above (Section IV), the outcome of BVD infection in pregnant cattle is very much stage dependent, and almost certainly influenced by strain type of the infecting virus (Baker, 1987). Infection of the fetus with so-called noncytopathic strains of BVD before the fetus is immune-competent can render the fetus permanently infected and immune-tolerant to the infecting strain of virus. It is now thought that postnatal exposure of immune tolerant animals to cytopathic strains of BVD, perhaps even attenuated vaccine strains, can lead to the development of mucosal disease (Bolin, 1985; Liess *et al.*, 1984; Radostits and Littlejohns, 1988). The requirement of a tolerant, susceptible population would explain the low morbidity (disease rate) that is characteristic of mucosal disease.

Earlier infection of the conceptus leads to abortion or developmental lesions of the central nervous system, as described previously (see Fig. 2). Some workers have also suggested that BVD infection of preimplantation bovine embryos can increase embryonic mortality and disrupt normal development (Archbald, *et al.*, 1979), but later work showed that there was little or no virus uptake by early bovine embryos, with or without their zonae pellucidae (Potter *et al.*, 1983). Furthermore, Singh (1987) stated that there was no effect on development of embryos after experimental exposure to BVD virus, or after collection from BVD-infected dams. It is not known if such early infection occurs naturally, but presumably, in a female that is viremic at the time of estrus/ovulation, or within the next few days following ovulation, the virus could have access to preimplantation embryos.

There is considerable debate as to the best means to control BVD's reproductive effects. Most control schemes rely on vaccination, but there is imperfect agreement on which animals to vaccinte, and when to vaccinate them. In general, vaccination of heifers is advocated immediately after their maternal (colostral) immunity has waned, that is, at about 6–8 months of age. Modified live virus vaccines can be used successfully in such animals, although congenitally infected (and therefore immunotolerant) individuals may "break" with mucosal disease when vaccinated. Killed vaccines are also available, and should be given at least twice in the first month, followed by annual revaccination.

F. Fungal Agents

Fungal agents usually cause sporadic (rather than epizootic) abortions. In cattle, the two most commonly incriminated genera include *Aspergillus* and *Mucor*. Spores from these molds are probably taken into the respiratory or alimentary tracts, from which they hematogenously travel to the uterus and placenta. The organisms apparently grow slowly in the placental tissues, developing a chronic and often severe placentitis. The fetus is usually aborted in late gestation, and is grossly normal, although in some cases, classic fungal skin lesions are seen. Diagnosis is based on histological examination of the placental and fetal tissues (especially skin) for fungal hyphae, or on fungal culture of these tissues (Kirkbride, 1984).

VII. Diagnostic Steps for Determining Causes of Abortion

A. Current Success Rate

Under ideal circumstances, the best diagnostic laboratories can only confirm a definitive diagnosis in about 25–35% of submitted cases. The reasons for this disappointing rate are many, and include the following:

1. Not all abortions are caused by infectious agents. Examples of noninfectious causes include twinning, genetically

defective fetuses, trauma, heat stress, improper drug use (e.g., corticosteroids), toxins, and probably nutritional deficiencies or excesses.

2. Even when the cause of abortion *is* infectious, the causative organism is often no longer viable in fetal tissues by the time abortion occurs. The processes of inflammation, abortion, and autolysis can destroy many microorganisms.

3. The use of serological profiles of the dam to determine the cause of abortion is fraught with uncertainty. As mentioned, some organisms are latent in the body of the dam for a long-period before abortion occurs. Therefore, the dam is likely to have a serological titer to these agents that will not show a classic "rise" between the time of abortion and the time of collection of a convalescent sample. Fetal serological tests are often negative, because the pathogen may have overwhelmed the fetus before it was capable of responding immunologically.

4. Many causes of abortion leave no "characteristic" lesions, that is, they kill the conceptus without grossly or historically changing the tissues in a recognizable way. Also, the conceptus is very often autolyzed (see discussion above), and subtle changes in tissues are obscured.

B. Enhancing the Probability of a Diagnosis

In order to have the greatest opportunity to establish a diagnosis of the cause of abortions, the diagnostic laboratory needs the following:

1. A thorough *history* of the problem. That is, which animals are affected (e.g., mature females or uniparous females)? At what state of gestation are they aborting?

Are they bred by natural service or artificial insemination? What is the diet of the aborting individual or her herd-mates? Is there animal traffic in and out of the property where the aborting animal is kept? What is the nature of that traffic? Against which diseases is the aborting female vaccinated, and when was she vaccinated for them? Was the aborting female visibly ill at any time before she aborted? The answers to these and other questions can be of considerable use to the diagnostician.

2. Appropriate *samples* from the individual and from the herd of origin. These samples include:

a. The *fetus:* submission of the entire fetus, on ice (but not frozen), is preferred by most laboratories. Freezing disrupts the architecture of the tissues, and makes interpretation of histology very difficult.

b. *Fetal tissue:* When the entire fetus cannot be submitted, aseptically taken tissues can be packaged in sterile containers, refrigerated, and sent to the laboratory. These tissues include lung, liver, kidney, and spleen (about a third of each organ's total mass).

c. *Placenta:* this is an extremely important tissue for the diagnostician, and often the one tissue that is not submitted. In ruminants, it is important to submit at least one cotyledon, in addition to intercotyledonary sections.

The above samples should be divided, with half of each tissue placed in sterile refrigerated containers and the other half cut in 6- to 8-mm-thick sections and placed in 10% formalin. Other samples that should be taken include:

d. *Dam's serum:* In spite of the fact that serological diagnosis is not always possible, it is sometimes useful to know what agents the dam's immune system has recently encountered. In some cases, interpretation is a matter of regulatory statute, as in the case of brucellosis, in which a positive serological titer is, by law, diagnostic. In any case, a single sample from a single aborting dam is often not diagnostic, whereas the serological *profile* of a subset of the herd (i.e., a randomly selected 10%) may be useful. In the latter case, a second sample from each cow, collected 2 weeks after the first, is usually advised.

e. *Fetus's serum:* As noted, the fetus can respond to some antigens, so it is sometimes rewarding to examine fetal serum for a possible antibody response to known pathogens. Positive findings, while not definitive, do indicate that the fetus was exposed *in utero* to the pathogen in question. If fetal lesions consistent with that pathogen are present, then a diagnosis can usually be made. In cases where fetal heart blood cannot be obtained, it is sometimes possible to aspirate thoracic and/or peritoneal fluid for the same serologic purpose.

References

Archbald, L. F., Fulton, R. W., Seger, C.-L., Al-Bagdadi, F., and Godke, R. A. (1979). *Theriogenology* **11,** 81–89.

Arthur, G. H., Noakes, D. E., and Pearson, H. (1989). "Veterinary Reproduction and Obstetrics," 6th ed. Balliere Tindall, London.

Baker, J. C. (1989). *J. Am. Vet. Med. Assoc.* **190,** 1449–1458.

Bolin, C. A., Thiermann, A. B., Handsaker A. L., and Foley, J. (1989). *Am. J. Vet. Res.* **50,**(1), 161–165.

Bolin, S. R. (1985). *Proc. U.S. Anim. Health Assoc.* **89,** 115–118.

Bouters, R., DeKeyser, J., Van dePlassche, M., van Aart, A., Brone, E., and Bonte, P. (1973). *Br. Vet. J.* **129,** 52–57.

Brownlie, J., Clarke, M. C., and Howard, C. J. (1989). *Res. Vet. Sci.* **46,**(3), 307–311.

Clark, B. L., Dufty, J. H., and Parsonson, I. M. (1977). *Aust. Vet. J.* **53,** 170–172.

Clark, B. L., Dufty, J. H., and Monsbourgh, M. J. (1974). *Aust. Vet. J.* **50,** 407–409.

Corbeil, L. B., Corbeil, R. R., and Winter, A. J. (1975). *Am. J. Vet. Res.* **36,** 403–406.

Corbeil, L. B., Schurig, G. D., Duncan, J. R., Corbeil, R. R., and Winter, A. J. (1981). *In* "The Ruminant Immune System. Advances in Experimental Medicine and Biology," vol. 137, pp. 729–743. Plenum Press, New York.

Cort, N., and Kindahl, H. (1986). *Acta Vet. Scand.* **27,**(2), 145–158.

Daels, P. F., Stabenfeldt, G. H., Hughes, J. P., Kindahl, H., and Odensvik, K. (1989). *In Proc. 34th Annu. Convention of the American Association of Equine Practitioners, San Diego,* pp. 169–171.

Done, J. T., Terlecki, S., Richardson, C., Harkness, J. W., Sands, J. J., Patterson, D. S. P., Sweasy, D., Shaw, I. G., Winkler, C. E., and Duffell, S. J. (1980). *Vet. Rec.* **106,** 473–479.

Dubey, J. P. (1984). *In* "Laboratory Diagnosis of Abortion in Food Animals" (C. Kirkbride, ed.), pp. 165–168. American Association of Veterinary Laboratory Diagnosticians, Madison, Wisconsin.

Ellis, W. A. (1986). *In* "Current Therapy in Theriogenology," 2nd ed. (D. Morrow, ed.), pp. 267–271. Saunders, Philadelphia.

Ellis, W. A., Songer, J. G., Montgomery, J., and Cassells, J. A. (1986). *Vet. Rec.* **118,** 11–13.

Fredriksson, G. (1984). *Acta Vet. Scand.* **25**(3), 365–377.

Gilbert, R.-O., Bosu, W. T. K., and Peter, A. T. (1986). *Proc. 14th World Congr. Diseases of Cattle, Dublin* **2,** 894–899.

Giri, S. N., Graham, T. W., Cullor, J., Keen, C. L., and Thurmond, M. C. (1989). *FASEB J.* **3**(4), A1137.

Hawk, H. W. (1971). *J. Anim. Sci.* **32**(Suppl. 1), 55–63.

Hellstrom, J. S., and Blackmore, D. K. (1979). *Proc. 2nd Int. Symp. Veterinary Epidemiology and Economics,* Canberra, Australia, pp. 214–219.

Houwers, D. J., and Richardus, J. H. (1987). *Zentralb. Bakteriol. Hygiene* **267**(1), 30–36.

Hussaini, S. N., Edgar, A. W., and Sawtell, J. A. A. (1986). *Res. Vet. Sci.* **41**(1), 131–132.

Iglesias, J. G., and Harkness, J. W. (1988). *Vet. Microbiol.* **16**(3), 243–354.

Kahrs, R. F. (1981). "Viral Diseases of Cattle," pp. 1–299. Iowa State University Press, Ames.

Kimsey, P. B. (1986). *In* "Current Therapy in Theriogenology," 2nd ed. (D. Morrow, ed.), pp. 260–263. Saunders, Philadelphia.

Kirkbride, C. (1984). *In* "Laboratory Diagnosis of Abortion in Food Animals" (C. Kirkbride, ed.), pp. 54–76, 106–110, 111–122, 133–139, 147–161. American Association of Veterinary Laboratory Diagnosticians, Madison, Wisconsin.

Liess, B., Orban, S., Frey, H.-R., Trautwein, G., Wiefel, W., and Blindow, H. (1984). *Zentralbl. Veterinarmed. B* **31**(9), 669–681.

Manthei, C. A. (1968). *In* "Abortion Diseases of Livestock" (L. Faulkner, ed.), pp. 80–94. C. C. Thomas, Springfield, Illinois.

McDermott, M. R., Clark, D. A., and Bienenstock, J. (1980). *J. Immunol.* **124**, 2536–2539.

Miller, J. M., and van der Maaten, M. J. (1987). *Am. J. Vet. Res.* **48**(11), 1555–1558.

Miller, J. M., and van der Maaten M. J. (1986). *Am. J. Vet. Res.* **47**(2), 223–228.

Miller, J. M., van der Maaten, M. J., and Whetstone, C. A. (1988). *Am. J. Vet. Res.* **49**, 1653–166.

National Research Council, Subcommittee on Brucellosis Research. (1977). "Brucella Research: An Evaluation," pp. 1–256. National Academy of Sciences, Washington, D.C.

Nicoletti, P. (1984). *In* "Laboratory Diagnosis of Abortion in Food Animals" (C. Kirkbride, ed.), pp. 20–23. American Association of Veterinary Laboratory Diagnosticians, Madison, Wisconsin.

Osburn, B. I. (1986). *In* "Current Therapy in Theriogenology," 2nd ed. (D. Morrow, ed.), pp. 203–204. Saunders, Philadelphia.

Parsonson, I. M., Clark, B. L., and Dufty, J. (1976). *J. Comp. Pathol.* **86**, 59–66.

Potter, M. L., Corstvet, R. E., Looney, C. R., Fulton, R.

W., Archbald, L. F., and Godke, R. A. (1983). *Am. J. Vet. Res.* **45**(9), 1778–1780.

Radostits, O. M., and Littlejohns, I. R. (1988). *Canad. Vet. J.* **29**, 513–528.

Rhyan, J. C., Stackhouse, L. L., and Quinn, W. J. (1988). *Vet. Pathol.* **25**(5), 350–355.

Roberts, S. J. (1986). "Veterinary Obstetrics and Genital Diseases (Theriogenology)," 3rd ed., published by the author (David and Charles, Inc., distributors), North Pomfret, Vermont.

Roeder, P. L., and Drew, T. W. (1984). *Vet. Rec.* **114**(13), 309–313.

Schwabe, C. W., Riemann, H. P., and Franti, C. E. (1977). "Epidemiology in Veterinary Practice," pp. 14–16. Lea and Febiger, Philadelphia.

Singh, E. (1987). *Theriogenology* **27**(1), 9–20.

Skirrow, S. Z., and BonDurant, R. H. (1990). *J. Am. Vet. Med. Assoc.* **196(6),** 885–889.

Skirrow, S. Z., and BonDurant, R. H. (1988). *Vet. Bull.* **58**(8), 591–603.

Storz, J. (1984). *In* "Laboratory Diagnosis of Abortion in Food Animals" (C. Kirkbride, ed.), pp. 34–43. American Association of Veterinary Laboratory Diagnosticians, Madison, Wisconsin.

Thawley, D. G., and Morrison, R. B. (1988). *J. Am. Vet. Med. Assoc.* **193**(2), 184–190.

Thiermann, A. B., Handsaker, A. L., Foley J. W., White, F. H., and Kingscote, B. F. (1986). *Am. J. Vet. Res.* **47,** 61–66.

Todd, J. D. (1976). *Dev. Biol. Stand.* **33,** 391–395.

Wilfert, C. M. (1984). *In* "Zinsser Microbiology," 18th ed. (W. Joklik, H. Willet, and D. Amos, eds.), p. 665. Appleton-Century-Crofts, Norwalk, Connecticut.

Winter, A. (1982). *J. Am. Vet. Med. Assoc.* **181**(10):1069–1073.

Index